Applied Palaeontology

Palaeontology, the scientific study of fossils, has developed from a descriptive science to an analytical science used to interpret relationships between earth and life history. *Applied Palaeontology* covers all aspects of palaeontology, although its principal focus is on the applied. It adopts a holistic, integrated approach, highlighting the key role of palaeontology in the study of the evolving earth, life and environmental processes.

After an introduction to fossils and how they are classified, each of the principal fossil groups is dealt with in detail, covering their biology, morphology, classification, palaeobiology and biostratigraphy. The latter half of the book focuses on the applications of fossils in the interpretation of earth and life processes and environments, including the events that control biodiversity. It concludes with case histories of how our knowledge of fossils is applied, in industry and elsewhere.

This will be a valuable reference for anyone involved in the applications of palaeontology, including earth and life science students and academics, professionals in petroleum, minerals, mining and engineering industries, palaeontologists, geologists and environmental scientists.

ROBERT WYNN JONES gained his B.Sc. in geological sciences at the University of Birmingham in 1979 and his Ph.D. at the University College of Wales, Aberystwyth in 1982. Since then he has worked as a palaeontologist in the oil industry, including the last 20 years with BP. His work has involved analysis of micropalaeontological samples and interpretation and integration of micro- and macropalaeontological data. He has worked on rocks and fossils from Proterozoic to Pleistocene, from sedimentary basins all around the world.

Dr Jones also maintains an active interest in academic research, especially in the study of foraminiferal taxonomy, palaeobiology, biostratigraphy and historical micropalaeontology. This interest has resulted in over 60 publications to date, including five books. It has also resulted in an honorary scientific associateship at the Natural History Museum, London and an honorary research fellowship at University College London.

Applied Palaeontology

Robert Wynn Jones

CAMBRIDGE UNIVERSITY PRESS
Cambridge, New York, Melbourne, Madrid, Cape Town,
Singapore, São Paulo, Delhi, Mexico City

Cambridge University Press
The Edinburgh Building, Cambridge CB2 8RU, UK

Published in the United States of America by
Cambridge University Press, New York

www.cambridge.org
Information on this title: www.cambridge.org/9781107407442

First published 2006
First paperback edition 2012

A catalogue record for this publication is available from the British Library

ISBN 978-0-521-84199-3 Hardback
ISBN 978-1-107-40744-2 Paperback

For my father, who unknowingly set me on a life-course when he showed me as a young boy a specimen of his favourite fossil, the brachiopod Rhynchonella . . .

To see a World in a grain of sand
And a Heaven in a wild flower
To hold Infinity in the palm of your hand
And Eternity in an hour.

William Blake, *Auguries of Innocence*

Gweled Nef ym mhlygion blodyn
Canfod Byd mewn un tywodyn
Dal mewn orig Dragwyddoldeb
Cau dy ddwrn am Anfeidroldeb.

A Welsh rendition of the above, by my great-grandfather, Thomas Gwynn Jones, *Caniadau*

Contents

Preface

Mankind has always been fascinated by fossils, by their beauty and their mystery, their charm and their strangeness, their mute testimony to lives and worlds lost unimaginably long ago. In prehistoric times, our forebears not only collected fossils, but evidently treated them as valued artefacts, as indicated, for example, by the discovery of an ammonite at an Upper Palaeolithic burial site in Aveline's Hole in Burrington in the West Country (Rahtz, 1993), and numerous different types of fossil at Cro Magnon sites in the Vézère valley in the Périgord region of France, truly the birthplace of European civilisation (many of which are now displayed in the magnificent Museum of Prehistory in Les Eyzies). The habit persisted both in so-called primitive and so-called advanced societies through historical times (Mayor, 2000).

Palaeontology, that is, the scientific study of fossils, may be said to have originated at least as long ago as the sixteenth century (Thackray, in Briggs and Crowther, 1990), and, obviously, continues to be practised to the present day. The earliest written observations on fossils were made by the German Bauer, or Agricola, in his book *De natura fossilum*, and the earliest illustrations by the Swiss Gesner in his book *De rerum fossilium lapidum et gemmarum*, both of which date from the sixteenth century. The usage by these and other early observers of the term 'fossil', from the Latin *fodere*, meaning 'to dig', pertained to literally anything dug up from the ground or mined, including what we would now classify as minerals, crystals and gemstones. The earliest interpretations as to the nature of what we would now accept as fossils were made by the Danish anatomist Stensen, or Steno, working in the Medici court in Florence, in his publications dating from the latter part of the seventeenth century (Cutler, 2003). Steno applied Descartes' 'method of doubt' and his own deductive logic to demonstrate that the so-called *glossopetrae* or 'tongue stones' much valued in medieval Europe for their supposed medicinal properties were in fact not the tongues of snakes turned to stone by St Paul, as was the superstition, but the fossilised equivalents of the shark's teeth he was familiar with from his dissection work. Elsewhere in his writings, Steno established three important principles of stratigraphy, namely the 'principle of superposition', the 'principle of original horizontality' and the 'principle of lateral continuity', such that he is regarded by many as the true founder of that science. Incidentally, in later life, he renounced science for religion, and was made a saint by Pope John Paul II!

There may be said to have been three, partially overlapping, areas or phases of subsequent palaeontological study: the descriptive; the synoptic; and the interpretive (Bowler, in Briggs and Crowther, 1990; Hoffman, in Briggs and Crowther, 1990; Thackray, in Briggs and Crowther, 1990; Valentine, in Briggs and Crowther, 1990; Jones, 1996). The emphasis through the three phases has shifted from the documentation of fossils to the analysis and interpretation of their relationship to evolving earth and life history and processes and environments, and their application to the elucidation thereof; from data acquistion and processing to interpretation and integration; from pure to applied. The descriptive phase began with the first descriptions of fossil species conforming to modern standards, made following the introduction of the binomial system for the naming of species by Linné, or Linnaeus, in the late eighteenth century. The synoptic phase has continued into the twenty-first century, with the establishment of higher-level taxonomic classification systems based on morphology and phylogeny, made following the publication of *On the Origin of Species by Means of Natural Selection* by Darwin (1859) in the late nineteenth century, and the advances in cladistics and molecular biology in the twentieth. The interpretive phase, ultimately resulting in the development of, and advances in, the applied sub-disciplines of palaeobiology and biostratigraphy, began with the establishment of the ordered succession of fossils in Great Britain, and the 'law of superposition' and the 'law of strata identified by organised fossils' by William ('Strata') Smith in the late eighteenth and early nineteenth centuries (Torrens, 2003); and by the publication also by Smith of the first geological map of Great Britain, 'the map that changed the world' (Winchester, 2001). The first application of biostratigraphy in the oil industry was by the Pole Josef Grzybowski in the late nineteenth and early twentieth centuries (Czarniecki, in Kaminski *et al.*, 1993). At a time when (micro)palaeontology was essentially in a stage of synthesis, it was he who first used the discipline in an analytical fashion to solve geological problems encountered in the oilfields of the eastern Carpathians (those around the village of Potok being the oldest still in production anywhere in the world). His contribution to biostratigraphy and also to palaeobiology has long been recognised and justly acclaimed in his own country, but is sadly seldom acknowledged in the west.

In the future, applied (micro)palaeontology will continue to play a vital role in exploiting the world's discovered petroleum and other mineral resources, and in exploring for undiscovered reserves. In view of the growing concern about the environment, applications in environmental science, and outwith the exploitative industries, are also likely to come to the fore.

A significant number of textbooks have been written about palaeontology, most focusing on pure rather than applied aspects.

This book covers all aspects of palaeontology, although its principal focus is on the applied. It attempts to adopt a holistic, integrated approach, highlighting the key role of palaeontology in earth and life science. It treats palaeontology not as an end in itself, but as a means to an end – of understanding earth and life history and processes, and global change. Its theme may be said to be that of 'fossils as recorders and indicators of global change'.

The following quotation from Erwin Schrödinger's book *What Is Life?* serves as a disclaimer:

> A scientist is . . . usually expected not to write on any topic of which he is not a master. This is regarded as a matter of *noblesse oblige*. For the present purpose I beg to renounce the *noblesse*, if any, and to be freed of the ensuing obligation . . .

Acknowledgements

I would like to acknowledge Cambridge University Press, Longman/Pearson, Poyser/A. & C. Black, and John Wiley & Sons for providing permission to reproduce figures; separate acknowledgements are given as appropriate in the text. Every reasonable effort has been made to contact all other copyright holders in this regard. To any whose rights I may have unintentionally infringed I offer my unreserved apologies.

I would also like to acknowledge the assistance and professionalism of Sally Thomas, Emily Yossarian, Vince Higgs, J. Bottrill, Anna Hodson and Wendy Phillips of Cambridge University Press in seeing the project through from conception to publication.

I would like to acknowledge the many and varied contributions of my geology teacher at Penglais School, Aberystwyth, namely Huw Spencer-Lloyd; my palaeontology and stratigraphy lecturers at the University of Birmingham, namely George Bennison, Russell Coope, Tony Hallam, Frank Moseley and Isles Strachan; my micropalaeontology professors at the University College of Wales, Aberystwyth, namely Jo Haynes and Robin Whatley; and my colleagues at BP, the Natural History Museum, University College London and around the world, namely the late Geoff Adams, Jordi Agusti, Nigel Ainsworth, Peter Andrews, John Athersuch, Haydon Bailey, Fred Banner, Deryck Bayliss, Joan Bernhard, Ray Bernor, Ian Boomer, Paul Bown, Barry Carr-Brown, Mike Charnock, Kevin Cooper, Phil Copestake, Andy Currant, the late Remmert Daams, Jill Darrell, Paul Davis, Louis de Bonis, Mikael Fortelius, John Frampton, Liam Gallagher, Tony Gary, Philip Gingerich, Matt Hampton, Andy Henderson, Nick Holmes, Jerry Hooker, David Horne, Mike Howarth, Wyn Hughes, Jake Jacovides, Clive Jones, the late Garry Jones, Dave Jutson, Mike Kaminski, Paul Kenrick, Eduardo Koutsoukos, Johanna Kovar-Eder, Nadezda Krstic, David Loydell, Norm MacLeod, Paul Marshall, Sue Matthews, Sebastian Meier, Giles Miller, Alex Mitlehner, Bob Morley, Noel Morris, John Murray, Bonnie O'Brien, Hugh Owen, Bernard Owens, Simon Parfitt, Simon Payne, Pete Rawson, Jelle Reumer, Fred Rogl, Lorenzo Rook, Brian Rosen, Mike Simmons, David Siveter, Andrew Smith, Nikos Solounias, the late Charles Stainforth, Mike Stephenson, Jean-Pierre Suc, Jon Todd, Jan van der Made, Janice Weston, Greg Wahlman, John Whittaker, the late Peter Whybrow, Mark Williams, Brent Wilson and Jeremy Young; and the invaluable assistance of Keith Greenwood, Dave Johnson, Mike Larby, Ashley Lawrence, Pat Randell and Aubrey Thomas in the drawing office, and of Eddie Murphy in the library.

On a more personal note, I would like to acknowledge the inspiration provided by Carlene Anderson, Laurie Anderson, Joan Armatrading, Devendra Banhart, Bach, Bjork, Jussi Björling, Billy Bragg, Bruch, Michael Buble, Maria Callas, The Calling, Laura Cantrell, Enrico Caruso, Johnny Cash, Eva Cassidy, Nick Cave and the Bad Seeds, Chopin, The Clancy Brothers, The Clash, Jimmy Cliff, Holly Cole, Copland, Elvis Costello, Jamie Cullum, Victoria de los Angeles, Donizetti, Elgar, Marianne Faithfull, Kathleen Ferrier, Aretha Franklin, Franz Ferdinand, Marvin Gaye, Bebel Gilberto, Philip Glass, Tito Gobbi, David Gray, Patty Griffin, Jimi Hendrix, Billie Holliday, Ives, Etta James, Karl Jenkins, Norah Jones, Joy Division, Keane, The Kings of Leon, Ladysmith Black Mambazo, Leighton, Mendelssohn, Robert Merrill, Van Morrisson, Mozart, The Muse, Willie Nelson, Sinead O'Connor, Orff, Pärt, Dolly Parton, The Pogues, Elvis Presley, Otis Redding, Steve Reich and Beryl Korot, Damien Rice, Paul Robeson, Rutter, Ravi Shankar, Frank Sinatra, Mindy Smith, Patti Smith, Bruce Springsteen, Kristi Stassinopoulou, Joss Stone, Stravinsky, Jesse Sykes and the Sweet Hereafter, Takemitsu, Tallis, Tavener, Tchaikovsky, Phil Thornton, Juliet Turner, Vaughan Williams, Verdi, Anne Sofie von Otter, Tom Waits, Walton, Gillian Welch, Paul Weller, Hayley Westenra, Brian Wilson, and all the artistes on my many Gregorian chant and Russian Orthodox Church music CDs.

Finally, I would like to acknowledge not only the technological assistance but also the love, support and, above all, forbearance of my wife Heather, and my sons Wynn and Gethin, no longer small. I really have finished now.

1 • Introduction

Palaeontology incorporates several, overlapping, areas, the principal of which are:

- Systematic palaeontology, or taxonomy, that is, the documentation, description and classification of new species and higher-level taxa of fossils
- Phylogenetics, that is, the analysis and interpretation of their evolutionary relationships
- Palaeobiology, that is, the documentation, analysis and interpretation of their relationships to evolving earth and life processes and environments, and their application to the elucidation thereof
- Biostratigraphy, that is, the documentation, analysis and interpretation of their ordered succession, their relationships to evolving earth and life history, and their application to the elucidation thereof.

The former two areas of study, systematics and phylogenetics, interfacing with life science, constitute pure palaeontology; the latter two, palaeobiology and biostratigraphy, interfacing not only with life science, but also with earth and environmental science and other disciplines, applied palaeontology (Fig. 1.1).

There is a logical progression from pure to applied aspects through the course of the book. The purer parts will perhaps be of most value to earth and life science undergraduate and postgraduate students and professionals in academia (and also to interested amateurs). The applied parts will be of value to professionals in industry, including not only applied palaeontologists, palaeobiologists and biostratigraphers, but also petroleum, minerals, mining and engineering geologists, environmental scientists and environmental archaeologists.

Chapter 2 ('Fossils and fossilisation') essentially deals with what fossils are, how they form, how they are collected, and how they are classified and identified, described and illustrated, and curated. It includes sections on fossils, on the fossilisation process (taphonomy), on collection of fossils, on classification and identification (systematic palaeontology or taxonomy), on description and illustration (palaeontography), and on preparation, conservation and curation (museology). The section on the fossilisation process contains sub-sections on the representativeness of the fossil record, and on exceptional preservation of fossil assemblages. The section on collection contains sub-sections on surface collection of macrofossils, and collection of surface samples for microfossil analysis, and contains a code of conduct. It also covers equipment and safety.

Chapter 3 ('Principal fossil groups') deals with the principal fossil groups. It includes sections on bacteria, plant-like protists or algae, animal-like protists or 'protozoans', plants, fungi, invertebrate and vertebrate animals, and trace fossils. The section on bacteria covers cyanobacteria and stromatolites. The section on plant-like protists or algae contains sub-sections on: dinoflagellates; silicoflagellates; diatoms; calcareous nannoplankton; calcareous algae; and Problematica (acritarchs and *Bolboforma*). The section on animal-like protists or 'protozoans' contains sub-sections on foraminiferans, radiolarians and calpionellids. The sub-section on foraminiferans covers, in some detail, '*Rhabdammina*' or 'flysch-type' or 'deep-water arenaceous foraminiferan (DWAF)' faunas, and larger benthic foraminiferans. The section on plants covers plant macrofossils, spores and pollen, and phytoliths. The coverage of plant macrofossils includes mosses and allied forms (bryophytes), club mosses, ferns, horsetails and allied forms (pteridophytes), seed-plants, that is, seed-ferns, tree ferns, conifers and allied forms (gymnosperms), and flowering plants (angiosperms). The section on fungi covers fungal spores and hyphae. The section on invertebrate animals contains sub-sections on sponges, archaeocyathans and stromatoporoids (poriferans), corals (cnidarians), brachiopods and bryozoans (lophophorates), bivalves (including, in some detail, rudists), gastropods, ammonoids, belemnites and tentaculitids (molluscs), trilobites, ostracods and

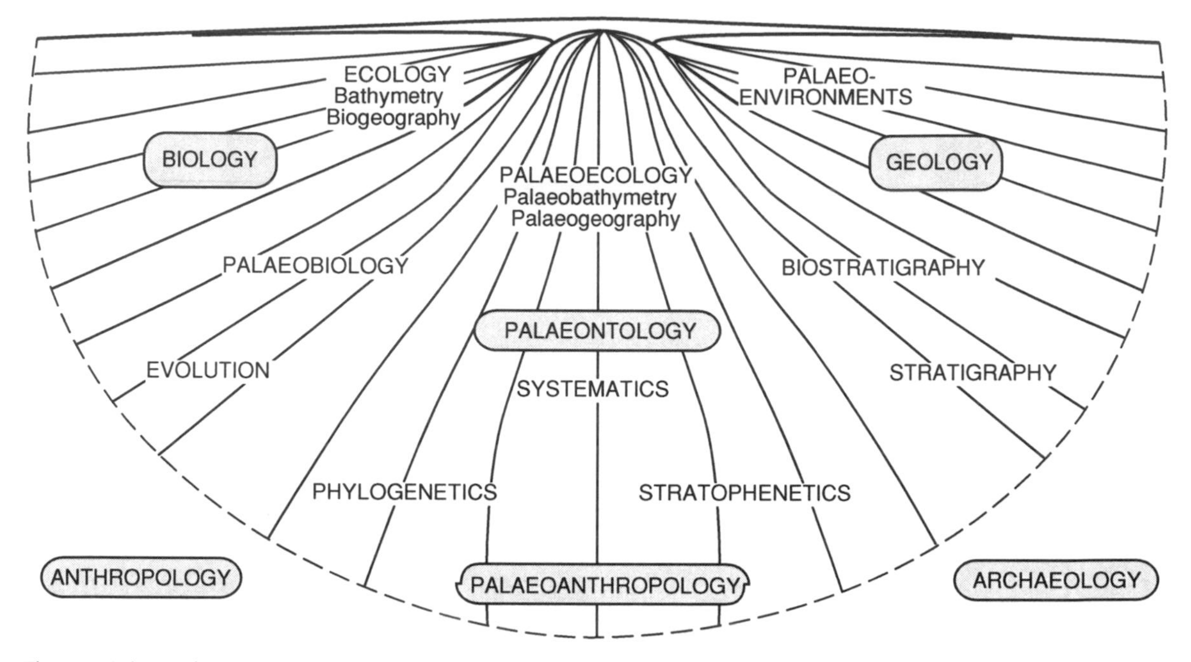

Fig. 1.1. Schematic representation of the relationship of palaeontology to biology and geology ('-ologies').

insects (arthropods), and crinoids and echinoids (echinoderms). It also includes sub-sections on 'ediacarians', 'small shelly fossils', tentaculitids, graptolites and chitinozoans. The section on vertebrate animals contains sub-sections on fish; and on amphibians, reptiles and birds, and mammals, collectively known as tetrapods. The sub-section on fish covers conodonts, ichthyoliths and otoliths, collectively colloquially known as 'fish bits'. Emphasis has been placed on the palaeobiological and biostratigraphic significance of the various groups, and their proven and potential applications in the interpretation of earth and life history and processes.

Chapter 4 ('Palaeobiology') deals with applications of fossils in the interpretation of earth and life processes (excluding evolution), and environments. It includes sections on palaeoecology or palaeoenvironmental interpretation, on discrimination of non-marine and marine environments, on palaeobathymetry, on palaeobiogeography, on palaeoclimatology, and on quantitative and other techniques. The section on palaeoenvironmental interpretation contains sub-sections on palaeoenvironmental interpretation on the basis of analogy and palaeoenvironmental interpretation on the basis of functional morphology. The sub-section on functional morphology covers life strategy, life position and feeding strategy. The section on palaeobathymetry contains sub-sections on marginal marine environments, shallow marine environments and deep marine environments. The sub-section on shallow marine environments covers shallow marine clastic and carbonate environments. The sub-section on deep marine environments covers: oxygen minimum zone sub-environments; submarine fan sub-environments; and hydrothermal vent and cold (hydrocarbon) seep environments, including, in some detail, observations on benthic foraminiferans associated with hydrocarbon seeps. The sections on palaeobiogeography and palaeoclimatology each contain sections on the Palaeozoic, the Mesozoic and the Cenozoic. The section on quantitative and other techniques contains sub-sections on palaeobathymetry, palaeobiogeography and palaeoclimatology.

Chapter 5 ('Key biological events in earth history') deals with the generalities and specifics of the evolutionary and extinction events and trends that have controlled past and present, and will control future, biodiversity on Earth. It includes general sections on evolution and extinction events, and on Proterozoic, Palaeozoic, Mesozoic and Cenozoic events. The section on evolution and extinction contains sub-sections on evolution and extinction. The sub-section on evolution covers evolutionary events and evolutionary biotas (including the Cambrian, Palaeozoic and Mesozoic–Cenozoic, or modern, evolutionary biotas). It also covers, in some detail, foraminiferal diversity trends through time. The sub-section on extinction covers mass extinction events (including causes, periodicity, selectivity and recovery). The section on the Proterozoic contains sub-sections on the origin of life (prokaryotes), on the evolution of complex life (eukaryotes), and of multicellularity, and on the Late Precambrian mass extinction. The section on the Palaeozoic contains sub-sections on the Cambrian evolutionary diversification, the evolution of reefs, the Early and Late Cambrian mass extinctions, the Ordovician evolutionary diversification, the evolution of vertebrates, the End-Ordovician mass extinction, the evolution of life on land, the Late Devonian mass extinction, the evolution of forests, and of flight, and the End-Permian mass extinction. The section on the Mesozoic contains sub-sections on the Mesozoic evolutionary diversification, the End-Triassic mass extinction, the evolution of flowering plants, and the Late and End-Cretaceous mass extinctions. The section on the Cenozoic contains sub-sections on the End-Palaeocene and End-Eocene mass extinctions, the evolution of grasses and grassland animals, the evolution of humans, and the Pleistocene and Holocene mass extinctions. It also covers, in some detail, firstly, new evidence for land mammal dispersal across the northern North Atlantic in the Early Eocene; and, secondly, aspects of the palaeogeography and palaeoclimate of the Oligocene–Holocene of the Old World, and consequences for land mammal evolution and dispersal – including that of *Homo sapiens*.

Chapter 6 ('Biostratigraphy and sequence stratigraphy') deals with applications of fossils in the interpretation of earth and life history. It includes sections on biostratigraphy, on biostratigraphic technologies, on allied disciplines, on stratigraphic timescales and on sequence stratigraphy. The section on biostratigraphy contains sub-sections on the Proterozoic, the Palaeozoic, the Mesozoic and the Cenozoic. The section on biostratigraphic technologies contains sub-sections on graphic correlation and on ranking and scaling. The section on allied disciplines contains sub-sections on chemostratigraphy, cyclostratigraphy, heavy minerals, magnetostratigraphy, radiometric dating and Quaternary dating methods. The sub-section on chemostratigraphy covers carbon, oxygen and strontium isotope stratigraphy, and trace element stratigraphy. The section on stratigraphic timescales covers global stratigraphic (boundary) sections and points (GSSPs). The section on sequence stratigraphy contains subsections on general and clastic sequence stratigraphy, carbonate sequence stratigraphy, mixed clastic–carbonate and carbonate–evaporite sequence stratigraphy, and seismic facies analysis. It also covers, in some detail, palaeontological characterisation of systems tracts.

Chapter 7 ('Case histories of applications of palaeontology') deals with how our knowledge of fossils and earth and life history is applied in industry and elsewhere. It includes sections on petroleum geology, mineral exploration and exploitation, coal mining, engineering geology, environmental science and archaeology. The section on petroleum geology contains sub-sections on the principles and practice of petroleum geology, and on applications of biostratigraphy and palaeobiology in petroleum exploration and in reservoir exploitation, each with case histories. The sub-section on the principles and practice of petroleum geology covers: play components (petroleum source-rocks and systems, reservoir-rocks and cap-rocks (seals) and traps), and stratigraphic control on their distribution; petroleum exploration; and reservoir exploitation (drilling, petrophysical logging and testing). It also covers, in some detail: palaeontological inputs into petroleum systems analysis; micropalaeontological characterisation of mudstone cap-rocks; and palaeontology and health, safety and environmental issues in the petroleum industry. The sub-section on applications of biostratigraphy and palaeobiology in petroleum exploration covers chronostratigraphy and palaeoenvironmental interpretation, and operational biostratigraphy, with accompanying case histories selected from the central and northern North Sea and the Middle East. The sub-section on applications in reservoir exploitation covers integrated reservoir description, and biosteering, with accompanying case histories selected from Cusiana field in Colombia, the Andrew field in the North Sea, the Sajaa field in Sharjah in the United Arab Emirates and the Valhall field in the North Sea. The section on mineral exploration and exploitation includes a sub-section on case histories from the La Troya mine in Spain (mineral exploration), and the Pitstone and East Grimstead quarries in the UK (mineral exploitation). The section on engineering geology includes a sub-section on case histories from the Channel tunnel, Thames barrage and 'Project Orwell' site investigations in the UK. The section on environmental science contains sub-sections on environmental impact assessment and environmental monitoring. The section on archaeology contains sub-sections on archaeostratigraphy and environmental archaeology, and case histories. The sub-section on environmental archaeology covers, in some detail, the palaeoenvironmental interpretation of the Pleistocene–Holocene of the British Isles, as determined by proxy benthic foraminiferal distribution data. The sub-section on case histories covers Westbury Cave, Somerset (Palaeolithic), Boxgrove, Sussex (Palaeolithic), Massawa, Eritrea (Palaeolithic), Goats Hole, Paviland, Gower (Palaeolithic), 'Doggerland', North Sea (Mesolithic), Mount Sandel, Coleraine, Co. Derry, Northern Ireland (Mesolithic), the Black Sea (Mesolithic/Bronze Age), the Tyrolean Alps (Neolithic), Littleton, Co. Tipperary, Ireland (Neolithic–Medieval), Skara Brae, Orkney (Neolithic) and the City of London (Medieval).

2 • Fossils and fossilisation

This chapter essentially deals with what fossils are, how they form, how they are collected, and how they are classified and identified, described and illustrated, and curated.

2.1 Fossils

The word 'fossil' derives from the Latin *fodere*, meaning 'to dig'. Its usage is in reference to any and all physical remains or other direct physical indications of past life. The physical remains of past life include not only what are often termed 'body fossils', that is, 'hard parts' of skeletons such as bones (Denys, in Briggs and Crowther, 2001) and shells (Meldahl, in Briggs and Crowther, 2001), but also, under exceptional circumstances, 'soft parts' of animals (Stankiewicz and Briggs, in Briggs and Crowther, 2001), parts of plants (van Bergen, in Briggs and Crowther, 2001), bacteria (Liebig, in Briggs and Crowther, 2001), and biomolecules such as fats, proteins and DNA (Briggs *et al.*, in Erwin and Wing, 2000; Collins and Gernaey, in Briggs and Crowther, 2001; Evershed and Lockheart, in Briggs and Crowther, 2001; Poinar and Paabo, in Briggs and Crowther, 2001; Jones, 2001). Other direct physical indications of past life include 'trace fossils' (see separate section in Chapter 3).

2.2 The fossilisation process (taphonomy)

The fossilisation process, whereby living individual organisms or communities are transformed after death into fossils or fossil assemblages, is termed taphonomy (Fursich, in Briggs and Crowther, 1990; Allison and Briggs, 1991; Donovan, 1991; Martin, 1999; Behrensmeyer *et al.*, in Erwin and Wing, 2000; Wilson, in Briggs and Crowther, 2001; Holz and Simoes, in Koutsoukos, 2005). In fact, a range of biological, physical and chemical processes is involved, from decomposition (Allison, in Briggs and Crowther, 2001), through disarticulation, fragmentation and transportation (Anderson, in Briggs and Crowther, 2001), to compaction, thermal alteration and dissolution on burial (Briggs, in Briggs and Crowther, 1990; Tucker, in Briggs and Crowther, 1990; Jones, 1996; McNeil *et al.*, 1996), and reworking. For a living organism to become fossilised requires that it can withstand the combined effects of these processes. In general, its chances of becoming fossilised are enhanced if it becomes buried immediately after death (and/or is otherwise protected from decomposition, perhaps by anoxic bottom conditions), and/or is and remains resistant to the destructive physical and chemical effects of the so-called diagenetic processes that enter into operation after burial (see also Sub-section 2.2.2). The rate of burial relative to that of destructive processes, and hence the likelihood of fossilisation, varies considerably from place to place, and indeed from time to time (Brett and Speyer, in Briggs and Crowther, 1990). However, fossilisation potential is generally best in marine environments. The processes by which living plants and animals become fossilised in terrestrial environments can be extremely complicated, and remain comparatively poorly understood. Preservation of fossils is typically in the form of mineralised hard parts such as bones and shells, or casts or moulds of shells, in the case of marine and terrestrial animals; and in the form of flattened two-dimensional impressions in the case of plants (but see also Sub-section 2.2.2).

Importantly, the palaeobiological and biostratigraphic usefulness of (micro)fossils can be impaired by alteration to the specific composition of the living community by natural taphonomic processes such as – selective – transportation, and diagenetic thermal alteration, dissolution and destruction (Jones, 1996; McNeil *et al.*, 1996; Murray and Alve, in Hart *et al.*, 2000; Reinhardt *et al.*, 2001; Ruiz *et al.*, 2003).

Readers interested specifically in the taphonomy of plants are referred to Scott, in Briggs and Crowther (1990), Spicer, in Briggs and Crowther (1990) and Gastaldo, in Briggs and Crowther (2001). Those interested specifically in the complex taphonomy of terrestrial vertebrates are referred to Andrews (1990), Behrensmeyer *et al.* (1992), Behrensmeyer and Kidwell (1993), Lynam (1994), Behrensmeyer, in Briggs and Crowther (2001) and Trueman *et al.* (2003).

2.2.1 *Representativeness of the fossil record*

In view of the destructive nature of the fossilisation and preservation process (see above), legitimate questions have been asked as to the likely completeness and representativeness of the fossil record, and as to the meaningfulness of attempts to analyse same (Holland, in Erwin and Wing, 2000; Foote, in Briggs and Crowther, 2001; Kidwell, in Briggs and Crowther, 2001; Smith, in Briggs and Crowther, 2001). It may be equally legitimately argued that sufficient exceptionally preserved fossil assemblages are now known, providing 'windows on the world' at every level, that this is no longer such a serious issue.

2.2.2 *Exceptional preservation of fossil assemblages*

As intimated above, a number of examples are now known from the fossil record of exceptionally preserved fossil assemblages or 'fossil-Lagerstatten' (Briggs, in Briggs and Crowther, 2001; Bottjer *et al.*, in Bottjer *et al.*, 2002; Selden and Nudds, 2004). The word 'Lagerstatte' derives from the German mining tradition, and its usage is in reference to deposits containing sufficient constituents of economic interest to warrant exploitation (Seilacher, in Briggs and Crowther, 1990). The nature of the preservation in fossil-Lagerstatten is exceptional in terms of quality and/or quantity. Exceptional quality preservation, or conservation – in 'Konservat-Lagerstatten' in marine environments typically arises through the process of obrution, whereby living communities become smothered and buried extremely rapidly, perhaps by an effectively instantaneous event (Brett, in Briggs and Crowther, 1990). The Burgess shale, the Hunsruck slate, and the Lithographic limestone are all interpreted examples of obrutionary stagnation deposits, in which preservation was enhanced by the protection from decomposition provided by anoxic bottom conditions (see also below). These examples are all characterised by exceptional quality soft-part as well as hard-part preservation, albeit only in two-and-a-half dimensions. Exceptional quality preservation in terrestrial environments typically arises through deep-freezing, pickling, or desiccation (Seilacher, in Briggs and Crowther, 1990). It can also arise through encapsulation in amber (Martinez-Delclos *et al.*, 2004; Poinar and Poinar, 2004), as in the case of Baltic amber (Schluter, in Briggs and Crowther, 1990; Janzen, 2002; Selden and Nudds, 2004) and Dominican amber (Poinar and Poinar, 1999; Poinar, in Briggs and Crowther, 2001). Exceptional quantity preservation – in 'Konzentrat-Lagerstatten' – typically arises through concentration by physical processes (Seilacher, in Briggs and Crowther, 1990). Concentration can also arise in so-called 'traps', such as the Pleistocene Rancho La Brea tar pit in Los Angeles in California (Selden and Nudds, 2004).

The most important of the exceptionally preserved fossil assemblages or fossil-Lagerstatten are those that provide insights into key biological events in earth history (see below; see also Chapter 5).

Proterozoic

Exceptionally preserved 'ediacarians' are known from the Late Precambrian of the Ediacara Hills in South Australia and elsewhere (see Sub-section 3.6.1). The 'ediacarian' biotas provide important insights into the evolution of multicellularity.

Palaeozoic

Exceptionally preserved Cambrian biotas are known from, for example, the Early Cambrian, Atdabanian/Botoman, *Nevadella* Zone, of Sirius Passet in Greenland (Conway Morris, 1998), and the Early Cambrian, Qiongzhusian, *Eoredlichia–Wutingaspis* Zone, Ya'anshan member of the Heinlinpu formation of Chengjiang in eastern Yunnan Province in southwest China, part of the South China micro-plate (Hou Xian-Guang and Bergstrom, 1997; Bergstrom, in Briggs and Crowther, 2001; Hagadorn, in Bottjer *et al.*, 2002; Hou Xian-Guang *et al.*, 2004). Further exceptionally preserved faunas are known from the Middle Cambrian, Albertan, Burgess shale of British Columbia in Canada (Conway Morris, in Briggs and Crowther, 1990; Gould, 1990; Briggs *et al.*, 1994; Conway Morris, 1998; Hagadorn, in Bottjer *et al.*, 2002; Selden and Nudds, 2004). These faunas provide important insights into the 'Cambrian evolutionary diversification', and information on rare groups. Although it is not as prolific or as famous as the Burgess shale biota, the Chengjiang biota is perhaps more important, as it is older (although not as old as the – comparatively less well-known – Sirius

Passet biota). The Chengjiang biota is characterised by hard-bodied calcareous algae, 'small shelly fossils', poriferans, cnidarians, brachiopods, phoronids, arthropods, vetulicolids and chordates, occasionally with their soft parts preserved, and, significantly, by soft-bodied nematomorphs, priapulids, chaetognathans and enigmatics.

Exceptionally preserved eurypterids and conodont parent animals are known from the 'Late' Ordovician Soom shale member of the Cedarberg formation of the area around Keurbos in southwestern Cape Province in South Africa (Aldridge *et al.*, in Briggs and Crowther, 2001; Selden and Nudds, 2004). The Soom shale fauna provides important insights into the 'Ordovician evolutionary diversification' and into the evolution of vertebrates.

Exceptional preserved conodont assemblages and articulated remains attributed to a polychaete worm are known from the 'Early' Silurian Birkhill shales formation of southern Scotland (Wilby *et al.*, 2003). Exceptionally preserved ostracods and starfish are known from serial thin-sections of nodules from the Herefordshire Lagerstatte from the heart of England (Sutton *et al.*, 2003). The ostracods are preserved with their soft parts intact, including their penes, as pruriently reported in the popular press in the UK! These Silurian faunas provide further important insights into the evolution of vertebrates, and/or information on rare groups.

Exceptionally preserved Devonian biotas are known from the Early Devonian Hunsruck slate of Hunsruck in the Rheinisches Schiefegebirge in Germany (Bartels *et al.*, 1998; Raiswell *et al.*, in Briggs and Crowther, 2001; Etter, in Bottjer *et al.*, 2002; Selden and Nudds, 2004), from the Early Devonian Rhynie chert of Rhynie in Scotland (Trewin, in Briggs and Crowther, 2001; Rice *et al.*, 2002; Anderson and Trewin, 2003; Kelman *et al.*, 2003; Trewin *et al.*, 2003; Selden and Nudds, 2004). The Rhynie biota of primitive land plants and rare invertebrate animals such as arthropods provides important insights into the evolution of life on land.

Exceptionally preserved Carboniferous biotas are known from, for example, the Late Carboniferous, Pennsylvanian of Mazon Creek in Illinois (Nitecki, 1979; Baird, in Briggs and Crowther, 1990; Schellenberg, in Bottjer *et al.*, 2002; Selden and Nudds, 2004). The Mazon Creek biota of land plants, invertebrate animals and primitive tetrapod vertebrates provides further important insights into the evolution of life on land, and into the evolution of forests.

Mesozoic

Exceptionally preserved Triassic faunas are known from, for example, the Middle Triassic Grès a *Voltzia* of France and the Middle Triassic of Monte San Giorgio in Switzerland (Etter, in Bottjer *at al.*, 2002; Selden and Nudds, 2004). The Grès a *Voltzia* and Monte San Giorgio faunas provide important insights into the 'Mesozoic evolutionary diversification' or 'Mesozoic marine revolution' in marginal and fully marine environments, respectively.

Exceptionally preserved Jurassic biotas are known from, for example, the Late Jurassic Morrison formation of the Mid-West of the USA (Selden and Nudds, 2004; Turner and Peterson, 2004), and the Late Jurassic, Tithonian lithographic limestone or 'Plattenkalk' of Solnhofen in Bavaria in Germany (Barthel, 1978; Barthel *et al.*, 1990; Viohl, in Briggs and Crowther, 1990; Etter, in Bottjer *et al.*, 2002; Selden and Nudds, 2004). The Morrison formation biota of land plants and animals, including spectacular dinosaurs, provides important insights into the 'Mesozoic evolutionary diversification' on land. The Solnhofen biota provides further insights into the 'Mesozoic evolutionary diversification' on the margins of the land, and important insights into the evolution of flight among reptiles, and in the early bird *Archaeopteryx*.

Cenozoic

Exceptionally preserved Cenozoic biotas are known from, for example, the Middle Eocene of Messel near Frankfurt in Germany (Franzen, in Briggs and Crowther, 1990; Schaal and Ziegler, 1992; Franzen, in Gunnell, 2001; Selden and Nudds, 2004). The Messel biota provides important insights into, among other things, the evolution of flight among mammals (bats).

2.3 Collection of fossils

Collection of fossils is required to constrain surface and subsurface geological mapping and correlation, and to build up collections for reference and for academic research purposes (Croucher and Woolley, 1982; Tucker, 1996; Goldring, 1999; Green, 2001). The roles of the palaeontologist in the field are to provide age assignments and palaeoenvironmental interpretations based on fossils, and to ensure that only appropriate ages and facies of sediments are sampled for palaeontological analysis, and that the stratigraphic and geographic location of every sample is recorded. Incidentally, it is worth noting that field age assignments can be made by means not only of macrofossils but also of macroscopically visible microfossils, such as the larger benthic foraminiferans or LBFs of the Permo-Carboniferous and 'Middle' Cretaceous through 'Middle' Tertiary of the Tethyan realm (author's unpublished observations).

Surface collection of macrofossils and collection of surface samples for microfossil analysis are discussed in turn below. Collection of subsurface samples for microfossil analysis is discussed in Section 7.1.

Code of conduct

Field collection of fossils has to be responsible and sustainable, so as to conserve or preserve what is a finite natural resource for future generations, preferably in place, or at least in publicly accessible institutions such as museums and universities (King, in Bassett *et al.*, 2001; Weighell, in Bassett *et al.*, 2001). Occasionally, though, it is necessary to undertake more drastic salvage operations, as when active quarries are becoming worked out (Thompson, in Bassett *et al.*, 2001). In the UK, fossil collecting in designated Sites of Special Scientific Interest or SSSIs is restricted to that for genuine and justifiable scientific purposes only, otherwise it would constitute an 'operation likely to damage' (OLD) the resource (King and Larwood, in Bassett *et al.*, 2001; MacFadyen, in Bassett *et al.*, 2001). Elsewhere in the UK, it is restricted by recommendation or voluntary code of conduct (Edmonds, in Bassett *et al.*, 2001; Munt, in Bassett *et al.*, 2001; Reid and Larwood, in Basssett *et al.*, 2001; Simpson, in Bassett *et al.*, 2001). In Germany, fossil collecting in so-called 'geotopes' ('parts of the geosphere clearly distinguishable from their surroundings in a geoscientific fashion') is restricted by nature conservation and by national monument protection legislation (Wuttke, in Bassett *et al.*, 2001).

The Geologists' Association 'geological fieldwork code' of conduct recommends the following actions (Robinson, in Bassett *et al.*, 2001): Firstly, 'Students should be encouraged to observe and record, and *not hammer indiscriminately*.' Secondly, 'Keep collecting to a minimum. Avoid removing *in situ* fossils, rocks and minerals unless they are *genuinely* needed for serious study.' Thirdly, 'For teaching purposes, the use of replicas is recommended. The collecting of actual specimens should be restricted to those localities where there is a plentiful supply, or to scree, fallen blocks and waste tips.' Fourthly, 'Never collect from walls or buildings. Take care not to undermine fences, walls, bridges or other structures.'

Equipment

The – more-or-less – technical equipment required or useful for the palaeontologist in the field is as follows: a global positioning satellite (GPS) system; a topographic map or aerial photographs or satellite images of the area of interest; a compass/clinometer; an altimeter; a range-finder; a pair of binoculars; a digital camera or video; a portable laptop computer on which to upload digital images; a portable solar panel with which to recharge electronic equipment; a measuring tape; a ×10 to ×20 magnifying glass or a pocket microscope; a bottle of dilute hydrochloric acid to test for carbonates; a sledge or 4-lb (2-kg) lump hammer; a 2-lb (1-kg) hammer; a set of chisels; a set of dental tools; a pickaxe; an entrenching tool; an auger; a supply of sample bags; a supply of indelible pens for labelling them; a waterproof notebook and a supply of pencils for recording observations.

Safety

Safety equipment should include clothing and footwear appropriate to the season and terrain; suncream; personal protection equipment, including a hard hat or climbing helmet, and goggles for use when hammering; sufficient food and water to see out an emergency, such as becoming benighted; fire-lighting equipment; a survival blanket; a torch (flashlight); a whistle, for attracting attention; and a first-aid kit.

Recommended safety procedures are as follows (from Goldring, 1999):

> Listen to the daily weather forecast (including wind direction), which may determine where it is prudent to work. Take account of the time and height of tides when planning coastal work. Write down each day your approximate route, working

area and time of return, and leave it for others to see. In worsening conditions, do not hesitate to turn back if it is still safe to do so. If you get lost, disabled, benighted, or cut off by the tide, stay where you are until conditions improve or until you are found. Supposed short cuts can be lethal.

Distress codes are as follows (also from Goldring, 1999):

On mountains: 6 long blasts, flashes, shouts or waves in succession, repeat(ed) at minute intervals. At sea: 3 short then 3 long, then 3 short blasts or flashes [Morse code for SOS],repeat(ed). Rescuers reply with 3 blasts or flashes repeated at minute intervals.

2.3.1 *Surface collection of macrofossils*

Macrofossils are generally sufficiently large to be seen in surface outcrops or in float. However, careful observation may be required in order that they may actually be seen. The angle of the sun is important in this regard. Early mornings and late afternoons, when the sun is low and the shadows long, are often the best times for searching for fossils. (Similarly, tilting slabs can cast shadows that throw previously unseen and unsuspected fossils into unexpected relief.) Intensive searching can commence once extensive searching has revealed a fossiliferous horizon. Hard rocks can be broken open using a lump hammer, or split along bedding planes using a hammer and chisel, in both cases carefully, so as not to damage specimens. Contained fossils are typically harder than containing rocks, and can be readily extracted. In the event that the fossils are softer than the rock, they can nonetheless still be extracted, carefully, using dental tools, a process often started in the field and finished in the laboratory (see Subsection 2.6.1). Especially fragile specimens such as long bones may need to be protected against damage during excavation – and/or transportation – by first being 'consolidated', that is, wrapped in hessian or jute (burlap) soaked in plaster of Paris and allowed to dry (Longbottom and Milner, in Whybrow, 2000; Milner, in Whybrow, 2000). Collecting fossils from certain hard rocks, such as massive limestones, can be effectively impossible. Specimens are probably better photographed than removed from these rocks. Soft rocks can be trenched and samples removed for laboratory preparation (see Subsection 2.6.1).

2.3.2 *Collection of surface samples for microfossil analysis*

The overall objectives of the fieldwork should be considered when determining the appropriate strategy for sampling. For example, if the objective is reconnaissance mapping, spot sampling might be all that is required, whereas if the objective is detailed logging, targeted or close systematic sampling would be required. As a general comment, the biostratigraphic or palaeoenvironmental resolution of the analytical results will depend as much on the sampling density as on the fossils themselves. Partly on account of this, and partly on account of the logistical effort and financial cost of mobilising field parties, it is always advisable to collect what might be thought of as too many rather than too few samples. However, any restrictions on access or sampling imposed by the landowner should be respected, as should the code of conduct (see above). The particular microfossil groups to be expected in the ages and environments of the rocks expected to be encountered should also be considered, together with any sampling requirements specific to those groups (see below).

Size of sample

The size of sample required depends to an extent on the fossil group targeted (see below). For example, for microfossils, it varies from approximately 1 cm^3 in the case of calcareous nannofossils to up to several kilograms in the case of conodonts.

Lithology

The lithology of sample required also depends to an extent on the fossil group targeted (see below). However, some generalisations can be made, as follows. The lithologies most likely to be productive for microfossils are fine-grained clastics such as shales and mudstones, especially where calcareous, and limestones such as lime mudstones, wackestones and packstones. Those least likely to be productive are coarse-grained clastics such as sandstones, limestones such as grainstones, rudstones and framestones, and altered dolomites. Those most likely to be unproductive are coarse-grained continental clastics, especially 'red beds', and evaporites. Weathered rocks of any lithology are unlikely to be productive for organic-walled microfossils, on account of the likelihood of oxidation, which, it is worth pointing out, can occur not only at the surface but also in the subsurface, for example, at the junction between permeable and impermeable beds or along joints. They are also less likely than unweathered rocks to be productive for calcareous microfossils, on account of the likelihood of leaching. Where weathering pervades some distance into the rock, unweathered samples should be obtained by digging, augering or trenching, using appropriate tools. Unweathered rocks can be recognised by their generally blocky rather than slabby, platy, fissile or earthy texture. If it is simply not possible to access unweathered rocks, because the effects of weathering have pervaded so deep, it is worth sampling any calcareous concretions that might be present, since experience has shown that these can be productive for calcareous microfossils. Thermally altered rocks of any lithology are less likely than unaltered rocks to be productive, particularly for organic-walled microfossils. The effects of thermal alteration can be either local or regional.

Sampling for specific microfossil groups

Calcareous microfossils are locally so abundant in rocks of the appropriate age-range and facies as to be rock-forming, as in the case of '*Globigerina*' or planktonic foraminiferal oozes. They are common in essentially all marine limestones and marls, especially in finer-grained ones, and even in indurated ones, which cannot be easily disaggregated and which are therefore best studied in thin-section, although they may be difficult to identify in altered dolomites. Calcareous microfossils are also common in essentially all marine calcareous mudstones, and, in the case of arenaceous foraminiferans, in non-calcareous mudstones. Even non-marine, lacustrine calcareous mudstones can contain calcareous microfossils, in the form of ostracods. The contained ostracods may be sufficiently large to be discernible on bedding planes with the aid of a hand-lens. Samples are best collected by chiselling along bedding planes rather than hammering, so as to avoid damage to specimens. Sandstones are generally poorly productive in terms of *in situ* calcareous microfossils. One large sample bag is generally sufficient to ensure recovery of calcareous microfossils, especially if the material is fresh and unweathered. It is invariably worth the effort ensuring that this is so!

Siliceous microfossils are locally so abundant in rocks of the appropriate age-range and facies as to be rock-forming, as in the case of diatomites, radiolarian cherts or radiolarites, and spiculites. Diatomites often resemble volcanic tuffs when weathered. Diatoms can be common not only in diatomites but also in siliceous mudstones, such as those of the Miocene of California, or in so-called 'opokas', such as those of the Miocene of Sakhalin. Radiolarians can be common not only in radiolarites but also in shales and in calcareous rocks of marine origin. Unfortunately, the silica of which diatoms is composed is an unstable variety (Opal-A), which converts to a more stable variety (cristobalite or Opal-CT) under the sort of pressure and temperature conditions encountered at burial depths of the order of 2 km, often resulting in the destruction of diagnostic morphological features. Even under these conditions, though, diatoms can be preserved, with their diagnostic morphological features intact, through recrystallisation, replacement – typically by pyrite or calcite – or entombment in concretions. Radiolarians are generally more robust, and more resistant to diagenetic alteration.

Phosphatic microfossils such as conodonts are at least locally common in most marine rocks of the appropriate age-range and facies. They are perhaps most common in limestones, especially bioclastic wackestones or packstones. The occurrence of macrofossils such as crinoids or brachiopods in a rock is an encouraging sign that it will be productive for conodonts. Cherts are also sometimes productive for conodonts on treatment with hydrofluoric acid. Conodonts are generally resistant to chemical attack, and also to diagenetic dolomitisation and thermal alteration. They can occasionally be seen on bedding planes with the aid of a hand-lens. They can be concentrated in lag deposits such as bone beds. The abundance of conodonts varies through time, such that sample sizes need to be adjusted accordingly. Ordovician faunas from the mid-continent of the USA contain abundant specimens, and samples need only be 0.5 kg. In contrast, Devonian faunas contain only rare specimens, and samples need to be several kilograms. The facies preference of conodonts also varies through time. Older conodonts are more common in shallower-water, younger ones in deeper-water deposits.

Organic-walled microfossils or *palynomorphs* are present to common in most clastic rocks of the appropriate age-range and facies that contain clay-sized particles and that have not been subject to excessive oxidation or thermal alteration (carbonates are generally poorly productive). Organic-walled microfossils can be extremely abundant, with up to 100 000 grains per gram present in some carbonaceous deposits, such that small samples are generally sufficient. Even individual conglomerate clasts can be analysed, in order to provide an indication of provenance of reworking. Organic-walled microfossils are prone to reworking on account of their small size and resistance to chemical attack. They are also susceptible to oxidation during and after deposition, at outcrop, in the laboratory, and in the sample storage facility. Oxidation is a particular problem in past and present arid environments (and has been observed to pervade to depths of up to 45 m, below the Permo-Carboniferous unconformity in the North Sea). Organic-walled microfossils are also susceptible to thermal alteration, and are unlikely be preserved in highly altered rocks.

Calcareous nannofossils are locally so abundant in marine rocks of the appropriate age-range and facies as to be rock-forming, as in the case of calcareous nannofossil oozes and chalks. They are common in essentially all marine limestones, marls and calcareous mudstones. Reefal sediments, though, should be avoided, as should recrystallised limestones and dolomites. Marine red beds, such as the *ammonitico rosso* of the circum-Mediterranean, should also be avoided, as they are likely to have been deposited below the calcite compensation depth, and to have had their original calcareous component destroyed by dissolution. Calcareous nannofossils are extremely small, so only about 1 cm^3 of sample is generally needed.

Contamination issues

Care must be taken during sampling, transportation and preparation to avoid contamination. During sampling, there are two main sources of contamination, namely reworking and stratigraphic admixture. Stratigraphic admixture arises when younger material is introduced among older. It can arise through natural agencies such as flowing or sloughing, or the filling of desiccation cracks, fissures or caves, or the development of a soil profile, or through ground-water percolation through porous rocks, or even airborne dust fallout. It can also arise artificially, for example, through the failure to clean hammers or other tools, or the use of cloth rather than plastic sample bags.

2.4 Classification and identification of fossils (systematic palaeontology or taxonomy)

2.4.1 Classification of fossils

Classification of organisms and of fossils by systematic biologists and palaeontologists is essentially based on a combination of biology, morphology, and evolutionary relationships or phylogeny (Fig. 2.1) (Futuyma, 1998; Purves *et al.*, 2004). Evolutionary relationships are in turn inferred from empirical observations on the fossil record, or stratophenetic analysis (Gingerich, in Briggs and Crowther, 1990), or from cladistic analysis (see below).

Classification based essentially solely on morphology is restricted to disarticulated plant and animal fossil groups such as leaves, spores and pollen, conodonts, and ichthyoliths (Aldridge, in Briggs and Crowther, 1990; Thomas, in Briggs and Crowther, 1990). The relationships between these groups, which are obviously all parts of organisms, and the actual parent organisms themselves are for the most part unknown, as they tend to occur in isolation from one another. In the absence of knowledge as to their actual relationships, classification of these groups is based on morphological form. Trace fossils are also classified according to 'form taxonomy' (Kelley, in Briggs and Crowther, 1990). Additional notable examples include evolutionary lineages arbitrarily subdivided into artificial 'species' on the basis of morphometrics (Harper and Owen, in Harper, 1999; Hughes, in Briggs and Crowther, 2001).

Cladistic analysis

Cladistic analysis is a methodology for phylogenetic reconstruction based on the number of morphological and/or molecular characters organisms have in

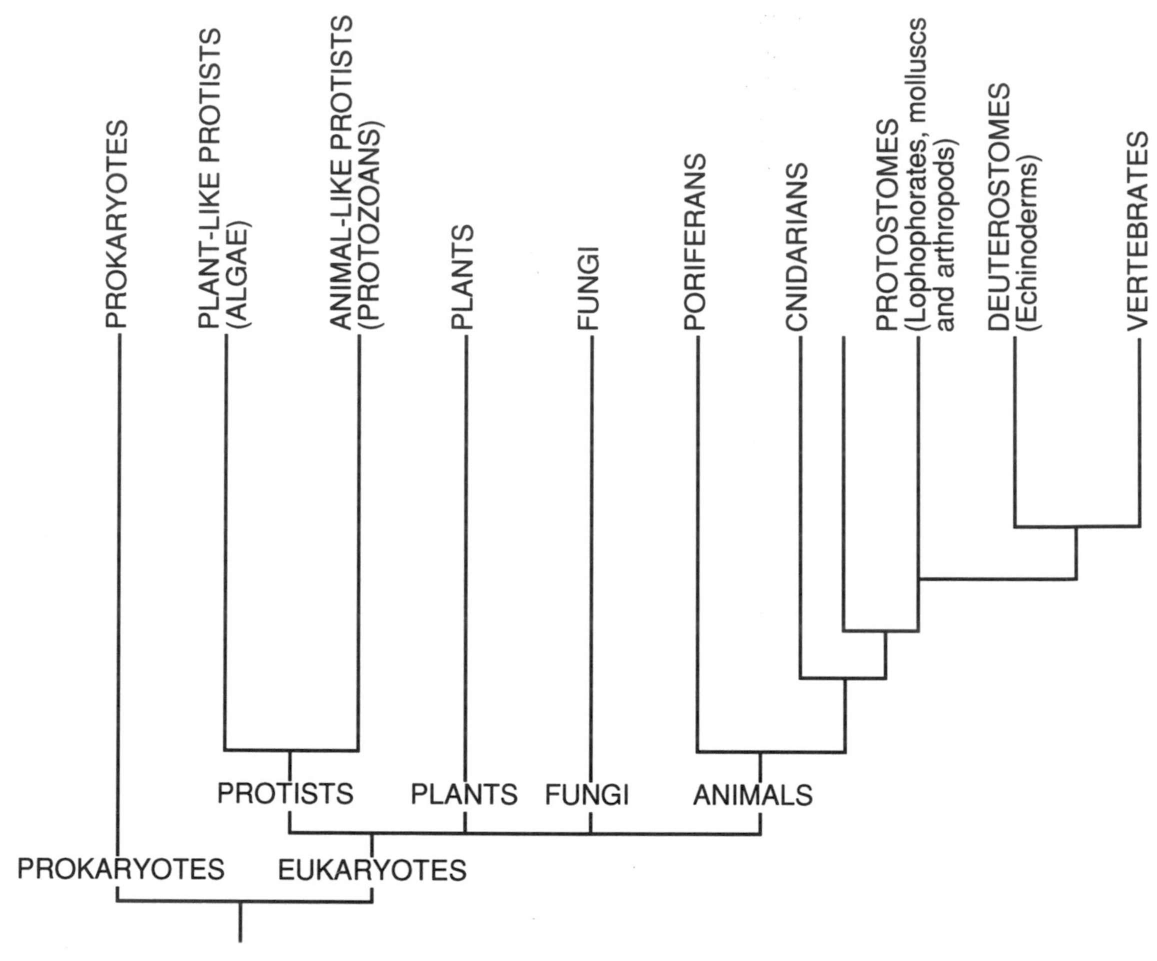

Fig. 2.1. Interpreted evolutionary relationships among selected high-level divisions of organisms ('cladogram'). The divisions are based on various criteria (see text). For example, the division between the prokaryotes and the eukaryotes is on the basis of the presence or absence of a nucleus.

common with one another (Fig. 2.1) (Hennig, 1979; Forey, in Briggs and Crowther, 1990; Forey *et al.*, 1992; Smith, 1994; Kitching *et al.*, 1998; Catzeflis, in Briggs and Crowther, 2001; Forey and Fortey, in Briggs and Crowther, 2001; Huenselbeck, in Briggs and Crowther, 2001; Norell, in Briggs and Crowther, 2001; Wilkinson, in Briggs and Crowther, 2001; Skelton and Smith, 2002). As such, in principle, it is objective and rigorous. In practice, subjectivity and bias can be introduced into cladistic analysis by the inappropriate selection and – conscious or unconscious – weighting of characters, such that an uncontentious consensus can be difficult to achieve, leastwise without 'parsimony analysis'. Critically, the differences between different groups of organisms that cladistic analysis identifies can only be used as meaningful measures of divergence, *and of the timing of divergence*, if they themselves are meaningful – that is, not of degree but of kind (Cooper *et al.*, in Briggs and Crowther, 2001; Smith and Peterson, 2002; Donaghue and Smith, 2004). Notwithstanding the above caveats, cladistic analysis is invaluable in its applicability to those groups whose evolutionary relationships cannot be inferred on the basis of empirical observations because they have a poor fossil record, for example, terrestrial mammals.

Bacteria

The hierarchical levels recognised by systematic biologists and palaeontologists in their attempts at classification are, from highest to lowest, domains, kingdoms, phyla, classes, orders, families, genera and species.

The two fundamental domains are those of the prokaryotes and the eukaryotes. The prokaryotes are diagnosed by the absence of, the eukaryotes by the presence of, nuclei.

The only kingdom of the domain of the prokaryotes that is important in the fields of palaeobiology and biostratigraphy is the bacteria, which contains the cyanobacteria (see also Section 3.1).

The kingdoms of the domain of the eukaryotes, all of which are important in the fields of palaeobiology and biostratigraphy, are the protists, the plants, the fungi and the animals (see below; see also appropriate sections in Chapter 3).

The unicellular to multicellular plant-like protists or algae and the multicellular plants are diagnosed by primary production through the process of photosynthesis, or autotrophism; the unicellular animal-like protists or 'protozoans', and the multicellular animals or 'metazoans', by secondary consumption, or heterotrophism. The fungi are diagnosed by chemo-organotrophism.

Plant-like protists or algae

The most important high-level divisions of the kingdom of the plant-like protists or algae are: pyrrhophytes, dinophytes or dinoflagellates; chrysophytes, especially silicoflagellates; bacillariophytes, or diatoms; haptophytes, especially coccolith-bearing or coccolithophore calcareous nannofossils; rhodophytes, or red algae; chlorophytes, or green algae; and charophytes, or stoneworts. Pyrrhophytes are diagnosed by golden-brown photosynthetic and accessory pigmentation caused by dinoxanthin and peridinin, by whip-like locomotory flagella and in some cases by calcareous shells; chrysophytes by golden-brown pigmentation, by flagella and by siliceous shells; bacillariophytes by yellowish pigmentation caused by carotenoids, and by siliceous shells or frustules; haptophytes by golden-brown pigmentation caused by chlorophyll and fucoxanthin, flagellar apparatuses known as haptonema, and calcareous shells; chlorophytes and charophytes by green pigmentation caused by chlorophyll, and usually by calcareous skeletons; and rhodophytes by red pigmentation caused by phycoerythrin, and by calcareous skeletons.

Animal-like protists or 'protozoans'

The most important high-level divisions of the kingdom of the animal-like protists or 'protozoans' are foraminiferans and radiolarians. Foraminiferans are diagnosed by the possession of granulo-reticulose pseudopodia, and an agglutinated arenaceous or secreted calcareous shell or test; radiolarians by a siliceous shell or scleracoma.

Plants

The most important high-level divisions of the kingdom of the plants are: bryophytes, including mosses and allied forms; pteridophytes, including clubmosses, ferns and horsetails; gymnosperms or seed-plants, including seed-ferns, tree ferns, conifers and allied forms; and angiosperms or flowering plants. Bryophytes are diagnosed by the absence of, and pteridophytes, gymnosperms and angiosperms by the presence of, vascular tissue. Pteridophytes are diagnosed by the absence of, and gymnosperms and angiosperms by the presence of, seeds. Gymnosperms are diagnosed by the absence of, and angiosperms by the presence of, flowers.

Fungi

As noted above, the fungi are diagnosed by chemo-organotrophism.

Animals

The most important high-level divisions of the invertebrate animals are: poriferans, especially sponges, archaeocyaths and stromatoporoids; cnidarians, especially corals; lophophorates, especially brachiopods and bryozoans; molluscs, especially bivalves, gastropods, ammonoids, belemnites and tentaculitids, arthropods, especially trilobites, ostracods and insects; and echinoderms, especially crinoids and echinoids. Poriferans are diagnosed by asymmetry; cnidarians by radial symmetry; lophophorates, molluscs, arthropods and echinoderms by bilateral to pentameral symmetry. Lophophorates, molluscs and arthropods are diagnosed by protostome development, of the mouth before the anus, in the embryo, and by external skeletons; echinoderms by deuterostome development, of the anus before the mouth, in the embryo, and by internal skeletons. Among the protostomes, lophophorates are diagnosed by the presence of specialised feeding apparatuses called lophophores; and arthropods by jointing, and by moulting or ecdysis.

The most important high-level divisions of the vertebrate animals are: fish; and tetrapods, including amphibians, reptiles and birds, and mammals. Fish are diagnosed by the absence of, and tetrapods by the presence of, limbs. Within the tetrapods, amphibians are diagnosed by the laying of eggs without shells; reptiles and birds by the laying of eggs with shells; and mammals by the presence of mammary glands.

2.4.2 *Identification of fossils*

Identification of fossils to higher hierarchical levels is possible through the use of this book, or those cited in the body of the text and reference list. Identification of fossils to species level may require the services of a specialist. (Most metropolitan natural history museums offer a specialist fossil identification service. This service is typically free to the general public, although a charge may be levied for private institutions.) It is worth noting in this context that palaeontologists and biologists define and identify species in slightly different ways. Biologists mainly use direct observation and/or molecular biology to define and identify 'biological species'. According to the 'biological species concept', a biological species is constituted of an interbreeding population capable of producing fertile progeny. In the absence of information on breeding behaviour, palaeontologists mainly use morphology to define and identify 'morphological species'.

2.5 Description and illustration of fossils (palaeontography)

2.5.1 *Description of fossils*

Formal descriptions of new species should include an explanation of the derivation of the chosen name, a synonymy, a diagnosis and a summary of the observed stratigraphic and palaeoecological distribution of the species, as well as a physical description of its size and characters (formal descriptions of higher-level taxa are not considered here). The synonymy is essentially a listing of previous records of the new species under old names, and is important in that it helps to establish the new species concept and its stratigraphic and palaeoecological range. The diagnosis is a succinct summary of how the new species differs from related species of the same genus, and is not the same as a physical description. The physical description should include comments on the amount of material available to the author of the species, and on any observed intraspecific variability. All of the specimens available to the author of the species constitute so-called syntypes. A single specimen from among the syntypes that well illustrates the characters of the species should be designated the holotype; the remainder paratypes. All type material should be deposited in publicly accessible collections for the purposes of consultation.

Readers interested in further details are referred to the *International Code of Botanical Nomenclature* (Greuter *et al.*, 2000) and the *International Code of Zoological Nomenclature* (Ride *et al.*, 1999). Incidentally, it is a rule under the *International Code of Zoological Nomenclature* that no name should give offence. I almost inadvertantly broke this rule myself once, by naming a new genus of foraminifer *Dyeria*, after my respected colleague Robin Dyer, but fortunately recognised my mistake just in time. I have heard it said, though, that another author once intentionally named a species of vertebrate with a particularly small brain and large mouth after an evidently less respected colleague. Fortunately, it is only a recommendation under the *Code* that naming should not be entered into with undue levity. I am told that BP's former chief palaeontologist Frank Eames spent a lifetime looking for a new species of the bivalve genus *Abra* that he could call *Abra cadabra*, but unfortunately died before his search bore fruit, forever leaving unanswered the question as to whether he would have got away with it or not. The classical scholars who published the following fossil names – abstracted from a website devoted to 'curiosities of biological nomenclature' – definitely, and deservedly, got away with it: *Dicrotendipes thanatogratus*, meaning 'Grateful Dead' for a chironomid (strictly not a fossil, but surely incapable of omission from any such list as this); *Arcticalymene viciousi*, *A. rotteni*, *A. jonesi*, *A. cooki* and *A. matlocki* for trilobites (Vicious, Rotten, Jones, Cook and Matlock were the mighty, mighty 'Sex Pistols'); *Norasaphus monroae* for a trilobite with an hourglass-shaped glabella; *Eucritta melanolimnetes*, meaning 'Creature from the Black Lagoon' for an amphibian; *Elvisaurus* for a dinosaur with a quiff-like crest; *Bambiraptor* for a small theropod; *Gojirasaurus*, meaning 'Godzilla', for a larger and rather more terrifying theropod; *Jenghizkhan* for a Mongolian tyrannosaurid; and *Mimatuta morgoth*, meaning, in Tolkien's elvish 'dark enemy of the world', for an evidently unattractive mammal.

2.5.2 *Illustration of fossils*

Illustration of fossils can be achieved by any one or more of a range of techniques, depending on the desired effect, including drawing, *camera lucida* drawing, analogue and digital optical photography, video photography, X-ray photography, and electron photomicrography (Claugher and Taylor, in Briggs and Crowther, 1990; Siveter, in Briggs and Crowther, 1990). Three-dimensional reconstruction can also be achieved.

2.6 Preparation, conservation and curation of fossils (museology)

The fundamental roles of a palaeontology department in a museum are to prepare, conserve and curate fossils, and to make them available for public exhibition and/or scientific research (Collins, 1995). The repositories of the principal museum palaeontological collections of the world are listed by Cleevely (1983), the repositories of palaeobiologically important collections, providing insights into key biological events in earth history, by Selden and Nudds (2004). British museums housing geological, including palaeontological, collections are listed by Nudds (1994). The scientific, cultural and financial value of museum natural history, including palaeontological, collections is discussed by Nudds and Pettitt (1997). The scientific content of the Natural History Museum in London is discussed by, among others, McGirr (1998); its history by Stearn (1998); and its magnificent Victorian Gothic architecture by Girouard (1981).

2.6.1 *Preparation of fossils*

Preparation of microfossils is discussed in detail by Aldridge, in Briggs and Crowther (1990), Hodgkinson, in Collins (1995), Jones (1996) and Green (2001), and summarised in the appropriate sections in Chapter 3. Preparation of macrofossils is discussed in detail by Croucher and Woolley (1982), Whybrow and Lindsay, in Briggs and Crowther (1990), Lindsay, in Collins (1995), and Wilson, in Collins (1995), and summarised below.

Mechanical preparation methods

Palaeontological preparators are able to extract macrofossils from rocks, or to finish the extraction process started in the field, by simple mechanical methods (see Sub-section 2.3.1). Automated tools, including percussive engraving pens, diamond or carborundum grinding wheels, and sandblasting or 'airbrasive' machines, are available to facilitate the final extraction.

Chemical preparation methods

Palaeontological prepators are also able to extract macrofossils from rocks by chemical methods, for example in the event that the fossils are so abundant and/or complex that they simply cannot be removed in a reasonable time-frame by simple mechanical methods reliant on manual dexterity. The chemicals used obviously have to be selected so as to be capable of disaggregation or dissolution of the rock without damage to the fossil, and without unmanageable hazard to the handler. They vary according to the chemistry of the fossil and rock. *Water* can be used to disaggregate soft, uncompacted mudstones, as it has the effect of causing the contained clay minerals to swell and disrupt the structure of the rock. Water is especially effective in combination with a detergent, which acts as a deflocculant. A similarly disruptive effect is caused by *hydrogen peroxide*, although it should be borne in mind that this is an unstable chemical that breaks down to release oxygen, which in turn constitutes a fire hazard. *Sequestrants* and *chelating agents* can also be used to disaggregate mudstones. Sequestrants include polyphosphates such as sodium hexametaphosphate. Chelating agents include ethylene diaminetetracetic acid, which can break down fossils as well as rocks, and which should therefore only be used under controlled conditions. *Acids* are widely used in the extraction of fossils from calcareous rocks such as limestones, and from siliceous rocks such as sandstones. Hydrochloric, acetic – or ethanoic – and other acids can be used to dissolve calcareous rocks, although they can also attack calcareous fossils and those with a calcareous component, including bones, and should therefore only be used under controlled conditions. Hydrofluoric acid can be used to dissolve siliceous rocks, although it can attack not only siliceous and calcareous – and indeed all other than organic-walled – fossils, but also glassware, and handlers, with potentially lethal consequences. It should therefore only be used after the provision of appropriate training and personal protection equipment, and even then only under the most rigorously controlled conditions. Other potentially hazardous chemicals are listed by Croucher and Woolley (1982).

2.6.2 Conservation of fossils

Macrofossils can be subjected to a process of artificial consolidation to minimise the risk of damage in the process of preparation, and to ensure conservation for subsequent study (Howie, in Collins, 1995). In the event that they do become damaged, they can be repaired by adhesion (Watkinson, 1987). The effects of the consolidants and adhesives used are reversible, so that no irreversible damage is done to the fossil, in the process of consolidation and adhesion. Commonly used consolidants include polyvinyl butyral resin and polybutylmethacrylate. Commonly used adhesives include polymethylmethacrylate. Polybutyl- and polymethylmethacrylate have the added advantage of being acid-resistant and therefore usable in the process of preparation by chemical methods as well as by physical or mechanical means.

Further conservation measures include the use of appropriate storage units in appropriate storage or display cabinets, in controlled environments free of fluctuations in temperature and, especially, relative humidity, and free of dust (Crowther, in Briggs and Crowther, 1990). Storage units (glass vials, cardboard boxes, wooden drawers etc.) are capable of being completely sealed, as appropriate. Storage and display cabinets are made of woods other than oak and birch, the hemicellulose component of which can on exposure to high temperatures and relative humidities yield acetic acid, which can in turn attack calcareous fossils. Relative humidity is kept in the range 45–55%, either by air-conditioning or the use of appropriate portable humidification or dehumidification equipment. Below this range, delicate, especially sub-Recent, fossils can dry out and crack. Above this range, pyritised fossils can oxidise and decompose. Still further specialised measures are required for the conservation of palaeobotanical material (Collinson, in Collins, 1995), sub-fossil bone (Shelton and Johnson, in Collins, 1995) and mummified humain remains (David and David, in Collins, 1995).

The serious issue of pest management in museums is discussed by Pinniger (2001).

2.6.3 Curation of fossils

Curation is nowadays generally computerised, although manual card index and other systems are still used as well as, or instead of, digital systems, and are still useful, not least as back-ups (Crowther, in Briggs and Crowther, 1990; Brunton, in Collins, 1995). The central register and/or computer database, and the labels accompanying every specimen, all contain as a minimum the museum number of the specimen, its scientific name, the name of its collector, and details of the geographical and stratigraphic location of its collection (as should any other, for example, geographical or stratigraphic, cross-referencing cataloguing systems).

3 • Principal fossil groups

This chapter includes sections on all of the high-level divisions of organisms identified in Chapter 2 that are important in the fields of palaeobiology and biostratigraphy, namely bacteria, plant-like protists or algae, animal-like protists or protozoans, plants, fungi, invertebrate animals and vertebrate animals. It also covers trace fossils.

As far as practicable, a standard format has been adopted. Areas covered include: biology, morphology and classification; palaeobiology; and biostratigraphy.

Emphasis has been placed on the palaeobiological and biostratigraphic significance of the various groups, and their proven and potential applications in the interpretation of earth and life history and processes. Other areas are covered only if they are relevant to these applications, and only in a level of detail commensurate with applicability. Fossil groups with minimal applicability have not been included.

More or less familiar names have been used wherever practicable, and an attempt has been made to keep the amount of scientific jargon to a necessary minimum.

Micropalaeontology and palynology

The groupings used in this book are essentially natural, reflecting evolutionary relationships, and the arrangement is in at least approximate order of evolutionary advancement. It is worth noting here, though, that an alternative, entirely artificial grouping of microfossils also exists, which is useful for practical purposes, as it is based on size range, composition, and required methods of collection, preparation and study. This treats together the calcareous microfossils, that is, the calcareous algae, the calcispheres or oligosteginids, *Bolboforma*, foraminiferans, calpionellids, ostracods and otoliths; the siliceous microfossils, that is, the silicoflagellates, diatoms and radiolarians; the phosphatic microfossils, that is, the conodonts and ichthyoliths; the organic-walled microfossils or palynomorphs, that is, the dinoflagellates, acritarchs, plant spores and pollen, and chitinozoans; and the calcareous nannofossils (Jones, 1996). The study of microfossils is termed micropalaeontology; that of organic-walled microfossils or palynomorphs, palynology.

Innovative applications of palynology

Palynology has a number of applications in the fields of petroleum geology, mineral exploration and exploitation, coal mining, and archaeology (see appropriate sections below). It also has applications in a number of other fields.

Forensic palynology. Forensic palynology has been used to test alibis, by comparing the pollen content in samples from suspects against that of the background at the scene of the crime, and that of their alleged whereabouts at the time of the crime; and to determine whether corpses have been moved after death, by comparing the pollen content in samples against the background at the scene of recovery (Mildenhall, 1990; Bryant *et al.*, in Jansonius and MacGregor, 1996; Jones, 1996). The first reported use was in Austria in 1959, in which case a palynologist was able to identify the source of the mud on the soles of the shoes of a murder suspect as an area of the Danube valley north of Vienna, close to where the body of the victim was eventually found (Bryant *et al.*, in Jansonius and MacGregor, 1996).

Forensic palynology has also been used to establish the provenance and travel history of illicit drugs (Mildenhall, 1990; Stanley, 1992; Bryant *et al.*, in Jansonius and MacGregor, 1996). In a recent case, cocaine seized by the New York Police Department turned out to contain not only *Lycopodium* spores and pollen from northern South America, probably introduced during initial processing of the coca leaves, but also *Tsuga canadensis* (hemlock) and *Pinus banksiana* (jack pine) pollen from New England, probably introduced during 'cutting', and a variety of pollen found in New York itself, probably introduced during storage.

Medical palynology. Many people suffer from allergic reactions to various types of airborne spores and pollen. Medical palynology aims to identify those types causing the most serious effects, and to alert particularly susceptible individuals to their presence through daily publication of a 'pollen count' (O'Rourke, in Jansonius and MacGregor, 1996).

Entomopalynology. Entomopalynology is the study of pollen associated with insects, and provides important information as to their feeding and migratory activities (Pendleton *et al.*, in Jansonius and MacGregor, 1996).

Melissopalynology. Melissopalynology is the study of pollen associated with honey, and provides important information as to the source (Jones and Bryant, in Jansonius and MacGregor, 1996). This is by no means as esoteric as it might sound, as some sources result in honey commanding a higher price than others, and obviously beekeepers are keen to site their hives as close as possible to these sources!

3.1 Bacteria

The only kingdom of the domain of the prokaryotes that is important in the fields of palaeobiology and biostratigraphy is the bacteria, which contains the cyanobacteria.

3.1.1 *Cyanobacteria and stromatolites*

Biology, morphology and classification

Biology

Bacteria are typically small, morphologically simple to comparatively complex and differentiated, unicellular organisms. They generate their energy by various means: some are photo-autotrophic; others, chemo-autotrophic. They reproduce by spores. In terms of ecology, bacteria live in every conceivable environment, and thrive in adverse and even extreme environments that other organisms are unable to tolerate (for example, superheated springs).

Cyanobacteria, formerly, erroneously, known as blue-green algae, are comparatively complex, differentiated forms up to 60 μm in diameter: some of them may even be multicellular (Carr and Whitton, 1973). They are essentially photo-autotrophic, that is, they generate energy through the process of photosynthesis (Whitton and Potts, 2000). They are thus restricted to the photic zone. Those of Shark Bay in Australia live in a restricted shallow marine lagoon, in hypersaline waters that the gastropods that would ordinarily graze them are unable to tolerate.

Morphology

Cyanobacteria characteristically form mats on supratidal flats or on the sea-floor, in order to maximise their exposure to the sunlight required for the process of photosynthesis to take place. The mats occasionally get buried by sediments, whereupon the cyanobacteria migrate through same to the new surface and colonise it, forming a fresh mat. Over time, a characteristically layered structure of cyanobacterial mats and sediment builds up. Layered rocks known as stromatolites are interpreted as having formed in this way (Fig. 3.1) (Walter, 1976; Awramik, in Briggs and Crowther, 1990; Bertrand-Sarfati and Monty, 1994; McNamara, 1997). The cyanobacteria themselves are typically lost in the process of fossilisation, while the layered structure is preserved (although occasional instances are known of the preservation of cyanobacterial filaments).

Classification

Classification of cyanobacteria and stromatolites is based essentially on morphology.

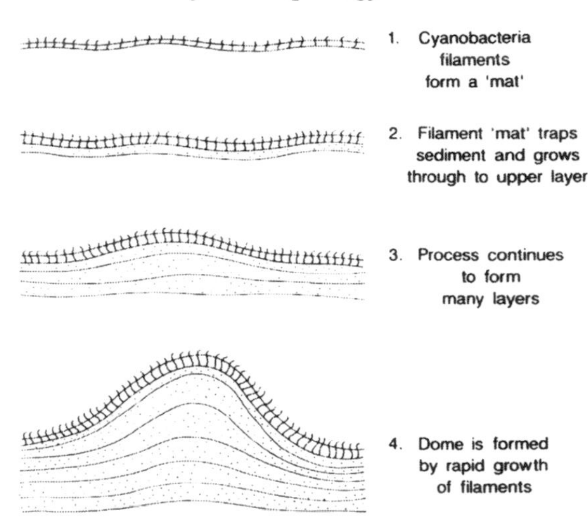

Fig. 3.1. Morphology of stromatolites. (Reproduced with permission from Doyle, 1996 *Understanding Fossils* . . . , © John Wiley & Sons Ltd.)

Palaeobiology

Fossil cyanobacteria are interpreted, essentially on the basis of analogy with their living counterparts, as having lived, and stromatolites as having formed, in a range of environments, and as having thrived in a range of adverse environments. For example, stromatolites are characteristic of the 'disaster biota' of the recovery interval following the End-Triassic mass extinction. They are also characteristic of the fossil 'cold seeps' of the Early Jurassic of the Neuquen basin of Argentina (Gomez-Perez, 2003).

Stromatolites were important constructors of reefs in the shallow marine environments of the Proterozoic and Palaeozoic (Pratt *et al.*, in Zhuravlev and Riding, 2001).

Following the Late Devonian mass extinction event, true skeletal reef-building organisms were effectively absent from latest Devonian, Famennian to Early Carboniferous, Dinantian marine ecosystems (Pickard, in Strogen *et al.*, 1996; Jian-Wen Shen and Webb, 2004). However, reefs of this age do exist, for example, the Miaomen reef in Guilin province, in which the reef core is composed of *Renalcis, Izhella, Paraepiphyton, Garwoodia,* and a *Keega*-like microbe (Jian-Wen Shen and Webb, 2004). Moreover, mud-dominated reef-like structures or 'Waulsortian mounds' of this age also exist. Microbial activity is considered to have played a significant role in the development of these structures, too. Evidence in support of this assertion is provided by the presence of stromatolitic and thrombolitic structures and of calcimicrobes such as *Renalcis. Girvanella* and *Renalcis* also characterise the latest Devonian of the Bolshoi Karatau in southern Kazakhstan and of the southern Urals (Cook *et al.*, in Zempolich and Cook, 2002; Zempolich *et al.*, in Zempolich and Cook, 2002). Early Carboniferous, Visean, Asbian–early Brigantian, reef-like build-ups of the Mullaghfin formation of the Kingscourt outlier on the margins of the Dublin basin are characterised for the most part by fine-grained sediments containing domal stromatolites, thrombolites etc. (Somerville *et al.*, in Strogen *et al.*, 1996).

Biostratigraphy

Stromatolites evolved in the Palaeoproterozoic, and have ranged through to the Recent. They diversified through the Neoproterozoic, but underwent something of a decline in the Riphean, from around 1000 Ma. Some authors have speculated that this decline was due to excessive grazing by early 'ediacarians' (see Walter and Heys, 1985; see also Subsection 5.2.3). However, other authors have hypothesised that it was brought about by environmental change associated with a series of glaciations, resulting in a so-called 'snowball Earth'. Incidentally, it has also been hypothesised that diversity promotes environmental stability, and therefore that low-diversity biotas, like those of the Proterozoic, are more susceptible to such external environmental factors than high-diversity biotas, like those of the Phanerozoic.

Cyanobacteria and stromatolites are known from the Zwartkoppie formation of the Onverwacht group of the Barberton supergroup, one of a number of Archaean greenstone belts in the north-eastern part of South Africa, dated to approximately 3300–3500 Ma, known from the predominantly volcanic Ventersdorp group, dated to 2642–2700 Ma, and from the Transvaal supergroup, dated to 2200–2642 Ma (MacRae, 1999). Banded ironstone formations or BIFs are also known from the Transvaal group, indicating that by the time these rocks were deposited there was a significant amount of oxygen in the ocean. (There may even have been enough oxygen to trigger the evolution of and to sustain complex, multicellular life, tantalising evidence of which is provided by enigmatic trace fossils resembling *Planolites* and *Rhizocorallium*.) Organic material of somewhat questionable origin has been found in the Witwatersrand supergroup, dated to 2700–2900 Ma. It has been hypothesised that this material originated either from bacteria or algae, or from lichen-like organisms (the observed columnar form as representing *in situ* growths, the particulate or 'fly-speck' form as dispersed spores). Whether or not this is the case, it is clear that whatever organism was responsible for the organic material was also somehow responsible for the observed concentration of gold in the organic material, for which the Witwatersrand is rather more famous.

The exceptionally slow rate of evolutionary turnover exhibited by the cyanobacteria renders them of limited use in biostratigraphy.

In my working experience in the petroleum industry, they have proved of limited use in stratigraphic subdivision and correlation in the following areas.

Proterozoic

The Precambrian, Riphean–Vendian of east Siberia. Here, the resolution of the existing inter-regional stromatolite- and microphytolite-based biostratigraphic zonation is poor (Jones, in Simmons, in press). Nonetheless, the Kamo group carbonate reservoir of the Yurubchen field on the Baikit arch has been dated as Early Riphean on microphytolite evidence. A similar unit in subsurface wells in the Prisayan–Yenisei area has been dated as Riphean on the basis of correlation with the Kamo group of Yurubchen. However, the underlying clastic unit in the wells in the Prisayan–Yenisei area has been dated as Middle Riphean, this time on the basis of correlation with the Dzhemkukan group of surface outcrops in the Patom Highlands dated as such on stromatolite and radiometric evidence.

Proterozoic or Palaeozoic

The 'Infracambrian' of the Middle East. Here, stromatolites characterise the Barut formation of north-east Iran (Jones, 2000). Thrombolites characterise the Ara group of the Huqf supergroup of Oman (Amthor *et al.*, 2000). Importantly, thrombolite build-ups form reservoirs in Oman (Amthor *et al.*, 2000).

Palaeozoic

The Early Carboniferous of Libya in North Africa. Here, *Collenia* marks the Visean M'Rar formation (Bellini and Massa, in Salem and Busrewil, 1980).

Mesozoic

The Early Cretaceous of the western margin of the British Isles. Here, the problematic organic-walled microfossil *Celyphus rallus* characterises the 'Wealden' facies (Jones, 1996). *Celyphus rallus* has recently been interpreted as a possible cyanobacterium (Batten, in Jansonius and MacGregor, 1996). Note, though, that it has also been interpreted as of fungal origin (Elsik, in Jansonius and MacGregor, 1996).

3.2 Plant-like protists or algae

3.2.1 *Dinoflagellates (pyrrhophytes)*

Biology, morphology and classification

Dinoflagellates, dinophytes or pyrrhophytes are an extant, Silurian?–Recent group of algae characterised by golden-brown pigmentation, by flagella, and in some cases by calcareous shells (note also that some dinoflagellates are characterised by siliceous shells, although only one of these, *Actiniscus*, is important in the fossil record, ranging from Oligocene to Recent). Their characteristic colour derives from photosynthetic and accessory pigments such as dinoxanthin and peridinin, and gave rise to the name 'pyrrhophytes', which translates as 'fire plants'. The name 'dinoflagellates' translates as 'whirling whips', and alludes to the whip-like flagella. The rather more prosaic ancient name, 'Animalcules which cause the Sparkling Light in Sea Water', alludes to the ability of some species to bioluminesce (blue-green).

Biology

The biology and ecology of modern dinoflagellates is comparatively well known. They achieve motility by movements of their longitudinal and transverse flagella. They are essentially autotrophic, that is, they produce food in the photic zone by means of photosynthesis (and indeed, with the diatoms, are the primary producers of food in the world's oceans). However, heterotrophic, predatory and parasitic, forms are also known. Living dinoflagellates may 'bloom' under particularly favourable, for example, seasonal, environmental conditions, resulting in 'red tides', highly toxic to other life-forms. Their reproductive cycle is complex, but in essence comprises a resting or cyst stage characterised by asexual reproduction, alternating with a motile or thecate stage characterised by sexual reproduction. Encystment of the thecate stage has been observed to occur not only after sexual reproduction but also in response to adverse environmental conditions. In general, only dinoflagellate cysts, sometimes, especially in the older literature, referred to as hystrichospheres, are found as fossils.

In terms of ecology, living dinoflagellates are more or less exclusively marine (Marret and Zonneveld, 2003).

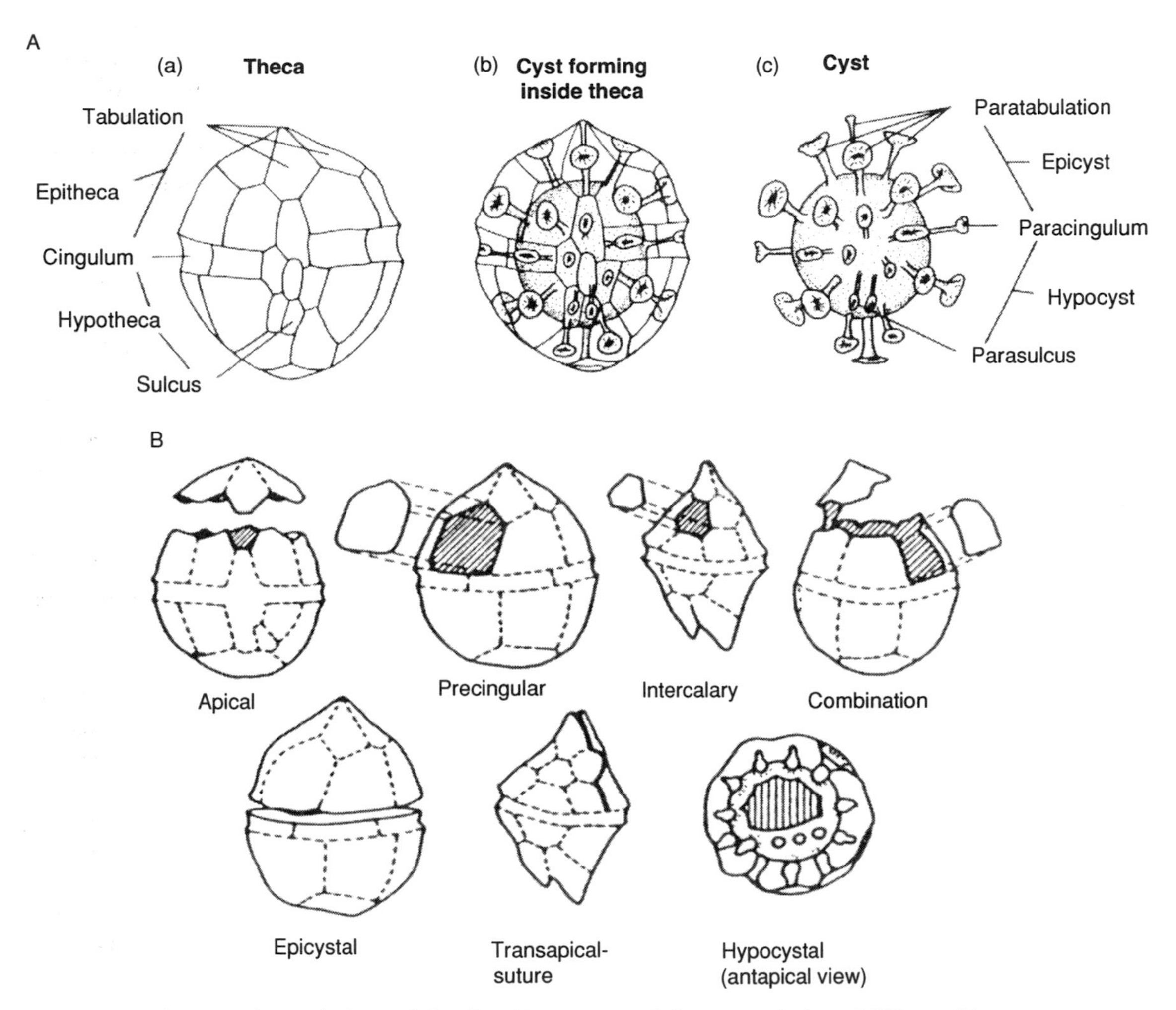

Fig. 3.2. (A) Morphology of dinoflagellates: cyst and theca morphology. (a) Theca; (b) cyst forming inside theca; (c) cyst. Cyst paratabulation parallels theca tabulation. (From Jones, 1996; reproduced with permission from Lipps, 1993.) (B) Morphology of dinoflagellates: archaeopyle types. (From Jones, 1996; in turn after Haq and Boersma, 1978.)

At the present time, there is comparatively little evidence of provincialism in dinoflagellate distribution. However, polar, sub-polar, intermediate, sub-equatorial and equatorial provinces can be discerned on the basis of the documented distributions of selected species, or on the basis of transfer functions based on these distributions.

Morphology

The basic morphology of dinocysts is determined by the cyst–theca relationship, and is described as proximate, cavate or chorate accordingly (Fig. 3.2). Both the cyst and the theca are constituted of variously arranged, and variously ornamented, plates, the arrangement referred to as paratabulation, in the case of the cyst, or tabulation, in the case of the theca. An excystment opening or aperture or archaeopyle is present in the wall of the cyst. This opening is capable of being closed by an operculum.

The so-called calcispheres or oligosteginids of the older literature are almost certainly calcareous dinoflagellates, although some lack obvious openings (Wendler and Willems, 2004; S. Meier, the Natural History Museum, personal communication). Calcispheres are characterised by a variously structured, and variously ornamented, hollow subspherical body (Fig. 3.3). They are found in micropalaeontological rather than in palynological preparations.

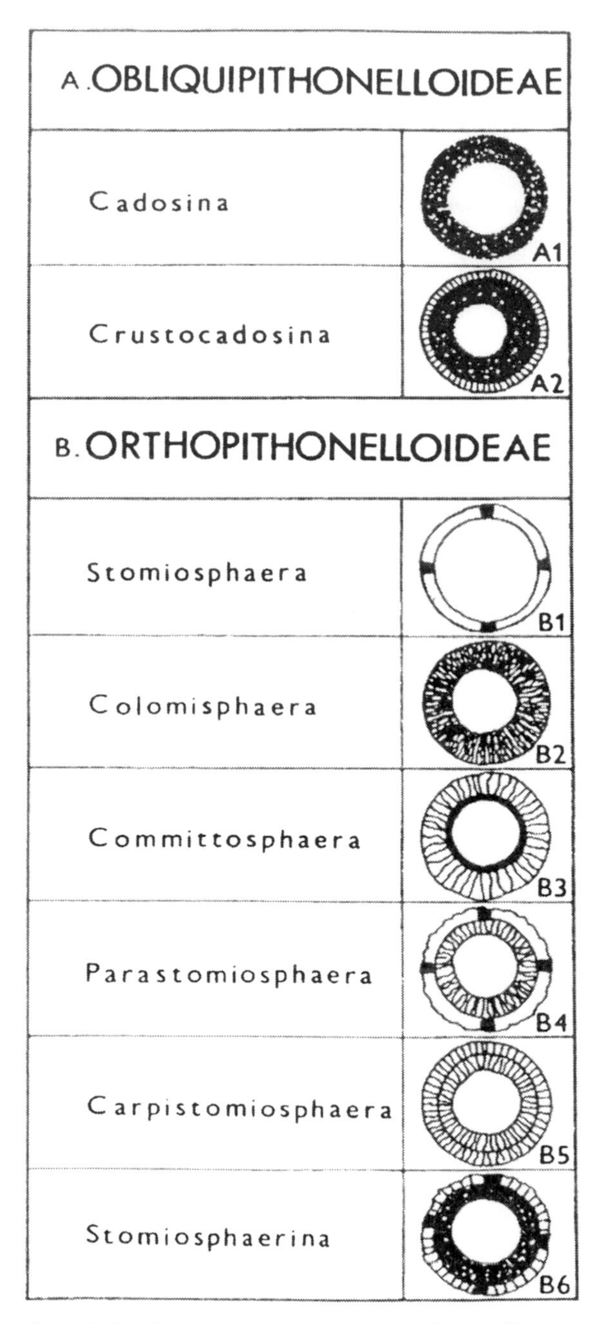

Fig. 3.3. Hard-part morphology of calcispheres. (From Jones, 1996; in turn after Rehanek and Cecca, 1993.)

Classification

Classification of dinoflagellates is based essentially on morphology, and on phylogeny inferred from empirical observations on the fossil record. Comparatively little molecular and cladistic analysis has thus far been undertaken on the group (Grzebyk *et al.*, 1998).

The position of the flagella, the nature of the (para)tabulation pattern, of the archaeopyle, and of the operculum, and the presence or absence of a calcareous shell, are all important classification criteria.

Two main sub-groups of non-calcareous dinoflagellates have been recognised on the basis of the nature of the (para)tabulation pattern, namely the gonyaulacoids and the peridinioids. The gonyaulocoids are autotrophic, and live in the photic zone in the late part of the water column. The peridinioids are heterotrophic, and live in areas of high productivity such as upwelling zones, ice margins or estuaries (Mudie and Rochon, 2001).

The calcareous dinoflagellates are distinguished by the presence of a calcareous shell. Their biology and ecology is still comparatively poorly known (Holl *et al.*, 1998; Meier and Willems, 2003; Vink, 2004). They appear to prefer areas of low productivity such as oceanic gyres, and at least *Pernambugia tuberosa, Calciodinellum elongatum* and *Scrippsiella regalis* appear to indicate oligotrophic conditions, although equally *Thoracosphaera heimii, Leonella granifera, Calciodinellum levantinum* and *Caracomia arctica* appear to indicate eutrophic conditions. At least in the Mediterranean, there appear to be distinct shallow, neritic and deep, oceanic water assemblages. There also appear to be distinct warm and cool water assemblages.

Palaeobiology

Most fossil dinoflagellates are interpreted, essentially on the basis of analogy with their living counterparts, as having been more or less exclusively marine, and planktonic, or, strictly, as they are able to achieve some motility, nektonic. Importantly, though, some have also been interpreted, essentially on the basis of associated fossil and sedimentary facies, as brackish and even non-marine.

Salinity has been inferred from transfer functions based on the documented distributions of living species.

Trophic levels have also been inferred from transfer functions, or from the ratio of heterotrophic to total – that is, heterotrophic plus autotrophic – species (see also below).

Palaeobathymetry

Within the marine realm, distinct dinoflagellate taxa have distinct ecological preferences. Some measure of depth and distance from shoreline is provided by transfer functions based on the documented distributions of living species.

Moreover, some measure of depth or distance from shoreline is provided by assemblage morphogroup composition, proximal environments being characterised by simple subspherical morphotypes, and distal environments by process-bearing morphotypes. For example, in the Eocene London clay of southern England, proximal/regressive assemblages are characterised by *Apectodinium* and *Wetzeliella*, and distal assemblages by *Areoligera* and *Spiniferites*. In the Miocene to Holocene, inner neritic assemblages are characterised by *Lejeunecysta* and *Selenopemphix*; inner to outer neritic assemblages by *Hystrichosphaeropsis obscura, Impagidinium, Multispinula, Nematosphaeropsis, Operculodinium israelianum, Polysphaeridium zoharyi, Spiniferites, Tectatodinium* and *Tuberculodinium vancampoe*; and outer neritic to oceanic assemblages by *Cleistosphaeridium diversispinosum, Lingulodinium machaerophorum, Operculodinium centrocarpum* and *Spiniferites mirabilis*.

Furthermore, some measure of depth, distance from shoreline and productivity is provided by the ratio between autotrophic gonyaulacoid and heterotrophic peridinioid dinoflagellates, as, for example in the Late Cretaceous, Cenomanian–Turonian of the western interior seaway of the USA (Harris and Tocher, 2003).

Fossil 'calcispheres' appear to have exhibited a preference for outer shelf to late slope environments, although they also appear more abundant in the Waulsortian mounds of the Carboniferous than in their level-bottom equivalents (Ahr and Stanton, in Strogen *et al.*, 1996). Significantly, where they occur, they tend to do so in flood abundance and to the virtual exclusion of other groups of calcareous plankton, indicating that they were able to thrive under conditions of environmental stress inimical to their competitors (?dysoxia).

Palaeobiogeography

Many fossil dinoflagellates appear to have had restricted or endemic biogeographic distributions, rendering them of use in the characterisation of palaeobiogeographic provinces, and in turn in the constraint of plate tectonic reconstructions. For example, in the Cretaceous, the 'Malloy suite' characterised the low- to moderate-latitude Tethyan realm, extending as far south as the Walvis Ridge in the South Atlantic, and the 'Williams suite' characterised the high-latitude Austral and Boreal realms. Fossil 'calcispheres' appear to have been restricted to the Tethyan realm.

Palaeoclimatology

The exacting ecological requirements and tolerances of many dinoflagellate species, and their rapid response to changing environmental and climatic conditions, render them useful in palaeoclimatology and climatostratigraphy, and in environmental archaeology (Dale and Dale, in Haslett, 2002).

Biostratigraphy

Molecular evidence indicates a Precambrian origin for the dinoflagellates (Moldowan *et al.*, in Zhuravlev and Riding, 2001). Indeed, 'possible dinoflagellates' have been recorded from the Neoproterozoic Wynniatt formation of Victoria Island in Arctic Canada, dated to between 721 and 1081 Ma (Butterfield and Rainbird, 1998), and questionable peridinioids from the Silurian (note also that the existence of zooxanthellates as long ago as the Ordovician is indirectly indicated by the occurrence of corals, with which the group presently has a symbiotic relationship). However, definite dinoflagellates do not appear in the rock record until the Triassic. The overall pattern of dinoflagellate evolution has been one of ever-increasing diversification through the Mesozoic and Cenozoic, with comparatively little loss of diversity other than that associated with the Late Cenomanian mass extinction (Fitzpatrick, in Hart, 1996) and, more especially, the End-Cretaceous mass extinction. Interestingly, dinoflagellates appear to have evolved (?iteratively) through the incorporation by a protist of a haptophyte, arising from the incorporation by a protozoan of an alga, arising in turn through the incorporation by a eukaryote of a cyanobacterium (Knoll, 2003)!

The rapid rate of evolutionary turnover exhibited by the dinoflagellates renders them of considerable use in biostratigraphy, at least in appropriate, marine, environments. Their usefulness is enhanced by their essentially unrestricted ecological distribution, and facies independence, a function of their pelagic habit. High-resolution biostratigraphic zonation schemes based on the dinoflagellates have been established, that have at least local to regional applicability. Biostratigraphic zonation schemes based on the 'calcispheres' have been established, that have at least local to regional applicability within a given palaeobiogeographic province.

In my working experience, dinoflagellates and 'calcispheres' have proved of particular use in stratigraphic subdivision and correlation, and/or in palaeoenvironmental interpretation, in the following areas.

Mesozoic

The Jurassic and Cretaceous of the North Sea. Here, dinoflagellates form part of the basis of the various regional biostratigraphic zonation schemes, calibrated against ammonite and other standards, and part of the basis of the calibration of the various regional- and reservoir-scale sequence stratigraphic schemes (Duxbury, in Jones and Simmons, 1999). They also form the basis of the identification of time-slices for regional and reservoir facies mapping purposes. Most importantly, they are of considerable use in reservoir characterisation and exploitation (author's unpublished observations).

Calcareous dinoflagellates and 'calcispheres' are of some use in stratigraphic subdivision and correlation in the chalk facies of the Late Cretaceous to Early Palaeocene of the North Sea (Jones, 1996). They are also of commercial importance as contributors to the formation of chalk reservoirs, as in the case of the several fields in the UK, Norwegian and Danish sectors (Jones, 1996; Fritsen *et al.*, 1999). Here, there may be some stratigraphic control on reservoir porosity, with the comparatively large and highly perforate *Thoracosphaera* only evolving in the Early Palaeocene, and thus being restricted to the uppermost part of the Chalk group (the Ekofisk formation).

The Cretaceous of northern South America and the Caribbean. Here, 'calcispheres' and dinoflagellates are of use in stratigraphic subdivision and correlation in the late Early–Late Cretaceous of the Campos basin of Brazil (Guardado *et al.*, in Edwards and Santogrossi, 1990). Dinoflagellates are important in reservoir characterisation and exploitation in the Late Cretaceous Guadalupe formation reservoir of the Cusiana field of Colombia (Jones *et al.*, in Koutsoukos, 2005; see also Section 7.1).

The 'Middle' Cretaceous of the Middle East. Here, dinoflagellates are of considerable use in the characterisation of the essentially Albian Burgan formation reservoir of the Raudhatain and Sabiriyah fields in North Kuwait (Al-Eidan *et al.*, 2001; author's unpublished observations). 'Calcispheres' are of use in stratigraphic subdivision and correlation, and to a lesser extent in palaeoenvironmental interpretation, in the Albian–Turonian of Iran (Jones, 1996).

Cenozoic

The Palaeogene of west Siberia. See Iakovleva and Kulkova (2003).

The Palaeogene of the Gulf coast of the USA. Here, the palaeobathymetry and sequence stratigraphy of the Late Eocene to Early Oligocene of the Gulf coast of the USA has been inferred from the diversity of dinoflagellate cysts, as measured using the unbiased Simpson index (SI), the range-through method, and de-trended correspondence analysis (DCA), so as to correct for sampling artefacts (Oboh-Ikuenobe and Jaramillo, in Prothero *et al.*, 2003). All the measures used reveal a peak in diversity associated with an interpreted maximum flooding surface in the Shubata clay of the Jackson group of the Late Eocene, and a trough in diversity associated with an interpreted sequence boundary at the boundary between the Forest Hill formation and the Mint Spring marl of the Vicksburg group of the Early Oligocene.

The Cenozoic of the North Sea. Here, dinoflagellates form part of the basis of the various regional biostratigraphic zonation schemes, loosely calibrated against planktonic foraminiferal and/or calcareous nannoplankton standards, and part of the basis of the calibration of the various regional- and reservoir-scale sequence stratigraphic schemes. They also form the basis of the identification of time-slices for regional and reservoir facies mapping purposes. Most importantly, they are of considerable use in reservoir characterisation and exploitation (author's unpublished observations).

The Cenozoic of eastern Paratethys. Here, dinoflagellates are of use in the stratigraphic subdivision and correlation of the Middle Eocene Kuma suite, Late Eocene Belaya alina suite and Oligocene–Early Miocene Maykop suite of the Caucasus (Zaporozhets, 1999). The Maykop suite is the principal petroleum source-rock of the Black Sea and Caspian. Dinoflagellates are also of use in palaeoenvironmental interpretation in the Black Sea and Caspian, enabling the discrimination of brackish and marine environments, calibrated against calcareous nannofossil, planktonic foraminiferal and/or oxygen isotopic data (Eaton, 1996; Mudie *et al.*, 2001). For example, impoverished assemblages of the cruciform species *Spiniferites cruciformis* and *Pyxidinopsis psilata* distinguish late glacial muds in the Black Sea associated with salinities of <7 ppt; *Lingulodinium machaerophorum–Spiniferites–Cymatiosphaera* assemblages, late glacial sapropels associated with salinities of 14–18 ppt; and diverse *Brigantedinium*–protoperidinoid–*Lingulodinium*–*Operculodinium* assemblages, post-glacial coccolith chalks associated with salinities of 18–20 ppt. Dinoflagellates are of particular use in the characterisation and exploitation of Early Pliocene 'productive series' reservoirs in the south Caspian (S. Lowe, unpublished observations).

The Cenozoic of northern South America and the Caribbean. Here, dinoflagellates are of considerable use in stratigraphic subdivision and correlation (de Verteuil and Johnson, 2002).

Other applications

Dinoflagellates are also useful in environmental science (see Section 7.5).

3.2.2 *Silicoflagellates (chrysophytes)*

Biology, morphology and classification

Silicoflagellates are an extant, Cretaceous–Recent, group of chrysophyte algae characterised by golden-brown pigmentation, by flagella, and by siliceous shells (Blome *et al.*, 1996). Their characteristic colour derives from photosynthetic pigments. They are mixotrophic, subsisting partly on the products of photosynthesis and partly on prey trapped in their pseudopodia. Silicoflagellate reproduction is probably asexual.

In terms of ecology, silicoflagellates are exclusively marine, and essentially planktonic, although they are also able to achieve some motility through whip-like movements of their flagella. They appear to be characteristically associated with cold, deep and/or upwelling water masses. *Bachmannocena* is interpreted as an indicator of upwelling.

In terms of hard-part morphology, silicoflagellates are characterised by siliceous skeletons (Fig. 3.4). The basic body-plan is a hollow ring. The interior may be traversed by bars. The exterior may bear spines.

Classification of silicoflagellates is based on details of morphology.

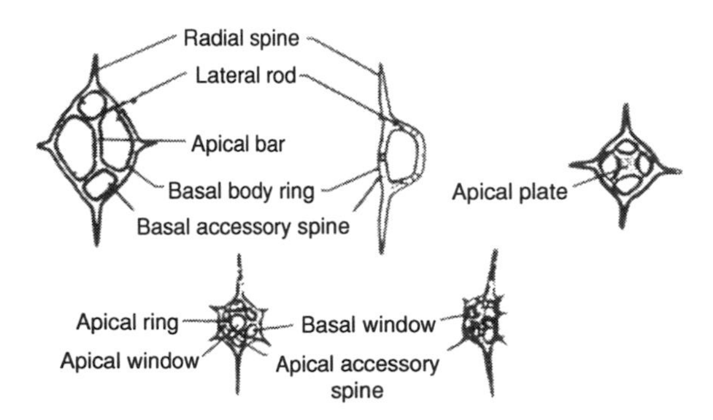

Fig. 3.4. Hard-part morphology of silicoflagellates. (From Jones, 1996; in turn after Haq and Boersma, 1978.)

Palaeobiology

Fossil silicoflagellates are interpreted, essentially on the basis of analogy with their living counterparts, as having been marine and planktonic.

Palaeobiogeography and palaeoclimatology

The ratio between the comparatively warm-water *Dictyocha* and the cool-water *Distephanus* provides some measure of palaeotemperature and palaeoclimate.

Biostratigraphy

Silicoflagellates range from Cretaceous to Recent. The moderate rate of evolutionary turnover exhibited by the silicoflagellates renders them of some use in biostratigraphy, at least in appropriate, marine, environments. Their usefulness is enhanced by their essentially unrestricted ecological distribution, and facies independence, a function of their habit. Biostratigraphic zonation schemes based on the silicoflagellates have been established, that have at least local to regional applicability.

In my working experience, silicoflagellates have proved of some use in stratigraphic subdivision and correlation in the following areas.

Cenozoic

The Palaeogene of the North Sea and north-west Europe. Here, the diatomaceous sediments of the Palaeocene–Eocene boundary section are marked and indeed independently zoned by silicoflagellates (Jones, 1996).

The Neogene of central Paratethys. Here, for example, the Badenian and Sarmatian regional stages of the Miocene are marked by particular silicoflagellate assemblages (Jones, 1996).

Other applications

Silicoflagellates are also useful in environmental science (see Section 7.5).

3.2.3 Diatoms (bacillariophytes)

Biology, morphology and classification

Biology

Diatoms or bacillariophytes are an extant, Jurassic–Recent, group of algae characterised by yellowish pigmentation, and by shells or frustules composed of opaline silica (Round *et al.*, 1990; Blome *et al.*, 1996; Stoermer and Smol, 1999). Their characteristic colour derives from carotenoids. They are autotrophic, that is, they manufacture food in the photic zone through the process of photosynthesis. They may form colonies. Diatom reproduction is primarily asexual, though sexual reproduction may take place intermittently.

In terms of ecology, diatoms occur in every aquatic, freshwater and marine, environment. Some are benthic; some planktonic. Benthic diatoms lack flagella, but are able to achieve some motility by simply sliding around on the substrate. Planktonic forms also lack flagella, and are unable to achieve any motility.

Although distinct marine diatom taxa have distinct ecological preferences, the group as a whole is characteristically associated with cold, deep-water masses and/or with upwelling systems. Diatoms may bloom under particularly favourable environmental conditions, resulting in the accumulation of siliceous mudstones, referred to in Russia as 'opokas', or diatomites. Siliceous mudstones are important petroleum source-rocks in the Miocene of the Monterey formation of California, in the Miocene of the Teradomari formation of Japan, and in the Miocene Pilsk formation of Sakhalin in the far east of the former Soviet Union. Diatomites are also economically important, on account of their peculiar properties of high porosity and absorbency and low density, and have a wide range of uses, most especially in filtration systems and as lightweight fillers (Harwood, in Stoermer and Smol, 1999). The world's largest deposits are in the Miocene of California. The Miocene was evidently a time of significantly enhanced biogenic production of silica (euphoniously known as 'the commotion in the ocean').

Morphology

In terms of hard-part morphology, diatoms are characterised by siliceous shells or frustules, comprising two valves, the larger, dorsal one being termed the epivalve and the smaller, ventral one the hypovalve (Fig. 3.5). The shells may be either radially or bilaterally symmetrical. Some of the shell surface is covered by perforations, which may be aligned as more or less radial striae separated by imperforate costae.

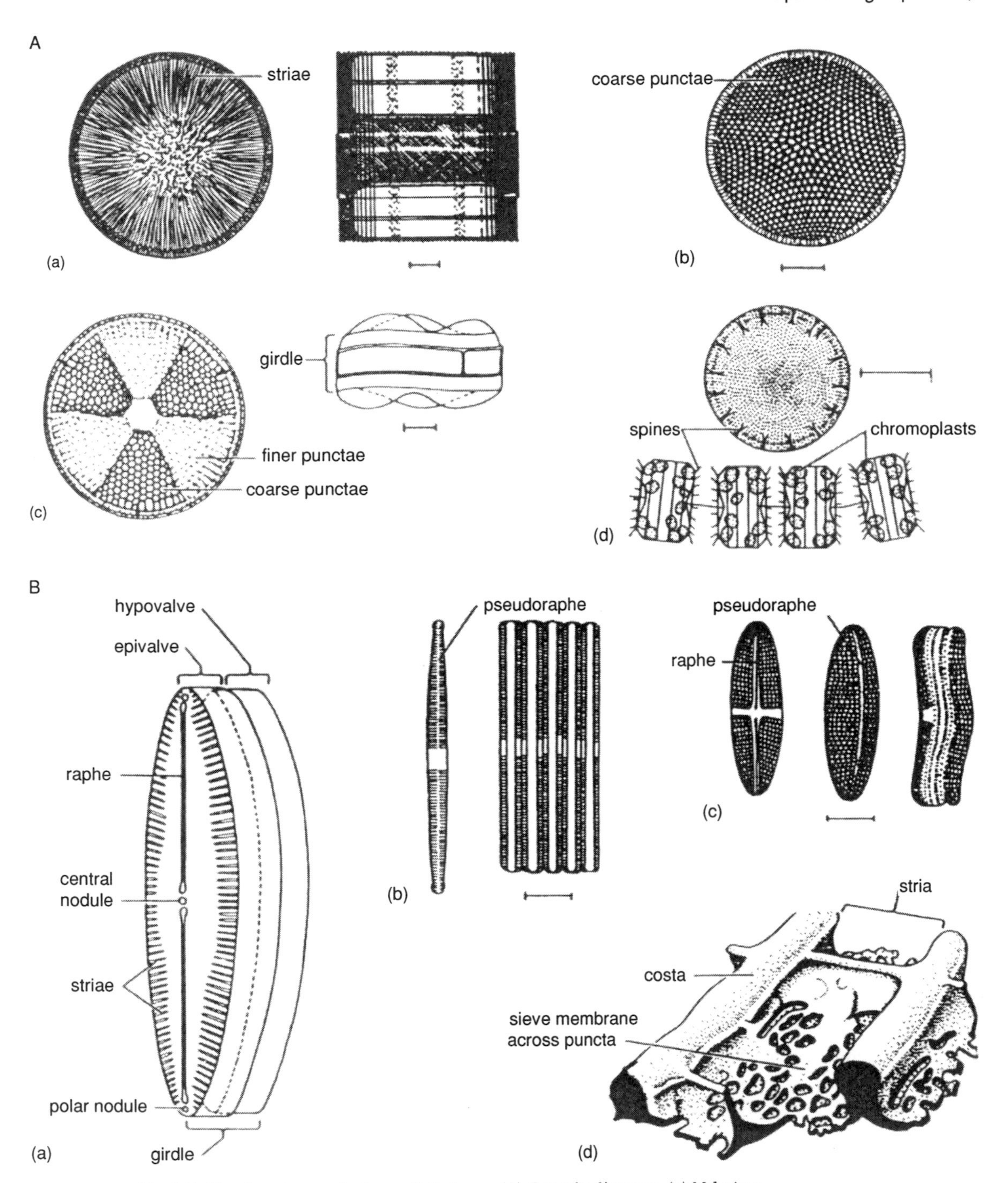

Fig. 3.5. Hard-part morphology of diatoms. (A) Centric diatoms. (a) *Melosira*; (b) *Coscinodiscus*; (c) *Actinoptychus*; (d) *Thalassiosira*. Scale bars 10μm. (B) Pennate diatoms. (a) *Pinnularia*, ×320; (b) *Fragillaria*; (c) *Achanthes*; (d) detail of puncta. Scale bars 10μm. (From Jones, 1996; in turn after Brasier, 1980.)

A longitudinal groove or raphe may be present in bilaterally symmetrical forms.

Classification

Classification of diatoms is based on details of morphology. The shape of the frustule is the most important criterion. Two sub-groups are recognised, the radially symmetrical centric diatoms, and the bilaterally symmetrical pennates.

Palaeobiology

Fossil diatoms are interpreted, essentially on the basis of analogy with their living counterparts, as having occupied a range of aquatic, freshwater and marine, benthic and planktonic environments (Cooper, in Stoermer and Smol, 1999; Denys and de Wolf, in Stoermer and Smol, 1999; Sancetta, in Stoermer and Smol, 1999; Snoejis, in Stoermer and Smol, 1999; Sullivan, in Stoermer and Smol, 1999). To oversimplify, centric diatoms are typically marine and planktonic, and pennates typically freshwater and benthic.

Palaeobathymetry

Transfer functions based on the environmental distributions and tolerances of diatoms are of considerable use in sea-level reconstructions (see also Sections 4.6 and 7.6).

Palaeobiogeography

In the Cenozoic, diatom evidence points toward the existence of a marine connection between Tethys and the Arctic Ocean in the Eocene, by way of the so-called Obik Sea or Turgai Strait (Radionova and Khokhlova, 2000).

Palaeoclimatology

The exacting ecological requirements and tolerances of many diatom species, and their rapid response to changing environmental, and especially climatic, conditions, render them extremely useful in palaeoclimatology and climatostratigraphy, and in archaeology (Boyd, in Neale and Brasier, 1981; Juggins and Cameron, in Stoermer and Smol, 1999; Kennington, in Haslett, 2002; see also Section 7.6).

Biostratigraphy

Diatoms range from Jurassic? to Recent. The moderate rate of evolutionary turnover exhibited by the diatoms renders them of some use in biostratigraphy, at least in appropriate, aquatic, environments. Unfortunately, their usefulness is compromised by their destruction in diagenesis, through the transformation of opaline silica into cristobalite and of cristobalite into quartz (Keller and Isaacs, 1985). Nonetheless, biostratigraphic zonation schemes based on the diatoms have been established, that have at least local to regional applicability. Important schemes applicable to the Neogene and Pleistogene of the North Pacific have been established by Barron and Gladenkov (1995), Gladenkov (1998), Motoyama and Maruyama (1998) and Yanagisawa and Akiba (1998) (Fig. 3.6).

In my working experience, diatoms have proved of some use in stratigraphic subdivision and correlation, and/or in palaeoenvironmental interpretation, in the following areas.

Mesozoic

The 'Mid' Cretaceous of the Sergipe–Alagoas basin in Brazil. See Koutsoukos and Hart (1990). Here, diatoms are characteristic of middle–outer shelfal environments.

Cenozoic

The Palaeogene of the North Sea. See Mitlehner (1994), Mitlehner, in Knox *et al.* (1996) and Bidgood *et al.*, in Jones and Simmons (1999). Here, diatoms form part of the basis of the regional biostratigraphic zonation, loosely calibrated against planktonic foraminiferal and/or calcareous nannoplankton standards. They therefore also form part of the basis of the calibration of the regional sequence stratigraphic scheme.

The Palaeogene–Neogene of Sakhalin in the former Soviet Union. Here, diatoms mark all the key horizons sampled in the Schmidt Peninsula in north Sakhalin, that is, the Machigarskaya, dated as Oligocene, the Tumskaya, dated as Oligocene–Early Miocene, the Pilskaya, dated as Middle Miocene, the Kaskadnaya, dated as Middle–Late Miocene, the Vengeriyskaya and Mayamrafskaya, dated as Late Miocene, and the Matitutskaya and Pomyrskaya, dated as Pliocene, and serve as the basis for time-slice definition for facies mapping purposes

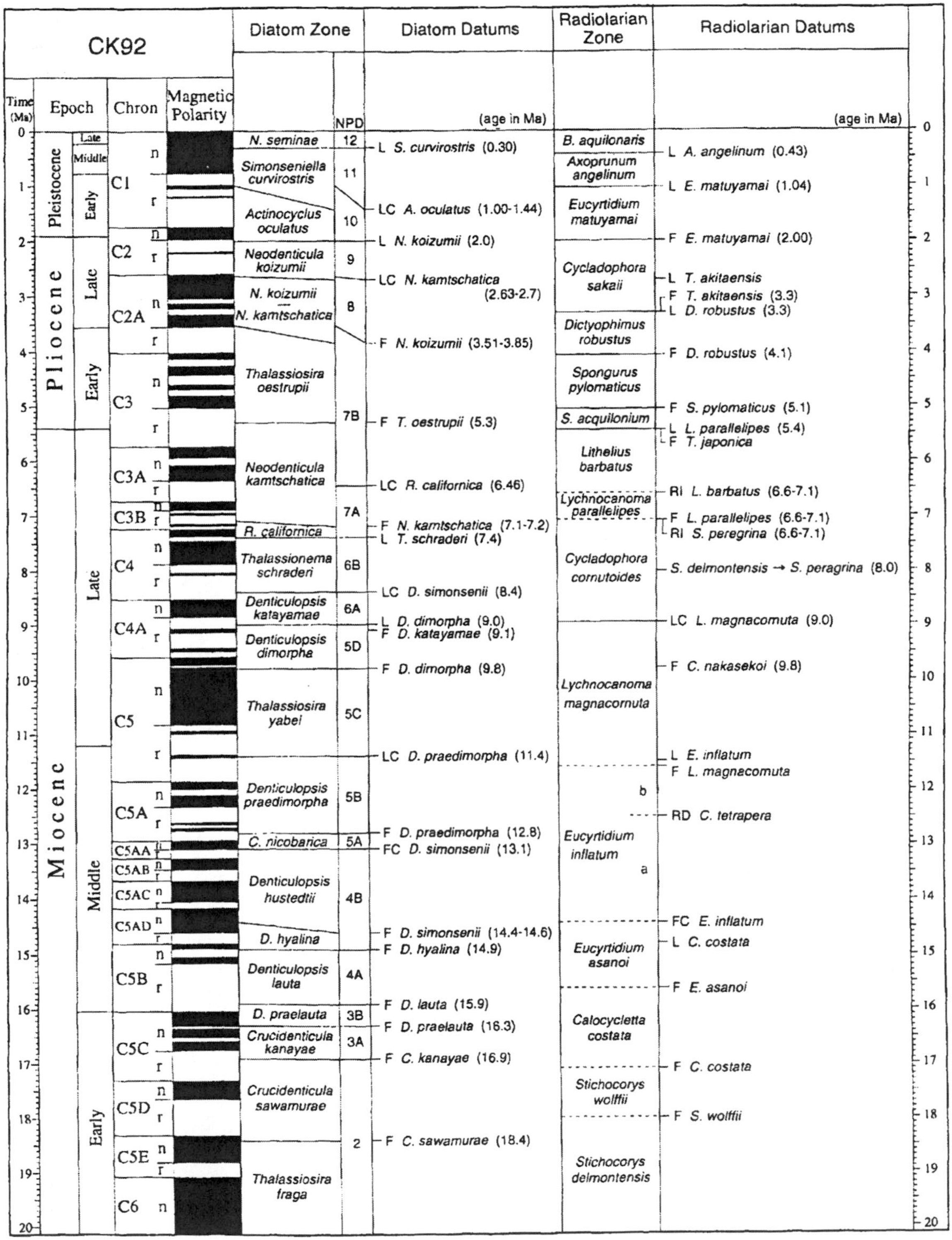

CK92

Time (Ma): 0, 1, 2, 3, 4, 5, 6, 7, 8, 9, 10, 11, 12, 13, 14, 15, 16, 17, 18, 19, 20

Epoch: Pleistocene (Late, Middle, Early); Pliocene (Late, Early); Miocene (Late, Middle, Early)

Chron / Magnetic Polarity: C1 (n, r); C2 (n, r); C2A (n, r); C3 (n, r); C3A (n, r); C3B (n, r); C4 (n, r); C4A (n, r); C5 (n, r); C5A (n, r); C5AA (n, r); C5AB (n, r); C5AC (n, r); C5AD (n, r); C5B (n, r); C5C (n, r); C5D (n, r); C5E (n, r); C6 (n)

Diatom Zone (NPD):
- *N. seminae* (12)
- *Simonseniella curvirostris* (11)
- *Actinocyclus oculatus* (10)
- *Neodenticula koizumii* (9)
- *N. koizumii* – *N. kamtschatica* (8)
- *Thalassiosira oestrupii* (7B)
- *Neodenticula kamtschatica* (7A)
- *R. californica*
- *Thalassionema schraderi* (6B)
- *Denticulopsis katayamae* (6A)
- *Denticulopsis dimorpha* (5D)
- *Thalassiosira yabei* (5C)
- *Denticulopsis praedimorpha* (5B)
- *C. nicobarica* (5A)
- *Denticulopsis hustedtii* (4B)
- *D. hyalina*
- *Denticulopsis lauta* (4A)
- *D. praelauta* (3B)
- *Crucidenticula kanayae* (3A)
- *Crucidenticula sawamurae* (2)
- *Thalassiosira fraga*

Diatom Datums (age in Ma):
- L *S. curvirostris* (0.30)
- LC *A. oculatus* (1.00-1.44)
- L *N. koizumii* (2.0)
- LC *N. kamtschatica* (2.63-2.7)
- F *N. koizumii* (3.51-3.85)
- F *T. oestrupii* (5.3)
- LC *R. californica* (6.46)
- F *N. kamtschatica* (7.1-7.2)
- L *T. schraderi* (7.4)
- LC *D. simonsenii* (8.4)
- L *D. dimorpha* (9.0)
- F *D. katayamae* (9.1)
- F *D. dimorpha* (9.8)
- LC *D. praedimorpha* (11.4)
- F *D. praedimorpha* (12.8)
- FC *D. simonsenii* (13.1)
- F *D. simonsenii* (14.4-14.6)
- F *D. hyalina* (14.9)
- F *D. lauta* (15.9)
- F *D. praelauta* (16.3)
- F *C. kanayae* (16.9)
- F *C. sawamurae* (18.4)

Radiolarian Zone:
- *B. aquilonaris*
- *Axoprunum angelinum*
- *Eucyrtidium matuyamai*
- *Cycladophora sakaii*
- *Dictyophimus robustus*
- *Spongurus pylomaticus*
- *S. acquilonium*
- *Lithelius barbatus*
- *Lychnocanoma parallelipes*
- *Cycladophora cornutoides*
- *Lychnocanoma magnacornuta*
- *Eucyrtidium inflatum* (b, a)
- *Eucyrtidium asanoi*
- *Calocycletta costata*
- *Stichocorys wolffii*
- *Stichocorys delmontensis*

Radiolarian Datums (age in Ma):
- L *A. angelinum* (0.43)
- L *E. matuyamai* (1.04)
- F *E. matuyamai* (2.00)
- L *T. akitaensis*
- F *T. akitaensis* (3.3)
- L *D. robustus* (3.3)
- F *D. robustus* (4.1)
- F *S. pylomaticus* (5.1)
- L *L. parallelipes* (5.4)
- F *T. japonica*
- RI *L. barbatus* (6.6-7.1)
- F *L. parallelipes* (6.6-7.1)
- RI *S. peregrina* (6.6-7.1)
- *S. delmontensis* → *S. peragrina* (8.0)
- LC *L. magnacomuta* (9.0)
- F *C. nakasekoi* (9.8)
- L *E. inflatum*
- F *L. magnacomuta*
- RD *C. tetrapera*
- FC *E. inflatum*
- L *C. costata*
- F *E. asanoi*
- F *C. costata*
- F *S. wolffii*

Fig. 3.6. Stratigraphic zonation of the Neogene and Pleistogene of the North Pacific by means of diatoms and radiolarians. (From Motoyama and Maruyama, 1998.)

(Gladenkov, 1999). Diatoms also mark key horizons sampled in the Pogranich area of east Sakhalin, that is, the Pilengskaya (Daekhurinskiy), dated as Oligocene, and the Borskaya (Uninskiy plus Daginskiy) (Gladenkov *et al.*, 2000). The Pilengskaya equates to the regional Oligocene *Rocella vigilans*, *Cavitatus rectus* and *Rocella gelida* Zones; the Borskaya to the Early Miocene *Thalassiosira praefraga*, *Thalassiosira fraga*, *Crucidenticula sawamurae*, *C. kanayae* and *Denticulopsis praelauta* Zones.

The Neogene of Angola in west Africa. Here, marine and freshwater diatoms proved of use in the stratigraphic subdivision and correlation of the Late Miocene part of the Malembo formation of the Pacassa and Veado wells in the offshore, Block 3 (Fourtanier and Seyve, 2001). Two zones, an early, *Thalassiosira sira* Zone and a late, *Nitzschia porteri* Zone, were recognised, as also at DSDP Site 362 on the Walvis Ridge to the south. These zones can be calibrated against the absolute timescale between 8.8 and 7.4 Ma. Significantly, this was a time of high-latitude cooling, as evidenced by the benthic foraminiferal oxygen isotopic record, and increased upwelling along the Benguela margin. As intimated above, freshwater as well as marine diatoms are found, contemporaneously transported by the palaeo-Congo.

Other applications

Living diatoms are extremely useful in environmental science (Hall and Smol, in Stoermer and Smol, 1999; Stevenson and Pan, in Stoermer and Smol, 1999; see also Section 7.5).

Incidentally, diatoms are also useful in forensic science (Peabody, in Stoermer and Smol, 1999). Their principal use in this field is in diagnosing death by drowning (from the diatom flora in the water in the lungs). Another use is in identifying diatomaceous earth used in safe construction, and thus the scene where a safe was broken open, or the suspect (from the presence of diatomaceous earth at the scene or on the suspect's skin or clothing).

3.2.4 Calcareous nannoplankton (haptophytes)

Biology, morphology and classification

Calcareous nannoplankton include coccoliths produced by living and fossil coccolithophore calcareous nannoplankton, and nannoliths produced by fossil non-coccolithophores, which latter are of uncertain taxonomic affinity (Green and Leadbeater, 1994; Winter and Siesser, 1994; Bown, 1998; Thierstein and Young, 2003; Young *et al.*, 2003).

Biology

Coccolithophores are an extant, Triassic–Recent, group of haptophyte algae characterised by golden-brown pigmentation, two flagella and an additional flagellar apparatus known as a haptonema, and calcareous shells. Their characteristic colour derives from chlorophyll and fucoxanthin. They are autotrophic, that is, they manufacture food in the photic zone through the process of photosynthesis. Their reproduction is predominantly asexual, but may incorporate a sexual phase.

In terms of ecology, living coccolithophores are more or less exclusively marine, and essentially planktonic, although they are also able to achieve some motility through whip-like movements of their flagella and haptonema.

As in the case of the planktonic foraminiferans, different sub-groups have different lateral distributions within the world's oceans, and different vertical distributions within the water column.

Coccolithophores may bloom under particularly favourable environmental conditions, resulting in the formation of calcareous nannoplankton oozes or chalks. Seasonal blooms are imageable on satellite data.

Morphology

In terms of hard-part morphology, coccolithophores are characterised by calcareous shells composed of rounded to elliptical calcareous plates or coccoliths (Fig. 3.7). Coccolithophore biomineralisation has recently been studied using novel atomic force microscopy (AFM) technology (Henriksen *et al.*, 2004).

In general, only isolated coccoliths are encountered in the fossil record, and coccolith–coccolithophore relationships are obscure.

The coccoliths themselves are, in section, subdivided into inner or proximal and outer or distal layers or shields. In plan view, they are subdivided into central and outer areas. The central area is very variable in nature, sometimes clear, sometimes traversed by bridges or bars, sometimes almost infilled

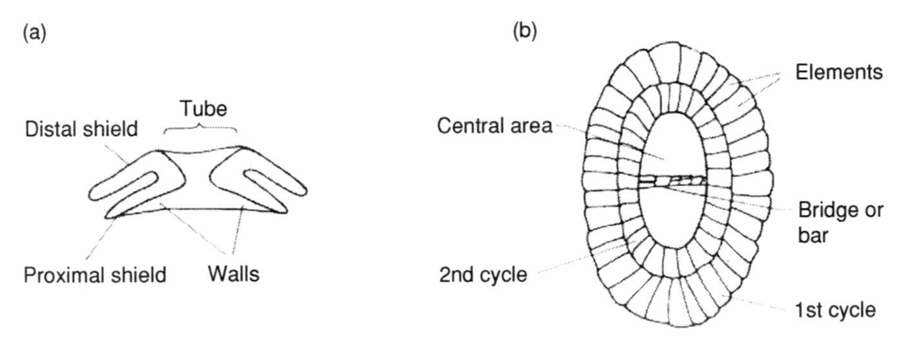

Fig. 3.7. Hard-part morphology of coccoliths. (a) Side view; (b) top view. (From Jones, 1996; reproduced with permission from Lipps, 1993.)

by complex arrangements of radial and tangential elements, sometimes sieve-like. The outer area is generally made up of two concentric rings or cycles of varying numbers of radially or obliquely oriented platelets or elements that may or may not bear rims or processes.

The morphology of nannoliths is more varied.

Classification

Classification of coccoliths and nannoliths is based on details of morphology.

Palaeobiology

Calcareous nannofossils are interpreted, essentially on the basis of analogy with living coccolithophores, as having been essentially marine and planktonic. Importantly, though, some forms are interpreted as having, or as having had, some brackish tolerance, for example *Braarudosphaera* and *Emiliania*.

Braarudosphaera is a long-ranging genus. In modern oceans it is found mainly in low-salinity nearshore waters, as in the Gulf of Maine and in the Gulf of Panama, and only rarely in the open ocean (Siesser *et al.*, 1992). It is also found on the oceanic side of the weak upwelling front off Portugal (Duarte-Silva *et al.*, 2004). Low-salinity 'events', such as heavy rainfall or increased meltwater, or upwelling, have had to be invoked to account for the ancient '*Braarudosphaera* chalks' observed in the Oligocene of the various DSDP and ODP Sites of the South Atlantic, which are clearly of open oceanic and not nearshore aspect (Bukry, 1974, 1978; Berger, in Talwani *et al.*, 1979; Parker *et al.*, 1985; Siesser *et al.*, 1992; Peleo-Alampay *et al.*, 1999): downslope transportation from a nearshore setting is another possible explanation (author's unpublished observations). Most recently, it has been postulated that the evident 'blooms' resulting in the chalks reflected rhythmic variations in the vigour of the South Atlantic gyre, and that their cessation towards the end of the Early Oligocene represented a response to an important shift in circulation, related to the opening of the Drake Passage (Kelly *et al.*, 2003). Interestingly, foraminiferal evidence indicates that benthic to thermocline oxygen and carbon isotope gradients were reduced during the blooms, a hallmark signature for strengthened upwelling. *Braarudosphaera* may have lived in the lower part of the euphotic zone, a favourable location for exploiting upwelling nutrients (see also Scarparo Cunha and Koutsoukos, 1998).

Palaeobathymetry

Different calcareous nannofossil sub-groups are interpreted as having had different vertical distributions within the water column. For example, ancient nannoconids are interpreted as characteristic of epipelagic environments.

Different calcareous nannofossil sub-groups are also interpreted as having had different horizontal distributions. Distributions along a 120-km-long proximal-to-distal transect in the Early Jurassic, Late Pliensbachian–Early Toarcian of the Umbria–Marche basin in central Italy have been analysed by Mattioli and Pittet (2004). Here, *Crepidolithus crassus* and *Schizosphaerella* spp. have been found to be most abundant in limestone facies at the proximal end of the transect, in proximity to the shallow marine

Latium–Abruzzi platform, indicating a tolerance of oligotrophic conditions. Species of *Lotharingius, Biscutum* and *Calyculus* have been found to be uniformly distributed along the depth transect, but to be most abundant in marl facies, indicating a preference for higher trophic levels (?and upwelling). *Mitrolithus jansae* has been found to be most abundant at the distal end of the transect, indicating a preference for deeper water: significantly, it is also abundant at times as well as in places of high sea level, such as the earliest Toarcian.

The abundance of calcareous nannofossils has been used as a measure of carbonate dilution or palaeo-productivity, as, for example, in the Early Cretaceous of the Vocontian basin in south-east France (Reboulet *et al.*, 2003). Calcareous nannofossils have also been used in the reconstruction of nutricline dynamics here (Herrle, 2003).

Abundance peaks of calcareous nannofossils have been widely used to characterise condensed sections in sequence stratigraphic interpretation. Note, though, that certain of these peaks have been observed to be associated with what benthic foraminiferal sedimentological data demonstrate to be allochthonous debris flows, such that they are recording allochthonous as well as autochthonous calcareous nannofossil species (Jones, 2003a). Thus, nannofossil abundance peaks identified in the absence of such benthic foraminiferal and sedimentological evidence should be regarded as of questionable reliability as indicators of anything meaningful!

Palaeobiogeography

Many calcareous nannofossils appear to have had restricted or endemic biogeographic distributions, rendering them of use in the characterisation of palaeobiogeographic provinces, and in turn in the constraint of plate tectonic reconstructions, in the Mesozoic (Roth, 1978), and in the Cenozoic (Ramsay and Funnell, in Hallam, 1973).

Biostratigraphy

Calcareous nannofossils evolved in the Triassic. They diversified through the Mesozoic, before sustaining severe losses in the End-Cretaceous mass extinction. Although they have staged some form of recovery from this event, they have not regained their pre-extinction diversity. Calcareous nannofossil diversity appears to exhibit cyclical change generally parallelling that of sea level, but perturbed by intermittent volcanic and impact events (Prokoph *et al.*, 2004).

The rapid rate of evolutionary turnover exhibited by the calcareous nannofossils renders them of considerable use in biostratigraphy, at least in appropriate, marine, environments. Their usefulness is enhanced by their essentially unrestricted ecological distribution, and facies independence, a function of their pelagic habit. High-resolution biostratigraphic zonation schemes based on the calcareous nannofossils have been established that have global applicability (Fig. 3.8). The schemes for the later Tertiary and Quaternary have been calibrated against the absolute chronostratigraphic timescale by astronomical tuning.

In my working experience, calcareous nannofossils have proved of particular use in stratigraphic subdivision and correlation in the following areas.

Mesozoic

The Jurassic–Cretaceous of the Middle East. Here, the lower, early–middle Oxfordian Hawtah member of the Hanifa formation of Saudi Arabia is marked by *Vekshinella stradneri* (Hughes, in Bubik and Kaminski, 2004). The regional Early Cretaceous, Berriasian K20 maximum flooding surface or MFS of Sharland *et al.* (2001), stratotypified in the Minagish formation of Kuwait, is marked by calcareous nannofossils of the KN52/KN53 Zone. The Hauterivian K40 MFS, stratotypified in the lower Zubair formation, is marked by calcareous nannofossils of the KN48/KN49 Zone. The Barremian K50 MFS, stratotypified in the middle Zubair formation, is marked by calcareous nannofossils of the KN46 Zone. The Late Cretaceous, Santonian K160 MFS, stratotypified in the Shargi member of the Fiqa formation of Oman, is marked by calcareous nannofossils of the *Marthasterites furcatus* Zone.

The Cretaceous of the North Sea. Here, calcareous nannofossils form part of the basis of the regional

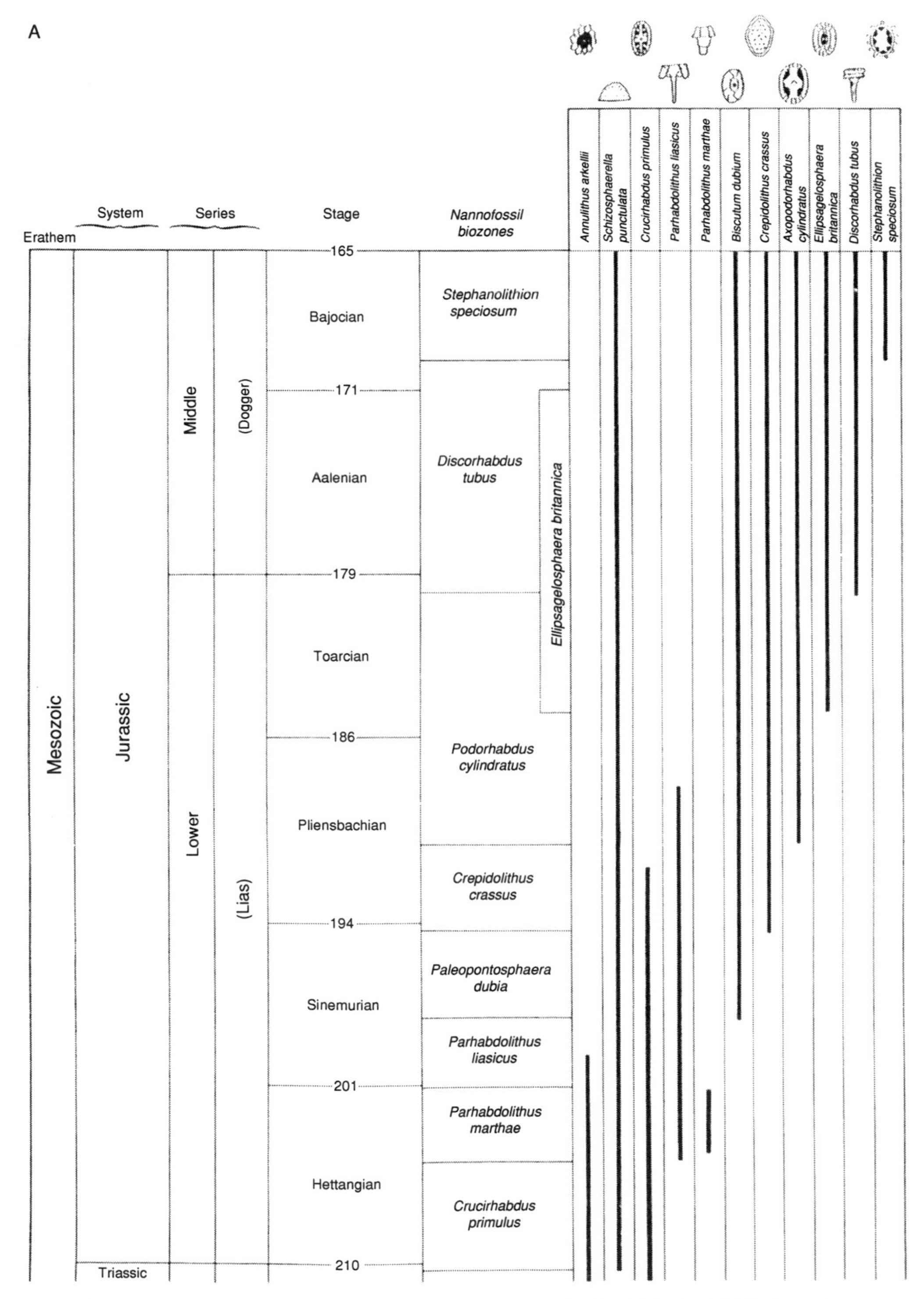

Fig. 3.8. Stratigraphic distribution of calcareous nannofossils. (A) Early–Middle Jurassic. (B) Middle–Late Jurassic. (C) Early Cretaceous. (D) Late Cretaceous. (E) Palaeogene. (F) Neogene–Pleistogene. (From Jones, 1996; reproduced with permission from Lipps, 1993.)

B

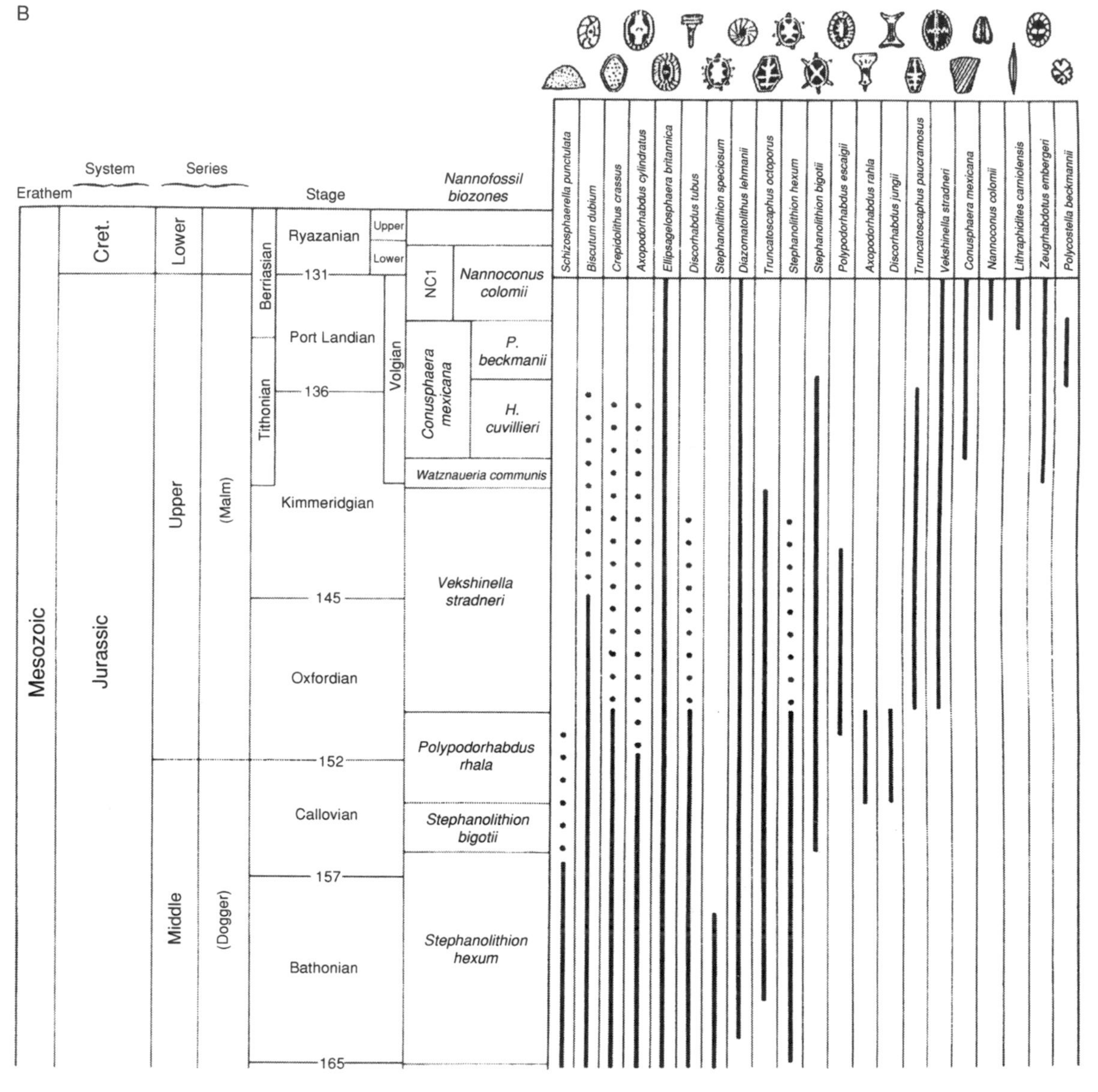

Fig. 3.8 *(cont.)*

biozonation, calibrated against ammonite and other standards. Importantly, they are also of use in the characterisation and exploitation of Late Cretaceous (to Early Palaeocene) Chalk group reservoirs in a number of fields in the British, Norwegian and Danish sectors (Shipp, in Jones and Simmons, 1999; Jones *et al.*, in Koutsoukos, 2005; see also Section 7.1). These reservoirs are actually composed of calcareous nannofossil oozes or chalks. The characteristics of certain of these reservoirs are controlled primarily by chalk composition and 'nannofabric', and only secondarily by diagenesis.

Cenozoic

The Cenozoic of the Middle East. Here, calcareous nannofossils proved important in dating the timing of movement of the Aksu thrust in the Isparta angle in south-west Turkey, as post-'mid' Pliocene, younger than previously accepted (Poisson *et al.*, in Akinci *et al.*, 2003).

C

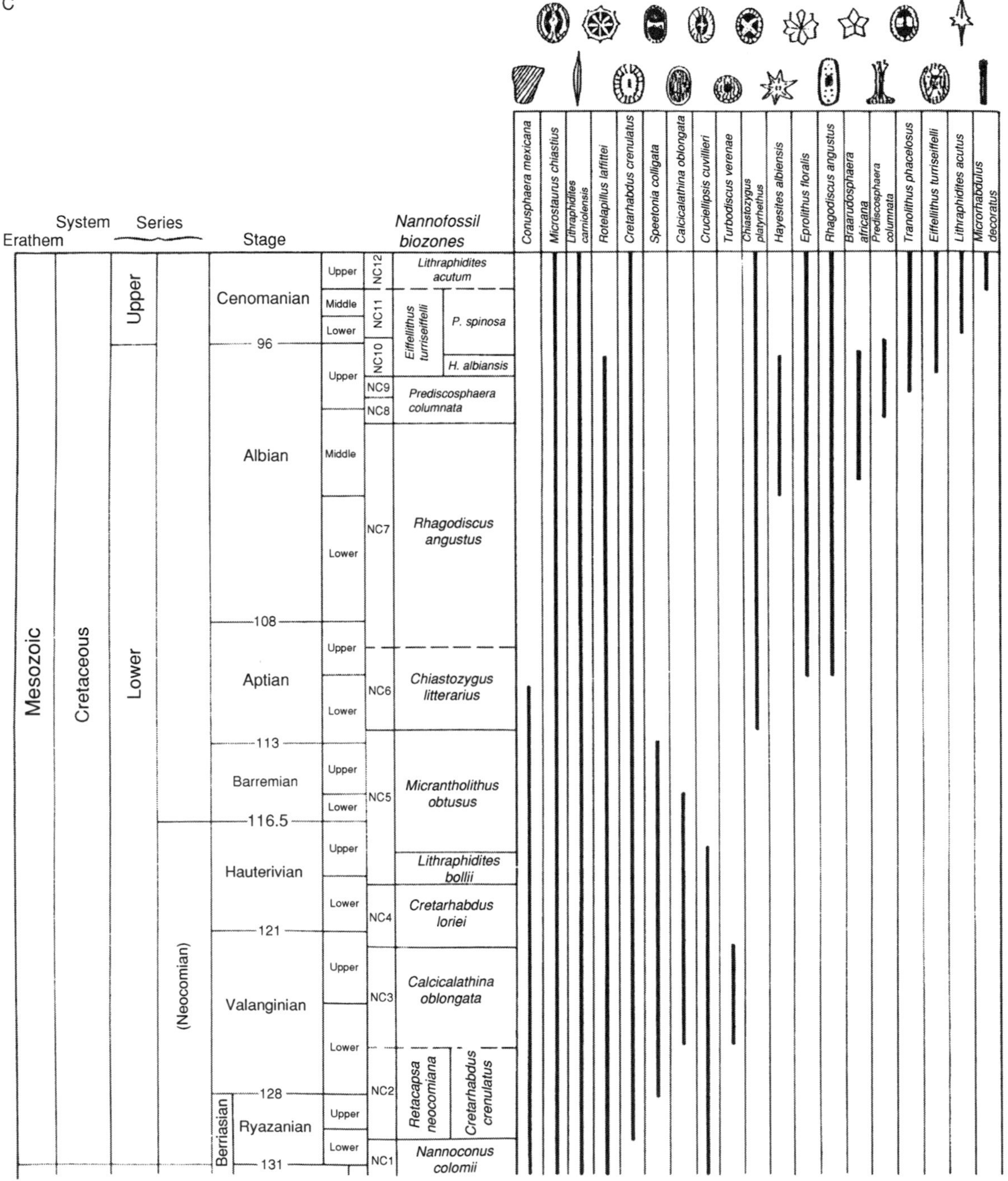
Erathem
System
Series
Stage
Nannofossil biozones
Conusphaera mexicana
Microstaurus chiastius
Lithraphidites carniolensis
Rotelapillus laffittei
Cretarhabdus crenulatus
Speetonia colligata
Calcicalathina oblongata
Cruciellipsis cuvillieri
Turbodiscus verenae
Chiastozygus platyrhethus
Hayesites albiensis
Eprolithus floralis
Rhagodiscus angustus
Braarudosphaera africana
Prediscosphaera columnata
Tranolithus phacelosus
Eiffellithus turriseiffelli
Lithraphidites acutus
Microrhabdulus decoratus
Mesozoic
Cretaceous
Upper
Lower
(Neocomian)
Cenomanian
Albian
Aptian
Barremian
Hauterivian
Valanginian
Berriasian
Ryazanian
96
108
113
116.5
121
128
131
Upper
Middle
Lower
NC12
NC11
NC10
NC9
NC8
NC7
NC6
NC5
NC4
NC3
NC2
NC1
Lithraphidites acutum
Eiffellithus turriseiffelli
P. spinosa
H. albiansis
Prediscosphaera columnata
Rhagodiscus angustus
Chiastozygus litterarius
Micrantholithus obtusus
Lithraphidites bollii
Cretarhabdus loriei
Calcicalathina oblongata
Retacapsa neocomiana
Cretarhabdus crenulatus
Nannoconus colomii

Fig. 3.8 (*cont.*)

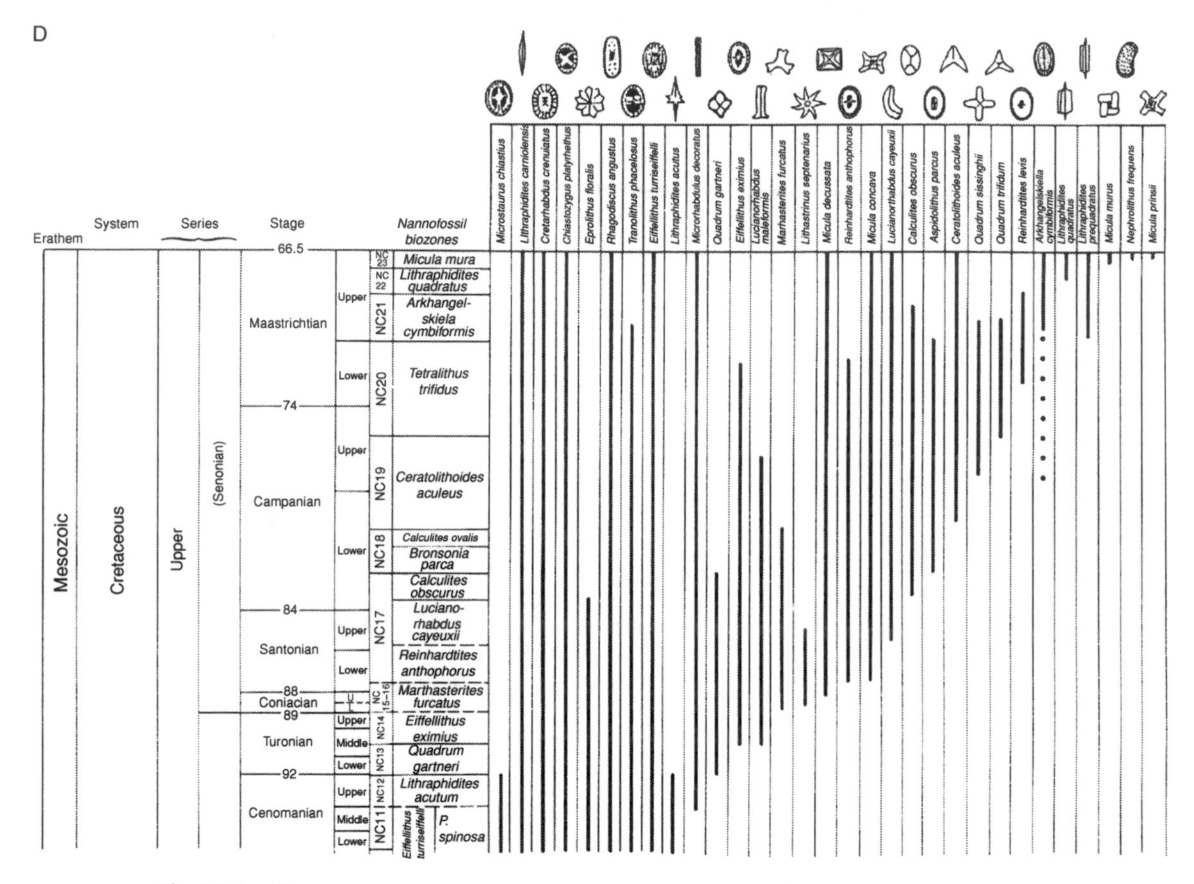

Fig. 3.8 (*cont.*)

The Cenozoic of northern South America and the Caribbean. See Flor *et al.* (1997). Here, specifically in the Neogene–Pleistogene of the Orinoco Delta and Columbus basin, offshore Trinidad, calcareous nannofossils form part of the basis of the regional biostratigraphic zonation scheme, part of the basis of the calibration of the regional sequence stratigraphic scheme, and part of the basis of the identification of time-slices for facies mapping purposes (author's unpublished observations).

The 'Mid'-Cenozoic of Angola in west Africa. See Grant *et al.* (2000). Here, calcareous nannofossils form part of the basis of the regional biozonation of the Oligocene–Miocene. They also form part of the basis of the calibration of the regional sequence stratigraphic framework (and the basis of the identification of time-slices for facies mapping purposes). Calcareous nannofossil abundance peaks in what benthic foraminiferal sedimentological data demonstrate to be autochthonous deposits have been used to characterise condensed sections.

The 'Mid'-Cenozoic of Paratethys. (Meszaros, 1992a, b; Nagymarosy and Voronina, 1993; Jones, 1996; Jones and Simmons, 1996; Jones and Simmons, in Robinson, 1997). Here, calcareous nannofossils are restricted in their development, because of the isolated geological evolution of the region. They are only periodically sufficiently well developed to enable ties to global biostratigraphic zonation schemes, at times of maximum flooding surfaces and normal marine connection to the world's oceans, as, for example, in the Maykopian (Oligocene–Early Miocene), Maeotian (Late Miocene) and Akchagylian (Late Pliocene).

E

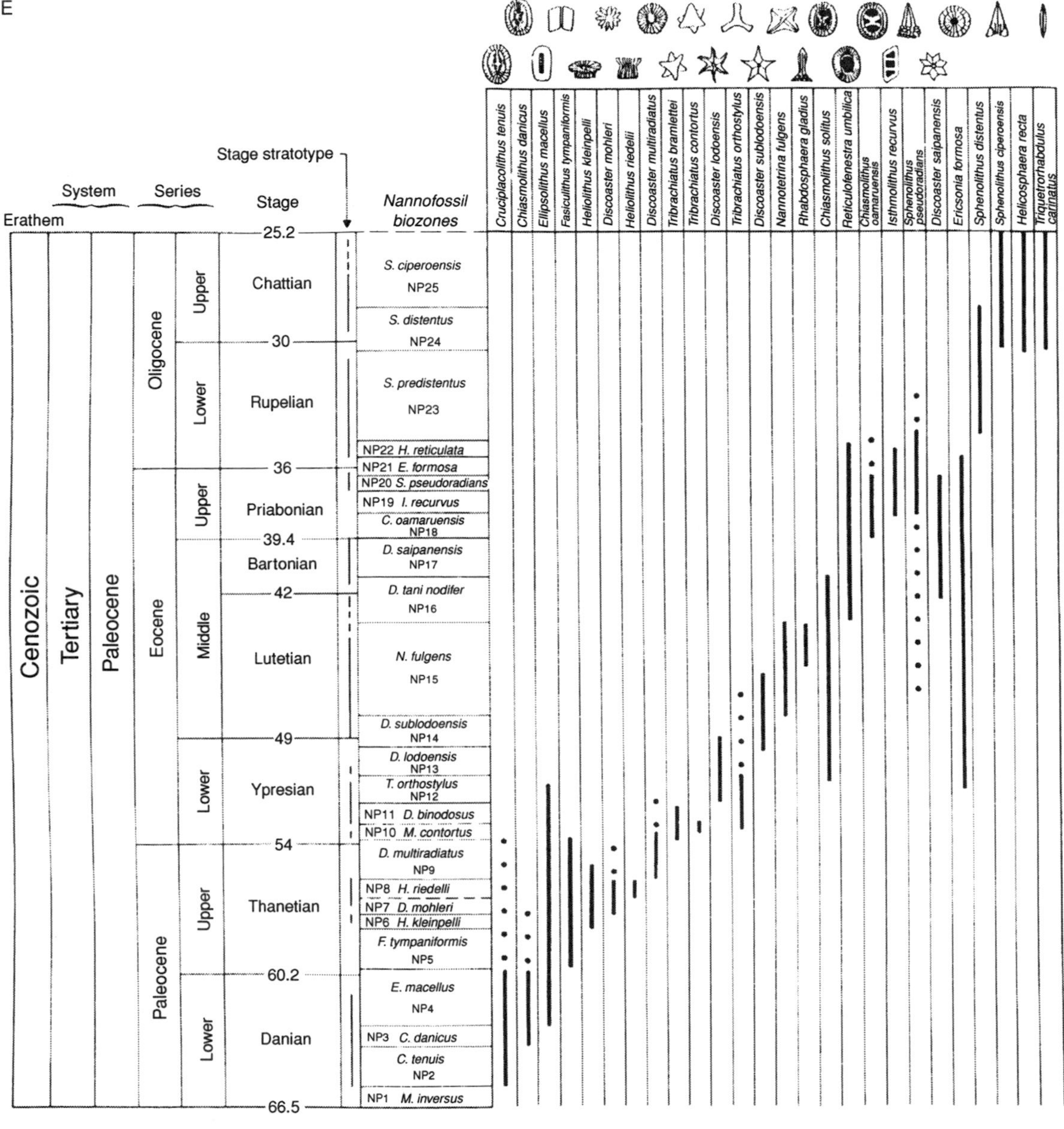

Fig. 3.8 (*cont.*)

The Neogene of the Pearl River Mouth basin, South China Sea. See Lusheng Huang (1997).

The Neogene–Pleistogene of the Gulf of Mexico. See Jones (1996). Here, calcareous nannofossils form part of the basis of the regional biostratigraphic zonation schemes, calibrated against the global standard, and part of the basis of the calibration of the various sequence stratigraphic schemes. They also form the basis of the identification of time-slices for facies mapping purposes.

Other applications

Calcareous nannofossils are useful in engineering geology (see Section 7.4). Calcareous nannofossil

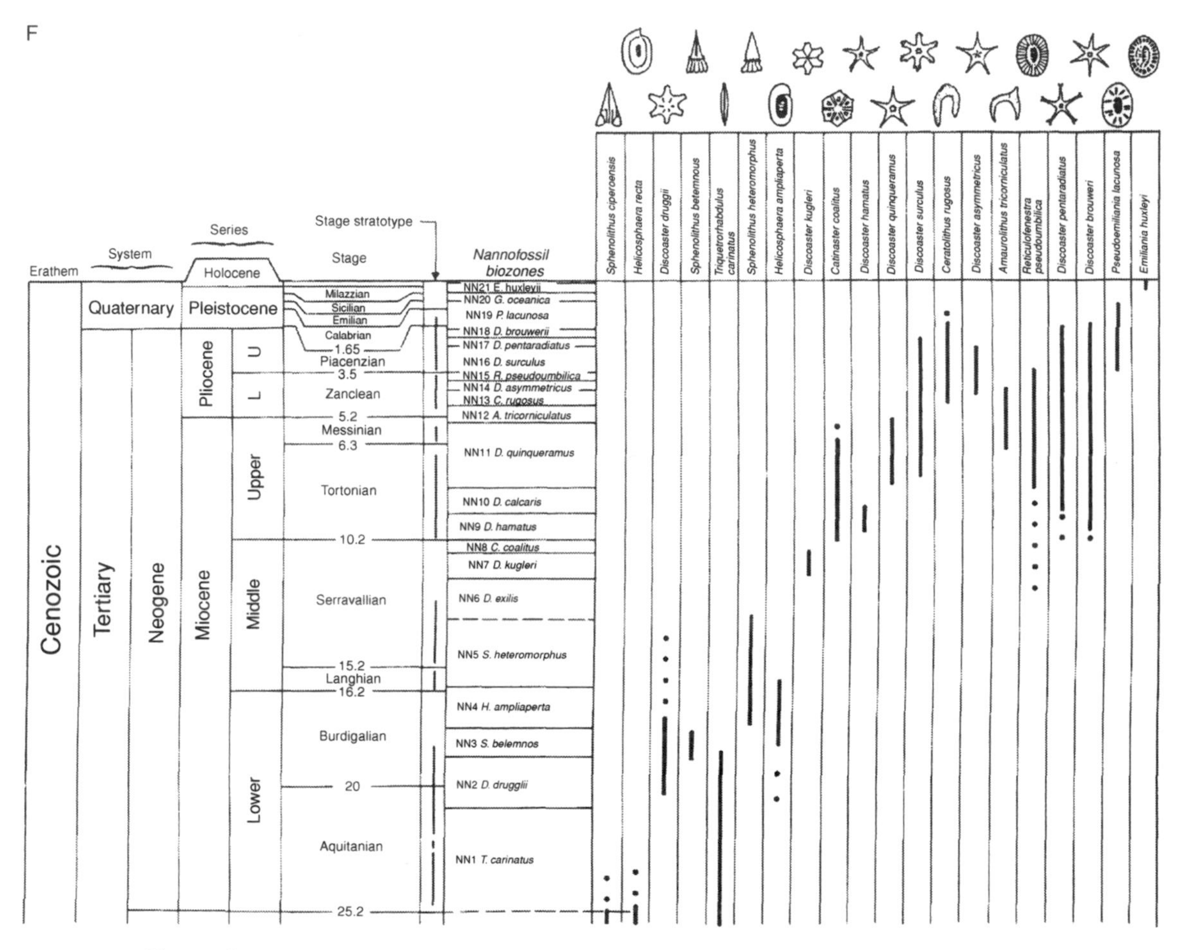

Fig. 3.8 *(cont.)*

chalks have peculiar engineering properties, controlled primarily by their 'nannofabrics'. For example, some are peculiarly dense on account of the close packing of the subcubical calcareous nannofossil *Micula* (author's unpublished observations).

3.2.5 *Calcareous algae (rhodophytes, chlorophytes and charophytes)*

Biology, morphology and classification

Biology

Calcareous algae are a disparate, extant, Precambrian–Recent group characterised by various types of pigmentation used in photosynthesis, and by calcareous skeletons (Toomey and Nitecki, 1985; Cole and Sheath, 1990).

In terms of ecology, they occupy a range of aquatic, freshwater and shallow marine, benthic environments. They are sufficiently abundant in some shallow marine environments to be important contributors to the formation of sediments. This is true both in tropical waters, where they constitute an important part of the chlorozoan group of grain types in carbonate sediments, and in temperate waters, where they constitute an important part of the foramol group (Lees *et al.*, 1969; Lees and Buller, 1972; Bosence, 1980). Rates of organic calcium carbonate production in temperate areas are comparable to those in tropical, non-reef areas, as in the Bahamas. However, rates are an order of magnitude lower than those in tropical reef areas.

Calcareous algae are especially characteristic of the temperate-water carbonates of the British Isles and elsewhere in western Europe, which are also known as 'maerls' (a Breton word), and which also, incidentally, are often misidentified as coralliferous, even on otherwise unimpeachably accurate Admiralty charts! Three main carbonate facies have been recognised between Roundstone and Clifden in Co. Galway in Connemara in the west of Ireland, namely algal gravels and beach sands ('coral strands'), composed primarily of the red alga *Lithothamnium*; algal–sponge–bryozoan–molluscan–ostracod sands; and molluscan sands (Lees *et al.*, 1969). Incidentally, very fine foraminiferal sands, velvety to the touch of the toes, are also found in Dog's Bay (author's unpublished observations). In Mannin Bay, the main facies are autochthonous *Lithothamnium corallioides* – and *Phymatolithon calcareum* – banks in water depths of 1–8 m; rippled algal gravels in exposed, high-energy areas; muddy algal gravels in partly exposed, intermediate-energy areas; and lime muds, in protected, low-energy areas (Bosence, 1980). Each facies is characterised by particular growth forms: the bank facies by unattached, interlocked, branched coralline red algae in growth positions, locally apparently bound together by filamentous algae (or the sea-grass *Zostera*); the clean algal gravel facies by densely branched sphaeroidal to ellipsoidal red algal rhodoliths formed by wave abrasion; the muddy algal gravel facies by open-branched morphotypes such as *Lithophyllum fasciculatum*; and the lime mud facies by epiphytes completely covering the corallines, and by oncolites trapped by filamentous algae. The main factor controlling overall facies distribution is coastal morphology, and its direct effects on wave and tidal energy, and indirect effects on turbulence and turbidity, in turn affecting light penetration.

Calcareous algae are also, incidentally, characteristic of the temperate-water carbonates of the 'Coralline Crag' of the Pliocene of East Anglia (Hodgson and Funnell, in Hart, 1987).

Morphology

In terms of hard-part morphology, calcareous algae are characterised by the possession of calcareous skeletons (Fig. 3.9). Growth forms observed in the Early Oligocene Gornyi Grad beds of northern Slovenia range from encrusting through lamellar and protuberant to arborescent; taphonomic features observed range from disarticulation to fragmentation, and also include encrustation and bio-erosion (Nebelsick and Bassi, in Insalaco *et al.*, 2000). Encrusting growth forms are rare in foraminiferal–coralline and grainstone facies, and common in coral facies. Arborescent growth forms are common only in grainstone facies.

Classification

Classification of calcareous algae is based on biology and morphology. Three sub-groups are recognised, namely the red algae or rhodophytes, the green algae or chlorophytes, and the stoneworts or charophytes.

Rhodophytes. Rhodophytes are characterised by red coloration caused by the photosynthetic pigment phycoerythrin, and by calcareous, calcitic skeletons. They are exclusively shallow marine, and photosynthesise within the euphotic and sub-euphotic zones in order to manufacture food. They reproduce both sexually and asexually, the structures involved in asexual reproduction being termed sporangia, those involved in sexual reproduction termed conceptacles. They have differentiated inner and outer cellular layers (hypothalli and perithalli respectively), the outer layers (perithalli) containing the conceptacles. The most important sub-group of red algae is the corallines, which includes encrusting, or crustose, and branching, or articulated, morphotypes. Encrusting morphotypes are characteristic of reef cores, branching morphotypes of reef complexes generally.

Chlorophytes. Chlorophytes are characterised by green pigmentation caused by chlorophyll, and usually, although not invariably, by calcareous, usually, although not invariably, aragonitic skeletons. They are both freshwater and shallow marine, and photosynthesise within the euphotic zone in order to manufacture food. They reproduce both sexually and asexually. In terms of hard-part morphology, they are characterised by cylindrical, branching or filamentous bodies sheathed by thalli, and often attached to the sea-floor by means of

A

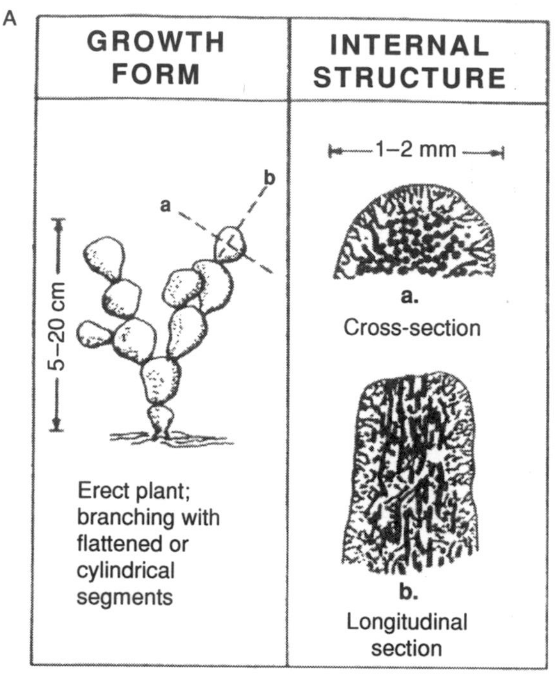

B

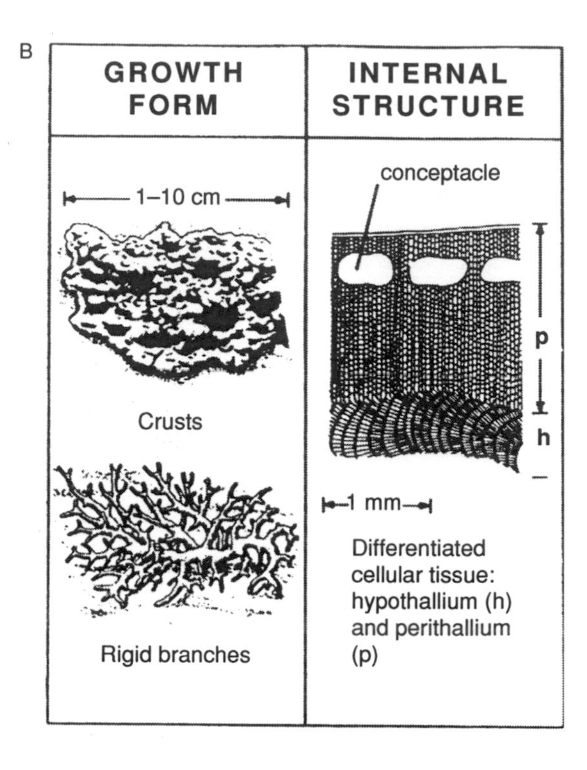

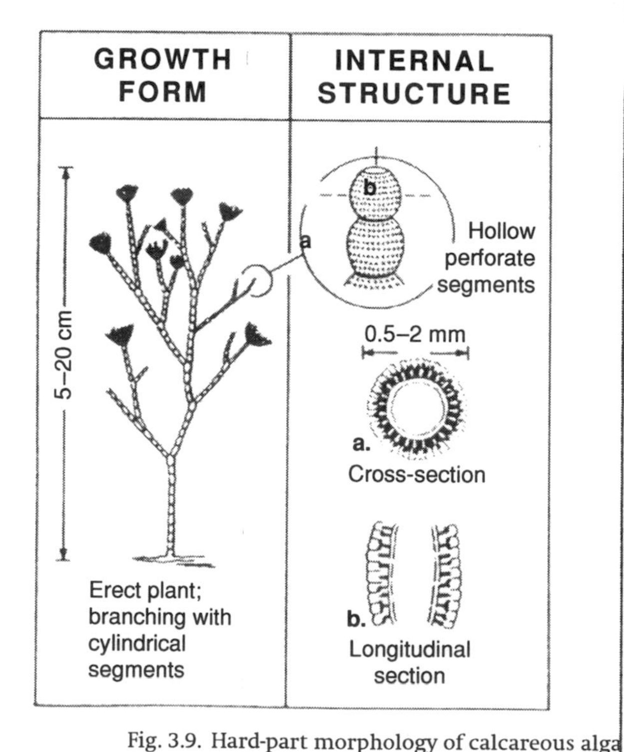

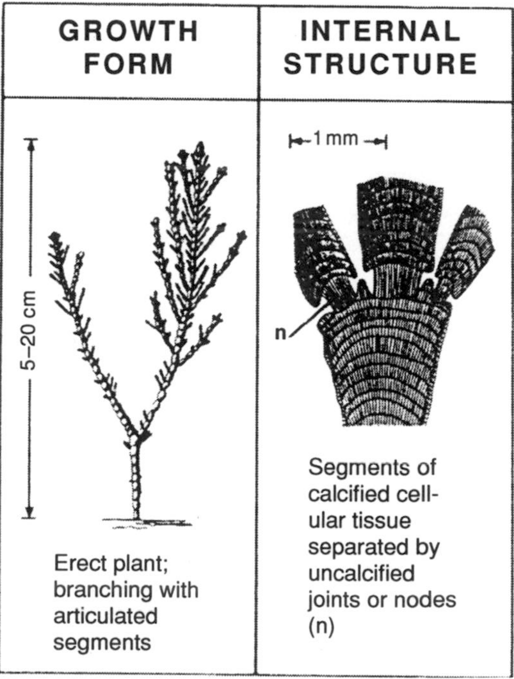

Fig. 3.9. Hard-part morphology of calcareous alga[e. (A) Green algae or chlorophytes.] Top, codiaceans; bottom, dasycladaceans. (B) Red algae or rhodophytes. Top, encrusting corallines; bottom, articulated corallines. (C) Charophytes. (From Jones, 1996; in turn after Haq and Boersma, 1978.)

C

Fig. 3.9 (*cont.*)

hold-fasts or rhizoids. The most important subgroups of the green algae are the codiaceans, also known as the udotaceans, and the dasycladaceans. Dasycladacean green algae are characteristic of sheltered back-reef lagoons, adjacent to reef cores.

Non-calcareous green algae such as free-living or colonial *Botryococcus* and *Pediastrum*, which commonly occur in palynological preparations, are essentially freshwater, although some species of *Botryococcus* appear to have some brackish tolerance (Batten and Grenfell, in Jansonius and MacGregor, 1996; van der Zwan and Brugman, in Jones and Simmons, 1999).

Charophytes. Charophytes are characterised by green pigmentation caused by chlorophyll, and by calcareous skeletons. They are essentially freshwater, although some appear to have some brackish tolerance, and instances of transportation into evidently marine environments are not unknown.

Palaeobiology

Fossil calcareous algae are interpreted, essentially on the basis of analogy with their living counterparts, as having occupied a range of aquatic, freshwater and shallow marine, habitats. Charophytes and non-calcareous green algae are essentially freshwater; calcareous green algae freshwater and shallow marine; and red algae shallow marine.

Charophytes are of considerable use in the palaeoenvironmental interpretation of freshwater environments, for example in the Oligocene 'molasse' of the Penzberg syncline in south Germany, where distinct deltaic, lagoonal, estuarine, riverine and lacustrine sub-environments have been identified (Reichenbacher *et al.*, 2004).

Palaeobathymetry

Calcareous algae photosynthesise within the photic zone in order to manufacture food. They are especially useful in palaeobathymetric interpretation, as specific taxa have specific light requirements and depth distributions even within the photic zone (light penetration being dictated primarily by water depth).

Calcareous algae were sufficiently abundant in some shallow marine environments as to have been important contributors to the construction of reefs, as, for example, in the Palaeozoic, Late Ordovician to Devonian and Late Carboniferous to Permian of the former Soviet Union (Kuznetsov, in Zempolich and Cook, 2002), the Late Devonian of the Canning basin of Western Australia (Wood, 2004), the Carboniferous to Triassic worldwide, the Cretaceous of the Middle East, and the Cenozoic of Papua

New Guinea in the Far East (author's unpublished observations).

In the Palaeozoic, the Late Devonian mass extinction caused reef ecosystems to collapse (or, in the case of the Canning basin of Western Australia, to reorganise), and resulted in a 'reef gap' in the rock record from latest Devonian, Famennian to Early Carboniferous, Dinantian (Pickard, in Strogen *et al.*, 1996). This time interval was characterised by the general absence of reef-building organisms and of reefs, but also by the extensive development of mud-dominated reef-like structures or mounds known as Waulsortian mounds or Waulsortian-like mounds. These structures are thought to have resulted partly from algal activity, and are distinguishable from their level-bottom equivalents partly in terms of elevated incidences of algae, indicating development within the photic zone (Lees and Miller, in Monty *et al.*, 1995; Jeffery and Stanton, in Strogen *et al.*, 1996).

Recognisably reef-like structures, characterised in part by the green alga *Koninckopora* and by the red alga *Ungdarella*, reappeared in the Asbian–early Brigantian, as in the case of the Mullaghfin formation of Ardagh on the margins of the Dublin basin; true reefs, characterised by the phylloid/ancestral coralline red alga *Archeolithophyllum*, in the late Brigantian (Somerville *et al.*, in Strogen *et al.*, 1996). Note that the Asbian–Brigantian was an important time of diversification of calcareous algae. Not only the red algae *Ungdarella* and *Archaeophyllum* but also the palaeoberesellid green alga *Kamaenella*, first appeared at this time, to rise to prominence in the reefs of the Late Carboniferous. Early Late Carboniferous, Bashkirian–early Moscovian carbonate build-ups are also characterised by *Dvinella/Donezella, Komia, Cuneiphycus* and phylloids (Wahlman, in Kiessling *et al.*, 2002). Late Late Carboniferous, late Moscovian–Gzhelian build-ups are characterised by erect udotacean phylloid algae, and, in deeper water, by encrusting red algae. Early Permian, Asselian–Sakmarian build-ups are characterised by 'baffling' erect phylloid algae, and 'binding' encrusting red algae. Middle–Late Permian build-ups such as the Capitan Reef of the Permian Basin of west Texas are characterised by similar, although more evolved, biotas. Late Carboniferous–Early Permian temperate-water, as opposed to tropical-water, build-ups, such as those of the north-western USA, Arctic Canada, the Barents Sea, and the Timan–Pechora basin, are characterised by *Palaeoaplysina*, phylloid algae and *Tubiphytes*.

The Asbian Urswick limestone formation of the southern Lake District in northern England has been subject to a semiquantitative sedimentological analysis (Horbury and Adams, in Strogen *et al.*, 1996). Cyclic sedimentation is evident. Cycle bases, interpreted as representing water depths of 10–20 m, are typically characterised by diverse algal assemblages, including *Coelosporella* and *Stacheoides*. The middles of cycles, representing water depths of 5–10 m, are characterised by *Kameana, Kamaenella, Epistacheoides* and *Ungdarella*. Cycle tops, representing water depths of 0–10 m, are characterised by *Koninckopora, Anatolipora* and *Polymorphcodium*. Note, though, that there are significant variations in allochem distribution according to palaeogeography as well as bathymetry.

The Asbian–Brigantian of the Burren, Buttevant and Callan areas of southern and western Ireland has been subject to a biostratigraphic and palaeobiological analysis based on foraminiferans and calcareous algae (Gallagher, in Strogen *et al.*, 1996). In terms of palaeoecology, the entire interval is characterised by cyclic platform sedimentation, with evidence for a change in the style of cyclicity, associated with deepening, at the Asbian/Brigantian boundary. The main contributors to carbonate production in the Asbian were the calcareous algae *Koninckopora, Ungdarella* and *Kamaenella*.

Palaeobiogeography

Many calcareous algae appear to have had restricted or endemic biogeographic distributions, rendering them of use in the characterisation of palaeobiogeographic provinces, and in turn in the constraint of plate tectonic reconstructions, in the Palaeozoic (Watkins and Wilson, 1989; Poncet, in McKerrow and Scotese, 1990).

Palaeoclimatology

Charophytes are of some use in palaeoclimatic interpretation, although the isotopic records that they reveal have to be interpreted with caution on account of possible disequilibrium effects (Andrews *et al.*, 2004).

Biostratigraphy

The oldest known rhodophytes are Mesoproterozoic, the oldest chlorophytes Neoproterozoic. The overall pattern of rhodophyte and chlorophyte evolution through the Phanerozoic has been one of diversification offset by loss of diversity associated with the various mass extinction events of that time. The Ordovician diversification of the calcareous algae is discussed by Nitecki *et al.*, in Webby *et al.* (2004).

The oldest known charophytes are Devonian, from the renowned Rhynie Lagerstatte of Aberdeenshire in north-east Scotland (Kelman *et al.*, 2003).

The moderate rate of evolutionary turnover exhibited by the calcareous algae renders them of some use in biostratigraphy, at least in appropriate environments. Unfortunately, the usefulness of many forms is compromised by their restricted bathymetric and biogeographic distribution and facies dependence. Nonetheless, biostratigraphic zonation schemes based in part on the calcareous algae have been established, that have at least local to regional applicability. Charophytes are of at least some biostratigraphic use in, for example, the non-marine facies of the Mesozoic–Cenozoic of Europe (Riveline *et al.*, 1996).

In my working experience, calcareous algae have proved of some use in stratigraphic subdivision and correlation, and perhaps more particularly in palaeoenvironmental and palaeobathymetric interpretation, in the following areas.

Palaeozoic

The Devonian–Permian of the former Soviet Union. Here, calcareous algae form part of the basis of the regional biozonation of the Carboniferous (Wagner *et al.*, 1996). Also here, calcareous algae contribute to the construction of reefs that in turn form subsurface petroleum reservoirs in the Late Devonian–Early Permian of the Peri-Caspian (Bliefnick *et al.*, in Ahr *et al.*, 2000), and surface outcrops and analogues in the Bolshoi Karatau and northern Tien-Shan mountains in Kazakhstan (Cook *et al.*, in Ahr *et al.*, 2000; Zempolich *et al.*, in Ahr *et al.*, 2000). *Tubiphytes* is characteristic of the shoal and mound facies of the Carboniferous, Visean to Serpukhovian; *Donezella, Archaeolithoporella* and phylloid algae of the Carboniferous, Bashkirian to Gzelian; *Tubiphytes, Archaeolithoporella*, phylloids and *Palaeoaplysina* of the Early Permian (Vennin *et al.*, in Zempolich and Cook, 2002; Zempolich *et al.*, in Zempolich and Cook, 2002). Calcareous algae also contribute to the construction of reefs that in turn form subsurface reservoirs in the Late Devonian of the Timan–Pechora basin in the north-eastern part of the Russian platform and in the Late Carboniferous–Early Permian of the Timan–Pechora basin and of the Russian and Norwegian sectors of the Barents Sea (Wahlman and Konovalova, in Zempolich and Cook, 2002; author's unpublished observations), and surface outcrops in the Shernyadeyyta and Kozhim areas of the Timan–Pechora basin and of Bjornoya and Spitsbergen in the Barents Sea (Wahlman and Konovalova, in Zempolich and Cook, 2002). *Tubiphytes* is characteristic of cooler or deeper water facies, *Palaeoaplysina* of warmer or shallower water facies. Significantly, neither occurs in the cold, deep-water facies of the Late Permian.

The Permian of the Middle East. Here, charophytes mark the Early Permian Unayzah formation in Saudi Arabia (Hill and El-Khayal, 1983). Various calcareous green and red algae mark the Late Permian–Early Triassic Khuff formation of Saudi Arabia and its correlatives elsewhere (Jones, 1996).

Mesozoic

The Mesozoic of the Middle East. Here, calcareous algae are of some use in stratigraphic subdivision and correlation, as in the dating of the Triassic 'elementary sequence' and reconstruction of the Isparta angle in south-west Turkey (Vrielynck, in Akinci *et al.*, 2003). They are also of use in the characterisation of the various formations of the Jurassic and Early Cretaceous of the Arabian plate (Jones, 1996). The Late Jurassic, late Kimmeridgian J90 MFS of Sharland *et al.* (2001), stratotypified in Member B of the Arab formation of Abu Dhabi, is marked by the calcareous alga *Clypeina jurassica*. The equivalent late part of the Surmeh formation of the Zagros Mountains of south-west Iran is also marked by *C. jurassica* (author's unpublished observations). Calcareous algae are also of considerable use in high-resolution palaeobathymetric and sequence stratigraphic interpretation, and in parasequence-scale reservoir characterisation (Banner and Simmons, in Simmons, 1994; author's unpublished observations). Calcareous algae are of considerable importance as contributors to the formation of platformal and peri-reefal carbonate reservoirs in the Mesozoic of the Middle East. The Late Jurassic Arab D reservoir of Saudi Arabia contains palaeoberesellids in abundance (Adams and Al-Zahrani, 2000). The Early Cretaceous Thamama group reservoirs of Abu Dhabi contain *Bacinella/Lithocodium* in abundance, indeed locally in sufficient abundance as to be rock-forming (author's unpublished observations). Here, reservoir quality is controlled primarily by depositional facies, and only secondarily by diagenetic factors, and the best reservoir quality is associated with *Bacinella/Lithocodium* boundstone formed in water depths of 30–50 m.

Cenozoic

The Cenozoic of Papua New Guinea. Here, calcareous algae are of particular use in the palaeoenvironmental and palaeobathymetric interpretation of the Oligo-Miocene Darai limestone formation (Jones, 1996).

The Neogene of Nigeria. Here, non-calcareous green algae are of particular use in palaeoenvironmental interpretation and reservoir characterisation (van der Zwaan, in Jones and Simmons, 1999).

The Neogene of Azerbaijan. Here, non-calcareous green algae are of particular use in palaeoenvironmental interpretation and in the characterisation and exploitation of Early Pliocene 'productive series' reservoirs (S. Lowe, unpublished observations).

3.2.6 *Acritarchs*

Biology, morphology and classification

Acritarchs are organic-walled microfossils of uncertain and probably disparate taxonomic affinity, some forms being at least arguably ancestral to dinoflagellates.

The biology and ecology of living acritarchs are poorly known.

In terms of morphology, acritarchs are characterised by a hollow spherical or subspherical body or vesicle (Fig. 3.10). Surface ornamentation in the form of spines or processes may be present. Excystment apertures may also be present. These appear analogous and homologous to those of dinoflagellates, which occur in the resting (cyst) stage of the life cycle, and facilitate the release of the motile cells that allow population growth.

Classification of acritarchs is based on details of morphology. Two main sub-groups are recognised, namely the smooth and unornamented sphaeromorphs, and the ornamented acanthomorphs.

Palaeobiology

Fossil acritarchs are interpreted, on the bases of analogy with their living counterparts, namely dinoflagellates, and associated fossils and sedimentary facies, as having been essentially marine, although arguably with some brackish tolerance, and probably planktonic.

Aggregates of individuals of the acritarch *Dilatisphaera laevigata* have been observed in the Silurian of the Downton Gorge section near Ludlow in the English Midlands (Mullins, 2003). These have been interpreted on the basis of analogy with living algae as having formed by faecal pelletisation, or by coagulation during phytoplankton blooms, or as a defence against predation.

Palaeobathymetry (shoreline proximity indication)

In the Silurian of the Welsh borderlands and the Devonian of western Canada, proximal, intermediate and distal assemblages are distinguishable on the basis of abundance, diversity, dominance and dominant morphogroup (Fig. 3.11) (Jones, 1996). In the Devonian of western Canada, proximal assemblages are characterised by simple subspherical forms, intermediate assemblages by thin-spined forms and distal assemblages by thick-spined forms.

Some measure of the favourability or otherwise of the environment for reproduction is provided by the proportion of specimens possessing excystment apertures.

Palaeobiogeography

Many acritarchs appear to have had restricted or endemic biogeographic distributions, rendering them of use in the characterisation of palaeobiogeographic provinces, and in turn in the constraint

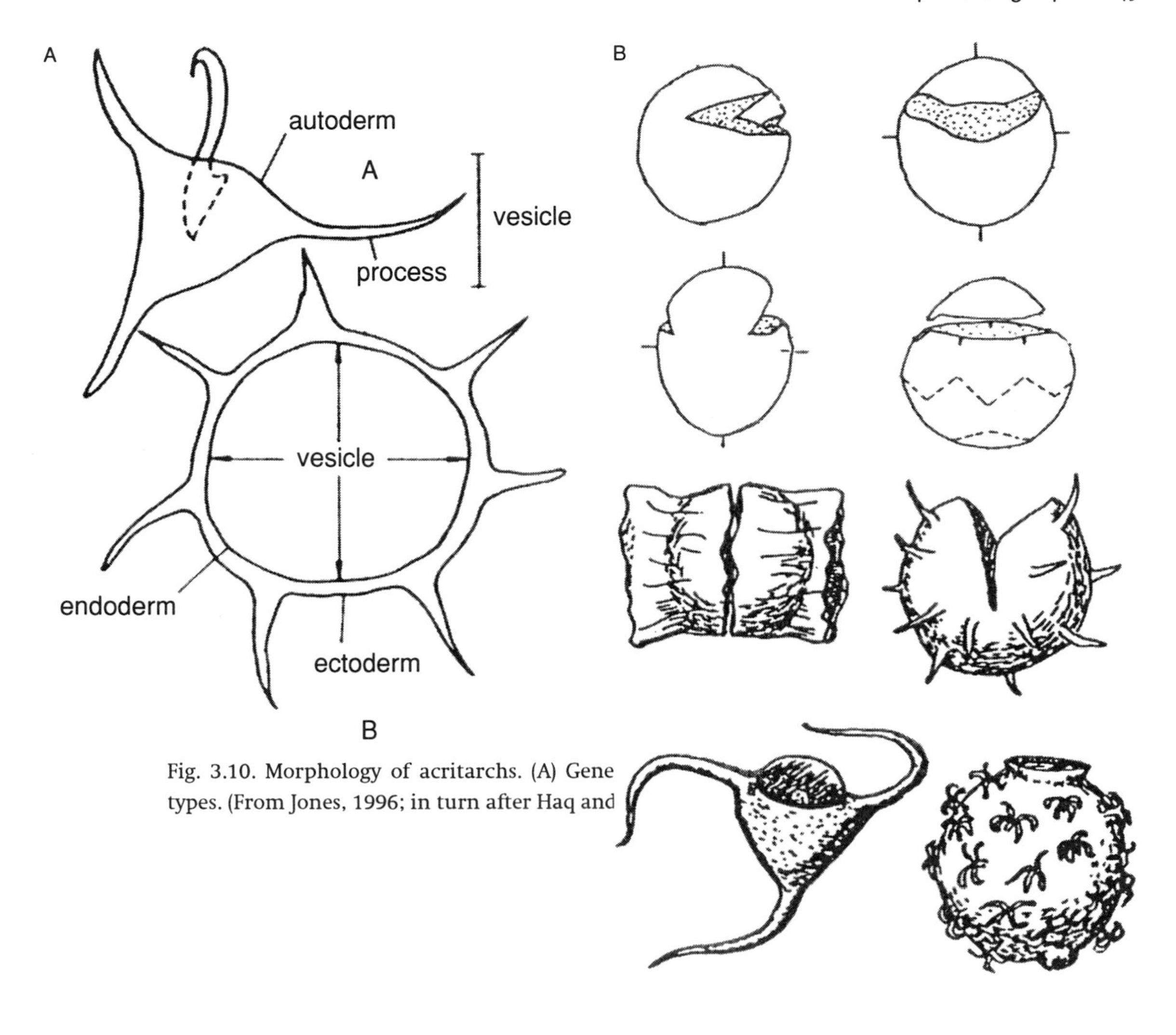

Fig. 3.10. Morphology of acritarchs. (A) Gene
types. (From Jones, 1996; in turn after Haq and

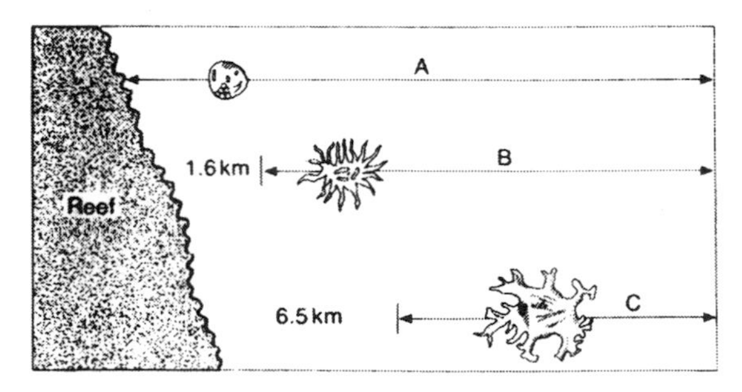

Fig. 3.11. Palaeoecological distribution of acritarchs. (A) Simple, spherical 'morphogroup'. (B) Thin-spined 'morphogroup'. (C) Thick-spined 'morphogroup'. (From Jones, 1996; reproduced with permission from Lipps, 1993.)

of plate tectonic reconstructions, in the Palaeozoic. Cramer and Diez (1978) recognised an American palynological unit and an African palynological unit (or *Coryphidium bohemicum* province) in the Early Ordovician, separated by an Appalachian oceanic gap; an African palynological unit or *Multiplicisphaeridium pilaris* province in the Arenig–Llanvirn; and a Baltic acritarch province and a 'Gondwanan' *Neoveryhachium carminae* facies in the Middle Silurian, late Llandovery–early Wenlock.

Biostratigraphy

Acritarchs evolved in the Proterozoic (Vidal and Knoll, 1982). The oldest known, sphaeromorph, forms are from the Palaeoproterozoic Gunflint chert of western Ontario. The oldest known acanthomorphs are from the Mesoproterozoic Lakhanda formation of east Siberia or the Neoproterozoic Doushantuo formation of south China (see also Yin Leiming, in Jin Yugan *et al.*, 1991). Incidentally, specimens from the Doushantuo formation are remarkably well preserved, in three dimensions (Zhou Chuanming *et al.*, 2001).

Acritarchs diversified in the Proterozoic, Riphean, before sustaining significant losses in the Vendian (Vidal and Moczydlowska Vidal, 1997; Moczydlowska, 2004). They recovered from this event, though, and underwent further adaptive radiations in the Cambrian, arguably in response to selection pressure exerted by the grazing activities of the newly evolved arthropods (Butterfield, cited in Conway Morris, 1998), and in the Ordovician (Vecoli and le Herisse, 2004; Servais *et al.*, in Webby *et al.*, 2004). They then sustained further significant losses in the End-Ordovician mass extinction (Vidal and Moczydlowska Vidal, 1997; Vecoli and le Herisse, 2004). However, they recovered from this event, too, and diversified again in the Silurian (Kaljo, in Hart, 1996; le Herisse, in Al-Hajri and Owens, 2000). They sustained severe losses in the Late Devonian mass extinction, from which they may be regarded as never having really recovered.

The moderate rate of evolutionary turnover exhibited by the acritarchs renders them of some use in biostratigraphy, at least in appropriate, marine environments. Their usefulness is enhanced by their essentially unrestricted ecological distribution, and facies independence, a function of their planktonic habit. Biostratigraphic zonation schemes based on the acritarchs have been established that have at least local to regional applicability.

In my working experience, acritarchs have proved of some use in stratigraphic subdivision and correlation, and to a lesser extent in palaeoenvironmental interpretation, in the following areas.

Proterozoic

The Late Precambrian, Vendian of the Middle East. Here, the Soltanieh dolomite formation of north-east Iran contains the 'small disc-like' fossils *Fermoria*, *Chuaria* and *Beltanella* (Jones, 2000). *Chuaria* has been recorded from the Upper Valdai series of the Russian platform, which has been independently dated as Vendian.

Proterozoic to Palaeozoic

The Late Precambrian, Vendian to Early Cambrian, ?Tommotian of east Siberia. Here, the Olkha formation of the Pri-Sayan area, the Kachergat formation and Ushakova or Ushakovskaya suite of the Pri-Baikal area, and the Bochugunor and Tir horizons of the Nepa–Botuobin area, have all been dated as Riphean–Vendian (Jones, in Simmons, in press). However, at least locally, these formations contain not only 'Assemblage 2' acritarchs effectively indistinguishable from those of the Kotlin horizon of eastern Europe, which is of Late Precambrian, Vendian age, but also 'Assemblage 3' acritarchs effectively indistinguishable from those of the Nemakit–Daldyn horizon of the Pri-Anabar area of north-east Siberia, which is of Early Cambrian, Nemakit–Daldynian age, and those of the Rovno and Lontova horizons of the Lublin slope of south-east Poland and elsewhere in eastern Europe, which are also of probable Nemakit–Daldynian age. Key species in common include *Baltisphaeridium cerinum*, *B. ciliosum*, *B. pilosiusculum*, *B. primarium*, *B. strigosum*, *Granomarginata prima*, *G. squamacea*, *Leiospheridia* spp. including *L. gigantea*, *L. pellucida* and *L. simplex*, *Micrhystridium tornatum*, *Origmatosphaeridium rubiginosum* and *Tasmanites tenellus*. The Lower Moty or Motskaya suite of a number of subsurface wells in the Angara–Lena and Nepa–Botuobin areas also contains 'Assemblage 2–3' acritarchs of Late Precambrian, Vendian to Early Cambrian, Nemakit–Daldynian, age. The Danilov horizon of the Pri-Lena area contains enriched 'Assemblage 3' acritarch assemblages similar to those of the L(y)ukati horizon of eastern Europe, which is of Early Cambrian, ?Tommotian, age.

In general, the resolution of the acritarch-based biozonation of the Vendian of east Siberia is comparatively poor, with only two zones recognisable over a time interval that could be as long as 60 Ma, from 605 to 545 Ma. Note in this context, though, that the Vendian in east Siberia could be as short as 20 Ma, if the interpretation of significant lack of representation of section through onlap onto an unconformity surface of 'Pre-Vendian' age is correct. According to this interpretation, which is supported by recent isotopic stratigraphic data, only the younger part of the Vendian (the Ediacar(i)an of authors) is represented in Siberia; and the older part (Varangerian) is entirely unrepresented. This interpretation provides a possible explanation for the apparent absence in Siberia of glaciogenic sediments of older Vendian (Varangerian) age, which are widespread elsewhere in the world, as on the Russian platform and in China, Canada, Scandinavia, Oman, Namibia and Australia.

Palaeozoic

The Cambrian–Silurian of the Middle East. Here, acritarchs form part of the basis of the biozonation of the Cambrian–Silurian of Saudi Arabia (le Herisse, in Owens *et al.*, 1995; Stump *et al.*, in Owens *et al.*, 1995; Al-Hajri and Owens, in Al-Hajri and Owens, 2000). They also form part of the basis of the calibration of the regional sequence stratigraphic framework (Sharland *et al.*, 2001). The regional Early Ordovician, Tremadoc O10 MFS of Sharland *et al.* (2001), stratotypified in the Mabrouk member of the Andam formation of Central Oman, is marked by acritarchs. The regional Early Ordovician, Tremadoc O20 MFS, stratotypified in the Barakat member of the Andam formation, is also marked by acritarchs.

The Silurian–Carboniferous of north Africa. Here, acritarchs form the basis of a number of regional biozonations, for example, of the Silurian–Devonian of the Polignac basin in the Algerian Sahara (Jardine and Yapaudjian, 1968), and the Devonian–Carboniferous of the Grand Erg Occidental in the Algerian Sahara (Lanzoni and Magloire, 1969). They are also of use in palaeoenvironmental interpretation on the northern Sahara platform (Vecoli, 2000).

3.2.7 *Bolboforma*

Biology, morphology and classification

Bolboforma is of uncertain taxonomic affinity. Nothing is known of its biology as it is extinct. In terms of hard-part morphology, it is characterised by a subspherical calcitic body with an opening at one end (Fig. 3.12). Its classification is based on morphology, especially on surface ornamentation.

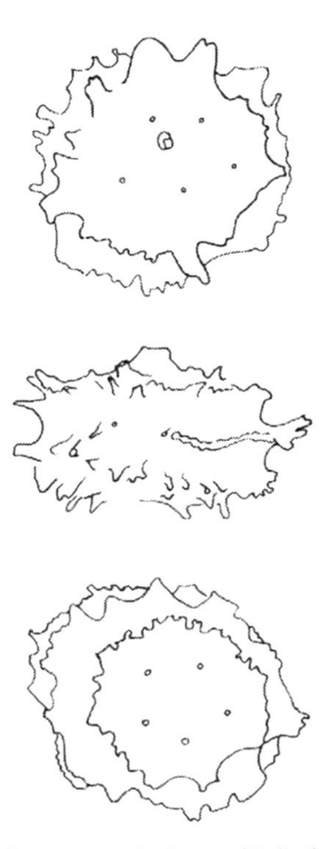

Fig. 3.12. Hard-part morphology of *Bolboforma*. (From Jones, 1996; in turn after Powell, 1986.)

Palaeobiology

Bolboforma is interpreted, essentially on the basis of associated fossils and sedimentary facies, as having been exclusively marine and planktonic.

It had a wide geographic distribution.

Biostratigraphy

Bolboforma evolved in the Eocene and became extinct in the Pliocene.

The moderate rate of evolutionary turnover exhibited by *Bolboforma* renders it of some use in biostratigraphy, at least in appropriate, marine environments. Its usefulness is enhanced by its essentially unrestricted ecological distribution, and facies independence, a function of its planktonic habit. Biostratigraphic zonation schemes based on *Bolboforma* have been established, that have at least local to regional applicability.

In my working experience, *Bolboforma* has proved of some use in stratigraphic subdivision and correlation in *the Oligocene–Miocene of central Paratethys* and in *the Miocene of the North Sea* (Jones, 1996).

In the Oligo-Miocene of central Paratethys, the Egerian–Eggenburgian regional stages are marked by *Bolboforma rotunda* and *B. spinosa*, the early Badenian by *B. reticulata*, the mid Badenian by *B. danielsi* and the late Badenian by *B. badensis* (Jones, 1996).

3.3 Animal-like protists or protozoans

3.3.1 *Foraminiferans*

Biology, morphology and classification

Foraminiferans are an extant, Cambrian–Recent, group of protozoans characterised by the possession of granulo-reticulose pseudopodia, and an agglutinated arenaceous or secreted calcareous shell or test (Jones, 1996; Sen Gupta, 1999; Holbourn *et al.*, 2004). Among their closest relatives are the living testate *Amoebae* or 'thecamoebans', which have little documented fossil record beyond the Pleistocene (but see comments on 'Biostratigraphy' below).

Foraminiferans were first noted, although not as such, as long ago as the first century BC, by Strabo, who came from Asia Minor, and travelled extensively around the ancient world, and who wrote, of his observations in Egypt, of what we now know to be *Nummulites gizehensis*:

> There are heaps of stone chips lying in front of the pyramids and among them are found chips that are like lentils both in form and size . . . They say that what was left of the food of the workmen has petrified and this is not improbable.

The first scientific description of foraminiferans was made in the eighteenth century, by Beccarius. The name 'foraminiferans', from the Latin for aperture-bearers, was first coined in the nineteenth century, by the great French naturalist Alcide Dessalines d'Orbigny.

Biology

The biology and ecology of modern foraminiferans is comparatively poorly known, partly because the group has historically received more attention from geologists than from biologists, and partly because of difficulties experienced in reproducing living conditions in laboratory culture experiments. Those species whose biology has been studied have been shown to feed by extruding protoplasm through their aperture(s) in the form of a pseudopodial net to capture food (typically diatoms, algae, bacteria and detritus): some are also able to achieve some motility through the use of their pseudopods. The reproductive cycle appears to be typically characterised by alternate asexual and sexual phases. Sexual reproduction provides a mechanism for mutation and evolution.

In terms of ecology, most species are benthic, some planktonic.

Living benthic foraminiferans occupy every conceivable brackish and marine benthic environment, from the high marsh to the abyssal plain, in all latitudes. Different taxa have different ecological, and especially bathymetric and biogeographic, distributions, such that the group is of considerable use in palaeoecological, and especially palaeobathymetric and palaeobiogeographic, interpretation.

The bathymetric distributions and zonations, and bathymetrically related morphological and other trends, of 'live' (as revealed by biologic staining techniques) and 'dead' modern benthic foraminiferans are well documented in the marine biological and biological oceanographic literature, and of use as proxies for the palaeobathymetric interpretation of ancient benthic foraminiferans (Jones, 1996). The distinction between 'live' and 'dead' is important in view of *post-mortem* transportation of tests (see Section 2.2).

Marginal-, shallow- and deep-marine environments can be distinguished in most cases (see section 'Palaeobiology' below, p. 60; see also Chapter 4). Many benthic foraminiferans accommodate algal photosymbionts, which essentially restrict them to the shallow marine photic zone (but see also Bernhard, 2003). High-resolution liquid chromatography can be used to detect the presence of chlorophyll, and hence to indicate photosymbiosis (Knight and Mantoura, 1985). Boron isotope composition can also be used as an indicator of photosymbiosis (Honisch *et al.*, 2003). Interestingly, the deep-marine benthic foraminiferan *Virgulinella fragilis*, from the oxygen-depleted, sulphide-enriched waters of the Cariaco basin off the north coast of Venezuela, accommodates diatom? endosymbionts tentatively identified as sulphide-oxidisers (Bernhard, 2003).

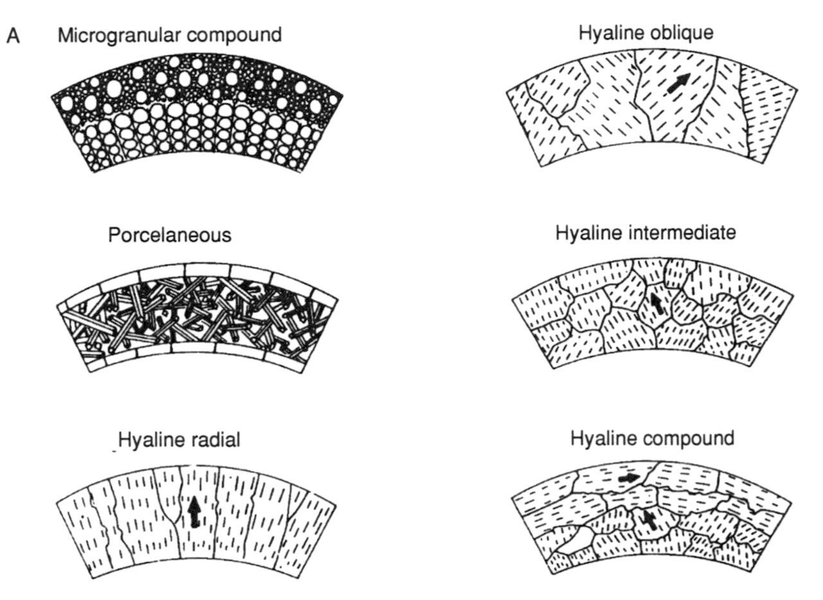

Fig. 3.13. Hard-part morphology of foraminiferans. (A) Wall structure. (B) Chamber arrangement and coiling mode. (From Jones, 1996; reproduced with permission from Haynes, 1981.)

Living planktonic foraminiferans are exclusively marine (see also section 'Globigerinides' below).

The ratio of planktonic to benthic foraminiferans tends to increase with increasing depth, so serves as some measure of depth (Jones, 1996). Note, though, that the trend is commonly reversed on the upper slope, presumably due to low predation and hence high benthic foraminiferal productivity in the oxygen minimum zone.

Morphology

In terms of hard-part morphology, foraminiferans are characterised by arenaceous or secreted calcareous shells or tests (Figs. 3.13–3.17). The simplest forms are undivided spheres or tubes with openings or apertures at one or both ends. More advanced forms are divided into chambers arranged in various ways. The final chambers communicate with the external world through apertures of various types (the preceding chambers with one another through intercameral openings or foramina). Internal chamber divisions of various types may be present. External ornamentation of various types may also be present.

Classification

Classification of foraminiferans is based essentially on morphology, in particular, in approximate (but by no means agreed!) order of importance, wall structure, chamber arrangement or coiling mode, and apertural form; and, to a lesser extent, on phylogeny inferred from empirical observations on the fossil record. Comparatively little molecular and cladistic analysis has thus far been undertaken on the group. Interestingly, though, what molecular work has been done has revealed the existence of true, molecular biological species within morphological 'species', as in the case of the planktonic foraminiferan *Orbulina universa* (de Vargas *et al.*, 1999; Norris, in Erwin and Wing, 2000; Norris and de Vargas, 2000). In this case, the 'cryptic' true species each have clearly distinct distributions, centred on areas of high, moderate and low nutrient concentration, as inferred from satellite images, in the Mediterranean, the Caribbean, and the Sargasso Sea and South Atlantic, respectively. The only morphological difference between them is their porosity, which had previously been used as a measure of temperature (porosity has also previously been used as a measure of palaeotemperature, as in the Cretaceous

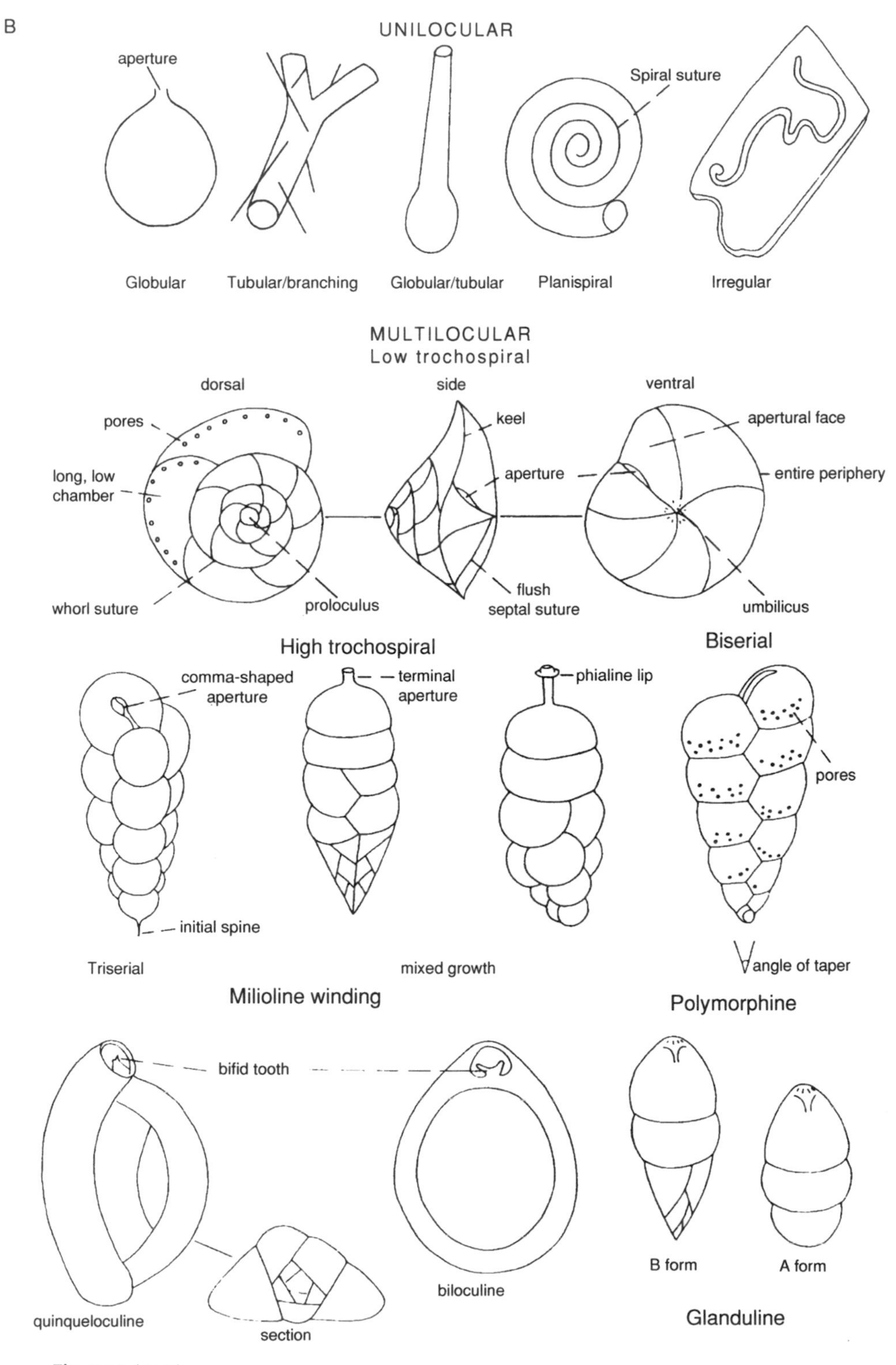
B
UNILOCULAR
aperture
Spiral suture
Globular
Tubular/branching
Globular/tubular
Planispiral
Irregular
MULTILOCULAR
Low trochospiral
dorsal
side
ventral
pores
keel
apertural face
long, low
chamber
aperture
entire periphery
whorl suture
proloculus
flush
septal suture
umbilicus
High trochospiral
Biserial
comma-shaped
aperture
terminal
aperture
phialine lip
pores
initial spine
Triserial
mixed growth
angle of taper
Milioline winding
Polymorphine
bifid tooth
B form
A form
biloculine
quinqueloculine
section
Glanduline

Fig. 3.13 *(cont.)*

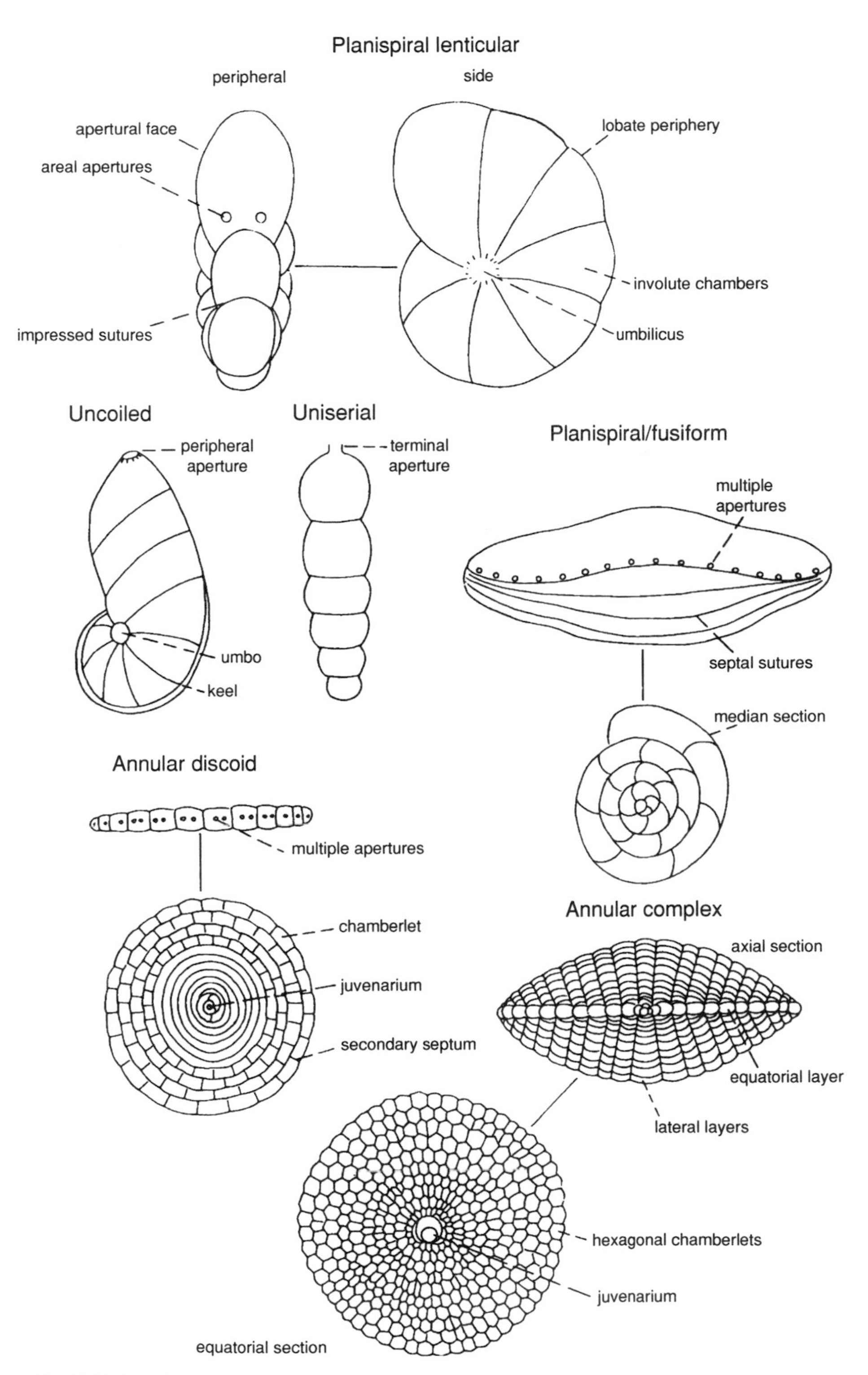
Planispiral lenticular
peripheral
side
apertural face
areal apertures
impressed sutures
lobate periphery
involute chambers
umbilicus
Uncoiled
peripheral aperture
umbo
keel
Uniserial
terminal aperture
Planispiral/fusiform
multiple apertures
septal sutures
median section
Annular discoid
multiple apertures
chamberlet
juvenarium
secondary septum
Annular complex
axial section
equatorial layer
lateral layers
hexagonal chamberlets
juvenarium
equatorial section

Fig. 3.13 *(cont.)*

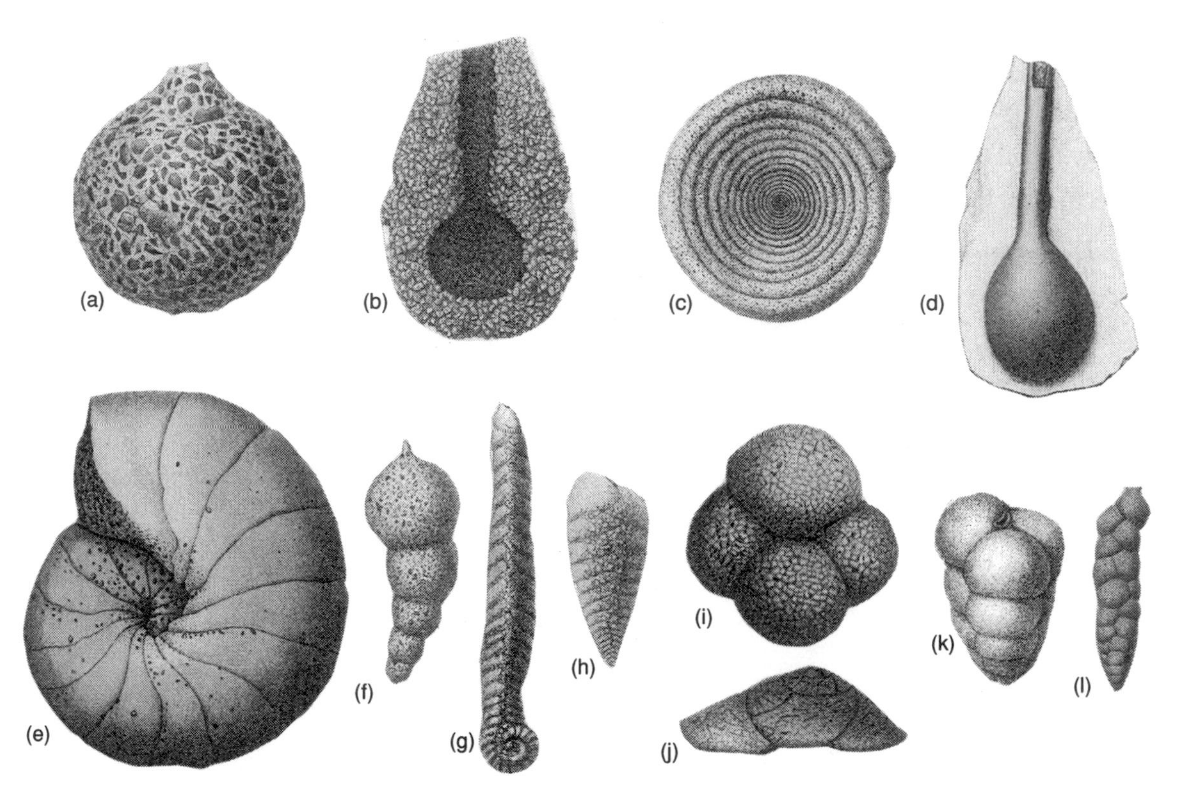

Fig. 3.14. Some representative modern foraminiferans: arenaceous foraminiferans. (a) *Saccammina sphaerica*, general view, North Atlantic, 'deep water', ×15; (b) *Hyperammina friabilis*, sagittal section, ×10; (c) *Ammodiscus anguillae*, general view, Culebra, West Indies, 780 m, ×15; (d) *Ammolagena clavata*, general view, Pernambuco, west Atlantic, 700 m, ×25; (e) *Cyclammina cancellata*, general view, North Atlantic, ×15; (f) *Hormosina globulifera*, general view, north-west Ireland, 1260 m, ×15; (g) *Spiroplectella earlandi*, general view, Raine Island, Torres Strait, 310 m, ×60; (h) *Spiroplectinella wrighti*, general view, Atlantic, ×35; (i) *Paratrochammina challengeri*, umbilical view, South Atlantic, 3800 m, ×25; (j) *Tritaxis challengeri*, lateral view, Culebra, West Indies, 780 m, ×50; (k) *Eggerella bradyi*, general view, Juan Fernández, South Pacific, 3650 m, ×25; (l) *Karrerulina conversa*, general view, South Atlantic, 4700 m, ×40. (Modified after Jones, 1994.)

Fig. 3.15. Some representative modern foraminiferans: porcelaneous foraminiferans or miliolides, nodosariides and buliminides. Miliolides: (a) *Cornuspira foliacea*, general view, West Indies, 780 m, ×15; (b) *Spirophthalmidium acutimargo*, general view, Pernambuco, West Atlantic, 1350 m, ×50; (c) *Sigmoilopsis schlumbergeri*, general view, north-west Ireland, 1260 m, ×50; (d) *Quinqueloculina seminulum*, general view, Skye, 90 m, ×50; (e) *Triloculina trigonula*, general view, north-west Ireland, 1260 m, ×40; (f) *Pyrgo murrhina*, general view, South Atlantic, 3800m, ×40; (g) *Marginopora vertebralis*, general view, Tongatabu, Friendly Islands, 36m, ×20; (h) *Peneroplis planulatus*, general view, Nares Harbour, Admiralty Islands, 32m, ×50; (i) *Alveolinella quoyi*, general view, ×15; (j) *Borelis melo*, sagittal section, coral sand, Bermuda. ×50. Nodosariides: (k) *Lenticulina anaglypta* general view, Torres Strait, 310m, ×35; (l) *Amphicoryna scalaris*, general view, New Zealand, 550 m, ×50; (m) *Dentalina flinti*, general view, north-west Ireland, 1260 m, ×16; (n) *Lagena sulcata*, general view, Kerguelen, 240m, ×60; (o) *Fissurina laevigata*, general and apertural views, ×70. Buliminides: (p) *Bulimina marginata* general view, west of Ireland, 3260 m, ×80; (q) *Uvigerina bradyana*, general view, Juan Fernández, South Pacific, 2750 m, ×40; (r) *Brizalina mexicana*, general view, west of Ireland, 3260 m, ×50; (s) *Cassidulina teretis*, general view, west of Ireland, 3260 m (?relict), ×75; (t) *Cassidulina subglobosa*, general view, Pernambuco, West Atlantic, 1350 m, ×60; (u) *Pleurostomella acuminata*, lateral and apertural views, Ki Islands, 238 m, ×70; (v) *Parafissurina lateralis*, general view, east of Shetland Islands, 128 m, ×75. (Modified after Jones, 1994.)

(a)
(b)
(c)
(d)
(e)
(f)
(g)
(h)
(i)
(j)
(k)
(l)
(m)
(n)
(o)
(p)
(q)
(r)
(s)
(t)
(u)
(v)

Fig. 3.16. Some representative modern foraminiferans. Robertinide: (a) *Hoeglundina elegans*, umbilical view, West Indies, 780 m, ×35. Rotaliides: (b) *Spirillina vivipara*, general view, Tahiti, 1240 m, ×100; (c) *Rosalina bradyi*, umbilical view, Hong Kong, 14m, ×75;

of the western interior seaway of the USA: Fisher *et al.*, 2003). The full implications of the ongoing work on the molecular biology of living foraminiferans for the palaeobiology of their fossil counterparts have yet to be experienced!

Five main sub-groups of foraminiferans have been recognised, essentially on the basis of wall structure. These are: the proteinaceous foraminiferans (allogromides); the arenaceous or agglutinating foraminiferans (textulariines of some authors; astrorhizides and lituolides of some others); the microgranular calcareous foraminiferans (fusulines or fusulinides of some authors); the porcelaneous calcareous foraminiferans (miliolines or miliolides of some authors); and the hyaline calcareous foraminiferans (rotaliines of some authors; nodosariides, buliminides, robertinides, rotaliides and globigerinides of some others).

A further, artificial but nonetheless practically useful, sub-group has also been recognised, namely the larger benthic foraminiferans or LBFs (see Box 3.2, p. 67).

Proteinaceous foraminiferans. Proteinaceous foraminiferans are an extant sub-group characterised by a proteinaceous wall structure. They have a poor fossil record (indeed, comparatively little is known about them generally).

Chamber arrangement in this sub-group ranges from simple to complex. Apertural form is generally simple, the position variable but often terminal.

Arenaceous or agglutinating foraminiferans. Arenaceous foraminiferans are an extant, Cambrian–Recent, sub-group characterised by an arenaceous wall structure, constructed out of randomly or specifically selected adventitious sand or other particles (including, incidentally, in at least one instance, microchondrules of extra-terrestrial origin). Chamber arrangement in this sub-group ranges from simple to complex. Apertural form is generally simple, the position variable but often terminal.

Living arenaceous foraminiferans are marine and benthic. They appear better able than their calcareous counterparts to tolerate conditions of reduced salinity, oxygen availability and alkalinity (though this may be in part a preservational phenomenon), and, proportionately, are most characteristic of marginal marine, and of deep marine, environments. Some, larger, species are interpreted as having accommodated algal symbionts, and as having been restricted to the photic zone. Some species have poor preservation potential.

The environmental interpretation of fossil arenaceous foraminiferal assemblages relies on analogy with their living counterparts. However, some arenaceous foraminiferans, constituting so-called '*Rhabdammina*' or 'flysch-type', or 'deep-water arenaceous foraminiferan (DWAF)' faunas, appear to have no counterparts, and are difficult to interpret palaeoenvironmentally, and in particular palaeobathymetrically.

Caption for fig. 3.16 (*cont.*) (d) *Buccella tenerrima*, umbilical view, Cape Frazer, ×50; (e) *Laticarinina pauperata*, umbilical view, Pernambuco, west Atlantic, 1350 m, ×35; (f) *Eponides repandus*, umbilical view, Pernambuco, west Atlantic, 1350 m, ×35; (f) *Eponides repandus*, umbilical view, Loch Scavaig, Scotland, 90 m, ×30; (g) *Oridorsalis umbonatus*, umbilical view, west coast of Patagonia, 240 m, ×75; (h) *Hyalinea balthica*, general view, west of Ireland, 3260 m, ×50; (i) *Cibicidoides wuellerstorffi*, umbilical view, New Zealand, 550 m, ×50; (j) *Elphidium crispum*, general view, Wednesday Island, 14 m, ×40; (k) *Elphidiella arctica*, general view, Smith Sound, 420 m, ×30; (l) *Melonis pompilioides*, general view, west of Ireland, 2886 m, ×75; (m) *Pseudorotalia schroeteriana*, umbilical view, China Sea, ×25; (n) *Calcarina spengleri*, umbilical view, Admiralty Islands, 32 m, ×30; (o) *Operculina complanata*, general view, Amboyna, 30 m, ×12; (p) *Operculinella cumingi*, general view, China Sea, ×18; (q) *Cycloclypeus carpenteri*, general view, Fiji, 420m, ×12; (r) *Heterostegina depressa*, general view, Admiralty Islands, 34 m, ×12. (Modified after Jones, 1994.)

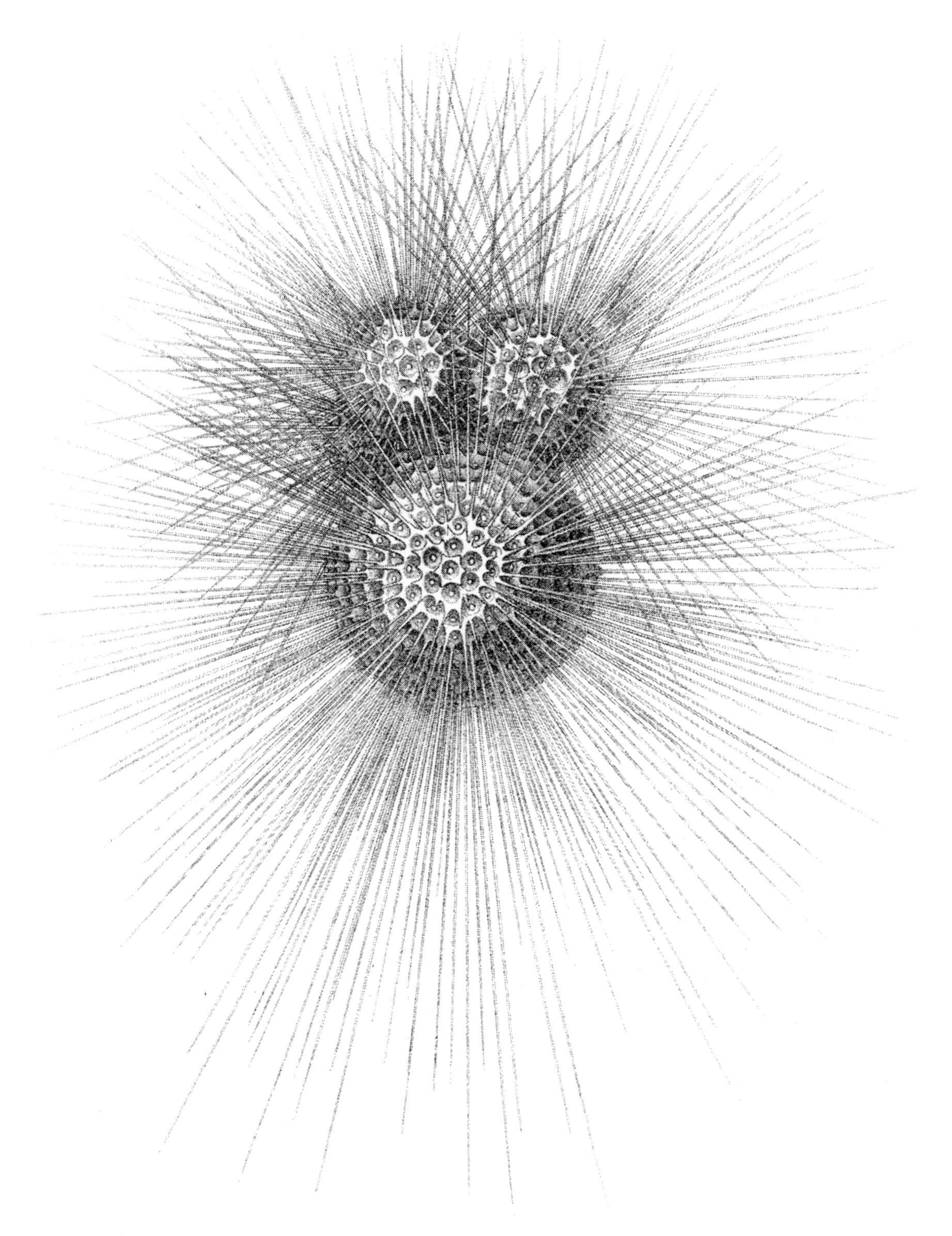

Fig. 3.17. Some representative modern foraminiferans: the globigerinide *Globigerina bulloides*, general view, plankton tow, ×200. (Modified after Jones, 1994.)

Box 3.1 Palaeoenvironmental interpretation of '*Rhabdammina*' faunas

Introduction

'*Rhabdammina*' faunas, first described from various classical turbidite or flysch localities in Europe, for example, the Gurnigel–Schlieren flysch in Switzerland, the Peira Cava flysch in France and the Macigno flysch in Italy, are composed almost exclusively of single-chambered or unilocular and uniserial multilocular agglutinating foraminiferans, which typically possess non-calcareous cements.

They have been interpreted on the basis of analogy with the arenaceous components of modern mixed foraminiferal faunas as associated with abyssal and hadal environments of deposition. A notable feature of these types of environment is that they occur below the calcite compensation depth, where secretion or preservation of calcareous tests or cements is inhibited.

Discussion

At the present time, there are considered to be a number of critical factors in the interpretation of '*Rhabdammina*' faunas, some of which are interrelated.

Geological age

There are few records known to the author of Jurassic or even earliest Cretaceous '*Rhabdammina*' or 'flysch-type' faunas, and none at all of any older. This is despite the fact that flysch deposits of these ages are well documented from various localities around the world. The inference is that arenaceous foraminiferans did not colonise environments characterised by flysch deposition until the later Cretaceous. Notably, some of the families most characteristic of '*Rhabdammina*' or 'flysch-type' faunas in later Cretaceous and younger deposits occur in carbonate shelf deposits of the Welsh borderlands in the Silurian, but not in coeval 'geosynclinal' deposits of mid-Wales.

Geological setting

Ancient '*Rhabdammina*' faunas occur in a variety of geological settings. These include both active margins such as the Alpine–Carpathian orogenic belt in Europe, parts of Trinidad in the Caribbean and parts of Borneo and Papua New Guinea in the Far East, and passive margins such as the Labrador and North Seas in the North Atlantic. One factor common to both active and passive margins characterised by '*Rhabdammina*' faunas is a restricted basin configuration brought about by structuration, in turn leading to restricted circulation and reduced oxygen concentration at or just below the sediment surface. These factors seem to favour the sustenance and/or preservation of non-calcareous arenaceous foraminiferans, and the exclusion and/or dissolution of their calcareous counterparts, especially at times of lowered sea level, when the input of organic material and hence the release of acids in early diagenetic decomposition reactions is at its highest. Restricted circulation and reduced oxygen concentration at or just below the sediment surface can also be brought about by salinity stratification associated with excess freshwater run-off from a major river system, which, incidentally, would also have the effect of eliminating the normal marine phytoplankton that, as – seasonal – phytodetritus, is known to constitute an important food source for at least some calcareous benthic species.

Water depth

Modern foraminiferal assemblages dominated by arenaceous foraminifera occur in every water depth from the marginal marine to the hadal (see above). However, true '*Rhabdammina*' faunas appear to be generally associated with deep rather than shallow environments. Certainly, many authors are of the opinion that the '*Rhabdammina*' faunas of the Alpine–Carpathian orogenic belt in Europe were associated with great depth. Some others regard depth as only one of a number of factors controlling the occurrence of the group, but nonetheless infer a minimum depth corresponding to wave-base (200 m).

Sedimentary regime

Somewhat self-evidently, '*Rhabdammina*' or 'flysch-type' faunas are associated with flysch or turbidites and hence submarine fan environments! In the case of the Miocene of Trinidad in the eastern Venezuelan basin, an association with brackish as well as turbid water, and with the proto-Orinoco delta, has been inferred.

Instances of both horizontal variation and vertical variation in microfaunal assemblage composition within fan complexes have been reported in the literature (see also sub-section on 'Submarine fans' in Chapter 4).

Other physico-chemical factors

A number of authors have intimated that calcium carbonate content might be a critical factor controlling the distribution of '*Rhabdammina*' faunas, and indeed there does appear to be an inverse correlation between the carbonate content of flysch deposits and the development of '*Rhabdammina*' faunas. However, the extent to which this apparent control is primary rather than secondary, that is, essentially a function of dissolution, is unknown (see Section 2.2).

Some authors have also intimated that the availability of iron (?and silica) and the ability to secrete a ferruginous (?and siliceous) cement might be a critical factor.

Dynamic factors

There appears to be a direct correlation between the development of flysch and of '*Rhabdammina*' or 'flysch-type' faunas and tectonoeustatically mediated sea level low-stands and associated systems tracts (see also Box 6.1).

Modified after Charnock and Jones, in Hemleben *et al.* (1990), Jones (1996), Charnock and Jones (1997), Jones, in Ali *et al.* (1998), Jones, in Jones and Simmons (1999), Jones (2001), Jones *et al.*, in Koutsoukos (2005) and Jones *et al.*, in Powell and Riding (in press).

Microgranular foraminiferans. Microgranular foraminiferans are an extinct, Devonian–Permian, sub-group characterised by an microgranular wall structure. Chamber arrangement ranges in this sub-group from simple to complex. Apertural form also ranges from simple to complex.

As they are extinct, the palaeoecology of the microgranular foraminiferans has had to be interpreted, in part from the ecology of the analogous extant porcelaneous foraminiferans, and in part from associated fossils and sedimentary facies. They are interpreted as having occupied a range of platformal and peri-reefal carbonate environments (Cozar and Rodriguez, 2003; Haig, 2003). Some, larger, species are interpreted as having accommodated algal symbionts. Depth appears to have been the main control on the distribution of microgranular foraminiferans. A number of microgranular foraminiferal biofacies have been recognised in the Permian of the southern Carnarvon basin in Western Australia that appear to be related to shallowing (Haig, 2003).

Porcelaneous foraminiferans. Porcelaneous foraminiferans are an extant, Carboniferous–Recent, sub-group characterised by a porcelaneous wall structure. Chamber arrangement in this sub-group ranges from simple to complex. Apertural form also ranges from simple to complex.

Modern porcelaneous foraminiferans are marine and benthic. They are characteristic of platformal and peri-reefal carbonate environments, and especially characteristic of restricted, hypersaline backreef lagoonal environments. The range of niches they occupy is wide. Some, larger, species accommodate algal symbionts. Others are associated with seagrasses. The environmental interpretation of ancient porcelaneous foraminiferal assemblages relies on analogy with their modern counterparts.

Hyaline foraminiferans. Hyaline foraminiferans are an extant, Devonian?–Recent, sub-group characterised by a hyaline wall structure (the nodosariides, buliminides, rotaliides and globigerinides are calcitic; the robertinides aragonitic). Chamber arrangement in this sub-group ranges from simple (unilocular to uniserial and enrolled planispiral in the nodosariides, planispiral to low trochospiral in the small rotaliides, high trochospiral in the buliminides) to complex ('orbitoidiform' in the large rotaliides). Apertural form also ranges from simple to complex.

Modern hyaline foraminiferans are marine. The environmental interpretation of ancient hyaline foraminiferans essentially relies on analogy with their modern counterparts.

Nodosariides (Devonian?–Recent) are marine and benthic, and characteristic of outer shelf and slope environments at the present time (but see comments in section on 'Palaeobiology' below). They are of comparatively little use in stratigraphic subdivision and correlation.

Buliminides (Jurassic–Recent) are marine and benthic, and characteristic of outer shelf and slope environments, and especially characteristic of upper slope oxygen minimum environments, such that they are useful indicators of oxygen minimum zone (OMZ) petroleum source-rock facies in modern and ancient environments. The dysoxic environment of the Late Cretaceous, Santonian–Maastrichtian of the upper Magdalena basin of Colombia was characterised by the buliminide *Siphogenerinoides*, which has been hypothesised to have been an opportunist, feeding on the phosphatised faeces of clupeoid fish (Martinez, 2003). Buliminides are of comparatively little use in stratigraphic subdivision and correlation.

Robertinides (Triassic–Recent) are marine and benthic, and characteristic of outer shelf and slope environments at the present time (but see comments in section on 'Palaeobiology' below). They are of comparatively little use in stratigraphic subdivision and correlation.

Rotaliides (Triassic–Recent) are marine and benthic, and live in every niche from marginal to deep marine. The range of niches they occupy is wide. Some, larger, species accommodate algal symbionts. Smaller rotaliides are of some use in the stratigraphic subdivision and correlation of the Mesozoic and Cenozoic of the Tethyan, Austral and Boreal realms (Jones, 1996).

Globigerinides (Middle Jurassic–Recent) are marine and planktonic. Many house algal photosymbionts, which essentially restrict them to the photic zone in the late part of the water column (Taylor, in Capriulo, 1990; Jones, 1996).

Living globigerinid globigerinides occupy shallow, epipelagic and intermediate, mesopelagic environments within the water column; globorotaliid globigerinides deep, bathypelagic environments. As a consequence of functional morphological adaptation to their respective environments, globigerinids are characterised by a higher ratio, and globorotaliids by a lower ratio, of surface area to volume. Note, incidentally, that the high ratio of surface area to volume in the globigerinids also has the effect of rendering them susceptible to solution. Living globigerinides are cosmopolitan (Zeitzschel, in Capriulo, 1990).

The environmental interpretation of fossil globigerinides relies not only on analogy with their living counterparts, but also on functional morphology, and on stable isotope chemistry. Note in this context that globorotaliid globigerinides appear, on the basis of stable isotope chemistry, to have occupied shallower environments, and globigerinid globigerinides deeper environments (within the water column) in the Palaeogene than they do today.

Thus, eutrophic epipelagic environments of the 'Middle'–Late Cretaceous are characterised, on the bases of all available lines of evidence, by unkeeled taxa such as the interpreted ecological generalist and opportunist *Guembelitria*, which bloomed under particularly favourable environmental conditions, and a number of – characteristically either small and spiral or large, ornate and serial – interpreted ecological specialists; mesotrophic mesopelagic environments by weakly keeled taxa such as *Praeglobotruncana* and *Dicarinella*; oligotrophic bathypelagic environments by strongly keeled taxa such as *Rotalipora*; and oxygen minimum zones by dysoxia-tolerant heterohelicids (Abramovich *et al.*, 2003; Keller *et al.*, 2003; Coccioni and Luciani, 2004; Keller, 2004; Keller and Pardo, 2004). The enriched $\delta^{13}C$ signatures of many interpreted specialist surface-dwelling taxa, such as *Rugoglobigerina*, *Pseudoguembelina*, *Planoglobulina* and *Racemiguembelina*, provide evidence of photosymbiosis. Significantly, many of these taxa are absent in areas characterised by expanded oxygen minimum zones. Also significantly, the interpreted surface-dwelling *Hedbergella planispira* appears tolerant of less than normal marine salinity.

Many fossil globigerinides had restricted or endemic biogeographic distributions, rendering them useful in the characterisation of palaeobiogeographic provinces in the Cretaceous (Anglada and

Randrianasolo, 1985) and Cenozoic (Funnell and Ramsay, in Hallam, 1973). Moreover, many had exacting ecological requirements and tolerances, and responded rapidly to changing environmental, and especially climatic, conditions, rendering them extremely useful in palaeoclimatology and climatostratigraphy (Chapman, in Culver and Rawson, 2000). Furthermore, their tests preserve stable isotope records of changing temperature and ice volume, rendering them extremely useful also in palaeotemperature and palaeoclimatic as well as in palaeobiogeographic interpretation, palaeobathymetric effects notwithstanding (see Section 4.5). Coiling direction ratios are also of use in palaeotemperature interpretation.

Globigerinides are extremely useful in the stratigraphic subdivision and correlation of the Cretaceous and Cenozoic worldwide (see also section on 'Biostratigraphy' below).

Palaeobiology

Fossil foraminiferans are interpreted, on the bases of analogy with their living counterparts, functional morphology, and associated fossils and sedimentary facies, as having occupied a range of brackish and marine, benthic and planktonic, environments, and as having assumed a variety of life positions and practised a variety of feeding strategies. Note, though, that some nodosariides and robertinides appear, on the basis of associated fossils and sedimentary facies, to have occupied shallower environments in the Jurassic than they do today, arguably becoming displaced by increased competition from evolving rotaliides. Note also that globorotaliid globigerinides appear to have occupied shallower environments, and globigerinid globigerinides deeper environments, in the Palaeogene than they do today (see also above).

In terms of functional morphology, living, and by inference fossil, epifaunal filter-feeding arenaceous benthic foraminiferans – representing 'morphogroup' A – are characterised by tubular to branching morphologies (Fig. 3.18) (Jones and Charnock, 1985; Jones, 1986; Jones, 1996; Jones, in Jones and Simmons, 1999; Jones *et al.*, in Powell and Riding, in press). Epifaunal detritus-feeding (including, incidentally, phytodetritus-feeding) arenaceous benthic foraminiferans – representing 'morphogroup' B – are characterised by globular to flattened, spiral morphologies. Infaunal deposit-feeding arenaceous benthic foraminiferans – representing 'morphogroup' C – are characterised by elongate, spiral to serial morphologies with low surface area to volume ratios. Epifaunal detritus-feeding calcareous benthic foraminiferans are characterised, like their arenaceous counterparts, by globular to flattened, biconvex or planoconvex morphologies (Corliss, 1985; Corliss and Chen, 1988; Corliss and Fois, 1991). Infaunal calcareous benthic foraminiferans are characterised, again like their arenaceous counterparts, by elongate morphologies.

Triangular cross-plots of arenaceous benthic 'morphogroups' A–C provide a means of distinguishing depth environments such as marginal marine, shelf to late slope, middle to early slope, and abyssal (Jones and Charnock, 1985; Jones, 1986; Jones, 1996), or sedimentary environments such as interturbiditic and hemipelagic (Jones, in Jones and Simmons, 1999), or submarine fan channel axis and levee/overbank (Jones *et al.*, in Powell and Riding, in press). Note also that the incidence of infaunal 'morphotypes' appears to be more or less proportional to the organic content of the substrate, or inversely proportional to the oxygen content, and is a key input in the calculation of the benthic foraminiferal oxygen index (Kaiho, 1994).

Arenaceous and calcareous LBFs are characterised by comparatively large sizes, complex internal structures, and high internal surface area to volume ratios, representing adaptations to the accommodation of algal photosymbionts in shallow marine environments (see also boxed sub-section on 'Larger benthic foraminiferans' above on pp. 67–71).

Planktonic foraminiferans are characterised by comparatively high external surface area to volume ratios, and often by projecting spines, representing adaptations to the reduction of the rate of sinking through the water column and the maintenance of neutral buoyancy (see also section on 'Globigerinides' above).

The organic linings of foraminiferal tests, found in palynological preparations, can be of use in palaeoenvironmental interpretation (Stancliffe, in Jansonius and MacGregor, 1996). Thermal and colour alteration to the organic components of

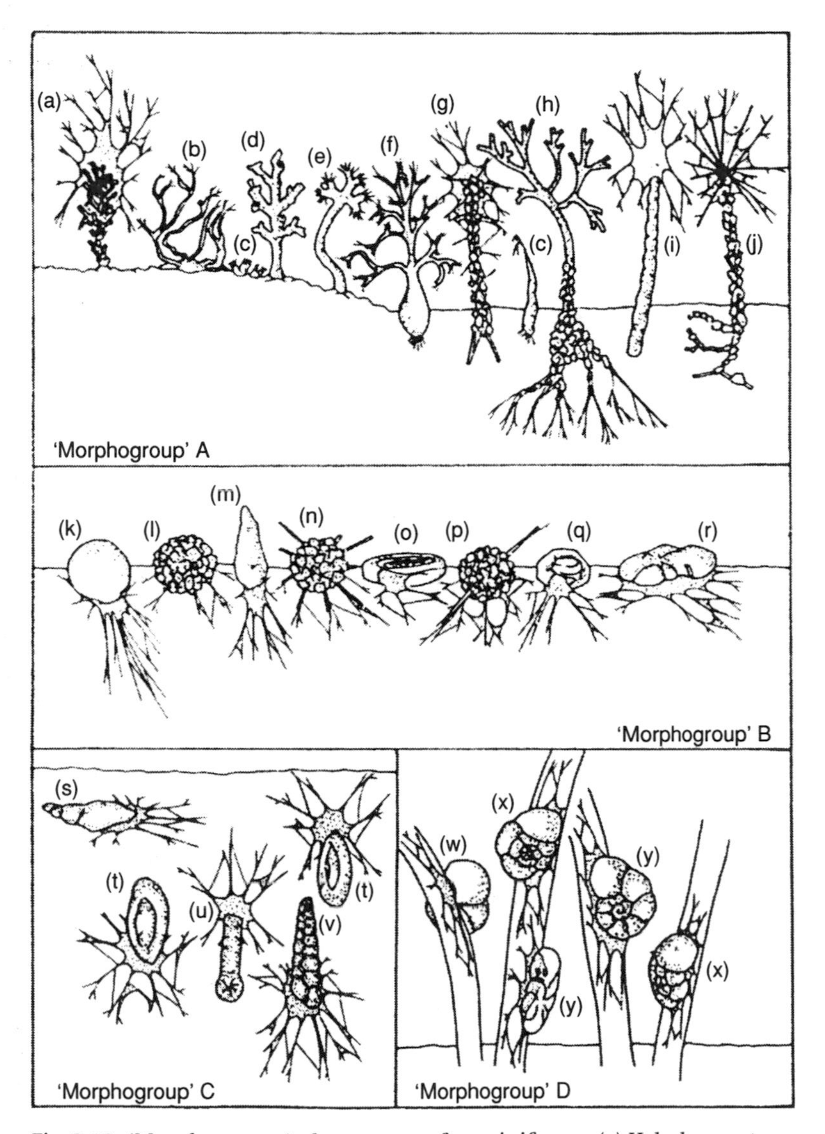

Fig. 3.18. 'Morphogroups' of arenaceous foraminiferans. (a) *Halyphysema tumanowiczii*; (b) *Saccodendron* sp.; (c) *Pelosina* sp.; (d) *Dendrophrya erecta*; (e) *Halyphysema* sp.; (f) *Pelosina arborescens*; (g) *Jaculella obtusa*; (h) *Notodendrodes antarctikos*; (i) *Bathysiphon* sp.; (j) *Marsipella arenaria*; (k) *Saccammina alba*; (l) *Saccammina sphaerica*; (m) *Hippocrepina* sp.; (n) *Saccammina sphaerica* var. *anglica*; (o) *Ammodiscus* sp.; (p) *Psammosphaera parva*; (q) *Glomospira* sp.; (r) *Labrospira jeffreysi*; (s) *Reophax subfusiformis*; (t) *Miliammina fusca*; (u) *Ammobaculites exiguus*; (v) textulariid; (w) *Trochammina pacifica*; (x) *Trochammina inflata*; (y) *Jadammina macrescens*. (From Jones, 1996; in turn after Jones and Charnock, 1985.)

foraminiferal tests provides a measure of thermal maturity brought about during burial and diagenesis (McNeil *et al.*, 1996; see also section on 'Palaeontological inputs into basin modelling' in Chapter 7).

Palaeobathymetry

As noted above, the bathymetric distributions and zonations of modern benthic foraminiferans are well documented, and of use as proxies for the palaeobathymetric interpretation of ancient benthic foraminiferans, with at least marginal, shallow and deep marine environments distinguishable in most basins (see also Section 4.3). Transfer functions based on the environmental distributions and tolerances of foraminiferans are of considerable use in sea level reconstructions (see also Sections 4.6 and 7.6).

Abundance and diversity trends, and bathymetric trends, have been used in the characterisation of condensed section and systems tracts in sequence stratigraphic interpretation (Jones *et al.*, in Jenkins, 1993; Jones, 1996; Jones, 2003a, b; see also Box 6.1). They have also been used in the identification of barriers and baffles in reservoir characterisation and in seal capacity analysis (Jones, 2003a, b; see also Section 7.1).

Palaeobiogeography

Many fossil foraminiferans appear to have had restricted or endemic biogeographic distributions, rendering them of use in the characterisation of palaeobiogeographic provinces, and in turn in the constraint of plate tectonic reconstructions, in the Palaeozoic, in the Mesozoic and in the Cenozoic.

In the Mesozoic, larger and smaller benthic foraminiferal distributions indicate the existence of generally distinct Austral, Boreal and Tethyan realms (LBFs, smaller benthic foraminiferans such as *Stensioeina*, and keeled planktonic foraminiferans such as globotruncanid globigerinides characterising the Tethyan realm). Note, though, that there was probably at least local or intermittent communication between these realms, particularly between the Boreal realm and Mesogean Tethys. Smaller benthic foraminiferal distributions also indicate the existence of two distinct provinces within the Boreal realm, namely the Boreal–Atlantic province, characterised by epistominids, and the Arctic province, characterised by arenaceous foraminiferans such as ammodiscids (Podobina, in Kaminski *et al.*, 1995; Westermann, 2000). Norway and Greenland would appear to have been in the Boreal–Atlantic province, and Alaska and Canada in the Arctic province, the latter here characterised by arenaceous foraminiferans such as haplophragmoidids, trochamminids and ataxophragmiids (author's unpublished observations). Kamchatka and Sakhalin, in the Far East of the former Soviet Union, would appear to have been in a separate, Boreal–Pacific, province, characterised by endemic rzehakinids (author's unpublished observations).

In the Cenozoic, Palaeocene–Eocene, smaller foraminiferal distributions point toward the existence of distinct shallow (Midway) and deep (Velasco) waters in the Tethyan and low-latitude Atlantic provinces (Berggren and Aubert, 1975; van Morkhoven *et al.*, 1986).

Palaeoclimatology

The exacting ecological requirements and tolerances of many foraminiferal species, and their rapid response to changing environmental and climatic conditions, render them useful in palaeoclimatology and climatostratigraphy, and in archaeology (see also Section 7.6).

Biostratigraphy

The oldest testate amoebae are from the Neoproterozoic Chuar group of the Grand Canyon (Porter and Knoll, 2000). Fossil testate amoebae have also been recorded, preserved in amber, from the Cenomanian of Schliersee in Bavaria in southern Germany (Schmidt *et al.*, 2004); *Centropyxis*, preserved in pyrite, from the Pliocene of the Pedernales field of eastern Venezuela (Jones, in Jones and Simmons, 1999). The fossil record of the testate amoebae is generally poor.

The oldest, arenaceous, foraminiferan, interpreted as having evolved from a Neoproterozoic allogromiid ancestor, is *Platysolenites*, from the earliest Cambrian, Nemakit–Daldynian to Tommotian of Newfoundland, Great Britain, Scandinavia, Estonia and Russia (McIlroy *et al.*, 2001). Taphonomic, teleological and ultrastructural studies suggest that ancient *Platysolenites* had a similar grade of organisation and mode of life to modern *Bathysiphon* (an epifaunal detritus-feeder), although morphological evidence, specifically the presence of an agglutinated proloculus, indicates that it was probably not directly related. Microgranular foraminiferans evolved in the Devonian, and diversified rapidly, before becoming extinct in the End-Permian mass extinction. Porcelaneous foraminiferans evolved at the end of the Palaeozoic, and diversified through the Mesozoic and Cenozoic. Among the hyaline foraminiferans, the nodosariides evolved in the ?Devonian, and the robertinides in the Triassic. Both diversified through the Jurassic and Early Cretaceous, and declined through the Late

Cretaceous and Cenozoic, possibly due to increasing competition from diversifying rotaliides, especially in shelfal environments. The buliminides evolved in the Jurassic, and diversified through the Cretaceous and Cenozoic. The rotaliides evolved in the Triassic, and diversified dramatically in the Cretaceous, before sustaining significant losses in the End-Cretaceous mass extinction. They recovered from this event, though, and in general have been radiating throughout the Cenozoic. The globigerinides, the planktonic foraminiferans, evolved at the beginning of the Middle Jurassic, coincidentally or otherwise immediately after the End-Toarcian mass extinction (BouDagher-Fadel *et al.*, 1997; Hart *et al.*, in Crame and Owen, 2002; Hart *et al.*, 2003). They diversified dramatically in the Cretaceous, before sustaining severe losses and indeed almost becoming extinct in the End-Cretaceous mass extinction. They staged a slow recovery from this event, and in general have been radiating through the Cenozoic.

The rapid rate of evolutionary turnover exhibited by at least the LBFs and planktonic foraminiferans renders them of considerable use in biostratigraphy, at least in appropriate, marine, environments. The usefulness of the planktonic foraminiferans is enhanced by their essentially unrestricted ecological distribution and facies independence, a function of their habit. A range of high-resolution biostratigraphic zonation schemes based on the planktonic foraminiferans has been established, that have inter-regional to global applicability (Fig. 3.19) (Jones, 1996).

In my working experience, foraminiferans have proved of particular use in stratigraphic subdivision and correlation, and in palaeoenvironmental interpretation, in the following areas (see also section on 'Larger benthic foraminiferans', on pp. 67–71).

Mesozoic

The Triassic–Cretaceous of the western margin of the British Isles. Here, foraminiferans form part of the basis of the regional biostratigraphic zonation, calibrated against ammonite and other standards (Ainsworth *et al.*, in Batten and Keen, 1989; Jones, 1996; Ainsworth *et al.*, in Underhill, 1998).

The Jurassic–Cretaceous of the North Sea. Here, foraminiferans form part of the basis of the regional biostratigraphic zonation, calibrated against ammonite and other standards, and part of the basis of the calibration of the regional sequence stratigraphic framework (Jones, 1996; Copestake, in Evans *et al.*, 2003; Copestake *et al.*, in Evans *et al.*, 2003; Husmo *et al.*, in Evans *et al.*, 2003; Surlyk *et al.*, in Evans *et al.*, 2003). Importantly, they are also of use in the characterisation and exploitation of a number of reservoirs (Morris, in Jones and Simmons, 1999).

The Jurassic–Cretaceous of the Middle East. Here, foraminiferans form part of the basis of the regional biostratigraphic zonation, calibrated against ammonite and other standards (Jones, 1996; Kuznetsova *et al.*, 1996). They also form part of the basis of the calibration of the regional sequence stratigraphic framework (Sharland *et al.*, 2001). The regional Early Cretaceous, 'mid' Aptian K80 MFS, stratotypified in the Bab member of the Shuaiba formation of the United Arab Emirates, is marked by the planktonic foraminiferan *Wondersella athersuchi*. The Late Cretaceous, early Cenomanian K120 MFS, stratotypified in member E of the Natih formation of Oman, is marked by *Favusella washitensis*. The early Turonian K140 MFS, stratotypified in member B of the Natih formation, is marked by planktonic foraminiferans of the *Whiteinella archaeocretacea* Zone. The Campanian K170 MFS, stratotypified in the Fiqa formation of the Haushi–Huqf area of Oman, is marked by planktonic foraminiferans of the *Globotruncana ventricosa* Zone.

The Cretaceous of northern South America and the Caribbean. Here, foraminiferans form part of the basis of the regional biostratigraphic zonation (Lamb, 1964). Indeed, many of the planktonic foraminiferal zones established here can be recognised worldwide (Saunders and Bolli, 1985). Foraminiferans are also of use here in palaeoenvironmental interpretation, as in Trinidad (Koutsoukos and Merrick, 1985) and Venezuela (Paredes *et al.*, 1994; Carrillo *et al.*, 1995).

Cenozoic

The Palaeogene of the North Sea. Here, foraminiferans form part of the basis of the regional biostratigraphic zonation, loosely calibrated against the calcareous nannofossil standard, and part of the basis of the calibration of the regional sequence

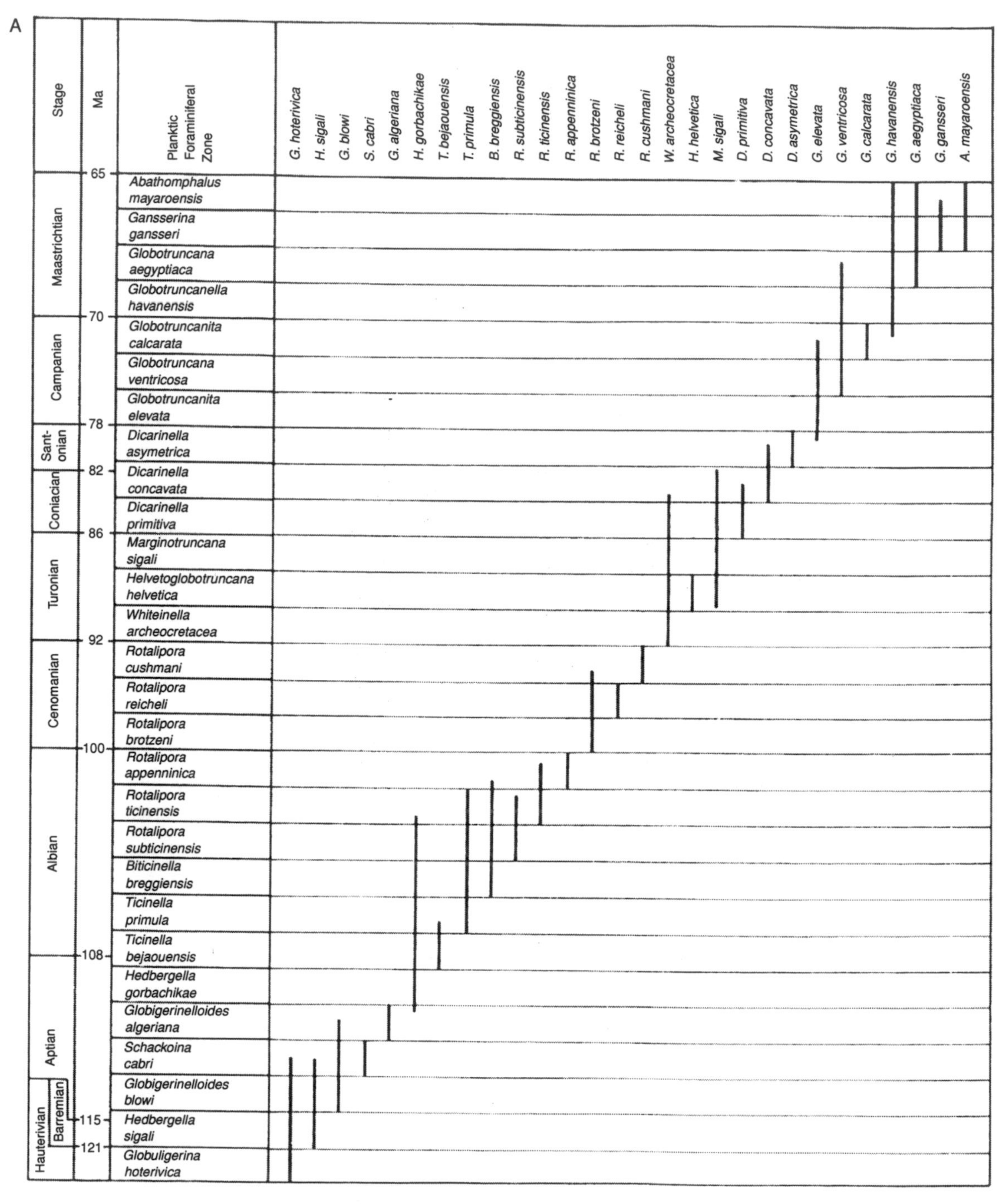

Fig. 3.19. Stratigraphic distribution of selected globigerinides. (A) Cretaceous. (B) Palaeogene. (C) Neogene–Pleistogene. (From Jones, 1996; reproduced with permission from Lipps, 1993.)

B

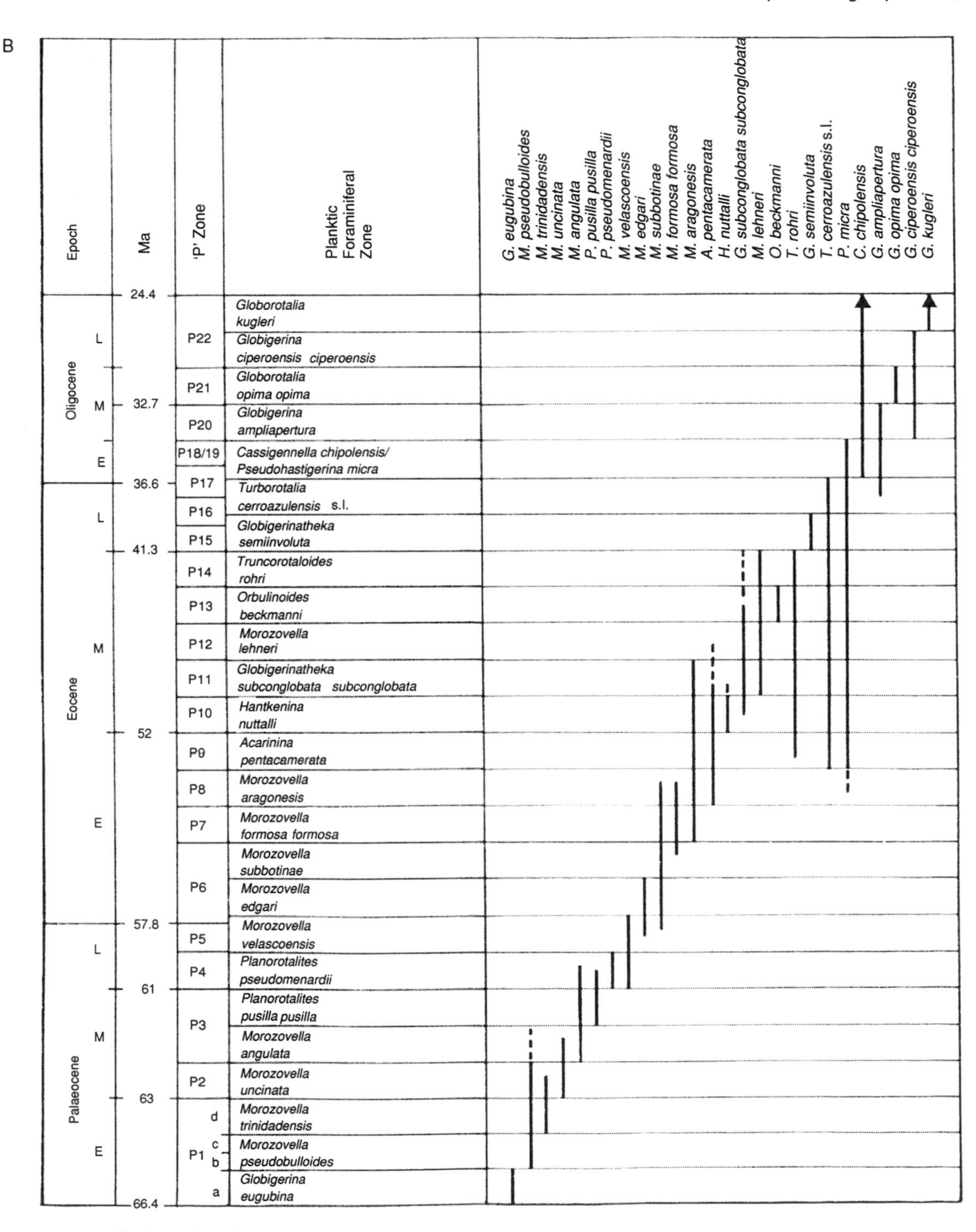

Fig. 3.19 (*cont.*)

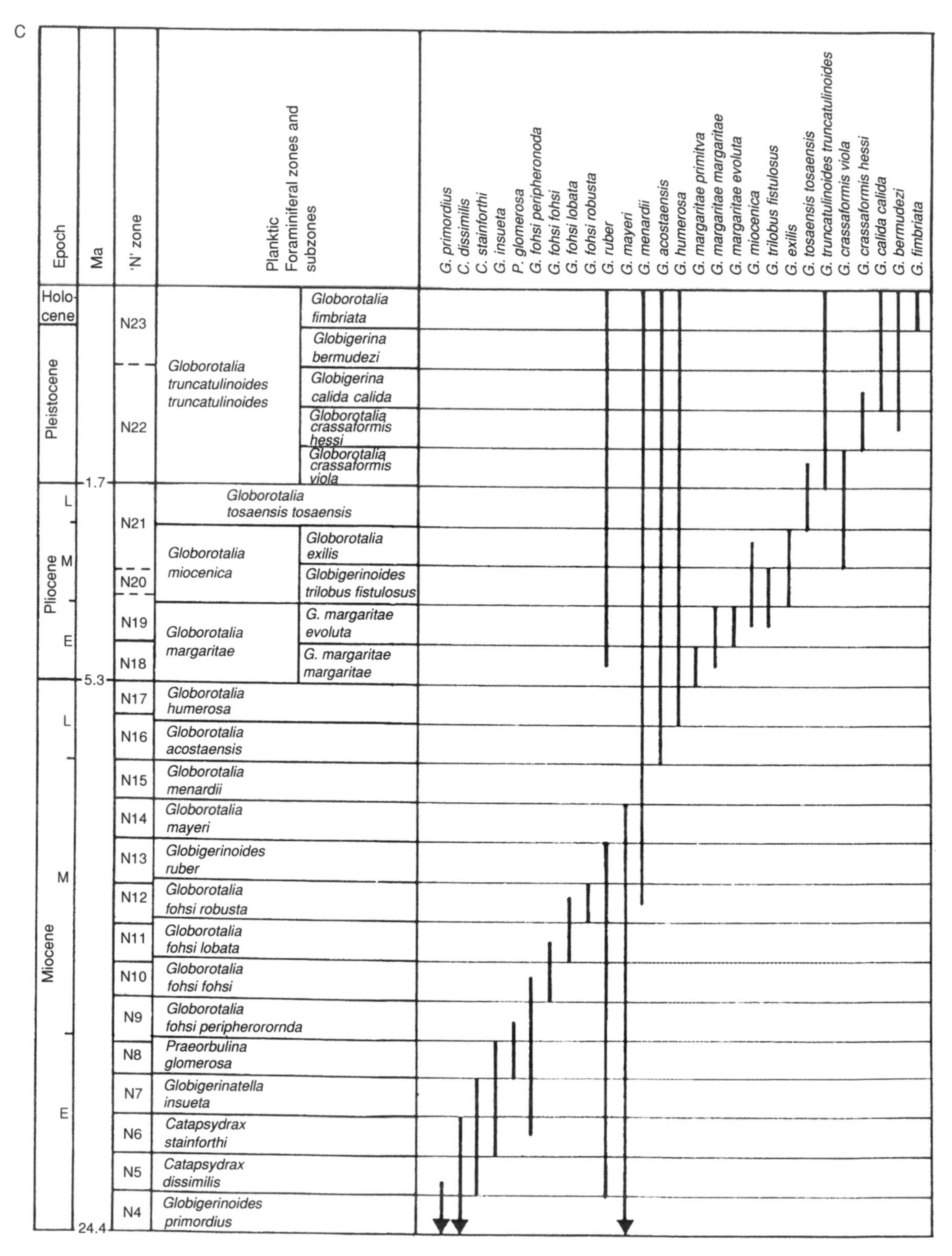
C
Epoch
Ma
'N' zone
Planktic Foraminiferal zones and subzones
G. primordius
C. dissimilis
C. stainforthi
G. insueta
P. glomerosa
G. fohsi peripheronoda
G. fohsi fohsi
G. fohsi lobata
G. fohsi robusta
G. ruber
G. mayeri
G. menardii
G. acostaensis
G. humerosa
G. margaritae primitva
G. margaritae margaritae
G. margaritae evoluta
G. miocenica
G. trilobus fistulosus
G. exilis
G. tosaensis tosaensis
G. truncatulinoides truncatulinoides
G. crassaformis viola
G. crassaformis hessi
G. calida calida
G. bermudezi
G. fimbriata
Holocene
Pleistocene
Pliocene
Miocene
L
M
E
1.7
5.3
24.4
N23
N22
N21
N20
N19
N18
N17
N16
N15
N14
N13
N12
N11
N10
N9
N8
N7
N6
N5
N4
Globorotalia truncatulinoides truncatulinoides
Globorotalia fimbriata
Globigerina bermudezi
Globigerina calida calida
Globorotalia crassaformis hessi
Globorotalia crassaformis viola
Globorotalia tosaensis tosaensis
Globorotalia miocenica
Globorotalia exilis
Globigerinoides trilobus fistulosus
Globorotalia margaritae
G. margaritae evoluta
G. margaritae margaritae
Globorotalia humerosa
Globorotalia acostaensis
Globorotalia menardii
Globorotalia mayeri
Globigerinoides ruber
Globorotalia fohsi robusta
Globorotalia fohsi lobata
Globorotalia fohsi fohsi
Globorotalia fohsi peripherorornda
Praeorbulina glomerosa
Globigerinatella insueta
Catapsydrax stainforthi
Catapsydrax dissimilis
Globigerinoides primordius

Fig. 3.19 (*cont.*)

stratigraphic framework, and of the identification of time-slices for facies mapping purposes (Charnock and Jones, in Hemleben *et al.*, 1990; Jones, 1996; Charnock and Jones, 1997; Ahmadi *et al.*, in Evans *et al.*, 2003; Jones *et al.*, in Evans *et al.*, 2003). Importantly, they are also of use in reservoir characterisation and exploitation in a number of fields (see also Section 7.1).

The Cenozoic of the Middle East. Here, foraminiferans form part of the basis of the regional biostratigraphic zonation (Jones and Racey, in Simmons, 1994; Jones, 1996). They also form part of the basis of the calibration of the regional sequence stratigraphic framework (Sharland *et al.*, 2001, 2004). The regional Late Palaeocene Pg10 MFS, stratotypified in the Shammar member of the Umm Er Radhuma formation of Saudi Arabia, is marked by the planktonic foraminiferan *Planorotalites pseudomenardii*. The late Middle Miocene Ng40 MFS, stratotypified in the Guri limestone member of the Mishan formation of south-west Iran, is marked by *Orbulina universa*.

The Cenozoic of the Sakhalin in the former Soviet Union. Here, foraminiferans form part of the basis of the regional biostratigraphic zonation, and part of the basis of the calibration of the regional sequence stratigraphic framework (Charnock and Salnikov, in Simmons, in press).

The Cenozoic of northern South America and the Caribbean. Here, foraminiferans form part of the basis of the regional biostratigraphic zonation (Lamb, 1964). Indeed, many of the planktonic foraminiferal zones established here can be recognised worldwide (Saunders and Bolli, 1985; Jones, in Ali *et al.*, 1998). Foraminiferans are also of use here in palaeoenvironmental interpretation, as in Venezuela (Paredes *et al.*, 1994; Moreno-Vasquez, 1995). Importantly, they are also of use in reservoir characterisation and exploitation, as in the case of the Pliocene Pedernales formation reservoir of the Pedernales field in eastern Venezuela (Jones *et al.*, in Jones and Simmons, 1999).

The 'mid' Cenozoic of Paratethys. See Popkhadze (1977), Jones and Simmons (1996) and Jones and Simmons, in Robinson (1997).

The 'mid'–late Cenozoic of the rift basins of the South Atlantic. Here, foraminiferans form part of the basis of the regional biostratigraphic zonation, for example in the Oligo–Miocene of Angola (Grant *et al.*, 2000), and the Plio-Pleistocene of Brazil (author's unpublished observations). They also form part of the basis of the calibration of the regional sequence stratigraphic framework. Indeed, foraminiferal abundance and diversity trends, and bathymetric trends, form part of the basis of the characterisation of the maximum flooding surfaces that constitute the bounding surfaces of the sequences.

The Neogene–Pleistogene of the Niger Delta in Nigeria. Here, foraminiferans are of particular use in the characterisation and exploitation of the Late Miocene reservoir of the Oso field (Armentrout *et al.*, in Jones and Simmons, 1999).

The Neogene–Pleistogene of the Gulf of Mexico. Here, foraminiferans are of particular use in reservoir characterisation and exploitation in the greater Mars area (O'Neill *et al.*, in Jones and Simmons, 1999).

Other applications

Foraminiferans are also useful in mineral exploration and exploitation, in engineering geology, and in environmental science (see Sections 7.4 and 7.5).

Box 3.2 Larger benthic foraminiferans

Biology, morphology and classification

Larger benthic foraminiferans (LBFs) constitute a disparate group ranging in age from Devonian to Recent: modern species belong to two orders, the porcelaneous miliolides and the hyaline rotaliides; ancient ones to four, the arenaceous foraminiferans and the microgranular fusulinides, as well as the porcelaneous miliolides and the hyaline rotaliides. They are grouped together not because they are taxonomically related but because they are of comparatively large size – being visible to the naked eye – and complex internal structure, and are best studied, and indeed in many cases can only be identified, in thin-section (by biometrics). Most, if not all, living LBFs host photosynthetic algal symbionts.

The ecological, bathymetric and biogeographic, distributions of modern LBFs are comparatively well documented (Hohenegger, 2000, 2004; Langer and Hottinger, 2000). Interestingly, the distributions are similar to those of zooxanthellate (z) corals.

Bathymetry

In terms of bathymetry, modern LBFs characterise shallow, warm waters of near-normal marine salinities (some tolerating slightly elevated salinities), and, commonly, platformal and/or peri-reefal carbonate environments (see also Section 4.3). As noted above, most, if not all, host photosynthetic algal symbionts, such that they are restricted to depths within the photic zone, that is, no deeper than 130 m in clear water, and shallower still in turbid or turbulent water (Taylor, in Capriulo, 1990; Hohenegger, 2004). Depth tolerances of individual species are dictated by the precise light requirements of their symbionts. Those with green or red algal or dinoflagellate symbionts, such as most of the miliolides, are restricted to the euphotic zone, and thus shallow, inner to middle platform, or back-reef – including sea-grass – and reef, sub-environments; those with diatom symbionts, such as most of the rotaliides, the sub-euphotic zone, and thus deeper, middle to outer ramp, or reef and fore-reef, sub-environments. Interestingly, the tests of modern rotaliides are more enriched in photosynthetic $\delta^{12}C$ (or impoverished in $\delta^{13}C$) than those of modern miliolides, arguably indicating a greater photosynthetic capacity (another possibility is that the rotaliides use $\delta^{12}C$-enriched photosynthetic carbon from their symbionts in test construction, whereas the miliolides use carbon from sea water). This is consistent with the observation that the rotaliides are characteristic of oligotrophic, and the miliolides of mesotrophic to eutrophic, environments, such that the rotaliides are more, and the miliolides less, reliant on their symbionts as a food source. In other words, the symbiosis may be obligate in the rotaliides and facultative in the miliolides. Incidentally, the tests of certain ancient rotaliide LBF species are more enriched in $\delta^{12}C$ (or impoverished in $\delta^{13}C$) than others, again arguably indicating a greater photosynthetic capacity. Coincidentally or otherwise, these species evolved and became extinct later than the others, arguably indicating that symbiosis played an important role in the evolution and extinction of the rotaliide LBFs.

Biogeography

In terms of biogeography, modern LBFs characterise three provinces, namely the Indo-(west) Pacific, the Mediterranean and the Caribbean, which last-named was contiguous with the (east) Pacific prior to the closure of the isthmus of Panama in the Pliocene.

Importantly, temperature tolerances of modern species fall within the range 14–40° C (generally between 0° and 35° N/S, but in early latitudes where cold-water currents flow northward along the western margins of the continental shelves in the southern hemisphere or southward along the eastern margins of the continental shelves in the northern hemisphere). A minimum summer sea-surface temperature of 18° C is apparently necessary for reproduction. (In common with most other foraminiferans, reproduction takes place in alternate sexual and asexual stages, with sexually produced individuals generally rare. Whether or not a planktonic habit is developed at any stage of the life cycle remains unclear.) LBF diversity appears to be at least in part a function of sea-surface temperature. A high proportion (94%) of high-diversity sites, a moderate proportion (65%) of moderate-diversity sites, and a low-proportion (47%) of low-diversity sites are located in essentially tropical waters characterised by temperatures >25° C. The diversity of LBFs can thus be used as a measure of sea-surface temperature, and, assuming uniformitarianism, palaeotemperature and palaeoclimate. However, it appears also to be a function of several other variables, namely: trophic regime (being highest in oligotrophic environments); evolution (being highest at times or in places of intense evolutionary activity); dispersal (being highest along routes of dispersal from loci of evolution); duration of studied interval and/or size of studied area (being highest at times or in areas of greatest niche availability); sampling artefact; and taxonomic artefact. Indeed, it may be more a function of any one or more of these variables than of palaeotemperature and palaeoclimate. Ideally, in order to take into account the effects of these variables, raw diversity data requires normalisation.

Palaeobiology

The environmental interpretation of ancient LBFs relies on analogy with their modern counterparts.

Palaeobathymetry

Thus, in terms of palaeobathymetry, ancient LBFs are interpreted as indicative of shallow, warm waters, and, commonly, platformal and/or peri-reefal carbonate environments. They are also interpreted as having hosted photosynthetic algal symbionts, and as having been restricted to depths within the photic zone. Those interpreted as having hosted green or red algal or dinoflagellate symbionts, such as the porcelaneous miliolides, and, by analogy, the microgranular fusulinides, are further interpreted as indicative of the euphotic zone, and thus shallow, back-reef and reef, sub-environments; those interpreted as having hosted diatom symbionts, such as the hyaline rotaliides, as indicative of the sub-euphotic zone, and thus deeper, reef and fore-reef, sub-environments (Jones, 1996). Microgranular LBFs were important contributors to the construction of reefs in the Late Carboniferous to Permian of the former Soviet Union (Kuznetsov, in Zempolich and Cook, 2002). Accumulations of certain hyaline LBFs, named nummulites ('coin-stones'), in the Palaeogene of the circum-Mediterranean are of commercial significance as petroleum reservoirs (Jones, 1996; Jones, 1997; Racey, 2001; Vennin *et al.*, 2003).

Palaeobiogeography

In terms of palaeobiogeography, many LBFs appear to have had restricted or endemic biogeographic distributions, rendering them of some use in the characterisation of palaeobiogeographic provinces, and in turn in the constraint of plate tectonic reconstructions, in the Palaeozoic (Kobayishi, 1997a, b; Katsumi Ueno, 2003), in the Mesozoic and in the Cenozoic.

In the Mesozoic, the Tethyan realm is characterised by the LBF *Orbitopsella* in the Pliensbachian; *Paracoskinolina (occitana), Spiraloconulus, Satorina* and *Pfenderina (salernitana)* in the Bathonian; *Alveosepta, Kurnubia, Parurgonia, Labyrinthina, Mangashtia* and *Everticyclammina* the Oxfordian–Kimmeridgian; *Anchispirocyclina, Choffatella (pyrenaica), Pseudocyclammina, Pfenderina* and *Valdanchella* in the Tithonian–Valanginian; *Valserina* and *Orbitolinopsis* in the Hauterivian–early Barremian; *Choffatella (decipiens), Palorbitolina, Orbitolinopsis* and *Palaeodictyoconus* in the late Barremian–early Aptian; *Orbitolina (Mesorbitolina), Coskinolinoides, Nummoloculina, Neoiraqia, Neorbitolinopsis, Pseudochoffatella, Archaealveolina, Barkerina, Simplorbitolina, Coskinolinella, Paracoskinolina (tunesiana)* and *Archaealveolina* in the late Aptian–middle Albian; *Neorbitolinopsis, Orbitolina, Neoiraqia* and *Valdanchella* in the late Albian–early Cenomanian; *Ovalveolina, Praealveolina* and *Cisalveolina* in the middle–late Cenomanian; and *Lacazina, Lamarmorella* and *Murgella* in the Santonian (Fourcade and Michaud, 1987; Cherchi, in Boriani *et al.*, 1989). Incidentally, *Orbitopsella* occurs in the Early Jurassic Rotzo member of the Calcari Grigi formation of the Venetian Alps, overlying the Triassic Dolomia Principale and underlying the Middle Jurassic *ammonitico rosso*, where it has been interpreted, on the bases of analogy with living soritids, functional morphology, and associated fossils and sedimentary facies, as having accommodated algal photosymbionts, and as having constituted part of a mesotrophic (?oligotrophic) community, associated with lithiotid bivalve 'reefs' (Fugagnoli, 2004). Interestingly, an apparent association between *Orbitopsella* and lithiotids has also been noted in the Oman and Zagros mountains (author's unpublished observations).

In the Cenozoic, the Indo-Pacific and Mediterranean provinces are characterised by LBFs. The commonality of LBFs between the Indo-Pacific and Mediterranean is a key indicator of the palaeo(bio)geographic and palaeoclimatic evolution of Eurasia over the critical Oligocene–Miocene time interval (see also Sub-section 5.5.8).

Biostratigraphy

The rapid rate of evolutionary turnover exhibited by the LBFs renders them of some use in biostratigraphy, at least in appropriate, marine, environments, although in practice, their usefulness is somewhat compromised by their restricted ecological, bathymetric and biogeographic distribution, and facies independence, a function of their habit. A number of biostratigraphic zonation schemes based on the LBFs have been established, that have local or regional applicability (Jones, 1996; see also Cariou and Hantzpergue, 1997 for the Mesozoic, and Cahuzac and Poignant, 1997, and Serra-Kiel *et al.*, 1998 for the Cenozoic). However, no inter-regional schemes have been established that can be correlated across palaeobiogeographic provinces.

Macroscopically visible LBFs are of use in stratigraphic subdivision and correlation, and in palaeoenvironmental interpretation, in the field, for example in the Permian of New Mexico and west Texas in the south-western USA, in the Cretaceous of the Middle East, in the Palaeogene of the Pyrenees, north Africa, the Middle East and northern South America, and in the Neogene of the Far East (author's unpublished observations).

In my working experience, LBFs have proved of particular use in stratigraphic subdivision and correlation, and in palaeoenvironmental interpretation, in the following areas.

Palaeozoic

The Late Palaeozoic, Devonian–Permian of the Russian platform in the former Soviet Union. Here, LBFs form the basis of the regional biostratigraphic zonation. They also form part of the basis of the calibration of the regional sequence stratigraphic framework (Makhlina, in Strogen *et al.*, 1996; Rukina, in Strogen *et al.*, 1996).

The Carboniferous of China. Here, as in the Tarim basin, the usefulness of LBFs has been enhanced through the use of graphic correlation (Groves and Brenckle, 1997).

The Carboniferous–Permian of the Middle East. See Jones (1996). Here, LBFs form part of the basis of the regional biostratigraphic zonation. They also form part of the basis of the calibration of the regional sequence stratigraphic framework. The Carboniferous of north-east Iran is marked by various LBFs (Vachard, in Wagner *et al.*, 1996). The Late Permian, Kazanian P20 MFS of Sharland *et al.* (2001), stratotypified in the Abadeh section in Iran, is marked by fusulinide LBFs of the *Neoschwagerina margaritae* Zone. The Late Permian, Tatarian P30 MFS, stratotypified in member C of the Khuff formation of Saudi Arabia, is marked by fusulinide LBFs of the *Colaniella parva* Zone. The Khuff formation of Oman also contains LBFs (Angiolini *et al.*, in Al-Husseini, 2004).

Mesozoic

The Jurassic–Cretaceous of the Middle East. Here, LBFs form part of the basis of the regional biostratigraphic zonation, calibrated against ammonite and other standards. They also form part of the basis of the calibration of the regional sequence stratigraphic framework (Sharland *et al.*, 2001). The Late Jurassic, late Kimmeridgian J80 MFS, stratotypified in member C of the Arab formation of Abu Dhabi, is marked by the arenaceous LBF *Kurnubia jurassica*. The Tithonian J110 MFS, stratotypified in the Arus member of the Hajar formation of southern Yemen, is marked by *Everticyclammina kelleri*. The Early Cretaceous, Barremian K60 MFS, stratotypified in the lower Kharaib formation of Abu Dhabi in the United Arab Emirates, is marked by *Palaeodictyoconus arabicus*. The early Aptian K70 MFS, stratotypified in the Hawar shale in the upper Kharaib formation of the United Arab Emirates, is marked by *Palorbitolina lenticularis*. The early Albian K90 MFS, stratotypified in marker limestone bed I of the Nahr Umr formation of Oman, is marked by *Mesorbitolina texana*. The middle Albian K100 MFS, stratotypified in marker limestone bed II of the Nahr Umr formation, is marked by *Mesorbitolina subconcava*. The late Albian K110 MFS, stratotypified in the Mauddud formation of Kuwait, is marked by *Orbitolina sefini*. The middle Cenomanian K130 MFS, stratotypified in member D of the Natih formation of Oman, is marked by the porcelaneous LBF *Praealveolina tenuis*. The Maastrichtian K180 MFS, stratotypified in the Simsima formation of northern Oman and the eastern United Arab Emirates, is marked by the hyaline LBF *Omphalocyclus macroporus*. In the field in the Zagros Mountains in south-west Iran, the upper part of the Dariyan formation, above the 'Kazhdumi tongue', is datable as within the range late Albian to early Cenomanian on the occurrence of the macroscopically visible LBF *Orbitolina ex gr. concava*, and the upper part of the Kazhdumi formation as latest Alban–earliest Cenomanian on the occurrence of *Orbitolina sefini* (author's unpublished observations). Importantly, LBFs are of use in reservoir characterisation, for example of the Late

Jurassic Arab formation reservoir of Ghawar field of Saudi Arabia (Hughes, 1996), and the Early Cretaceous Shuaiba formation reservoir of Shaybah field in Saudi Arabia (Hughes, 2000) (see also Hughes, in Hart *et al.*, 2000).

Cenozoic

The Cenozoic of the Middle East. Here, LBFs form part of the basis of the regional biostratigraphic zonation, loosely calibrated against the calcareous nannofossil standard. They also form part of the basis of the calibration of the regional sequence stratigraphic framework (Sharland *et al.*, 2001, 2004). The regional Early Eocene Pg20 MFS, stratotypified in the Midra shale member of the Dammam formation of Saudi Arabia, is marked by the LBFs *Nummulites globulus* and *Coskinolina balsillei*. The Early Oligocene Pg30 MFS, stratotypified in the Taqa formation of Oman, is marked by *N. fichteli*. The 'Middle' Oligocene Pg40 MFS, stratotypified in the Bajawan formation of the Kirkuk group of Iraq, is marked by *N. fichteli*, *Lepidocyclina* and *Praerhapidionina delicata*. The Late Oligocene Pg50 MFS, stratotypified in the Anah formation of the Kirkuk group, is marked by *Miogypsinoides complanatus*. The early Early Miocene Ng10 MFS, stratotypified in the Euphrates formation of Iraq, is marked by *Miogypsina globulina*. The late Early Miocene Ng20 MFS, stratotypified in the Jeribe formation of Iraq, is marked by *Borelis melo curdica*. The early Middle Miocene Ng30 MFS, stratotypified in the Lower Fars formation of Iraq, is marked by the presence of *Miogypsina* and the absence of *Borelis melo curdica*. The late Middle Miocene Ng40 MFS, stratotypified in the Guri limestone member of the Mishan formation of south-west Iran, is marked by *Miogypsina* and *Flosculinella bontangensis*. In the field in the Zagros Mountains in south-west Iran, the lower Asmari formation is datable as Oligocene on the occurrence of the macroscopically visible hyaline LBF *Eulepidina elephantiana*, and the Middle Asmari as Miocene on the occurrence of *Miogypsina* (author's unpublished observations).

The Cenozoic of northern South America and the Caribbean. See Robinson (1993).

3.3.2 Radiolarians

Biology, morphology and classification

Radiolarians are an extant, Cambrian–Recent, group of protozoans characterised by a – usually – siliceous shell or scleracoma (Blome *et al.*, 1996; de Wever *et al.*, 2002).

Biology

The biology and ecology of living radiolarians is well known. They are essentially heterotrophic, that is, consumers rather than producers, entrapping prey in pseudopodial networks. However, some are autotrophic, housing symbiotic algae that produce food for them by photosynthesis. Radiolarian reproduction, as far as is known, is exclusively asexual.

In terms of ecology, radiolarians are exclusively marine and planktonic, and, as a group, are cosmopolitan (Zeitzschel, in Capriulo, 1990). In the Sea of Japan, shallow environments, from 40 to 120 m within the water column, are characterised by *Spirocyrtis, Spongodiscus, Lipmanella* and juvenile *Larcopyle*; shallow intermediate environments, from 40 to 300 m, by *Ceratocyrtis, Spongotrochus* and adult *Larcopyle*; deep intermediate environments, from 160 to 300 m by *Ceratospyris*; and deep environments, from 1000 to 2000 m, by *Cycladophora* and *Actinomma* (Itaki, 2003). Still deeper-water taxa such as *Cornutella, Cyrtoptera* and *Peripyramis* are not found in the Sea of Japan, because of the shallow sill depth.

Although, as intimated above, distinct radiolarian taxa have distinct ecological preferences, the group as a whole is characteristically associated with cold, deep-water masses and/or upwelling systems. Radiolarians may bloom under particularly favourable environmental conditions, resulting in the accumulation of radiolarian oozes or cherts.

Morphology

In terms of hard-part morphology, radiolarians are characterised by a shell or scleracoma comprising a hollow tangential lattice or a series of concentric lattices supported by radial spines (Fig. 3.20). The overall shape is generally spherical or conical.

Classification

Historically, three main sub-groups of radiolarians have been recognised on the basis of skeletal

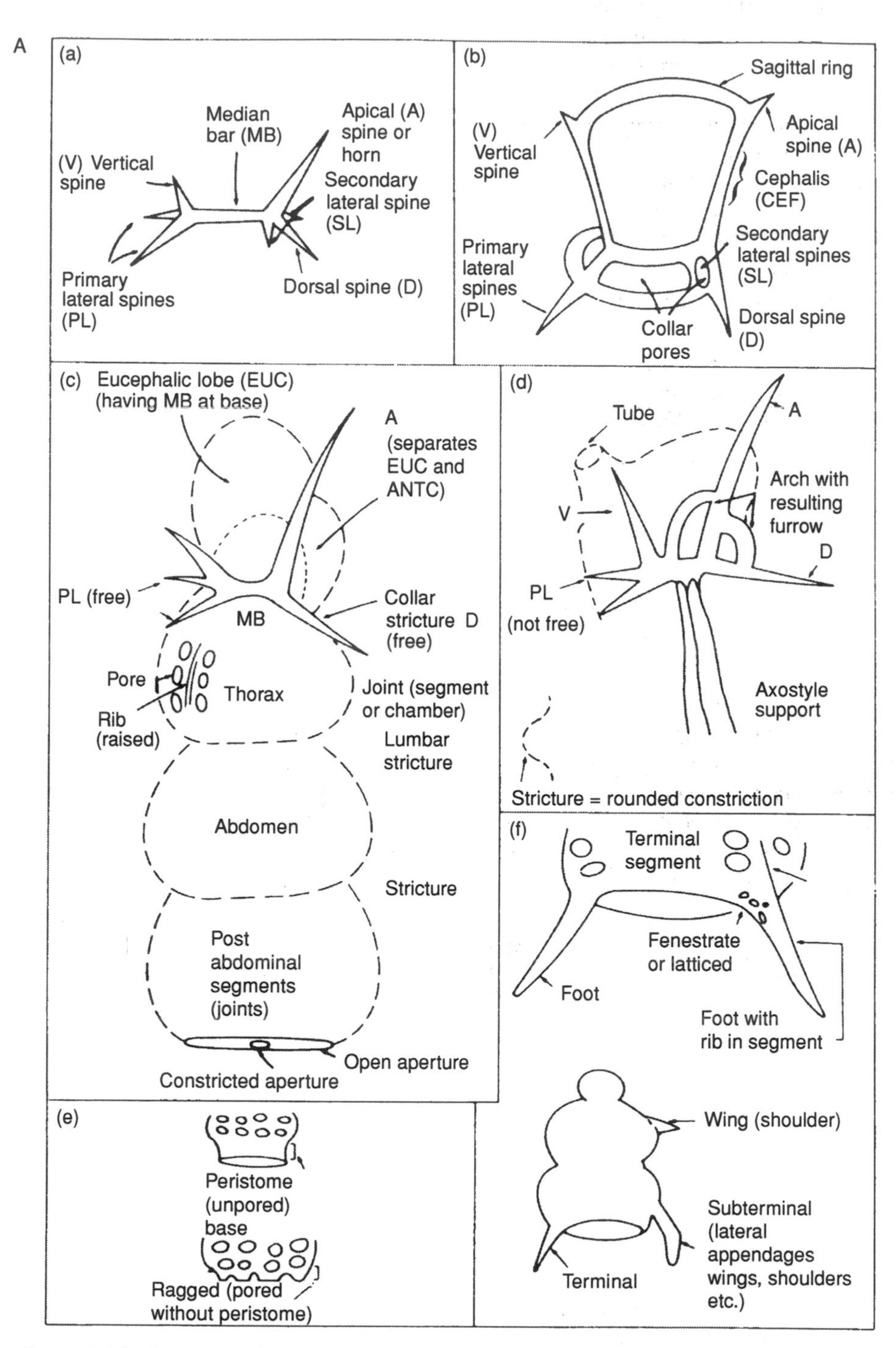

Fig. 3.20. Hard-part morphology of radiolarians. (A) Nasselarians. (a–b, d) Spicules; (c) cephalic, thoracic, abdominal and post-abdominal segments; (e–f) nature of terminations; (g–h) details of segment morphology; (i–j) details of pore morphology. (B) Spumellarians. (a) Spicules; (b) cortical and medullary shells and interconnecting structures; (c) details of pore morphology; (d) discoid/lenticular spumellarian; (e) 'larcoid' spumellarian; (f) development of spongy inter-arm 'patagium'; (g) ellipsoidal spumellarian. (From Jones, 1996; reproduced with permission from Lipps, 1993.)

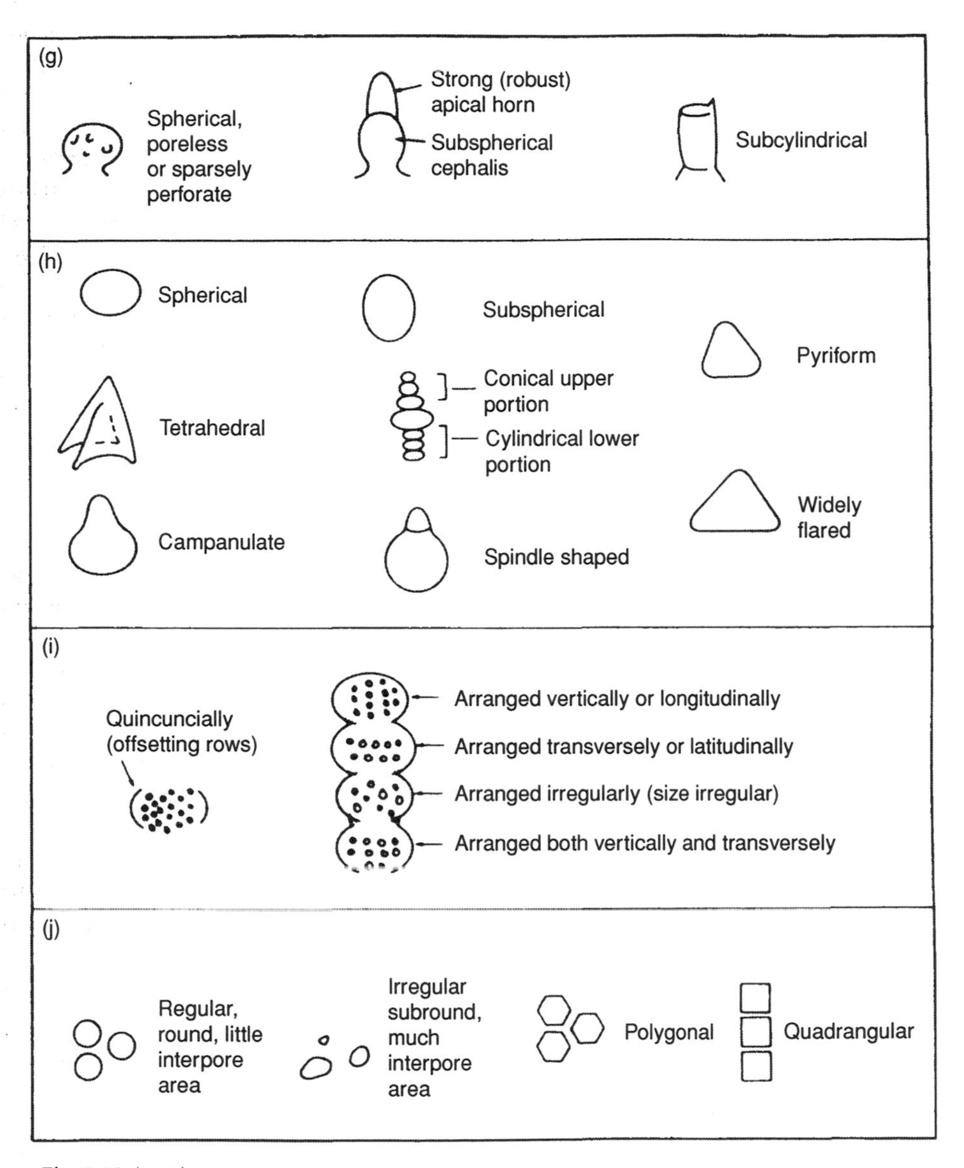

Fig. 3.20 (*cont.*)

composition and morphology, namely, polycystines, acantharians and phaeodarians. However, these sub-groups are probably polyphyletic, and therefore the historical classification is in need of revision, the better to reflect phylogeny. Unfortunately, as yet no clear consensus has been reached as to a revised classification.

Polycystines. Polycystines are characterised by siliceous skeletons. Two further sub-groups of polycystines are in turn recognised on the basis of morphology, namely, nasselarians and spumellarians. *Nassellarians* are characterised by sphaeroidal to discoidal morphologies, and the absence of algal symbionts, and are in turn characteristic of mesopelagic and bathypelagic environments intermediate to deep within the water column (and underlying sediments). *Spumellarians* are characterised by conical to fusiform morphologies, adaptations aimed at reducing sinking rate, and the

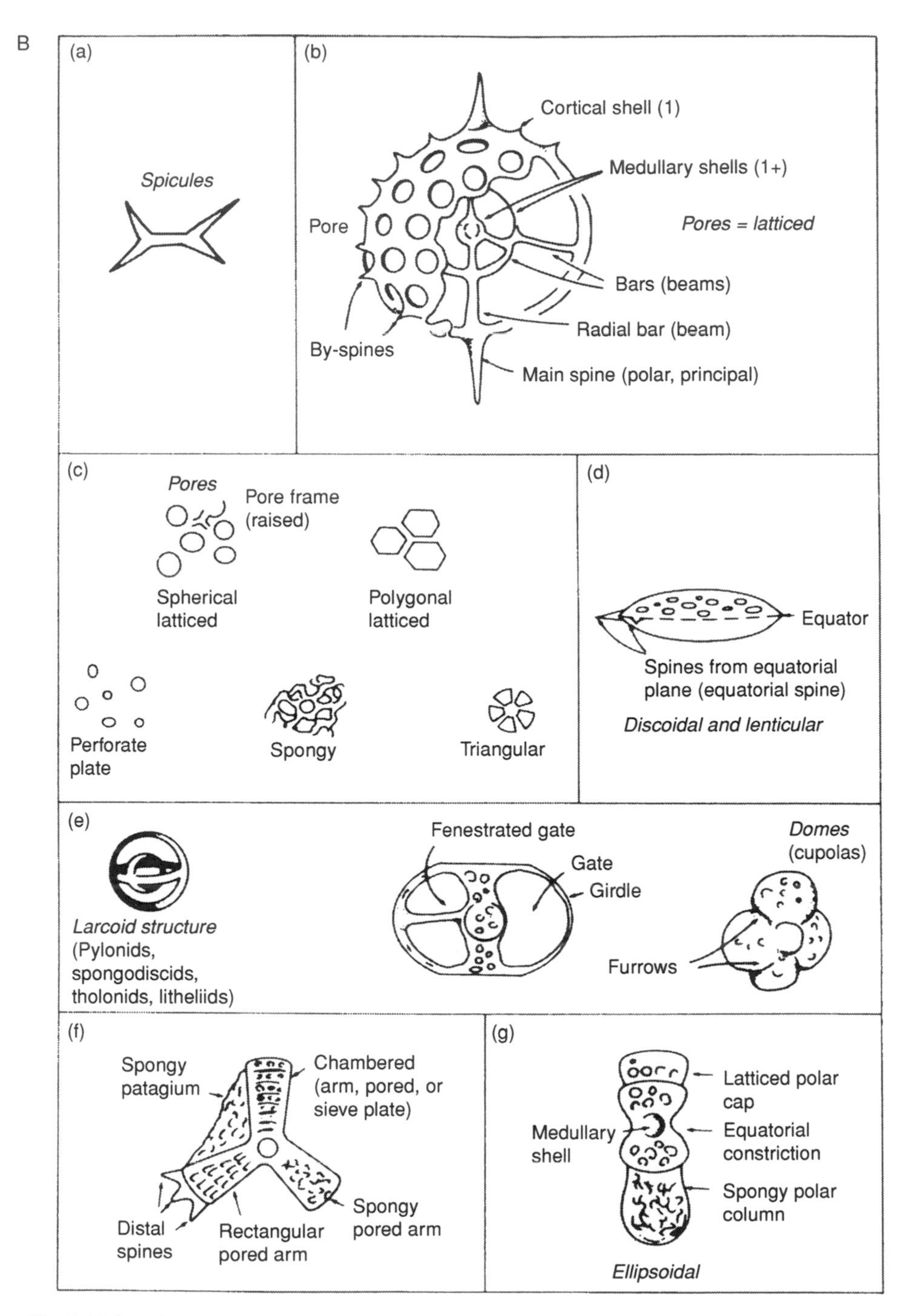

Fig. 3.20 *(cont.)*

presence of photosymbionts, and are characteristic of epipelagic environments shallow within the water column, and within the photic zone. Symbiosis is observable using epifluorescence microscopy (EFM) (Takahashi *et al.*, 2003). (Importantly from the point of view of palaeobiology, nasselarians are disproportionately susceptible to dissolution, and commoner in water column than sea-bottom sediment samples.)

Acantharians. Acantharians are characterised by strontium sulphate skeletons.

Phaeodarians. Phaeodarians are characterised by organic skeletons strengthened with up to 20% opaline silica.

Palaeobiology

Fossil radiolarians are interpreted, essentially on the basis of analogy with their living counterparts, as having been exclusively marine and planktonic. They bloomed under particularly favourable environmental conditions, resulting in the accumulation of cherts and siliceous limestones.

Radiolarians are further interpreted as associated with upwelling, as in the Early Carboniferous of the Culm basin of Germany (Gursky, in Strogen *et al.*, 1996), and the Miocene of the Pungo River formation of North Carolina (Palmer, 1988). In the Early Carboniferous of the Culm basin, westward-driven nutrient-rich currents from the Palaeo-Pacific favoured high radiolarian productivity, resulting in the widespread deposition of cherts and siliceous limestones, associated with phosphorites and, possibly significantly, volcanics. This was terminated by changes to oceanic circulation and by increased clastic input associated with the collision between Laurasia and Gondwana in the Hercynian orogeny (named after the Harz Mountains).

Radiolarians may also be associated with eutrophication resulting from riverine input rather than upwelling, as in the Oligo-Miocene of Angola in west Africa (author's unpublished observations).

Palaeobathymetry

Fossil spumellarians and nasselarians are interpreted, essentially on the basis of analogy with their living counterparts, as having been epipelagic to mesopelagic, and bathypelagic, respectively (see also Funnell, in Hallam, 1967).

Palaeobiogeography

Many radiolarians appear to have had restricted or endemic biogeographic distributions, rendering them of use in the characterisation of palaeobiogeographic provinces, and in turn in the constraint of plate tectonic reconstructions, as in the Palaeozoic, in the Mesozoic, and in the Cenozoic.

In the Mesozoic, radiolarian distributions point toward the development of three palaeobiogeographic provinces, namely: a southern or Austral province characterised by *Praeparvicingula* and *Parvicingula*; a central, Tethyan province, characterised by high abundances and diversities of pantanellids and by *Mirifusus, Andromeda, Acanthocircus dicranocanthos* and *Ristola*; and a northern or Boreal province characterised by *Praeparvicingula, Parvicingula, Stichocapsa* and, in its southern part only, by *Ristola* (Vishnevskaya, in Crasquin-Soleau and Barrier, 1998a; Pessagno *et al.*, in Mann, 1999).

In the Cenozoic, radiolarian evidence points toward the existence of a marine connection between Tethys and the Arctic Ocean in the Eocene, by way of the so-called Obik Sea or Turgai Strait (Radionova and Khokhlova, 2000).

Palaeoclimatology

Radiolarians are of use in palaeoclimatic as well as in palaeobiogeographic interpretation (Dyer, in Neale and Brasier, 1981; Haslett, in Haslett, 2002). For example, in the Middle–Late Miocene of the Equatorial Pacific, collosphaerids have been interpreted, on the basis of analogy with their living counterparts, as having been essentially tropical to temperate, and a specific variant of the theoperid *Stichocorys delmontensis* as an indicator of cooler conditions characterised by enhanced upwelling (Dyer, in Neale and Brasier, 1981). Data on the distributions of the respective cool- and warm-water markers, taken together, indicate a significant cooling within the Late Miocene.

Biostratigraphy

Radiolarians evolved in the Cambrian. They underwent dramatic diversifications in the Ordovician (Noble and Danielian, in Webby *et al.*, 2004), and again in the Mesozoic and Cenozoic. The diversification in the Apto-Albian appears to have been associated with 'oceanic anoxic event I' (Danielian *et al.*, 2003).

The moderate rate of evolutionary turnover exhibited by the radiolarians renders them of some use in biostratigraphy, at least in appropriate, marine, environments. Their usefulness is enhanced by their essentially unrestricted ecological distribution, and facies independence, a function of their planktonic habit. Biostratigraphic zonation schemes based on the radiolarians have been established, that have at least local to regional applicability. An important scheme applicable to the Cenozoic of the tropics has been established by Sanfilippo and Nigrini (1998). An important scheme applicable to the Neogene–Pleistogene of the North Pacific has been established by Motoyama and Maruyama (1998) (Fig. 3.6). Independent schemes for the Late Neogene–Pleistogene of the tropics, the Atlantic, the Indian Ocean and the Pacific have been reviewed, and correlated, by Haslett (2004).

In my working experience, diatoms have proved of some use in stratigraphic subdivision and correlation, and/or in palaeoenvironmental interpretation, in the following areas.

Mesozoic

The Middle Jurassic to earliest Cretaceous of the Middle East. Here, radiolarians are critical to the dating and reconstruction of the Isparta angle of south-west Turkey (Vrielynck, in Akinci *et al.*, 2003). Dating is with reference to the 'unitary association zonation' of Baumgartner *et al.* (1995).

The Late Jurassic–Early Cretaceous of the North Sea. See Copestake *et al.*, in Evans *et al.* (2003). Here, radiolarians form part of the basis of the regional biozonation, calibrated against the ammonite standard.

The Late Jurassic–Cretaceous of the Timan–Pechora basin, the Urals and west Siberia in the former Soviet Union. See Amon *et al.* (1997), De Wever and Vishnevskaya (1997) and Vishnevskaya, in Crasquin-Soleau and Barrier (1998a). Here, interestingly, the Early Kimmeridgian and Middle Volgian of the Timan–Pechora basin contain markers for the *Parvicingula vera* Zone, and the *P. haeckeli* or *P. papulata* Zones, respectively, and indicate a tentative correlation with the Kimmeridge clay formation of the North Sea, which is marked by abundant *P. jonesi*.

The 'mid' Cretaceous of the Sergipe–Alagoas basin in Brazil. See Koutsoukos and Hart (1990). Here, spumellarians are characteristic of middle shelfal to upper bathyal, and nasselarians of upper bathyal – including oxygen minimum zone – environments.

Cenozoic

The Cenozoic of the European platform. See de Wever and Popova (1997).

The Palaeogene of the North Sea. See Ahmadi *et al.*, in Evans *et al.* (2003) and Jones *et al.*, in Evans *et al.* (2003). Here, radiolarians form part of the basis of the regional biozonation, loosely calibrated against the calcareous nannoplankton standard.

The Palaeogene of the Caribbean. Here, three radiolarian zones have been recognised in the Late Eocene of the Bath Cliff section in Barbados, namely, in ascending stratigraphic order, the *Cryptoprora ornata* Zone, the *Calocyclas bandyca* Zone, and the '*Carpocanistrum*' *azyx* Zone (Saunders *et al.*, 1984).

The Palaeogene–Neogene of central Paratethys. Here, specifically in the Moravian part of the Carpathian foredeep, certain of the regional stages of the Oligocene–Miocene of central Paratethys are marked by particular radiolarian zones, for example, the Miocene, Badenian by the *Dorcadospiris alata* Zone (Slama, in Thon, 1983).

The Neogene of Sakhalin in the former Soviet Union. See Popova (1993).

3.3.3 Calpionellids

Biology, morphology and classification

Calpionellids are an extinct group of uncertain taxonomic affinity, although they exhibit some morphological similarities with modern tintinnids, which are ciliates (Jones, 1996). Nothing is known directly about their biology, as they are extinct. In terms of morphology, they are characterised by open-ended flask-like bodies or loricae (Fig. 3.21). Their classification is based on details of morphology, including overall shape, and the form of the lorical collar.

Palaeobiology

Fossil calpionellids are interpreted, essentially on the basis of associated fossils and sedimentary facies, as having been marine and pelagic. They appear to have exhibited a preference for open oceanic environments. They exhibited a wide geographic distribution in low to mid latitudes.

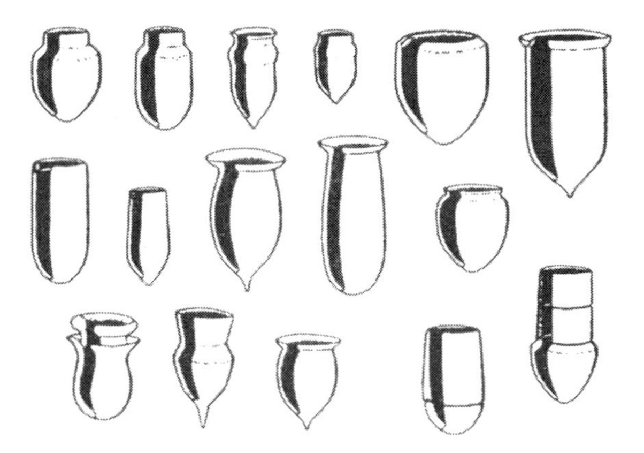

Fig. 3.21. Hard-part morphology of calpionellids. (From Jones, 1996; in turn after Haq and Boersma, 1978.)

Biostratigraphy

Calpionellids evolved in the latest Jurassic (Tithonian) and became extinct in the earliest Cretaceous (Neocomian, ?Barremo-Aptian).

The rapid rate of evolutionary turnover exhibited by the calpionellids renders them of considerable use in biostratigraphy, at least in appropriate, deep marine, environments, in low latitudes. Unfortunately, their usefulness is somewhat compromised by their restricted bathymetric and biogeographic distribution. Nonetheless, high-resolution biostratigraphic zonation schemes based on the calpionellids have been established, that have at least regional applicability.

In my working experience, calpionellids have proved of particular use in stratigraphic subdivision and correlation in the *latest Jurassic to earliest Cretaceous of the Middle East*. Here, the regional earliest Cretaceous, Berriasian K10 MFS of Sharland *et al.* (2001), stratotypified in the Rayda formation of the Oman Mountains, is marked by *Calpionella alpina*.

3.4 Plants

3.4.1 *Plant macrofossils, spores and pollen, and phytoliths*

Biology, morphology and classification

Biology

Plants are an extant, Silurian–Recent, group of autotrophic multicellular organisms. They photosynthesise in order to manufacture food (Stewart and Rothwell, 1993; Taylor and Taylor, 1993; Thomas and Spicer, 1995; Iwatsuki and Raven, 1997; Niklas, 1997; Jones and Rowe, 1999; Willis and McElwain, 2002; Kenrick and Davis, 2004; Ingrouille and Eddie, in press).

In terms of ecology, plants are essentially terrestrial to aquatic. In terms of biogeography, there is evidence of marked provincialism in present-day plant distributions, related primarily on land to climate and hence to latitude or altitude, and on oceanic islands to dispersal and hence to prevailing winds and currents. There is also evidence of similar provincialism in the geological past (see section on 'Palaeobiology' below).

Morphology

In terms of morphology, plants are typically characterised by roots that bind them to the substrate, by branched stems, and by leaves that facilitate the processes of photosynthesis and transpiration. They may or may not possess vascular systems.

Plant preservation potential tends to be poor, as the organic components of which they are composed tend to oxidise during diagenesis. However, instances of exceptional preservation, even of the cellular structure, are not uncommon. Cellular structure is observable using the novel technique of incident-light darkfield microscopy (Thomas *et al.*, 2004).

Classification

Classification of plants is based on morphology and phylogeny.

Four sub-groups of plant macrofossils are recognised, namely: the bryophytes (mosses and allied forms); the pteridophytes (club mosses, ferns and horsetails); the gymnosperms (seed-plants, that is, seed-ferns, tree-ferns, conifers and allied forms); and the angiosperms (flowering plants) (Figs. 3.22–3.24).

A further two sub-groups of plant microfossils are also recognised, namely the phytoliths (Fig. 3.25) and the spores and pollen (Fig. 3.26).

Bryophytes (mosses and allied forms). Bryophytes are an extant, Silurian–Recent, group of non-vascular plants. Their living representatives include the mosses, liverworts and hornworts. Fossil occurrences of these fragile plants are rare. The oldest known forms are from interpreted wet mudflats of

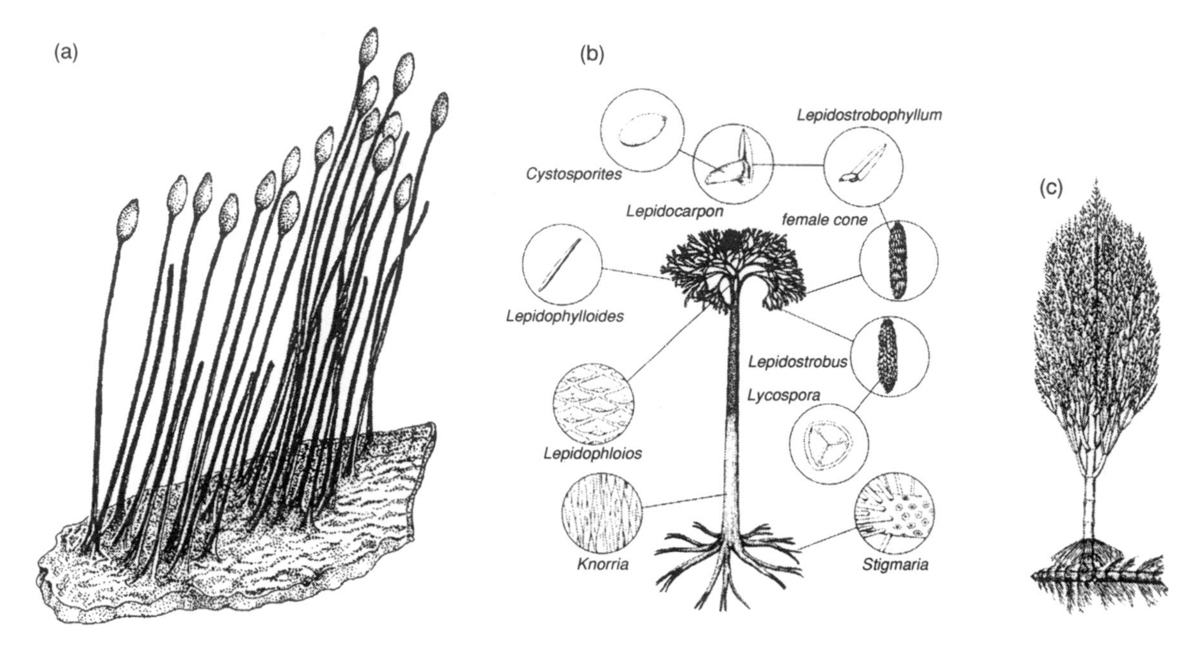

Fig. 3.22. Morphology of plants: bryophytes and pteridophytes. (a) *Sporogonites* (bryophyte), Early Devonian; (b) *Lepidodendron* (pteridophyte, club-moss), Carboniferous; (c) *Calamites* (pteridophyte, horsetail), Carboniferous. (Modified after Benton and Harper, 1997.)

Devonian age. More or less fully terrestrial forms are known from the Silurian onwards. The slow rate of evolutionary turnover exhibited by the bryophytes renders them of limited use in biostratigraphy. In my working experience, they have not proved of particular use.

Pteridophytes (club-mosses, ferns, horsetails and allied forms). Pteridophytes are an extant, Silurian–Recent, group of non-seed vascular plants. Their living representatives include the club-mosses or lycophytes, ferns or filicinophytes, horsetails or sphenophytes, and psilophytes. Fossil occurrences of these plants or of their spores are comparatively common, fossil ferns, for example, being represented by *Pecopteris, Lobatopteris* and *Polymorphopteris*. Ferns formed understoreys in the conifer-dominated forests of the Early Cretaceous (Skelton, 2003).

The ancestral pteridophyte is *Cooksonia*, from the Silurian, a small plant characterised by the presence of spore sacs at the ends of its branched stems, and by the absence of leaves. The ancestors of the club-mosses, namely the zosterophyllophytes, and the ancestors of the ferns and horsetails and indeed all of the other vascular plants, namely the trimerophytes, evolved in the Devonian, as did the rhyniophytes. Many new groups appeared in the Carboniferous, including spectacular giant club-mosses and horsetails, that came to form luxuriant swamp-forests, resounding with insects, in the palaeo-tropics. The swamp-forests of the 'coal measures' of the Carboniferous would have shared certain structural, though not taxonomic, characteristics with those of the lower reaches of the River Mississippi of the modern era.

The giant club-mosses of the Carboniferous 'coal measure' swamp-forests were characterised by spirally arranged leaf bases – the scars left on the trunk and branches when the leaves were shed. They are exemplified by *Lepidodendron* (the 'scale tree', the remains of which have, bizarrely, been mistaken for fish). The trunk of *Lepidodendron* is known to have reached a height of 36 m, before it branched into a crown of leafy shoots to give a possible total height of 45 m. The leafy shoots bore needle-like leaves

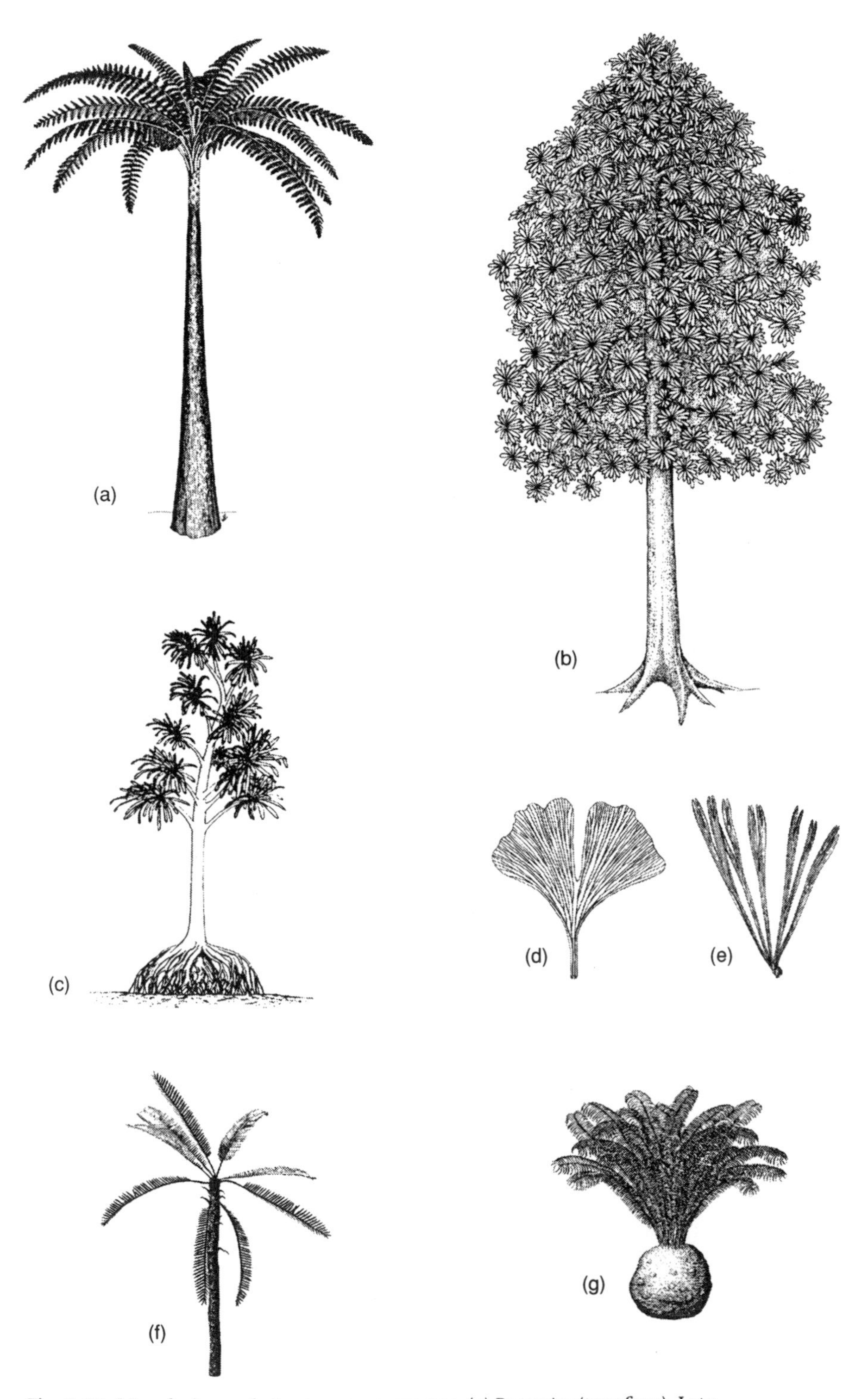

Fig. 3.23. Morphology of plants: gymnosperms. (a) *Psaronius* (tree-fern), Late Carboniferous, North America; (b) *Glossopteris* (seed-fern), Late Permian, Australia; (c) *Cordaites* (conifer); (d) *Ginkgo biloba* (ginkgo), modern; (e) *Sphenobaiera paucipartita* (ginkgo), Jurassic; (f) *Leptocycas gigas* (cycad), Late Triassic, North America; (g) Bennettitalean (cycadeoid), Cretaceous, North America. (Modified after Benton and Harper, 1997.)

Fig. 3.24. Some representative plants. Pteridophytes: Club-mosses: (a) *Lepidodendron aculeatum*, stem, Late Carboniferous, Sunderland, Tyne-and-Wear, ×0.75; (b) *Lepidodendron sternbergi*, branch, Late Carboniferous, Coseley, West Midlands. ×1; (c) *Sigillaria mamillaris*,

and pendant spore-producing cones assigned to the form genera *Flemingites* and *Lepidostrobus* respectively. *Sigillaria* was similar to *Lepidodendron*, but distinguished by strong vertical ribs bearing leaf scars. *Lepidodendron* and *Sigillaria* roots are assigned to the form genus *Stigmaria*, characterised by spirally arranged circular scars which once bore rootlets. Dwarf herbaceous club-mosses grew in the undergrowth alongside the 'coal measure' giants, and are interpreted as having given rise to the club-mosses of the present day, exemplified by *Lycopodium* and *Selaginella*.

The giant horsetails or equisetids of the margins of the Carboniferous 'coal measure' swamp-forests were characterised by tapered, jointed trunks bearing whorls of branches that in turn bore whorls of leaves (incidentally, the trunks were hollow in life, although they often became filled with sediment on death and burial, so as to produce a natural 'pith cast' of the internal structure). The giant equisetids are exemplified by *Calamites*, which reached heights of 12–15 m. Equisetid leaves are represented by the form genera *Annularia* and *Asterophyllites*, spore-producing cones by *Calamostachys* and *Palaeostachya*. Dwarf equisetids grew alongside the giants, as exemplified by *Sphenophyllum*, characterised by whorls of wedge-shaped leaves.

The slow rate of evolutionary turnover exhibited by the pteridophytes renders them of limited use in biostratigraphy. However, in my working experience, I have found them to be of some use in *the Palaeozoic of the Middle East* and in *the Mesozoic of the Middle East, north Africa and South America* (see section on 'Biostratigraphy' below). I have also found pteridophyte spores to be of use.

Gymnosperms (seed-plants). Gymnosperms are an extant, Devonian–Recent, group of seed-bearing vascular plants. Their living representatives include the conifers or coniferophytes; cycads or cycadophytes; ginkgos or ginkgophytes, also known as maiden-hairs; and gnetophytes. Fossil occurrences of these plants, their extinct relatives, or their spores, are comparatively common.

The ancestral gymnosperm evolved, from a trimerophyte pteridophyte, in the Devonian (Beck, 1988). It was probably functionally bisexual, containing megaspores with attached microspores, presumably shed together, facilitating both self- and cross-fertilisation (Marshall and Hemsley, 2003).

Conifers and allied forms, including tree-ferns, seed-ferns and cordaites, diversified in the Carboniferous, apparently on comparatively high,

Caption for fig. 3.24 (*cont.*) stem, Late Carboniferous, Darton, near Barnsley, Yorkshire, ×0.5; (d) *Stigmaria ficoides*, rootstock, Late Carboniferous, Dudley, West Midlands, ×0.5. Ferns: (e) *Pecopteris polymorpha*, foliage, Late Carboniferous, Radstock, Somerset, ×1. Horsetails: (f) *Calamites suckowi*, pith cast, Late Carboniferous, Gosforth, near Newcastle, ×0.5; (g) *Annularia stellata*, foliage, Late Carboniferous, Clandown, Radstock, Somerset, ×0.75; (h) *Asterophyllites equisetiformis*, foliage, Late Carboniferous, Radstock, Somerset, ×1; (i) *Sphenophyllum emarginatum*, foliage, Late Carboniferous, Forest of Dean, Gloucestershire, ×1. Gymnosperms: Seed-ferns: (j) *Trigonocarpus* sp., seed, Late Carboniferous, Stevenston, Ayrshire, ×1; (k) *Neuropteris gigantea*, frond, Late Carboniferous, Coseley, West Midlands, ×0.5; (l) *Alethopteris serli*, frond, Late Carboniferous, Newcastle-on-Tyne, ×0.5; (m) *Mariopteris nervosa*, frond, Late Carboniferous, Coal Measures, Netherton, near Dudley, West Midlands, ×1; (n) *Sphenopteris alata*, frond, foliage, Late Carboniferous, Radstock, Somerset, ×1. Cordaites: (o) *Cordaites angulostriatus*, leaf, Late Carboniferous, Cameron, Somerset, ×0.75. Ginkgos: (p) *Ginkgo huttoni*, leaf, Middle Jurassic, Bajocian, Middle Deltaic ('Estuarine') series, Cayton Bay, Yorkshire, ×1. Angiosperms: (q) *Magnolia lobata*, internal mould of seed, Early Eocene, London clay, Isle of Sheppey, Kent, ×2. (Modified after Natural History Museum, 2001a, b, c.)

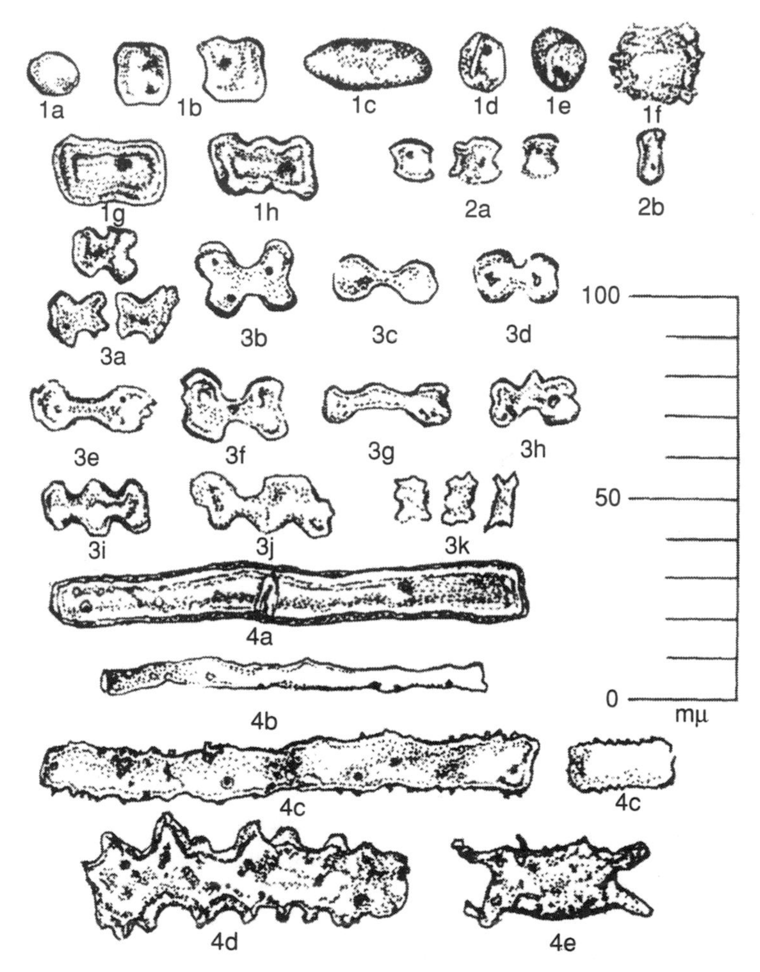

Fig. 3.25. Morphology of phytoliths. (1), Festucoid; (2), chloridoid; (3), panicoid; (4), elongate. (Reproduced with permission from Pearsall, 2000.)

dry ground rather than in the low-lying swamp-forest. They dominated the forests of the Early Cretaceous (Skelton, 2003). Seed-ferns and cordaites are extinct; tree-ferns and conifers are still extant. Seed-ferns, or pteridospermophytes, reproduced by means of seeds, and in this respect, and in lacking spore sacs or sporangia, are quite distinct from true ferns. Some specimens have been found with seeds still attached. Seed-fern seeds are exemplified by the form genus *Trigonocarpus*; fronds by the form genera *Neuropteris, Laveineopteris, Macroneuropteris, Alethopteris, Mariopteris, Paripterus* and *Sphenopteris*. The parent plants probably reached heights of up to 10 m. Seed-ferns became extinct in the Cretaceous. Cordaites reached heights of up to 30 m, with sword-shaped leaves up to 1 m long. The cones of these plants, one sort producing pollen, the other seeds, are assigned to the form genus *Cordaitanthus*. Fossil tree-ferns resembled living forms, and reached similar heights, of up to 10 m. They are exemplified by *Psaronius*, which resembled the living *Marattia*. Fossil conifers also resembled living forms. They are exemplified by the Early Cretaceous, 'Wealden' form *Pseudofrenolepis*.

Cycads and ginkgos evolved in the Carboniferous (cycads from a group of seed-ferns called the Trigonocarpales). Cycads and ginkgos diversified and became widely distributed in the Cretaceous (Skelton, 2003), as did the cycad-like cycadeiods, which had what

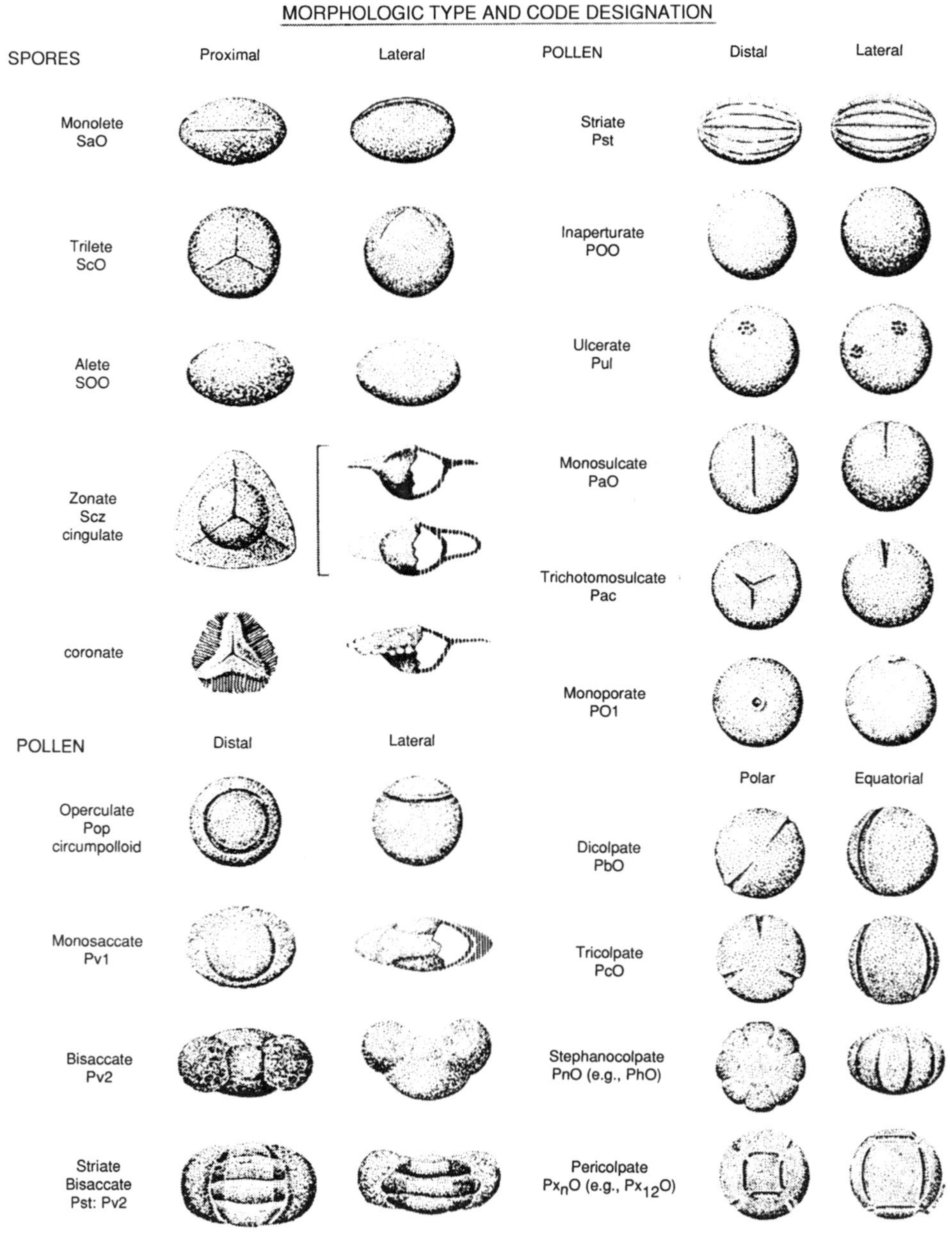

Fig. 3.26. Morphology and descriptive terminology of spores and pollen. Coding is according to the so-called 'Shell scheme'. (From Jones, 1996; in turn after Traverse, 1988.)

looked like flowers, but were in fact apparatuses for seed dispersal rather than pollination. Ginkgos grew throughout the northern hemisphere until the end of the Cretaceous, when they all but disappeared, leaving only a single species alive today, in the forests of western China. In the Cretaceous, they appear to have grown along river banks (Skelton, 2003).

Gnetophytes evolved in the Cretaceous. They are represented by three living genera, *Gnetum, Ephedra*

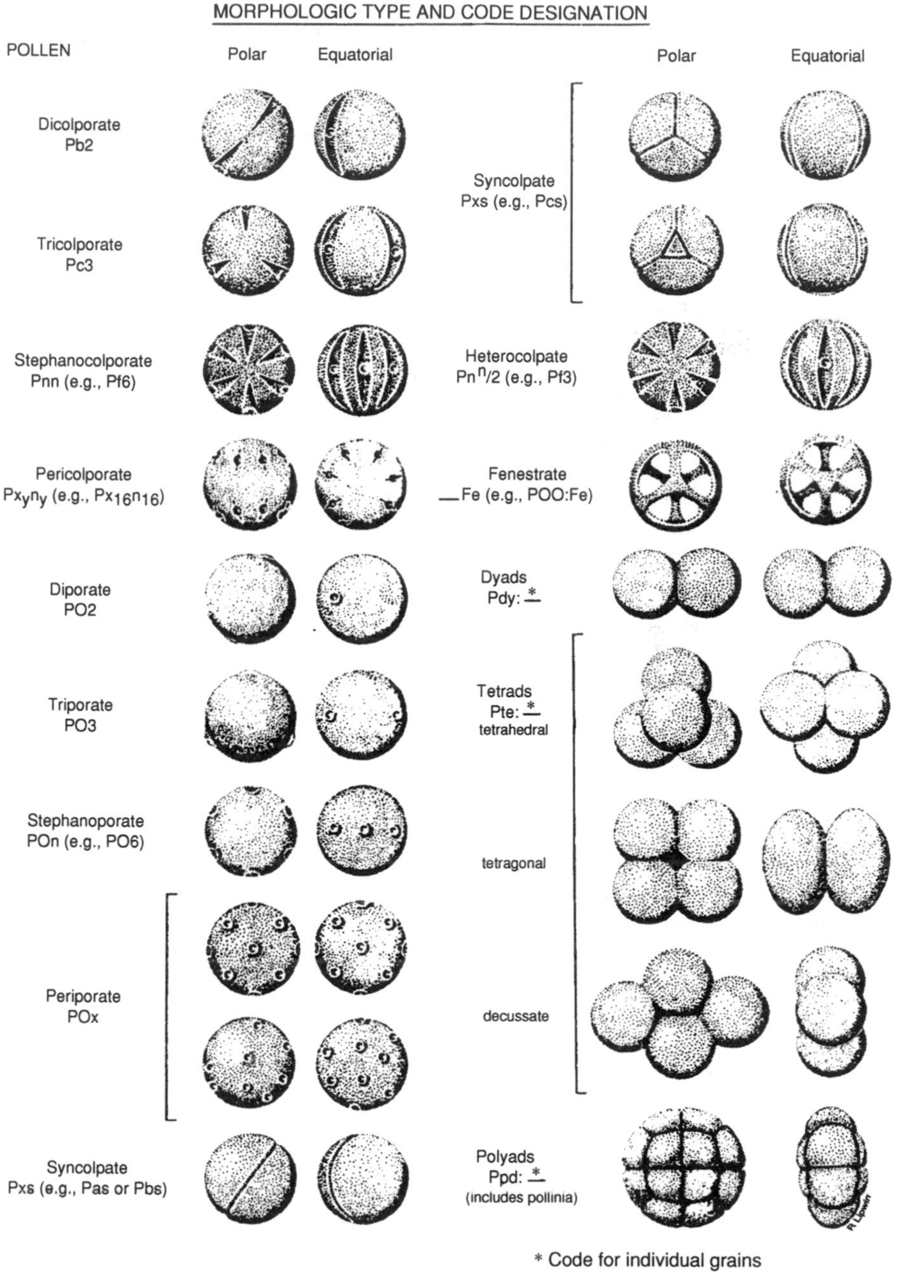

MORPHOLOGIC TYPE AND CODE DESIGNATION

POLLEN

Polar
Equatorial

Dicolporate
Pb2

Tricolporate
Pc3

Stephanocolporate
Pnn (e.g., Pf6)

Pericolporate
Px_yn_y (e.g., $Px_{16}n_{16}$)

Diporate
PO2

Triporate
PO3

Stephanoporate
POn (e.g., PO6)

Periporate
POx

Syncolpate
Pxs (e.g., Pas or Pbs)

Polar
Equatorial

Syncolpate
Pxs (e.g., Pcs)

Heterocolpate
$Pn^n/2$ (e.g., Pf3)

Fenestrate
—Fe (e.g., POO:Fe)

Dyads
Pdy: *

Tetrads
Pte: *
tetrahedral

tetragonal

decussate

Polyads
Ppd: *
(includes pollinia)

R Lipow

* Code for individual grains

Fig. 3.26. (*cont.*)

and *Welwitschia*. The last named is an unusual plant that grows in the Namib Desert of Namibia, where its only source of water is in the form of fog that condenses on its leaves, which may be as much as 10 m in length. *Ephedra* is also indicative of arid conditions.

The moderate rate of evolutionary turnover exhibited by the gymnosperms renders them of some use in biostratigraphy. In my working experience, gymnosperms have not proved of particular use, although their pollen has (see below).

Angiosperms (flowering plants). Angiosperms are an extant, Cretaceous–Recent, group of flowering vascular plants. Fossil occurrences are comparatively common, generally as branches, leaves, seeds and pollen, and less so as flowers, although examples of flowers are known from the 'Middle' Cretaceous Orapa deposits of South Africa (MacRae, 1999), and examples of *Magnolia*-like flowers are known from the 'Middle' Cretaceous of North America. Angiosperms evolved in the Cretaceous, from seed-ferns (Beck, 1976; Hughes, 1976; Taylor and Hickey, 1995; Friis, 2003). The oldest known are from the Early Cretaceous of America, Russia, and Jixi in eastern Heilongjiang in north-east China (Ge Sun and Dilcher, 2002). Angiosperms diversified through the Cretaceous (Lupia *et al.*, in Culver and Rawson, 2000; Cantrill and Poole, in Crame and Owen, 2002). They also dispersed through the Cretaceous (Brenner, in Beck, 1976), displacing the ginkgophytes from their riparian habitats, and then the conifers from their mires (Skelton, 2003). By the beginning of the Cenozoic, angiosperms had become distributed throughout the world other than in the highest latitudes and altitudes. Angiosperms are the most successful plants to have evolved, possibly at least in part on account of mutually beneficial relationships with a number of important groups of pollinators, including insects and, more rarely, birds or bats (Crepet, in Briggs and Crowther, 2001). Their living representatives include some 250 000 out of a total of 300 000 plant species.

The moderate rate of evolutionary turnover exhibited by the angiosperms renders them of some use in biostratigraphy. In my working experience, angiosperms have not proved of particular use, although their pollen has. The angiosperm derivative oleanane has also proved useful, as a geochemical biomarker for the Cenozoic.

Phytoliths. Phytoliths are siliceous deposits occurring within cells or within or between cell walls in certain plants. They are typically subspherical or festucoid or chloridoid, dumb-bell-shaped or panicoid, or elongate in morphology (Fig. 3.25). Phytoliths are produced entirely in non-marine environments. They can be extremely common in terrestrial and freshwater sediments such as palaeosols and peats. They typically occur without their parent plants, but can be typed back to them. Phytoliths provide invaluable information regarding vegetation and climate in the source area. Thus, the vegetation and vegetation history of the great plains of North America – especially the rise and spread of grasslands – during the Late Eocene to Early Miocene has been reconstructed using phytoliths (Stromberg, 2004; see also Smith and White, 2004). Phytoliths also provide invaluable information regarding the diet of animals, and the domestication of plant species by ancient humans at archaeological sites (Piperno, 1988; Pearsall, 2000; see also Section 7.6).

Spores and pollen. Spores and pollen are produced by land plants during their reproductive cycle (Harley *et al.*, 2000). The relationship of most modern spores and pollen to their parent plants is comparatively well documented. Spores, which may or may not be differentiated into micro- or mio-spores and macro- or mega-spores, are produced by evolutionarily primitive bryophytes and pteridophytes during the asexual or sporophyte phase of their reproductive cycle. The succeeding sexual phase takes place after the disseminated spores have arrived at a suitable substrate and germinated into a gametophyte plant. The suitability or otherwise of the substrate is determined essentially by the availability or absence of moisture, which is the medium for gamete fusion. Pollen is produced by evolutionarily advanced gymnosperms and angiosperms during the asexual phase of their reproductive cycle. The succeeding sexual phase (pollination) is generally facilitated not by water but by wind in the case of the gymnosperms, and by insects, birds or bats in the case of the angiosperms. Greater volumes of pollen are produced by wind-pollinated gymnosperms than by angiosperms.

In terms of ecology, spores and pollen are produced almost exclusively in non-marine environments (pollen produced by sea-grasses in marine environments has no protective exine coating, presumably on account of the negligible risk of desiccation, and hence has low preservation potential). The vast majority, probably approximately 99%, come to rest within 1 km of the source, although, as the grains are designed to be easily transportable, some may be transported by wind and/or water into marine environments. Some bisaccate pollen has even been recorded ingested by benthic foraminiferans living in abyssal environments hundreds of kilometres offshore. The concentration of spores and pollen in a sample generally varies in inverse proportion to the distance from source. In marine environments, the concentration generally varies in inverse proportion to distance from shoreline. In the tropical Tertiary, fern spores are especially indicative of vegetated river banks, and mangrove pollen of coastal swamps.

In terms of morphology, spores and pollen are characterised by isolated subspherical bodies, sometimes distorted by diagenesis. The precise shape can be described as prolate or oblate depending on whether the polar axis is longer or shorter than the equatorial axis. In the case of spores, the polyad arrangement commonly observed in modern sporangia is seldom seen. However, lesions or laesurae may be present on proximal faces, and are presumed to indicate the positions of former attachment sites. Spores can be described as monolete, trilete or zonate depending on the nature of these lesions, if present; and alete if not. In the case of pollen grains, germinal apertures may be present on the distal faces. Pollen grains can be described as colpate, operculate, porate, sulcate or ulcerate depending on the nature of these apertures, if present; and inaperturate if not. Those bearing air sacs designed to aid dispersal are described as saccate. Various qualifiers describe the surface sculpture of spores and pollen.

Classification of ancient spores and pollen (into form genera) is based on details of morphology. The relationship between the form taxonomy of spores and pollen and the formal taxonomy of their parent plants is discussed by Hughes and Moody-Stuart (1966) and Muller (1981).

In terms of palaeobiological significance, the concentration of spores and pollen provides some measure of distance from source, and, in marine environments, distance from shoreline. Moreover, spores and pollen, like phytoliths, provide invaluable information regarding vegetation and climate in the source area. Furthermore, again like phytoliths, spores and pollen also provide invaluable information regarding the diet of animals, and the domestication of plant species by ancient humans (Pearsall, 2000; see also Section 7.6).

In terms of biostratigraphic significance (see also appropriate section below), in my working experience, spores and pollen have proved of particular use in a number of areas.

Palaeobiology

Fossil plants are interpreted, essentially on the bases of analogy with their living counterparts, and functional morphology, as having been essentially terrestrial, and as having occupied a range of habitats in all other than the highest latitudes, and at all other than the highest altitudes (Niklas, in Erwin and Wing, 2000).

Hardiness, moisture needs and salinity tolerance have been inferred from transfer functions based on the documented distributions of living species.

Climate has been inferred from functional morphology and leaf physiognomy; tropical rain-forest from a high incidence of 'drip tips' etc. (Spicer, in Culver and Rawson, 2000).

Palaeobiogeography

Many fossil plants appear to have had restricted or endemic biogeographic distributions, rendering them of use in the characterisation of palaeobiogeographic provinces, and in turn in the constraint of plate tectonic reconstructions, in the Palaeozoic, in the Mesozoic and in the Cenozoic (Morley, 2003).

In the Palaeozoic, the floras of the Late Silurian, Devonian, Carboniferous and Permian were characterised by greater or lesser degrees of endemism (Cleal and Thomas, in Cleal, 1991; Wnuk, in Wnuk and Pfefferkorn, 1996). Plant distributions over this time-interval point toward the existence of generally separate Angara, Euramerica, Cathaysia and Gondwana realms. In the Carboniferous–Permian, the southern hemisphere Gondwana realm was characterised by *Glossopteris* (Plumstead, in Hallam, 1973; Jones, 1996). Interestingly, in the Permian, part of the equatorial Euramerica realm was also characterised by *Glossopteris*, specifically *Glossopteris anatolica*, interpreted as an immigrant from Gondwana, where it is known from India, Patagonia and Antarctica (Archangelsky and Wagner, 1983). Also in the Permian, the plant macrofossils and microfossils of Oman were of essentially tropical aspect, and similar to those of south China, supporting the so-called

'Pangaea B' plate reconstruction (Berthelin *et al.*, 2003).

In the Mesozoic, plant distributions point toward the existence of separate Angara, Euramerica and Gondwana realms in the Triassic (Barnard, 1973). Plant and pollen distributions point toward the existence of separate *Aquilapollenites* and *Normapolles* provinces in Euramerica, and separate *Galeacornea–Constantisporis*, India and *Nothofagus* provinces in Gondwana, in the Cretaceous. Gondwana was characterised by the parent plant of *Classopollis* pollen in the Early Cretaceous (author's unpublished observations).

In the Cenozoic, Palaeocene–Eocene plant distributions point toward the palaeogeographic and palaeoclimatic evolution of the North Atlantic (see Box 5.2). In the Oligocene–Holocene, plant distributions point toward the existence, and the palaeogeographic and palaeoclimatic evolution, of Tethyan and Paratethyan realms (Jones, in Agusti *et al.*, 1999; Akhmetiev, 2000; Hoorn *et al.*, 2000; Mishra, 2000; Morley, 2000; Saxena, 2000; Utescher *et al.*, 2000; Sanyal *et al.*, 2004; see also Sub-section 5.5.8).

Palaeoclimatology

As intimated above, plants are important in palaeoclimatic as well as in palaeobiogeographic interpretation (see also Greenwood, in Briggs and Crowther, 2001; McElwain, in Briggs and Crowther, 2001; Upchurch, in Briggs and Crowther, 2001; da Rosa Alves and Guerra-Sommer, in Koutsoukos, 2005; Dutra, in Koutsoukos, 2005; Kurschner, in Gerhard *et al.*, 2001).

In the Palaeozoic, the leaf physiognomy of Cathaysian gigantopterids has been used in palaeoclimatic interpretation (Glasspool *et al.*, 2004). Leaf margin analysis (LMA) and climate leaf analysis multivariate program (CLAMP) analytical techniques have been applied.

In the Mesozoic, in the Triassic, plant distributions point toward the development of forests at high palaeo-latitudes, even in Antarctica, at 70–75° S (Cuneo *et al.*, 2003). The forests appear to have been associated with a time of at least local if not global warming, resulting in a prolonged growing season, as indicated by the occurrence of the frost-sensitive cycad *Antarcticycas*, and by tree-ring analysis, respectively. Interestingly, during this evident climatic amelioration, the *Glossopteris* flora of the Carboniferous–Permian came to be replaced by a – seasonally deciduous – *Dicroidium* flora. In the latest Cretaceous, plant distributions point toward the development of forests apparently at extraordinarily high palaeo-latitudes, even in Arctic Alaska and Canada, at 75–85° N (Skelton, 2003). These ecosystems evidently thrived under light and temperature regimes that do not exist in the present world, such that they cannot be interpreted solely on the basis of the principle of uniformitarianism. Also in the Cretaceous, CLAMP analytical techniques have been applied in palaeoclimatic interpretation (Skelton, 2003). The analytical results are generally in good agreement with those derived from computer modelling. However, serious discrepancies exist in continental interiors, where the computer models predict lower mean annual temperatures. The cause of this mismatch is not understood, although it has been hypothesised that the flora of the time was somehow capable of buffering the extreme effects of the continental climate.

In the Cenozoic, plant microfossils have been used in the palaeoclimatic interpretation of the Eocene of northern Tanzania (Jacobs and Herendeen, 2004). The vegetation was dominated not by tropical rain-forest but rather by caesalpinioid legume woodland, similar to the modern miombo. The annual precipitation is interpreted as having been in the range 600–800 mm, similar to that of today. Most of the precipitation is interpreted as having occurred in the wet season, although a small but significant proportion is interpreted as having occurred in the dry season, indicating a more equable distribution than that of today. Plant microfossils have been used in the palaeoclimatic interpretation of the Eocene/Oligocene transition of the Gulf coast of the USA (Oboh-Ikuenobe and Jaramillo, in Prothero *et al.*, 2003; Yancey *et al.*, in Prothero *et al.*, 2003), and plant macrofossils in the palaeoclimatic interpretation of the Eocene/Oligocene transition of the Pacific north-west (Myers, in Prothero *et al.*, 2003). Plant macrofossils have been used in the palaeoclimatic interpretation of the Miocene Shanwang flora of China (Ming-Mei Liang *et al.*, 2003). LMA, CLAMP and CoA (coexistence approach) analytical techniques have been applied, and the results compared and contrasted. Plant macrofossils have also been used in the palaeoclimatic interpretation of the Miocene–Pliocene of Australasia (Gallagher *et al.*, 2003; Pole, 2003). In New Zealand, the Early Miocene, characterised by the southern beech, *Nothofagus*, has been interpreted as having been ever-wet and cool, but frost-free (Pole, 2003). The Middle Miocene, characterised initially by eucalypts and palms, and then by casuarinaceans, chenopods and asteraceans, has been interpreted as having been initially seasonally dry, and subject to bushfires, and then dry. The Late Miocene, characterised by *Nothofagus* and *Sphagnum*, has been interpreted as having been wet and cool again. In south-east Australia, the Pliocene has been interpreted as having been dry (Gallagher *et al.*, 2003). The climatic changes have been interpreted as having been caused by oscillations in the positions of the sub-tropical high pressure system or sub-tropical front.

The exacting ecological requirements and tolerances of many plant – and associated pollen – species, and their rapid response to changing environmental, and especially climatic, conditions, render them especially useful in Quaternary palaeoclimatology and climatostratigraphy, and in archaeology (see also Section 7.6).

Biostratigraphy

Bryophytes and pteridophytes evolved in the Silurian, and gymnosperms in the Devonian. Pteridophytes diversified and came to dominate the plant kingdom in the Carboniferous, before sustaining significant losses at the time of the glaciation in the Late Carboniferous. Pteridophytes did stage some form of recovery from this event, although, ultimately, gymnosperms became the dominant elements in the flora (DiMichele and Phillips, in Hart, 1996). Indeed, gymnosperms remained the dominant elements in the flora throughout most of the Mesozoic. Angiosperms evolved in the Cretaceous, and came to dominate the plant kingdom by the beginning of the Cenozoic.

Fossil spores and pollen first appeared in the Ordovician or Silurian, and range through to the Cenozoic, their Ordovician diversification being discussed by Steemans and Wellman, in Webby *et al.* (2004). Different sub-groups made their appearances at different times. In general, spores and pollen are useful in the stratigraphic subdivision and correlation of marine and non-marine sediments of the appropriate age in all other than the highest (palaeo-) latitudes.

The moderate rate of evolutionary turnover exhibited by the plants renders them of some use in biostratigraphy, at least in appropriate, non-marine and adjacent marine, environments. Biostratigraphic zonation schemes based on plant macrofossils and microfossils have been established, that have at least local to regional applicability, for example, in the Silurian–Devonian of the 'Old Red Sandstone continent' (Richardson and MacGregor, 1986), in the Carboniferous of Euramerica (Smith and Butterworth, 1967; Riley, 1993; Somerville, in Strogen *et al.*, 1996; Owens, in Jansonius and MacGregor, 1996; McLean and Davies, in Jones and Simmons, 1999; Owens *et al.*, 2004), and in the Permian of Gondwana (Stephenson and McLean, 1999). In the Permian of Gondwana, the Morupule formation of the Ecca group of the Karoo supergroup of Botswana is dated as Early Permian, Aktastinian, and correlated with the 3a Microfloral Zone of the Karoo basin of adjacent South Africa and the *Striatopodocarpus fusus* Zone of the Collie basin of Western Australia.

In my working experience, plant macrofossils and microfossils have proved of some use in stratigraphic subdivision and correlation, and/or in palaeoenvironmental interpretation, in the following areas.

Palaeozoic

The Silurian–Permian of north Africa. Here, spores and pollen form the basis of a number of regional biozonations, for example, of the Silurian–Carboniferous of the Polignac or Illizi basin in the Algerian Sahara (Jardine and Yapaudjian, 1968; Attar *et al.*, 1980), the Devonian–Carboniferous of the Grand Erg Occidental in the Algerian Sahara (Lanzoni and Magloire, 1969), and the Devonian–Carboniferous of the Timimoun basin in the Algerian Sahara (Alem, 1998).

The Silurian–Permian of the Middle East. Here, spores and pollen form part of the basis of the biozonation of the Silurian–Permian of, for example, Saudi Arabia (Owens *et al.*, 1995; Al-Hajri *et al.*, 1999; Al-Hajri and Owens, 2000; Stephenson *et al.*, 2003b). They also form part of the basis of the calibration of the regional sequence stratigraphic framework (Sharland *et al.*, 2001). The regional Late Silurian, Pridoli S20 maximum flooding surface, stratotypified in the Tawil formation of Saudi Arabia, is marked by palynomorphs of the D4B Zone. The Early Devonian, Pragian D10 MFS, stratotypified in the Qasr limestone member of the Jauf formation, is marked by palynomorphs of the DB4A Zone. The Middle Devonian, Emsian D20 MFS, stratotypified in the Hammamiyat limestone member of the Jauf formation, is marked by miospores of the D3B Zone. The Late Devonian, Famennian, Strunian D30 MFS is widely marked by *Retispora lepidophyta*. The Early Carboniferous, Visean C10 MFS, stratotypified in the Berwath formation of Saudi Arabia, is marked by miospores of the RT Zone. The Carboniferous of north-east Iran is marked by various spores (Vachard, in Wagner *et al.*, 1996). The Late Permian, Tatarian P30 MFS, stratotypified in member C of the Khuff formation of Saudi Arabia, is marked by palynomorphs of the *Hamipollenites insolitus* Zone. The Late Permian, Tatarian P40 MFS, stratotypified at a higher level in member C of the Khuff formation of Saudi Arabia, is also marked by palynomorphs.

Plant macrofossils, including the *Marattia*-like *Qasimia* and the pecopterid *Gemellitheca*, mark the Late Carboniferous?–Early Permian Unayzah formation of Qasim province in central Saudi Arabia and the Gomaniimbrik formation of the Hazro inlier in south-eastern Turkey (Hill and El-Khayal, 1983; Hill *et al.*, 1985; Wagner *et al.*, 1985). *Sigillaria persica* marks the Late Carboniferous?–Early Permian Chal-i-Sheh member of the Faraghan formation of the Zagros Mountains of south-west Iran (author's unpublished observations).

The Devonian–Carboniferous of the Russian platform in the former Soviet Union. Here, plant macrofossils and microfossils form the bases of regional biozonations, calibrated against that of the 'Old Red Sandstone continent' (Wagner *et al.*, 1996).

Mesozoic

The Triassic–Jurassic of the North Sea. Here, spores and pollen form part of the basis of a regional biostratigraphic zonation, calibrated against ammonite and other standards (Goldsmith *et al.*, in Evans *et al.*, 2003; Husmo *et al.*, in Evans *et al.*, 2003).

The Triassic–Cretaceous of the western margin of the British Isles. Here again, spores and pollen form part of the basis of a regional biostratigraphic zonation, calibrated against ammonite and other standards (Ainsworth *et al.*, in Batten and Keen, 1989; Jones, 1996; Ainsworth *et al.*, in Underhill, 1998). The plant macrofossil *Weichselia*, a fern, marks the 'Wealden' facies of the Weald basin in Sussex (Stokes and Webb, 1824). It was originally described in a paper memorably entitled 'Description of some fossil vegetables of the Tilgate Forest . . . '.

The Triassic–Cretaceous of the Middle East. Here, the Early Triassic, Scythian Tr10 MFS of Sharland *et al.* (2001), stratotypified in the Khartam member of the Khuff formation of Saudi Arabia, is marked by palynomorphs of the *Densoisporites playfordi* Zone in the United Arab Emirates. The Early Triassic, Scythian Tr30 MFS, stratotypified in the Mirga Mir formation of Iraq, is marked by palynomorphs of the *Aratripollenites paenulatus* Zone in the United Arab Emirates. The Late Triassic, Norian Tr80 MFS, stratotypified in the Sarki formation of Iraq, is marked by palynomorphs in Kuwait. Spores and pollen are of considerable use in the characterisation of the essentially Albian Burgan formation reservoir of the Raudhatain and Sabiriyah fields in north Kuwait (Al-Eidan *et al.*, 2001; author's unpublished observations). *Weichselia* marks the Early – to 'Middle' – Cretaceous 'Wealden' facies of the Wasia formation in central Saudi Arabia (El-Khayal, 1985).

The Early Cretaceous 'Wealden' facies of Gondwana. Here, spores and pollen form part of the basis of the biozonation of the rift basins of eastern South America and West Africa (Jones, 1996). *Weichselia* marks the Early Cretaceous 'Wealden' facies of the 'continental intercalaire' of the Ahnet basin in Algeria in north Africa (Follot, 1952) and the Barranquin formation of Venezuela in South America (Gonzalez de Juana *et al.*, 1980), elsewhere in Gondwana.

Cenozoic

The Palaeogene of the North Sea. Here, spores and pollen form part of the basis of a regional biostratigraphic zonation, loosely calibrated against planktonic foraminiferal and/or calcareous nannoplankton standards (Jones, 1996; Jones *et al.*, in Evans *et al.*, 2003).

The Cenozoic of northern South America and the Caribbean. See Pocknall *et al.*, in Goodman (1999), de Verteuil and Johnson (2002) and Rull (2002). Here, spores and pollen are important in reservoir characterisation and exploitation in the Eocene Mirador formation reservoir of the Cusiana field of Colombia (author's unpublished observations). Phytoliths are important in palaeoenvironmental interpretation in Amazonia (author's unpublished observations).

The 'Mid'–Late Cenozoic of the Black Sea and Caspian in eastern Paratethys. Here, spores and pollen form part of the basis of a regional biostratigraphic zonation (Jones, 1996; Jones and Simmons, 1996; Jones and Simmons, in Robinson, 1997). They are also important in reservoir characterisation and exploitation in the Pliocene 'productive series' reservoirs of fields in the south Caspian (author's unpublished observations).

The Neogene–Pleistogene of Nigeria. Marginal, shallow and deep marine environments of the Niger Delta can be differentiated on the basis of the ratio of marine dinoflagellates to terrestrially derived spores and pollen, and marginal marine sub-environments can be differentiated on the basis of palynofacies. Also here, spores and pollen are important in reservoir characterisation and exploitation in the Oso and EA fields (Armentrout *et al.*, in Jones and Simmons, 1999; van der Zwan and Brugman, in Jones and Simmons, 1999).

Other applications

Plants, and spores and pollen are also useful in coal mining (see Section 7.3).

3.5 Fungi

3.5.1 *Fungal spores and hyphae*

Biology, morphology and classification

Fungi belong in a kingdom of their own, being chemo-organotrophic rather than either autotrophic or heterotrophic, as in the case of the plants and animals respectively (Madigan *et al.*, 2003). They are somewhat similar to certain plant-like protists or algae, but differ in lacking photosynthetic pigments. They are also somewhat similar to certain animal-like protists or protozoans, but differ in producing spores.

In terms of ecology, a few fungi are marine, and some are freshwater, but the majority are terrestrial, and live in soil or on dead plant matter such as leaf litter, playing an important role in the recycling of nutrients. A large number are parasitic on land plants, and a smaller number are parasitic on land animals.

In terms of morphology, fungi are characterised by a somewhat plant-like cellular structure, growth form and overall appearance (Fig. 3.27). The congregation of individual, branching, filaments or hyphae forms a so-called mycelium. Some hyphae bear asexual spores that disperse to allow the colonisation of new niches. Others bear sexual spores. Fungal spores are highly resistant to physical and even chemical attack, and have high preservation potential. Note, though, that they may be destroyed by the excessive oxidation involved in the processing of coals into palynological preparations (and preserved only in thin-sections and acetate peels).

Classification of fungi is based on morphology. Two main sub-groups are recognised, namely, the unicellular yeasts, and the multicellular moulds, mushrooms and allied forms.

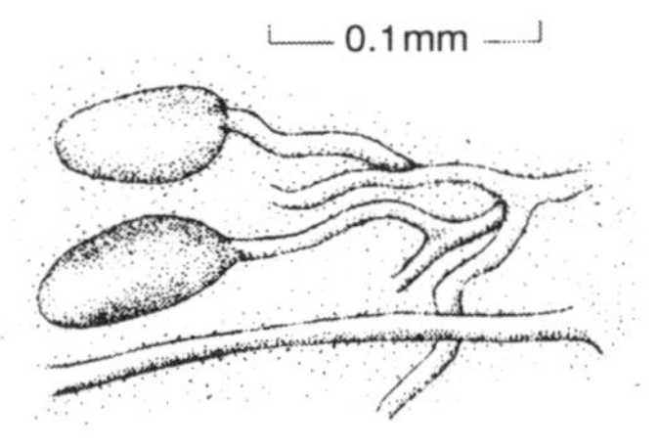

Fig. 3.27. Morphology of fungi. The fossil fungus *Palaeomyces* from the Early Devonian Rhynie Chest, showing branching hyphae terminated by enlarged vesicles. (Reproduced with permission from Benton Harper, 1997 *Basic Palaeontology*, © Longman/Pearson Education Ltd.)

Classification of fungal spores, hyphae and other microscopic remains found in palynological preparations ('fungal palynomorphs'), into form genera, is based on details of morphology (Elsik *et al.*, 1983).

Palaeobiology

Fossil fungal remains are interpreted, on the basis of analogy with living fungi, as having been essentially terrestrial (Taylor and Taylor, in Briggs and Crowther, 2001).

The occurrence of fungal remains is often indicative of the development of a palaeosol. It is worth noting in this context that outcrop samples can be contaminated by fungi living in the soil at the surface.

The superabundant occurrence of fungal remains may be indicative of the existence of significant amounts of dead vegetation, and thus of an ecological catastrophe. Note in this context that there was a pronounced 'fungal spike' following the End-Permian mass extinction (Steiner *et al.*, 2003).

Biostratigraphy

Recognisable fungal remains range from Carboniferous to Recent.

The slow rate of evolutionary turnover exhibited by the fungi renders them of limited use in biostratigraphy (Elsik, in Jansonius and MacGregor, 1996).

In my working experience in the petroleum industry, fungal spores and fruiting bodies have proved of some use in stratigraphic correlation, and/or in palaeoenvironmental interpretation, in the following areas.

Cenozoic

The Oligo-Miocene of offshore Angola. Here, apparently paradoxically, terrestrially derived, fungal spores and hyphae, and other spores and pollen, occur to the virtual exclusion of marine palynomorphs and other microfossils in deep marine submarine fan reservoir sandstones (author's unpublished observations). They are interpreted as representing hyperpycnal flows of dense, sediment-laden, fresh water, originating in river mouths.

The Pliocene of offshore Azerbaijan. Here, fungal spores and hyphae again occur to the virtual exclusion of other palynomorphs in lacustrine reservoir sandstones (author's unpublished observations). They are interpreted as representing the intermittent development of palaeosols during periods of lake desiccation associated with phases of arid climate.

3.6 Invertebrate animals

Readers interested in further details of the biology and palaeontology of invertebrates are referred to Willmer (1990), Raff, in Anderson (2001) and Taylor and Lewis (in press).

Palaeobiology

Fossil invertebrates are interpreted, essentially on the bases of analogy with their living counterparts, and functional morphology, as having occupied a range of habitats (Savazzi, 1999).

3.6.1 *'Ediacarians'*

Biology, morphology and classification

'Ediacarians' are an enigmatic, extinct, Precambrian–?Cambrian, morphologically diverse and probably polyphyletic group of primitive soft-bodied metazoans. Over 100 species have now been described from all around the world, the best-documented localities being in the Ediacara Hills of South Australia, Namibia and the adjacent western part of South Africa, the East European platform, the Russian platform, the Siberian platform, the Charnwood Forest area of Leicestershire in the East Midlands of England, Carmarthenshire in South Wales, Co. Wexford in Ireland, and south-east Newfoundland. Preservation, usually as two-dimensional impressions, and occasionally in three dimensions, is surprisingly good, with even fine details of morphology discernible (Narbonne, 2004). The possible processes responsible for such exceptional preservation are discussed by Gehling (1999) and Bottjer, in Bottjer *et al.* (2002).

Nothing is known directly of the biology of 'ediacarians', as they are extinct.

In terms of morphology, 'ediacarians' are characterised by a soft body with a high surface area to volume ratio and a radial or bilateral symmetry (Fig. 3.28).

Classification of 'ediacarians' is problematic, as a range of alternatives is possible based on the poorly

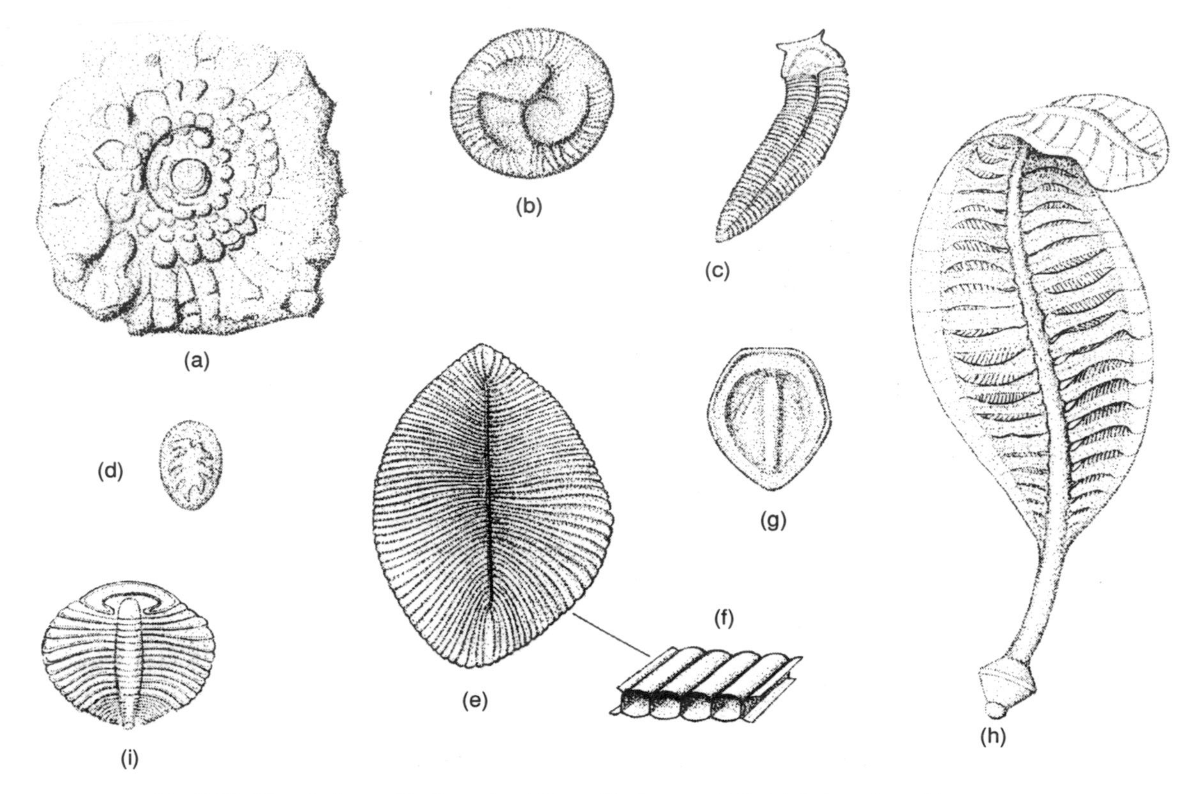

Fig. 3.28. Morphology of 'ediacarians' (from Clarkson, 1998). (a) *Cyclomedusa*; (b) *Tribrachidium*; (c) *Spriggina*; (d) *Praecambridium*; (e, f) *Dickinsonia*; (g) *Paruancorina*; (h) *Charniodiscus*; (i) soft-bodied trilobite.

understood evolutionary or phylogenetic relationships and on the difficult-to-interpret morphology of the group. Historically, many authors classified them alongside more familiar groups of metazoans such as sea-pens, jellyfish and corals (cnidarians), on the basis of at least superficial morphological similarity. In contrast, more recently, and more radically, Seilacher and his collaborators have classified them essentially entirely separately from the metazoans, in an extinct group of 'hopeful monsters' of their own, the 'vendozoans' or 'vendobionts' (Seilacher, 1992; Seilacher, in Bengtson, 1994). Seilacher and his collaborators interpreted their 'vendozoans' as unique in terms of the so-called 'pneu' structure – somewhat similar to that of corrugated cardboard – observed in many forms, which could have conferred both rigidity and flexibility. Most recently, Fedonkin, in Briggs and Crowther (1990) has reverted to a conservative classification of the 'ediacarians' as metazoans. Fedonkin recognises two sub-groups, the one characterised by radial, the other by bilateral symmetry. His radially symmetrical sub-group contains *Charnia* and *Ediacaria*, which exhibit certain similarities to sea-pens and to jellyfish, respectively. His bilaterally symmetrical sub-group contains both segmented forms such as *Dickinsonia*, which exhibits certain similarities to corals, and also smooth forms which exhibit certain similarities to turbellarian flatworms (platyhelminthes). The segmented bilaterian *Spriggina* is of uncertain, although arguably annelid (segmented worm) or arthropod affinity.

Palaeobiology

'Ediacarians' are interpreted, essentially on the basis of associated fossils and sedimentary facies, as having been marine. They are also interpreted as having lived in shallow marine environments, some as part of the sessile (?or motile) epifaunal benthos, as 'sediment stickers', 'flat recliners' or 'erect elevators', like living sea-pens (Seilacher, in Bengtson, 1994);

others as part of the infaunal benthos (Runnegar, 1982, 1995); still others, the medusioids, as part of the pelagos, like living jellyfish. Most 'ediacarians' are interpreted as having fed by filter-feeding. It has been postulated, and widely accepted, that they were able to obtain oxygen and sustenance directly from sea water through diffusion (which process would have been facilitated by their high surface area to volume ratio), thereby obviating any need for complex respiratory or digestive organs. It has also been postulated, although not widely accepted, that they hosted photosynthetic algae, maintaining an autotrophic existence in a so-called 'garden of Ediacara' (McMenamin, 1986, 1998). There do not appear to have been many, if any, scavenging or predatory 'ediacarians' (although many modern turbellarian flatworms, to which smooth bilaterians exhibit certain similarities, do exhibit these feeding behaviours).

As intimated above, as a group, 'ediacarians' appear to have had an essentially cosmopolitan palaeobiogeographic distribution. However, the bilaterally symmetrical sub-group appears to have had a provincial or endemic distribution, in equatorial palaeo-latitudes (Waggoner, 2003). Parsimony analysis of endemism (PAE) points toward separate 'ediacarian' provinces centred on Avalonia, the White Sea and Namibia.

Biostratigraphy

'Ediacarians' first appear in the rock record in the Late Vendian, or Ediacarian, immediately above Early Vendian, or Varangerian, tillites dated to approximately 620–650 Ma (Fedonkin, in Briggs and Crowther, 1990). The abundance and diversity of the 'ediacarians' already in the Vendian is arguably indicative of a still older origin, perhaps in the Riphean. One indication that this was the case comes from the observation that stromatolites began to decline in the Riphean, from around 1000 Ma, which some authors have suggested was due to excessive grazing by 'ediacarians' (Walter and Heys, 1985).

'Ediacarians' appear to disappear at the end of the Vendian, although morphologically similar forms are known from the Cambrian (Conway Morris, 1993). The extinction may be more apparent than real, though, and an artefact of generally poor preservation in the Cambrian, in turn resulting from an increase in atmospheric oxygen and/or in scavengers and predators. Perhaps significantly, one of the better-known 'ediacarian-like' organisms of the Cambrian, *Xenusion*, is characterised by modification to the basic body-plan to incorporate spines interpreted as a defence against predation.

In my working experience in the petroleum industry, 'ediacarians' have proved of stratigraphic use in the *Precambrian, Vendian of the Middle East*. Here, the occurrence of *Spriggina*, with or without associated *Charnia*, *Dickinsonia* and unidentified medusioid forms, has enabled a correlation between the Hormuz salt formation of south-west Iran and the Rizu and Esfordi formations of north-east Iran, radiometrically dated to 595–715 ± 120 Ma (Jones, 2000).

3.6.2 'Small shelly fossils'

Biology, morphology and classification

'Small shelly fossils' or 'SSFs' are an extinct, Precambrian–Early Cambrian, morphologically diverse and probably polyphyletic group of metazoans.

Nothing is known directly of the biology of 'SSFs', as they are extinct. However, exceptional instances of preservation of soft parts are known that have enabled inferences to be drawn not only as to biology but also as to ontogeny and phylogeny (Wilson and Ratcliffe, 2003).

In terms of morphology, 'SSFs' are characterised, as their name suggests, by small size (typically less than 1 cm) and a mineralised skeleton (Fig. 3.29). They appear to be the first group to have successfully experimented with skeletal mineralisation, although there is some doubt as to whether their skeletons represent primary constructions or secondary replacements. 'SSFs' can be composed of carbonate, either with or without an organic component, or of phosphate, and are usually extracted from containing rocks by acid treatment, such that there is a bias towards acid-resistant forms among the group.

Classification of 'SSFs' is based essentially exclusively on morphology and composition (their evolutionary or phylogenetic relationships remaining poorly understood). Two main sub-groups are recognised: calcareous and phosphatic.

The calcareous sub-group is exemplified by *Cloudina* and *Anabarites*. Its representatives are typically

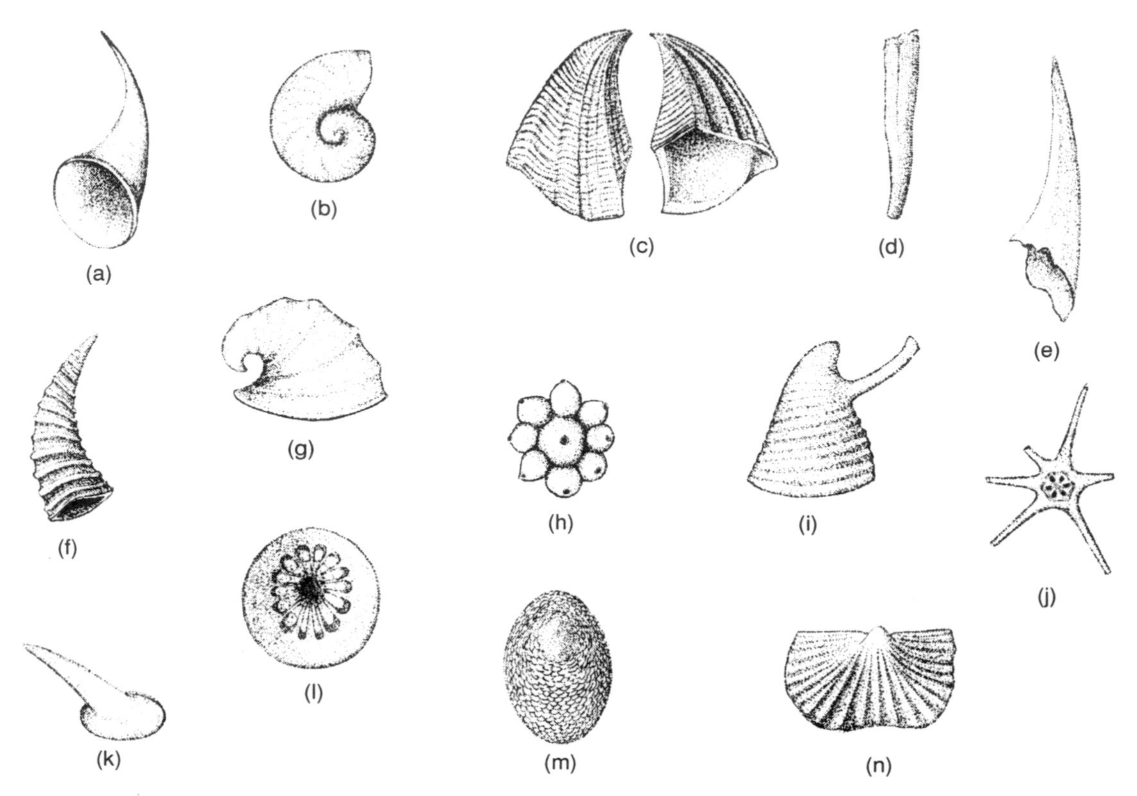

Fig. 3.29. Hard-part morphology of 'small shelly fossils' (from Clarkson, 1998). (a) *Fomitchella*; (b) *Aldanella*; (c) *Tommotia*; (d) *Anabarites*; (e) *Proteroherzina*; (f) *Lapworthella*; (g) *Coreospira*; (h) *Chancelloria*; (i) *Yochelcionella*; (j) (k) *Cambroclavus*; (l) *Mobergella*; (m) *Canopoconus*; (n) orthid brachiopod.

parallel-sided or slightly flared, straight, curved or coiled tubes open at one end, and appear to represent complete organisms. *Anabarites* is further characterised by an unusual, triradial symmetry, and arguably evolved from the similarly triradial 'vendobionts'.

The phosphatic sub-group includes the hyolithans, which are also tubes open at one end, with the opening capable of being closed by a lid or operculum, and which also appear to represent complete organisms. The hyolithans are further characterised by bilateral symmetry, and arguably evolved into the molluscs. The phosphatic sub-group also includes what are evidently the remains of disarticulated larger organisms such as *Halkieria* and *Microdictyon*. Articulated specimens of *Halkieria* have recently been discovered in Siberia, and in the Early Cambrian, Atdabanian Sirius Passet Lagerstatte of north Greenland, which exhibit sclerite structures that have led to their interpretation as part of the ancestral lineage leading to the annelids and brachiopods (Conway Morris and Peel, 1990, 1995). Note, though, that *Halkieria* has also been interpreted as a mollusc (Vinther and Nielsen, 2004). Articulated specimens of *Microdictyon*, with their soft parts preserved as impressions, have recently been discovered in the Qiongzhusian (Atdabanian to Botomian) Chengjiang Lagerstatte of south China, which show that the usually disarticulated phosphatic elements are actually the sites of attachment of numerous pairs of unjointed legs or lobopods (Hou Xian-Guang *et al.*, 2004). In the respect of possessing unjointed lobopods, they resemble living appendiculate worms or onychophorans such as *Peripatus*, that are currently interpreted on molecular evidence as primitive arthropods (Purves *et al.*, 2004).

Palaeobiology

'SSFs' are interpreted, essentially on the basis of comparison with associated fossils and sedimentary facies, as having been shallow marine. They are further interpreted on the basis of hard- and soft-part functional morphology as having lived as part of the sessile (?or motile) epifaunal or semi-infaunal benthos, and, at least in the case of the tubular morphotypes such as *Cloudina, Anabarites* and hyolithans, as having fed by passive filter-feeding (Burzin *et al.*, in Zhuravlev and Riding, 2001; Kouchinsky, in Zhuravlev and Riding, 2001; Zhuravlev, in Zhuravlev and Riding, 2001). Orthothecimorph hyoliths are interpreted on the basis of empirical observation as having lived in an upright position (Landing, 1993); hyolithomorph hyolithans on the basis of experimental observation as having lived in a prone position, orienting themselves into currents in order to feed by means of keels (Marek *et al.*, 1997). Importantly, *Cloudina* has been recorded in low-relief boundstone 'reefs' in Namibia (Germes, 1983), *Anabarites* in 'mounded aggregations' elsewhere (Droser and Xing Li, in Zhuravlev and Riding, 2001), and orthothecimorph and hyolithomorph hyoliths in association with archaeocyathan reefs in Siberia (Pratt *et al.*, in Zhuravlev and Riding, 2001). Also importantly, *Cloudina* has been observed to have been attacked by boring parasites or predators, leading to the suggestion that its mineralised skeleton evolved as a measure to counter such attack (Bengtson and Yue, 1992).

'SSFs' as a group appear to have had an essentially cosmopolitan palaeobiogeographic distribution. However, *Cloudina* appears to have been essentially endemic to proto-Gondwana, and *Anabarites* to proto-Laurasia (McMenamin, 1982).

Biostratigraphy

Calcareous 'SSFs' first appeared in the latest Precambrian, below the horizon marked by the carbon isotope excursion used as a proxy for the Precambrian/Cambrian boundary: *Cloudina* in Shaanxi, China and in Oman; and *Anabarites* in Mongolia. However, 'SSFs' are generally regarded as characterising the earliest Cambrian, Nemakit–Daldynian to Botoman. They did not appear until the base of the Nemakit–Daldynian in Siberia, which must now be accepted as defining the base of the Cambrian, on the basis of correlation, by means of the marker trace fossil *Phycodes pedum*, with the Precambrian/Cambrian boundary stratotype section in south-east Newfoundland. They disappeared in the Botoman, at around the level of the so-called 'Sinsk (anoxic) event' at the *Bergeroniellus micmacciformis–Erbiella* Zone/*B. gurarii* Zone boundary (Zhuravlev and Wood, 1996).

The phosphatic hyolithans first appeared in the Tommotian, and disappeared in the Botoman (Pratt *et al.*, in Zhuravlev and Riding, 2001); in Britain, they have been recorded in the Hartshill formation of Warwickshire in the West Midlands. *Halkieria* and *Microdictyon* first appeared in the Atdabanian, and may more accurately be regarded to be representative of the evolutionary diversification of that, rather than of any earlier, time. Non-'SSF' phosphatic fossils such as bradoriide and phosphatocopide arthropods also first appeared in the Atdabanian (Cook and Shergold, 1984; Melnikova *et al.*, 1997).

The rapid rate of evolutionary turnover and short range exhibited by 'SSFs' renders them of use in biostratigraphy, at least in appropriate, marine, environments. Thus, the early part of the Meishucunian in south China, is marked by *Anabarites* and *Protohertzina*, and the middle part by *Paragloborilus* and *Siphogonuchites* (Hou Xian-Guang *et al.*, 1996).

In my working experience in the petroleum industry, SSFs have proved of use in the following areas.

Proterozoic

The Late Precambrian of the Middle East. On the Arabian plate, the Precambrian/Cambrian boundary has recently been placed at the base of the A4 carbonate of the Ara group of Oman (Amthor *et al.*, 2003; Al-Husseini *et al.*, 2003). A negative $\delta^{13}C$ excursion at the base of the A4 carbonate appears to correlate with that observed at the boundary elsewhere. An ash bed at a stratigraphically slightly higher position within the A4 carbonate has yielded a U–Pb age of 542 ± 0.6 Ma, and an ash bed at a slightly lower stratigraphic position within the A3 carbonate has yielded a U–Pb age of 542.6 ± 0.3 Ma (incidentally, in turn indicating that the carbonate-evaporite cycles were deposited in hundreds of thousands of years). Regional MFS Cm10 of Sharland *et al.* (2001), picked on what was interpreted on petrophysical log evidence to be a 'thin shale' at the base of the A4 carbonate in the Birba-1 well, actually corresponds to the higher of the two ash beds. There are candidate additional high-frequency MFSs associated with the A0, A1, A2, A3, A5 and A6 carbonates. In Oman, *Cloudina* occurs exclusively in the Precambrian A1, A2 and A3 carbonates; and *Namacalathus* in the Precambrian A2 and A3 carbonates. Supposed hyolithans have been recorded in the Soltanieh dolomite formation of north-east Iran, that may in fact also be *Cloudina* (J. Amthor, Petroleum Development Oman, personal communication).

Palaeozoic

The Early Cambrian of east Siberia. Here, subdivision and correlation of the Nemakit–Daldynian is possible using 'SSFs' (Jones, in Simmons, in press). The early part is marked by the anabaritid *Anabarites trisulcatus*, the late part by the siphogonuchitid *Purella antiqua*. In the Irkutsk Amphitheatre area, the lower Moty suite is marked by unspecified anabaritids.

3.6.3 *Poriferans: sponges*

Biology, morphology and classification

Sponges are an extant, Precambrian–Recent, group of poriferans characterised by a primitive, essentially cellular, structure (Hartman *et al.*, 1990; Rentner and Keupp, 1991; Bergquist, in Anderson, 2001; Debrenne and Reitner, in Zhuravlev and Riding, 2001; Hooper and van Soest, 2002). Famously, they are able to deconstitute and then reconstitute themselves at the cellular level!

The biology and ecology of living sponges is well known. All are aquatic, and almost all marine, constituting part of the sessile benthos, and obtaining their nutriment through filter-feeding. Most are shallow marine; some deep marine, some of these associated with cold seeps, as on the Barbados accretionary prism (Jollivet *et al.*, 1990).

In terms of morphology, sponges comprise a cup-shaped, discoidal or flattened body with a simple central cavity, sometimes with additional simple or complex cavities in its sides (Fig. 3.30). The sides are densely perforated to allow the passage of inhalant currents and food, the top open to allow the passage of exhalant currents and waste. Specially modified collar cells assist the circulation of the currents by cracking their whip-like flagella, and amoeboid cells fulfil digestive, and also reproductive, functions (there being no differentiated tissues and organs). The skeleton is usually composed of an organic material called spongin, strengthened by calcareous or siliceous spicules. Sometimes it is effectively entirely calcareous (or calcified). Essentially only the articulated or disarticulated calcareous and siliceous skeletons are ever found as fossils. The exceptionally fine detailing in the siliceous construction of the 'Venus's flower basket', *Euplectella aspergillum* makes it one of the most exquisitely beautiful organisms on Earth, and unquestionably my favourite exhibit in the Natural History Museum in London.

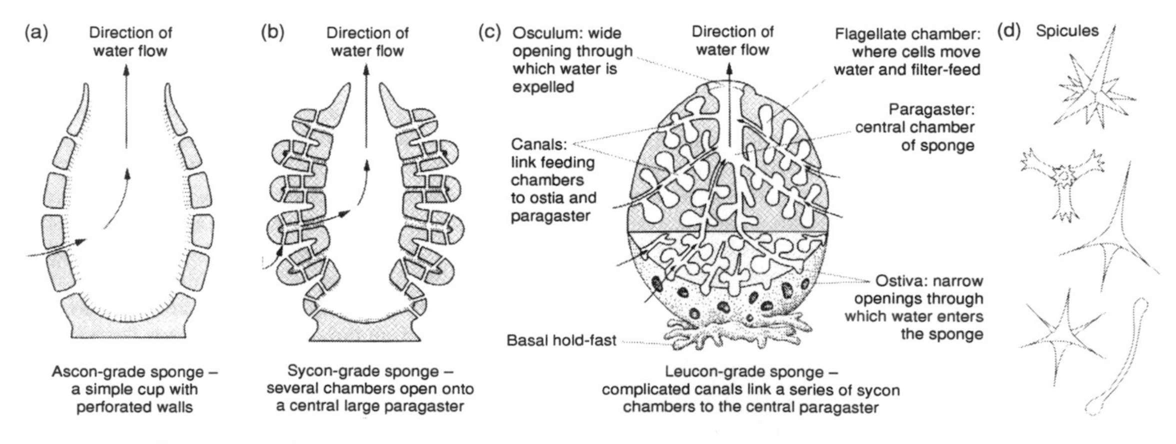

Fig. 3.30. Hard-part morphology of sponges (from Milsom & Rigby, 2004).

Classification of sponges is based on morphology and composition (phylogenetic relationships remaining unclear). Three sub-groups are recognised: the common sponges, or demosponges; the siliceous sponges, or hexactinellids; and the calcareous sponges, or calcareans.

Palaeobiology

Fossil sponges are interpreted, essentially on the basis of analogy with their living counterparts, as having been mostly shallow marine. They are also interpreted as having been active filter-feeders (Zhuravlev, in Zhuravlev and Riding, 2001).

Sponges were important contributors to the construction of reefs in the Palaeozoic, less so in the Mesozoic, and not at all in the Cenozoic (their decline appearing to mirror the rise of the scleractinian corals with algal photosymbionts). Cup-shaped and discoidal morphotypes colonised soft substrates, and often built substantial structures. Flattened morphotypes encrusted hard substrates.

Early Ordovician, Tremadoc reefs of south China are characterised by lithistid sponges (Li *et al.*, 2004). *Solenopora*, recently reclassified as a chaetitid sponge rather than a red alga, is characteristic of the Early Carboniferous, Asbian–early Brigantian, reef-like structures of the Mullaghfin formation of the Kingscourt outlier on the margins of the Dublin basin in Ireland (Somerville *et al.*, in Strogen *et al.*, 1996).

Early Late Carboniferous, Bashkirian–early Moscovian carbonate reefs are characterised by chaetitid sponges; late Late Carboniferous, late Moscovian–Gzhelian reefs by calcareous sponges, and Early Permian, Asselian–Sakmarian reefs by 'baffling' calcareous sponges and heliosponges (Wahlman, in Kiessling *et al.*, 2002). Middle–Late Permian build-ups such as the Capitan reef of the Permian basin in west Texas are characterised by biotas similar to, although more evolved than, those of the Early Permian (Wahlman, in Kiessling *et al.*, 2002). The cores of the Late Permian, Changhsingian reefs of Laolongdon, near Chongqing in Sichuan province in China are also characterised by sponges (Yang Wanrong, in Jin Yugan *et al.*, 1991).

Importantly, sponges are also characteristic of Permian temperate-water, as opposed to tropical, reefs (Beauchamp and Baud, 2002).

Palaeobiogeography

Sponges as a group appear to have had an essentially cosmopolitan palaeobiogeographic distribution.

Palaeoclimatology

Sponges are of some use in palaeoclimatology (Hughes and Thayer, in Gerhard *et al.*, 2001).

Biostratigraphy

Sponges first appeared in the Precambrian, Ediacarian, although they only rose to prominence in the Cambrian, Atdabanian (Debrenne and Reitner, in Zhuravlev and Riding, 2001). They diversified in the Ordovician (Carrera and Rigby, in Webby *et al.*, 2004). They remained important through the remaining part of the Palaeozoic, but became progressively less so through the Mesozoic and Cenozoic. There are at least 5000 living species.

In my working experience in the petroleum industry, sponges have proved of some use in stratigraphic subdivision and correlation, and/or in palaeoenvironmental interpretation, in the following areas.

Palaeozoic

The Late Palaeozoic of the former Soviet Union. Sponges are characteristic of the platform and peri-reefal facies of the Devonian to Permian of the Bolshoi Karatau Mountains of southern Kazakhstan, and of the southern Urals (Cook *et al.*, in Ahr *et al.*, 2000; Zempolich *et al.*, in Ahr *et al.*, 2000; Cook *et al.*, in Zempolich and Cook, 2002; Zempolich *et al.*, in Zempolich and Cook, 2002). These surface outcrops are important analogues for coeval subsurface petroleum reservoirs in the north Caspian, peri-Caspian and Volga–Urals basins. Sponge spicules form cold, deep-water accumulations in the Late Permian, Kungurian–Tatarian of the Barents Sea that in turn form petroleum reservoirs (Bruce and Toomey, in Vorren *et al.*, 1993; Ehrenberg *et al.*, 2001; author's unpublished observations). Outcrop analogues are known from Spitsbergen (Worsley and Aga, 1986; Ehrenberg *et al.*, 2001).

Mesozoic

The Late Jurassic of the North Sea. Influxes of *Rhaxella* and other, unidentified, sponge spicules constitute useful correlative datums in reservoir

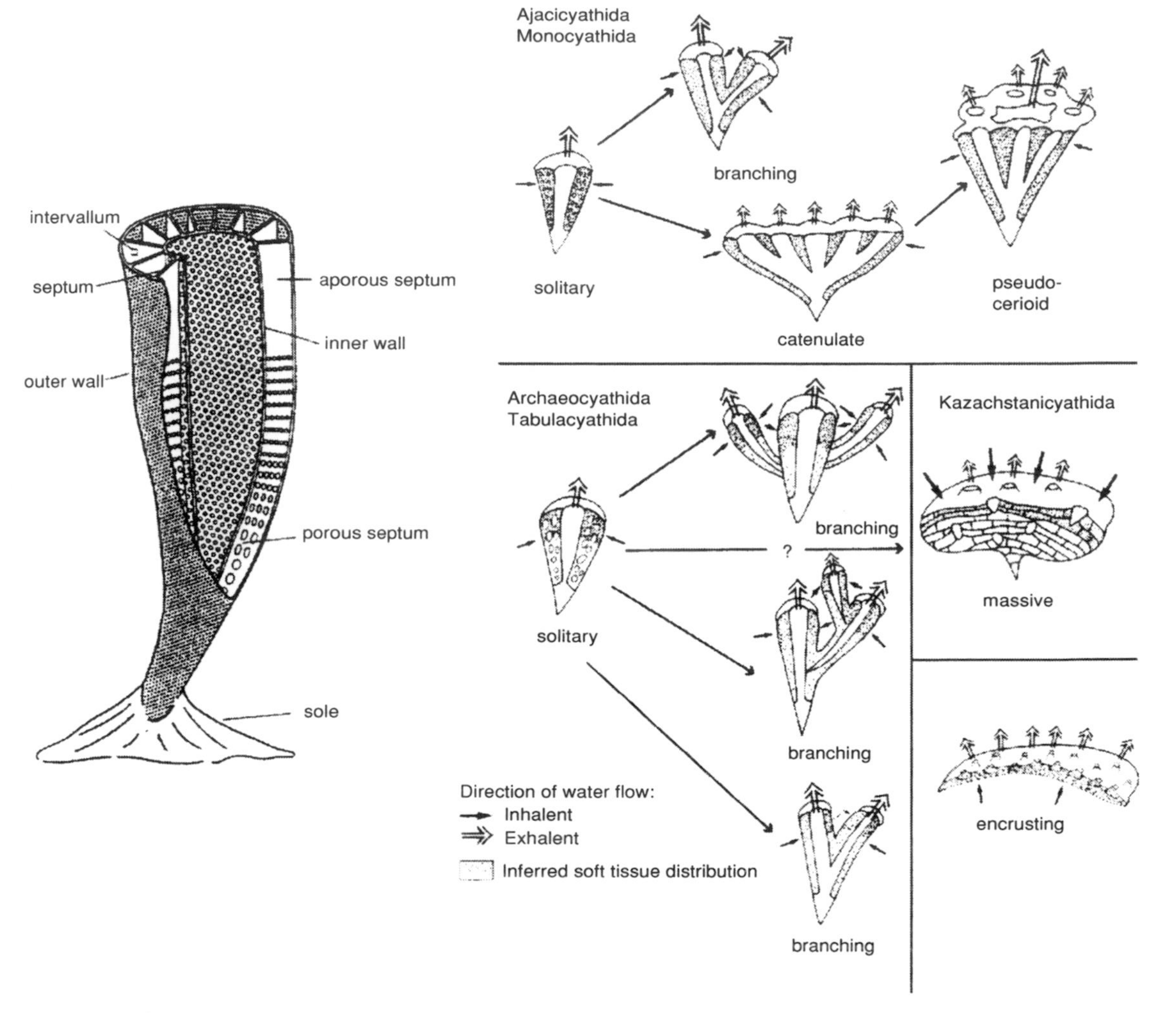

Fig. 3.31. Hard-part morphology of archaeocyathans. (Reproduced with permission from Benton and Harper, 1997 *Basic Palaeontology*, © Longman/Pearson Education Ltd.)

sections in many fields (Jones, 1996; Morris *et al.*, in Jones and Simmons, 1999).

Cenozoic

The Neogene of central Paratethys. Here, specifically in the Carpathian foredeep of Moravia, the Karpatian and Badenian regional stages are marked by particular sponge spicule assemblages (Riha, in Thon, 1983).

3.6.4 Poriferans: archaeocyathans

Biology, morphology and classification

Archaeocyathans are an extinct, Cambrian, group of calcareous fossils of uncertain – although almost certainly poriferan – affinity (Debrenne and Reitner, in Zhuravlev and Riding, 2001).

Nothing is known directly of the biology of the archaeocyathans, as they are extinct.

In terms of hard-part morphology, the simplest, solitary forms of archaeocyathan are characterised by a conical morphology, more complex, modular forms by a branching, colonial or encrusting habit (Fig. 3.31). In fact, the morphology of even the simplest forms is moderately complex, and is better described as cone-in-cone. The concentrically arranged inner and outer cones are separated by radially arranged vertical septa that partition the intervening space or intervallum into chambers or loculi.

Horizontal tabulae and dissepiments may also partition the intervallum. The inner and outer cones are densely perforate, the tabulae perforate, the dissepiments imperforate, the septa either perforate or imperforate. The inner cone circumscribes the so-called central cavity, which is open at the top and closed at the bottom.

Classification of archaeocyathans is based essentially on morphology (their evolutionary or phylogenetic relationships remaining poorly understood). Two main sub-groups are recognised: the regular archaeocyathans, which lack dissepiments; and the irregular archaeocyathans, which possess dissepiments. The regular group, to which most archaeocyathans belong, contains the ajacicyathids and monocyathids. The irregular group contains the archaeocyathanids, tabulacyathids and kazachstanicyathids. Complex, modular growth appears to have evolved independently among advanced representatives of both groups.

Palaeobiology

Archaeocyathans are interpreted, essentially on the basis of associated fossils and sedimentary facies, as having been shallow marine (Burzin *et al.*, in Zhuravlev and Riding, 2001). They are further interpreted, essentially on the basis of functional morphology, as having lived embedded in soft substrates, in the case of the ajacicyathids, or held fast on hard substrates, in the case of the archaeocyathans, thus forming part of the sessile benthos, and as having fed by filter-feeding (LaBarbera, in Briggs and Crowther, 1990; Savarese, 1992; Wood *et al.*, 1992; Burzin *et al.*, in Zhuravlev and Riding, 2001).

Archaeocyathans apparently lived in colonies, and at times, and in places, especially when in association with stromatolite-producing cyanobacteria, were sufficiently abundant as to be rock- or reef-forming (Pratt *et al.*, in Zhuravlev and Riding, 2001). Some are interpreted as having lived in reef cavities or crypts, probably because adjacent substrates were covered by microbial biofilms that prevented colonisation. Archaeocyathan–stromatolite reefs are typically patch-reefs, between 10 and 30 m in diameter and 3 m in thickness, and exhibit a dominantly outward growth form that has been interpreted as indicating optimisation of exposure to sunlight and facilitation of photosynthesis. At least one example is known to the author of what appears to be an archaeocyathan–stromatolite barrier-reef complex, that in the Tommotian–Botoman extended from the Anabar shield to the Aldan shield in Siberia, and now is exposed along the Lena, Aldan and Uchur rivers (Jones, in Simmons, in press). This reef complex foundered in the Botoman–Toyonian, coincident with a major archaeocyathan extinction. Importantly from the petroleum geological point of view, the restricted back-barrier lagoon on the landward, south-western, side of this reef complex was characterised by the precipitation of evaporites, that now constitute the cap-rocks to underlying reservoirs in the Irkutsk Amphitheatre.

Archaeocyathans as a group appear to have had an essentially cosmopolitan palaeobiogeographic distribution, although only in low palaeo-latitudes. They are known from Siberia to the Altai Mountains and Urals in the former Soviet Union, and from North America, in the proto-Laurasian province, and from north Africa and South Australia, in the proto-Gondwanan province.

Biostratigraphy

Archaeocyathans first appeared at the base of the Tommotian, and diversified and dispersed in the Botoman. They underwent a major extinction in the Botoman–Toyonian, and declined thereafter, with only a few genera recorded from the Middle Cambrian, only one from the Late Cambrian, and none after the end of the Cambrian.

The rapid rate of evolutionary turnover and short range exhibited by the archaeocyathans renders them of considerable use in biostratigraphy, at least in appropriate, marine, environments. Unfortunately, their usefulness is somewhat compromised by their restricted bathymetric and biogeographic distribution and facies dependence. Nonetheless, high-resolution biostratigraphic zonation schemes based on the archaeocyathans have been established, that have at least local to regional applicability. For example, three regional archaeocyathan zones are recognisable in the Atdabanian of Australia, all with an average duration or resolution of less than 1 Ma (Zhuravlev and Riding, in Zhuravlev and Riding, 2001). Moreover, some stratigraphic significance can be assigned simply on the basis of observation of archaeocyathan morphology, since older forms tend to be solitary, and younger ones modular.

In my working experience, archaeocyathans have proved of particular use in stratigraphic subdivision and correlation in *the Early Cambrian of east Siberia*. Here, seven regional archaeocyathan zones are recognisable in the Tommotian–Atdabanian, all with an average duration or resolution of less than 1 Ma (Zhuravlev and Riding, in Zhuravlev and Riding, 2001). In the Irkutsk Amphitheatre area, the middle and upper Moty suites are marked by unspecified archaeocyathans; the Usolye suite by *Ajacicyathus*, *Timulocyathus* and *Aldanicyathus* (author's unpublished observations).

3.6.5 *Poriferans: stromatoporoids*

Biology, morphology and classification

Stromatoporoids are an extinct, Palaeozoic–Mesozoic, ?Cambrian–?Cretaceous group of colonial calcareous sponge- or coral-like fossils of uncertain – although probably poriferan – affinity (Stearn, 1975).

Nothing is known directly of the biology of stromatoporoids, as they are extinct.

In terms of hard-part morphology, stromatoporoids are characterised by an essentially flat and encrusting, or domal and hemispherical, or bulbous and columnar, body or coenosteum formed from horizontal layers supported by vertical pillars (Fig. 3.32). They are further characterised by internal canals that open to the external world through radiating fissures or astrorhizae atop raised areas on the surface of the body. The canals and astrorhizae are thought to represent the components of a water circulation system. The reported occurrence of spicules in some stromatoporoids, together with the suggestion that the calcareous component of the skeleton is essentially secondary, and the result of precipitation of carbonate within a primary framework of spongin, strongly suggests an affinity with sponges.

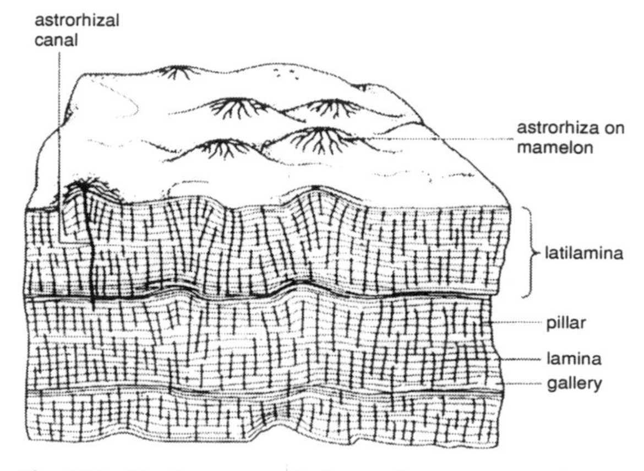

Fig. 3.32. Hard-part morphology of stromatoporoids. (Reproduced with permission from Benton and Harper, 1997 *Basic Palaeontology*, © Longman/Pearson Education Ltd.)

Classification of stromatoporoids is essentially based on details of the morphology, observable only in thin-section.

Palaeobiology

Palaeobathymetry

Stromatoporoids are interpreted, on the basis of associated fossils and sedimentary facies, as having been mostly shallow marine. They are also interpreted as having been active filter-feeders, creating their own currents, the better to utilise dispersed resources (LaBarbera, in Briggs and Crowther, 1990; Zhuravlev, in Zhuravlev and Riding, 2001).

Stromatoporoids were important contributors to the construction of reefs, including core frameworks, in the Ordovician to Devonian of northern Eurasia and North America, the Late Ordovician of south China, and the Late Devonian of the Canning basin of Western Australia. The Late Devonian mass extinction caused the Canning basin reef ecosystem to reorganise rather than to collapse (Wood, 2004).

Palaeobiogeography

Stromatoporoids as a group appear to have had an essentially cosmopolitan palaeobiogeographic distribution, although only in low to moderate palaeolatitudes (Nestor, in McKerrow and Scotese, 1990; Stock, in McKerrow and Scotese, 1990; Talent *et al.*, in Wright *et al.*, 2000; Webby *et al.*, in Wright *et al.*, 2000; Kiessling, 2001).

Biostratigraphy

Stromatoporoids first appeared in the rock record in the Cambrian, Botoman, almost certainly evolving from a soft-bodied and/or sponge-like ancestor (Debrenne and Reitner, in Zhuravlev and Riding, 2001). They diversified in the Ordovician (Webby, in Webby *et al.*, 2004), before sustaining losses in the

Late Ordovician mass extinction, taking much of the Silurian to recover (Nestor and Stock, 2001). They then diversified again in the Devonian, before being almost eliminated in the Late Devonian mass extinction, taking until the Jurassic to stage a partial recovery. They appear to have finally disappeared during the End-Cretaceous mass extinction. Note, though, that some living so-called calcified sponges exhibit stromatoporoid grades of organisation (Wood, in Simonetta and Conway Morris, 1991).

In my working experience, stromatoporoids have proved of use in stratigraphic subdivision and correlation, and/or in palaeoenvironmental interpretation, in the following areas.

Palaeozoic

The Devonian of Canada. The stratigraphic subdivision and correlation of the Devonian of Arctic and western Canada by means of stromatoporoids is discussed by Stearn (1997). The earliest Devonian is marked by the stromatoporoid *Parallelostroma*; the *Dehiscens* to *Gronbergi* Conodont Zones of the lower Blue Fiord formation by *Gerronostroma, Stictostroma* and *Stromatoporella perannulata*; the *Inversus* to *Partitus* Conodont Zones of the upper Blue Fiord formation by *Plectostroma salairicum, Stromatopora* and *Anostylostroma*; the late Eifelian Elm Point limestone by *Stromatoporella* and *Strictostroma cf. foraminosum*; the early Givetian, Early *Varcus* Conodont Zone by *Neosyringostroma logansportense* and *Actinostroma tyrelli*; the middle Givetian of the Dawson Bay formation of Manitoba by a transitional fauna; the late Givetian and early Frasnian of the Beaverhill Lake group by *Arctostroma contextum*; the middle Frasnian by *'Ferestromatopora' parksi*; the late Frasnian of the Canadian Rockies and the Great Slave Lake area by *Trupetostroma saintjeani*; and the Famennian by *Labechia palliseri* and *Stylodictyon sinense*.

Mesozoic

The Jurassic of the Middle East. Here, the first occurrences of *Shuqraia zuffardi, Dehornella crustans, Actinostromarianina praesalvensis* and *Parastromatopora libani* mark the early Oxfordian Musandam group 'e' or Beni Zaid limestone of the Oman Mountains, and the Shuqra Limestone of Yemen (Toland, in Simmons, 1994). The first occurrences of *Promillepora pervinquieri, Actostroma damesini, Steineria somaliensis* and *Shuqraia hudsoni* mark the middle-late Oxfordian Musandam group 'f' or *Valvulinella* limestone of the Oman Mountains, and the Hanifa formation of Saudi Arabia (Toiland, in Simmons, 1994). *Cladocoropsis* marks the Hanifa formation of Saudi Arabia (Hughes, in Bubik and Kaminski, 2004). The first occurrences of *Burgundia trinorchii, Actinostromarianina lecompti* and *Burgundia ramosa* mark the early Kimmeridgian Arab 'D' of eastern Saudi Arabia, Bahrain, Qatar, the United Arab Emirates, the Oman Mountains and Yemen (Toland, in Simmons, 1994). *Shuqraia zuffardi, Burgundia trinorchii, Actinostromarianina lecompti* and *Burgundia ramosa* occur in the late Kimmeridgian–Tithonian Asab oolite of the eastern United Arab Emirates (Toland, in Simmons, 1994).

3.6.6 Cnidarians: corals

Biology, morphology and classification

Corals are an extant, Ordovician–Recent group of cnidarians (Hinde, in Anderson, 2001; Spalding *et al.*, 2001). Specifically, they are anthozoans, related to the jellyfish and jellyfish-like hydrozoans, scyphozoans and cubozoans, and diagnosed from them by their calcareous skeletons with hexamerous to octomerous symmetry.

Two sub-groups of anthozoans are recognised, namely, the hexacorals or zoantharians, diagnosed by hexamerous symmetry, and the octocorals or alcyonarians, diagnosed by octomerous symmetry.

Only the hexacorals are considered here, as only they are important as fossils. Indeed, only the extant scleractinian corals and their extinct relatives the tabulate and rugose corals are considered. A number of minor extant and extinct groups are not considered, as they are not important as fossils.

The octocorals are important contributors to the construction of reefs in modern times, but were not in ancient times. Familiar representatives of the octocorals include the sea fan *Gorgonia* and the organ-pipe coral *Tubipora*.

Biology

The biology and ecology of living, scleractinian, corals is extremely well known, from extensive

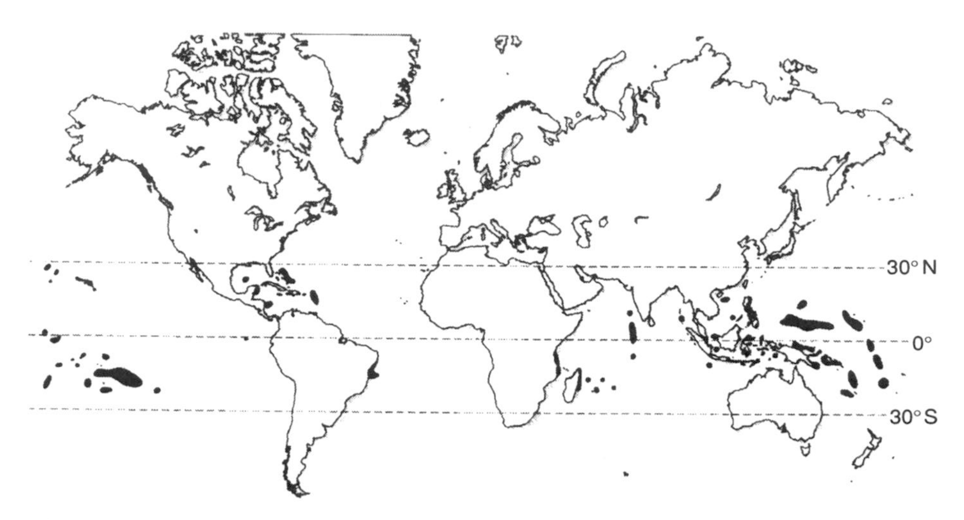

Fig. 3.33. Biogeographic distribution of living corals (from Milsom and Rigby, 2004).

studies on the Great Barrier Reef of Australia and elsewhere (Spalding *et al.*, 2001).

Some are solitary; others colonial, often with individual animals adapted to fulfil specialised functions. Colonial corals exhibit varying degrees of integration between individual animals, with some even entirely lacking hard divisions between them, and possessing digestive and nervous systems shared between them. High degrees of integration appear to be associated with the presence of photosynthetic algal symbionts or zooxanthellae.

Coral animals typically capture food by means of – stinging – tentacles, and both ingest it and excrete waste products through a centrally located mouth. However, as intimated above, some are partly mixotrophic rather than purely heterotrophic, and rely to an extent on the products of photosynthetic algal symbionts as sources of food. The symbiont-bearing corals are referred to as zooxanthellate or z-corals, and the non-symbiont-bearing ones as non-z-corals.

Corals grow by daily, monthly and annual increments. Living corals exhibit approximately 360 daily increments per year (they do not grow during breeding or under adverse environmental conditions), while Devonian fossils exhibit approximately 400 daily increments per year, which has been interpreted as indicating that there were actually more days in the year then, because the Earth was rotating faster (Wells, 1963).

Corals reproduce sexually.

In terms of ecology, living corals are exclusively marine, and essentially benthic, epifaunal and sessile, although they also have planktonic or medusoid developmental stages in their life cycles, enabling them to disperse and to colonise new niches. They live at most depths and latitudes, although most, and all forms with photosynthetic algal symbionts, live in depths of less, and often much less, than 500 m, and in temperatures of more than 10 °C. Pennular forms with photosynthetic algal symbionts appear adapted to living in the very deepest parts of the photic zone (Gill *et al.*, 2004)

Coral reefs. Practically all coral reefs are found in depths of less than 100–150 m, and temperatures of more than 16–17 °C (Fig. 3.33) (Wells, in Hedgpeth, 1957; Wells, in Hallam, 1967). This is essentially on account of the partial reliance of reef-constructing, or hermatypic, corals on food produced by photosymbionts, which can only live in the photic zone. The main areas of development of modern coral reefs are in the Indo-Pacific and Atlantic–Caribbean biogeographic provinces (Rosen, in Stoddart and Yonge, 1971; Rosen, 1988; Rosen, in Matteucci *et al.*, 1994; Spalding *et al.*, 2001). Coral diversity is higher in the Indo-Pacific than in the Atlantic–Caribbean. In fact, in detail, it is highest in, and decreases in proportion to increased distance from, south-east Asia, and it is conspicuously low on the western margins of the Americas and west Africa. Diversity has been used as a measure of temperature (Rosen, in Stoddart and Yonge, 1971; Fraser and Currie, 1996), and, by inference, palaeotemperature (Rosen, in Agusti *et al.*, 1999). However, it is also influenced by other factors, such as distance to, and dispersal from, the locus of evolution (Wilson and Rosen, in Hall and Holloway, 1998), and, significantly, depth (Wells, in Hedgpeth, 1957). A depth zonation of coral species and/or morphotypes is often evident on remote sensing satellite images and/or visual inspection by scuba-diving (Kowalik *et al.*, in Harris and Kowalik, 1994; Spalding *et al.*, 2001).

Some coral reefs, chiefly constructed by *Lophelia*, are found in cold, deep marine environments. Interestingly, there is some evidence to indicate that *Lophelia* is associated with 'cold seeps' (of hydrocarbons). In the Porcupine area off the west coast of Ireland in the North Atlantic, geophysical or seismic profiles have revealed the existence of a number of reefs up to 200 m high and 2000 m across, that have been demonstrated by dredging and gravity coring to be chiefly constructed by *Lophelia pertusa* (Coles *et al.*, 1996; identification by my colleague Brian Rosen, recently retired from the Natural History Museum, London). The *Lophelia* reefs here have in turn been demonstrated by the results of geochemical analysis to be characterised by higher than average interstitial hydrocarbon concentrations, interpreted to be related to seepage associated with underlying geological faults. However, it has not yet been demonstrated that *Lophelia*, or the associated fauna, including other corals such as *Desmophyllum cristagalli*, is actually metabolising the seeping hydrocarbon.

Morphology

In terms of hard-part morphology, the more or less cylindrical calcitic or aragonitic building (corallite or corallum) in which the individual coral animal lives possesses an external wall or theca, and storeys internally walled by radially arranged septa, and, sometimes, concentrically arranged dissepiments, and floored by tabulae (Fig. 3.34). A central strengthening rod or columella may also be present. Considerable morphological variation is exhibited by all the elements of the construction. Thus, the overall external form of solitary corals can range from flattened or patellate through discoid and trochoid to cylindrical; that of colonial corals from cateniform, or chain-like, through fasciculate, or finger-like, to massive. The growth mode of the massive colonial corals also exhibits considerable variation.

Classification

Classification of corals is based on a combination of morphology and phylogeny (evolutionary relationships). Three main sub-groups are recognised that are important as fossils, namely, the tabulate corals, the rugose corals, and the scleractinian corals.

Tabulate corals. Tabulate corals are an extinct, Ordovician–Permian, group of colonial corals.

Nothing is known directly of the biology of the tabulate corals, as they are extinct.

In terms of hard-part morphology, tabulate corals are characterised by the presence of numerous closely spaced tabulae, and the absence of septa and dissepiments. They are further characterised by walls perforated by mural pores. Tabulate corals are exemplified by the honeycomb-like *Favosites*, the chain-coral *Halysites*, and the sun-coral *Heliolites*, all from the Silurian.

Rugose corals. Rugose corals are an extinct, Ordovician–Permian, group of solitary to colonial corals.

Nothing is known directly of the biology of rugose corals, as they are extinct.

In terms of hard-part morphology, rugose corals are characterised by the possession of six primary septa or protosepta (the cardinal, surrounded by two alars or laterals, and the counter-cardinal, surrounded by two counter-laterals), numerous secondary septa or metasepta in between the primaries (specifically in the four spaces in between the cardinal and two alars or laterals, and in between the counter-cardinal and two counter-laterals), numerous tertiary septa in between the secondaries, and dissepiments. They are further characterised by the

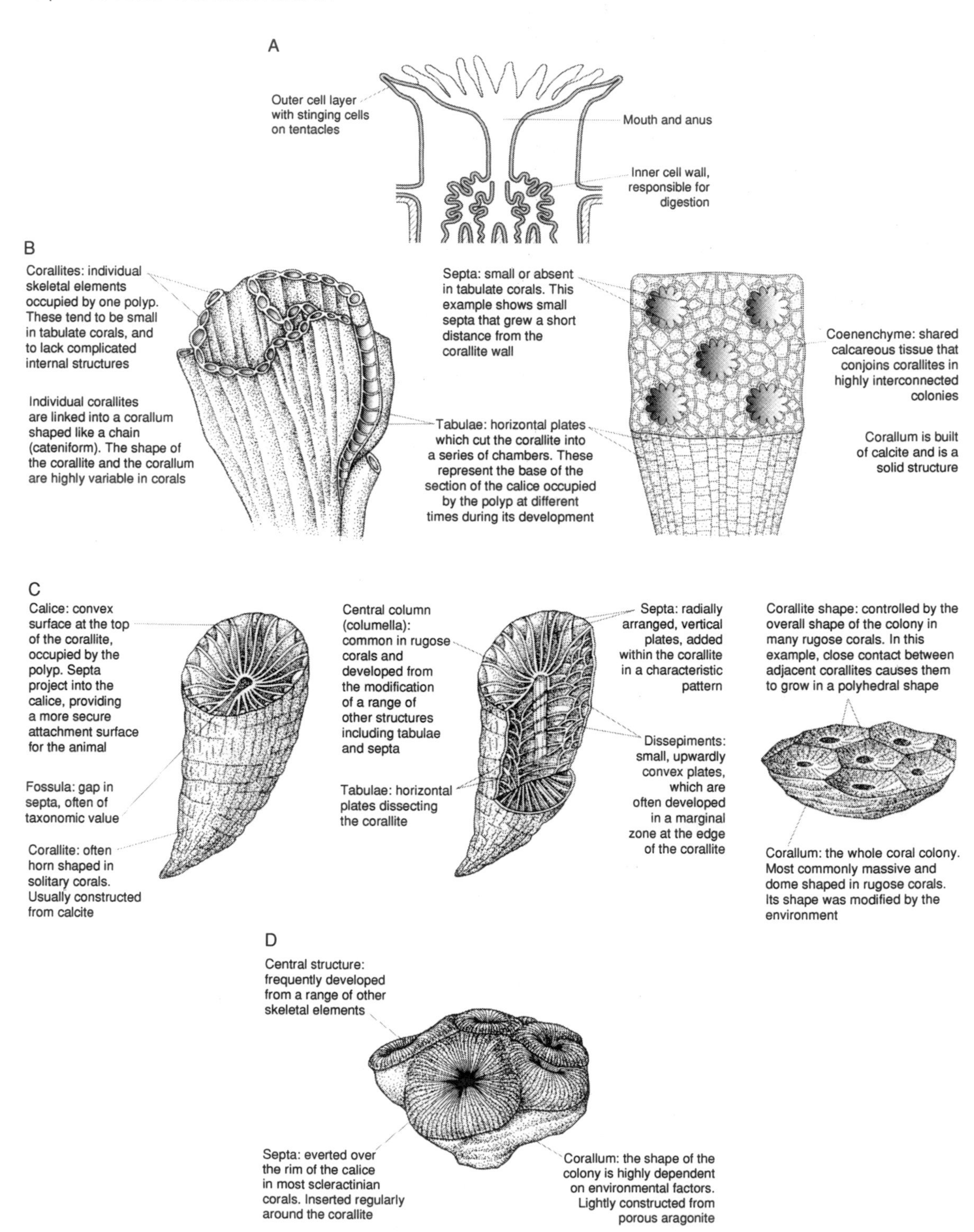

Fig. 3.34. Hard-part morphology of corals (from Milsom and Rigby, 2004). (A) Generalised morphology. (B) Tabulate corals. (C) Rugose corals. (D) Scleractinian corals.

possession of numerous tabulae, and by a cup-shaped depression or calice, to house the polyp. The septa are reflected externally by longitudinal grooves separated by ridges, the tabulae by transverse ridges. Rugose corals are exemplified by the solitary ceratoid or horn-shaped *Zaphrentis*, and the cylindrical *Dibunophyllum*, from the Carboniferous.

Scleractinian corals. Scleractinian corals are an extant, Triassic–Recent, group of solitary to colonial corals.

The biology of the scleractinian corals is well known (see above).

In terms of hard-part morphology, scleractinian corals are characterised by the presence of six primary septa; secondary septa or metasepta in all six spaces between the primaries; tertiary septa in between the secondaries; and dissepiments; and by the absence of tabulae. They are further characterised by anchoring base-plates and other attachment surfaces, and by porous, aragonitic constructions combining light weight with strength. The latter characteristics render the scleractinian corals important contributors to the construction of reefs. Scleractinian corals are exemplified by the colonial *Thamnastraea*, from the Jurassic–Cretaceous, and the stag's horn-coral *Acropora*, from the Pleistocene–Recent, the latter of which is presently an important reef-forming coral in the Indo-Pacific.

Palaeobiology

Fossil corals are interpreted, essentially on the bases of analogy with their living counterparts, functional morphology, and associated fossils and sedimentary facies, as having been marine, and as having lived principally as part of the sessile epifaunal benthos.

Fossil rugose corals are interpreted, on the basis of functional morphology, as having lived in a variety of positions (Neuman, 1988). The most prevalent appears to have been ambitopic, or recumbent on the sea-floor, concave side upward, with growth essentially vertical, giving rise to the stepped or roughened appearance that gives the group its name.

Palaeobathymetry

Fossil corals are interpreted as having been typically shallow marine (Wells, in Hallam, 1967). They are further interpreted on the basis of functional morphology, specifically on the degree of integration, as either having housed algal photosymbionts, or not, and thus as having been restricted to the photic zone, or not (Coates and Jackson, 1987). Many fossil corals appear to have had restricted bathymetric distributions within the photic zone, rendering them of use in the characterisation of palaeobathymetric zones (Perrin *et al.*, in Bosence and Allison, 1995). Depth-related morphological trends may also be evident.

It has been observed that corals seldom occur in close association with rudist bivalves in the shallow marine environments of the Cretaceous. On the basis of this observation, it has been hypothesised that the corals were displaced or excluded from these environments by competition from the rudists (Kauffman and Johnson, 1988). However, it has also been counter-hypothesised that the two simply had different ecological preferences or tolerances, especially as regards substrate (Skelton *et al.*, 1997).

Corals were important contributors to the construction of reefs in shallow marine environments to a lesser extent in the Palaeozoic, and to a greater extent in the Mesozoic and Cenozoic, coincident with the rise of the scleractinian corals with algal photosymbionts (Rosen, in Culver and Rawson, 2000). Palaeozoic corals were not generally particularly successful reef constructors, at least partly as they appear to have lacked structures that allowed anchorage or provided stability. However, corals were important contributors to the construction of reefs in the Ordovician to Devonian of the former Soviet Union (Kuznetsov, in Zempolich and Cook, 2002), in the Early Ordovician of south China (Li *et al.*, 2004), and in the Late Devonian of the Canning basin of Western Australia (although calcareous algae and stromatoporoids were more important here). Moreover, fasciculate rugose corals are characteristic of the Early Carboniferous, Brigantian boundstone reef of the Mullaghfin formation of Ardagh on the margins of the Dublin basin (Somerville *et al.*, in Strogen *et al.*, 1996); and rugose and tabulate corals of the Late Carboniferous, Bashkirian–early Moscovian seamount reefs of Japan (Wahlman, in Kiessling *et al.*, 2002). Note in this context that colonial rugose corals had just begun to radiate again in the Carboniferous after recovering from the effects of the Late Devonian mass extinction (Kossovaya, in Hart, 1996).

Palaeobiogeography

Many fossil corals appear to have had restricted or endemic biogeographic distributions, rendering them of use in the characterisation of palaeobiogeographic provinces, and in turn in the constraint of plate tectonic reconstructions, in the Palaeozoic, in the Mesozoic, and in the Cenozoic.

In the Cenozoic, coral distributions point toward the palaeogeographic and palaeoclimatic evolution of the Tethyan and Paretethyan realms (Rosen, 1988; Rosen, in Matteuci *et al.*, 1994; Wilson and Rosen, in Hall and Holloway, 1998; Rosen, in Agusti *et al.*, 1999; see also Sub-section 5.5.8).

Palaeoclimatology

Corals are of use in palaeoclimatic as well as in palaeobiogeographic interpretation (Parrish, 1998). For example, *Cladophora caespitosa* has been used in palaeoclimatic interpretation in the Pleistocene of the Mediterranean (Peirano *et al.*, 2004). Sclerochronology shows highest mean annual growth rates in warm phases.

High-resolution – decadal – interpretation based on oxygen isotopic compositions of corals has historically been difficult, on account of the suppression of signal by non-climatic noise (Lough, 2004). However, process studies are under way to assess what is being measured.

Biostratigraphy

Coral-like organisms or coralimorphs first appeared in the Cambrian (Zhuravlev and Riding, 2001). However, recognisable tabulate and rugose corals did not appear until the Ordovician. These sub-groups diversified through the Ordovician (Webby *et al.*, in Webby *et al.*, 2004), before sustaining losses in the End-Ordovician mass extinction. They recovered from this event, though (Kaljo, in Hart, 1996), and diversified again in the middle part of the Palaeozoic, before sustaining severe losses in the Late Devonian mass extinction. The tabulate corals never really recovered from this event, and finally disappeared in the End-Permian mass extinction. As noted above, the rugose corals staged some form of recovery from the Late Devonian mass extinction in the Carboniferous (Kossovaya, in Hart, 1996), but they, too, disappeared in the End-Permian mass extinction. Recognisable scleractinian corals first appeared in the Triassic, although arguably ancestral corals that exhibit scleractinian patterns of septal insertion have been described from the Ordovician of the Southern Uplands of Scotland (Scrutton and Clarkson, 1990). Scleractinian corals diversified in the Mesozoic and again in the Cenozoic, and are the dominant sub-group, and the only stony sub-group, at present. Interestingly, their rise through the Mesozoic–Cenozoic mirrors the decline of the essentially Palaeozoic stromatoporoids. There are some 3000 living species.

The moderate rate of evolutionary turnover exhibited by the corals renders them of some use in biostratigraphy, at least in appropriate, marine, environments (Fig. 3.35). Unfortunately, their usefulness is compromised by their restricted bathymetric and biogeographic distribution and facies dependence.

Nonetheless, they are of some use in stratigraphic subdivision and correlation, for example, in the British Isles. Here, corals form part of the basis of the biozonation of the Carboniferous Limestone (Fig. 3.36) (Vaughan, 1905; Riley, 1993; Jones and Somerville, in Strogen *et al.*, 1996).

In my working experience, corals have proved of some use in stratigraphic subdivision and correlation, and/or in palaeoenvironmental interpretation, in the following areas.

Palaeozoic

The Ordovician–Permian of the former Soviet Union. Here, corals form part of the basis of the regional biozonation of the Carboniferous (Wagner *et al.*, 1996). Tabulate and rugose corals are especially characteristic of the post-Late Devonian mass extinction platform and peri-reefal facies of the Late Devonian to Early Permian of the Bolshoi Karatau Mountains of southern Kazakhstan, and of the southern Urals (Cook *et al.*, in Zempolich and Cook, 2002; Vennin *et al.*, in Zempolich and Cook, 2002; Zempolich *et al.*, in Zempolich and Cook, 2002). These surface outcrops are important analogues for subsurface petroleum reservoirs in the North Caspian, peri-Caspian and Volga–Urals basins.

The Carboniferous of Egypt. Here, corals are characteristic of the Moscovian–Gzelian of Egypt (Izart *et al.*, 1998).

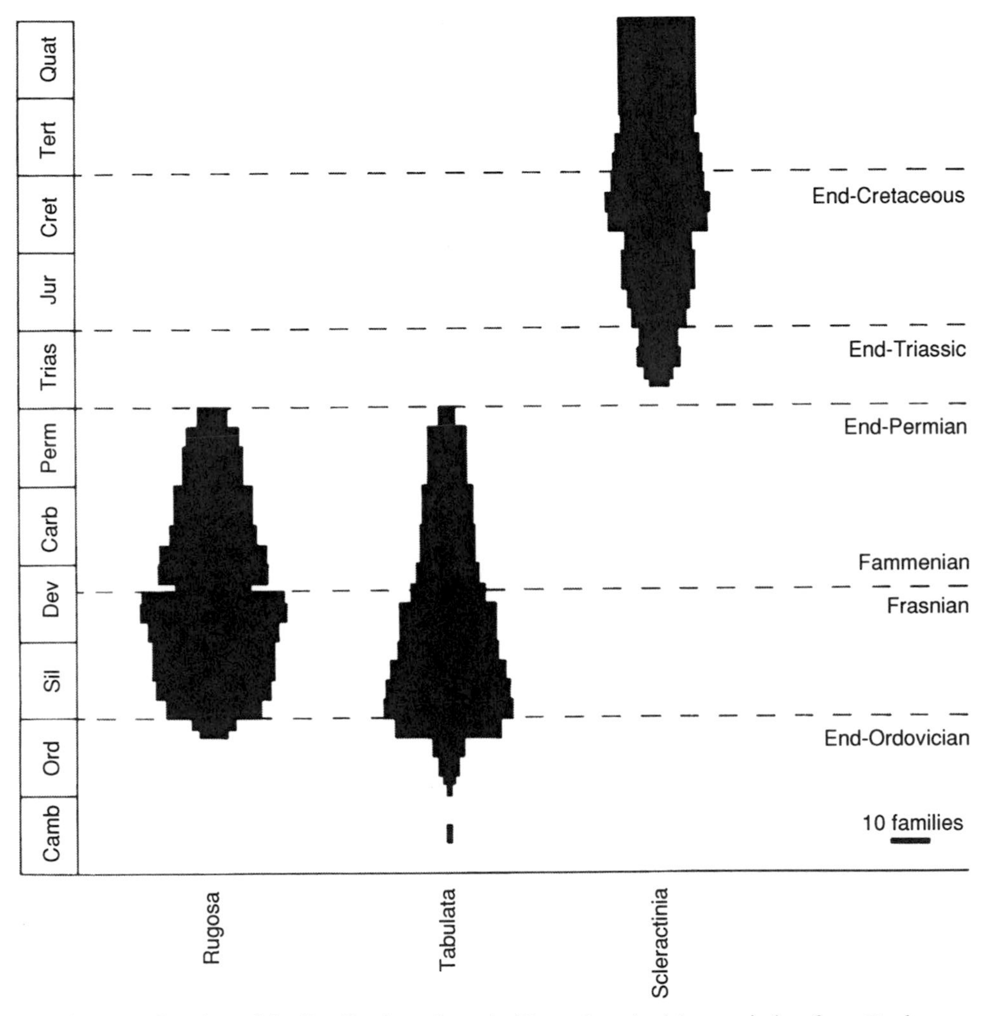

Fig. 3.35. Stratigraphic distribution of corals. (Reproduced with permission from Doyle, 1996 *Understanding Fossils*..., © John Wiley & Sons Ltd.)

The Carboniferous of the Middle East. Here, the Carboniferous of Iraq and of north-east Iran is marked by various corals (Monod and Weisbrod, in Wagner *et al.*, 1996; Vachard, in Wagner *et al.*, 1996).

Mesozoic

The Jurassic of north Africa. Here, *Myriophyllia rastelliana* and *Antiguastrea flandrini*, of Oxfordian–Kimmeridgian age, occur in reworked clasts in Palaeogene conglomerates in the Djurdjura Range in northern Algeria (Kotanski *et al.*, 1988). Coeval shelf limestones in Almeria, Sicily, Calabria, Sardinia and Corsica point to a widespread carbonate platform in the circum-Mediterranean in the Late Jurassic.

The Jurassic–Cretaceous of the Middle East. Here, corals are characteristic of the Jurassic Surmeh formation of the Zagros Mountains in south-west Iran and Musandam limestone of the Oman Mountains, and the peri-reefal facies of the Early Cretaceous Minagish formation of Kuwait (author's unpublished observations). They are also characteristic of the Late Cretaceous, late Campanian–Maastrichtian of the Oman/United Arab Emirates border region (Baron-Szabo, 2000).

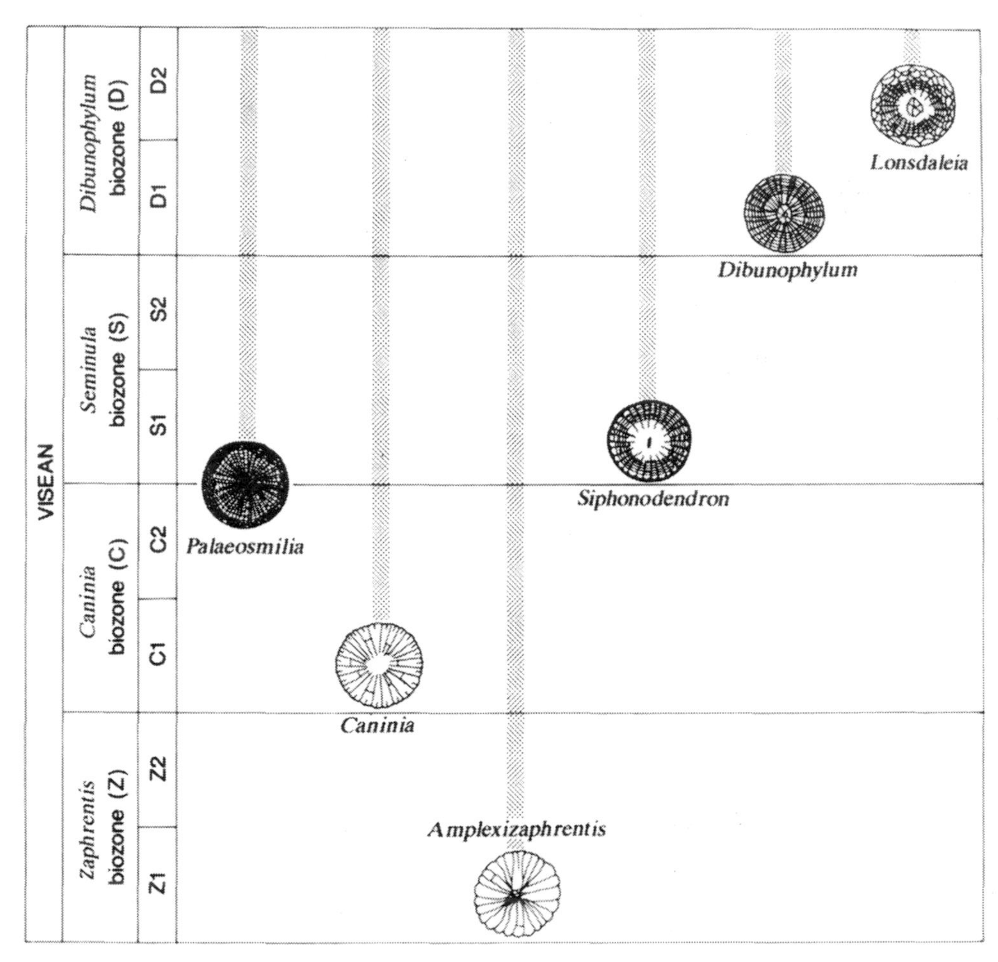

Fig. 3.36. Stratigraphic zonation of part of the Carboniferous Limestone of the Avon area of Great Britain by means of corals and brachiopods. (Reproduced with permission from Doyle, 1996 *Understanding Fossils . . .* , © John Wiley & Sons Ltd.)

Cenozoic

The Oligo-Miocene of the Middle East. Here, corals are characteristic of the peri-reefal facies of the Oligo-Miocene Asmari formation of the Zagros Mountains and Oman Mountains (author's unpublished observations). The platy coral *Colpophyllia stellata* is especially characteristic of the exhumed Asmari formation reef on the outskirts of the city of Al Ain on the Oman/United Arab Emirates border in the Oman Mountains (identification by my friend and colleague Brian Rosen, formerly of the Natural History Museum, London). A shoaling-upward succession of solitary corals and *Leptoseris–Stylophora* and *Porites*–Faviidae assemblages characterises the Asmari-equivalent Qom formation of Abadeh in north-east Iran (Schuster and Wielandt, 1999).

3.6.7 Lophophorates: brachiopods

Biology, morphology and classification

Brachiopods, colloquially known as 'lamp-shells', on account of the similarity of certain forms, when inverted from their growth position, to early Roman oil-lamps, are an extant, Cambrian–Recent group of lophophorates characterised by bivalved shells (MacKinnon *et al.*, 1991; Copper and Jin, 1996; MacKinnon *et al.*, 1996; Brunton *et al.*, 2001; Doherty, in Anderson, 2001; Carlson and Sandy, 2001). They have been interpreted as having evolved from halkierids (Holmer *et al.*, 2002).

Biology

The biology and ecology of living brachiopods is well known. They feed by passive filter-feeding, drawing in water and food particles suspended therein through lateral inhalant apertures, filtering out the food particles through a structure called a lophophore in the mantle cavity, and pumping out the waste water through a median exhalant siphon, all of which are situated along the anterior margin (all of the other internal organs are situated at the posterior end). The arrangement of the lophophore into a series of coils, filaments that close the gaps between the coils, and cilia that close the gaps between the filaments, is such as to allow little if anything to escape.

Living brachiopods are exclusively marine, occupying a range of marginally to fully marine environments worldwide, from the intertidal to the hadal, and from the equator to the poles. They are perhaps especially characteristic of deep or cold waters, where they are locally abundant, not least because they are gregarious. Most are epifaunal, and live attached directly or indirectly by means of a stalk or pedicle to the substrate, while some are unattached; few are infaunal. None possesses any significant capacity for motility, although some achieve a form of motility by being attached to floating substrates (and at least some, if not all, have a planktonic developmental stage, enabling them to disperse and to colonise new niches).

Morphology

In terms of hard-part morphology, brachiopods are characterised by a secreted calcitic or horny chitinophosphatic bivalved shell for the protection of their soft parts (Figs. 3.37–3.38). The valves are termed pedicle or ventral, and brachial or

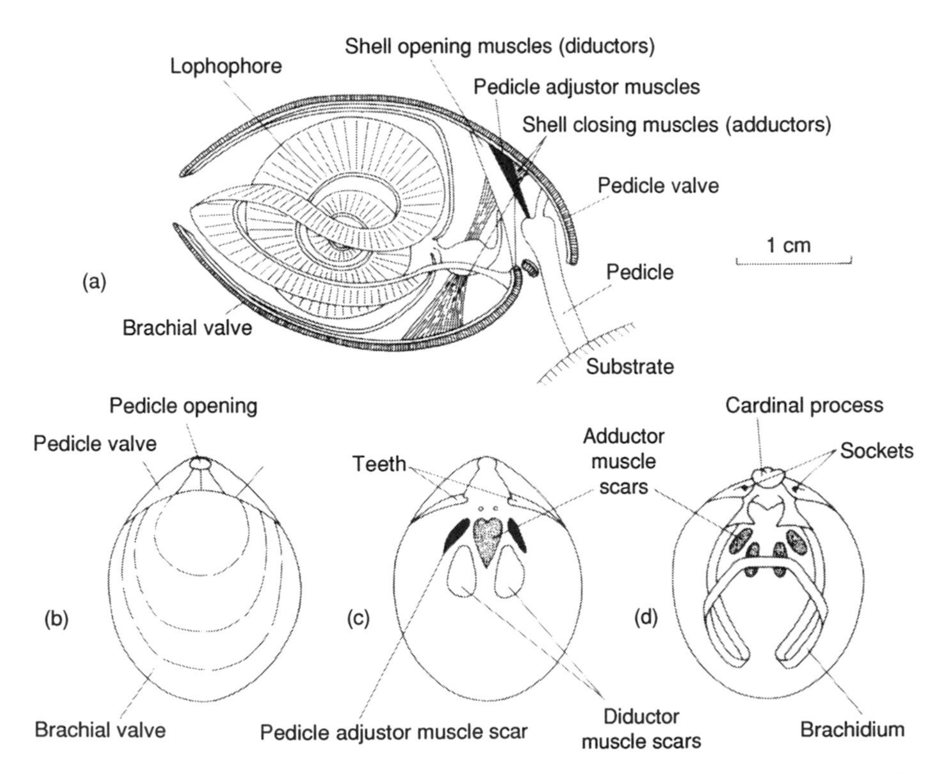

Fig. 3.37. Hard-part morphology of brachiopods. (a) Cross-section through a generalised living brachiopod; (b) exterior; (c) interior of the dorsal or pedicle valve; (d) interior of the ventral or brachial valve. (Modified after Black, 1989.)

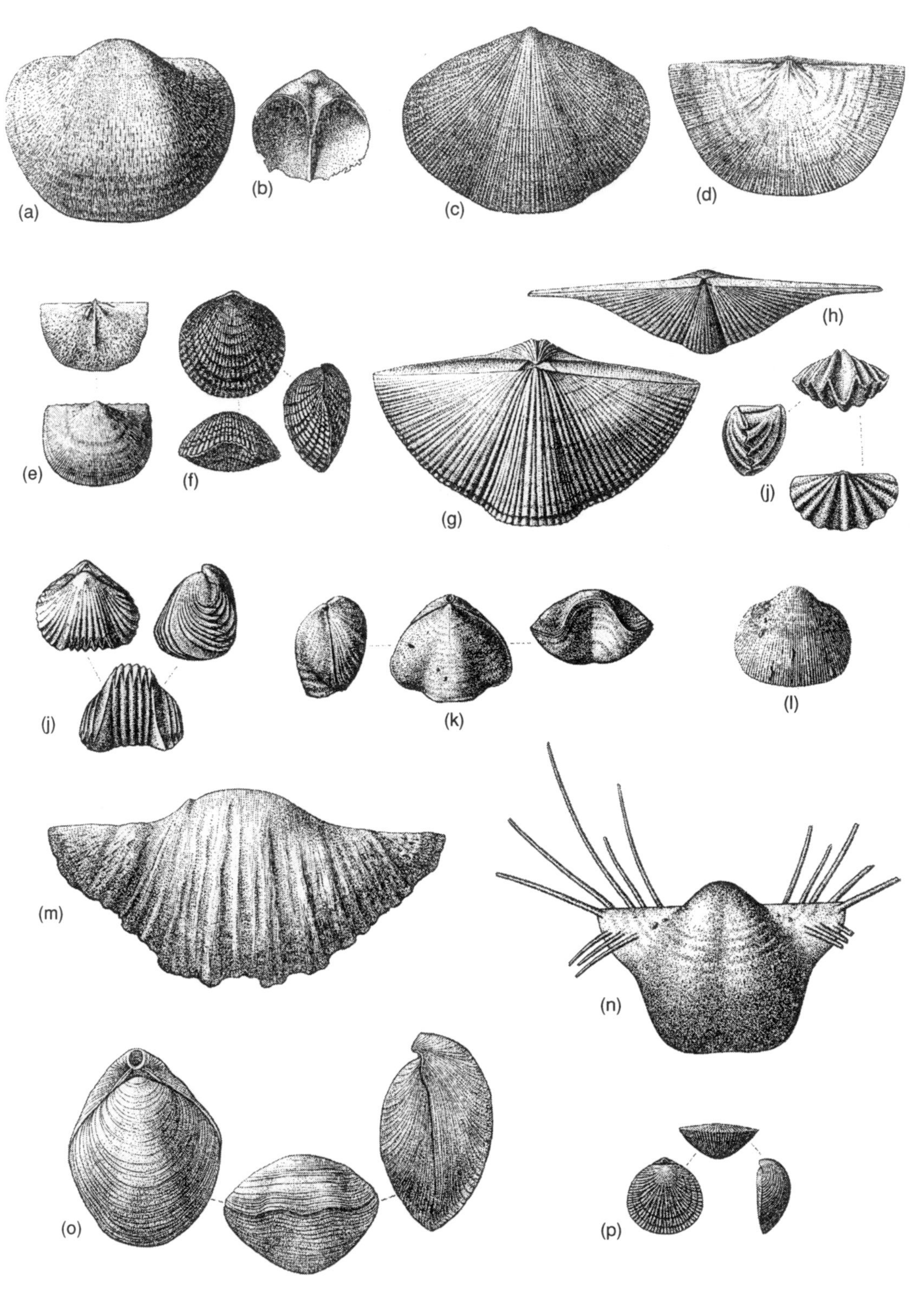

Fig. 3.38. Some representative brachiopods. Orthide: (a) *Schizophoria resupinata*, Early Carboniferous, Visean, Narrowdale Hill, near Longnor, Staffordshire, ×0.5. Pentamerides: (b) *Pentamerus oblongus*, Silurian, Llandovery, Norbury, Shropshire, ×0.75;

dorsal. The ventral valve is generally convex, while the dorsal valve can be either convex or concave. The ventral valve is also larger than the dorsal valve, and overlaps it, most markedly at the posterior end, where the pedicle protrudes. The valves are articulated together at the posterior hinge-line. They open and close along the anterior commissure. Opening of the valves is effected by the spring-like action of ratcheted diductor muscles; closing, by contraction of adductor muscles. The inner surface of the shell bears scars at the sites of attachment of these muscles, and may also exhibit evidence of lophophore supports. The outer surface may bear concentric growth rings, or radial corrugations.

In brachiopods, the plane of symmetry runs across the valves from the anterior to the posterior. In contrast, in bivalves, it runs between the valves.

Classification

Classification of brachiopods is currently in a state of flux. Historically, two main sub-groups have been recognised on the basis of morphology, namely, the inarticulate brachiopods, without, and the articulate brachiopods, with, articulating structures. However, it has recently been shown that both of these sub-groups are polyphyletic, and therefore that the historical classification is in need of revision, the better to reflect phylogeny. Unfortunately, as yet no clear consensus has been reached as to a revised classification. One suggestion has been to recognise two main sub-groups essentially on the basis of shell composition, perceived to be a conservative feature, namely, the lingulate brachiopods, with chitinophosphatic shells, and the calciate brachiopods, with calcitic shells.

Lingulate brachiopods. Lingulate or linguliform brachiopods in turn include the well-known lingulinides; and the poorly known paterinides, acrotretides, siphonotretides and discinides. Only the lingulids are considered here.

Lingulinides, which range from Cambrian to Recent, are characterised by spatulate valves. They are exemplified by the 'living fossil' *Lingula*, which burrows in tidal mudflats in China and Japan, and appears to have done so since the Cambrian. Incidentally, its Japanese name refers to its resemblance to a classical Japanese guitar. Somewhat less prosaically, its Chinese name translates as 'sea bean-sprout', and refers to its edible pedicle! Lingulids with preserved pedicles have recently been discovered in *Trypanites* borings in tabulate corals and stromatoporoids in the Ordovician and Silurian of eastern Canada (Tapanila and Holmer, 2004).

Calciate brachiopods. Calciate brachiopods in turn include the poorly-known inarticulate or craniiform craniopsides, trimerellides and craniides,

Caption for fig. 3.38 (*cont.*) (c) *Costistricklandia lirata*, Silurian, Ledbury, Herefordshire, ×0.75. Strophomenides: (d) *Strophomena grandis*, Ordovician, Caradoc, Cheney Longville, Shropshire, ×1; (e) *Rugosochonetes hardrensis*, Early Carboniferous, Visean, Wrexham, Clwyd, ×1.5. Atrypide: (f) *Atrypa reticularis*, Silurian, Wenlock, Dudley, West Midlands, ×1. Spiriferides and spiriferinides: (g) *Spirifer striatus*, Early Carboniferous, Derbyshire, ×0.5; (h) *Cyrtospirifer extensus*, Late Devonian, Delabole, Cornwall. ×0.75; (i) *Spiriferellina cristata*. Late Permian, Magnesian Limestone, Humbleton Hill, near Sunderland, Tyne-and-Wear. Rhynchonellide: (j) *Tetrarhyncia tetraedra*, Early Jurassic, middle Lias, Tilton, Leicestershire, ×1. Athyridide: (k) *Composita ambigua*, Early Carboniferous, Settle, North Yorkshire, ×1. Productides: (l) *Productus productus*, Beresford Hall, near Longnor, Staffordshire, ×1; (m) *Gigantoproductus giganteus*, Early Carboniferous, Visean, Llangollen, Clwyd, ×0.5; *Horridonia horrida*, Late Permian, Magnesian Limestone, Bishop Auckland, Co. Durham, ×1. Terebratulides: (o) *Terebratula maxima*, Pliocene, Coralline Crag, Orford, Suffolk, ×0.5; (p) *Terebratulina lata*, Late Cretaceous, Turonian, *Terebratulina lata* Zone, middle Chalk, Hooken Cliff, near Beer, Devon, ×3. (Modified after the Natural History Museum, 2001a.)

intermediate chileides, kutorginides and obolellides, and indeterminate thecideidinides; and the well-known articulate or rhynchonelliform orthides, pentamerides, strophomenides, atrypides, spiriferides and spiriferinides, athyridides, productides, and terebratulides. Only the articulate calciates are considered here.

Orthides (Cambrian–Permian) are characterised by biconvex shells and straight hinge-lines and simple cardinal processes, and by the absence of calcified lophophore supports. They are exemplified by *Orthis* and *Schizophoria*.

Pentamerides (Middle Cambrian–Devonian) are characterised by biconvex shells with curved hinge-lines, and by the presence of calcified lophophore supports termed cruralia and spiralia. They are exemplified by *Pentamerus, Stricklandia* and *Costistricklandia*.

Strophomenides (Ordovician–Permian or Triassic) are characterised by concavo-convex shells with complex cardinal processes, and by the absence of calcified lophophore supports. They are exemplified by *Strophomena, Billingsella, Chonetes* and *Rugosochonetes*.

Atrypides (Middle Ordovician–Devonian) are characterised by biconvex shells, and by dorsoventrally directed spiralia. They are exemplified by *Atrypa*.

Spiriferides and spiriferinides (Middle Ordovician–Permian) are characterised by laterally extended hinge-lines, and by laterally directed spiralia. They are exemplified by *Spirifer, Cyrtospirifer, Spiriferina, Spiriferellina* and *Tylothyris*.

Rhynchonellides (Middle Ordovician–Recent) are characterised by biconvex shells with curved hinge-lines, and by the presence of calcified lophophore supports termed crurae. They are exemplified by *Rhynchonella, Eocoelia* and *Tetrarhyncia*.

Athyridides (Late Ordovician–Jurassic) are characterised by biconvex shells with short, straight hinge-lines, and by posterolaterally directed spiralia. They are exemplified by *Athyris, Cleothyridina* and *Composita*.

Productides (Devonian–Triassic) are characterised by planoconvex, often aberrant, shells, and by the absence of calcified lophophore supports. They are exemplified by *Productus, Gigantoproductus, Horridonia* and *Richthofenia*.

Terebratulides (Devonian–Recent) are characterised by biconvex shells with short, curved hinge-lines, and by the presence of calcified lophophore supports termed brachidia. They are exemplified by *Terebratula* and *Terebratulina*.

Incidentally, some calciates, as well as lingulates, are recognised in popular culture. For example, the pentameride *Pentamerus oblongus* is sufficiently abundant in the Silurian of Shropshire to be rock-forming, with the rock popularly known as 'government rock' in reference to the supposed resemblance of the brachiopod remains to the government bench-marks on prisoners' uniforms! More prosaically, the spiriferide *Cyrtospirifier extensus* from the Devonian of Delabole in Cornwall, is popularly known as the 'Delabole butterfly', while some other spiriferides are known as 'stone swallows' in China, and some rhynconellides as 'little doves' in the Jura Mountains of central Europe.

Palaeobiology

Fossil brachiopods are interpreted, essentially on the basis of analogy with their living counterparts, as having been marine. They are also interpreted, on the basis of associated fossils and sedimentary structures, as having been more typically shallow marine than their living counterparts, although deeper marine fossil forms are also known, for example from the Mesozoic of Alpine Europe (Ager, 1965). They are further interpreted, essentially on the basis of functional morphology, as having lived principally as part of the sessile epifaunal or semi-infaunal filter-feeding benthos (Rudwick, 1961, 1964; Grant, 1966; Bassett, 1984; Vogel, in Racheboeuf and Emig, 1986; Ushatinskaya, in Zhuravlev and Riding, 2001). Rudwick's work elegantly demonstrated the function of the often-observed zigzag morphology of the commissure, which effectively increases the intake area, while maintaining the same gape and particle size restriction.

Brachiopods were sufficiently abundant in shallow marine environments in parts of the Palaeozoic, when they dominated the low-level, filter-feeding benthos, as to be rock-forming, and to contribute to the construction of reefs, as in the Early Permian, Asselian–Sakmarian (Wahlman, in Kiessling *et al.*, 2002).

In the Early Carboniferous, Visean of the areas around Loughs Carra, Corrib and Mask, and Galway Bay in western Ireland, shallow marine brachiopods occur in distinct biofacies apparently related to lithofacies (Harper and Jeffrey, in Strogen *et al.*, 1996). Here, the sandy limestone facies of the Carra–Mask unit is variously dominated by *Rhipidomella, Schizophoria, Leptagonia, Rugosochonetes, Krotovoa, Echinoconchus, Tylothyris, Punctospirifer* or *Spiriferellina*; the muddy limestone facies of the Kiltullagh Bridge unit by *Composita, Schizophoria, Brochocarina, Minythyra* and *Cleothyridina*, and the pure limestone of the Burren unit by large linoproductids.

Palaeobathymetry

Many fossil brachiopods appear to have had restricted bathymetric distributions, rendering them of use in the characterisation of palaeobathymetric zones. For example, in the Early Silurian, Llandoverian of the margins of the Anglo-Welsh basin, *Lingula, Eocoelia, Pentamerus, Costistricklandia* and *Clorinda* characterise successive benthic assemblage zones from shallowest to deepest (Cocks, in Hallam, 1967; Orr, in Briggs and Crowther, 2001). Similar zones are recognisable in the Silurian of south China (Wang *et al.*, 1987; Fortey and Cocks, 2003). Here, the shallowest, *Lingula*, zone is interpreted as intertidal, and the deepest, *'Stricklandia'*, zone as sub-tidal, below the lower limit for photosynthesis and reef-building activity.

Palaeobiogeography

Many fossil brachiopods appear to have had restricted or endemic biogeographic distributions, rendering them of use in the characterisation of palaeobiogeographic provinces, and in turn in the constraint of plate tectonic reconstructions, in the Palaeozoic and in the Mesozoic.

In the Palaeozoic, the cold-adapted *Hirnantia* brachiopod fauna appears to have had an unrestricted, pandemic or cosmopolitan distribution during the Late Ordovician, late Ashgillian, Hirnantian glaciation; and warm-adapted taxa essentially cosmopolitan distributions during the succeeding Early Silurian, Llandoverian deglaciation (Boucot and Johnson, in Hallam, 1973; Sheehan and Coorough, in McKerrow and Scotese, 1990). The cold-adapted opportunistic *r*-strategist *Pachycyrtella omanensis* was the dominant taxon in the early stages of the Permian deglaciation in Oman (Angiolini, in Brunton *et al.*, 2001; Angiolini *et al.*, 2003a, b; Angiolini *et al.*, in Al-Husseini, 2004). The *Pachycyrtella omanensis* community was succeeded by a more diverse and stable warm-adapted *Reedoconcha permixta–Punctocyrtella spinosa* community, also known from India, Afghanistan, the Himalayas and Thailand (the so-called 'Westralian province', possibly part of a still larger 'Indoralian province', also including eastern Australia), in the later stages of the deglaciation. It is possible that this apparent ecological succession may actually have been prompted by a 'catastrophic physical perturbation' that wiped out the *Pachycyrtella omanensis* community.

Biostratigraphy

Brachiopods first appeared in the rock record in the Early Cambrian; lingulates at the base of, and calciates within, the Tommotian (Ushatinskaya, in Zhuravlev and Riding, 2001). They diversified in the Ordovician (Harper *et al.*, in Webby *et al.*, 2004), before sustaining significant losses in the End-Ordovician mass extinction (Bassett *et al.*, in Crame and Owen, 2002; Harper and MacNiocaill, in Crame and Owen, 2002). They recovered from this event, though, and diversified again in the Silurian and Devonian, before sustaining significant losses again in the Late Devonian mass extinction. They recovered from this event, too, and diversified yet again in the Carboniferous and Permian, before sustaining severe losses in the End-Permian mass extinction. They may be said never to have really recovered from this event, and to have been in general decline ever since. There are some 12 000 fossil but only 350 living species.

Some authors have suggested that the post-Palaeozoic decline of the brachiopods is due to predation pressure. However, there are few records of predation or bio-erosion on living brachiopods, despite the fact that some, such as those from South Georgia, live in communities alongside a full range of predatory taxa (Harper *et al.*, 2003). Moreover, brachiopods have been observed to be invariably rejected by potential predators such as fish and asteroids in feeding trials (Doherty, in Anderson, 2001). This may be because they make use of chemical defences to deter predators.

The post-Palaeozoic decline of the brachiopods is perhaps more likely due to competition from the bivalve molluscs, whose rise appears to mirror it. Brachiopods would probably have been less able to disperse, and to colonise infaunal niches, than bivalves.

The moderate rate of evolutionary turnover exhibited by the brachiopods renders them of some use in biostratigraphy, at least in appropriate, marine, environments (Fig. 3.39). Unfortunately, their usefulness is compromised by their restricted ecological distribution and facies dependence, a function of their benthic habit.

Nonetheless, they are of some use in stratigraphic subdivision and correlation. For example, *Eocoelia* is extremely useful in the Silurian, where it forms the basis of a biozonation of the Llandovery to Wenlock that has inter-regional applicability throughout the 'Silurian cosmopolitan province' referred to above. Brachiopods are also useful in the Carboniferous of Britain, where they form part of the basis of the biozonation of the Carboniferous Limestone (Vaughan, 1905; Riley, 1993), and are locally important in the identification of marine bands in the 'Coal Measures', as in Northumberland and Durham. Also in Britain, *Terebratula lata* forms part of the basis of the biozonation of the Late Cretaceous Chalk.

In my working experience, brachiopods have proved of some use in stratigraphic subdivision and correlation, and/or in palaeoenvironmental interpretation, in the following areas.

Palaeozoic

The Palaeozoic of the Middle East. Here, the Cambrian Member C of the 'Mila formation', otherwise known as the Haft Tanan formation, of southwest Iran, is marked by *Billingsella* (Setuhdenia, 1975). The Carboniferous of north-east Iran is marked by various brachiopods (Vachard, in Wagner *et al.*, 1996). The regional Middle Devonian, Emsian D20 MFS of Sharland *et al.* (2001), stratotypified in the Hammamiyat Limestone member of the Jauf formation of Saudi Arabia, is marked by *Hysterolites ex gr. hystericus, Anathyris, Sulcathyris arabica, Fascistropheodonta, Schizophoria, 'Athyris' cf. undata, Areostrophia* and *?Devonaria*. The Early Permian, Sakmarian P10 MFS, stratotypified in shales immediately below the Haushi Limestone member of the lower Gharif formation of Oman, is marked by *Reedoconcha permixta* and *Punctocyrtella spinosa*. Member 1 of the overlying Khuff formation is characterised by *Celebetes manarollai* (Angiolini *et al.*, in Al-Husseini, 2004). The Khuff formation also contains *Derbyia cf. diversa, Neochonetes (Nongtaia) arabicus, Bilotina yanagidai, Linoproductus* aff. *kaseti, Juresania omanensis, Perigeyerella raffaelae, Acritosia* sp., *Vediproductus* sp., *'Dielesma'* sp., *Magniplicatina* sp. and *Globosobucina* sp. (Angiolini *et al.*, in Al-Husseini, 2004).

The Late Palaeozoic of the former Soviet Union. Here, brachiopods are useful in the stratigraphic subdivision and correlation of the Devonian in the Timan–Pechora basin in the north-eastern part of the Russian platform (author's unpublished observations), and of the Carboniferous throughout the former Soviet Union (Wagner *et al.*, 1996). They are also characteristic of the platform and peri-reefal facies of the Devonian to Permian of the Bolshoi Karatau Mountains of southern Kazakhstan, and of the southern Urals.

Mesozoic

The Jurassic–Cretaceous of the Middle East. Here, *Daghanirhyncia* marks the Middle Jurassic, Unit c, *Somalirhyncia* the Late Jurassic, Unit e (Beni Zaid Limestone), and *Septirhyncia* the Late Jurassic, Unit f, of the Musandam Limestone (Hudson and Chatton, 1959). *Somalirhyncia* also marks the Oxfordian Hanifa formation of Saudi Arabia (Hughes, in Bubik and Kaminski, 2004). Incidentally, it also appears to mark the Late Jurassic in east and north Africa (author's unpublished observations). *Pseudogibbithyris arabica* marks the Late Cretaceous, Maastrichtian, Qahlah formation of Jebel Huwayyah in the Oman Mountains (Owen, in Howarth, 1995). In the field in the Zagros Mountains of south-west Iran, the Jurassic Surmeh formation is marked by *Septirhyncia azaisi*, the Cretaceous, Aptian Early Dariyan formation and 'Kazhdumi Tongue' by *Terebratula sella* (author's unpublished observations).

3.6.8 Lophophorates: bryozoans

Biology, morphology and classification

Bryozoans, also known as ectoprocts, or, colloquially, as 'sea-mosses', are an extant, Ordovician–Recent group of lophophorates characterised – with the solitary exception of *Monobryozoon* – by a colonial habit (Herrera Cubilla and Jackson, 2000; Doherty, in Anderson, 2001; Carlson and Sandy, 2001).

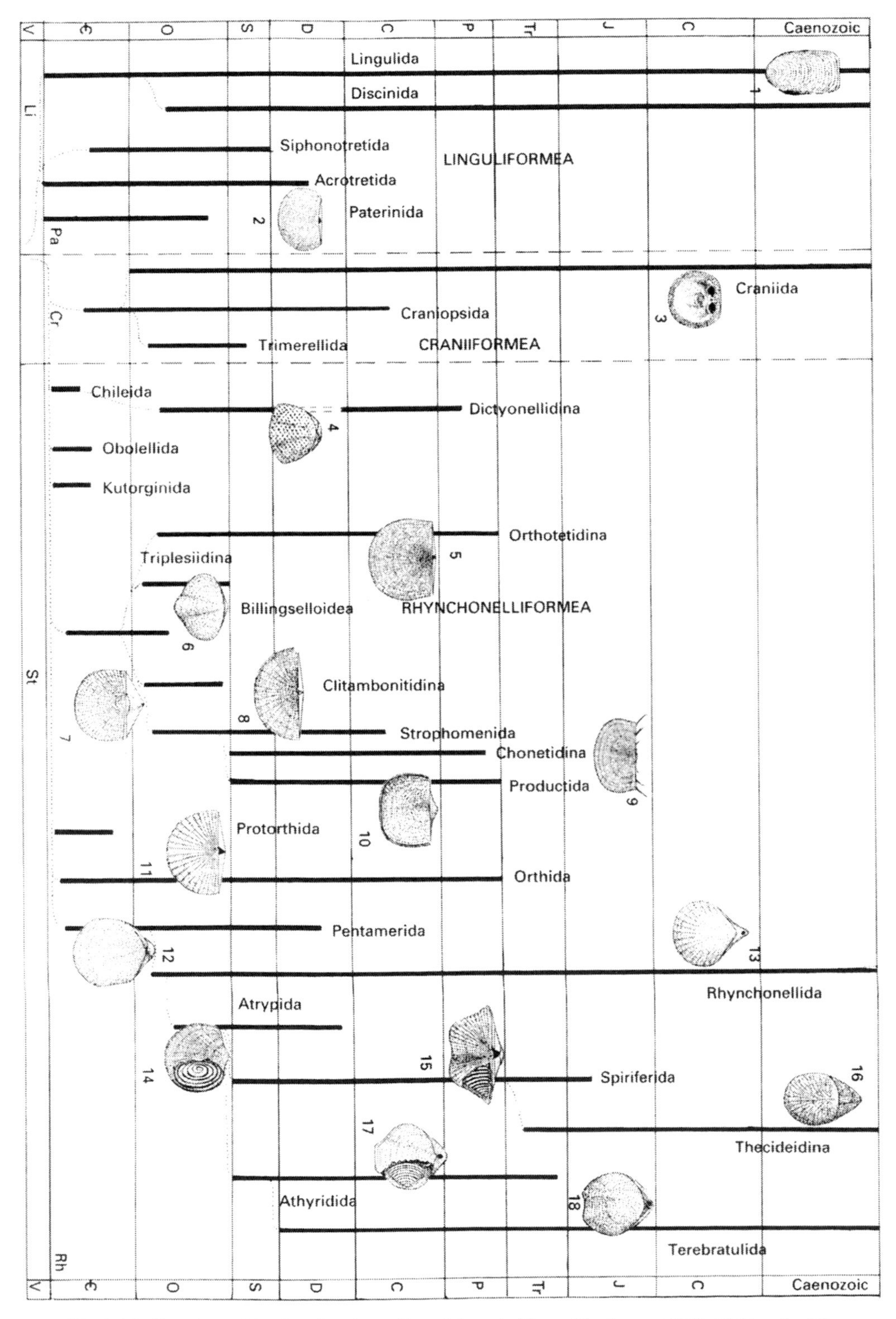

Fig. 3.39. Stratigraphic distribution of brachiopods (from Clarkson, 1998). (1) *Lingula*; (2) *Micromitra*; (3) *Crania*; (4) *Dictyonella*; (5) *Orthotetes*; (6) *Triplesia*; (7) *Vellamo*; (8) *Eoplectodonta*; (9) *Chonetes*; (10) *Dictyoclostus*; (11) *Hesperorthis*; (12) *Pentamerus*; (13) *Plicirhyncia*; (14) *Atrypa*; (15) *Cyrtospirifer*; (16) *Thecidea*; (17) *Composita*; (18) *Stiphrothyris*.

Biology

The biology and ecology of living bryozoans is well known. Bryozoan colonies superficially resemble coral colonies, although individual animals or zooids are typically less than 1 mm in diameter, and in this respect are rather more reminiscent of phoronids. As in the case of the corals, some individual bryozoan animals within the colony are adapted to fulfil specialised functions such as protection or reproduction, although most serve simply in the gathering of food. Like the brachiopods, the bryozoans feed by passive filter-feeding by means of a lophophore. They appear to make use of chemical defences known as bryotoxins to deter predators from feeding on them. One British species, *Alcyonidium diaphanum*, commonly known as sea chervil, is said to cause an irritating skin condition known to North Sea fishermen as 'Dogger Bank itch' (Naylor, 2003).

In terms of ecology, most living bryozoans are marine, although some are brackish or freshwater. Almost all of the marine forms are comparatively shallow marine, although a few are associated with deep marine 'cold seeps', as on the Barbados accretionary prism (Jollivet *et al.*, 1990). Almost all living bryozoans are benthic, epifaunal and sessile, although some are motile, and one Antarctic species is pelagic, forming floating colonies. Attached sessile epifaunal bryozoans characteristically exhibit encrusting, or rooted, or upright and branching growth; unattached motile forms concentric growth that enables them to roll around. Encrusting forms decrease, and upright forms increase, in abundance with depth, such that the ratio between the two provides a measure of depth, and, by inference, palaeodepth (evolutionary effects notwithstanding). Moreover, uniserial upright forms increase, and multiserial upright forms decrease, in abundance with depth below about 1000 m, such that the ratio between the two provides a measure of great depth or palaeodepth. Further details of the relationship between growth form and depth are provided by Amini *et al.* (2004).

Bryozoans are sufficiently abundant in some shallow marine environments as to be important contributors to the formation of sediments. This is especially true in temperate – as opposed to tropical – waters, where they constitute an important part of the foramol group of grain types in carbonate sediments (Lees and Buller, 1972). Ancient as well as modern bryozoans appear especially characteristic of temperate-water carbonates.

Morphology

In terms of morphology, individual bryozoan animals are characterised by a protective gelatinous or chitinous membrane or by a calcareous exoskeleton or zooecium of tubular or box-like construction (Fig. 3.40).

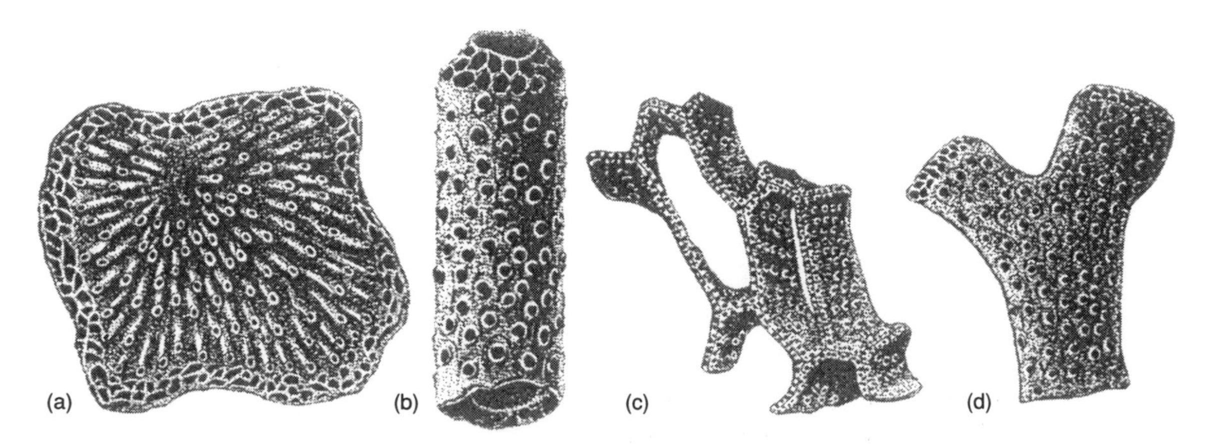

Fig. 3.40. Hard-part morphology of bryozoans. (a) Encrusting (turbulent water); (b) erect (calm water); (c) retiform (turbulent water); (d) bifoliate (calm water). (From Jones, 1996; in turn after Haq and Boersma, 1978.)

Bryozoan colonies can be described as encrusting, stoloniferous or arborescent. Colonial form appears to vary according to habitat.

Classification

Historically, three main sub-groups of bryozoans have been recognised essentially on the basis of morphology, namely, the phylactolaemates, the stenolaemates and the gymnolaemates. However, recent molecular evidence indicates that these sub-groups are probably polyphyletic, and therefore that the historical classification is in need of revision, the better to reflect phylogeny (Snell, 2004).

Phylactolaemates. The phylactolaemates are an extant, ?Cretaceous–Recent, group characterised by non-calcified tubular zooecia, and thus possess a poor fossil record, apparently from Jurassic or Cretaceous to Recent. They are exclusively freshwater.

Stenolaemates. The stenolaemates are an extant, Ordovician–Recent, group, characterised by tubular zooecia only partially filled by zooids enclosed by membranous sacs (with an exosaccal space or coelom remaining). The stenolaemates in turn include the extant, Ordovician–Recent, cyclostomates, and the extinct trepostomates, cystoporates, cryptostomates and fenestrates. The cyclostomates are exclusively marine and epifaunal, and predominantly shallow marine and epilithic, being both abundant and diverse on hard substrates, although it remains to be discovered whether their larvae are actually selective of substrate, as apparently is the case among the gymnolaemates (Hayward and Ryland, 1985).

Gymnolaemates. The gymnolaemates are an extant, Ordovician–Recent group, in turn including the Ordovician–Recent ctenostomates and the Jurassic–Recent cheilostomates. Most are marine, although some are brackish or freshwater. The ctenostomates are characterised by bud-like, tubular zooecia that are not calcified as such, although they may be bored into calcareous structures. The cheilostomates are characterised by box-like zooecia.

Palaeobiology

Fossil bryozoans are interpreted, essentially on the basis of analogy with their living counterparts, as having been mostly shallow marine (Smith, in Bosence and Allison, 1995). However, some appear to have been associated with deep marine vents, as in the Carboniferous of Newfoundland (von Bitter *et al.*, 1992).

Bryozoans were sufficiently abundant in some shallow marine environments as to have been important contributors to the construction of reefs, or bryoherms, especially in the Palaeozoic (Taylor and Allison, 1998).

For example, bryozoans are characteristic of the Early Ordovician, Tremadoc reefs of South China (Li *et al.*, 2004), and the – cool-water – Late Ordovician, Ashgill mud mounds of Libya (Buttler *et al.*, 2004).

Fenestrate bryozoans are characteristic of the initial, and large cylindrical and vase-shaped fenestrates of the final, stages of development of the extensively studied 'Muleshoe Mound', a Waulsortian mound with a relief in excess of 100 m in the Early Carboniferous, Tournaisian, early to late *Typicus* to *Anchoralis–Latus* Conodont Zone, Nunn and Terra Blanca members of the Lake Valley formation of the Sacramento Mountains of New Mexico (Kirkby and Hunt, in Strogen *et al.*, 1996). Fenestrate, ramose and encrusting bryozoans are characteristic of simple, Waulsortian-like mounds in the slightly older, early *Typicus* Conodont Zone Alamagordo member (Ahr and Stanton, in Strogen *et al.*, 1996), which differ in their early relief and lesser degree of environmental differentiation from crest to flank (Lees and Miller, in Monty *et al.*, 1995). Importantly, bryozoans are much less characteristic of the level-bottom equivalents of the Waulsortian and Waulsortian-like mounds (Ahr and Stanton, in Strogen *et al.*, 1996).

Encrusting bryozoans are characteristic of the Early Carboniferous, Brigantian algal boundstone reefs of the Mullaghfin formation of the Kingscourt outlier on the margins of the Dublin basin (Somerville *et al.*, in Strogen *et al.*, 1996). Bryozoans are also characteristic of the Early Carboniferous, Brigantian peri-reefal or platformal carbonates of the Burren, Buttevant and Callan areas of southern and western Ireland (Gallagher, in Strogen *et al.*, 1996), and of Late Carboniferous, late Moscovian–Gzhelian reefs generally (Wahlman, in Kiessling *et al.*, 2002).

'Baffling' fenestellid and ramose bryozoans and 'binding' fistuliporid bryozoans are characteristic of Permian reefs generally (Wahlman, in Kiessling *et al.*, 2002); for example, the Late Permian, Changhsingian reefs of Laolongdon, near Chongqing in Sichuan province in China (Yang Wanrong, in Jin Yugan *et al.*, 1991).

Fenestrate bryozoans are characteristic of the deeper water facies of Late Carboniferous–Early Permian temperate-, as opposed to tropical-, water build-ups, such as those of the north-western USA, Arctic Canada, the Barents Sea, and the Russian platform (Wahlman, in Kiessling *et al.*, 2002); for example, the latest Carboniferous, Gzhelian to earliest Permian, Asselian part of the Kozhim bank in the Timan–Pechora basin (Wahlman and Konovalova, in Zempolich and Cook, 2002).

Palaeobiogeography

Many bryozoans appear to have had restricted or endemic biogeographic distributions, rendering them of use in the characterisation of palaeobiogeographic provinces, and in turn in the constraint of plate tectonic reconstructions, in the Palaeozoic (Ross and Ross, in McKerrow and Scotese, 1990; Tuckey, in McKerrow and Scotese, 1990), and in the Cenozoic (Lagaaij and Cook, in Hallam, 1973).

Biostratigraphy

Bryozoans first appeared in the Ordovician; the stenolaemates in the Arenigian, the gymnolaemates in the Ashgillian (McKinney and Jackson, 1991; McKinney and Taylor, 2001). They diversified in the Ordovician (Taylor and Ernst, in Webby *et al.*, 2004), before sustaining significant losses in the End-Ordovician mass extinction. They recovered from this event, though, and diversified again, before sustaining severe losses in the End-Permian mass extinction. The stenolaemates may be said to have been in general decline ever since this event, and their only living representatives are the cyclostomates. The gymnolaemates, in contrast, recovered from the End-Permian mass extinction, and have been diversifying throughout the Mesozoic to Cenozoic. In fact, in detail, the diversity of the one constituent group, the ctenostomates, appears to have remained more or less constant through the Cretaceous to Recent, while that of the other, the cheilostomates, has expanded almost exponentially over that same time interval (Taylor, in Culver and Rawson, 2000). Cheilostomates are especially abundant in the shallow marine facies of the Late Cretaceous to Early Palaeocene Danskekalk or Danish Chalk of the Danish North Sea (author's unpublished observations). There are some 16 000 fossil and 5000 living species.

The moderate rate of evolutionary turnover exhibited by the bryozoans renders them of some use in biostratigraphy, at least in appropriate, aquatic, environments. Unfortunately, their usefulness is compromised by their restricted ecological distribution and facies dependence, a function of their benthic habit.

In my working experience, bryozoans have proved of some use in stratigraphic subdivision and correlation, and/or in palaeoenvironmental interpretation, in the following areas.

Palaeozoic

The Ordovician–Permian of the former Soviet Union. Here, bryozoans form part of the basis of the regional biozonation of the Carboniferous (Wagner *et al.*, 1996). Bryozoans are characteristic of the Ordovician–Permian peri-reefal and platform facies of the surface outcrops of the southern Urals, the Bolshoi Karatau Mountains of southern Kazakhstan, and elsewhere (Wagner *et al.*, 1996; Ahr *et al.*, 2000; Zempolich and Cook, 2002). These surface outcrops are important analogues for subsurface petroleum reservoirs in the Volga–Urals, peri-Caspian and north Caspian basins. Bryozoans form cool- or deep-water reefs in the Early Permian, Sakmarian–Artinskian of the Barents Sea that in turn form petroleum reservoirs (Bruce and Toomey, in Vorren *et al.*, 1993; author's unpublished observations).

Mesozoic

The Late Cretaceous of the Middle East. Here, seven encrusting species characterise the Campanikan–Maastrichtian Qahlah formation of Jebel Huwayyah in the Oman Mountains (Taylor, in Howarth, 1995).

Cenozoic

The Miocene of the Black Sea. Here, the recovery of *Schizoporella* from deep-water dredge samples from the Andrusov Ridge in the eastern Black Sea showed the sea-bed sediments there to be of probable Miocene age, and to have been deposited in water depths of no more than 200 m (Jones, 1996). In view of the present-day bathymetry of the Andrusov Ridge, approximately 2000 m, there exists the possibility that the dredged bryozoans might be allochthonous. Interestingly in this context, shallow marine bryozoan limestones of late Middle to early Late Miocene, Sarmatian–Maeotian age have been recorded slumped into coeval deep marine environments on the Taman peninsula, immediately east of where the Black Sea meets the Sea of Azov (Rostovtseva and Soloveva, 1999). There remains the possibility, though, that the dredged bryozoans might be autochthonous, and might relate to a time of sea level low-stand, such as the latest Miocene, Pontian (Messinian).

3.6.9 Molluscs: bivalves

Molluscs include a number of sub-groups that are important in the fields of palaeobiology and biostratigraphy, namely the bivalves, gastropods, ammonoids, belemnites and tentaculitids, which are discussed in turn below. Molluscs also include a number of sub-groups that are not so important in these fields, namely the heliconelloids, rostroconchs, polyplacophorans, tergomyans, scaphopods and aplacophorans, which are not discussed. There are somewhere between 50 000 and 150 000 living species.

Biology, morphology and classification

Bivalves, also known as lamellibranchs or pelecypods, are an extant, Cambrian–Recent group of molluscs characterised, as their name suggests, by a bivalved shell (Healy, in Anderson, 2001).

Biology

The biology and ecology of living bivalves is well known (Healy, in Anderson, 2001).

Most bivalves feed by passive filter-feeding, drawing in water and food particles suspended therein through an inhalant siphon, filtering out the food particles through modified gill filaments termed ctenidia, passing them on to the labial palps for sorting and transferring them to the mouth, and pumping out the waste water through an exhalant siphon. On the Crati submarine fan of the Ionian Sea, there is a significantly higher proportion of filter-feeding morphotypes in areas affected by turbidity currents than in unaffected areas (Colella and di Geronimo, 1987). Some bivalves feed by detritus-feeding, using ciliary action generated by tentaculate structures termed proboscides to pass particles encapsulated in mucus to the palps and to the mouth. A few are actively carnivorous, sucking in small prey animals through the inhalant siphon; and one, the ship-worm *Teredo*, feeds on wood pulp converted into sugar by the action of bacterial cellulase.

Significantly, a few bivalves feed by chemosynthesis or chemosymbiosis. For example, living methane-oxidising '*Bathymodiolus*-like' seep-mussels and the vesicomyid clam *Calyptogena* feed by chemosynthesis at the Bush Hill and other 'cold seep' sites in the Gulf of Mexico (Brooks *et al.*, 1989; MacDonald *et al.*, 1989; Callender *et al.*, 1990). Living *Thyasira sarsi* harbours sulphide-oxidising chemosymbionts among its gill filaments at a 'pockmark' site characterised by active methane seepage in Block 15/25 in the UK sector of the North Sea, and in the Skagerrak (Dando, in Austen *et al.*, 1991; Dando, in Bussmann *et al.*, 1994); and fossil *T. micheloti* from the Middle Miocene, Badenian Grund formation of Austria has also been interpreted as having harboured chemosymbionts under dysoxic conditions, and, incidentally, as having produced *Chondrites*-like traces (Pervesler and Zuschin, 2002). At least some modern lucinids also harbour sulphide-oxidising chemosymbionts (Taylor and Glover, in Harper *et al.*, 2000); and comparative morphology suggests that their ancient counterparts also did, as long ago as the Silurian (Reid and Brand, 1986). And ancient *Congeria* has also been interpreted as having harboured chemosymbionts, in the Late Miocene, Pannonian of the Pannonian basin in central Paratethys (Harzhauser and Mandic, 2004). It certainly appears to characterise times of dysoxic and sulphidic bottom and interstitial water, in contrast to the extreme *r*-strategists *Mytilopsis* and *Sinucongeria*, which characterise times of oxic water, and which produce 'boom' populations at these times (?in turn triggered by low-frequency orbital forcing).

A few bivalves, namely the giant Indo-Pacific clams of the family Tridacnidae, are mixotrophic rather than purely heterotrophic, and rely to an extent on the products of photosynthetic algal

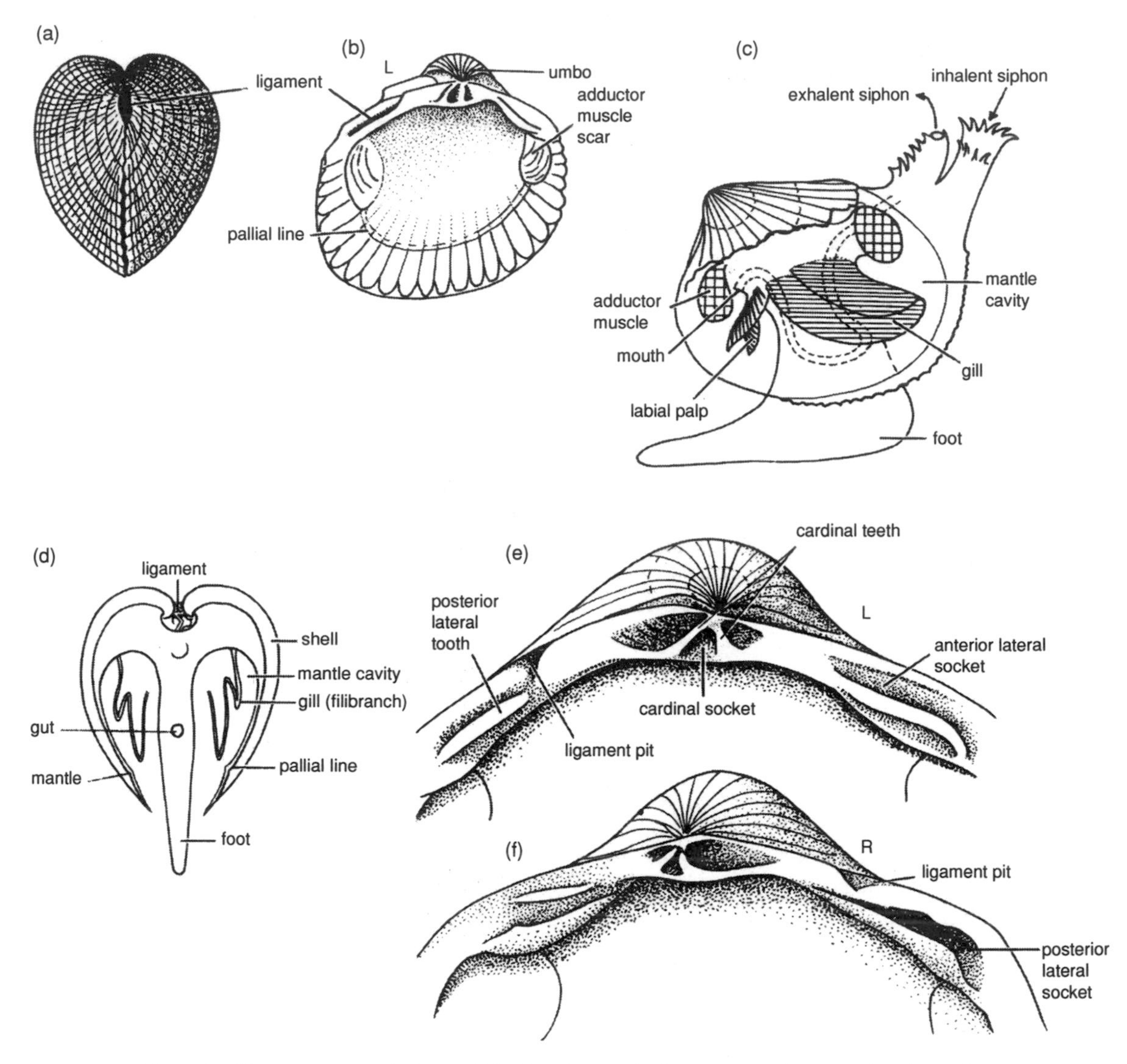

Fig. 3.41. Hard-part morphology of bivalves (from Clarkson, 1998). (a) Posterior view of shell of the edible cockle *Cerastoderma edule*, a living heterodont; (b) interior of left valve; (c) lateral view of living animal in life position; (d) vertical section of living animal; (e) cardinal region of left valve; (f) cardinal view of right valve. L, left; R, right.

symbionts housed in the mantle and siphonal tissues as sources of food (in this respect recalling the zooxanthellate corals). Ancient lithiotids have also been interpreted as having harboured photosymbionts (Fraser, 2001, 2003). This interpretation is essentially based on empirical observations of the arrangement of lithiotids in growth position in 'bouquets', maximising exposure not to current, as would be expected from filter-feeders, but to light; and of their rapid growth rates, calculated from sclerochronology.

In terms of ecology, bivalves are entirely aquatic in habit, and occur in freshwater and, more commonly, brackish and marine environments. Virtually all are

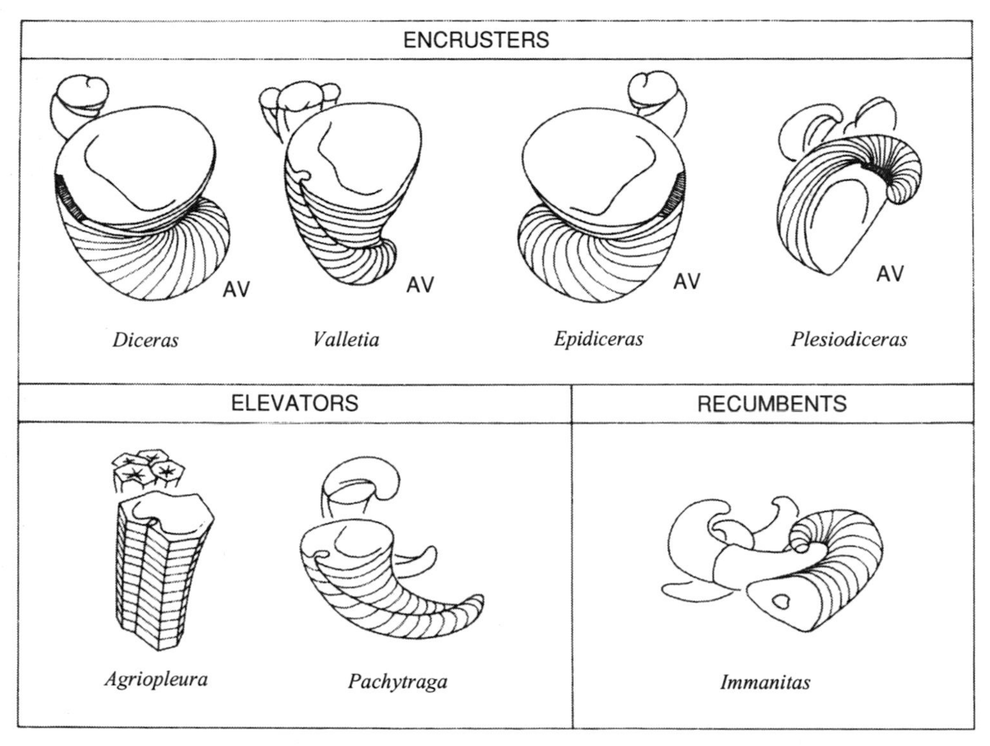

Fig. 3.42. Hard-part morphology of rudist bivalves. (Reproduced with permission from Doyle, 1996 *Understanding Fossils* . . . , © John Wiley & Sons Ltd.)

benthic, although some, such as scallops, are nektobenthic, that is, at least capable of free swimming. Many, if not most, are infaunal, excavating their burrows with a muscular so-called 'foot', normally housed within the shell, but protruded from the anterior margin as and when required. Some are epifaunal. Epifaunal forms can be either entirely free-living, although generally with little if any capacity for motility, or attachable to (and detachable from) the substrate by means of an organic thread or byssus, or permanently attached to the substrate by means of a secreted cement, and hence entirely sedentary or sessile. The diversity of living bivalves is highest in tropical and temperate latitudes, and tails off towards the poles (Crame, in Culver and Rawson, 2000; Crame, in Harper *et al.*, 2000; Jabklonski *et al.*, in Harper *et al.*, 2000).

Morphology

In terms of hard-part morphology, bivalves are characterised by secreted aragonitic or calcitic bivalved shells, held together at the dorsal margin by various types of hinge (Figs. 3.41–3.43). The shells serve to protect the 'soft parts' of the animals within (some shell layers are insoluble in acid, possibly as a defence against attack by predatory gastropods). The valves are opened and closed by various types of muscle, enabling the animals within to perform vital functions. The valves can be either more or less mirror images of one another, or equivalve; or not, or inequivalve. Bivalve shells can be ornamented with radially or concentrically arranged ribs or costae, or with spines for support or for protection against predators. They may also be marked by concentric growth lines.

The overall form of bivalve shells differs from sub-group to sub-group, reflecting both evolutionary inheritance and ecological adaptation.

Classification

The classification of bivalves is essentially based on hard-part morphology and on phylogeny inferred

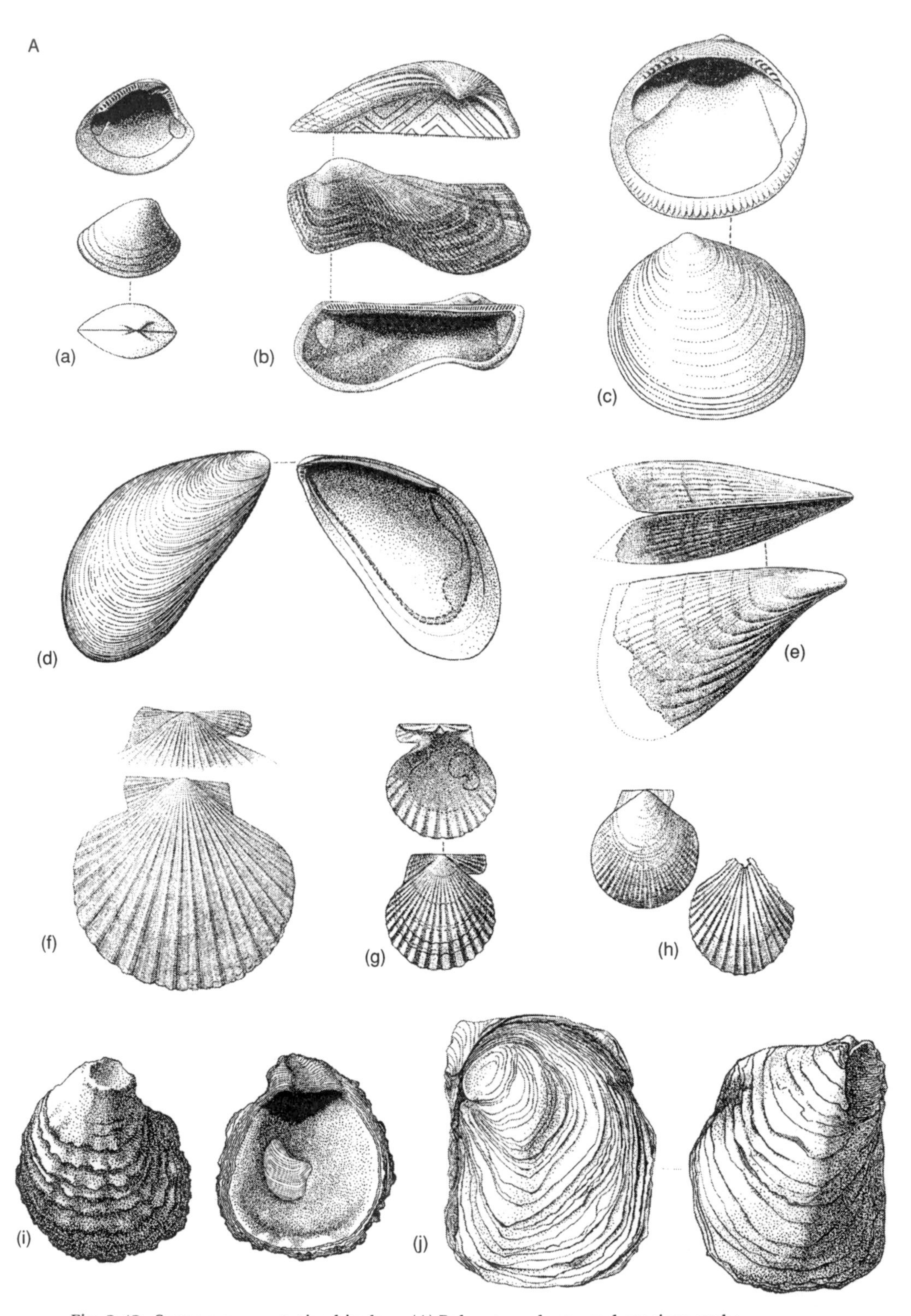

Fig. 3.43. Some representative bivalves. (A) Palaeotaxodonts and pteriomorphs. Palaeotaxodont: (a) *Nucula similis*, Middle Eocene, Barton beds, Barton, Hampshire (internal view, top); Alum Bay, Isle of Wight (external and dorsal views, bottom). ×1.5.

Caption for fig. 3.43 (*cont.*) Pteriomorphs: (b) *Arca biangula*, Middle Eocene, Bracklesham beds, Bracklesham Bay, Sussex, ×0.75; (c) *Glycymeris glycymeris*, Pliocene, Red Crag, Walton-on-Naze, Essex, ×0.75; (d) *Mytilus edulis*, Pliocene, Red Crag, Sutton, Suffolk, ×0.75; (e) *Pinna affinis*, Early Eocene, London clay, Fareham, Hampshire, ×0.5; (f) *Aequipecten opercularis*, Pliocene, Coralline Crag, Sutton, Suffolk, ×0.75; (g) *Chlamys recondita*, Middle Eocene, Barton beds, Barton, Hampshire, ×1.5; (h) *Pecten duplicatus*, Early Eocene, London clay, Haverstock Hill, London, ×1; (i) *Ostrea edulis*, Pliocene, Coralline Crag, Suffolk, ×0.5; (j) *Exogyra latissima*, Early Cretaceous, lower Greensand, Hythe, Kent, ×0.5. (B) Palaeoheterodonts and heterodonts. Palaeoheterodonts: (a) *Trigonia costata*, Middle Jurassic, Inferior Oolite, Dundry, Somerset, ×0.75; (b) *Myophorella (Trigonia) clavellata*, Late Jurassic, Oxfordian (Corallian) *Trigonia clavellata* beds, near Osmington, Dorset, ×0.75. Heterodonts: (c) *Venus casina*, Pliocene, Red Crag, Sutton, Suffolk, ×1; (d) *Lucina concinna*, Late Jurassic, Kimmeridge clay, Weymouth, Dorset, ×1; (e) *Cerastoderma edule*, Pliocene, Red Crag, Sutton, Suffolk, ×0.75; (f) *Mya truncata*, Pliocene, Coralline Crag, Ramsholt, Suffolk, ×0.75; (g) *Corbula regulbiensis*, Late Palaeocene, Thanet sands, Herne Bay and Richborough, Kent, ×2.

from molecular data (Campbell, in Harper *et al.*, 2000; Steiner and Hammer, in Harper *et al.*, 2000). The form of the gills is also an important criterion.

Six main sub-groups are recognised, namely: the palaeotaxodonts; the cryptodonts; the pteriomorphs; the palaeoheterodonts; the heterodonts; and the anomalodesmatans.

Palaeotaxodonts. The palaeotaxodonts are a primitive and ancient, yet extant, Ordovician–Recent, group characterised by so-called taxodont hinges and by equivalve shells; and also by so-called protobranch gills. Most are marine, benthic and infaunal. Palaeotaxodonts are exemplified by *Nucula*, which, incidentally, is characteristically deep marine.

Cryptodonts. The cryptodonts are an extant, Ordovician–Recent, group characterised by toothless or dysodont hinges. Most are marine, benthic and infaunal. Cryptodonts are exemplified by *Solemya*.

Pteriomorphs. The pteriomorphs are an extant, Ordovician–Recent, group characterised by hinges either with no teeth or with only one or two so-called cardinal teeth (and occasionally by further, smaller teeth situated anteriorly and/or posteriorly), by inequivalve shells; and by filibranch gills. They are brackish to marine, benthic and epifaunal, typically byssally attached or cemented to the substrate, and atypically motile rather than sessile benthic, or nektobenthic. Pteriomorphs are exemplified by the arcoids *Arca* (the ark shell), *Anadara* and *Glycymeris* (the dog cockle); the mytiloids *Mytilus* (the edible mussel) and *Pinna*; and the pterioids *Aequipecten* (the queen scallop), *Chlamys*, *Pecten* (the scallop), *Ostrea* (the oyster), *Exogyra* and *Inoceramus*.

Incidentally, *Aequipecten opercularis* has been reported serving as a mobile substrate for foraminiferans and other epibionts around the coasts of the British Isles, as in Cardigan Bay, Morecambe Bay and the Firth of Clyde, and off Plymouth in Devon and Falmouth in Cornwall. *Aequipecten opercularis* and other pectinids, and indeed ostreids, fulfil similar functions elsewhere in the world, as in the Gullmar Fjord in Sweden, in the Rio de Arosa in Spain, in the Osterschelde in Holland, and, further afield, in Newfoundland.

In parts of the rock record, ostreids or oysters are sufficiently abundant as to be rock-forming, as for example in the case of the Late Cretaceous, Maastrichtian '*Ostrea*' *lunata* bed of the chalk of Norfolk on the east coast of England. *Gryphaea* is also exceptionally abundant, as for example in the case of the Early Jurassic, lower Lias of Lyme Regis in Dorset on the south coast of England, characterised by *G. arcuata* or *incurva* (the 'devil's toenail'), through to the Late Jurassic, Oxfordian Oxford clay along the length of its outcrop from Weymouth in Dorset on the south coast to Scarborough in Yorkshire on the east coast, characterised by *G. dilatata*. Importantly, *Gryphaea* played a pivotal role in early evolutionary studies.

Palaeoheterodonts. The palaeoheterodonts are an extant, Ordovician–Recent, group characterised by heterodont to schizodont dentition. They are freshwater to marine. Palaeoheterodonts are exemplified by the heterodont modiomorphioid *Modiolopsis*, the freshwater unionoid *Unio*, and the schizodont trigonoids *Trigonia*, *Myophorella* and *Neotrigonia*.

Heterodonts. The heterodonts are an extant, Ordovician–Recent, group characterised by heterodont, degenerate desmodont or pachydont hinges, and by the development of elongate siphons. They are marine, benthic and either epifaunal, sessile and attached, or burrowing or infaunal, or boring or cryptic. Heterodonts are exemplified by the heterodont-hinged veneroids *Venus*, *Lucina*, *Cerastoderma* (the edible cockle), *Mactra* and *Tellina* (the tellen); by the desmodont myoids *Mya* (the sand-gaper), *Corbula*, *Pholas* and *Teredo* (the ship-worm); and by the pachydont hippuritoids or rudists *Hippurites*, *Diceras* and *Radiolites* (See Box 3.3 on pp. 131–33).

Anomalodesmatans. The anomalodesmatans are an extant, Ordovician–Recent group characterised by desmodont dentition, and by strongly modified shells. They, too, are marine, benthic and deeply infaunal. Anomalodesmatans are exemplified by the pholadomyoid *Pholadomya*. Oblique ribs in the pholadomyoids appear to represent a functional adaptation toward shell strengthening.

Palaeobiology

Fossil bivalves are interpreted, essentially on the bases of analogy with their living counterparts, and associated fossils and sedimentary facies, as having been freshwater, brackish or marine. Freshwater examples include *Unio* and allied forms from the Late Carboniferous 'Coal Measures' of Great Britain, and *Pseudunio* and allied forms from the Early Cretaceous 'Wealden' facies of the Weald and Wessex basins in the south-east of England. Marine examples associated with interpreted fossil seeps include lucinids from the Late Cretaceous, Campanian Pierre shale formation of Colorado in the USA and the Late Jurassic, Oxfordian Terres Noires formation of south-eastern France (Gaillard *et al.*, 1992), and *Cryptolucina* and allied forms from Early Cretaceous limestone mounds in the Wollaston Forland in north-east Greenland (Kelly *et al.*, in Harper *et al.*, 2000).

Fossil bivalves are further interpreted, essentially on the basis of functional morphology, as having been either benthic or nektobenthic; benthic forms as either free-living or motile, or byssally attached or cemented, and sessile. In detail, seven main life strategies are adopted by living or are inferred to have been adopted by ancient bivalves, namely: shallow infaunal benthic; deep infaunal benthic; byssally attached sessile epifaunal benthic; cemented sessile epifaunal benthic; motile epifaunal benthic; nektobenthic; and boring or cryptic (Fig. 3.44) (Stanley, 1970). Each life strategy is represented by a particular functional morphological adaptation.

Shallow infaunal. Proven and potential shallow infaunal morphotypes, exemplified by *Glycimeris*, are characterised by equivalve shells, and by equally sized adductor muscles. They are also often characterised by a robust construction, with strong hingement and ribbing or other ornamentation. The oblique ribs in the trigonoids of the Palaeozoic–Mesozoic, and, especially, in the nuculoids and veneroids of the Cenozoic, have been interpreted as representing an adaptation toward enhanced efficiency in burrowing (Checa and Jimenez-Jimenez, 2003).

Deep infaunal. Deep infaunal morphotypes, exemplified by the sand-gaper *Mya*, are characterised by elongated, streamlined and smooth equivalve shells. The burrowing foot may be sufficiently large as to require to be accommodated by a gape in the anterior margin of the shell; the feeding siphons sufficiently large as to require to be accommodated by a gape in the posterior margin, and a corresponding indentation or pallial sinus in the pallial line in the interior of the shell.

Byssally attached epifaunal. Byssally attached epifaunal morphotypes, exemplified by mussels such as *Mytilus*, are characterised by wedge-shaped shells with reduced anteriors and anterior adductor muscles. They are also occasionally characterised by a so-called byssal gape.

Cemented epifaunal. Cemented epifaunal morphotypes, exemplified by the oyster *Ostrea*, are characterised by markedly inequivalve shells, by a flattened to irregular attachment surface, occasionally, in the case of so-called xenomorphic forms, reflecting that of the substrate. They are also characterised by a single large adductor muscle.

Motile epifaunal. Free-living epifaunal morphotypes, exemplified by the extinct *Gryphaea*, are also characterised by markedly inequivalve shells, specifically by a relatively large and convex early valve resting on the – soft – substrate (sometimes with spines attached to it to stop it sinking), and a smaller, lid-like upper valve.

Nektobenthic. Nektobenthic morphotypes, exemplified by scallops such as *Pecten*, are characterised by relatively large, inequivalve shells, and by ear-like extensions of the hinge-line. The commonly observed corrugations on the surface of the valve may serve to reduce turbulent flow.

Boring or cryptic. Boring or cryptic morphotypes, exemplified by the wood-boring ship-worm *Teredo*, and by the rock-boring *Lithophaga* and *Pholas*, are characterised by elongated shells with rasping ornamentation. Importantly, traces left by rock-borers provide evidence of hardgrounds.

The pronounced shell ornamentation observed across functional morphological groupings may

Infaunal shallow burrowers

Glycimeris

Equivalved, adductor muscles of equal sizes and commonly with strong external ornament.

Infaunal deep burrowers

Mya

Elongated valves, often lacking teeth and with permanent gape and a marked pallial sinus.

Epifaunal with byssus

Mytilus

Elongate valves with flat ventral surface and reduction of both the anterior part of the valve and the anterior muscle scar. Attached by thread-like byssus.

Epifaunal with cementation

Ostrea

Markedly differently shaped valves, sometimes with crenulated commissures; large single adductor muscle.

Unattached recumbents

Gryphaea

Markedly differently shaped valves sometimes with spines for anchorage or to prevent submergence in soft sediment.

Swimmers

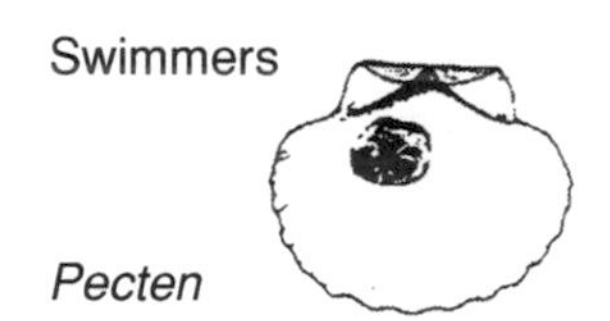

Pecten

Valves dissimilar in shape and size with very large, single adductor muscle and commonly with hinge line extended as ears.

Borers and cavity dwellers

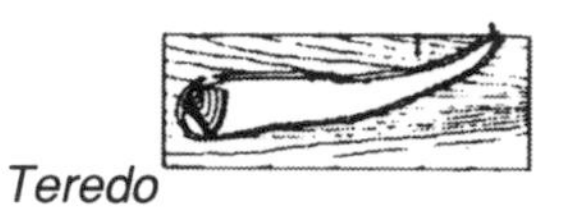

Teredo

Elongate, cylindrical shells with strong, sharp external ornament; cavity dwellers commonly grow in dimly lit conditions following the contours of the cavity.

Fig. 3.44. Functional morphology of bivalves. (Reproduced with permission from Benton Harper, 1997 *Basic Palaeontology*, © Longman/Pearson Education Ltd.)

represent a common adaptation to predator deterrence (Stone, 1998).

Palaeobathymetry

Fossil bivalves can be used as indicators of palaeobathymetry, or proximity to shoreline, in marine environments (McAlester and Rhoads, in Hallam, 1967). It is important to note, though, that certain living bivalves appear to have their distributions controlled less by bathymetry than by other factors, such as temperature. Thus, *Venericardia borealis* occurs in shallow water in high latitudes and in deep water in early latitudes (in the north-west Atlantic), indicating that it is a stenothermal, eurybathyal species.

Palaeobiogeography

Many fossil bivalves appear to have had restricted or endemic biogeographic distributions, rendering them of use in the characterisation of palaeobiogeographic provinces, and in turn in the constraint of plate tectonic reconstructions, in the Palaeozoic, in the Mesozoic and in the Cenozoic.

In the Palaeozoic, specifically in the Early Ordovician, bivalves appear to have been endemic to Gondwana (Cope, in Crame and Owen, 2002). In the Middle Ordovician, bivalves as a whole were more widespread, although sub-groups continued to exhibit endemic or provincial distributions, for example, pteriomorphs and nuculoids in low latitudes, and heteroconchs to mid to high latitudes. In the Late Ordovician, bivalves as a whole exhibited cosmopolitan distributions, encompassing for the first time the carbonate shelf facies of Laurentia and Baltica, although the pteriomorphs continued to remain restricted to low latitudes.

In the Mesozoic, bivalve distributions can be used to characterise Tethyan and Boreal realms, Boreal–Atlantic, Boreal–Pacific, and Arctic provinces within the Boreal realm, and Chukotka–Canada and Greenland–North Siberia sub-provinces within the Arctic province (author's unpublished observations). The Early Triassic bivalve *Claraia* and the Middle–Late Triassic *Monotis*, which latter, incidentally, may have been pseudoplanktonic or nektonic in habit, are widespread in the Tethyan realm, from the Mediterranean to the Indo-Pacific (Kummel, in Hallam, 1973; Westermann, in Hallam, 1973). The large, aberrant Early Jurassic, Pliensbachian–Toarcian bivalve *Lithiotis* and the allied forms *Cochlearites, Lithioperna, Gervilleioperna* and *Mytiloperna* exhibited essentially pan-Tethyan distributions, from Oregon and California states on the west coast of North America in the west, through southern Europe, north Africa and the Middle East, to East Timor in the east (Fraser, 2000, 2001, 2003; Fraser and Bottjer, 2001).

In the Cenozoic, bivalve distributions indicate the development of three palaeobiogeographic provinces in the north-east Pacific, namely the Gulf of Alaska province to the north, the Pacific North-West province centred on British Columbia and Vancouver in Canada and Washington State and Oregon in the conterminous USA, and the California province to the south (Addicott, in Armentrout, 1981). The dispersal of the bivalve *Astarte* from the Arctic into the North Pacific between 4.8 and 5.5 Ma implies that the Bering Strait had opened by this time (Marincovich and Gladenkov, 1999; Gladenkov *et al.*, 2002).

Palaeoclimatology

Also in the Cenozoic, bivalve distributions point toward the palaeogeographic and palaeoclimate evolution of the Tethyan and Paratethyan realms (Piccoli *et al.*, 1991; Popov, 1995; Harzhauser *et al.*, 2003; see also Sub-section 5.5.8).

Biostratigraphy

Bivalves evolved in the Cambrian (Taylor, 1996; Kouchinsky, in Zhuravlev and Riding, 2001). They underwent significant diversification in the Ordovician, by the end of which all of the known sub-groups and life strategies had become established (Cope, in Crame and Owen, 2002; Cope, in Webby *et al.*, 2004). Indeed, they continued to diversify until the end of the Permian, when they sustained losses in the End-Permian mass extinction, although not to the same extent as the brachiopods. They recovered from this event, and radiated, to take over from the brachiopods as the dominant filter-feeding group in the marine environment, in the so-called 'Mesozoic Revolution' (Harper and Skelton, 1993). The group as a whole may be regarded as having been continuously diversifying through the Cenozoic, although locally, as in the US Atlantic and Gulf coastal plains, sustaining severe losses over the Eocene/Oligocene transition, characterised by falling sea level and cooling climate (Campbell and Campbell, in Prothero *et al.*, 2003; Dockery and Lozouet, in Prothero *et al.*, 2003). The evident shift toward efficiency in burrowing through the Cenozoic has been interpreted as an evolutionary response to an increasing diversity of durophagous predators or an accelerating rate of sediment reworking (Checa and Jimenez-Jimenez, 2003).

The moderate rate of evolutionary turnover exhibited by the bivalves renders them of some use in biostratigraphy, at least in appropriate environments (Fig. 3.45). Unfortunately, their usefulness is somewhat compromised by their facies

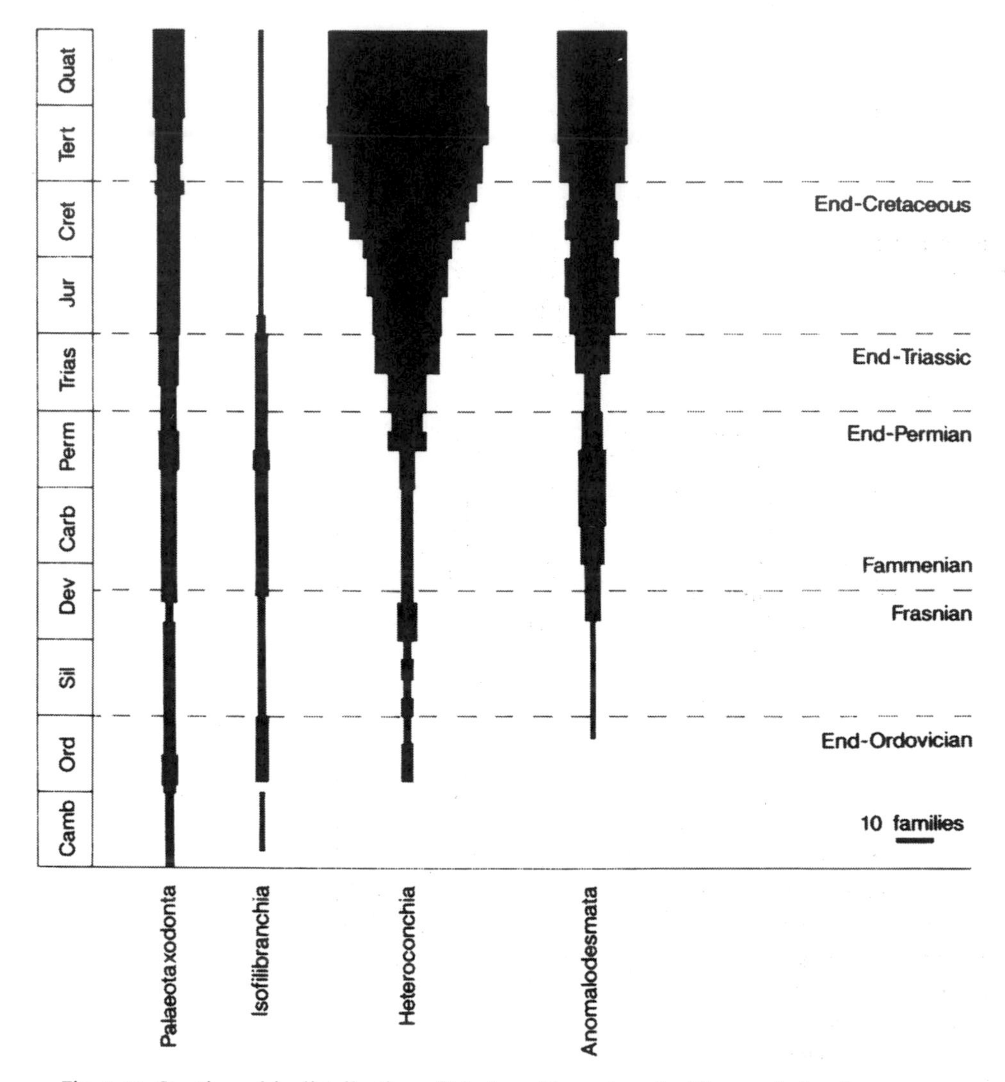

Fig. 3.45. Stratigraphic distribution of bivalves. (Reproduced with permission from Doyle, 1996 *Understanding Fossils*, © John Wiley & Sons Ltd.)

dependence. Nonetheless, they are of local use, especially in the absence of more suitable stratigraphic index fossils.

In the Palaeozoic, non-marine bivalves form the basis of biozonation of the Carboniferous 'Coal Measures' of Great Britain (Trueman, 1933; Ramsbottom *et al.*, 1978; see also Section 7.3).

In the Mesozoic, *Inoceramus labiatus* and *'Ostrea' lunata* form part of the basis of biozonation of the Late Cretaceous Chalk of Great Britain. Incidentally, inoceramid debris is locally sufficiently abundant in the Late Cretaceous of the adjoining North Sea basin as to be rock-forming (and even, in the case of the Turonian of Bruce field, to contribute to the formation of a petroleum reservoir). Inoceramids and other bivalves form part of the basis of biozonation of the Late Cretaceous of the western interior seaway of the USA (Eaton, in Nations and Eaton, 1991). Species of *Lopha, Pycnodonte, Exogyra* and *Ostrea* form part of the basis of the biozonation of the Late Cretaceous Austin

Chalk of Texas. I have on my desk before me as I write a specimen of the splendidly named *Exogyra ponderosa* from the Austin Chalk.

In the Cenozoic, bivalves are useful in the stratigraphic subdivision and correlation, and in the palaeoenvironmental interpretation, of the Eocene London clay and associated deposits, and Bracklesham and Barton beds. They are especially useful in the stratigraphic subdivision and correlation, and in the palaeoenvironmental interpretation, of the essentially Pleistocene 'Crags' and raised-beach deposits of the coasts of the British Isles, and indeed of the Pleistocene throughout north-west Europe. Interestingly, the classical subdivisions of the Cenozoic (Palaeocene, Eocene, Oligocene, Miocene, Pliocene, Pleistocene, Holocene) are based on Lyell's estimates of the proportions of still-extant bivalves in successive stages in the marginal basins of north-west Europe.

In my working experience, bivalves have proved of particular use in the following areas.

Palaeozoic

The Carboniferous of the former Soviet Union. Here, non-marine bivalves form the basis of a regional biozonation (Wagner *et al.*, 1996).

Mesozoic

The Triassic–Cretaceous of the Middle East. Here, the Early Triassic, Scythian Tr30 MFS of Sharland *et al.* (2001), stratotypified in the Mirga Mir formation of Iraq, is marked by *Halobia*. The Middle Triassic, Ladinian Tr50 MFS, stratotypified in the Late Geli Khana formation of Iraq, is marked by *Daonella*. The Late Triassic, Carnian Tr60 MFS, stratotypified in the Early Kurra Chine formation of Iraq, is marked by pelagic bivalves. The Early Jurassic part of the Musandam limestone in the Oman Mountains is marked by the occurrence of *Lithiotis* (Hudson and Chatton, 1959). The Middle Jurassic part of the Musandam limestone is marked by *Eligmus* (Hudson and Chatton, 1959); the Middle–Late Jurassic Sargelu and Naokelekan formations of the Gotnia Basin of south-west Iran and adjoining Iraq and Kuwait by the pelagic '*Posidonia*' or *Bositra* (author's unpublished observations). The Late Jurassic part of the Musandam limestone of the Oman Mountains is marked by the 'proto-rudist' *Diceras (Eodiceras)* (Hudson and Chatton, 1959). The Late Cretaceous, Campanian–Maastrichtian Qahlah and Simsima formations of the Oman Mountains are marked by inoceramids (Morris, in Howarth, 1995). In the field in the Zagros Mountains of south-west Iran, the Early Triassic Kangan formation is marked by *Claraia aurita* and *C. clarai elegans* (Johnson, in Neale and Brasier, 1981). The Early–Middle Triassic Khaneh Kat formation is marked by *Halobia parthanensis* and *Pseudomonotis ovata* (author's unpublished observations). The Early Jurassic part of the Surmeh formation is marked, like the Early Jurassic part of the Musandam limestone of the Oman Mountains, by *Lithiotis*; the Late Jurassic part, like the Late Jurassic part of the Musandam limestone, by *Eodiceras* (author's and other unpublished observations, based in part on identifications by Noel Morris of the Natural History Museum, London). The Kazhdumi formation is marked by the 'Middle' Cretaceous, Albian–Cenomanian *Ceratostraeon flabellatum (Exogyra flabellata)*, *Plagiostoma hoperi*, *Neitheia dutrigei*, *Corbiculopsis*, *Pterotrigonia (Scabrotrigonia)*, ? Lucinidae, ?*Sphaera*, Carditidae, ?Arcticidae, Cardiidae, Inoceramidae, Tellinacea and the *Eligmus*-like genus *Naiadina* (Collignon, 1981; author's unpublished observations, based on identifications by Noel Morris of the Natural History Museum, London). The '*Exogyra* marl' or Ahmadi member of the Sarvak formation is marked by the 'Middle' Cretaceous, Cenomanian–Turonian *Exogyra conica* and also by *Inoceramus labiatus*; the *Lopha* limestone member of the Gurpi formation by the Late Cretaceous, Campanian–Maastrichtian *Lopha dichotoma* (unpublished observations).

Cenozoic

The Cenozoic of Sakhalin. Bivalves, especially species of *Yoldia*, have proved of particular use in the stratigraphic subdivision and correlation of the Cenozoic of Sakhalin, and also of Kamchatka, Chukotka and Japan (Gladenkov, in Herman, 1974; Savitskii *et al.*, 1979; Gladenkov, 1980; Gladenkov, in Ikebe and Tsuchi, 1984; Shainyan *et al.*, 1990; Gladenkov, 1999; Gladenkov *et al.*, 2000; Charnock and Salnikov, in Simmons, in press). Incidentally, bivalves have also proved of use in the stratigraphic subdivision and correlation of the Cenozoic of the western coast of North America on the opposite side of the North Pacific, as, for example, in California, Oregon and Washington states and the Gulf of Alaska (Addicott, in Armentrout, 1981; Hickman, in Prothero *et al.*, 2003; Nesbitt, in Prothero *et al.*, 2003; Prothero, in Prothero *et al.*, 2003; Squires, in Prothero *et al.*, 2003).

The Oligocene–Pleistocene of Paratethys. Here, bivalves have proved of particular use in stratigraphic subdivision and correlation, and/or in palaeoenvironmental interpretation, in freshwater, brackish and marine environments (Zhizhchenko, 1959; Laliyev, 1964; Steininger *et al.*, 1976; Paramonova *et al.*, 1979; Semenenko, 1979; Rogl and Steininger, in Brenchley, 1984; Baldi, 1986; Zubakov and Borzenkova, 1990; Rogl, 1998; Rogl, in Agusti *et al.*, 1999; Vrsaljko, 1999; Mandic and Steininger, 2003). For example, it has been possible to demonstrate through the application of a high-resolution biostratigraphic zonation based on bivalves (Paramonova *et al.*, 1979) that the Sarmatian succession drilled in wells in the Rioni basin in Georgia on the east coast of the Black Sea is stratigraphically rather than tectonically thickened, thus contributing significantly to the structural model of the basin. It has also been possible through the use of a combination of biostratigraphic and palaeobiological data and interpretations to produce a series of detailed maps of the palaeogeography and, in particular, of the dispositions of freshwater, brackish and marine facies, of the Black Sea and Caspian, by time-slice. This has in turn proved of use in reconstructing the uplift history of the Caucasus, and implications for the provenance, and thus the reservoir quality, of the sediments in the Black Sea and Caspian.

The Miocene of the Middle East. The Terbol formation of the Aleppo area in north-western Syria is marked by the Early–Middle Miocene *Chlamys albina, C. submalvinae* and *Flabellipecten larteti* (Jones, in Agusti *et al.*, 1999). In the field in the Zagros Mountains of south-west Iran, the upper Asmari and Gachsaran formations are marked by the Early Miocene *Ostrea latimarginata*; the Mishan formation by the Middle Miocene *Antigona granosa, Chlamys pusio, C. senatoria, Lithophaga lithophaga, Ostrea digitata* and *O. virleti* (unpublished observations).

The Miocene–Pleistocene of the Caribbean. The bivalve *Anadara (Grandiarca) patricia* occurs in the ?Caparo clay member of the 'post-Springvale' or Talparo formation of the Freeport area in the western part of the northern basin of Trinidad (Maury, 1925), and in the upper part of the Morne l'Enfer formation of the Brighton area of the southern basin (Saunders and Kennedy, in Saunders, 1968), and enables their correlation. The upper Morne l'Enfer formation is characterised palynologically by a distinct acme of *Grimsdalea magnaclavata*, which has been dated as Late Pliocene, 2.1–2.5 Ma (Pocknall *et al.*, in Goodman, 1999). The abundance of *G. magnaclavata*, which is a type of coastal palm pollen, has been correlated with the warm climate obtaining at that time (Pocknall *et al.*, in Goodman, 1999). The upper part of the Talparo formation is characterised palynologically by the alder pollen *Alnipollenites verus*, which has been interpreted as Pleistocene (Pocknall *et al.*, in Goodman, 1999). The appearance of *A. verus* in the eastern Venezuelan basin has been correlated with the time of the elevation of the basin catchment area to a sufficient height to support Andean montane vegetation, or to the cool climate obtaining at that time (Pocknall *et al.*, in Goodman, 1999). On the basis of this independent palynological evidence, the Trinidadian range of *Anadara (Grandiarca) patricia* would appear to be Late Pliocene to Pleistocene, rather than Late Miocene, as was interpreted by Rutsch (1942). The Caribbean range would appear to be Late Miocene to Pleistocene, based on additional records from elsewhere. The species occurs, locally in rock-forming abundance, in the Cercado formation of the Rio Cano area of the northern Dominican Republic, which is demonstrably of Late Miocene, global standard calcareous nannoplankton zone NN11 age (Saunders *et al.*, 1986). It also occurs in the early part of the Cumana formation of Anzoategui in north-eastern Venezuela, which is of Pleistocene, *Globorotalia truncatulinoides* Zone, age (Hunter, 1978).

Incidentally, another species of *Anadara (Grandiarca), A. (G.) waringi* occurs around Point Paloma on the Manzanilla coast in the eastern part of the northern basin of Trinidad, in what was originally interpreted as 'a lignitic eastern phase . . . equivalent . . . to the western Springvale' (Maury, 1925). Recent remapping indicates that the Springvale formation is reduced or absent in this area, however, and that the beds with *waringi* should be reinterpreted as representing the Telemaque member of the Manzanilla formation, and also, possibly, in the case of those cropping out near the mouth of the Oropouche River, the ?Comparo Road Beds member of the Talparo formation (unpublished observations).

Anadara (Grandiarca) is useful for palaeobiological as well as biostratigraphic interpretation. In the Dominican Republic, *A. (G.) patricia* occurs in association with a *Larkinia–Mytilus–Melongena* assemblage, interpreted as indicative of brackish water conditions (Saunders *et al.*, 1982, 1986). In Trinidad, it occurs in association with the extinct benthic foraminifer *Miliammina telemaqensis* (Saunders and Kennedy, in Saunders, 1968), the extant counterparts of which live in mangrove swamps (Saunders, 1958); and also with interpreted delta-top lignites, or naturally sintered lignites distinguished by their brick-red coloration and known locally as porcelanites (author's unpublished observations). Indeed, a still more specific habitat for fossil *Anadara (Grandiarca)* is suggested on the basis of analogy with the evidently closely related Recent species *A. (G.) grandis*, namely an intertidal flat in a mangrove swamp (Maury, 1922; Olsson, 1932, 1961; J. Todd, the Natural History Museum, personal communication). Interestingly, *A. (G.) grandis* is restricted to the Pacific. Note in this context that the Caribbean and Pacific were connected through the Strait of Panama until as recently as the middle part of the Pliocene, when the connection was broken by the emergence of the Isthmus of Panama.

Box 3.3 Rudists

Biology, morphology and classification

Rudists are an important, extinct, Late Jurassic–Cretaceous, sub-group of heterodonts, characterised the attachment of one valve and the uncoiling of the other, and by accretion along the entire mantle margin thereof, and hence the construction of aberrant, essentially tubular, shells, often of large size, with lid-like upper valves (Fig. 3.42) (Anonymous, 1991; Masse and Skelton, 1998; Hofling and Steuber, 1999). Requienid rudists were attached by the left valve; caprotinids, caprinids, radiolitids and hippuritids by the right valve.

Palaeobiology

Palaeobathymetry

Rudists are interpreted, essentially on the bases of associated fossils and sedimentary facies, functional morphology, and analogy with the living 'fan-mussel' *Pinna* of the Great Pearl Bank Barrier in the Middle East Gulf, as having been shallow marine, benthic, and epifaunal, either embedded in or reclining or recumbent on the sediment surface, clinging to or encrusting it, or elevated above it (Skelton and Gili, in Anonymous, 1991; Ross and Skelton, 1993; Skelton *et al.*, 1995; Hughes; 1997; Skelton, 2003). It has been hypothesised that they harboured algal symbionts (Kauffman and Sohl, 1974; Vogel, 1975). Worked examples of rudist-based palaeobathymetric interpretation and palaeocommunity reconstruction in the peri-tidal carbonate environments of the Urgonian (late Barremian–early Aptian) of Provence in south-east France are provided by Masse *et al.* (2003) and Fenerci-Masse *et al.* (2004).

Rudists were locally sufficiently abundant in shallow marine environments as to have constructed so-called 'reefs', with elevator morphotypes providing the framework, and encrusters and recliners the binding, as in the circum-Mediterranean and Middle East, and also the southern USA, Central and northern South America, and the Gulf of Mexico and Caribbean at the western end of Tethys (Fig. 3.46). Note, though, that it is questionable whether these constructions actually represent reefs in the conventionally accepted sense, as they are essentially sediment-supported rather than bound, and constratal or low-relief rather than superstratal or high-relief, and probably did not form wave-resistant frameworks (Gili *et al.*, 1995): it is perhaps more appropriate to visualise them as 'meadows' (Skelton, 2003). Nonetheless, it is unquestionable that the volume of carbonate produced by the rudist

Fig. 3.46. Rudist reef, Late Cretaceous, Saiwan, Oman Mountains.

'factory', if I might mix industrial and agricultural metaphors, was significant (Steuber, in Insalaco *et al.*, 2000).

Rudist 'reefs' are of considerable commercial importance as petroleum reservoirs in the Cretaceous of the Middle East (Frost *et al.*, in Harris, 1983; Alsharhan and Nairn, in Simo *et al.*, 1993; Burchette, in Simo *et al.*, 1993; Alsharhan, 1995; Aqrawi *et al.*, 1998; Sadooni and Alsharhan, 2003; see also Chapter 7): for example in the Neocomian Ratawi formation of the Wafra field in the partitioned neutral zone between Kuwait and Saudi Arabia (Longacre and Ginger, in Lomando and Harris, 1988); the Aptian Shuaiba formation of Bu Hasa field in Abu Dhabi in the United Arab Emirates (Alsharhan, 1987), and Shaybah field in the Rub al-Khali on the border between Saudi Arabia and the United Arab Emirates (Skelton *et al.*, in Hofling and Steuber, 1999; Hughes, 2000); the Albian Mauddud formation of the Raudhatain and Sabiriyah fields in north Kuwait (Fig. 3.47) (Jones *et al.*, in Bubik and Kaminski, 2004); the Cenomanian Mishrif formation of the Fateh field in the Middle East Gulf off the coast of Dubai in the United Arab Emirates (Jordan *et al.*, in Roehl and Choquette, 1985; Videtich *et al.*, 1988); and the Albian Mauddud and Cenomanian Mishrif formations of Fahud field in north-western Oman

Fig. 3.47. Photomicrograph of rudist debris, Mauddud formation reservoir, north Kuwait.

(Harris and Frost, 1984). Importantly, a not insignificant proportion of the porosity, or storage capacity, in these and other reservoirs is in rudist or rudist debris facies: radiolitid debris is particularly porous, although it only occurs in reservoirs of Aptian and later age, as the radiolitids had not evolved earlier. Interestingly, rudists are recognisable in cores, in computerised tomography (CT) scans of cores, and on image logs, from some reservoirs (Hughes *et al.*, 2003; author's unpublished observations).

Rudist 'reefs' are also of considerable importance as hosts to mineral deposits, for example in the Early Cretaceous, Urgonian (Barremian–Aptian) of the La Troya lead–zinc mine in the Basco-Cantabrian basin in the Basque country of Spain (Lunar *et al.*, in Gibbons and Moreno, 2002; see also Chapter 7).

Palaeobiogeography

The rudists exhibited essentially pan-Tethyan distributions, and indeed their occurrence serves to distinguish the Tethyan from the Boreal realm (Kauffman, in Hallam, 1973; Kauffman and Sohl, 1974; Masse, 1985; Philip, 1985; Scott, 1990; Ozer, in Simo *et al.*, 1993; Scott, in Simo *et al.*, 1993; Yurewicz *et al.*, in Simo *et al.*, 1993; Masse and Gallo Maresca, 1997; Steuber and Loser, 2000; Masse *et al.*, 2002). Some sub-groups exhibited much more restricted distributions, related to continental plate reconfigurations, such that their occurrence serves to distinguish, for example, Afro-Arabian and American–Caribbean provinces within the Tethyan realm (Kauffman and Sohl, 1974). Interestingly, rudists from Peru exhibit affinities with those of the Caribbean, the two areas having been contiguous prior to the closure of the Strait of Panama (Philip and Jaillard, 2004).

Incidentally, the most northerly occurrences of a rudist known to the author is that of *Durania borealis* in the Late Cretaceous, Cenomanian part of the Hibernian greensand of Portmuck, near Islandmagee in Co. Antrim in Northern Ireland, some specimens of which are housed in the Ulster Museum in Belfast. This species is colloquially known as the 'cornetto animal' by the sweet-toothed Northern Irish, of whom my wife is one, in reference to its resemblance to the ice-cream cone of that name!

Biostratigraphy

Rudists evolved, probably from a diceratid rootstock characterised by incipient uncoiling, in the Late Jurassic, and became extinct by the end of the Cretaceous. Their extinction appears to have been incremental rather than instantaneous. The caprinids appear to have became extinct at the Cenomanian/Turonian boundary (see also Sub-section 5.4.4).

The moderate rate of evolutionary turnover exhibited by the rudists renders them of some use in biostratigraphy. Their usefulness is enhanced by their having had their stratigraphic ranges calibrated against either ammonite or strontium isotope standards (Philip, 1998; Steuber, 2003).

In my working experience, rudists have proved of particular use in the following areas.

The Cretaceous of the Middle East. Here, the Aptian Shuaiba formation, the Albian Nahr Umr formation, the Cenomanian Natih formation and the Campanian–Maastrichtian Qahlah and Simsima formations of the Oman Mountains are all marked by rudists (Simmons and Hart, in Hart, 1987; Skelton *et al.*, in Robertson *et al.*, 1990; Morris and Skelton, in Howarth, 1995; van Buchem *et al.*, 1996; Borgomano *et al.*, 2002; Pittet *et al.*, 2002; van Buchem *et al.*, 2002a, b). The stratigraphic usefulness of the rudists in the Cretaceous of the Middle East is enhanced by their having had their stratigraphic ranges calibrated through graphic correlation against composite standards (Scott, in Robertson *et al.*, 1990).

Importantly, rudists are of use in the characterisation of the Early Cretaceous Shuaiba formation reservoir of Shaybah field in Saudi Arabia (Hughes, 2000). Here, rudists are recognisable in cores, in CT scans of cores and on image logs (Hughes *et al.*, 2003).

The Late Cretaceous of north Africa. Here, the Turonian of Algeria is marked by the rudist *Praeradiolites* (Majoran, 1987).

3.6.10 Molluscs: gastropods

Biology, morphology and classification

Gastropods are a diverse, extant, Cambrian–Recent, group of molluscs characterised by an asymmetrical univalved shell (Healy, in Anderson, 2001).

Biology

The biology and ecology of most living gastropods is well known. The gastropod, or 'stomach-foot', animal has a large muscular 'foot', with the aid of contractions of which it is able to achieve at least a modicum of motility. Other soft parts include a head with sensory tentacles, eyes and a feeding apparatus; a digestive system; a functionally advanced respiratory system comprising gills or lungs; and a sexual or asexual reproductive system. Development may embody a planktonic larval stage, enabling dispersal and the colonisation of new niches.

In terms of ecology, gastropods live in terrestrial and aquatic, freshwater, brackish and marine, environments: some terrestrial forms aestivate in the summer months, to prevent desiccation. Almost all are benthic; most epifaunal, and either motile or sessile; some infaunal. Most terrestrial gastropods feed by browsing on vegetation. Some marine forms adopt a similar strategy, and feed by grazing on algae or by scavenging for detritus, often by means of a retractable mucus string. Some others are active predators, capable of boring through the shells of their prey with rasping teeth and acidic secretions; and, remarkably, cone-shells and allied forms are actually able to hunt by injecting prey with poison darts modified from marginal teeth. A few gastropods are filter-feeders, for example, the turritellids; a few others, parasitic.

Morphology

In terms of hard-part morphology, gastropods are characterised by layered, commonly attractively ornamented and coloured, calcareous, shells, typically coiled either in a plane spire or in a sinistral or dextral conical spire or conispire or turreted spire or trochospire, around a central rod or columella or a central cavity or umbilicus (Fig. 3.48). They are further characterised by an opening or aperture at the distal end, capable of being closed by a lid or operculum for protective purposes. The gastropod animal lives in a body or mantle cavity within the shell, which, in contrast to that of the ammonoid or coleoid cephalopod, the ammonite or belemnite (see below), is not internally partitioned.

Classification

Classification of gastropods is based on morphology, especially that of the feeding apparatus or radula and of the gills or ctenidia, and on phylogeny. Three main sub-groups are recognised by palaeontologists, namely the prosobranchs, the opisthobranchs and the pulmonates (a further sub-group is recognised by molecular biologists, namely the prosobranch-like heterobranchs).

Prosobranchs. Prosobranchs are an extant, Cambrian–Recent, group characterised by the primitive condition of torsion, with posteriorly positioned heads, and by simple gills; and are typically aquatic, and atypically terrestrial. Two sub-groups of prosobranchs are recognised, namely the basal prosobranchs or archaeogastropods, and the advanced prosobranchs or coenogastropods. In turn, three further sub-groups of archaeogastropods are recognised by molecular biologists and increasingly by palaeontologists, namely the patellogastropods, with docoglossate radulae, and the vetigastropods and the neritimorphs, with rhipidoglossate radulae. Two further sub-groups of coenogastropods are recognised by palaeontologists, namely the mesogastropods and neogastropods (three by molecular biologists, namely the architaenioglossans and the neotaenioglossans, with taenioglossate radulae, and the neogastropods, with stenoglossate or toxoglossate radulae). Archaeogastropods are exemplified by *Bellerophon*, *Platyceras* and the familiar limpet *Patella*; mesogastropods by *Turritella* (the 'auger shell'), *Nerinea* and the cowrie *Cypraea*; neogastropods by the cone-shell *Conus*.

Most prosobranchs graze or, by inference, grazed on algae, or scavenge or scavenged for detritus, although some, mesogastropods and, notably, neogastropods, are or were active carnivores.

Some Palaeozoic platycerid archaeogastropods have been interpreted as having been harboured by crinoids and even to have fed on their faeces (Bowsher, 1955). Note, though, that this has been disputed, most recently by Lindstrom and Peel (2003),

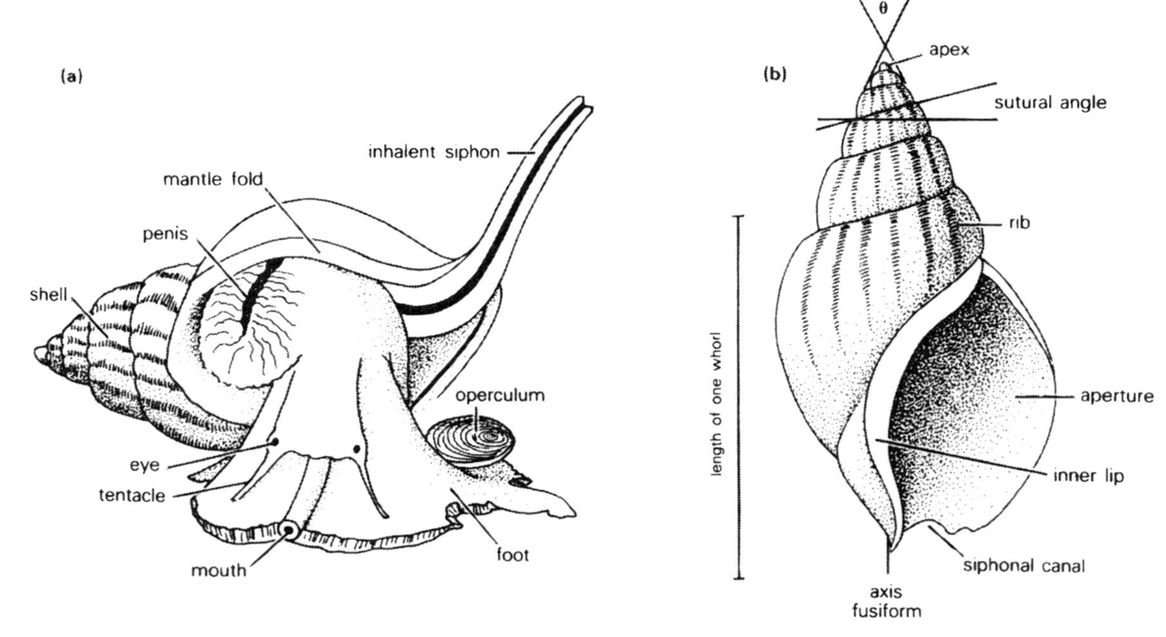

Fig. 3.48. Hard-part morphology of gastropods: general morphology. (a) General view of living *Buccinum*, ×1; (b) general view of shell. (Modified after Clarkson, 1998.)

who have reinterpreted the association as rather less, shall we say, intimate and more incidental, while acknowledging the attendant benefit to the gastropod in the area of avoidance of predation.

Some Tethyan Mesozoic nerineoidean mesogastropods were as large as coeval rudist bivalves, and it has been hypothesised, as it has been in the case of the rudists, that they harboured algal symbionts (Skelton and Wright, 1987). However, it has also been hypothesised on the basis of observations of their distributions that they were tolerant of stressed peritidal as well as sub-tidal environments, where they may have fed by grazing on algae (Morris and Taylor, in Culver and Rawson, 2000).

Opisthobranchs. Opisthobranchs are an extant, Carboniferous ?–Recent, group characterised by the advanced condition of detorsion, and further characterised by the secondary acquisition of bilateral symmetry. Two further sub-groups of opisthobranchs are recognised, namely, the pteropods, also known as euthecosomatans or, colloquially, as 'sea-butterflies', with, or secondarily without, and the nudibranchs or 'sea-slugs', essentially without, shells.

Pteropods are an important, extant, Palaeocene–Recent, group of opisthobranchs (see also comments on 'Biostratigraphy' below). The biology and ecology of living pteropods is comparatively poorly known (Be and Gilmer, in Ramsay, 1977; Lalli and Gilmer, 1989; Janssen, 2003). However, it is known that they are exclusively marine and nektonic, capable of free swimming by movements of a specially adapted 'foot' or 'wing', and that they feed principally on phytoplankton and zooplankton. They live in both epipelagic and bathypelagic environments, and exhibit a preference for open oceanic environments. The majority live in the circum-global warm-water biogeographic province, and only a few in the northern and southern cold-water provinces (some species, such as *Limacina helicina*, exhibiting bipolar distributions). On death, pteropods descend to the sea-floor, often in such numbers as to form 'pteropod oozes', principally in low to moderate latitudes. In terms of hard-part morphology, pteropods are characterised by conical, arcuate or trochospirally coiled shells (Fig. 3.49). The shells are composed of aragonite, such that they are only preserved above the aragonite compensation depth of approximately

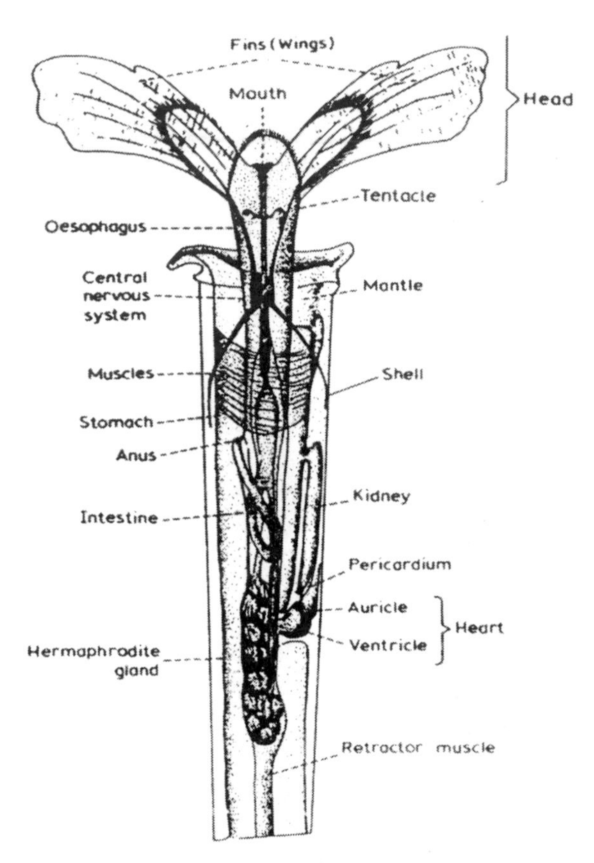

Fig. 3.49. Morphology of gastropods: pteropods. (From Jones, 1996; in turn after Haq and Boersma, 1978.)

2000–3000 m. Classification of pteropods is based essentially on morphology.

Pulmonates. Pulmonates are an extant, Jurassic–Recent, group also characterised by detorsion, with anteriorly positioned heads, and by gills modified into lungs; and are typically terrestrial, and atypically secondarily aquatic. They are further characterised by being hermaphroditic, i.e. not differentiated into males and females. Two further sub-groups of pulmonates are recognised by palaeontologists, namely the basommatophores with, and the stylommatophores with only vestigial, or without, shells (a further sub-group is recognised by molecular biologists, namely the systellommatophores). Basommatophores are exemplified by the freshwater snails *Lymnaea* and *Planorbis*; stylommatophores by the land-snails *Helix* and *Pupilla*, and by the slugs.

Palaeobiology

Fossil gastropods are interpreted, essentially on the bases of analogy with their living counterparts, and associated fossils and sedimentary facies, as having occupied a range of aquatic and terrestrial environments. Thusly interpreted freshwater gastropods include *Viviparus* (the 'river snail') from the Pleistocene of the Swanscombe archaeological site in Kent, and elsewhere. Terrestrial gastropods or land-snails, which, importantly, provide evidence as to substrate, ground cover and disturbance, include *Spermodea, Acanthinula, Aegopinella, Pupilla* and *Vallonia* from the Pleistocene of the Boxgrove archaeological site, and elsewhere (Preece and Bates, in Roberts and Parfitt, 1999; see also Section 7.6). Incidentally, calcitic plates attributed to the slug *Deroceras/Limax* also occur at the Boxgrove archaeological site.

Fossil gastropods with complete apertural margins are interpreted on the basis of analogy as having been herbivorous; those with siphonal canals as either actively herbivorous or carnivorous; those with wide apertures as having been carnivorous, and as having swallowed their prey whole.

Thick-shelled, patelliform or low-spired, fossil morphotypes are interpreted, on the basis of functional morphology, as indicating high-energy epifaunal environments. Thin-shelled morphotypes are similarly interpreted as indicating low-energy, often freshwater, epifaunal environments; high-spired morphotypes, infaunal environments.

Palaeobiogeography

Many fossil gastropods appear to have had restricted or endemic biogeographic distributions, rendering them of use in the characterisation of palaeobiogeographic provinces, and in turn in the constraint of plate tectonic reconstructions, in the Palaeozoic, Mesozoic and Cenozoic.

In the Palaeozoic, three palaeobiogeographic provinces are characterised by gastropods, namely the Old World, Eastern Americas and Malvinokaffric provinces (Blodgett *et al.*, in McKerrow and Scotese, 1990).

In the Mesozoic, the Tethyan realm is characterised by nerineoidean, acteonellid, pseudomelaniid, cassiopid, campanilid, strombid, cypraeid and chilodont gastropods (Sohl, 1987; Morris and Taylor, in Culver and Rawson, 2000).

In the Cenozoic, gastropod distributions point toward the opening of the Bering Strait in the Late Miocene to Pliocene (Durham and McNeil, in Hopkins, 1967; Vermeij, 1991), and its closure in the Pleistocene, resulting in the evolution of separate populations in the North Atlantic and North Pacific (Reid, 1996); and also the closure of the Strait of Panama in the Pliocene, resulting in the evolution of separate populations in the eastern Atlantic and western Pacific (Vermeij, 1978).

Palaeoclimatology

In the Cenozoic, gastropod distributions point toward the palaeogeographic and palaeoclimatic evolution of the Tethyan and Paratethyan realms (Piccoli *et al.*, 1991; Esu, in Agusti *et al.*, 1999; Amitrov, 2000; Harzhauser *et al.*, 2002, 2003; see also Subsection 5.5.8).

Gastropod distributions and turnovers also point toward the palaeoclimatic evolution of the North Pacific over the critical transition from Eocene to Oligocene (Oleinik and Marincovich, in Prothero *et al.*, 2003; Squires, in Prothero *et al.*, 2003). Palaeocene–Middle Eocene, unnamed to Tejon stage gastropod faunas of the west coast of the USA are of cosmopolitan, warm-water aspect, especially in the late Early–early Middle Eocene, Capay and Domengine stages, coincident with the global climatic optimum; Late Eocene, Galvinian stage faunas, interestingly, including a number of species indicative of 'cold-seep' and 'whale-fall' habitats of cooler-and/or deeper- water aspect; and Early Oligocene, Matlockian stage, and Late Oligocene, Juanian stage, faunas of cool-water aspect (Squires, in Prothero *et al.*, 2003). Moreover, Middle Eocene gastropod faunas from the Katalla district of the Gulf of Alaska are of cosmopolitan, warm-water aspect; Late Eocene faunas of mixed cosmopolitan and North Pacific temperate-water aspect; Early Oligocene faunas of mixed cosmopolitan and North Pacific temperate- and cold-water aspect; and Late Oligocene assemblages of North Pacific cold-water aspect (Oleinik and Marincovich, in Prothero *et al.*, 2003). Furthermore, Middle Eocene faunas from the Tigil and Palana districts of the Kamchatka Peninsula on the other side of the Bering Sea in the far east of the former Soviet Union are of mixed cosmopolitan and North Pacific temperate-water aspect; Late Eocene and Early Oligocene faunas of mixed cosmopolitan and North Pacific temperate- and cold-water aspect (Oleinik and Marincovich, in Prothero *et al.*, 2003).

The exacting ecological requirements and tolerances of many gastropod species, and their rapid response to changing environmental and climatic conditions, render them especially useful in Quaternary palaeoclimatology and climatostratigraphy, and in archaeology (Preece and Bates, in Roberts and Parfitt, 1999; see also Section 7.6).

Biostratigraphy

Gastropods evolved in the Cambrian, probably from ancestors with a monoplacophoran grade of organisation. They diversified in the Ordovician (Fryda and Rohr, in Webby *et al.*, 2004), and by the Carboniferous had colonised freshwater, and possibly also terrestrial, as well as marine environments. They sustained losses in the End-Ordovician, Late Devonian, and, especially End-Permian mass extinctions, but recovered from all of these events (Erwin and Hua-Zhang, in Hart, 1996). They have been diversifying from the Triassic, Jurassic or Cretaceous to the present day, that is, through much of the Mesozoic and, especially, Cenozoic, sustaining only comparatively minor losses in the End-Cretaceous and End-Eocene mass extinctions (Kelley and Hansen, in Hart, 1996; Morris and Taylor, in Culver and Rawson, 2000). Terrestrial forms diversified most dramatically during the sea-level low-stands at the end of the Jurassic and at the end of the Cretaceous (Morris and Taylor, in Culver and Rawson, 2000). Marine forms diversified during the mid-Cretaceous, coincident with a prolonged sea-level high-stand, and also, significantly, with a diversification of predators within and outwith the group (Morris and Taylor, in Culver and Rawson, 2000).

Interestingly, it was a study of the land-snail *Poecilizontes* from the Pleistocene of Bermuda that first led to the development of the 'punctuated equilibrium' model of evolution, involving long periods of stasis interrupted by short periods of rapid evolutionary change, interpreted as the result of allopatric speciation at times of isolation of populations by glacioeustatically mediated sea-level rise (Gould, 1969; Eldredge and Gould, in Schopf, 1972).

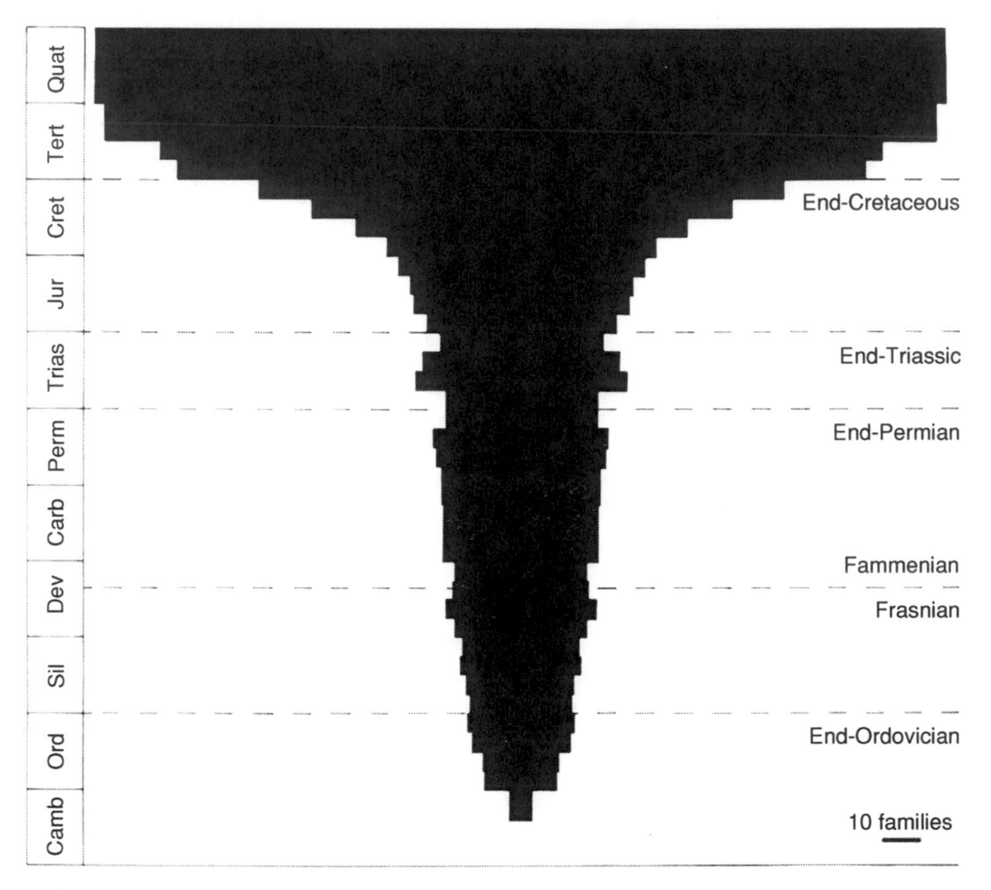

Fig. 3.50. Stratigraphic distribution of gastropods. (Reproduced with permission from Doyle, 1996 *Understanding Fossils*, © John Wiley & Sons Ltd.)

A similar study of snails from the Pliocene of Lake Turkana in the East African Rift Valley in Kenya later led to similar interpretations (Williamson, 1981).

The moderate rate of evolutionary turnover exhibited by the gastropods renders them of some use in biostratigraphy, at least in appropriate environments (Fig. 3.50). Unfortunately, the potential usefulness of most is somewhat compromised by their restricted ecological distribution and facies dependence, a function of their benthic habit. However, the usefulness of some, pteropods, is enhanced by their essentially unrestricted ecological distribution, and facies independence, a function of their nektonic habit. Biostratigraphic zonation schemes based on pteropods have been established, that have at least local to regional applicability, as in the Cenozoic of the onshore parts of the North Sea basin in north-west Europe (Jones, 1996; Janssen, 2003).

In my working experience, gastropods have proved of some use in stratigraphic subdivision and correlation, and/or in palaeoenvironmental interpretation, in the following areas.

Palaeozoic

The Devonian of the Middle East. In north-western Saudi Arabia, the Tawil formation is marked by the Devonian bellerophontid *Plectonotus derbyi* (Boucot *et al.*, 1988).

Mesozoic

The Jurassic and Cretaceous of the Middle East. In central Saudi Arabia, the Middle–Late Jurassic Dhruma, Tuwaiq Mountain and Hanifa formations are characterised by 35 species of gastropods representing the families Euomphalidae, Ataphridae, Pseudomelaniidae, Coelostylinidae, Procerithidae,

Nerineidae, Purpurinidae, Aporrhaidae, Naticidae, Acteonidae, Retusidae and Akeridae (Fischer *et al.*, 2001). The faunas are indicative of a range of platform and peri-reefal, and in particular platform interior or back-reef lagoonal, carbonate environments; their palaeobiogeographic affinities are with those of the Levant, Sinai and Europe, and, to a lesser extent, north and east Africa. In the Oman Mountains, the Late Jurassic part of the Musandam limestone is characterised by *Nerinea* (Hudson and Chatton, 1959). In the Zagros Mountains of south-west Iran, the upper, Late Jurassic, part of the Surmeh formation of the Zagros is also characterised by *Nerinea* (author's unpublished observations). The 'Middle' Cretaceous, Albian–Cenomanian Kazhdumi formation is characterised by Turritellidae, Aporrhaidae (?*Drepanochilus*) and Naticidae (?*Gyrodes*) (Collignon, 1981; author's unpublished observations, based in part on identifications by Noel Morris, formerly of the Natural History Museum, London).

Cenozoic

The Oligocene–Pleistocene of Paratethys. Gastropods are also of considerable use in palaeoenvironmental interpretation in this area, for example in the determination of palaeosalinity based on analogy, on associated fossils and sedimentary facies, and on isotope records. Terrestrial environments are variously characterised by the land-snail *Helix* and by *Potamius conicus*; non-marine, freshwater aquatic environments by *Lymnaea, Planorbis* and *Viviparus*, and by *Potamides hartbergensis* assemblages; marginal marine, brackish-water environments by *Granulolabium bicinctum* assemblages; and marine environments by the pteropod *Spirialis*, and by *Potamides disjunctus* assemblages. The presence of *Potamides disjunctus* assemblages in the late Sarmatian *Mactra* Zone of the St Margharethen–Zollhaus section in the Vienna basin in Austria indicates that conditions there were fully marine rather than brackish, as had previously been hypothesised by Papp (1954) (Harzhauser and Piller, 2001; Harzhauser and Kowalke, 2002; Latal *et al.*, 2004). The occurrence of the land-snail *Helix* in the late Sarmatian of Ciscaucasia is important in that it constrains not only palaeoenvironmental and palaeo(bio)geographic interpretation, but also the timing of uplift in the Greater Caucasus, and attendant consequences for the provenance and petroleum reservoir potential of sediments in the Black Sea and Caspian basins to the south (author's unpublished observations).

3.6.11 Molluscs: ammonoids

Biology, morphology and classification

Ammonoids are an extinct, Late Palaeozoic–Mesozoic, Devonian–Cretaceous, group of cephalopod ('head-foot') molluscs, characterised by large heads, and by tubular to planispirally coiled external shells (House, 1993; Landman *et al.*, 1996; Oloriz and Rodriguez-Tovar, 1999; Monks and Palmer, 2002). Their most immediate living relatives are coleoid and nautiloid cephalopods. Belemnite coleoids are important in the fields of palaeobiology and biostratigraphy, and are discussed below. Nautiloids are relatively unimportant, and are not discussed. Note, though, that fossil forms are known, from, for example, the Middle Jurassic, Callovian of southern Yemen; the Late Jurassic, Oxfordian of Saudi Arabia; the Late Cretaceous, Maastrichtian of the Oman Mountains; the Early Palaeocene, Danian of the Brazos River section in Texas in the south-western USA; and the Early Eocene London clay of the south-east of England.

Biology

Nothing is known directly of the biology of the ammonoids, as they are extinct (but see below).

Morphology

In terms of external hard-part morphology, ammonoids are characterised by a more or, in the case of the aberrant 'heteromorph' ammonites, less, regularly coiled shell comprising a number of volutions or whorls; the inner and outer layers or laminae of the shell composed of the prismatic form of aragonite, the middle one of the tabular or nacreous form, otherwise known as 'mother-of-pearl' (Fig. 3.51). Internally, the shell is supported, and separated into a series of chambers, by simple to complex walls or septa, reflected on the external surface as sutures. The chambers communicate with one another through openings in the septa. The ammonite animal is also thought to have been capable of filling the earlier chambers with, and emptying them of, fluid, for the purpose of buoyancy regulation. It is also thought to have lived in the last chamber, and to have been capable

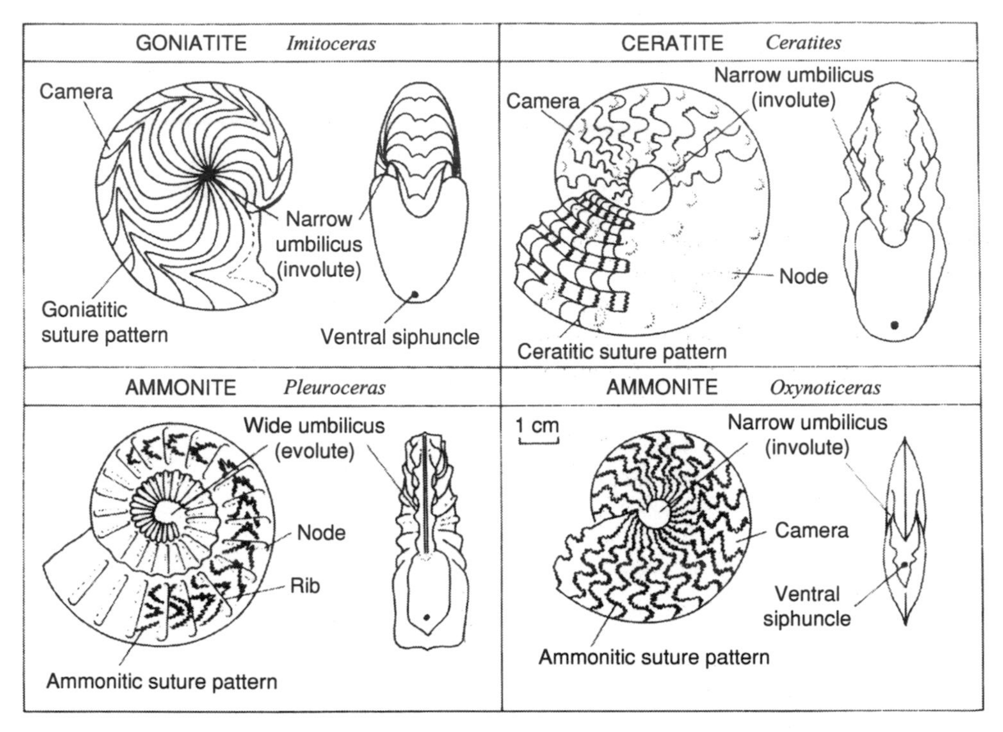

Fig. 3.51. Hard-part morphology of ammonoids. (Reproduced with permission from Doyle, 1996 *Understanding Fossils*, © John Wiley & Sons Ltd.)

of closing it off from the outside world by means of a pair of flaps or aptychi. Interestingly, the aptychi are composed of calcite, and thus capable, unlike the rest of the shell, of being preserved below the aragonite compensation depth, as in the case of the Jurassic *ammonitico rosso* of the circum-Mediterranean.

The increasing intricacy of the form of the septa through time probably represents a, progressive, adaptation to a pelagic habit. The development of ornamentation, most commonly in the form of ribs, knobs or tubercles, and spines, on the external surface of the shell, may have fulfilled a further protective function or helped to maintain stability. Another possibility is that the ornamentation, perhaps in combination with colour, unfortunately bleached out in the process of fossilisation, was used for camouflage from predators, or for sexual display.

Classification

Classification of ammonoids is based on morphology and phylogeny. Three main sub-groups are recognised: the goniatites, the ceratites and the ammonites.

Goniatites. Goniatites are an extinct, Late Palaeozoic, Devonian–Permian, group of ammonoids, thought to have evolved from orthocone nautiloids or bactritids.

In terms of hard-part morphology, goniatites are diagnosed by the simple to goniatitic form of their sutures.

Classification of goniatites and allied forms is based on morphology, and on phylogeny inferred from empirical observations on the fossil record. Three sub-groups are recognised: the anarcestids, the clymeniids and the goniatitids. The ancestral anarcestids, which range from Devonian to Permian, are characterised by ventral siphuncles and by simple sutures; the clymeniids, which are restricted to the Late Devonian, by dorsal siphuncles and by goniatitic sutures; and the goniatitids, which range from Carboniferous to Permian, by ventral siphuncles and by goniatitic sutures consisting of eight lobes.

Ceratites. Ceratites are an extinct, latest Palaeozoic–earliest Mesozoic, Carboniferous–Triassic, group of ammonoids.

In terms of hard-part morphology, ceratites are diagnosed by the goniatitic to ceratitic form of their sutures.

Classification of ceratites is based on morphology, and on phylogeny inferred from empirical observations on the fossil record. Two sub-groups are recognised, the prolecanitids and the ceratitids. The ancestral prolecanitids, which range from Carboniferous to Permian, are characterised by goniatitic to ceratitic sutures; the ceratitids, which are essentially Triassic, by ceratitic sutures.

Ammonites. Ammonites are an extinct, Mesozoic, Triassic–Cretaceous, group of ammonoids. They take their name from the hellenised version of the ancient Egyptian god Amun, depicted on Cyrenean coins and in sculptures as having a head with curling rams' horns. Interestingly, ammonites or 'saligrams', from the Jurassic Spiti shale of Salagrama in Nepal, on the banks of the Gandaki River, are regarded as holy in the Hindu religion to this day, on account of their resemblance to the chakra, or 'wheel of time', depicted held in one of the four hands of Vishnu (the other hands holding a clam, a lotus flower and a mace). More interestingly still, sacred Sanskrit poetry identifies them as having been made by the activities of marine worms! In other cultures, ancient legends identified ammonites as serpents that had been turned to stone. Their abundance in Whitby in Yorkshire on the north-east coast of England led to the incorporation of stylised 'snakestones' into the town's ancient coat of arms, and also, ultimately, to the naming of *Hildoceras*, after the seventh-century Anglo-Saxon abbess St Hilda, who lived there. The suffix- *ceras* is the Greek for 'horn'.

In terms of hard-part morphology, ammonites are diagnosed by the ammonitic form of their sutures.

Classification of ammonites is based on morphology, and on phylogeny inferred from empirical observations on the fossil record, and is complicated by iterative evolution and homoeomorphy. Four subgroups are recognised: the phylloceratids, the lycoceratids, the ammonitids and the ancyloceratids. The ancestral phylloceratids, which range from Triassic to Cretaceous, are thought to have evolved from ceratites, and are characterised by sutures with distinctive phylloid or leaf-like saddles and lituid lobes, and by smooth, involute, commonly compressed, shells; the lycoceratids, which range from Jurassic to Cretaceous, by evolute shells; the ammonitids, which range from Triassic to Cretaceous, by ventral siphuncles, and, commonly, ornamented shells; the ancyloceratids, which are essentially restricted to the Cretaceous, by heteromorphy.

The term 'ammonite' is hereafter used in reference essentially only to the ammonitids, unless otherwise explicitly stated.

Palaeobiology

Goniatites, ceratites and ammonites are interpreted, essentially on the basis of analogy with their living counterparts, namely coleoids and nautiloids, and of associated fossils and sedimentary facies, as having been marine and nektonic (Kennedy and Cobban, 1976). It has been argued that coleoids are more appropriate analogues than nautiloids (Jacobs and Landman, 1993).

Ammonites are interpreted to have fed on plankton or slow-swimming animals in the water column, in the case of nektonic forms, or on slow-moving animals – such as foraminiferans, ostracods and other crustaceans, bryozoans, corals and brachiopods – on the sea-floor, in the case of nektobenthic forms (Keupp, 2000). Some forms may have been detrital scavengers, still others active predators on other ammonites. The abundance of ammonites has been used as a measure of carbonate dilution or palaeoproductivity in the Early Cretaceous, Valanginian of the Vocontian basin in south-east France (Reboulet *et al.*, 2003).

Palaeobathymetry

In the Late Cretaceous of Tunisia, the Scaphitidae appear to have occupied shallow marine environments, and the Desmocerataceae and Tetragonitiaceae deep marine environments (Goolaerts *et al.*, 2004). Sea-level changes have been inferred on the basis of ammonite distributions, for example in the Early Cretaceous of the Kopet Dagh basin in north-east Iran (Raisossadat, 2003).

Palaeobiogeography

Many ammonoids appear to have had restricted or endemic biogeographic distributions, rendering

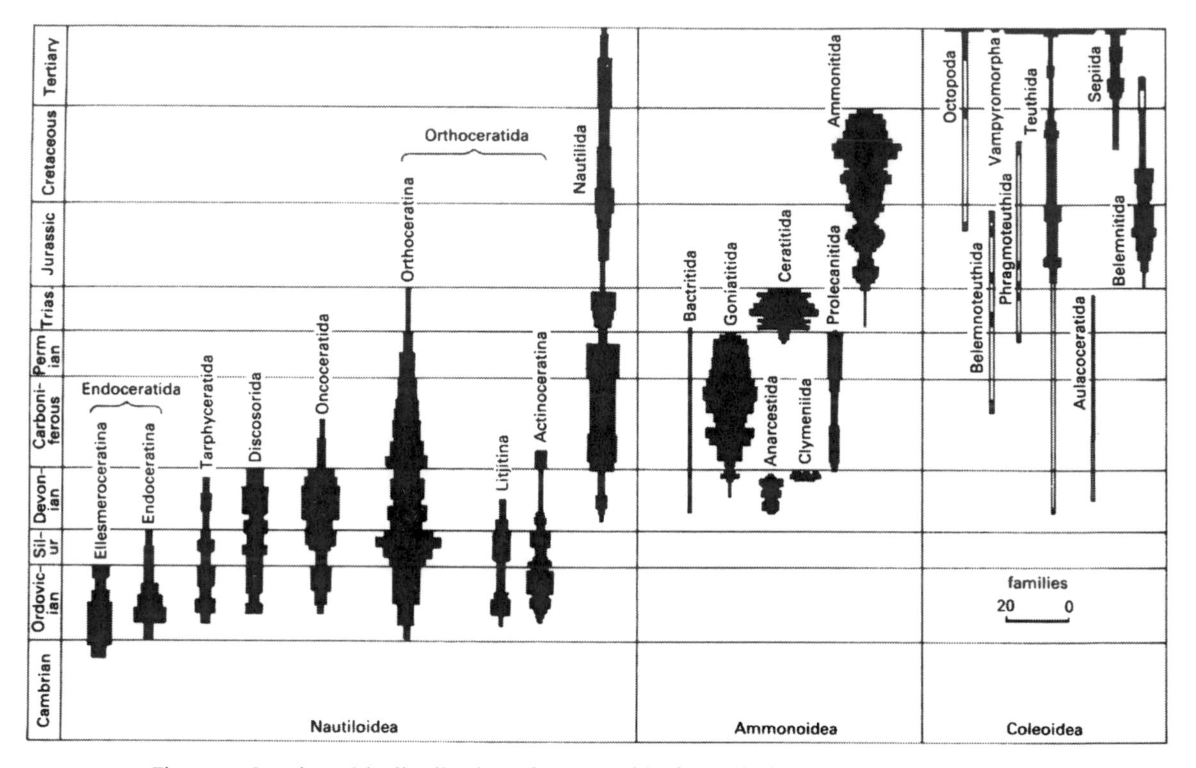

Fig. 3.52. Stratigraphic distribution of ammonoids (from Clarkson, 1998).

them of use in the characterisation of palaeobiogeographic provinces, and in turn in the constraint of plate tectonic reconstructions, in the Late Palaeozoic and in the Mesozoic.

In the Mesozoic, ammonite distributions can be used to characterise low-latitude Tethyan, and high-latitude southern, or Austral, and northern, or Boreal, realms (Kennedy and Cobban, 1976). Ammonites are typically characteristic of only one of these realms, for example, the phylloceratids of the Boreal realm. Note, though, that the lycoceratids are essentially cosmopolitan (Rawson, in Casey and Rawson, 1973). Ammonite distributions can also be used to characterise Boreal–Atlantic, Boreal–Pacific, and Arctic provinces within the Boreal realm, the last-named also encompassing the Chukotka–Canada sub-province, characterised by an essentially endemic fauna. Importantly, connection and dispersal between these provinces, and indeed between the Boreal and sub-Boreal–Tethyan realms, was facilitated by marine transgressions (Sey and Kalacheva, 1983). Ammonite distributions point towards the existence of such connections intermittently through the late Middle–Late Jurassic, Callovian, Oxfordian and Kimmeridgian, and in the *Kilmovi, Panderi*, early and late *Nikitini* and *Nodiger* Zones of the latest Jurassic, Volgian (Hantzpergue *et al.*, in Crasquin-Soleau and Barrier, 1998b). Ammonite distributions also point towards the existence of a connection between the Boreal and sub-Boreal–Tethyan realms in the form of a longitudinal strait in the area west of the Urals in the Early Cretaceous, generally characterised by southward flow of Boreal waters in the Neocomian and by northward flow of Tethyan waters in the Apto-Albian; of a disconnection in the Albo-Cenomanian; and of a reconnection in the area of Turgai in the Turonian (Baraboshkin *et al.*, 2003). The Turgai Strait connected the peri-Tethyan basins, the West Siberian Boreal basins, and the western interior seaway of the USA.

Tectonic as well as eustatic effects were an important factor in ammonite palaeobiogeography, for

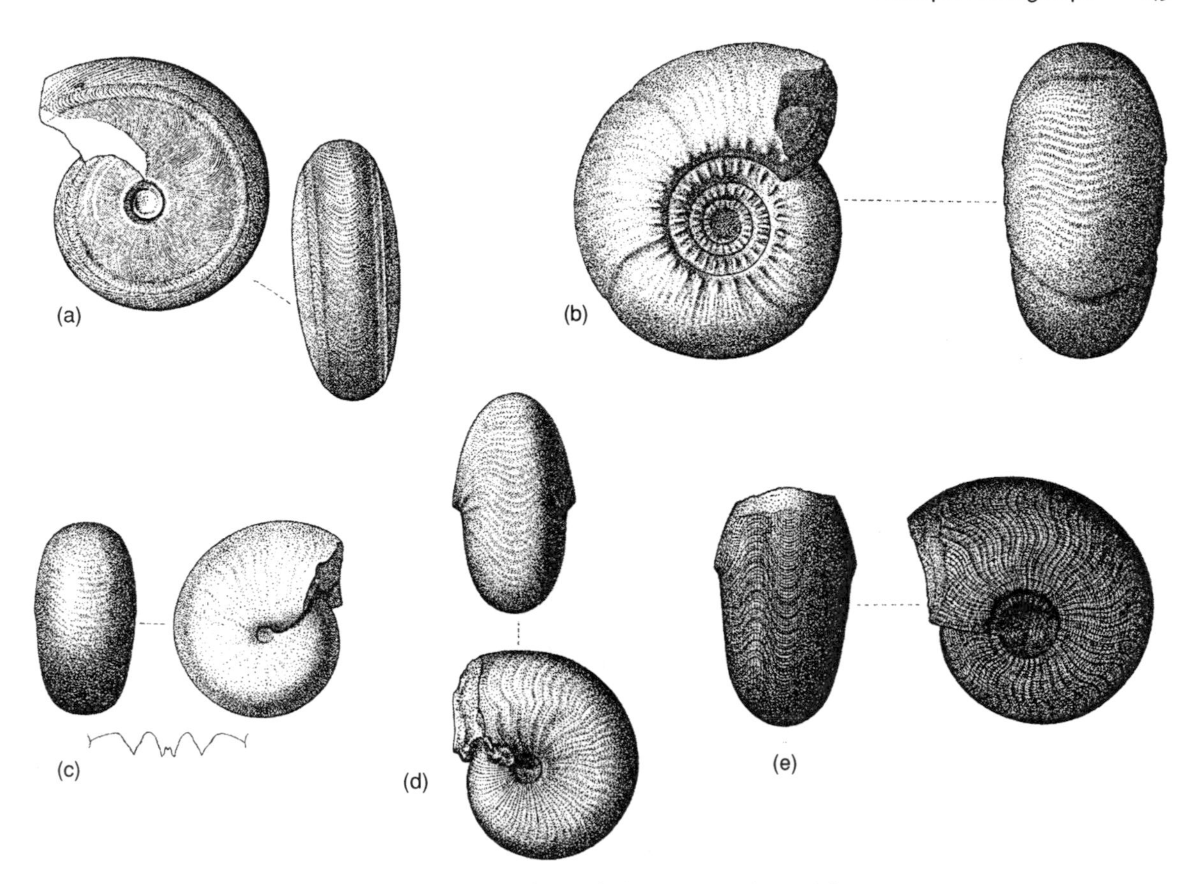

Fig. 3.53. Some stratigraphically useful goniatites. (a) *Reticuloceras bilinigue*, Late Carboniferous, Namurian (Marsdenian), *Reticuloceras* Zone R2, Hebden Bridge, North Yorkshire, ×1.5; (b) *Gastrioceras carbonarium*, Late Carboniferous, Westphalian A, *Gastrioceras* Zone G2, Churnet Vally, near Leek, Staffordshire, ×1.5; (c) *Beyrichoceras obtusum*, Early Carboniferous, Visean, *Beyrichoceras* Zone B, Bowland, near Clitheroe, Lancashire, ×1; (d) *Homoceras diadema*, Late Carboniferous, Namurian (Chokierian–Alportian), *Homoceras* Zone H, West Yorkshire, ×1.5; (e) *Reticuloceras reticulatum*, Late Carboniferous, Namurian (Kinderscoutian), *Reticuloceras* Zone R1, Hebden Bridge, North Yorkshire, ×1.5. (Modified after the Natural History Museum, 2001a.)

example those associated with the rifting of the south and equatorial Atlantic and the drifting apart of Africa and South America (Kogbe and Mehes, 1986; Rawson, in Culver and Rawson, 2000).

Biostratigraphy

As intimated above, goniatites evolved in the Early Devonian. The group then underwent a dramatic diversification in the Middle and Late Devonian before almost becoming extinct in the Late Devonian mass extinction. Nonetheless, they recovered from this event, and underwent renewed dramatic radiation in the Carboniferous, before finally becoming extinct in the End-Permian mass extinction. The rapid rate of evolutionary turnover exhibited by the goniatites renders them of considerable use in biostratigraphy, at least in appropriate, marine, environments (Figs. 3.52–3.53). Their usefulness is enhanced by their essentially unrestricted ecological distribution, and facies independence, a function of their nektonic habit. High-resolution biostratigraphic zonation schemes based on the

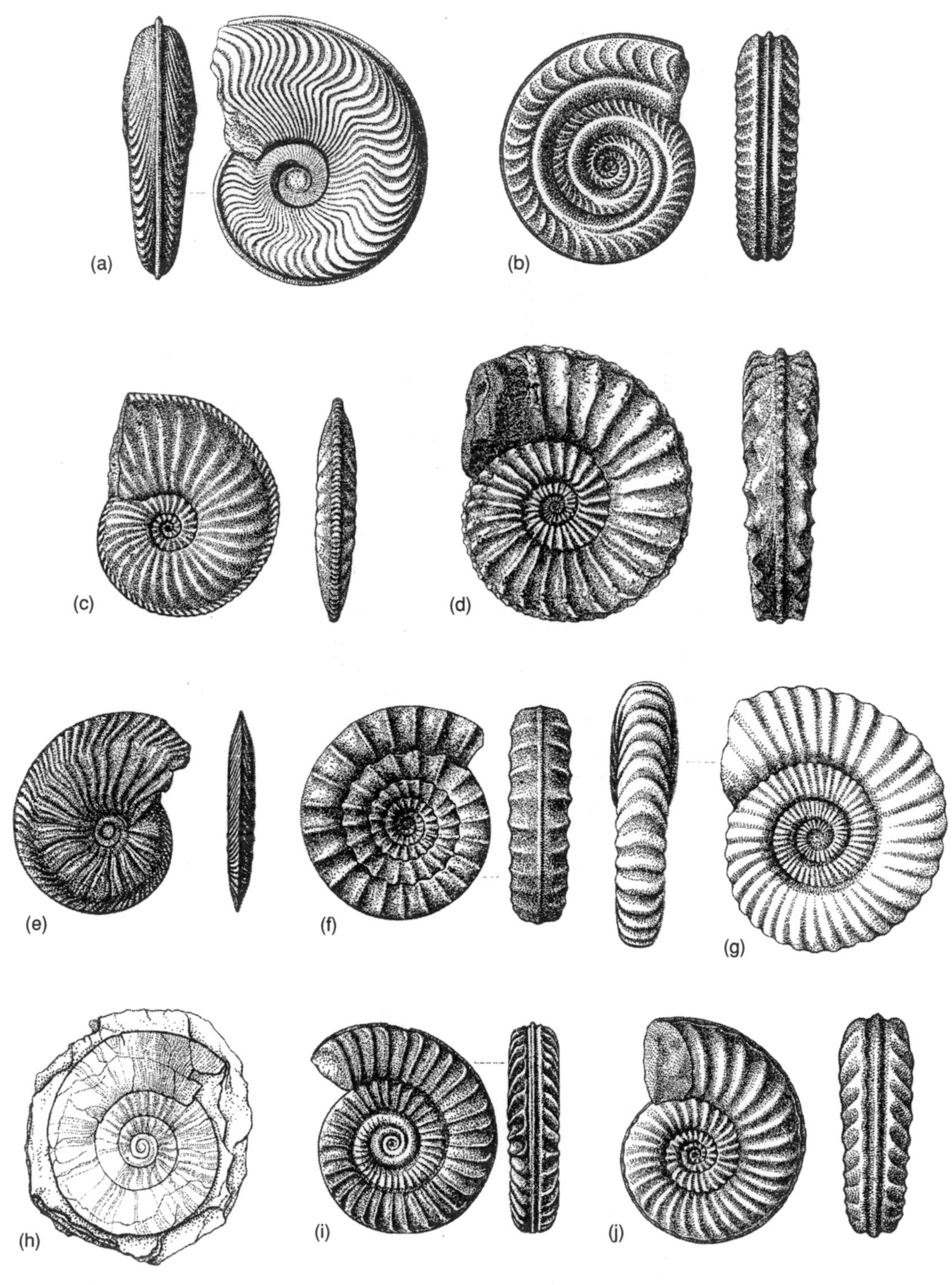

Fig. 3.54. Some stratigraphically useful ammonites. Early Jurassic. (a) *Harpoceras falciferum*, nominate taxon of the Toarcian *Harpoceras falciferum* Zone, Upper Lias, Lincoln, ×0.75; (b) *Hildoceras bifrons*, nominate taxon of the Toarcian *Hildoceras bifrons* Zone,

goniatites have been established, that have been calibrated against standard conodont schemes, or by orbital tuning, and that have regional to essentially global applicability (see, for example, Becker and House, in Bultynck, 2000; see also Chapter 6). Anarcestids and clymeniids form the bases of the biozonations of the Middle and Late Devonian respectively of the so-called Rhenish facies of western Europe, including the British Isles. Goniatitids form part of the basis of the biozonation of the Late Devonian and Carboniferous of the British Isles, including the Carboniferous 'Coal Measures' (Prentice and Thomas, 1960; Ramsbottom *et al.*, 1978; Riley, 1993; see also Section 7.3).

Ceratites are essentially restricted to the Triassic. The rapid rate of evolutionary turnover exhibited by ceratites renders them of considerable use in biostratigraphy, at least in appropriate, marine, environments (Fig. 3.52). Their usefulness is enhanced by their essentially unrestricted ecological distribution, and facies independence, a function of their nektonic habit.

Ammonites evolved in the Triassic. The group then underwent rapid diversification through the Jurassic and Cretaceous, before becoming extinct in the End-Cretaceous mass extinction. The recovery of the desmoceratacean ammonites from the effects of the Late Cretaceous, Campanian extinction in Sakhalin in the far east of Russia is discussed by Yazykova, in Hart (1996). The rapid rate of evolutionary turnover exhibited by the ammonites renders them of considerable use in biostratigraphy, at least in appropriate, marine, environments (Figs. 3.52, 3.54–3.57). Unfortunately, their usefulness is somewhat compromised by their restricted biogeographic distribution (see above). However, high-resolution biostratigraphic zonation schemes based on the ammonites have been established, that have regional to inter-regional applicability within a given palaeobiogeographic province (see Chapter 6); and, moreover, the independent schemes established for the various provinces have been calibrated against one another, at least for the most part (Hantzpergue *et al.*, in Crasquin-Soleau and Barrier, 1998b; Grocke *et al.*, 2003). Individual ammonite zones or sub-zones can have a duration or resolution of as little as 200 000 years! This level of temporal resolution is sufficient to enable the identification of even relatively minor unconformities or sequence boundaries. Significantly, it is also sufficient to provide a meaningful framework within which to assess at least some of the effects of climatic change, as inferred principally from isotope geochemistry (Grocke *et al.*, 2003).

Ammonites were first used for stratigraphic subdivision and correlation in the Jurassic of the Swabian and Franconian Alb of southern Germany, the extension of the Jura Mountains of France

Caption for fig. 3.54 (*cont.*) Upper Lias, Whitby, Yorkshire, ×0.75; (c) *Amaltheus margaritatus*, nominate taxon of the late Pliensbachian (Domerian) *Amaltheus margaritatus* Zone, Middle Lias, Ilminster, Somerset, ×0.5; (d) *Pleuroceras spinatum*, nominate taxon of the late Pliensbachian (Domerian) *Pleuroceras spinatum* Zone, Middle Lias, Down Cliff, near Bridport, Dorset, ×0.75; (e) *Oxynoticeras oxynotum*, nominate taxon of the Sinemurian *Oxynoticeras oxynotum* Zone, Lower Lias, Cheltenham, Gloucestershire, ×0.75; (f) *Echioceras raricostatum*, nominate taxon of the Sinemurian *Echioceras raricostatum* Zone, Lower Lias, Radstock, Somerset, ×0.75; (g) *Uptonia jamesoni*, nominate taxon of the early Pliensbachian (Carixian) *Uptonia jamesoni* Zone, Lower Lias, Munger Quarry, Paulton, Somerset, ×0.33; (h) *Psiloceras planorbis*, nominate taxon of the Hettangian, *Psiloceras planorbis* Zone, Lower Lias, Watchet, Somerset, ×0.75; (i) *Arnioceras semicostatum*, nominate taxon of the Sinemurian *Arnioceras semicostatum* Zone, Lower Lias, Robin Hood's Bay, Yorkshire, ×1.5; (j) *Asteroceras obtusum*, nominate taxon of the Sinemurian, *Asteroceras obtusum* Zone, Lower Lias, Lyme Regis, Dorset, ×1.5. (Modified after the Natural History Museum, 2001b.)

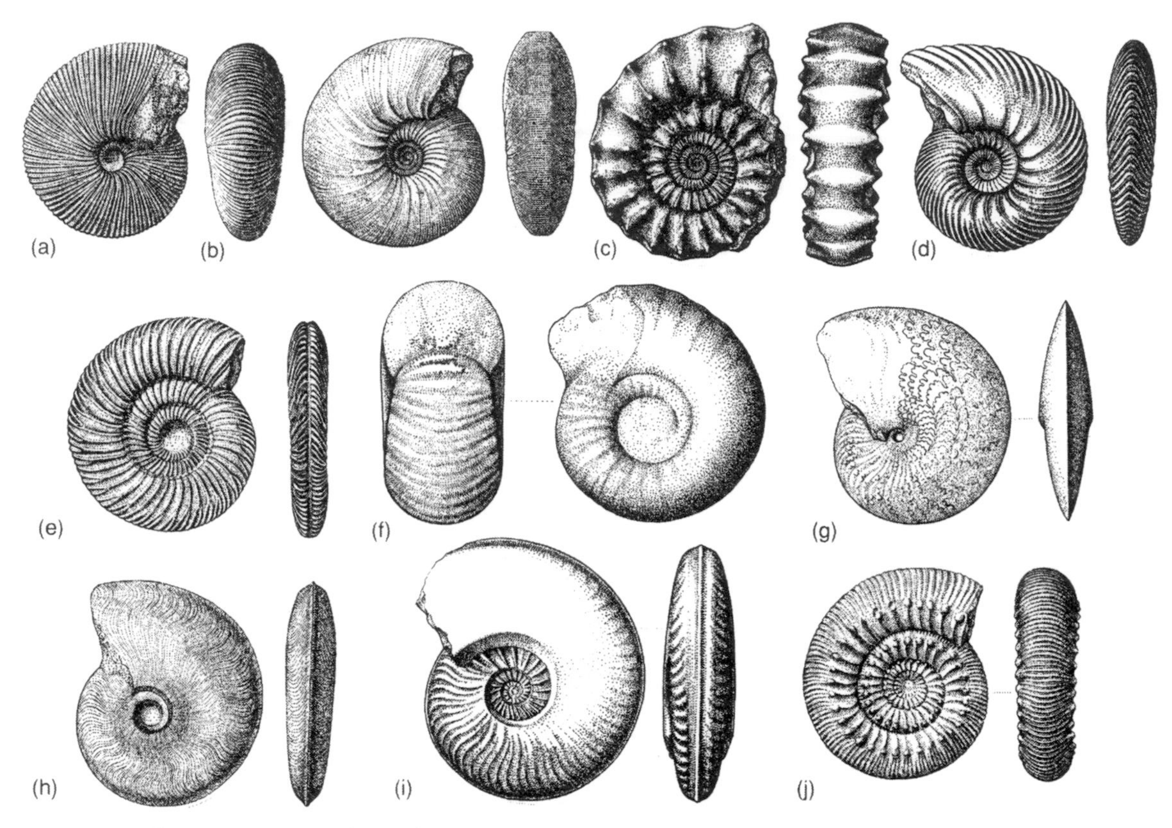

Fig. 3.55. Some stratigraphically useful ammonites: Middle Jurassic. (a) *Macrocephalites macrocephalus*, nominate taxon of the Callovian *Macrocephalites macrocephalus* Zone, Upper Cornbrash, Scarborough, Yorkshire, ×0.66; (b) *Sigaloceras calloviense*, nominate taxon of the Callovian *Sigaloceras calloviense* Zone, Kellaways rock, Kellaways, Wiltshire, ×0.75; (c) *Peltoceras athleta*, nominate taxon of the Callovian *Peltoceras athleta* Zone, Oxford clay, Eye, Cambridgeshire, ×0.75; (d) *Quenstedtoceras lamberti*, nominate taxon of the Callovian, *Quenstedtoceras lamberti* Zone, Oxford clay, Ashton Keynes, Wiltshire, ×0.66; (e) *Parkinsonia parkinsoni*, nominate taxon of the Bathonian *Parkinsonia parkinsoni* Zone, Upper Inferior oolite, Sherborne, Dorset, ×0.5; (f) *Tulites subcontractus*, nominate taxon of the Bathonian *Tulites subcontractus* Zone, Great Oolite, Fuller's Earth rock, Somerset, ×0.75; (g) *Clydoniceras discus*, nominate taxon of the Bathonian *Clydoniceras discus* Zone, Lower Cornbrash, Closeworth, Somerset, ×0.75; (h) *Leioceras opalinum*, nominate taxon of the Bajocian *Leioceras opalinum* Zone, Lower Inferior oolite, Bridport, Dorset, ×0.75; (i) *Ludwigia murchisonae*, nominate taxon of the Bajocian *Ludwigia murchisonae* Zone, Lower Inferior oolite, Beaminster, Dorset, ×0.5; (j) *Stephanoceras humphresianum*, nominate taxon of the Bajocian *Stephanoceras humphresianum* Zone, Middle Inferior oolite, Dundry, Somerset, ×0.5. Modified after the Natural History Museum (2001b).

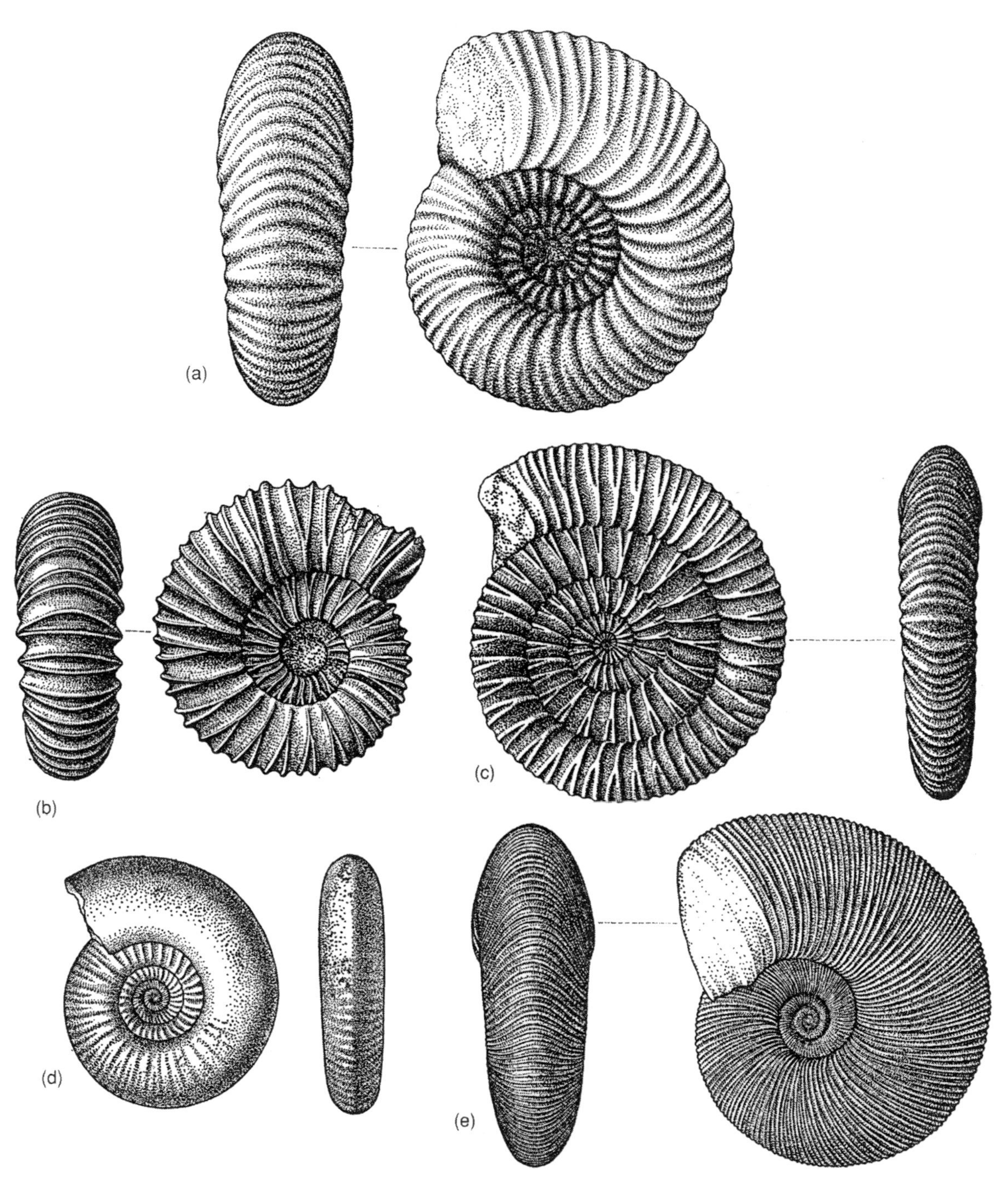

Fig. 3.56. Some stratigraphically useful ammonites: Late Jurassic. (a) *Titanites giganteus*, nominate taxon of the Portlandian *Titanites giganteus* Zone, Upper Portland beds (Portland Stone), Tisbury, Wiltshire, ×0.33; (b) *Pavlovia pallasioides*, nominate taxon of the Kimmeridgian *Pavlovia pallasioides* Zone, Hartwell clay, Hartwell, Buckinghamshire, ×0.75; (c) *Crendonites (Glaucolithes) gorei*, nominate taxon of Portlandian *Glaucolithes gorei* Zone, Lower Portland beds, Portland, Dorset. ×0.5; (d) *Pictonia baylei*, nominate taxon of the Kimmeridgian *Pictonia baylei* Zone, Lower Kimmeridge clay, Wootton Bassett, Wiltshire, ×0.5; (e) *Pectinatites pectinatus*, nominate taxon of the Kimmeridgian *Pectinatites pectinatus* Zone, Upper Kimmeridge clay, Swindon, Wiltshire, ×1. (Modified after the Natural History Museum, 2001b.)

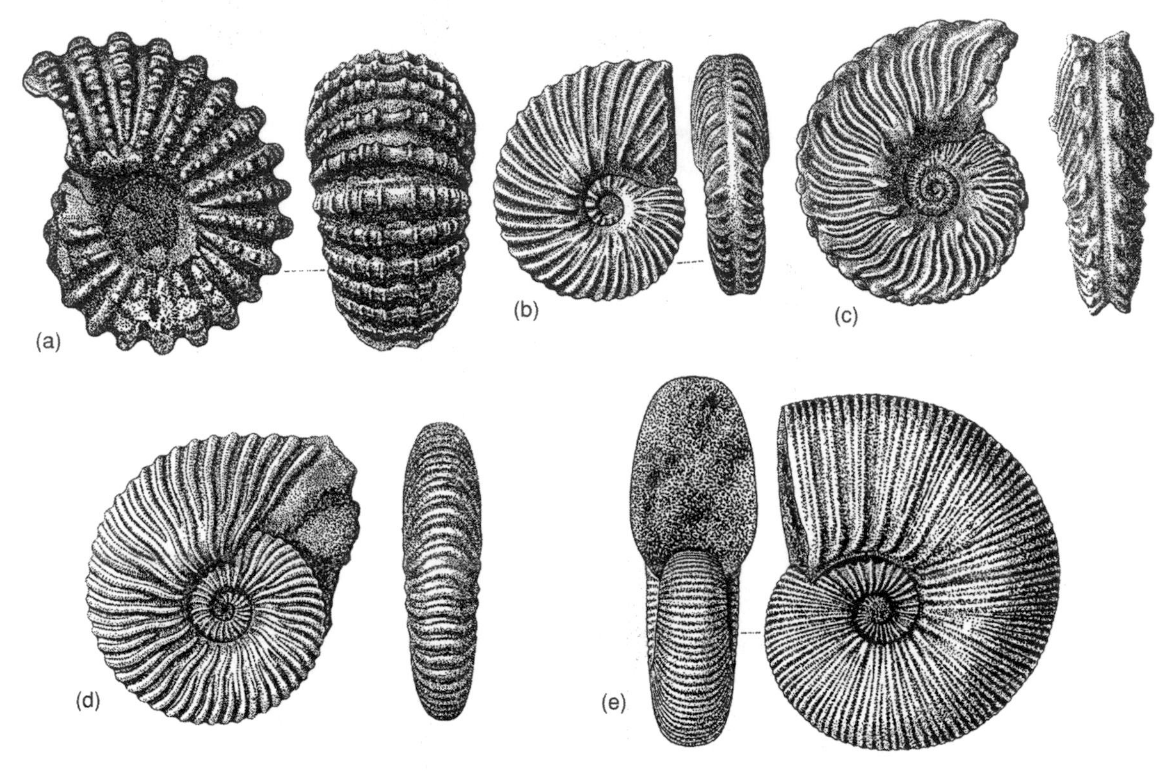

Fig. 3.57. Some stratigraphically useful ammonites: Early Cretaceous. (a) *Douvilleiceras mammillatum*, nominate taxon of the early Albian *Douvilleiceras mammillatum* Zone, Lower Greensand/Gault junction, Folkestone, Kent, ×1; (b) *Hoplites dentatus*, nominate taxon of the middle Albian *Hoplites dentatus* Zone, Lower Gault, Folkestone, Kent, ×0.75; (c) *Euhoplites lautus*, nominate taxon of the middle Albian *Euhoplites lautus* Zone, Lower Gault, Folkestone, Kent, ×1; (d) *Deshayesites forbesi*, nominate taxon of the early Aptian *Deshayesites forbesi* Zone, Lower Greensand, Atherfield, Isle of Wight, ×0.75; (e) *Parahoplites nutfieldensis*, nominate taxon of the late Aptian *Parahoplites nutfieldensis* Zone, Lower Greensand, Nutfield, Surrey, ×0.33. (Modified after the Natural History Museum, 2001b.)

and Switzerland from which the Jurassic takes its name, as long ago as the nineteenth century, by the German palaeontologists Oppel and Quenstedt. Ammonites form the bulk of the basis of the biozonation of the marine facies of the Jurassic and, to a slightly lesser extent, the Cretaceous of the British Isles, especially in the classical areas on the south and east coasts of England (Callomon and Cope, in Taylor, 1996; Hesselbo and Jenkyns, in Taylor, 1996; Rawson and Wright, in Hesselbo and Parkinson, 1996).

In my working experience, goniatites, ceratites and ammonites have proved of particular use in stratigraphic subdivision and correlation in the following areas.

Palaeozoic

The Devonian–Carboniferous of north Africa. In the Ahnet basin of Algeria, the Middle Devonian is marked by *Sobolewia, Parodiceras, Agoniatites, Tornoceras* and *Pleurotomia*, the Late Devonian by *Pharciceras*, and the Early Carboniferous by *Munstericeras, Pericyclus, Nautellipsites* and *Beyrichoceras* (Follot, 1952). In the adjacent Timimoun basin of

Algeria, successive horizons in the Devonian, Famennian to Carboniferous, Visean are marked by *Gonioclymenia, Gattendorfia, Acrocanites, Merocanites* and *Dimorphoceras* (unpublished observations).

The Devonian–Carboniferous of the former Soviet Union. Here, goniatites are useful in the stratigraphic subdivision and correlation of the Devonian in the Timan–Pechora basin in the north-eastern part of the Russian platform (House *et al.*, in Insalaco *et al.*, 2000; author's unpublished observations), and of the Carboniferous throughout the former Soviet Union (Wagner *et al.*, 1996). In the Timan–Pechora basin, goniatites provide part of the basis of the identification of the time-slices for palaeogeographic and facies mapping, which has important consequences for the depositional modelling of the basinal source-rock.

The Carboniferous–Permian of the Middle East. Here, the Carboniferous of north-east Iran is marked by various goniatites (Vachard, in Wagner *et al.*, 1996). The Early Permian, Sakmarian P10 MFS of Sharland *et al.* (2001), stratotypified in shales immediately below the Haushi limestone member of the lower Gharif formation of Oman, is marked by *Metalegoceras*. Member 2 of the overlying Khuff formation is characterised by *Pseudohalorites arabicus* (Angiolini *et al.*, in Al-Husseini, 2004).

Mesozoic

The Triassic–Jurassic of the Barents Sea. In the Barents Sea, *Keyserlingites robustus* marks a MFS in the Early Triassic, Spathian; *Karangatites evolutus, Lenotropites caurus, Anagymnotoceras varium* and *Frechites lacqueatus* successive MFSs in the Middle Triassic, Anisian; and *Tsvetkovites varius* a MFS in the Middle Triassic, Ladinian. Many of the MFSs are imageable on seismic data (Rasmussen, in Vorren *et al.*, 1993). In Spitzbergen, ammonites mark the Early–Middle Jurassic, Toarcian–Aalenian.

The Triassic–Cretaceous of the Middle East. In the Zagros Mountains of south-west Iran, the Triassic is marked by ceratites (unpublished observations). The Surmeh formation is marked by the Bathonian ammonite *Dhrumaites* (originally described from the upper part of the Middle Dhruma formation of Saudi Arabia), and by the early Kimmeridgian *Torquatisphinctes* and *Virgataxioceras* (unpublished observations, based in part of identifications by Noel Morris and Hugh Owen from the Natural History Museum, London). The Garau is marked by the Tithonian–Berriasian *Berriasella* and the late Valanginian *Neocomites neocomiensis, N. similis, Olcostephanus psilostomus, O. radiatus* and *O. salinarius*; the Fahliyan by the Tithonian–Berriasian *Spiticeras*. The lower part of the Dariyan is marked by the late Barremian *Turkmeniceras*, the early Aptian *Dufreynoia*, and the undifferentiated Aptian *Cheloniceras* and *Deshayesites*; the lower part of the Kazhdumi, the 'Kazhdumi tongue' between the lower and upper parts of the Dariyan, by the Aptian *Colombiceras* and *Parahoplites*; the upper part of the Kazhdumi by the late Albian to early Cenomanian *Hamites, Knemiceras* (including *K. attenuatum, K. syriacum* and *K. uhligi*) and *Venezoliceras*. The Sarvak is marked by the late Albian *Idiohamites turgidus, Mortoniceras albense* and *Puzosia*, and the Cenomanian *Acanthoceras*; the Ilam by the Santonian *Texanites* cf. *oliveti*; the Gurpi also by the Santonian *Texanites* cf. *oliveti*, and by the Maastrichtian *Pachydiscus* and *Sphenodiscus*.

The regional Early Jurassic, Toarcian J10 MFS of Sharland *et al.* (2001), stratotypified in the Marrat formation of central Saudi Arabia, is marked by ammonites of the *Hildoceras bifrons* Zone. The Middle Jurassic, Bajocian J20 MFS, stratotypified in the lower Dhruma formation, is marked by ammonites of the *Witchellia laeviuscula* Zone. The Bathonian J30 MFS, stratotypified in the Middle Dhruma, is marked by ammonites of the *Zigzagoceras zigzag* Zone. The Callovian J40 MFS, stratotypified in the Hisyan member of the upper Dhruma, is marked by ammonites of the *Erymnoceras coronatum* Zone. The Late Jurassic, Oxfordian J50 MFS, stratotypified in the lower Hanifa formation of offshore Qatar, is marked by ammonites of the *Perisphinctes plicatilis* Zone. The Early Kimmeridgian J60 MFS, stratotypified in the upper Hanifa formation of the United Arab Emirates, is marked by ammonites elsewhere. The Late Kimmeridgian J70 MFS, stratotypified in the Kilya member of the Naifa formation of southern Yemen, is marked by ammonites of the *Hybonoticeras beckeri* Zone. The Tithonian J110 MFS, stratotypified in the Arus

Member of the Hajar formation, is marked by ammonites of the *Micracanthoceras microcanthum* Zone. The Early Cretaceous, 'mid' Aptian K80 MFS, stratotypified in the Bab member of the Shuaiba formation of the United Arab Emirates, is marked by ammonites of the *Martinioides* Zone. The middle Albian K100 MFS, stratotypified in 'marker limestone bed II' of the Nahr Umr formation of Oman, is marked by *Knemiceras dubertreti*. The Late Cretaceous, Maastrichtian K180 MFS, stratotypified in the Simsima formation of northern Oman and the eastern United Arab Emirates, is marked by ammonites of the *Pachydiscus neubergicus* Zone.

The Jurassic–'Middle' Cretaceous of north and east Africa. The Early–Middle Jurassic of a north–south transect from the Traras Mountains to the Sahara in western Algeria is characterised by local to regional as well as inter-regional to global ammonite markers, for example, the *Margarita* Horizon in the Early Jurassic, late Sinemurian; the *Tropidoceras* Horizon in the early Pliensbachian, Carixian; the *Portisi* Horizon in the late Pliensbachian, Domerian; the *Mirabile* Horizon in the early Toarcian; the *Sublevinsoni* and *Alticarinata* Horizons in the middle Toarcian; the *Haplopleuroceras* Horizon in the Middle Jurassic, late Aalenian–early Bajocian; the *Sauzei* Horizon in the early Bajocian; the *Ermoceras* Horizon in the late Bajocian; and the *Oraniceras* Horizon in the early Bathonian (Elmi *et al.*, in Crasquin-Soleau and Barrier, 1998b). Also here, palaeobathymetric and other evidence points toward rapid subsidence and deepening associated with rifting in the Traras Mountains in the Toarcian.

The Nara formation of central Tunisia is marked by Early Jurassic, early Toarcian, *Polymorphum* Zone, and *Falcifer* or *Serpentinum* Zone, by middle Toarcian, *Bifrons* Zone, *Sublevensoni* Sub-Zone, and by late Toarcian, *Variabilis* or *Thouarsense* Zone ammonites; by Middle Jurassic, late Aalenian *Concavum* Zone and early Bajocian, *Discites* to *Humphresianum* Zone ammonites; by late Bajocian, *Niortense* Zone ammonites; by early Bathonian, *Zigzag* and *Aurigena* Zone and middle Bathonian, *Progracilis* Zone ammonites; by late Bathonian, *Breveri* to *Retrocostatum* Zone ammonites; by Late Jurassic, middle Callovian, *Anceps* Zone ammonites; by late Callovian, *Athleta* Zone ammonites; and by early Oxfordian, *Mariae* and *Cordatum* Zone, undifferentiated middle Oxfordian and late Oxfordian, *Bimammatum* Zone ammonites (Soussi *et al.*, in Crasquin-Soleau and Barrier, 2004). The major unconformities interpreted between the Sinemurian and Toarcian (UC1), between the Toarcian and late Aalenian (UC2), and between the Bathonian and middle Callovian (UC3), are further interpreted to be related to rifting.

The Antalo limestone of the Danakil horst and Afar depression in Eritrea is marked by the Late Jurassic, late Callovian ammonite *Peltoceras*, and the middle–late Oxfordian *Epimayaites falcoides et sp.*, *Mayaites* and *Perisphinctes* (Sagri *et al.*, in Crasquin-Solea and Barrier, 1998a).

The Early Cretaceous, latest Albian, *Dispar* Zone of Algeria is marked by the ammonites *Mortoniceras, Stoliczkaia, Hysteroceras* and *Knemiceras*; the earliest Cenomanian, *Mantelli/Cantianum* Zone, *Saxbi* Sub-Zone by *Mantelliceras saxbi, Sharpeiceras* and *Cunningtoniceras* (Majoran, 1987).

The Middle Jurassic–Cretaceous of the former Soviet Union. The Middle–Late Jurassic to earliest Cretaceous ammonite stratigraphy of Transcaucasia and the Mangyshlak peninsula in the peri-Caspian is discussed by Rostovtsev (1985), Vinks (1988) and Gaetani *et al.*, in Crasquin-Soleau and Barrier (1998b); the Late Jurassic of the Volga basin by Dain and Kuznetsova (1971) and Hantzpergue *et al.*, in Crasquin-Soleau and Barrier (1998b); the Late Jurassic–Cretaceous of the Russian platform by Vishnevskaya *et al.* (1999); the Early Cretaceous of the Mangyshlak peninsula by Kopaevich *et al.* (1999); the Early and Late Cretaceous of the Crimea, north Caucasus, peri-Caspian and Russian platform by Baraboshkin, in Crasquin-Soleau and Barrier (1998b) and Baraboshkin *et al.* (2003); and the Late Cretaceous of the Urals and west Siberia by Amon and de Wever, in Roure (1994). The calibration of the ammonite stratigraphy of the Russian platform with the standard north-west European scheme is discussed by Grocke *et al.* (2003).

The Middle Jurassic–Cretaceous sequence stratigraphy of the Russian platform, calibrated against ammonite stratigraphy, is discussed by Sahagian *et al.* (1996). The Middle–Late Jurassic sequence stratigraphy of the Nyurolskaya depression in west Siberia, calibrated against ammonite stratigraphy, is discussed by Pinous *et al.* (1999b). The Early

Cretaceous, Neocomian sequence stratigraphy of the 'productive complex' of the Priobskoe oil field and elsewhere in west Siberia, calibrated against ammonite stratigraphy, is discussed by Pinous *et al.* (1999a, 2001). Recovered ammonites indicate definite development of the *Neotollia, Temnoptychites insolitus* and *Polyptychites michalskyi* Zones of the early Valanginian; the *Dichotomites* Zone of the late Valanginian; the *Homolsomites bojarkensis* Zone of the early Hauterivian; and the *Speetoniceras inversum* Zone of the late Hauterivian. Importantly, the zonal resolution is sufficient as to allow detailed maps of reservoir distribution through time to be drawn.

The latest Jurassic–Cretaceous of northern South America. A wide range of ammonite zones is recorded in the latest Jurassic, Kimmeridgian to Cretaceous in Colombia (Petters, 1954, 1955), and in the Cretaceous of Venezuela (Maync, 1949; Gonzalez de Juana *et al.*, 1980; Renz, in Reyment and Bengtson, 1981; Vahrenkamp *et al.*, in Simo *et al.*, 1993; unpublished observations). Zonal marker taxa common to both countries include *Cheloniceras* and *Oxytropidiceras* in the Late Aptian–Early Albian, *Coelopoceras* in the Cenomanian or Turonian, *Texanites* in the Santonian, and *Sphenodiscus* in the Maastrichtian.

The Cretaceous of the Gulf Coast and western interior seaway of North America. The stratigraphic usefulness of the ammonites of the Late Cretaceous, Cenomanian of the Black Hills of north-eastern Wyoming and south-eastern Montana is enhanced by graphic correlation (Fisher, in Mann and Lane, 1995).

The 'Middle'–Late Cretaceous of west Africa. The Madiela formation of Gabon is marked by the Early Cretaceous, late Aptian ammonite *Prodeshayesites consorboides* and the early Albian *Douvilleiceras mammillatum*; the middle–late Albian of Gabon, Ivory Coast and Nigeria by *Oxytropidoceras, Diploceras, Hysteroceras, Mortoniceras, Elobiceras, Neoharpoceras* and *Stoliczkaia*; the Late Cretaceous, Cenomanian of Angola and Gabon and the Odukpani formation of Calabar by *Acanthoceras, Acompsoceras, Turrilites, Euhystrichoceras, Forbesiceras* and *Sharpeiceras*; the Eze–Aku group of Nigeria by early Turonian vascoceratids and pseudotissotids; the Turonian of Nigeria and of Cameroon by hoplitoids, mammitids and advanced pseudotissotids; the undifferentiated Turonian of Nigeria, Senegal and Brazzaville by *Peroniceras, Gauthiericeras, Solgerites, Forresteria, Reginaites* and *Tissotia*; the Coniacian of Gabon by *Peroniceras, Gauthiericeras* and barroisiceratids; the higher Senonian of Libreville by *Texanites* and *Hoplitoplacenticeras*; the Santonian of Nigeria, Douala and Vosso in Zaire by *Texanites*; the early Campanian of Nigeria by *Menabites, Krapadites* and *Eupachydiscus*; the late Campanian and Maastrichtian of Nigeria by *Libycoceras, Sphenodiscus, Baculites* and *Cirroceras* (Kogbe and Mehes, 1986; Meister *et al.*, 1992; Pascal *et al.*, in Simo *et al.*, 1993).

3.6.12 Molluscs: belemnites

Biology, morphology and classification

Belemnites are an extinct, Mesozoic, Jurassic–Cretaceous, group of coleoid cephalopod molluscs, characterised by vestigial internal shells. They take their scientific name from the Greek word *belemnon*, meaning 'dart', referring to their characteristic shape, which also led to the folk belief that they were fossilised thunderbolts. They also have a popular name, again referring to their shape: 'devil's fingers'!

Biology

Little is known directly of the biology of the belemnites, as they are extinct, and there are only occasional instances of the preservation of their soft parts as fossils (Doyle and Shakides, 2004).

However, the soft parts that have on those occasions been preserved include ink sacs similar of those of living coleoids such as sepiids or cuttlefish, octopods or octopus, and teuthids or squid. Thus, inferences as to the biology of the belemnites can be drawn from that of these, evidently related, living animals (the living 'paper Nautilus' *Argonauta* is also related).

Chitinous hook- or pincer-like structures called arm-hooks have also been preserved, that have been interpreted as having been attached to the belemnite animal's ten tentacles, and as having been used to grasp prey animals, or mates!

Belemnite prey is interpreted to have included other molluscs, crustaceans and small fish. Predators are interpreted to have included fish such as sharks and marine reptiles such as ichthyosaurs and marine crocodiles. Interestingly, some specimens of fossil sharks and ichthyosaurs have been found with their stomachs containing common belemnite arm-hooks, but only rare skeletons or rostra, possibly indicating either that only the head and tentacles of the belemnite were eaten, or that the whole animal was swallowed and any unpalatable parts regurgitated.

Morphology

In terms of preservable hard-part morphology, belemnites are characterised by a skeleton or rostrum of conical shape, like a cuttlefish bone, which is interpreted as having been located internally and posteriorly, with the pointed end or guard directed posteriorly, and the opposite end or pro-ostracum anteriorly (Fig. 3.58).

The guard is solid and composed of radially arranged crystals of calcite, with their long axes aligned perpendicular to the surface. It may also be characterised by concentrically arranged structures that have been interpreted as growth rings.

The pro-ostracum, representing a reduced body chamber, tends to break off after death, and is seldom found in the fossil record.

The guard and pro-ostracum were joined in life in the mid-section of the rostrum by the phragmacone, that fitted into a conical depression or alveolus at the anterior end of the guard.

The phragmacone is essentially hollow, although it is characterised by closely spaced internal transverse walls or septa that divide it into a series of narrow chambers. The chambers are in communication through openings in the septa that form a so-called siphuncular tube or siphuncle along the ventral margin.

The belemnite animal is interpreted to have lived with its body in the last chamber, and with its head and tentacles protruding.

The rostrum is interpreted as having acted as a counter weight to the head and tentacles at the anterior end of the animal, and as having enabled it to maintain an even keel.

Classification

Classification of belemnites is essentially based on details of morphology, such as overall shape and size, surface ornamentation (in the form of grooves etc.), and growth characteristics. The shape is particularly important, and is variously described as cylindrical, conical or hastate (spear-shaped) in ventral or dorsal outline, symmetrical or asymmetrical in lateral profile, and quadrate through pyriform and circular to elliptical in section. The shape of the apex is described as attenuate, acute, obtuse or mucronate, depending on the angle it subtends.

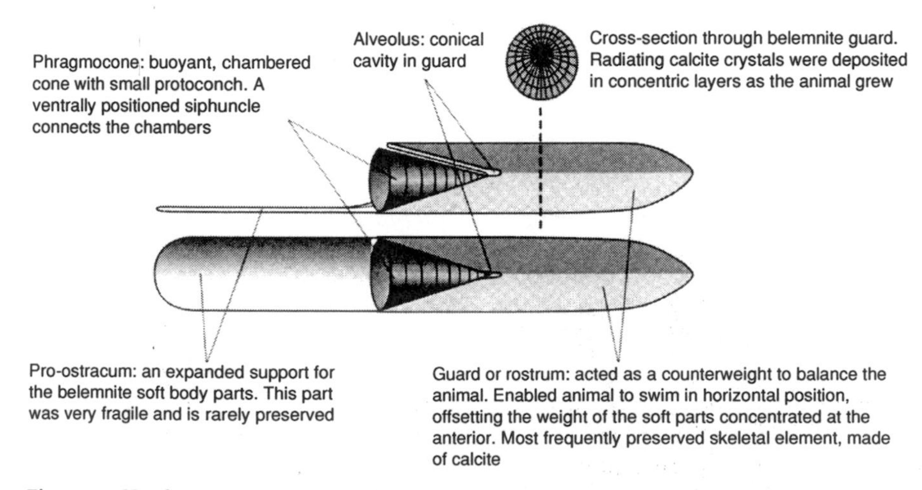

Fig. 3.58. Hard-part morphology of belemnites (from Milsom and Rigby, 2004).

Palaeobiology

Belemnites are interpreted, essentially on the bases of analogy with their living counterparts, and of associated fossils and sedimentary facies, as having been marine and nektonic, horizontally balanced in the water.

Palaeobathymetry

Belemnite rostra commonly exhibit preferred orientation, enabling inference as to the activity and direction of palaeo-currents, not necessarily always in shallow water (see below).

Mass accumulations, or 'belemnite battlefields', may be related to mass mortality events taking place after mating, as observed in living coleoids; to other catastrophic events; to concentration by predation, either by regurgitation (see above) or by the death of the predator; or to concentration by sedimentological processes, such as winnowing in shallow water, and gravity flow in deep water (Doyle and MacDonald, 1993).

Palaeobiogeography

Belemnites appear to have evolved in the Mediterranean region of Tethys, and in the Early and Middle Jurassic were essentially restricted to that region, with only isolated occurrences elsewhere (Stevens, in Hallam, 1973). By the Late Jurassic, though, they had extended their range to cover much of the world. Nonetheless, even at this time, many belemnites had restricted or endemic biogeographic distributions, rendering them of use in the characterisation of palaeobiogeographic provinces, and in turn in the constraint of plate tectonic reconstructions. Thus, at this time, the Boreal realm is characterised by the belemnitids and cylindroteuthids; and the Tethyan realm by the docoelitids, duvaliids, *Belemnopsis* and *Hibolites*. Moreover, Arctic and Boreal–Atlantic provinces are characterised within the Boreal realm; and Mediterranean, Ethiopian and Indo-Pacific provinces within the Tethyan realm. In the Early Cretaceous, the Boreal realm is characterised by cylindroteuthids and newly evolved oxyteuthids; the Tethyan realm by duvaliids, *Belemnopsis, Hibolites* and the newly evolved *Curtohibolites, Mesohibolites, Neohibolites* and *Parahibolites*; and the Austral realm by dimitobelids. By the Late Cretaceous, the cylindroteuthids and oxyteuthids were extinct, and the (sub-)Boreal realm was characterised by the newly evolved *Actinocamax, Belemnella, Belemnitella* and *Gonioteuthis*.

Palaeoclimatology

Importantly, belemnites are of considerable use in palaeotemperature and palaeoclimatic as well as in palaeobiogeographic interpretation, palaeobathymetric complications notwithstanding (see Section 4.5). This is on account of their enhanced resistance to chemical change during diagenesis, and their consequent ability to preserve intact geochemical signals such as oxygen – and strontium – isotope ratios. Indeed, the 'Pee Dee' belemnite standard is that against which such measurements are calibrated.

Biostratigraphy

True belemnites evolved in the Jurassic, and became extinct during the End-Cretaceous mass extinction. The arguably ancestral aulacocerids ranged from ?Devonian to Jurassic.

The moderate rate of evolutionary turnover exhibited by the belemnites, and their generally good preservation, even in diagenetically altered or even metamorphic rocks, renders them of some use in biostratigraphy, at least in appropriate, marine, environments (Fig. 3.59). Note, though, that their usefulness is somewhat compromised by the restricted, regional biogeographic distribution they exhibited in the earliest and again in the latest stages of their evolution.

In Great Britain and Northern Ireland, belemnites form part of the basis of the biozonation of the Late Cretaceous Chalk, with *Gonioteuthis quadrata* and *Belemnitella mucronata* marking the middle and late parts of the Campanian, respectively, and *Belemnella lanceolata* the early part of the Maastrichtian (Wood, 1988; Christensen, 1995).

In my working experience, belemnites have proved of use in the Late Cretaceous of the Crimea, north Caucasus, peri-Caspian, Russian platform, Urals and west Siberia in the former Soviet Union (Amon and de Wever, in Roure, 1994; Amon *et al.*, 1997; Baraboshkin *et al.*, 2003).

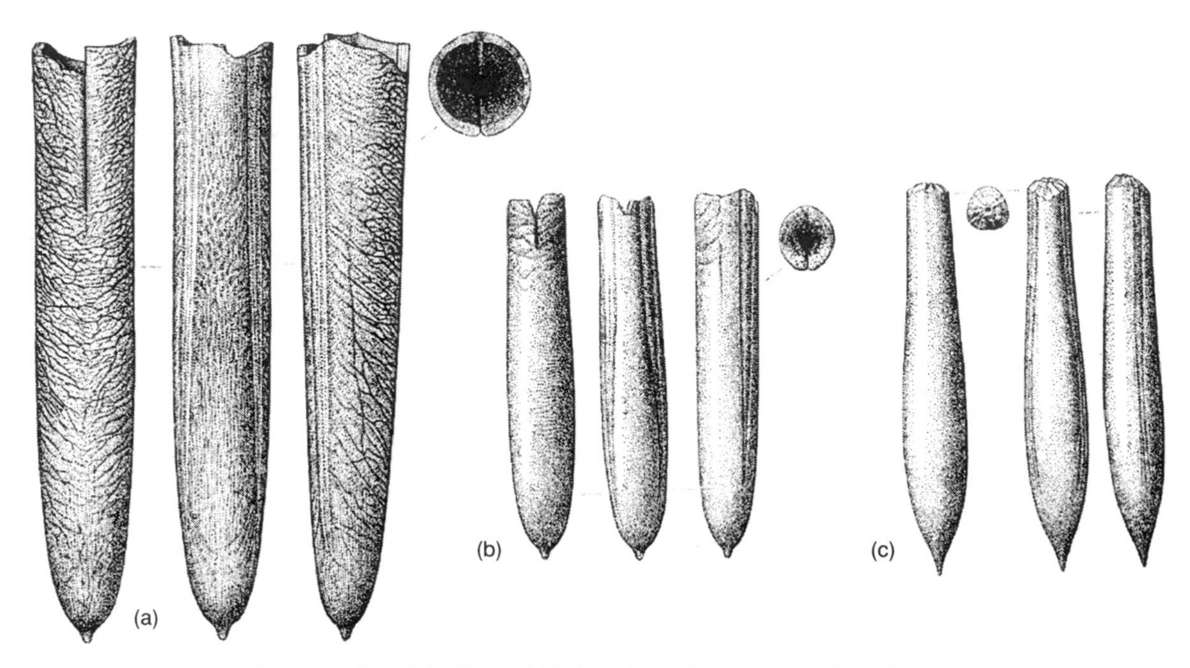

Fig. 3.59. Some stratigraphically useful belemnites. (a) *Actinocamax plenus*, Late Cretaceous, upper Cenomanian/lower Turonian, *Plenus* Marl, West Cliffs, Dover, Kent, ×0.75; (b) *Gonioteuthis quadrata*, nominate taxon of the Late Cretaceous, Campanian *Gonioteuthis quadrata* Zone, Upper Chalk, East Harnham, near Salisbury, Wiltshire, ×0.75; (c) *Belemnitella mucronata*, nominate taxon of the Late Cretaceous, Campanian *Belemnitella mucronata* Zone, Upper Chalk, Whitlingham, Norfolk, ×0.75. (Modified after the Natural History Museum, 2001b.)

3.6.13 *Molluscs: tentaculitids*

Biology, morphology and classification

Tentaculitids are an extinct, Ordovician?–Devonian, group of uncertain, although arguably molluscan, affinity.

Little is known of the biology of tentaculitids, as they are extinct, and their soft parts are only exceptionally preserved, as in the Early Devonian Hunsruck slate Lagerstatte of the Rhenish massif in Germany (Bartels *et al.*, 1998). However, it is known from high-magnification X-ray examination of slabs of Hunsruck slate that they possessed tentacles with which to capture food, and guts, and, arguably, siphuncles, suggesting a relationship with the cephalopod molluscs.

In terms of hard-part morphology, tentaculitids are characterised by elongate, conical shells variously adorned with transverse rings and transverse and longitudinal ribs.

Classification of tentaculitids is based on details of morphology, and on phylogeny inferred from empirical observations on the fossil record.

Palaeobiology

Tentaculitids have historically been interpreted, on the basis of associated fossils and sedimentary facies, as having been marine and nektonic. However, it is also possible to interpret them, on the basis of functional morphological analogy with the microconchids, as having been benthic, and either cemented to a hard substrate or vertically embedded in a soft substrate, with their feeding apparatuses extended into the overlying water column (Bartels *et al.*, 1998).

They had a wide geographic distribution in low to moderate latitudes (central and southern Europe, north Africa, South Africa, central Asia and China).

Biostratigraphy

Tentaculitids evolved in the ?Ordovician or Silurian. They became extinct in the Late Devonian mass extinction, sustaining severe losses in the late *Rhenana* Zone of the Frasnian, and dying out in the *Triangularis* Conodont Zone of the Fammenian (Schindler, in Kauffman and Walliser, 1990).

The rapid rate of evolutionary turnover exhibited by the tentaculitids renders them of considerable use in biostratigraphy, at least in appropriate, marine, environments, and in low to moderate latitudes.

High-resolution biostratigraphic zonation schemes based on the tentaculitids have been established for the Devonian, that have widespread applicability, especially in central Europe, and that have been calibrated against conodont standards (Boucek, 1964; Lutke, in House, 1979; Alberti, 1983). In the Barrandian area of the Czech Republic, the Lochkovian is represented by two zones; the Pragian by three; the Zlichkovian by four; the Daleian by three; the Eifelian by three; and the Givetian by one (Holcova, in Bubik and Kaminski, 2004).

In my working experience, tentaculitids have proved on particular use in *the Devonian of north Africa*, for example in the Ahnet basin, Ougarta Hills and elsewhere in Algeria (Follot, 1952; Le Fevre, 1971; Alberti, 1983), in Libya (Bellini and Massa, in Salem and Busrewil, 1980), and in Morocco (Alberti, 1983).

3.6.14 Arthropods: trilobites

Arthropods include three sub-groups that are important in the fields of palaeobiology and biostratigraphy, namely the trilobites, ostracods and insects, which are discussed in turn below. There are thought to be several million living species, the overwhelming majority of which are insects (although there are also something of the order of 100 000 living species of other arthropods).

Biology, morphology and classification

Trilobites are an extinct, Palaeozoic, Cambrian–Permian, group of trilobitomorph arthropods characterised by three-lobed external skeletons (Mickulic, 1990; Whittington, 1992; Levi-Setti, 1993; Fortey, 2000; Lane *et al.*, 2003).

Their most immediate relatives would appear to be the chelicerates, including the eurypterids or sea-scorpions, the arachnids or sea-spiders, spiders and scorpions, and the xiphosurans or horseshoe-crabs. None of these related groups is especially important in the fields of palaeobiology and biostratigraphy, and none is discussed. Note, though, that fossil forms are known. For example, fossil eurypterids are known from the Silurian of Lesmahagow in Lanarkshire in the Midland Valley of Scotland and from the Carboniferous 'Granton shrimp bed' of Granton near Edinburgh in Lothian, arachnids from the Carboniferous 'Coal Measures' of Dudley in the West Midlands, and xiphosurans from the 'Coal Measures' of Coseley in the West Midlands, all in Britain. Arachnids and xiphosurans are also known from the Late Jurassic Solenhofen limestone of Bavaria in southern Germany, and arachnids from Eocene–Miocene amber deposits of the Dominican Republic, and Oligocene lacustrine deposits of Florissant in Colorado.

Biology

Little is known directly of the biology of trilobites as they are extinct, and there are only rare instances of the preservation of their soft parts as fossils.

However, from these, we know that trilobites had a pair of jointed, apparently sensory, antennae protruding forwards from the head, and rows of pairs of jointed limbs on either side of the body, three on the head, and one on every segment of the body and tail.

Moreover, inferences as to the biology of the trilobites can be drawn on the basis of analogy with their closest living relatives, the horseshoe-crab *Limulus*, which is distributed along the Atlantic coast of North America from the Bay of Fundy southwards, and related genera distributed along the coasts of Japan and Indonesia (Anderson, in Anderson, 2001; Tanacredi, 2001; Shuster *et al.*, 2003).

Thus, it is inferred that the lower parts of the limbs, below the joints, were used in movement, in paddling about on the sea bottom or swimming, life-saving breaststroke style, in the water column, and in food gathering; the upper parts, above the joints, with gill filaments or 'books', in respiration (and specially modified clasping fore-limbs, only found in males, in reproduction).

It is further inferred that the preferred habitat was at least marginal marine, and the life strategy to burrow through the sediment on the bottom of the sea in search of worms and molluscs to eat (or, if available, as in aquaria in certain laboratories, dog biscuits).

Incidentally, because the requisite patterns of muscular contraction are antagonistic, *Limulus* is unable to move and eat at the same time, in this if in no other respect putting me in mind of the former US President Gerald Ford!

Morphology

In terms of preservable hard-part morphology, trilobites are characterised by an essentially calcitic external skeleton covering at least the exposed dorsal region (Figs. 3.60–3.61). The venter may simply have been protected by a soft skin-like membrane with poor preservation potential.

The skeleton is divisible into three, both from anterior to posterior, and from side to side. From anterior to posterior, the parts are the head or cephalon, body or thorax, and tail or pygidium; from side to side, lateral lobes or pleurae on either side of a central axial lobe interpreted as having housed the digestive system. Trilobites in which the pygidia are smaller than the cephala are described as micropygous; those in which the pygidia and cephala are of equal size as isopygopus; those in which the pygidia are larger than the cephala as macropygous.

The head, body and tail are all in turn subdivided into segments. Those of the head and tail were fused, while those of the body were free, enabling at least a modicum of motility (some forms were able to roll into a ball for protection like a modern woodlouse, and some of these possessed sophisticated

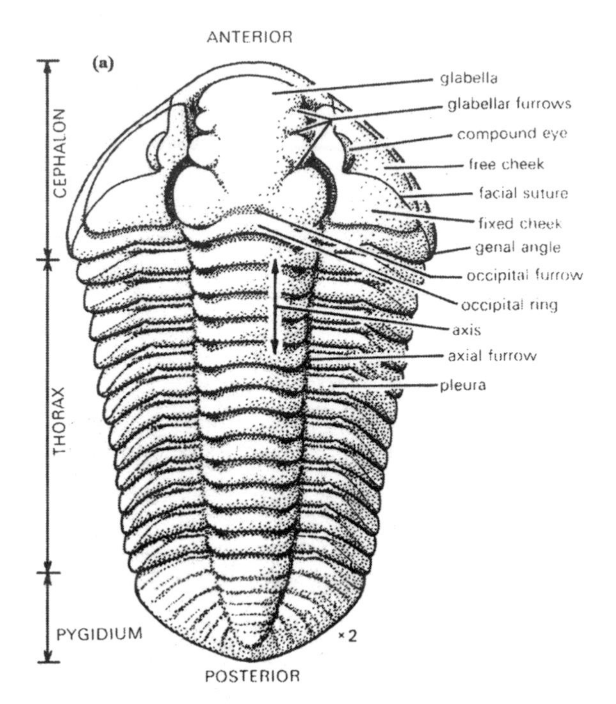

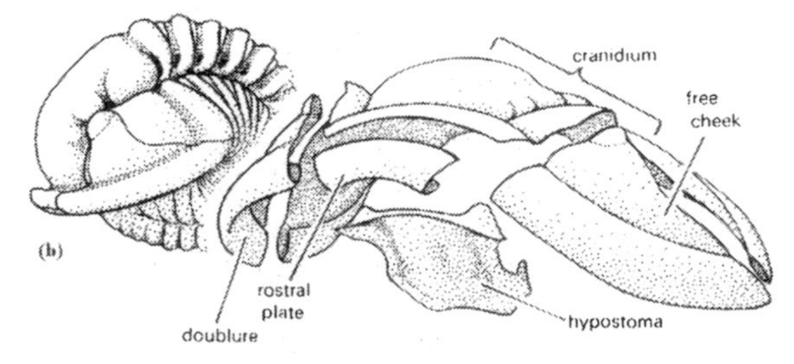

Fig. 3.60. Hard-part morphology of trilobites. (a) Dorsal view of *Calymene*; (b) lateral view of enrolled specimen, and oblique frontal view of cephalon exploded along suture lines, so as to show relationships of regions. (Modified after Black, 1989.)

Fig. 3.61. Some representative trilobites. Redlichiides: (a) *Olenellus lapworthi*, Lower Cambrian, Kinlochewe, Ross-shire, ×1.5; (b) *Paradoxides davidus*, Middle Cambrian, St Davids, Dyfed, ×0.5. Agnostide: (c) *Lotagnustus trisectus*, Late Cambrian, Malvern Hills, ×3. Corynexochides: (d) *Scutellum costatum* (tail), Middle Devonian, Newton Abbot, Devon, ×0.75; (e) *Bumastus (Illaenus) barriensis*, Silurian, Wenlock, Dudley, West Midlands, ×0.5. Ptychopariide: (f) *Olenus gibbosus*, Late Cambrian, Dolgellau, Gwynedd, ×1.25. Lichide: (g) *Leonaspis deflexa*, Silurian, Wenlock, Dudley, West Midlands, ×2. Phacopides: (h) *Remopleurides girvanensis*, Ordovician, Caradoc, Girvan, Ayrshire, ×1.5; (i) *Trinucleus fimbriatus*, Ordovician, Caradoc, Builth Wells, Powys, ×2; (j) *Ampyx linleyensis*, Ordovician, Llanvirn, Linley, Shropshire, ×1.5; (k) *Onnia gracilis*, Ordovician, Caradoc, Onny Valley, near Winstantow, Shropshire, ×1.25. Phacopides: (l) *Phacops accipitrinus* (head), Late Devonian, Shirwell, near Barnstaple, Devon, ×1; (m) *Dalmanites myops*, Silurian, Wenlock, Dudley, West Midlands, ×1. Phacopides: (n) *Acaste downingiae*, Silurian, Wenlock, Dudley, West Midlands, ×1.5; (o) *Chasmops extensa* (head), Ordovician, Caradoc, Onny Valley, near Winstantow, Shropshire, ×0.75; (p) *Calymene blumenbachi*, Silurian, Wenlock, Dudley, West Midlands, ×1; (q) *Trimerus delphinocephalus*, Silurian, Wenlock, Dudley, West Midlands, ×1. Proetide: (r) *Phillipsia gemmulifera* (tail), Early Carboniferous, Bowland, near Clitheroe, Lancashire, ×2. (Modified after the Natural History Museum, 2001a.)

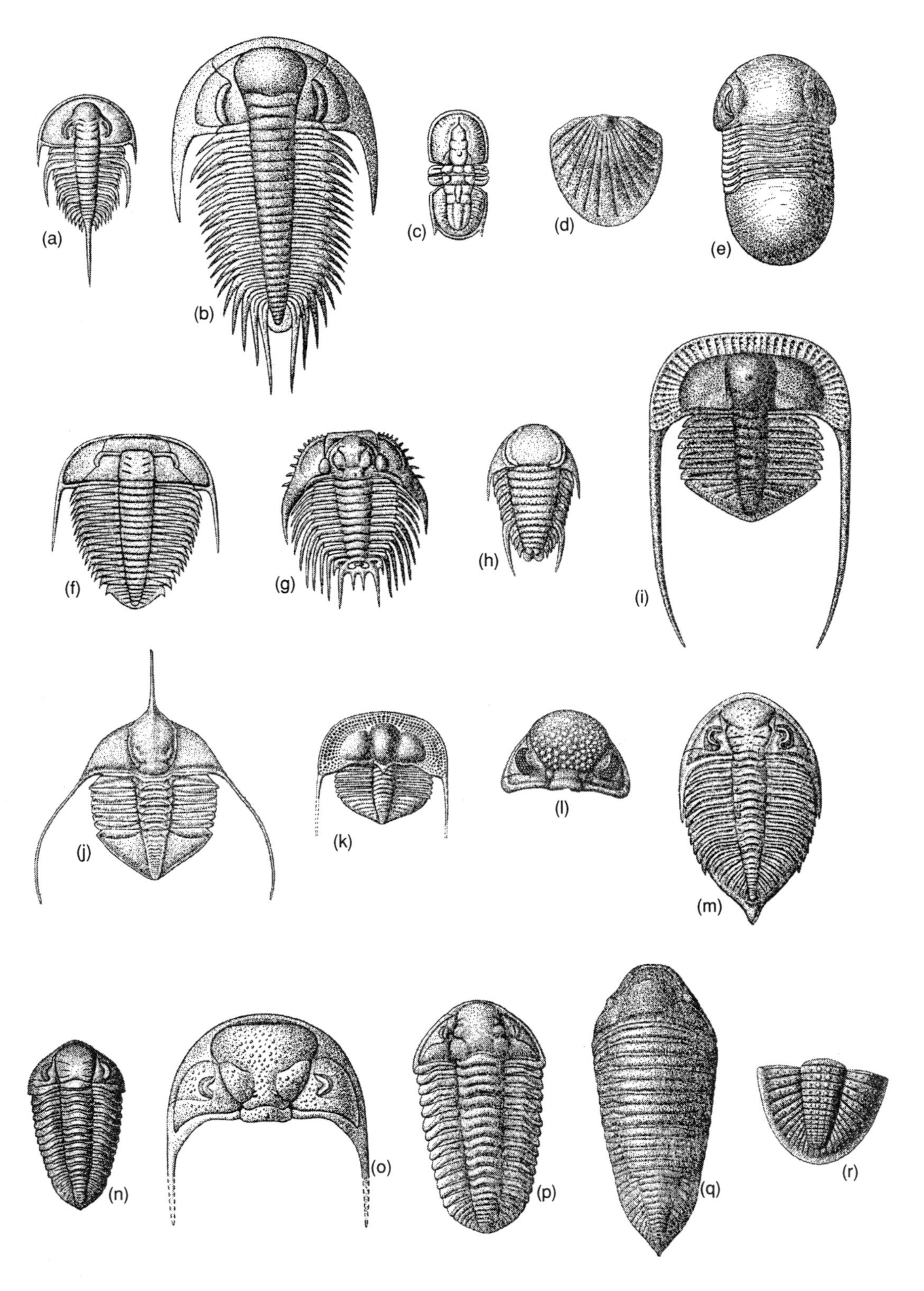
(a)
(b)
(c)
(d)
(e)
(f)
(g)
(h)
(i)
(j)
(k)
(l)
(m)
(n)
(o)
(p)
(q)
(r)

coaptative structures capable of locking the skeleton in the rolled position).

The head is also subdivided from side to side into fixed and free cheeks on either side of a central glabella. The fixed and free cheeks are separated by a suture, which is thought to have facilitated the process of moulting or ecdysis, which must have taken place periodically to accommodate the growth of the animal (in fact, there would appear to have been between six and eight moult stages). There are essentially five types of suture: ventral, lateral, proparian, gonatoparian and opisthoparian.

The eye, if present (it is absent in many forms that are interpreted to have lived in the shade or in the dark, in burrows or below the light-floor), is present on the inner margin of the free cheek, adjacent to the fixed cheek (Clarkson, 1979). It is compound, that is, it consists of a number of separate lenses. Primitive, simple types of eye, characterised by numerous small lenses composed of individual crystals of calcite arranged with their long axes aligned perpendicular to the surface, are categorised as holochroal; more advanced, complex, types, characterised by fewer, larger lenses arranged in discrete rows, as schizochroal. Depending on the type of eye possessed, trilobites would have had either comparatively poor vision, this presumably being all that they required, or comparatively good vision. Some, such as *Encrinurus*, evolved eyes mounted on stalks, presumably enabling them to cover themselves in soft sediment on the sea-floor for camouflage, and yet still see out. Others evolved complex arrangements that probably gave them all-round vision, like a modern fly, with some even able to correct for spherical aberration, or to estimate distances. Interpreted pelagic forms evolved downward-directed eyes. Incidentally, it has been suggested that the evolution of the trilobite eye, indeed of eyes generally, was driven by predation pressure at the time of the 'Cambrian explosion' (A. Parker, 2003; see also Sub-section 5.3.1).

Interpreted sensory organs other than eyes include various types of canals, pits, tubercles, spines and bristles or setae on the surface of the skeleton, for example the cephalic fringes of the trinucleids and harpids.

A so-called hypostome is found on the underside of the head, and is thought to have been some form of support for the mouth parts. There are three types of hypostome: conterminant, impedent and natant.

Classification

Classification of trilobites is based on morphology, and on phylogeny inferred from empirical observations on the fossil record. Nine sub-groups are recognised, namely: the redlichiides, the agnostides, the corynexochides, the ptychopariides, the naraoiides, the lichides, the asaphides, the phacopides and the proetides.

Redlichiides. The redlichiides, which range from Early to Middle Cambrian, are characterised by numerous thoracic segments, much reduced pygidia, and spinose morphology. They are exemplified by *Redlichia, Olenellus* and *Paradoxides*.

Agnostides. The agnostides (Cambrian–Ordovician) are characterised by minute size, and by much reduced thoraces, with only two segments. They are also blind, and are thus interpreted as probably having occupied deep marine environments below the light-floor. They are widely distributed, and are thus interpreted as having been pelagic. They are exemplified by *Agnostus* and *Lotagnostus*.

Corynexochides. The corynexochides (Cambrian–Middle Devonian) are characterised by subequal cephala and pygidia and by moderately numerous thoracic segments, by conterminant to impedent hypostomes, and by opisthoparian sutures. They are exemplified by *Olenoides, Scutellum, Illaenus* and *Chuangia*.

Ptychopariides. The ptychopariides (Cambrian–Devonian) are characterised by natant hypostomes or by conterminant or impedent hypostomes derived from the natant condition. They are exemplified by *Olenus, Shumardia* and *Harpes*, which last-named lacks eyes, but possesses a well-developed sensory cephalic fringe, and is interpreted as a burrower.

Naraoiides. The naraoiides (Middle Cambrian) are characterised by soft bodies, lacking thoracic segments altogether. They are exemplified by *Naraoia*, which was originally described from the Burgess shale of British Colombia (as a branchiopod).

Lichides. The lichides (Middle Cambrian–Middle Devonian) are characterised by conterminant hypostomes, and by spinose morphology. They are exemplified by *Lichas* and *Leonaspis*, which latter has been assigned by some authors to a separate group, the odontopleurides.

Asaphides. The asaphides (Late Cambrian–Silurian) are characterised by conterminant to impedent hypostomes derived from the natant condition, and by ventral sutures. They are exemplified by *Asaphus*, by *Remopleurides*, by interpreted pelagic forms such as *Cyclopyge*, the amphipod- or isopod-like *Opipeuter* and *Nileus*, and by *Trinucleus, Ampyx* and *Onnia*.

Phacopides. The phacopides (Ordovician–Devonian) are characterised by holochroal to schizochroal eyes, and by dominantly proparian sutures. They are exemplified by *Phacops*, by *Dalmanites*, with its distinctive telson-like tail, and by *Acaste, Chasmops, Calymene, Trimerus* and *Cheirurus*.

Proetides. The proetides (Ordovician–Permian) are characterised by large hypostomes, large glabellae, holochroal eyes and opisthoparian sutures. They are exemplified by *Proetus, Bathyurus* and *Phillipsia*.

Palaeobiology

Trilobites are interpreted, essentially on the bases of analogy with their living counterparts, namely horseshoe-crabs, and associated fossils and sedimentary facies, as having been at least marginal marine, and either benthic, nektobenthic, or nektonic.

Trilobites are also interpreted, essentially on the basis of functional morphology, as having been benthic and infaunal, as in the case of the smooth 'illaenimorph' morphotypes, to epifaunal and motile, as in the case of the tuberculate 'phacomorph' morphotypes, to pelagic (Fortey and Owens, in McNamara, 1990). In fact, a further five trilobite morphotypes have been recognised, namely 'atheloptic', 'marginal/cephalic spines', 'miniaturization', 'pitted fringe' and 'olenimorph' (Fortey and Owens, in McNamara, 1990). Articulated *Asaphus (Asaphus) raniceps* body fossils have recently been discovered in *Thalassinoides* burrows (Cherns *et al.*, 2004).

Trilobites are further interpreted, again essentially on the basis of functional morphology, as having practised a variety of feeding strategies, including predation or scavenging, detritus-feeding and filter-feeding (Fortey and Owens, 1999; Hughes, in Zhuravlev and Riding, 2001). Interpreted predators or scavengers are characterised by claw-like spines at the end of their legs, rigidly attached conterminant or impedent hypostomes with posterior forks, and expanded anterior glabellar folds; and are exemplified by *Colpocoryphe, Dindymene, Illaenopsis* and *Ormathops*. Interpreted detritivores are characterised by detached hypostomes, that may have functioned as scoops; and are exemplified by *Shumardia* and other generalised ptychopariides (and, incidentally, by the *Cruziana* trace-maker). Interpreted filter-feeders are characterised by elevated heads and thoraces, often flanked by extended genal spines; and are exemplified by *Ampyx* and other generalised trinucleids.

Many trilobites bore spines either as a defence against predators, or as an adaptation to reduce the sinking rate within the water column in the case of pelagic forms, or to maintain stability on, or to facilitate burrowing into, the sea-floor, in the case of benthic forms.

Palaeobathymetry

In general, shallow marine benthic environments are characterised by low-diversity assemblages of sighted, and typically large, trilobites, including *Flexicalymene, Merlinia* and *Neseuretus* (and the locomotion or feeding trace *Cruziana*), in association with algae, crinoids and gastropods; intermediate depth benthic environments by high-diversity assemblages, including *Selenopeltis, Geragnostus, Chasmops* and *Remopleurides*, in association with brachiopods and nautiloids; deep-marine environments by moderate diversity assemblages of blind, and typically small, forms including *Ampyx, Tretaspis, Shumardia* and trinucleids; and pelagic environments by forms with downward-directed eyes, including *Cyclopyge*.

A general model of sighted trilobites in illuminated, shallow-water environments and blind ones in dark, deep-water environments was proposed by Clarkson, in Hallam (1967), and slightly modified by

Lamont, in Hallam (1967), to take into account his observations of blind trilobites in interpreted dark, shallow-water environments such as algal glades. The general model is complicated by the further observations that blind and sighted species often occur together, in interpreted deep-water as well as shallow-water environments. Interestingly, similar observations have been made on Recent arthropods. Importantly, though, while sighted Recent arthropods from below 200 m have normal-looking eyes specially adapted to the perception of dim light (eyes, incidentally, rejoicing in the wonderful, Wagnerian name of 'Dammerungsaugen'), those from below 500 m have distinctly degenerate eyes, possibly sensitive to bioluminescence.

The general model of trilobite distribution is further complicated by evident substrate as well as depth control (Fortey and Cocks, 2003). Taking this into account, specifically in the Early Ordovician of Laurentia, intertidal environments are characterised by bathyurids, shallow marine limestones by illaenids and cheirurids, deep marine limestones by nileids, and deep marine graptolitic shales by olenids. In the Middle Silurian of Gondwana, marginal to shallow marine sandstones are characterised by *Acaste* and *Trimerus*, shallow marine limestones by *Proetus* and *Warburgella*, shallow marine shales by *Dalmanites* and *Raphiophorus*, deep marine limestones by *Radnoria* and *Cornuproteus*, and deep marine graptolitic shales by *Delops* and *Miraspis*.

Palaeobiogeography

Many trilobites appear to have had restricted or endemic biogeographic distributions, rendering them of use in the characterisation of palaeobiogeographic provinces, and in turn in the constraint of plate tectonic reconstructions (Palmer, in Hallam, 1973; Whittington, in Hallam, 1973; Cocks and Fortey, 1982; Cocks and Fortey, in McKerrow and Scotese, 1990; Turvey, in Crame and Owen, 2002; Fortey and Cocks, 2003; Javier Alvaro, in Servais *et al.*, 2003; Hou Xian-Guang *et al.*, 2004; Pour, 2004).

In the Cambrian, there were three trilobite provinces, a Laurentia–Baltica–Avalonia province, characterised by olenellids, a Siberia province, characterised by bigotinids, and a Gondwana province, characterised by redlichiides. The interpreted pelagic agnostides exhibited essentially cosmopolitan distributions. Throughout most of the Ordovician, there were four trilobite provinces, a Laurentia province, characterised by bathyurids, a Baltica–Avalonia province, characterised by megalaspids and asaphids, a low-latitude Gondwana province, characterised by *Selenopeltis* and dikelokephalinids, and a high-latitude Gondwana province, characterised by calymenaceans and dalmanitaceans. At this time, Baltica–Avalonia, of which England and Wales were a part, was separated from Laurentia, of which Scotland was a part, by the Iapetus Ocean, a substantial body of deep water that constituted an insuperable barrier to the dispersal of shallow-water benthic trilobites. Baltica–Avalonia was also separated from Gondwana by the Rheic Ocean. Towards the end of the Ordovician, trilobite provincialism generally decreased, and cosmopolitanism increased, as Baltica–Avalonia and Laurentia drifted into one another, thereby uniting Great Britain, albeit as part of a larger northern continent, and as said continent and Gondwana drifted towards one another. Warm-water elements became extinct, and a cosmopolitan cold-water fauna evolved, at the time of the Late Ordovician, Hirnantian glaciation. Cosmopolitanism remained the norm through the succeeding Silurian deglaciation and transgression. In the Devonian, provincialism was re-established. At this time, the Malvino-Kaffric province, centred on South America, the Falklands or Malvinas, and South Africa, became established for the first time. It is difficult to discern any palaeobiogeographic patterns in the interval from the Late Devonian mass extinction to the End-Permian mass extinction.

Biostratigraphy

Trilobites evolved in the Early Cambrian, Atdabanian. They underwent a dramatic diversification through the remainder of the Cambrian and Ordovician (Adrain *et al.*, in Webby *et al.*, 2004), before sustaining severe losses in the End-Ordovician mass extinction. Nonetheless, they recovered from this event, only to sustain severe losses again in the Late Devonian mass extinction. They finally became extinct in the End-Permian mass extinction.

The rapid rate of evolutionary turnover exhibited by the trilobites renders them of considerable use

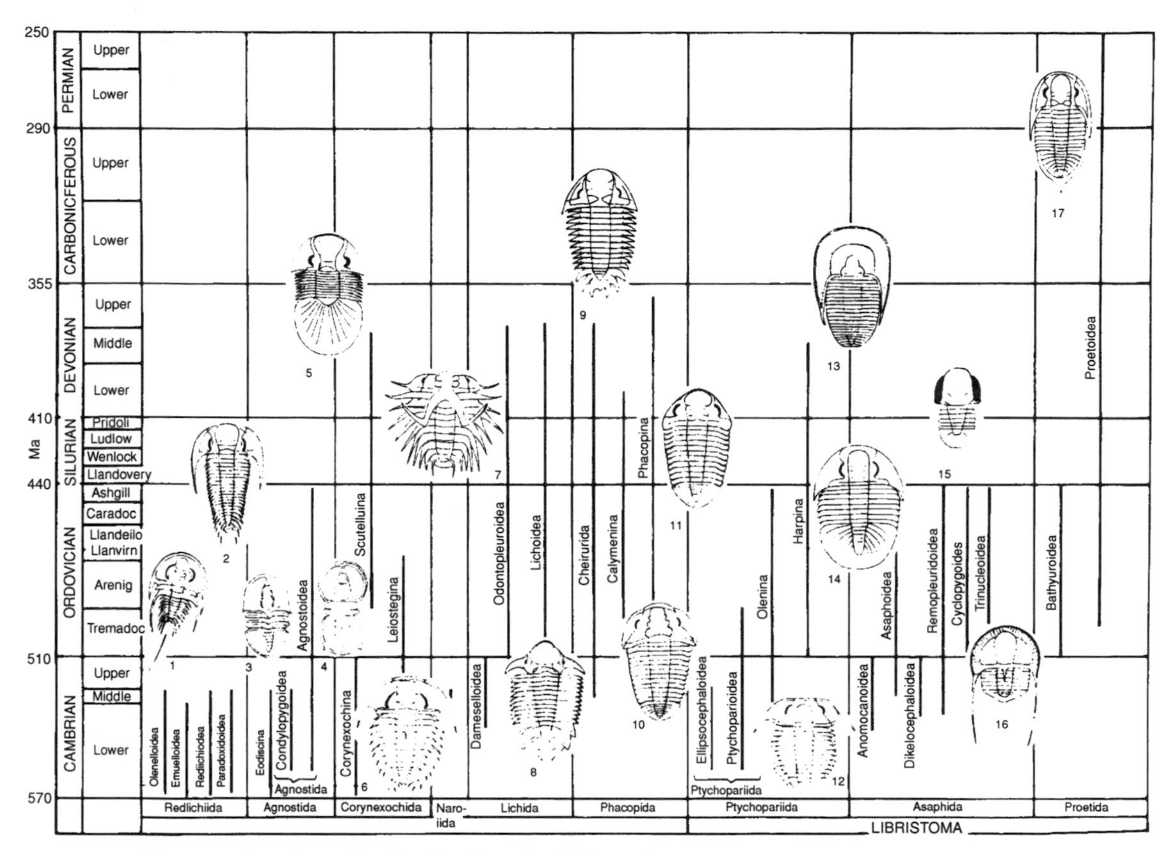

Fig. 3.62. Stratigraphic distribution of trilobites (from Clarkson, 1998). (1) *Paedumias*; (2) *Paradoxides*; (3) *Eodiscus*; (4) *Agnostus*; (5) *Scutellum*; (6) *Kootenia*; (7) *Dicranurus*; (8) *Trochurus*; (9) *Cheirurus*; (10) *Calymene*; (11) *Acaste*; (12) *Olenus*; (13) *Harpes*; (14) *Ogygiocaris*; (15) *Cyclopyge*; (16) *Trinucleus*; (17) *Paladin*.

in biostratigraphy, at least in appropriate, marine, environments (Figs. 3.62–3.63). High-resolution biostratigraphic zonation schemes based on the trilobites have been established, that have widespread applicability (see also Chapter 6). The zonal resolution is capable of further enhancement through the use of graphic correlation technology, as in the Late Cambrian of Pennsylvania (Loch and Taylor, in Mann and Lane, 1995). Interestingly, trilobites, for example, those of the Ordovician *Teretiusculus* shale of Wales, played a pivotal role in the development of the 'phyletic gradualism' model of evolution (Sheldon, 1997).

In England and Wales, olenellids, paradoxidids and agnostids are useful for stratigraphic subdivision and correlation in the Cambrian. *Merlinia* is locally useful in the Ordovician, and *Calymene* in the Silurian. Disarticulated pygidia of *Merlinia* are so conspicuous around Carmarthen in south-west Wales as to have gone down in local legend as butterflies petrified by a spell cast by the magician Merlin (or, in Welsh, Myrddin, as, in mutated form, in Caer-Fyrddin, anglicised into Carmarthen); hence the unusual name. *Calymene blumenbachi* is so conspicuous in the Wenlock limestone around the town of Dudley in the West Midlands as to have acquired the local name of the 'Dudley bug', and even to have had its image incorporated into the town's coat of arms. *Calymene (Neseuretus) ramseyensis* is sufficiently abundant as to lend its name to the *Calymene* ashes of the Arenig of the type area in north Wales.

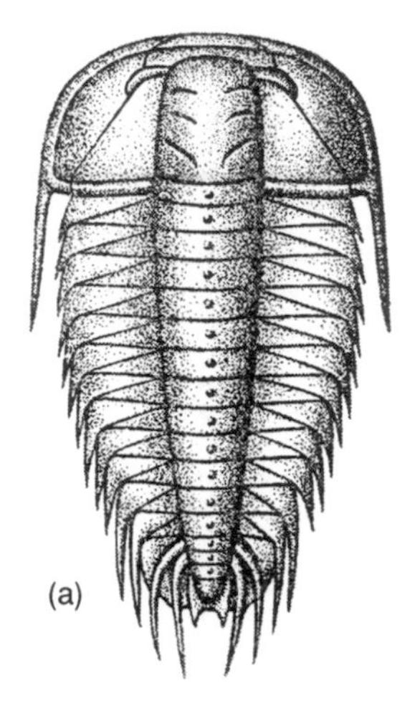

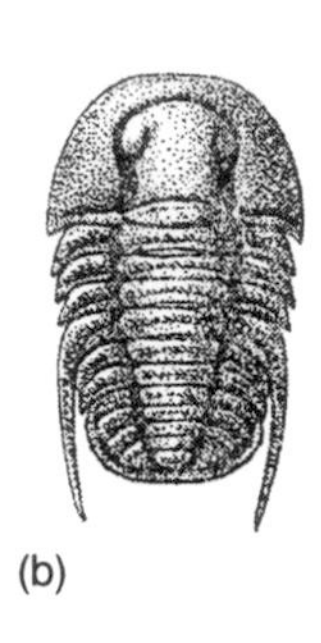

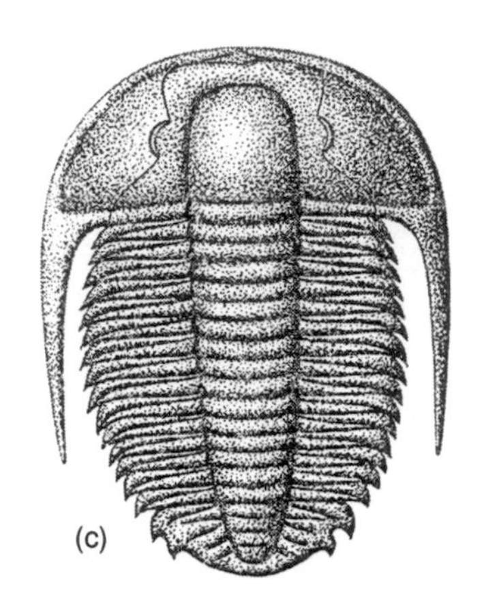

Fig. 3.63. Some stratigraphically useful trilobites. (a) *Parabolina spinulosa*, nominate taxon of the Cambrian, Merioneth, *Parabolina spinulosa* Zone, Dolgellau, Gwynedd, ×2; (b) *Shumardia pusilla*, nominate taxon of the Ordovician, Tremadoc, *Shumardia pusilla* Zone, Sheinton, Shropshire, ×6; (c) *Angelina sedgwickii*, nominate taxon of the Ordovician, Tremadoc, *Angelina sedgwickii* Zone, Tremadoc, Gwynedd, ×1. (Modified after the Natural History Museum, 2001a.)

In my working experience, trilobites have proved of particular use in stratigraphic subdivision and correlation in the following areas.

The Cambrian of the Irkutsk amphitheatre area of east Siberia. Here, the Bulay suite is marked by the Early Cambrian, Botoman trilobite *Bergeroniellus*; the Litvinsev suite by the Middle Cambrian, Mayan (Menevian) *Solenopleura* (Jones, in Simmons, in press).

The Cambrian–Silurian of the Middle East. Here, the regional Middle Cambrian, Menevian Cm20 MFS of Sharland *et al.* (2001), stratotypified in the Seydisehir formation of south-east Turkey, is marked by *Solenopleuropsis*. The regional Middle Cambrian, Dolgellian or Changshanian Cm30 MFS, stratotypified in the Al-Bashair member of the Andam formation of Oman, is marked by *Chuangshania kursteni*. The regional Early Devonian, Pragian D10 MFS, stratotypified in the Qasr limestone member of the Jauf formation of Saudi Arabia, is marked by trilobites.

In south-west Iran, the Hormuz salt formation is marked by poorly age-diagnostic 'Infracambrian' trilobites (Jones, 2000). The unassigned Cambrian is marked by late Early–early Middle Cambrian *Redlichia*, late Middle Cambrian *Irania*, Late Cambrian *Chuangia, Farsia* and *Afghanocare*, and latest Cambrian to earliest Ordovician *Saukia*. The upper member, C of the 'Mila formation' ('Haft Tanan limestone') is marked by Middle Cambrian *Dorypygella, Solenoparia, Nisusia* and *Paradoxides*, and Late Cambrian *Euradagnostus, Agnostus* and *Loganellus*; the Ilebeyk Formation by Late Cambrian *Saukia iranicus, Plectotrophia, Saratogia latefrons, Idahoia, Calvinella, Baltagnostus, Meeria, Lotagnostus, Pseudagnostus* and *Chuangia*; the lower member of the Zard Kuh formation by Early Ordovician *Dikelokephalina* cf. *asiatica*; and the upper member by Late Ordovician or Silurian *Hysterolenus*. In north-east Iran, the Lalun formation is marked by late Early–early Middle Cambrian *Redlichia*. This indicates a correlation with part of the unassigned Cambrian of south-west Iran.

3.6.15 *Arthropods: ostracods*

Biology, morphology and classification

Ostracods, also known as ostracodes, are an extant, Cambrian–Recent, group of crustacean arthropods characterised by a bivalved carapace (Whatley and Maybury, 1990; McKenzie and Jones, 1993; Greenaway, in Anderson, 2001). Their most immediate relatives are the cirripedes, or barnacles, the phyllocarid malacostracans, the eumalacostracans, including the isopods, amphipods and decapods, and a number of other, minor groups. None of these groups is particularly important in the fields of palaeobiology and biostratigraphy, and none is discussed. Note, though, that fossil forms are known. For example, some spectacular decapods (the group represented today by the shrimps, prawns, crabs and lobsters), are known from Carboniferous 'Granton shrimp bed' of Granton near Edinburgh in Lothian in Scotland, the Cretaceous Greensand, Gault and Chalk, and the early Tertiary London clay of various localities along the south coast of England; and conchostracans from the Triassic Mercia mudstone group of Worcestershire in the West Midlands. Decapods and isopods are known from the Late Jurassic Solenhofen limestone of Bavaria in southern Germany, and isopods from Eocene–Miocene amber deposits in the Dominican Republic. The crustacean coprolite (faecal pellet!) *Favreina* has proved a useful stratigraphic marker in the Triassic and Oligo-Miocene in parts of the Middle East, and also in the Jurassic–Cretaceous of the Gulf of Paria between Venezuela and Trinidad.

Incidentally, ostracods derive their name from the Greek for 'pot-sherd'. Coincidentally, it may be the case that the stylised 'water bug' on a Pueblo pot made by the Mogollon people of New Mexico as much as a thousand years ago is the oldest image of an ostracod on record. It certainly bears a strong resemblance to *Chlamydotheca*, which still lives in freshwater habitats in the Central American region, and grows to a size discernible to the keen observer.

Note in this context that ostracods are mainly microscopic, and typically treated as microfossils. Note also, though, that some grow to a sufficiently large size to be discernible to the naked eye, and indeed even to the underwater video camera, the free-swimming myodocopide *Gigantocypris* being among the stars of David Attenborough's excellent recent series of television documentaries entitled *The Blue Planet*.

Biology

The biology – even cytology – of modern ostracods is well known.

As a group, ostracods are omnivorous, some sophisticated species even containing chloroplast symbionts. In turn, they are preyed upon by larger organisms, including, in fresh water, trout (my old professor, Robin Whatley, once caught a trout with what he estimated to be 150 000 specimens of *Herpetocypris reptans* in its belly, whereupon he tied a fly in imitation of same, and caught a further seven trout with it!).

Ostracods grow by periodic moulting or ecdysis of the external skeleton, with the various moult stages, of which there are typically nine, termed instars. The population age structure, that is the proportion of juveniles to adults, can provide information as to *post-mortem* modification of life assemblages, or taphonomy, although obviously a high incidence of juveniles might simply reflect a high mortality rate rather than hydrodynamic sorting. The percentage of disarticulated carapaces also provides a measure of the extent of taphonomic processes. Other taphonomic effects are discussed by Jones (1996) (see also Section 2.2).

Ostracods usually reproduce sexually, although in some species only females occur, and these reproduce asexually or parthenogenetically, with further females, and only females, hatching from unfertilised eggs. In sexually reproducing species, males and females may have different valve outlines designed to accommodate differing genitalia, to facilitate copulation, or, in the case of some Palaeozoic palaeocopids, to accommodate eggs or young (in brood pouches). These are thus said to exhibit sexual dimorphism, precocious forms of which may be evident in juveniles. Ostracod eggs usually hatch in the spring, although they are also able to lie dormant under unfavourable conditions. They can withstand a certain amount of desiccation, enabling the ostracods to inhabit temporary water bodies.

Interestingly, some species of ostracod bioluminesce. It is said that during the Second World War, Japanese soldiers and sailors kept cultures of bioluminescent ostracods to enable them to read their maps or instruments by night without being observed and giving away their position.

In terms of ecology, ostracods exhibit a wide variety of life strategies (Fig. 3.64). Almost all are aquatic, although at least two, *Mesocypris* and *Terrestricythere*, are more or less terrestrial, living in moist leaf litter, and others may have been overlooked (Horne *et al.*, 2004). The aquatic forms have colonised

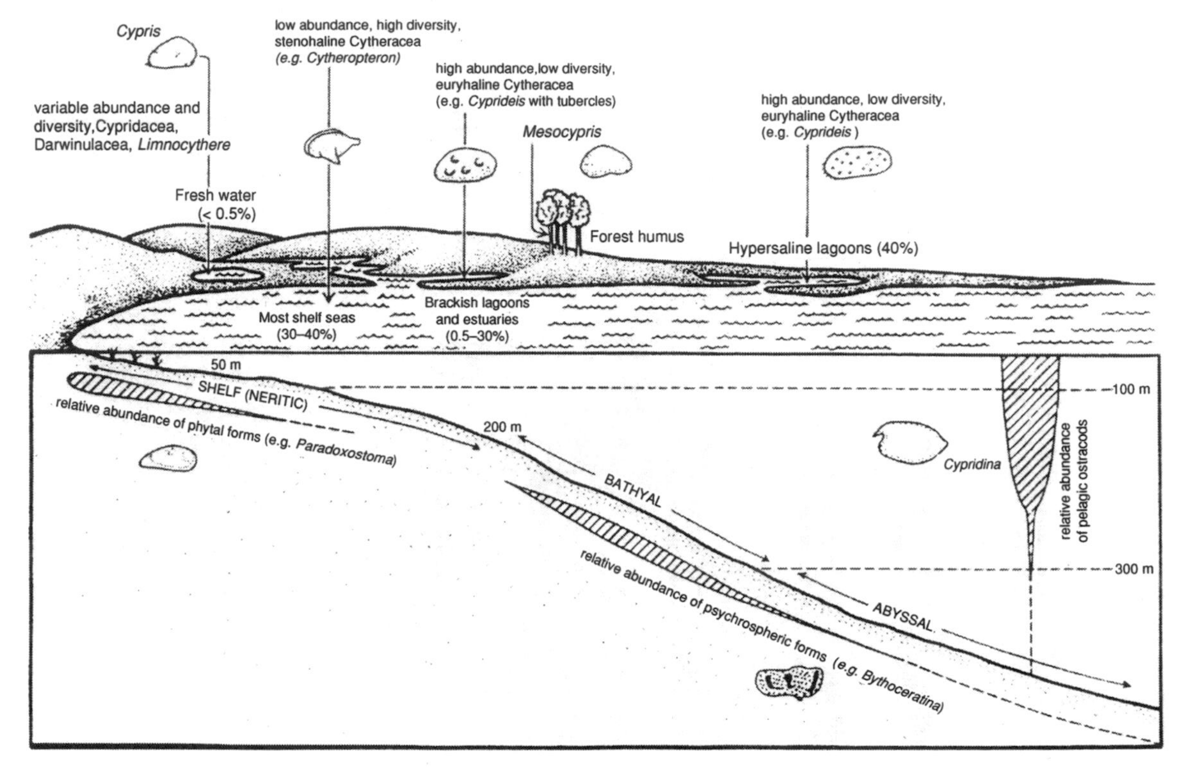

Fig. 3.64. Ecological distribution of living ostracods. (From Jones, 1996; in turn after Brasier, 1980.)

every conceivable niche in freshwater, brackish and marine environments, including freshwater springs and associated tufas. Both benthic and nektonic forms are known, although the latter have only a poor, or at least a poorly documented, fossil record (Colin and Andreu, in Whatley and Maybury, 1990). Both epifaunal and infaunal forms are known among the benthos. Epifaunal forms are attached by silk-like threads secreted by the spinneret gland (also found in spiders).

Non-marine forms tend to be thin-shelled and unornamented, as, for example, in the case of *Darwinula* (Jones, 1996).

Marginal marine forms subjected to wide variations in salinity tend to develop variable surface ornamentation in the form of nodes or torosities, as in the case of the 'estuarine' form *Cyprideis torosa*.

Shallow marine forms from the turbulent zone tend to be robust, thick-shelled and highly ornamented, as in the case of *Aurila* and *Loxoconcha*. Those from the photic zone typically possess eye tubercles and ocular sinuses. Forms associated with plant or phytal habitats or substrates tend to be elongate in morphology, leastwise if living on leaves, as in the case of *Paradoxostoma*, less so if living on stems or roots, as in the case of *Xestoleberis*. Forms associated with soft substrates tend to develop broad bases in the form of ventral tumidities or alae (literally, 'wings'), as in the case of *Cytheropteron*.

Deep marine forms tend to be thin-shelled and unornamented, as in the case of *Krithe*, although some thick-shelled and ornamented forms that dispersed from shallow into deep water and retained these features as useful preadaptations are also known. Deep marine forms tend also to be blind, although those that dispersed from shallow into deep water and secondarily lost their sight nonetheless retained their ocular sinuses and eye tubercles.

Deep marine ostracod assemblages associated with turbidite sediments off southern Australia exhibit anomalously high abundance and diversity

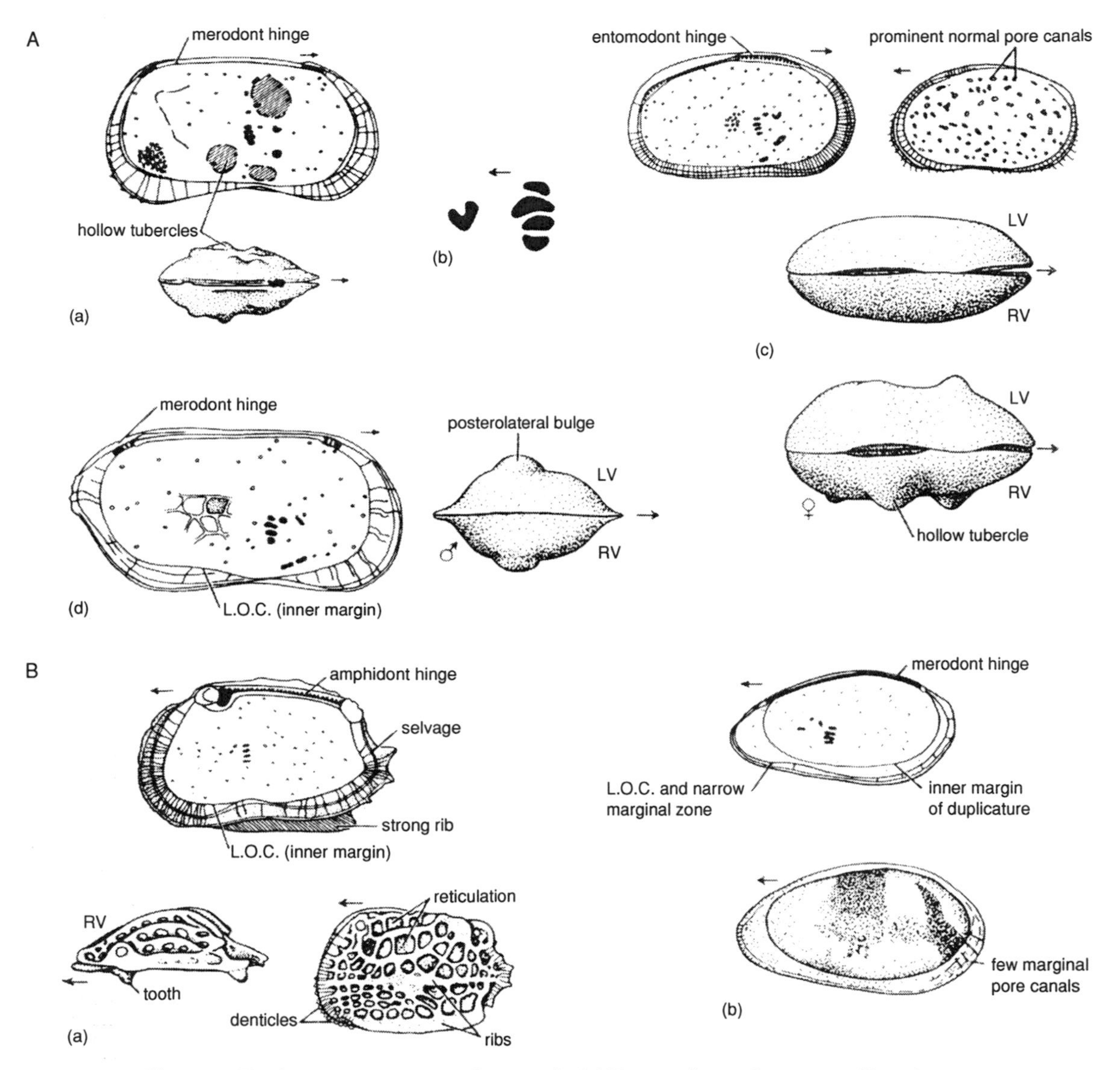

Fig. 3.65. Hard-part morphology of ostracods. (A) Non-marine cytheraceans: (from Jones, 1996; in turn after Brasier, 1980). (a–b) *Limnocythere* (b – detail of muscle scar); (c) *Cyprideis*; (d) *Cytherura*. Magnifications are in the range $\times 27$–$\times 88$. Arrows are directed toward the anterior. (B) Shallow marine cytheraceans: (a) *Quadracythere*; (b) *Paradoxostoma*. Magnifications are in the range $\times 40$–$\times 62$. Arrows are directed toward the anterior. (C) Deep marine cytheraceans: (a) *Cytheropteron*; (b) *Krithe*; (c) *Bythoceratina*. Magnifications are in the range $\times 20$–$\times 50$. Arrows are directed toward the anterior. LV, left valve; RV, right value; LOC, line of commissure. (From Jones, 1996; in turn after Brasier, 1980.)

due to admixture of allochthonous shallow and autochthonous deep marine species (Drapala, in McKenzie and Jones, 1993). They also exhibit bimodal size distribution, because of size differences between typically small shallow and large deep marine species.

Deep marine forms associated with hydrothermal vents are characterised by pore clusters thought

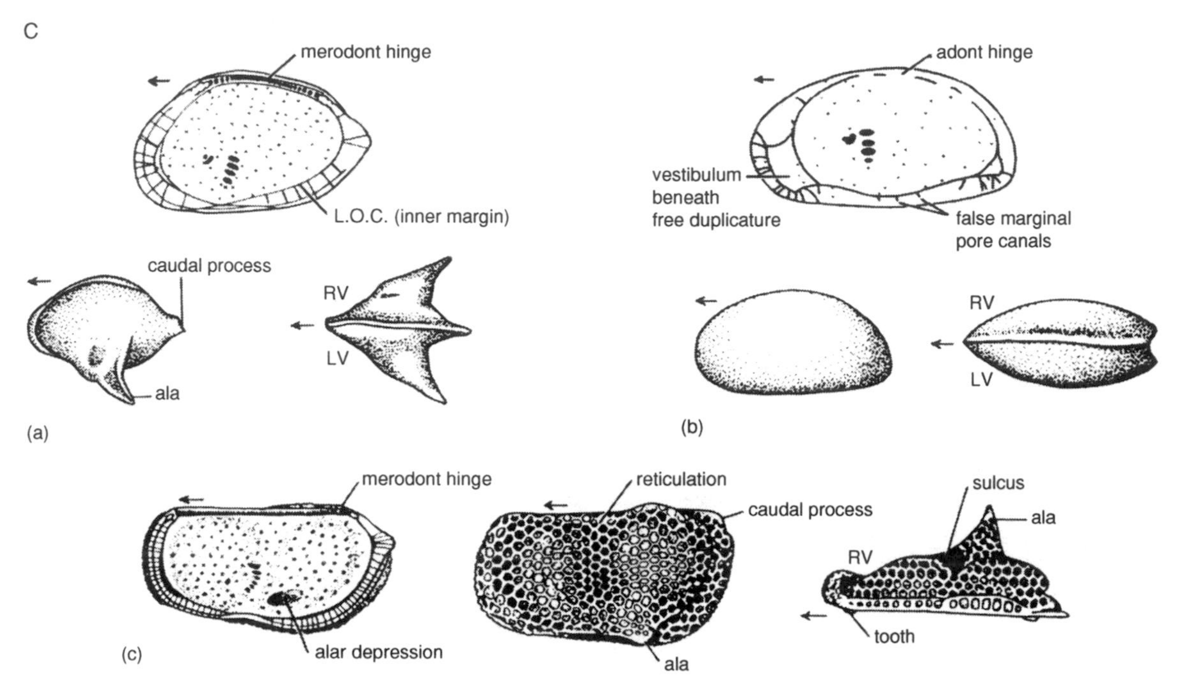

Fig. 3.65. (*continued*)

to represent adaptations to absorption of dissolved matter through the epidermis, to endosymbiosis, or to the absorption of oxygen, as in the case of the eucytherines (van Harten, in McKenzie and Jones, 1993).

Morphology

In terms of morphology, ostracods are characterised by soft bodies with a head, a thorax and an abdomen, together typically bearing seven pairs of limbs or appendages variously specialised for locomotion, feeding or other vital functions (Fig. 3.65). The head typically bears eyes. Other sensory appendages or setae protrude through pores in the shell.

Occasional instances of soft-part preservation are documented, for example from the Early Cretaceous of Brazil (Bate, 1972; Smith, 2000).

However, it is generally only the protective calcareous bivalved shell or carapace that is preserved in the fossil record, or its disarticulated valves. The valves are held together at the dorsal margin by various types of hinge arrangement. They are opened and closed by various types of muscle, whose imprints are preserved as scars. The imprints of the eyes are preserved as internal ocular sinuses and external eye tubercles.

Classification

Classification of ostracods is based on morphology, and on phylogeny inferred from empirical observations on the fossil record. Six sub-groups are recognised: the bradoriides, the leperditicopides, the palaeocopides, the myodocopides, the platycopides and the podocopides.

Bradoriides. The bradoriides, which range from Cambrian to Ordovician, are characterised by chitino-phosphatic rather than calcareous compositions (and may not be true ostracods at all).

Leperditicopides. The leperditicopides (Cambrian or Ordovician to Devonian) are characterised by thick, smooth valves, and by several hundred adductor muscle scars.

Palaeocopides. The palaeocopides, which range from Ordovician to Permian or Triassic (Crasquin-Soleau *et al.*, 2004), are characterised by thick valves with tuberculate lobes, and, in some cases, a brood pouch and/or a frill, or vellum, around the ventral margin.

Myodocopides. The myodocopides (Ordovician–Recent) are characterised generally by thin, smooth valves, and, in some cases, a so-called rostral incisure, through which natatory or swimming appendages protrude or protruded.

Platycopides. The platycopides (Ordovician–Recent) are characterised by elongate ovate valves, the right overlapping the left, and by 10–18 adductor scars arranged in rows or in a rosette.

Podocopides. The podocopides (Ordovician–Recent), to which most modern species belong, are characterised by an arched dorsal margin, a complex hinge, and generally by four to five adductor scars arranged in a single row. Four further main sub-groups of podocopides are recognised, namely the essentially freshwater cypridaceans and darwinulaceans, the freshwater and marine cytheraceans and the marine bairdiaceans.

Palaeobiology

Fossil ostracods are interpreted, essentially on the bases of analogy with their living counterparts, and associated fossils and sedimentary facies, as having occupied a range of terrestrial and aquatic, freshwater, brackish and marine, environments (Jones, 1996). They appear to have first colonised freshwater environments in the Carboniferous (Williams *et al.*, 2003). Fluctuation in the amount of terrigenous input, in turn related to delta growth, is interpreted to have been the most important factor controlling ostracod distribution in the Early Carboniferous Yoredale series of northern England (Masurel, 1989).

Ostracods have been used as indicators of the palaeo-chemistry, and in particular the salinity and alkalinity, concentration and turbidity, and organic matter content and oxygenation, of freshwater lakes, such as those of the Pleistocene–Holocene of the East African Rift Valley (Cohen, 2003). The abundance, diversity and specific composition of assemblages, key species, and ornamentation, are all guides.

Ostracods have also been used as indicators of the degree of oxygenation of the seas of the Devonian–Carboniferous of western Europe (Lethiers and Whatley, 1994). The percentage of filter-feeding morphotypes appears to have been inversely proportional to the oxygenation. Significantly, the percentage of filter-feeding ostracods is highest, and the oxygenation apparently lowest, in the Frasnian (i.e. at the time of the Late Devonian mass extinction).

Ostracod shell chemistry, and especially the ratio between strontium and calcium, and the ratio between calcium and magnesium, is useful as an indicator of water chemistry and salinity, and can be used to distinguish non-marine and marine environments (Chivas *et al.*, 1985; Lister, in Colin *et al.*, 1987; Jones, 1996).

The interpreted hypersaline species *Simeonella brotzenorum* is an indicator of the 'Carnian salinity crisis' of the Levant (Gerry *et al.*, in Whatley and Maybury, 1990).

Palaeobathymetry

Fossil ostracods are interpreted, essentially on the bases of analogy with their living counterparts, and associated fossils and sedimentary facies, as having occupied a range of depth habitats in marine environments.

As in the case of their living counterparts, fossil marginal marine forms tend to be torose; shallow marine robust, thick-shelled and highly ornamented; and deep marine forms thin-shelled and unornamented, and also blind rather than sighted. Fossil deep marine forms interpreted as associated with oxygen minimum zones include filter-feeding platycopines such as *Cytherella*, which manage to obtain sufficient oxygen by virtue of the greater volume of water which they circulate across their respiratory surface in the course of their normal feeding behaviours.

Paraparchitids, and bairdiids, have been interpreted as indicative of shallow marine environments in the Late Devonian–Early Carboniferous of the Nanbiancun section of Guilin in Guangxi province in China (Wang Shangqi and Becker, in Jin Yugan *et al.*, 1991). (Paraparchitids have also been interpreted as indicative of shallow marine environments in the Permian of the mid-continent of the USA: Kaesler *et al.*, in Whatley and Maybury, 1990).

The nektonic? myodocopides *Bolbozoe* and *Entomis* have been interpreted as indicative of 'mid-water' environments in the Silurian of France and Bohemia (Perrier *et al.*, 2004). They are associated with bivalves, cephalopods and graptolites.

Spinose podocopides have been interpreted as indicative of deep marine environments in the Late Devonian–Early Carboniferous of the Shizhiling section of Guilin in Guangxi province in China (Wang Shangqi and Becker, in Jin Yugan *et al.*, 1991). They have also been interpreted as indicative of deep marine environments elsewhere in south-east Asia, in Eurasia, especially around Thuringia, and in north Africa (Becker and Bless, in Whatley and Maybury, 1990).

Ostracod assemblages associated with a chemosynthetic community in the Early Carboniferous on the Port au Port peninsula, Newfoundland are dominated by the paraparchitacean *Chamishella* (Dewey, in McKenzie and Jones, 1993).

Palaeobiogeography

Many fossil ostracods appear to have had restricted or endemic biogeographic distributions, rendering them of use in the characterisation of palaeobiogeographic provinces, and in turn in the constraint of plate tectonic reconstructions, in the Palaeozoic, in the Mesozoic, and in the Cenozoic.

In the Palaeozoic, in the Cambrian, *Petrianna* is characteristic of north Greenland in Laurentia (in the olenellid trilobite realm), *Cambria* of Siberia (in the bigotinid trilobite realm), and *Cambria, Paracambria, Auriculatella, Chuanbeiella* and *Shangsiella* of China in Gondwana (in the redlichiid trilobite realm) (Siveter *et al.*, 1996; Williams *et al.*, 1996; Melnikova *et al.*, 1997; Floyd *et al.*, 1999; Williams *et al.*, in Servais *et al.*, 2003). Interestingly, *Isoxys* is found in Laurentia, in Siberia and in Gondwana.

In the Mesozoic, ostracod distributions point towards generally distinct Austral, Boreal and Tethyan realms (Babinot and Crumiere-Airaud, in Whatley and Maybury, 1990; Kristan-Tollmann, in Whatley and Maybury, 1990; Luger, 2003; Seeling *et al.*, 2004). Note, though, that there was also at least local or intermittent communication between these realms. The presence of the essentially Tethyan ostracod *Orthonotocythere* in the South Atlantic in the Early Cretaceous, Barremian indicates a connection between these areas at this time, possibly through the interior rifts of Africa or South America (Davison *et al.*, 2004). Incidentally, the Tethyan to sub-Boreal foraminiferan *Lilliputianella globulifera (Hedbergella maslakovae)* also occurs in the South Atlantic in the Barremian to Aptian (author's unpublished observations).

In the Cenozoic, Palaeocene–Eocene ostracod distributions point toward the – ?continued – existence of a trans-Saharan seaway linking Tethys and the South Atlantic (Luger, 2003).

Palaeoclimatology

Also in the Cenozoic, ostracod distributions point towards the palaeogeographic and palaeoclimatic evolution of the Old World in the Oligocene–Holocene (Benson, in Whatley and Maybury, 1990).

The exacting ecological requirements and tolerances of many ostracod species, and their rapid response to changing environmental and climatic conditions, render them especially useful in Quaternary palaeoclimatology and climatostratigraphy, and in archaeology (Athersuch *et al.*, 1989; Henderson, 1990; Whittaker, in Roberts and Parfitt, 1999; Griffiths and Holmes, 2000; Meisch, 2000; Boomer, in Haslett, 2002; Whittaker *et al.*, 2003; see also Section 7.6).

Biostratigraphy

Ostracods appeared first in the Cambrian, Atdabanian (Siveter and Williams, 1997), or Ordovician if the bradoriides are excluded. They diversified in the Ordovician, by the end of which all of the major sub-groups had appeared. They sustained significant losses in the End-Ordovician mass extinction, when the bradoriides disappeared. The recovery of the remaining sub-groups is discussed by Swain, in Hart (1996). The leperditicopides became extinct at the time of the Late Devonian mass extinction; the palaeocopides either in the Permian, at the time of the End-Permian mass extinction, or in the Triassic, at the time of the End-Triassic mass extinction. The remaining sub-groups diversified during the Mesozoic and Cenozoic, sustaining some losses in the Late Triassic, Pliensbachian, Bathonian, 'Purbeckian', Barremian, Turonian, Maastrichtian, Late Eocene and

Late Miocene (Babinot and Crumiere-Airaud, in Whatley and Maybury, 1990; Horne *et al.*, in Whatley and Maybury, 1990; Whatley, in Whatley and Maybury, 1990).

The moderate rate of evolutionary turnover exhibited by the ostracods renders them of some use in biostratigraphy, at least in appropriate environments. Unfortunately, their usefulness is compromised by their restricted ecological distribution and facies dependence, a function of their benthic habit (see above). Nonetheless, biostratigraphic zonation schemes based on the ostracods have been established, that have at least local to regional applicability (Jones, 1996).

In my working experience, ostracods have proved of use in stratigraphic subdivision and correlation, and/or in palaeoenvironmental interpretation, in the following areas:

Palaeozoic

The Palaeozoic of the former Soviet Union. In the Cambrian of east Siberia, purported ostracods (in context, if anything, bradoriides) have been recorded from the lower Moty suite of the Ust Kut and Verkhnechona fields of the Nepa–Botuobin area of east Siberia, from around the same level as the first anabaritids (Nemakit–Daldynian), but below the levels of the first archaeocyathans (Tommotian) and trilobites (Atdabanian) (author's unpublished observations). If correctly identified, these would be the oldest ostracods on record, the oldest to date being from the Atdabanian of Siberia or Qiongzhusian of China. Unfortunately, it is unclear from the photographs in the archives, the quality of which is insufficient to allow them to be reproduced here, whether they are correctly identified or not (G. Miller and J. E. Whittaker, the Natural History Museum, D. J. Siveter, University of Leicester and M. Williams, British Geological Survey, Nottingham, personal communications), and there would appear to be no surviving samples. In the Devonian–Carboniferous of the Russian platform, ostracods form the basis of a regional biozonation, the usefulness of which is enhanced by its having been calibrated against the conodont standard (Wagner *et al.*, 1996).

The Devonian of Algeria in north Africa. Here, the Emsian–Givetian of the Ougarta Hills is characterised by ostracods (Le Fevre, 1971). Marginal marine clastic eco-zones are characterised by beyrichids; open marine carbonate eco-zones are characterised by diverse palaeocopides and podocopides (bairdiocyprids, krausellids, pachydomellids, beecherellids, acronotellids, bufinids, aechminids, kirkbyellids, hollinids, paraparchitids, arcyzonids and scrobiculids). A total of 13 eco-zones have been recognised, whose distributions appear to be controlled primarily by lithology (limestone vs. silt and shale, grain size, mineralogical composition, and trace element geochemistry).

The Devonian of Canada. Here, ostracods form the basis of a biozonation applicable in open and restricted marine environments (Braun, in Whatley and Maybury, 1990). Five biozones are recognised in the Emsian–Eifelian, four in the Givetian, six in the Frasnian, and four in the Famennian.

The Devonian–Permian of the Middle East. Here, the regional Late Devonian, Famennian, Strunian D30 MFS of Sharland *et al.* (2001), stratotypified in the black shales of the Koprulu formation of Zap in Hakkari province in south-east Turkey, is marked by the ostracod *Cryptophyllus diatropus*. Ostracod evidence indicates that the black shales were subtidal (and also, incidentally, that the underlying red sandstones of the Yiginli formation were supratidal, and the overlying limestones of the Belek littoral). The Permian Khuff formation of Oman is marked by diverse ostracods, including *Bairdia omanensis, Cavellina boomeri, C. gerryi, C. huqfensis, Sulcella arabica, Langaia hornei, Sargentina woutersi, Carinaknightina braccinii, Jordanites lordi, Hollinella (Hollinella) benzartii* and *H. (H.) martensi*, which are only known from Oman (Crasquin-Soleau *et al.*, 1999; Angiolini *et al.*, in Al-Husseini, 2004).

The Carboniferous–Permian of Texas. See Melnyk and Maddocks (1988a, b).

Mesozoic

The Triassic–Cretaceous of the western margin of the British Isles. Here, ostracods form part of the basis of the regional biostratigraphic zonation, calibrated against ammonite and other standards (Ainsworth *et al.*, in Batten and Keen, 1989; Jones, 1996; Ainsworth *et al.*, in Underhill, 1998). Importantly, they form the principal basis of the regional biozonation of the non-marine earliest Cretaceous 'Wealden' facies.

The Jurassic–Cretaceous of the North Sea. Here again, ostracods form part of the basis of the regional biozonation, calibrated against ammonite and other standards (Jones, 1996; Copestake, in Evans *et al.*, 2003; Copestake *et al.*, in Evans *et al.*, 2003; Husmo *et al.*, in Evans *et al.*, 2003; Surlyk *et al.*, in Evans *et al.*, 2003).

The Jurassic–Cretaceous of the Middle East. Here, the Late Cretaceous, Coniacian K150 MFS of Sharland *et al.* (2001), stratotypified in the Laffan formation of the United Arab Emirates, is marked by ostracods of the *Ovocytheridea* AUR1496 Zone.

The Early Cretaceous 'Wealden' facies of Gondwana. Here, ostracods form part of the basis of the biozonation of the rift basins of eastern South America and west Africa (Jones, 1996; Grosdidier *et al.*, in Jardine *et al.*, 1996; Braccini *et al.*, 1997; Bate, in Cameron *et al.*, 1999). Interestingly, they also form part of the basis of the dating of coeval sections in China, and have thus contributed to the understanding of the evolutionary significance of the important vertebrate fossils, including primitive birds, recently discovered there (Pang Qiqing and Whatley, in Whatley and Maybury, 1990).

Cenozoic

The Mid–Late Cenozoic of eastern Paratethys. Here, ostracods form part of the basis of the regional biostratigraphic zonation, especially in the marginal to non-marine environments of the Neogene–Pleistogene. They are also of considerable use in palaeoenvironmental interpretation, again, especially in the marginal to non-marine environments of the Neogene–Pleistogene. By analogy with the modern Black Sea and Caspian, non-marine environments are characterised by *Aglaiocypris, Candona, Candoniella, Cyclocypris, Cypria, Eucypris, Ilyocypris, Pseudostenocypria* and *Zonocypris*, marginal marine environments by *Cyprideis* (salinity tolerance range 2–14 ppt) and more or less 'normal' marine environments by *Loxoconcha* (5–14 ppt), *Bakunella, Caspiolla* and *Cytherissa* (11–14 ppt) and *Graviacypris* (12–13 ppt) (Jones, 1996; Jones and Simmons, 1996; Jones and Simmons, in Robinson, 1997). The regional Middle Miocene to Pleistocene Sarmatian, Maeotian, Pontian, Kimmerian, Akchagylian, Apsheronian, Bakunian and Khazarian stages of Azerbaijan and the south Caspian are all individually distinguishable, and indeed in many cases internally divisible, on the basis of ostracods (author's unpublished observations, based in part on identifications by Dilara Mamedova, formerly of the Geological Institute of Azerbaijan, Baku).

This level of stratigraphic resolution renders ostracods of considerable use in Azerbaijan and the south Caspian, not only in regional basin modelling and in reservoir characterisation (see below), but also in well-site geohazard investigation, specifically, in the discrimination of different soil horizons with different engineering properties (see Section 7.4). Importantly, the essentially Early Pliocene Kimmerian reservoir interval is internally divisible into nine zones, corresponding to the lithostratigraphic units constituting the 'productive series', namely the Kalin, Pre-Kirmaky Sand, Kirmaky, Post-Kirmaky Sand, Post-Kirmaky Clay, Pereriva, Balakhany, Sabunchi and Surakhany suites. The upper part of the Surakhany suite is locally further subdivisible into at least three zones, a lower one marked by *Ilyocypris* and *Candoniella*, a middle one marked by *Limnocythere*, and an upper one marked by *Cyprideis* and *Candona* (the Balakhany suite is potentially also further subdivisible). The occurrence in the *Ilyocypris–Candoniella* Zone of the upper part of the Surakhany suite of *Ilyocypris* and *Candoniella*, the latter of which is known to be tolerant of freezing and of desiccation, indicates a cold, freshwater environment, and possibly one such as that of the modern ephemeral ponds or 'takyrs' of the salt deserts or 'solonchaks' of the terraces of the Amu Darya and Syr Daria (such an environment is also indicated by the palynological 'steppe index'). In contrast, the occurrence in the *Cyprideis–Candona* Zone of *Cyprideis* indicates a comparatively warm, brackish-water environment. Thus, salinity and temperature seem to co-vary, although in actuality salinity is controlled by the balance between evaporation and precipitation rather than temperature *per se*. The relationship between salinity, temperature and the water level of the, then, Caspian lake, remains unclear. Note in this context also that ostracod abundance appears to be correlated with oxicity as well as salinity.

The Mid–Late Cenozoic of northern South America, the Caribbean and the Gulf of Mexico. Here, the Oligocene–Holocene is marked by five primary and seven secondary ostracod zones in shallow marine environments, and five primary and two secondary zones in brackish environments (Jones, 1996). The primary zones in shallow marine environments are, in ascending stratigraphic order, the *Pokornyella saginata laresensis* Total-Range Zone, of Oligocene–early Early Miocene age (as determined by calibration against the global standard planktonic foraminiferal zonation), the *Procythereis? deformis* Total-Range Zone, of late Early–early Late Miocene age, the *Coquimba congestocostata* Partial-Range Zone, of late Late Miocene age, the *Radimella confragosa* Partial-Range Zone, of Early–Middle Pliocene age, and the *Radimella wantlandi* Total-Range Zone, of Late Pliocene–Holocene age. The primary zones in brackish environments are, again in ascending stratigraphic order, the *Cyprideis aff. ovata* Partial-Range Zone, of Early Miocene age, the *Cyprideis pascagoulensis–aff. ovata* Concurrent-Range Zone, of early Middle Miocene age, the *Cyprideis pascagoulensis* Partial-Range Zone, of late Middle Miocene–early Early Pliocene age, the *Cyprideis subquadraregularis* Partial-Range Zone, of mid Early Pliocene age, and the *Cyprideis salebrosa* Total-Range Zone, of late Early Pliocene–Holocene age. Incidentally, the distinction between shallow marine and brackish environments is made mainly on the basis of analogy between fossil ostracod species and their living congeners (van den Bold, 1974). Note, though, that in the Gulf of Paria, between Venezuela and Trinidad, there is a clear relationship between ostracod assemblages and lithology, as well as salinity. Here, shelly sands are characterised by *Bairdia*, silty muds by *Orionina* and clayey muds by *Cativella*, *Cushmanidea* and *Cytheropteron*.

Other applications

Ostracods are also useful in environmental science (see Section 7.5).

3.6.16 Arthropods: insects

Biology, morphology and classification

Insects are an extant, Devonian–Recent, group of hexapod uniramian arthropods characterised by the possession of six unbranched or uniramous limbs (Gullan and Cranston, 2000; Hales, in Anderson, 2001; Rasnitsyn and Quicke, 2002; Grimaldi and Engel, in press). Their most immediate relatives are the proturan, collembolan and dipluran hexapods, including the springtails, and the onychophoran and myriapod uniramians, including the appendiculate worms, millipedes and centipedes. None of these related groups is especially important in the fields of palaeobiology and biostratigraphy, and none is discussed. Note, though, that fossil forms are known.

Biology

The biology and ecology of living insects is extremely well known (Gullan and Cranston, 2000).

The phenomenal diversity of morphological form exhibited by the group is thought to be a consequence of functional adaptation to a correspondingly wide range of ecological niches.

In terrestrial environments, insects are found in and on the ground, feeding on roots, leaf litter, dung and carrion, and either directly or indirectly, on fungi (through cultivation or 'farming' in the case of termites and leafcutter ants). Some are social, such as the termites, ants, bees and wasps. Others have developed complex relationships with plants. In some cases these relationships are mutually beneficial, and the insects and plants are ultimately interdependent on one another. In other cases the relationships clearly either benefit the insect more than the plant, as in the case of feeding, boring or gall formation; or may be said to benefit the plant more than the insect, as in the case of inadvertent pollination, or seed dispersal, in the course of feeding.

In aquatic environments, insects are found in, on and around fresh-, brackish and sea water.

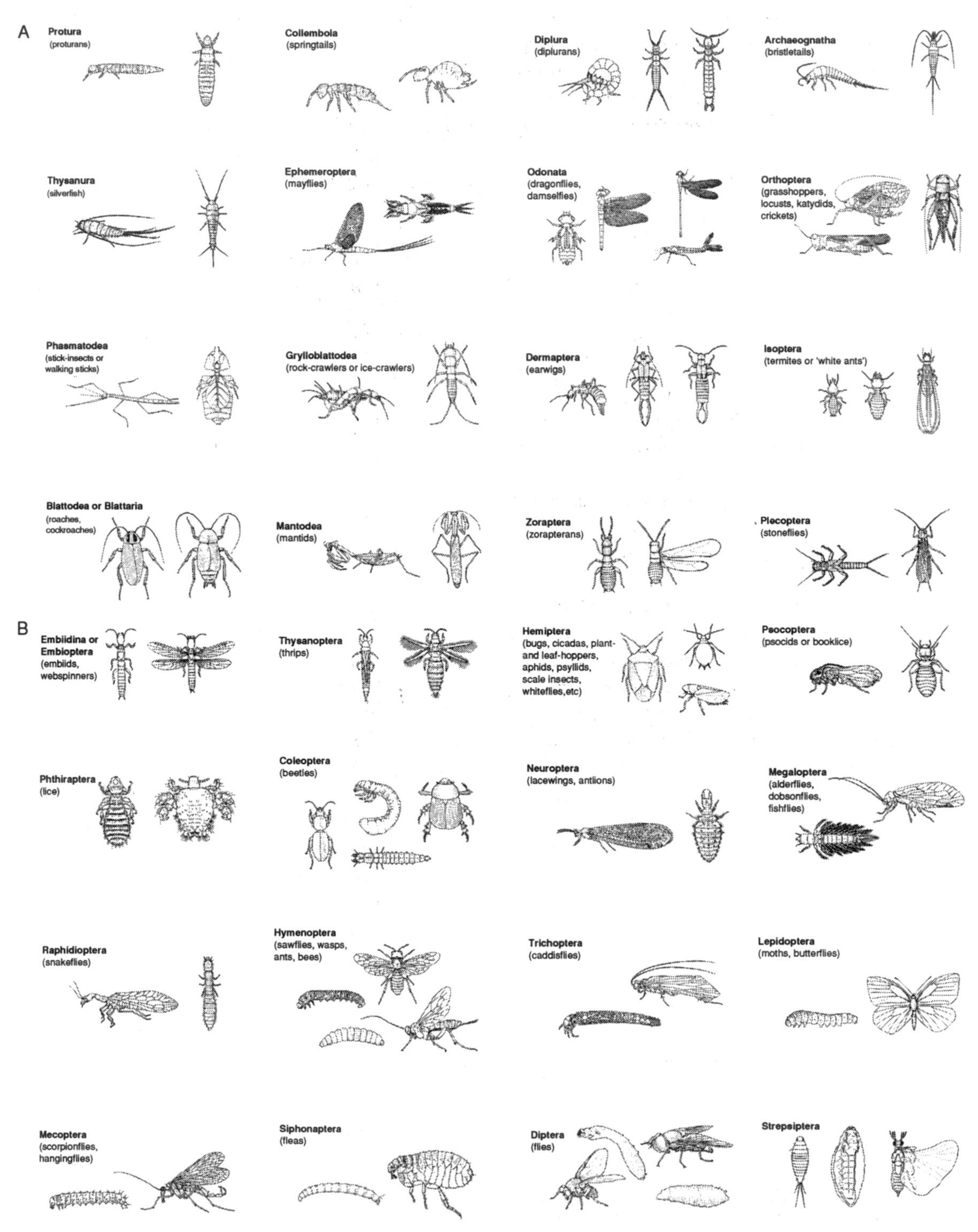

Fig. 3.66. Morphology of insects. (A) Protura to Plecoptera. (B) Embioptera to Strepsiptera. (Modified after Gullan and Cranston, 2000.)

Morphology

In terms of hard-part morphology, insects are characterised by external skeletons composed of two layers of cuticle, the outer of which is sclerotinised for strength (Fig. 3.66). The skeletons comprise heads, bodies with six legs, and tails (and also, in most cases, wings).

Preservation of insects in ancient, pre-Quaternary sediments is generally poor, owing to the poor long-term preservation potential of cuticle. However, it is good in Lagerstatten and in other exceptional deposits, such as fossil resin or amber, or tufa, from which a reasonably complete record can be constructed (Ponomarenko, 1985; Barthel *et al.*, 1990; Poinar and Poinar, 1999; Glover and Robertson, in Akinci *et al.*, 2003; Meyer, 2003; Martinez-Delclos *et al.*, 2004).

Incidentally, the insects from which dinosaur DNA was supposedly extracted in Spielberg's blockbuster *Jurassic Park* were from amber from the Dominican Republic, which is actually of Eocene–Miocene age, and post-dates the End-Cretaceous extinction of the dinosaurs by at least 20 million years!

Classification

Classification of insects is based on morphology and on phylogeny inferred from cladistic analysis of molecular data (Gullan and Cranston, 2000). The pattern of veining on the wings is an important morphological classification criterion, and is clearly imageable using novel non-contact surface laser scanning technology (Bethoux *et al.*, 2004).

Two main sub-groups of insects are recognised, the wingless apterygotes and the winged pterygotes.

Two further sub-groups of apterygotes are recognisable; the archaeognaths or bristletails, and the thysanurans or silverfish.

Nine further sub-groups of pterygotes are recognisable: the ephemeropterans, the odonatans, the neopterans, the paraneopterans, the neuropteridans, the coleopterans, the antiliophorans, the amphiesmenopterans and the hymenopterans.

The *ephemeropterans* in turn include the mayflies.

The *odonatans* include the dragonflies and damselflies.

The *neopterans* include the isopterans or termites, the blattodeans or roaches, the mantodeans or mantids, the orthopterans or crickets, katydids, grasshoppers and locusts, the phasmatodeans or stick-insects, the embidiinans or web-spinners, the grylloblattodeans or rock-crawlers, the dermapterans or earwigs, the plecopterans or stoneflies, and the zorapterans, which have no familiar name.

The *paraneopterans* include the thysanopterans or thrips, the hemipterans or bugs, the psocopterans or booklice, and the phthirapterans or parasitic lice.

The *neuropteridans* include the neuropterans or lacewings and antlions, the raphidiopterans or snakeflies, and the megalopterans or alderflies, dobsonflies and fishflies.

The *coleopterans* include the beetles.

The *antiliophorans* include the dipterans or flies, the mecopterans or hangingflies and scorpionflies, and the siphonapterans or fleas.

The *amphiesmenopterans* include the trichopterans, or caddisflies, and the lepidopterans, or butterflies and moths.

The *hymenopterans* include the ants, bees, wasps and sawflies.

Palaeobiology

Fossil insects are interpreted, essentially on the bases of analogy with their living counterparts, and functional morphology (Labandeira, 1997), as having been terrestrial, aquatic and aerial.

The Triassic insect faunas from the Molteno formation of the Karoo group of South Africa, dominated by beetles, are interpreted as representing a range of habitats, including mature and immature *Dicroidium* riparian forest (bordering channels), also with cockroaches, bugs and dragonflies; *Dicroidium* woodland (on the open flood-plain), also with bugs and cockroaches; *Sphenobaiera* woodland (adjacent to lakes on the flood-plain), also with cockroaches, bugs, dragonflies, crickets and moths; *Heidiphyllum* thicket (on flood-plains or channel sand-bars, or in areas of high water table), also with cockroaches and bugs; *Equisetum* marsh (on flood-plains), also with bugs; and fern/*Ginkgophytopsis* meadow (on sand-bars in braided rivers), with no insect faunas (MacRae, 1999). The constituent species are further interpreted as herbivores, carnivores, omnivores and, importantly, pollinators (of gymnosperms, excluding glossopterids, extinct by this time).

Palaeoclimatology

The exacting ecological requirements and tolerances of many insect species, and their rapid response to changing environmental, and especially climatic, conditions, render them especially useful in Quaternary palaeoclimatology and climatostratigraphy, and also in archaeology (Elias, 1994; Bidashko *et al.*, 1995; Coope, in Culver and Rawson, 2000; Ashworth, in Gerhard *et al.*, 2001; Cohen, 2003; see also Section 7.6). Coleopterans or beetles are especially useful in these fields, as their strongly sclerotinised wing-cases or elytra have uncommonly good preservation potential. My former lecturer Russell Coope thus developed an 'inordinate fondness' for them; as, incidentally, no doubt for his own reasons, did God, as J. B. S. Haldane once famously remarked.

Biostratigraphy

Wingless insects first appeared in the Devonian, as exemplified by the Rhynie chert of Aberdeenshire in Scotland (Labandeira and Sepkoski, 1993; Jarzembowski and Ross, in Hart, 1996). Primitive winged, dragonfly-like, insects, unable to fold their wings against their bodies, appeared next, in the Early Carboniferous; many more advanced groups in the Late Carboniferous and Permian (Jarzembowski and Ross, in Hart, 1996). Most of the remaining groups, including the lepidopterans or butterflies and allied forms, and hymenopterans or bees and allied forms, that are now important plant pollinators, appeared in the Jurassic through Cretaceous, more or less coincident with the radiation of the flowering plants (Labandeira and Sepkoski, 1993; Ross *et al.*, in Culver and Rawson, 2000). The last remaining group, the siphonapterans or fleas, appeared in the Cenozoic, coincident with the radiation of the mammals that they parasitise.

The overall pattern of insect evolution has been one of ever-increasing diversification, with little or no evident loss of diversity associated with any of the main mass extinction events, other than, arguably, the End-Permian mass extinction (Jarzembowski and Ross, in Hart, 1996). All nine sub-groups of pterygotes are found as fossils in the Late Jurassic Solenhofen limestone of Bavaria in southern Germany (Ponomarenko, 1985; Barthel *et al.*, 1990); one sub-group of apterygotes and six of pterygotes in the Eocene–Miocene amber deposits of the Dominican Republic, together with collembolans and myriapods (Poinar and Poinar, 1999); and all nine sub-groups of pterygotes in the Oligocene lacustrine deposits of Florissant in Colorado, together with myriapods (Meyer, 2003). In Britain, diverse insects are found in the Late Carboniferous 'Coal Measures', and in Early Cretaceous and Oligocene continental deposits on the Isle of Wight.

The rapid rate of evolutionary turnover exhibited by the insects renders them potentially useful in biostratigraphy (Fig. 3.67). However, their usefulness is somewhat limited in practice by their poor preservation potential.

In my working experience, insects have proved of particular use in *the Late Miocene, late Sarmatian of the Stavropol arch in central Ciscaucasia in the former Soviet Union* (author's unpublished observations). Here, insects with lowland habitat preferences indicate the precise location and timing of the initiation of uplift in the central part of the Greater Caucasus. This has important consequences not only for the structural geology, but also for the petroleum geology of the area, insofar as it influences the provenance and thus the quality of potential reservoir sandstones.

Other applications

Living insects are useful in forensic science (Erzinclioglu, 2000; Gullan and Cranston, 2000). Their principal use in this field is in determining the time of death, from an assessment of the stage of advancement of the 'ecological succession' that results from changes in the attractiveness of the corpse to different groups through time (complicated by changes to the extent of exposure of the corpse, movement of the corpse, etc). Blowflies and houseflies are typically the first insects to visit a corpse, and oviposit eggs or drop live larvae onto it, that develop into maggots over a fixed period of time determined essentially by ambient temperature. Certain types of dermestid beetle larvae and adults only appear later, followed by cheese-skipper larvae. Fruit flies and hoverflies appear later still, after the body has started to desiccate, and tineid moths latest of all, only after the body has completely desiccated.

3.6.17 Echinoderms: crinoids

Echinoderms include two sub-groups that are important in the fields of palaeobiology and biostratigraphy, namely the crinoids and echinoids, which are

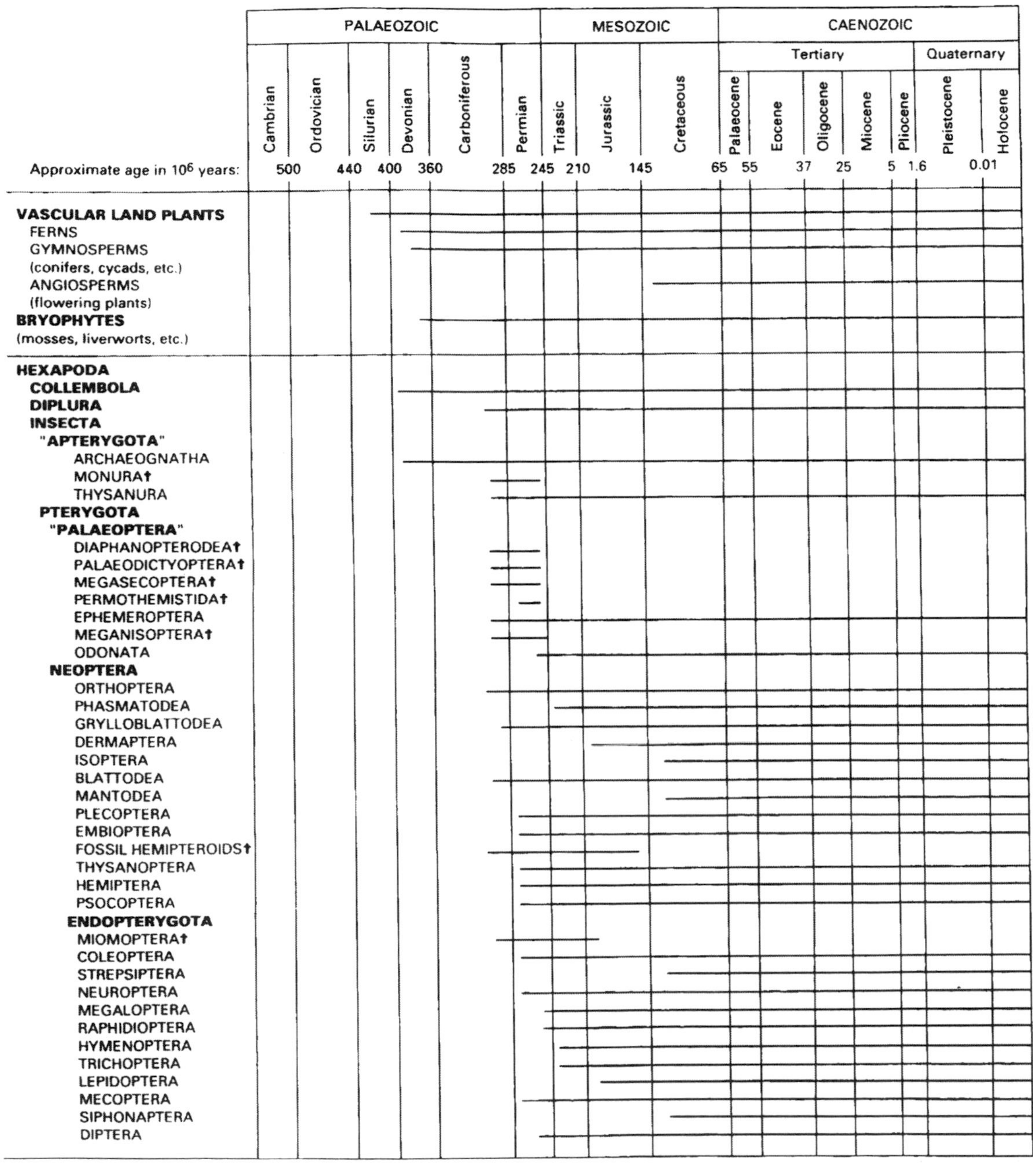

Fig. 3.67. Stratigraphic distribution of insects. (Reproduced with permission from Gullan and Cranston, 2000.)

discussed in turn below. There are some 13 000 fossil and 6600 living species of echinoderms.

Biology, morphology and classification

Crinoids, which, rather misleadingly, take their scientific name from the Greek for 'lily-like', are an extant, Ordovician–Recent, group of pelmatozoan echinoderms, typically characterised by the possession of a stem-like stalk (Waters and Maples, 1997; Hess *et al.*, 1999; Byrne, in Anderson, 2001). Their living representatives include the sea-lilies or isocrinids, and the feather-stars or comatulids. Their living relatives include the starfish or asteroids, and the brittle-stars or ophiuroids; their fossil relatives the eocrinoids, 'cystoids' and blastoids. None of these related groups is especially important in the fields of palaeobiology and biostratigraphy, and none is discussed. Note, though, that fossil forms are known. For example, the ophiuroid *Palaeocoma (Ophioderma) egertoni* is known from the Early Jurassic, middle Lias so-called 'starfish bed' of Down Cliff near Bridport in Dorset on the south coast of England.

The morphologically similar extinct carpoids or homalozoans are actually not closely related, but belong in a separate group, the calcichordates, arguably ancestral to the chordates or vertebrates (Jefferies, 1968, 1975, 1986, 1997; Jefferies, in House, 1979; Philip, 1979; Jollie, 1982; Gee, 1996; Jefferies *et al.*, 1996).

Biology

The biology and ecology of living crinoids is comparatively poorly known, possibly at least in part because they typically live either in open water or on the bottom of the ocean and are difficult to study by scuba-diving. However, it is known that the crinoid animal lives in a protective cup-like structure or calyx. It is also known that the animal feeds by filter-feeding, drawing food into its centrally located, upwardly directed mouth by means of tube-feet and cirri on its arms or brachials, and excreting waste through an adjacent, but separate, anus. In fact, crinoids lack distinct circulatory and excretory systems, gas exchange and the removal of waste being accomplished by coelomic fluids and coleomocytes; and also lack centralised nervous systems. Crinoids reproduce sexually, although it is worth noting that many asteroids and ophiuroids, and indeed also many holothurians, reproduce asexually, essentially by cloning.

In terms of ecology, living crinoids are exclusively marine, and either benthic, with a planktonic developmental stage, enabling dispersal and the colonisation of new niches, or, more commonly, nektonic or nektobenthic, with a benthic developmental stage.

Benthic forms are exclusively epifaunal. The stalked sea-lilies or isocrinids are high-level epifaunal and sessile, attached to the sea-floor by means of a hold-fast. The stalkless feather-stars or comatulids are low-level epifaunal, and either sessile, attached to the sea-floor by means of modified hook-like cirri, or motile, achieving motility through the use of their arms and cirri, rather than their tube-feet. As intimated above, some crinoids, typically sea-lilies, are deep marine, although some, typically feather-stars, are shallow marine. Some authors have attributed the relative rarity of living crinoids in shallow marine environments to predation by fish (Meyer, 1985; Oji, 1996). Others have attributed the relative rarity of fossil crinoids in the shallow marine environments of the Mesozoic and Cenozoic to predation by fish (Meyer and Macurda, 1977). Interestingly, even some Palaeozoic crinoids show signs of having been predated by fish with specialised crushing teeth (while others show signs of having been parasitised, possibly by myzostomid polychaetes).

Nektonic or nektobenthic forms, exemplified by the stalkless feather-star *Antedon*, are able to achieve a swimming motion through reportedly graceful, ballerina-like movements of alternate arms (although they spend much of their time at sites on the substrate that are favourable for feeding). Incidentally, *Antedon* occurs around the British Isles (Naylor, 2003).

Morphology

In terms of hard-part morphology, crinoids are characterised by internal skeletons or tegmens secreted of calcite, and constituted of hundreds of individual plates or ossicles of different shapes and sizes, each characterised by pentameral or five-rayed symmetry (Fig. 3.68). On death, the soft tissues holding the skeleton together decay, and it usually disaggregates, such that the normal mode of preservation is as isolated plates. The plates that form the calyx are flattened in section and polygonal in outline, and fit together side by side like pieces of a jigsaw, while those that form the stalk, known as columnals, are disc-like, and rounded, elliptical, pentagonal or stellate in outline, and fit together one on top of the other.

Classification

The classification of crinoids is based on morphology, and on phylogeny inferred in the main from

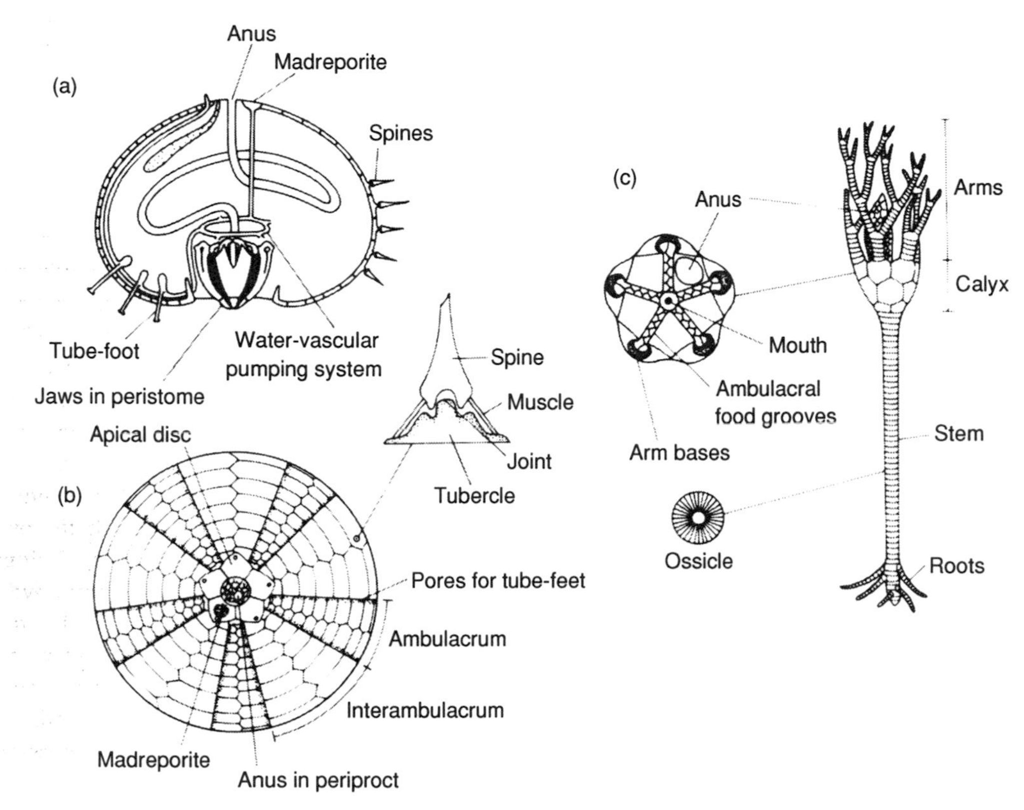

Fig. 3.68. Hard-part morphology of crinoids and echinoids: general morphology. (a–b) Echinoid; (c) crinoid. (Modified after Black, 1989.)

empirical observations on the fossil record (Simms, in Hess *et al.*, 1999). Four sub-groups are recognised, namely the comparatively well-known cladiids and camerates; and the less well-known disparids and hybocrinids. The echmatocrinids are currently excluded from the crinoids.

Cladiids. The cladiids, which range from Ordovician to Recent, are characterised by fused and rigid to flexible structures. They include the 'stem-group' cladiids or dendrocrinids and cyathocrinids (Ordovician–Permian), characterised by completely fused and rigid structures; the flexibilians (Ordovician–Permian), characterised by only partially fused and more or less flexible structures; and the articulates (Triassic–Recent), characterised by flexible structures. The articulates in turn include the encrinids (Triassic), holocrinids (Triassic), millericrinids (Triassic–Recent) and cyrtocrinids (Jurassic–Recent), as well as the familiar isocrinids (Triassic–Recent) and comatulids (Triassic–Recent).

Camerates. The camerates (Ordovician–Permian) are characterised by only partial fusion (of the proximal ossicles of the brachials with the theca), and by disproportionately long stalks. They include the diplobathrids and monobathrids.

The observed evolutionary trends appear to represent progressive adaptations to more efficient filter-feeding, especially the acquisition of the ability to flex and thus orient the feeding apparatus into the flow of current (euphoniously, rheophily).

Palaeobiology

Most fossil crinoids are interpreted, essentially on the basis of analogy with their living counterparts, as having been marine, and either nektonic, nektobenthic or benthic.

They are further interpreted, on the bases of functional morphology, and associated fossils and sedimentary facies, as having been more typically benthic than their living counterparts.

However, the Jurassic *Seriocrinus*, with its unusual distribution of weight and flexibility, has been interpreted, partly on the basis of functional morphology, and partly on the basis of empirical observation, as having lived suspended upside-down from floating logs (Seilacher *et al.*, 1968; Haude, in Jangoux, 1980); and the Jurassic *Pentacrinites* has since been interpreted as similarly pseudoplanktonic (Simms, 1986).

Moreover, Silurian scyphocrinitids have been interpreted as wholly planktonic, and as having lived suspended upside-down from floating holdfasts (Haude *et al.*, 1994).

Furthermore, the Jurassic–Cretaceous *Saccocoma*, which has secondarily lost its stalk, has been interpreted as nektonic, like the living *Antedon* alluded to above. Interestingly in this context, well-preserved specimens show projections called 'Schwimmplatten' on the axillary and on some of the brachial plates (Fig. 3.69) (Jaekel, 1918).

The similarly stalkless, although larger, Cretaceous *Uintacrinus* and *Marsupites* have been interpreted as nektonic by some authors, but as benthic by some others. The latest interpretation is that they were benthic, although possibly with planktonic developmental stages, and that they lived with their calices floating like icebergs in a sea of soft, chalky substrate, and, to mix metaphors, with their arms outstretched in the form of a bowl to collect food (Hess, in Hess *et al.*, 1999). Interestingly in this context, *Uintacrinus* is especially characteristic of the Late Cretaceous, Santonian Niobrara Chalk of the Uinta Mountains in Utah and of contiguous Colorado and Kansas in the American Mid-West, where it was first discovered by the dinosaur collector Marsh, and both *Uintacrinus* and *Marsupites* of the Santonian part of the Chalk of Great Britain and Northern Ireland.

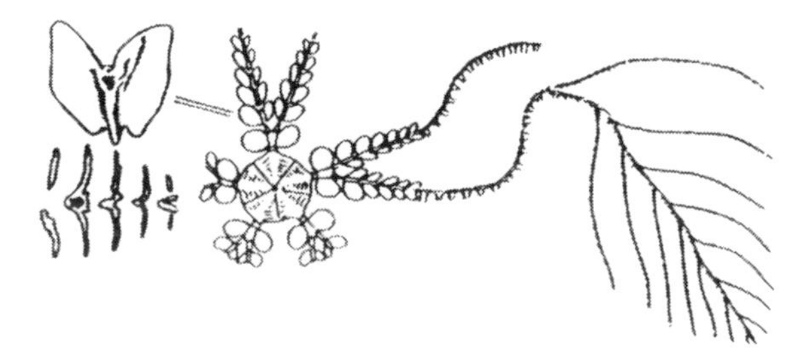

Fig. 3.69. Hard-part morphology of crinoids and echinoids: the pelagic crinoid *Saccocoma*. Magnification ×5. (From Jones, 1996; reproduced with permission from Bignot, 1985.)

Palaeobathymetry

Fossil benthic crinoids are interpreted, essentially on the basis of associated fossils and sedimentary structures, as having been more typically shallow marine than their living counterparts. In parts of the Palaeozoic, they appear to have congregated together to form 'crinoid gardens' in shallow marine environments, locally in sufficient abundance as ultimately to be rock-forming. Indeed, in parts of the Palaeozoic, crinoids were important contributors to the construction of reefs and reef-like structures in shallow marine environments. For example, they are characteristic of the Early Carboniferous, Brigantian platformal and peri-reefal facies of the Burren, Buttevant and Callan areas of southern and western Ireland (Gallagher, in Strogen *et al.*, 1996). They are also characteristic of the initial stages of development of the Early Carboniferous, Tournaisian 'muleshoe mound', a Waulsortian mound in the Sacramento Mountains of New Mexico (Kirkby and Hunt, in Strogen *et al.*, 1996).

Palaeobiogeography

Many crinoids appear to have had restricted or endemic biogeographic distributions, rendering them of some use in the characterisation of palaeobiogeographic provinces, and in turn in the constraint of plate tectonic reconstructions, in the Palaeozoic (Lane and Sevastopulo, in McKerrow and Scotese, 1990, Webster, in Mawson *et al.*, 2000; Lefebvre and Fatka, in Servais *et al.*, 2003).

Biostratigraphy

Crinoids first appeared in the Middle Cambrian, if the echmatocrinids are included, or Early Ordovician, if they are excluded (Hess and Ausich, in Hess *et al.*, 1999). They diversified through the Ordovician (Sprinkle and Guensberg, in Webby *et al.*, 2004), but sustained significant losses in the End-Ordovician mass extinction. They recovered from this event, though, and diversified again through the Silurian and Devonian, but sustained further significant

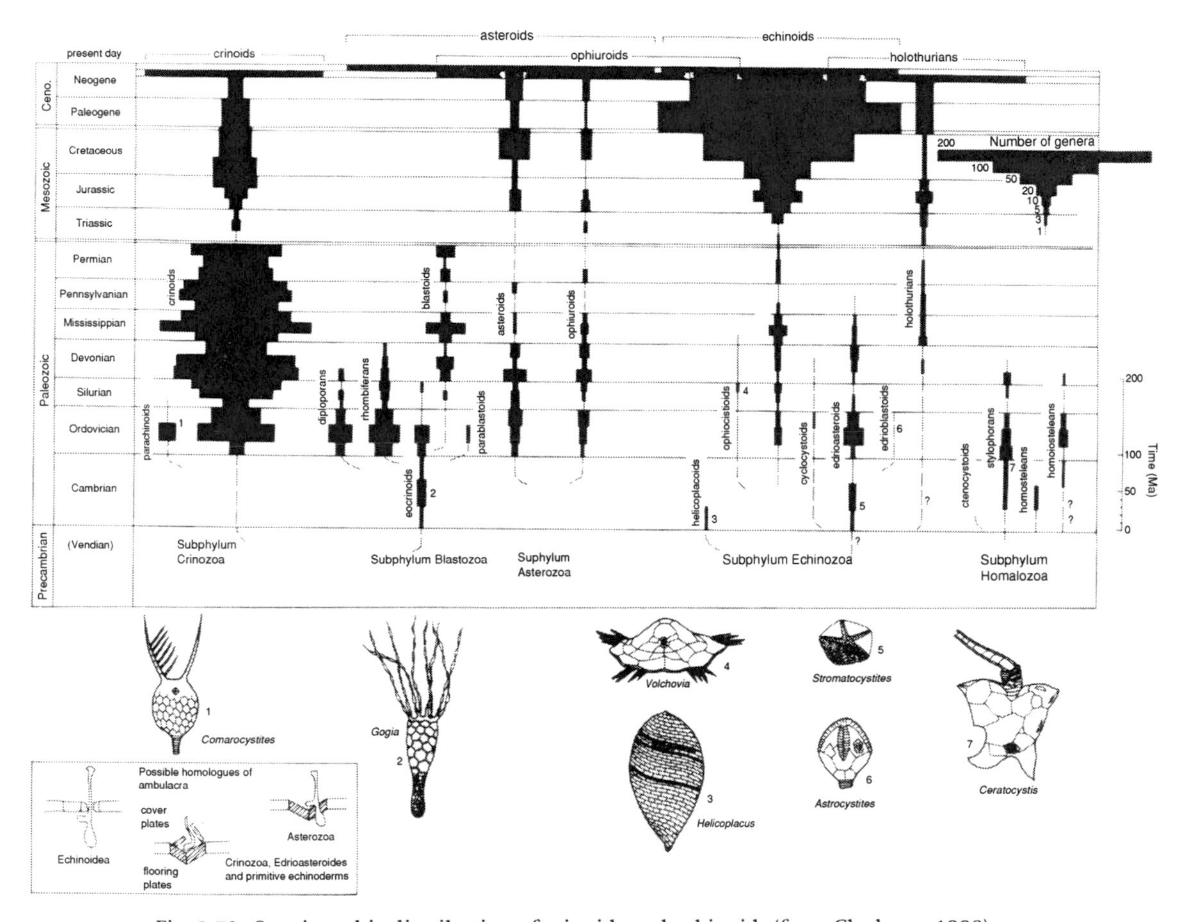

Fig. 3.70. Stratigraphic distribution of crinoids and echinoids (from Clarkson, 1998).

losses in the Late Devonian mass extinction. They recovered from this event, too, and underwent renewed radiation in the Carboniferous and Permian, before almost becoming extinct in the End-Permian mass extinction. They did stage some form of recovery from this event, and did diversify through the Triassic and Jurassic, but never attained the prominence they had in the Palaeozoic, in the Mesozoic. They may be said to have been in decline since the Mesozoic, possibly due either to increased predation by fish, especially in shallow marine environments, or to increased competition from echinoids. The perceived decline may be more apparent than real, though, and an artefact of preservation. There are some 600 living species of crinoids.

The moderate rate of evolutionary turnover exhibited by the crinoids renders them of some use in biostratigraphy, at least in appropriate, marine, environments (Figs. 3.70–3.71).

Crinoids are of use in, for example, the Ordovician and Cretaceous of the UK. Here, crinoid columnals are of use in the Ordovician (Donovan, 1995) and Cretaceous (Rasmussen, 1961); crinoids in the Late Cretaceous, where they form part of the basis of the biozonation of the Chalk, *Uintacrinus socialis* and *Marsupites testudinarius* marking the middle and late parts of the Santonian, respectively. Note also, incidentally, that exquisitely preserved specimens of *Pentacrinites fossilis* occur in the Early Jurassic, lower Lias of Charmouth in Dorset (Simms, in Hess *et al.*, 1999).

Crinoids are also of use in, for example, the Early Carboniferous of North America, the Permian of Australia, and the Triassic of Europe.

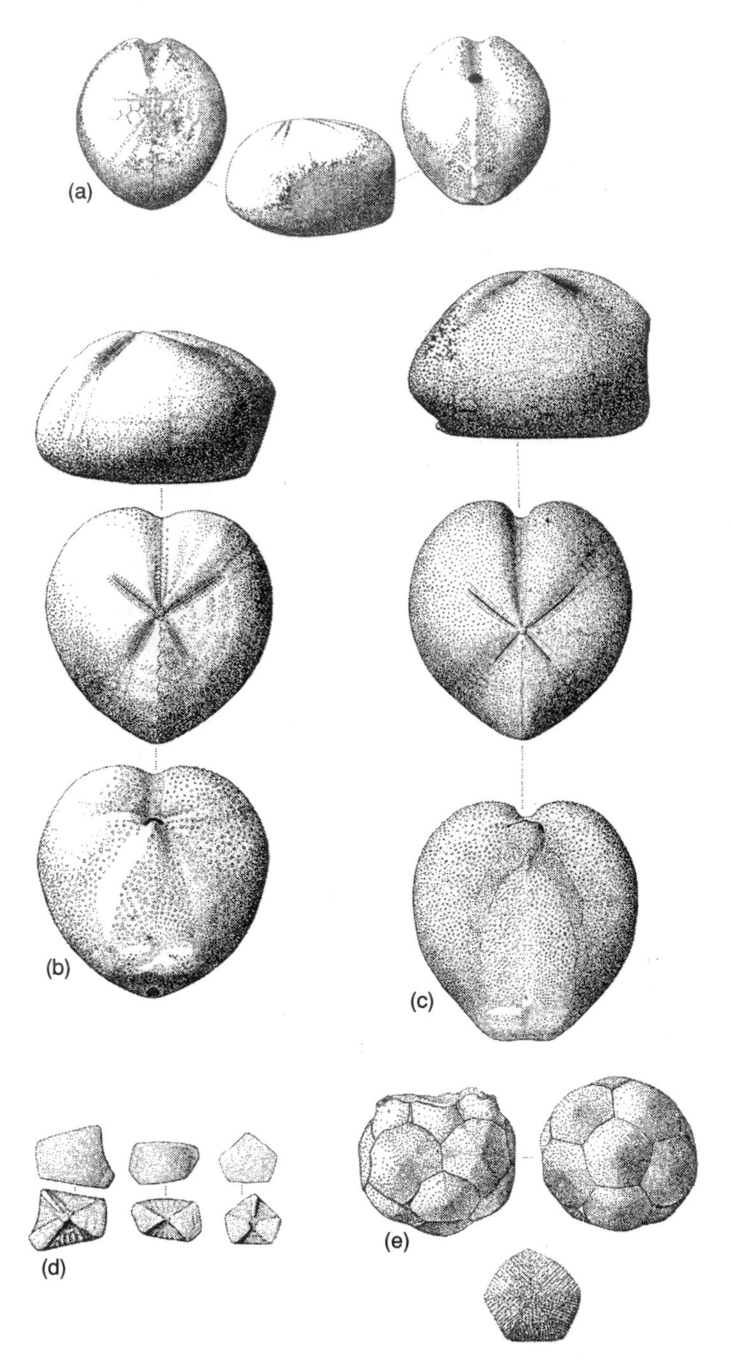

Fig. 3.71. Some stratigraphically useful crinoids and echinoids. Echinoids: (a) *Sternotaxis (Holaster) planus*, nominate taxon of the Late Cretaceous, Coniacian *Holaster planus* Zone, Upper Chalk, east of Dover, Kent, ×1; (b) *Micraster cortestudinarium*, nominate taxon of the Late Cretaceous, Coniacian *Micraster cortestudinarium* Zone, Upper Chalk, Chatham, Kent, ×1; (c) *Micraster coranguinum*, nominate taxon of the Late Cretaceous, Santonian *Micraster coranguinum* Zone, Upper Chalk, Gravesend, Kent, ×1. Crinoids: (d) *Uintacrinus socialis*, nominate taxon of the Late Cretaceous, Santonian *Uintacrinus socialis* Zone, Upper Chalk, Kent, ×2; (e) *Marsupites testudinarius*, nominate taxon of the Late Cretaceous, Santonian *Marsupites testudinarius* Zone, Upper Chalk, southern England, ×1.

In my working experience, crinoids have proved of some use in stratigraphic subdivision and correlation, and/or in palaeoenvironmental interpretation, in the following areas.

Palaeozoic

The Ordovician–Permian of the former Soviet Union. Crinoid columnals are of use in stratigraphic subdivision and correlation in the Ordovician–Permian of the former Soviet Union (Stukalina, 1988; Ausich *et al.*, in Hess *et al.*, 1999). Crinoids are of use in palaeoenvironmental interpretation in the platformal and peri-reefal facies of the Devonian–Permian of the Bolshoi Karatau Mountains of southern Kazakhstan, and of the southern Urals. Crinoids form cool- or deep-water reefs in the Early Permian, Sakmarian–Artinskian of the Barents Sea that in turn form petroleum reservoirs (Bruce and Toomey, in Vorren *et al.*, 1993; author's unpublished observations).

Mesozoic

The latest Jurassic–earliest Cretaceous of the Middle East. Here, the pelagic crinoid *Saccocoma* forms a useful marker for the latest Jurassic to earliest Cretaceous of the deep marine environments of the Gotnia Basin, centred on south-west Iran, adjacent parts of Iraq, and Kuwait (Jones, 1996). Its usefulness is enhanced by its ease of identification in thin-section. Incidentally, *Saccocoma* also characterises coeval deep marine deposits elsewhere, from Russia in the east to Cuba in the west.

3.6.18 Echinoderms: echinoids

Biology, morphology and classification

Echinoids are an extant, Ordovician–Recent, group of eleutherozoan echinoderms, characterised by the absence of a stalk (Waters and Maples, 1997; Byrne, in Anderson, 2001). Their living representatives include the sea urchins or 'true' echinoids, the heart-urchins or spatangoids and allied forms, the sand-dollars or clypeasteroids, and the slate-pencil-urchins, or cidaroids. Their living relatives include the holothurians or sea cucumbers; their fossil relatives the edrioasteroids. Neither of these related groups is especially important in the fields of palaeobiology and biostratigraphy, and neither is discussed. Note, though, that holothurian sclerites have some as yet unrealised biostratigraphic potential.

Biology

The biology and ecology of living echinoids is well known. The echinoid animal feeds by detritus-feeding, drawing food into a modified mouth, or 'Aristotle's lantern', situated on the underside or oral side, and excreting waste through an anus usually situated on the topside or aboral side. In fact, like crinoids, echinoids lack distinct circulatory and excretory systems; and also lack centralised nervous systems. Also like crinoids, echinoids reproduce sexually (bizarrely, echinoid gonads are edible, and indeed are apparently considered a gourmet delicacy in some cultures).

In terms of ecology, living echinoids are exclusively marine and benthic, with a planktonic developmental stage enabling dispersal and the colonisation of new niches (apparently capable of being replaced by a brooding stage under unfavourable conditions). Many are deep marine, although evidently ultimately of shallow marine origin (Smith, 2004). Some are epifaunal; some infaunal.

Epifaunal forms are either motile, achieving motility through use of their tube-feet (which also, incidentally, appear to play a role in respiration), and feed by scavenging for detrital food particles strewn about the sea-floor, or by grazing on algae; or sessile.

Infaunal forms feed on detrital food particles disseminated through the bottom sediment. They often exhibit functional morphological adaptations to an infaunal habit, such as elongation, and migration of the mouth from the underside to the anterior, and of the anus from the topside to the posterior, resulting in the superimposition of bilateral symmetry onto the basic pentameral plan.

Morphology

In terms of hard-part morphology, echinoids are characterised by internal skeletons or tests secreted of calcite, and constructed of hundreds of individual plates, arranged according to an intricate plan, in alternate rows of perforate ambulacrals, that allow the tube-feet to protrude, and imperforate inter-ambulacrals, five of each (Fig. 3.68). The 'Aristotle's lantern' is strengthened by high-magnesium calcite, and is capable of being used for burrowing. The amount of bio-erosion caused by the burrowing activities of echinoids can be significant.

The test is covered with spines, which in some cases are poisonous, for protection (hence the name, which derives from the Greek for 'spiny skin'), and which, on death, detach, leaving tubercles.

In outline, it is rounded, with an oral–aboral axis of symmetry, in the case of the so-called regular echinoids, to heart-shaped, with an anterior–posterior axis of symmetry, in the case of the so-called irregular echinoids.

Classification

Classification of echinoids is currently in something of a state of flux. Historically, two main sub-groups have been recognised on the basis of morphology, namely, the regular echinoids and the irregular echinoids (see above). However, it has recently been demonstrated by cladistic analysis of molecular data that both of these sub-groups are polyphyletic, and therefore that the historical classification is in need of revision, the better to reflect phylogeny. Unfortunately, as yet no clear consensus has been reached as to a revised classification. One suggestion has been to recognise three main sub-groups, namely the poorly known perischoechinoids, and the well-known cidaroids and euechinoids. Only the cidaroids and euechinoids are considered here.

Cidaroids. The cidaroids, which range from Devonian to Recent, are characterised by regular morphology. They are exemplified by *Cidaris*, and by the more-or-less familiar living slate-pencil-urchin *Heterocentrotus*.

Euechinoids. The euechinoids (Late Triassic–Recent) are characterised by regular to irregular morphology. They are exemplified by the more-or-less familiar regular sea urchins or 'true' echinoids, such as *Echinus* and *Diadema*; irregular heart-urchins or spatangoids, such as *Echinocardium* and *Micraster*, and allied forms, such as the tragically named *Disaster*; and irregular sand-dollars or clypeasteroids, such as *Clypeaster* and *Scutellum*.

Many of the observed evolutionary trends appear to represent progressive adaptations to an infaunal habitat.

Palaeobiology

Palaeobathymetry

Fossil echinoids are interpreted, on the basis of analogy with their living counterparts, as having been marine and benthic. They are further interpreted, on the basis of associated fossils and sedimentary structures, as having been more typically shallow marine than their living counterparts. Note, though, that demonstrable deep marine fossil echinoids are known, for example from fossil 'cold seeps' (Gaillard *et al.*, 1992). Note also that fossil echinoid diversity appears to be highest during times of transgression, at least in the Cretaceous (Smith and Jeffery, in Culver and Rawson, 2000).

Seven echinoid biofacies have been recognised in the Early Miocene of Egypt, which have in turn been used to identify sedimentary facies and cycles of transgression and regression (Kroh and Nebelsick, 2003). The *Phyllacanthus* biofacies represents a shallow-marine carbonate environment; the *Clypeaster martini* biofacies a shallow marine, high-energy environment; the *Parascutella* biofacies a shallow marine, high-energy environment characterised by storm winnowing of sand-dollars; the Mixed biofacies a shallow marine, low- to moderate-energy environment; the Cidaroid–Echinacea biofacies a deeper, low-energy environment, characterised by a highly structured habitat and corresponding diversity of regular and irregular sea urchins; the Spatangoid biofacies a deep, low-energy, environment characterised by diverse burrowers; and the Transported biofacies a deep marine, high-energy environment characterised by transportation of material from up-dip.

Palaeobiogeography

Many echinoids appear to have had restricted or endemic biogeographic distributions, rendering them of some use in the characterisation of palaeobiogeographic provinces, especially in the Cenozoic (Jones, in Agusti *et al.*, 1999; Carter, in Prothero *et al.*, 2003; see also Sub-section 5.5.8).

Biostratigraphy

Echinoids first appeared in the Ordovician (Guensberg and Sprinkle, in Zhuravlev and Riding, 2001). They diversified through the Ordovician (Sprinkle and Guensberg, in Webby *et al.*, 2004), and continued to do so through the Silurian and Devonian, when the cidaroids evolved, before sustaining significant losses in the Late Devonian mass extinction or in the Carboniferous. They recovered from this event, though, and underwent renewed radiation in the Permian, before sustaining further significant losses in the End-Permian mass extinction. They recovered from this event, too, and underwent a dramatic diversification in the Mesozoic, when the euechinoids evolved, at which time they overtook the crinoids as the dominant group of echinoderms (Smith and Jeffery, in Culver and Rawson, 2000). They were little if at all affected by the End-Cretaceous mass extinction, and indeed underwent renewed radiation in the Cenozoic. There are some 900 living species of echinoids.

In theory, the moderate rate of evolutionary turnover exhibited by the echinoids renders them of some use in biostratigraphy, at least in appropriate, marine, environments (Figs. 3.70–3.71).

Unfortunately, however, in practice, their usefulness is compromised by their restricted ecological distribution and facies dependence, a function of their benthic habit. Moreover, their potential usefulness in the Palaeozoic is further compromised by their poor preservation potential at this time, prior to the innovation of the fusion of the plates in the Mesozoic.

Nonetheless, echinoids are of use in, for example, the Late Cretaceous of Great Britain and Northern Ireland, where they form part of the basis of the biozonation of the Chalk. Here, *Holaster planus* marks the late part of the Turonian, *Micraster cortestudinarium* the early part of the Coniacian, *M. coranguinum* the late part of the Coniacian and the early part of the Santonian, and *Offaster pilula* the early part of the Campanian.

In my working experience, echinoids have proved of some use in stratigraphic subdivision and correlation, and/or in palaeoenvironmental interpretation, in the following areas.

Mesozoic

The Cretaceous of the Middle East. Here, 'unit o' of the Musandam limestone of Musandam Peninsula is marked by the Early Cretaceous, Aptian echinoid *Heteraster musandamensis* (Hudson and Chatton, 1959). 'Member c' of the Natih formation of the Oman Mountains is marked by nine species of Late Cretaceous, Cenomanian echinoids, although with one, *Coenholectypus larteti*, dominant (Smith *et al.*, 1990). The assemblage has been interpreted as representing a lagoonal facies landward of a rudist reef or bioclastic shoal. The Simsima formation of the Oman Mountains is marked by no fewer than 45 species of Maastrichtian echinoids (Smith, in Howarth, 1995). These have been grouped into assemblages or biofacies largely on the basis of inferred feeding strategies, and in turn used to identify sedimentary facies. Back-reefal lagoonal facies are characterised by epifaunal grazers such as *Noetlingaster* and *Hattopsis*, semi-infaunal selective deposit-feeders such as *Hemipneustes*, infaunal selective deposit-feeders such as *Mecaster* and *Faujasia*, and non-selective deposit-feeders such as *Petalobrissus*. Reefal facies are characterised by epifaunal hardground grazers such as *Echinotiara, Phymechinus* and *Goniopygus*. Peri-reefal facies are characterised by epifaunal grazers such as *Glyphopneustes, Mimiosalenia* and *Goniopygus*; browsers such as *Hattopsis*, and generalists; infaunal generalists such as *'Globator'* and *Conulus*; and selective and non-selective deposit-feeders. Fore-reef facies are characterised by infaunal generalists such as *'Globator'*, selective deposit-feeders such as *Faujasia* and non-selective deposit-feeders such as *Petalobrissus*. Off-reefal facies are characterised by epifaunal grazers, and infaunal selective deposit-feeders such as *Proraster*. In the field in the Zagros Mountains in the south-west of Iran, the lower part of the Dariyan formation is marked by the occurrence of the Aptian echinoid *Heteraster oblonga* (author's unpublished observations). The Kazhdumi formation is marked by the late Albian–early Cenomanian *Emiritia cf. libanoticum, Trochodiadema cf. dhofarensis, Anorthopygus aff. orbicularis, Douvillaster cf. husseini* and *Epiaster cf. blanckenhorni* (Collignon, 1981; author's unpublished observations, based in part on identifications by Andrew Smith of the Natural History Museum, London). The Sarvak formation is marked by the

Cenomanian *Anorthopygus orbicularis* and *Caenholectypus serialis* (author's unpublished observations).

Cenozoic

The Oligocene–Miocene of the Middle East. In the field in the Zagros Mountains in the south-west of Iran, the '*Brissopsis* beds' of the Pabdeh formation are marked by the Oligocene echinoids *Brissopsis cf. biarritzensis, Echinolampas* and *Schizaster*, the Mishan formation by the Miocene *Echinolampas jacquemonti* (author's unpublished observations).

3.6.19 Graptolites

Biology, morphology and classification

Graptolites, which take their name from the Greek for 'writing in the rocks', are an extinct, Cambrian–Carboniferous, group which have conventionally always been interpreted by palaeontologists as deuterostome hemichordates (Palmer and Rickards, 1991). This interpretation is essentially based on the morphological similarity of the extinct graptolites to the extant pterobranchs, which have been classified as deuterostome hemichordates by most biologists (Stocker, in Anderson, 2001). Note, though, that pterobranchs have also been classified as protostome lophophorates by some biologists (Purves *et al.*, 2004).

Biology

Nothing is known of the biology of the graptolites, as they are extinct. However, inferences can be drawn on the basis of analogy with what would appear to be their closest fossil and living relatives, the solitary to colonial pterobranchs *Rhabdopleura*, species of which are distributed around the Atlantic and Mediterranean coasts of Europe (van der Land, in Costello *et al.*, 2001), and *Cephalodiscus*. Thus, it is inferred that the mode of life was either sessile, as in *Rhabdopleura*, or involved a degree of motility, as in *Cephalodiscus*. It is further inferred that the preferred habitat was deep marine, benthic and epifaunal, the feeding strategy to gather floating food particles with feather-like tentacles, and the reproductive strategy asexual.

Morphology

In terms of hard-part morphology, graptolites are characterised by skeletal frameworks, or rhabdosomes, that are thought to have housed colonies of up to 5000 individual animals, or zooids, in a complex system of internal tubes (Fig. 3.72) (Zalasiewicz

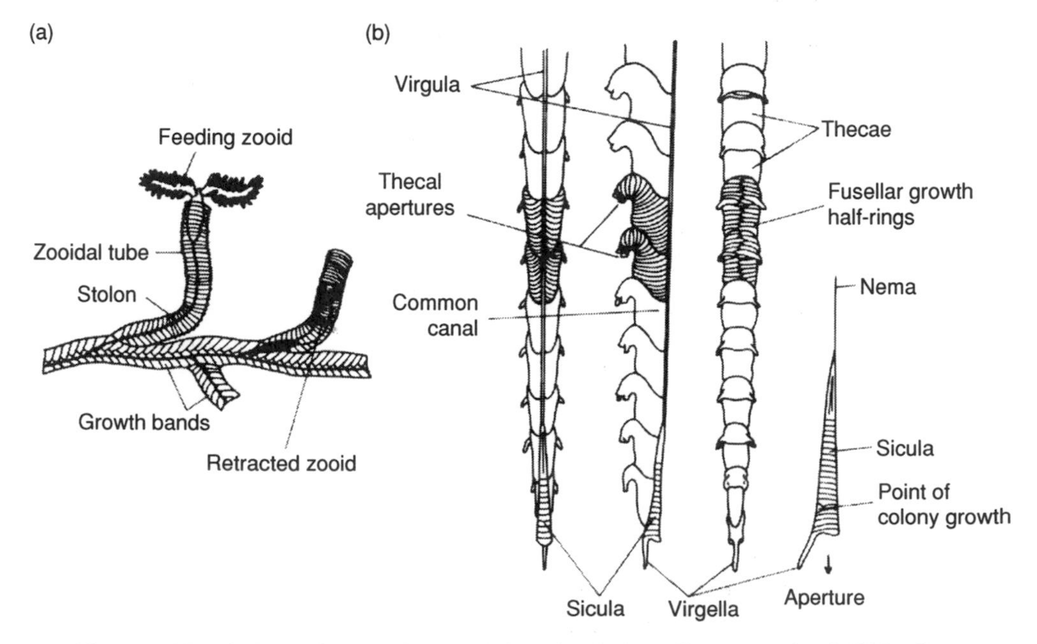

Fig. 3.72. Morphology of graptolites. (a) Living *Rhabdopleura* (for comparison); (b) fossil *Monograptus*. (Modified after Berry, in Boardman *et al.*, 1987.)

et al., 2000). The skeletal frameworks are composed of two layers of collagen, an inner layer of rings or fusellae and an outer layer of wraps of cortical tissue believed to have been secreted by special glands and laid down in a manner reminiscent of the bandages around a mummy. The skeletal frameworks are of varying types. The general arrangement is of an initial sicula, which may or may not bear a spine or virgella, succeeded by a varying number of tubular branches or stipes, in turn bearing a varying number of side-branches or thecae, opening to the outside world through apertures. The stipes contain a central thread- or rod-like nema or virgula. The arrangement of the stipes is either horizontal; or directed the same way as the sicula ('down'), or pendent; or directed the opposite way from the sicula ('up'), or reclined; or fused back-to-back, or scandent. *Isograptus*, with two reclined stipes, has an outline uncannily reminiscent of 'Batman'. Graptolite soft parts are only exceptionally preserved (Loydell *et al.*, 2004).

Classification

Classification of graptolites is based on morphology, and on phylogeny inferred from empirical observations on the fossil record (Palmer and Rickards, in Palmer and Rickards, 1991; Rickards, in Palmer and Rickards, 1991).

Eight sub-groups have been recognised: the well-known dendroids and graptoloids, and the poorly known dithecoids, archaeodendrids, stolonoids, cameroids, tuboids and crustoids. Only the dendroids and graptoloids are considered here.

Dendroids. The dendroids, which range from Middle Cambrian to Carboniferous, are characterised by numerous stipes, and by thecae differentiated into two types, interpreted as male and female. They are exemplified by *Dendrograptus* and *Dictyonema*.

Graptoloids. The graptoloids (Ordovician–Early Devonian) are characterised by less numerous stipes, and in some cases only one, and by undifferentiated thecae, which has led to their having been interpreted as hermaphroditic. They are exemplified by dichograptids (Ordovician), such as *Dichograptus* and *Tetragraptus*; by glyptograptids (Ordovician), such as *Glyptograptus*; by dicranograptids (Ordovician), such as *Dicranograptus*, *Dicellograptus* and *Nemagraptus*; by diplograptids (Ordovician–Silurian), such as *Diplograptus* and *Climacograptus*; and by monograptids (Silurian–Early Devonian), such as *Monograptus*, *Cyrtograptus* and *Monoclimacis*. The main evolutionary trend observed in the graptoloids is towards a reduction in the number of stipes.

Palaeobiology

Palaeobathymetry

Fossil graptolites are interpretable, essentially on the basis of analogy with their living counterparts, namely *Rhabdopleura* and *Cephalodiscus*, as having been deep marine and benthic, and either sessile or motile. However, some are interpreted, on the bases of functional morphology and associated fossils and sedimentary facies, as having been essentially pelagic rather than benthic (and benthic forms as more typically shallow marine than their living counterparts).

Specifically, some dendroids are interpreted as having been benthic and sessile (Rigby and Fortey, in Palmer and Rickards, 1991). These are envisaged as having had the form of small shrubs, with a specially strengthened proximal or sicular end held fast or rooted to the sea bottom, and the distal stipes branching out into the overlying water column. Other dendroids, though, are interpreted as having lived suspended upside-down from floating seaweed, that is, as having been pseudoplanktonic.

Graptoloids have been interpreted as planktonic by some authors (Bulman, 1964; Rickards, 1975), but as nektonic by some others (Kirk, 1969, 1972). Observations on the behaviour of life-size models in columns of water indicated that they would have required some means of auto-mobility, or at least some means of increasing buoyancy, possibly involving the deployment of stored gas or fat, in order to stop themselves simply falling to the bottom (Rigby and Rickards, 1989; Rigby, 1991). The latest interpretation is that they were nektonic, achieving a swimming motion through undulating movements of lateral extensions to the muscular cephalic shields of individual zooids (Melchin and DeMont, 1995). Whether they were planktonic or nektonic, graptoloids were well adapted to a pelagic habit by virtue of their long nemas, hook-like thecae and net-like overall forms, all of which helped to reduce their sinking rate and maintain their position within the water column. Interestingly, some show evidence of having been parasitised.

Palaeobiogeography

Many graptolites appear to have had restricted or endemic biogeographic distributions, rendering them of some use in the characterisation of palaeobiogeographic provinces, and in turn in the constraint of plate tectonic reconstructions (Berry, in Hallam, 1973; Skevington, in Hallam, 1973; Berry and Wilde, in McKerrow and Scotese, 1990; Finney and Chen, in McKerrow and Scotese, 1990; Rickards *et al.*, in McKerrow and Scotese, 1990; Cooper *et al.*, 1991).

In the Ordovician, there were two graptolite provinces, a Gondwanan province, characterised by *Didymograptus*, and a Laurentian province, characterised by isograptids, cardiograptids and oncograptids.

Biostratigraphy

Dendroid graptolites first appeared in the Cambrian, and ranged through to the Carboniferous (Rickards, in Palmer and Rickards, 1991).

Graptoloids evolved, from dendroids, in the Ordovician (Rickards, in Palmer and Rickards, 1991). They diversified through the Ordovician (Cooper *et al.*, in Webby *et al.*, 2004), but then sustained significant losses in the End-Ordovician mass extinction. They recovered from this event, though, and diversified again in the Silurian, especially in the Llandovery and Ludlow (Berry, in Hart, 1996; Kaljo, in Hart, 1996). However, they then appear to have gone into a decline, and became extinct during the Devonian.

The rapid rate of evolutionary turnover exhibited by the graptolites renders them of considerable use in biostratigraphy, at least in appropriate, deep marine environments (Figs. 3.73–3.74). Indeed, high-resolution biostratigraphic zonation schemes based on the graptolites have been established, that have essentially global applicability (see Chapter 6). The zonal resolution is sufficiently high to facilitate the construction of detailed chronostratigraphic diagrams, for example, in the Ordovician–Silurian of the Welsh basin (Woodcock *et al.*, in Hesselbo and Parkinson, 1996). It is also capable of further enhancement through the use of graphic correlation technology, for example in the Ordovician of the Appalachians (Grub and Finney, in Mann and Lane, 1995) and in the Silurian of central Nevada and elsewhere (Kleffner, in Mann and Lane, 1995).

In Great Britain, *Dictyonema* is useful for stratigraphic subdivision and correlation in the Ordovician, Tremadoc; *Didymograptus* in the Arenig and Llanvirn; *Glyptograptus* in the Llandeilo; *Nemagraptus* in the Llandeilo–Caradoc; *Climacograptus, Dicranograptus* and *Pleurograptus* in the Caradoc; and *Dicellograptus* and *Glyptograptus* again in the Ashgill; *Akidograptus, Monograptus* and *Monoclimacis* in the Silurian, Llandovery; *Cyrtograptus, Monograptus* again and *Pristiograptus* in the Wenlock; and *Pristiograptus* again, *Monograptus* yet again and *Bohemograptus* in the Ludlow. Here, graptolites are locally sufficiently abundant as to lend their name to the containing rocks, as in the case of the Tremadoc *Dictyonema* Band of the west of Portmadoc, and the Llanvirn *(Didymograptus) Bifidus* Beds of the area around Cader Idris in southern Snowdonia, both in north Wales.

In my working experience, graptolites have proved of particular use in stratigraphic subdivision and correlation in the following areas.

The Ordovician and Silurian of the Middle East. Here, the regional Middle Ordovician, Llanvirn O30 MFS of Sharland *et al.* (2001), stratotypified in the Hanadir member of the Qasim formation of Saudi Arabia, is marked by graptolites of the *Didymograptus murchisoni* Zone. The regional Late Ordovician, Caradoc O40 MFS, stratotypified in the Ra'an member of the Qasim formation, is marked by graptolites of the Caradoc *Dicranograptus clingani* Zone. The regional Early Silurian, Llandoverian, Aeronian S10 MFS, stratotypified in the Qusaiba member of the Qalibah formation of Saudi Arabia, is marked by graptolites of the *Monograptus convolutus* Zone. The associated Rhuddanian transgressive systems tract is characterised by the extensive development of petroleum source-rocks (Luning *et al.*, 2000).

In the Zagros Mountains of south-west Iran, the upper member of the Zard Kuh formation around Zard Kuh and Kuh-e Dinar is marked by the occurrence of the Late Ordovician–Silurian graptolites

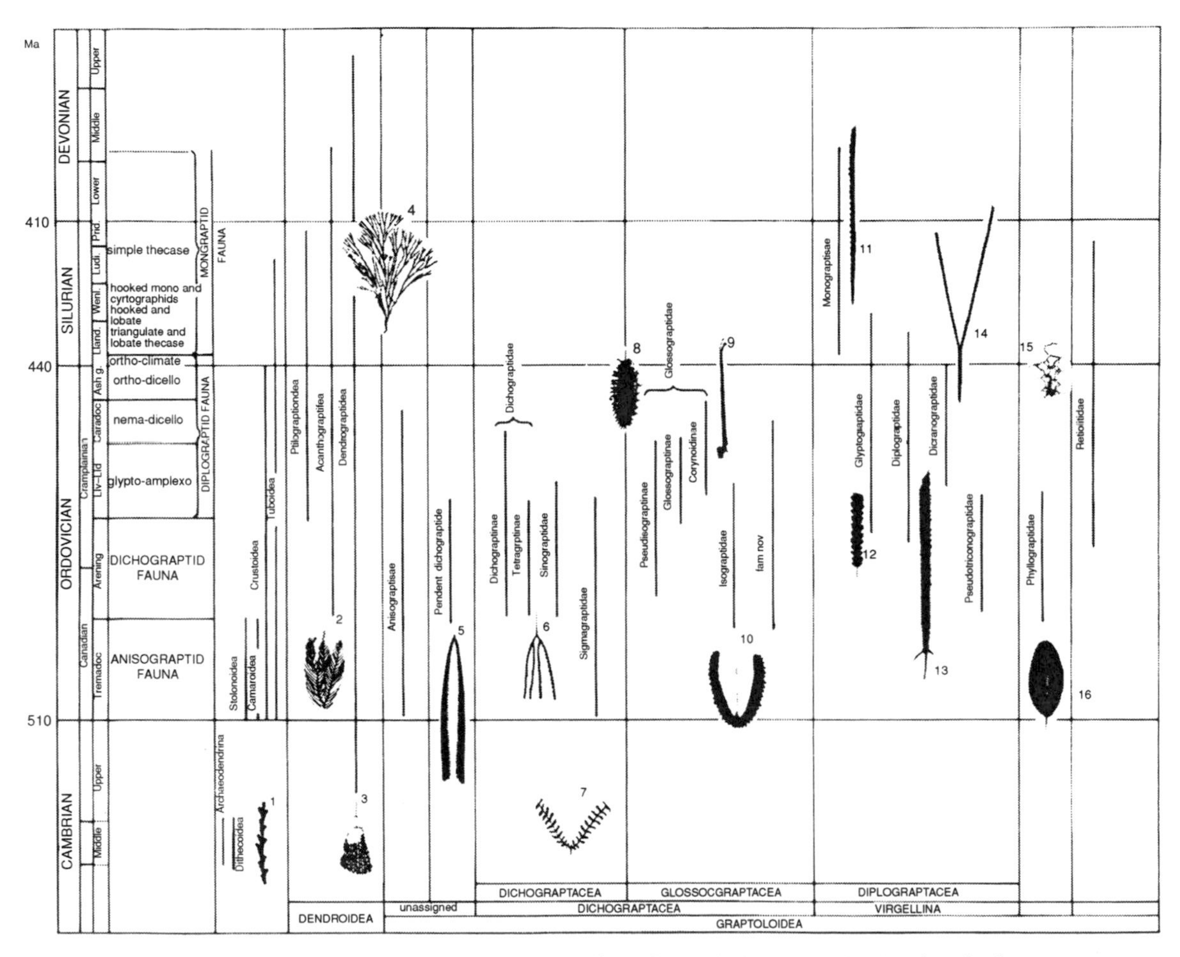

Fig. 3.73. Stratigraphic distribution of graptolites (from Clarkson, 1998). (1) *Dithecodendrum*; (2) *Ptilograptus*; (3) *Rhabdinopora*; (4) *Dendrograptus*; (5) *Didymograptus*; (6) *Tetragraptus*; (7) *Sinograptus*; (8) *Paraglossograptus*; (9) *Corynoides*; (10) *Isograptus*; (11) *Monograptus*; (12) *Glyptograptus*; (13) *Orthograptus*; (14) *Dicranograptus*; (15) *Orthoretiolites*; (16) *Phyllograptus*.

Didymograptus cf. extensus, Temnograptus, Schizograptus, Cyrtograptus and *Monograptus* (Jones, 2000). Unnamed correlative rocks cropping out at Kuh-e Faraghun and Kuh-e Gahkum in the Bandar Abbas hinterland contain the markers for the Late Ordovician *Persculptograptus persculptus* to Early Silurian *Stimulograptus sedgwickii* Zones (Rickards *et al.*, 2000).

The Silurian of north Africa. The Silurian of the Ahnet basin in Algeria and of the Murzuq basin and of Tripolitania in Libya is marked by graptolites (Follot, 1952; Bellini and Massa, in Salem and Busrewil, 1980). The petroleum source-rocks in the Murzuq basin are dated as Llandovery, Rhuddanian on graptolite evidence, indicating that they correlate with those of the Middle East (Luning *et al.*, 2000).

3.6.20 Chitinozoans

Biology, morphology and classification

Chitinozoans are an extinct, Cambrian–Carboniferous, group of organic-walled microfossils of uncertain affinity, although it has been argued that they are probably metazoans, and possibly pre-prosicular graptolites (Jones, 1996).

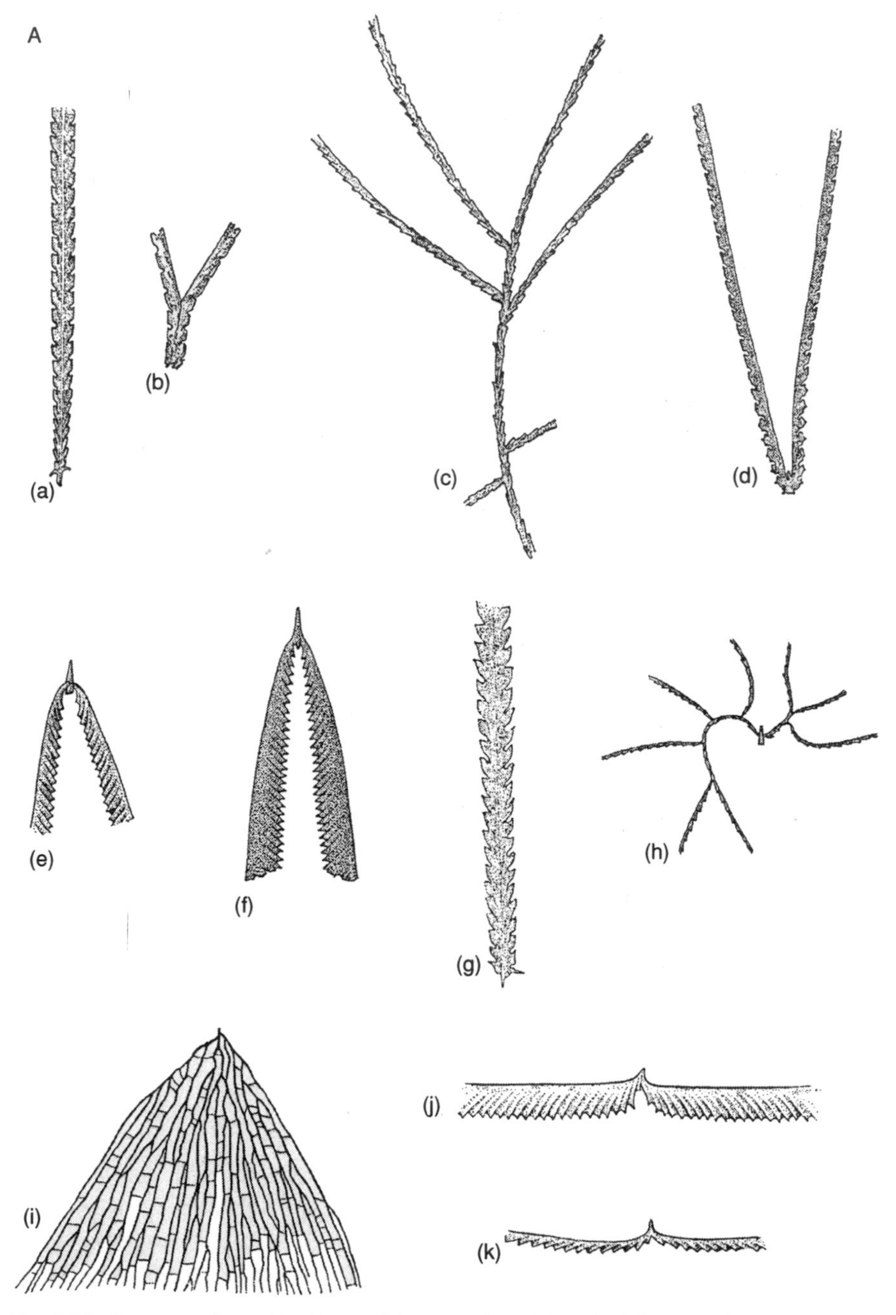

Fig. 3.74. Some stratigraphically useful graptolites. (A) Ordovician: (a) *Climacograptus wilsoni*, nominate taxon of the Caradoc (Soudleyan) *Climacograptus wilsoni* Zone, Moffat, Dumfriesshire, ×2; (b) *Dicranograptus clingani*, nominate taxon of the Caradoc (Longvillian–Actonian) *Dicranograptus clingani* Zone, Moffat, Dumfriesshire, ×2; (c) *Pleurograptus linearis*, nominate taxon of the Caradoc (Onnian) *Pleurograptus linearis* Zone, Moffat, Dumfriesshire, ×2; (d) *Dicellograptus anceps*, nominate taxon of the Ashgill *Dicellograptus anceps* Zone, Moffat, Dumfriesshire, ×2; (e) *Didymograptus bifidus*, nominate taxon of the Llanvirn, *Didymograptus bifidus* Zone, Aberdaron, Gwynedd, ×2;

B

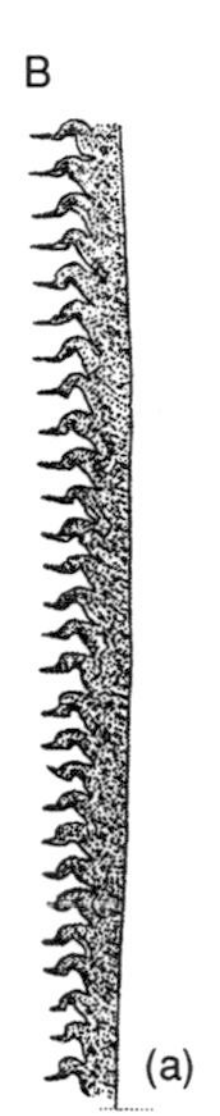

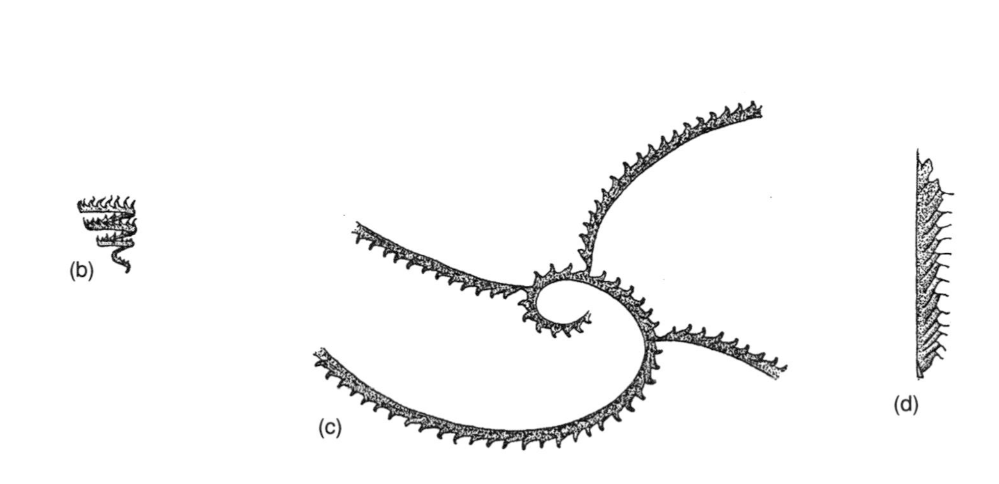

Caption for fig. 3.74 (*cont.*) (f) *Didymograptus murchisoni*, nominate taxon of the Llanvirn *Didymograptus murchisoni* Zone, Abereiddy Bay, Dyfed, ×1; (g) *Glyptograptus teretiusculus*, nominate taxon of the Llandeilo *Glyptograptus teretiusculus* Zone, Pwllheli, Gwynedd, ×2; (h) *Nemagraptus gracilis*, nominate taxon of the Llandeilo–Caradoc (Costonian) *Nemagraptus gracilis* Zone, Moffat, Dumfriesshire, ×2; (i) *Dictyonema flabelliforme*, nominate taxon of the Tremadoc *Dictyonema flabelliforme* Zone, Ffestiniog, Gwynedd, ×1; (j) *Didymograptus hirundo*, nominate taxon of the Arenig *Didymograptus hirundo* Zone, Skiddaw, Cumbria, ×2; (k) *Didymograptus extensus*, nominate taxon of the Arenig *Didymograptus extensus* Zone, Lleyn, Gwynedd, ×2.
(B) Silurian: (a) *Monograptus sedgwickii*, nominate taxon of the Llandovery (Fronian) *Monograptus sedgwickii* Zone, Stockdale, Cumbria, ×2; (b) *Monograptus turriculatus*, nominate taxon of the Llandovery (Telychian) *Monograptus turriculatus* Zone, Stockdale, Cumbria, ×2; (c) *Cyrtograptus murchisoni*, nominate taxon of the Wenlock (Sheinwoodian) *Cyrtograptus murchisoni* Zone, Builth Wells, Powys, ×2; (d) *Monograptus leintwardinensis* nominate taxon of the Ludlow (Leintwardinian) *Monograptus leintwardinensis* Zone, Leintwardine, Herefordshire, ×2.

Nothing is known directly of the biology of chitinozoans, as they are extinct.

In terms of morphology, chitinozoans are characterised by hollow flask-shaped bodies or vesicles (Fig. 3.75). The broad main chamber narrows into a tube, culminating in an opening or aperture, which is capable of being closed by an operculum, at the oral end. The aboral end may be carinate, or may possess appendices. Various other types of surface ornament may be present. Chitinozoans may be siphonate. They may also be colonial, attached at the oral end by means of an annulated tube in turn attached to the operculum, and at the aboral end by a copula.

Classification of chitinozoans is based on details of morphology. The nature of the operculum is the most important criterion. Forms with simple opercula are grouped in the Simplexoperculati, and forms with complex opercula in the Complexoperculati.

Palaeobiology

Fossil chitinozoans are interpreted, essentially on the basis of associated fossils and sedimentary facies, as having been marine and pelagic. They are further interpreted as typically deep marine.

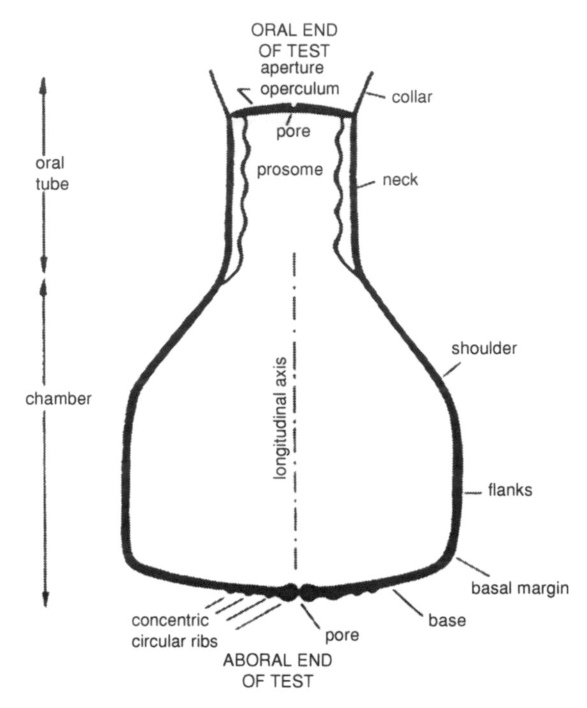

Fig. 3.75. Morphology of chitinozoans. (From Jones, 1996; in turn after Haq and Boersma, 1978.)

Biostratigraphy

Chitinozoans evolved in the Cambrian, diversified in the Ordovician (Paris *et al.*, in Webby *et al.*, 2004) and Silurian, declined in the Devonian, and became extinct in the Carboniferous.

The rapid rate of evolutionary turnover exhibited by the chitinozoans renders them of considerable use in biostratigraphy, at least in appropriate, deep marine, environments. Unfortunately, their usefulness is somewhat compromised by their restricted bathymetric distribution. Nonetheless, high-resolution biostratigraphic zonation schemes based on the chitinozoans have been established, that have widespread applicability (Vandenbroucke, 2004).

In my working experience, chitinozoans have proved of particular use in stratigraphic subdivision and correlation in the following areas.

Palaeozoic

The Early Palaeozoic of the Middle East. Here, chitinozoans form part of the basis of the biozonation of, for example, the Ordovician–Silurian of Saudi Arabia (Owens *et al.*, 1995; Jones, 1996; Al-Hajri and Owens, 2000).

The Early Palaeozoic of north Africa. Here, chitinozoans form the basis of a number of regional biozonations, for example, of the Silurian–Devonian of the Illizi and Polignac basins in the Algerian Sahara (Benoit and Taugourdeau, 1961; Jardine and Yapaudjian, 1968; Massa and Moreau-Benoit, 1976; Jones, 1996).

3.7 Vertebrate animals

Readers interested in further details of the palaeontology of vertebrates are referred to Carroll (1988), Colbert and Morales (1991), Prothero and Schoch (1994), Kocsis (2002), Pough *et al.* (2002) and Benton (2005). Those interested in tetrapods are referred to Schultze and Trueb (1991), Ahlberg and Milner (1994), Fraser and Sues (1994), Clack (2002) and Warren and Turner (2004).

Palaeobiology

Fossil vertebrates are interpreted, essentially on the bases of analogy with their living counterparts, and functional morphology, as having occupied a range of habitats (Radinsky, 1987; Thomason, 1995; Rowe, 2000).

Palaeobiogeography

Many fossil vertebrates appear to have had restricted or endemic biogeographic distributions, rendering them of some use in the characterisation of palaeobiogeographic provinces, and in turn in the constraint of plate tectonic reconstructions, in the Palaeozoic, in the Mesozoic and in the Cenozoic.

In the Palaeozoic, the occurrence of essentially identical vertebrate assemblages in the Permian of China, in the Karoo basin of South Africa and in the Russian Urals points toward the complete assembly of the super-continent of Pangaea by this time (Lucas, 2001).

In the Mesozoic, the occurrence of essentially identical tetrapods and/or tetrapod trackways in the Triassic of the Cuyana basin of Argentina in South

America, and of the Karoo basin in South Africa, points toward these areas having been contiguous at this time, within the western part of Gondwana (Marsicano and Barredo, 2004). Studies using *Q*- and *R*-mode cluster analysis of vertebrate microfossil assemblages from the Late Cretaceous of Alberta in Canada have revealed two clusters corresponding to interpreted biogeographically distinct communities in the north and south of the study area (Brinkman *et al.*, 2004). *Paratarpon* is more abundant in the north; *Adocus* in the south.

In the Cenozoic, the occurrence of vertebrates with Orinoco basin affinities in the Socorro and Urumaco formations of the Falcon basin in western Venezuela indicates that the palaeo-Orinoco discharged into the Caribbean rather than into the Atlantic in the Oligocene to Middle or Late Miocene (Diaz de Gamero, 1996).

Palaeoclimatology

Importantly, vertebrates are of considerable use in palaeoclimate as well as in palaeobiogeographic interpretation (see Sub-section 5.5.8).

In the Cenozoic, analysis of non-avian tetrapods, using a computerised, geographic information system (GIS) analogue database on worldwide Recent distributions, enables reconstruction of the palaeoclimate of the Middle Eocene of the Messel Lagerstatte in Germany (Markwick, in Crame and Owen, 2002). Results compare well with those based on other proxies of climate. The abundance of 'cold-blooded', or, more correctly, ectothermic, taxa appears to be directly proportional to, and an accurate measure of, temperature. Overall diversity, though, appears to be a function not only of temperature, but also of other variables, notably the preferred adaptive strategy for procuring energy, which differs from one group of organisms to another.

3.7.1 Fish macrofossils, conodonts, ichthyoliths and otoliths

Biology, morphology and classification

Fish are an extant, Cambrian–Recent, group of vertebrates diagnosed by the absence of limbs (Long, 1995; Maisey, 1996; Forey, 1997; Gayet *et al.*, 2003). They are thought to have evolved from a primitive lancelet-like chordate such as *Pikaia*.

Biology

The biology and ecology of living fish is well known, and will be more or less familiar. Like all vertebrates, fish are characterised by comparatively advanced respiratory, digestive, circulatory, excretory and sexual reproductive systems, and by highly advanced central and peripheral nervous systems, including a unique sensory apparatus situated in the lateral line on the side.

In terms of ecology, fish live in every conceivable aquatic, freshwater, brackish and marine habitat, extracting oxygen from water by means of gills. Significantly, some, like lungfish, could even be characterised as semi-terrestrial in habit, and extract oxygen by means of lungs. Others, like the 'flying fish' *Exocetus*, are even capable of brief periods of flight, which is a sight to behold. However, all are ultimately dependent on water in one form or another for their survival and reproduction.

Some are nektonic, and are active swimmers, living in the water column, and characterised by a streamlined shape, powerful musculature and stabilising dorsal, lateral and caudal fins on the back, side and tail, respectively; some nektobenthic, living on or just above the bottom; some benthic, living on the bottom, often in a shallow burrow, and characterised by flattening and by migration of the eyes to the upper or dorsal side. Most of the active swimmers are active predators, and are thus typically further characterised by sharp teeth.

Morphology

In terms of hard-part morphology, fish – and indeed tetrapods – are characterised by an internal skeleton comprising a skull, backbone and ribcage, usually made of bone (Fig. 3.76). Sometimes the skeleton is made of gristle or cartilage, which has poorer preservation potential.

Classification

The classification of fish is based on a combination of morphology, and phylogeny inferred both from empirical observations on the fossil record, and from cladistic analysis of molecular data.

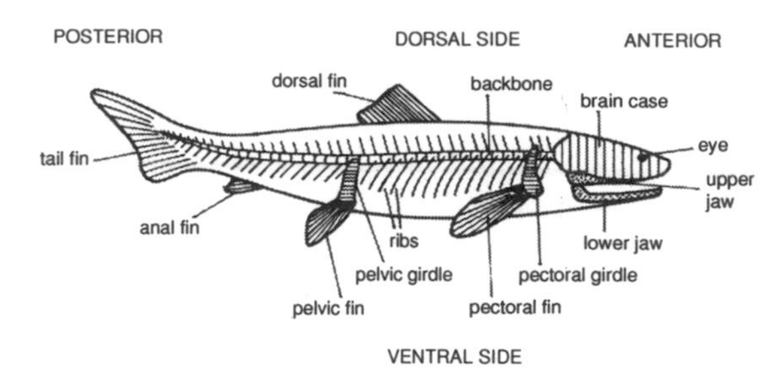

Fig. 3.76. Morphology of fish: general view of skeleton. (Modified after Black, 1989.)

Two main sub-groups of fish macrofossils have been recognised, namely the jawless fish or agnathans, and the jawed fish or gnathostomates. The jawed fish include the armoured fish or placoderms; the spiny sharks or acanthodians; the sharks and allied forms, or chondrichthyans; and the bony fish or osteichthyans, in turn including the ray-finned fish or actinopterygians, and the lobe-finned fish or sarcopterygians.

A further three sub-groups of fish-derived microfossils have also been recognised in micropalaeontological preparations, namely the conodonts, ichthyoliths and otoliths.

Jawless fish or agnathans. Jawless fish are an extant, Late Cambrian–Recent group of fish. They are characterised, as the name suggests, by the absence of a jaw. They are exemplified by the fossil ostracoderms, including the thelodonts, heterostracans, anaspids, galeaspids, pituriaspids, and armoured agnathans or osteostracans, the last-named characterised by protective bony head-shields. They are also exemplified by the living, parasitic, marine myxiniforms or hagfish, and the freshwater petromyzontiforms or lampreys, the latter characterised by a head with a large circular opening or cyclostome full of rasping teeth in place of a mouth with jaws, and also by a notochord in place of a backbone, and by a soft body. Incidentally, I once chanced across a dead lamprey while fishing on the River Ystwyth in mid-Wales, and will confess to having been somewhat startled by its alien appearance, and by what my fanciful youthful imagination perceived as green blood: I could not conceive why King Henry would have wanted to eat even a single one, never mind the 'surfeit' that supposedly killed him!

Jawed fish: armoured fish or placoderms. Armoured fish are an extinct, Silurian to Devonian group, thought to have evolved from the jawless fish, by modification of gill supports or branchial arches into jaws. They are characterised, as their name indicates, by an armoured head-shield and body plates. They are exemplified by the ptychtodontids, pseudopetalichthyids, acanthothoracids, rhenanids, petalichthyids, antiarchs, phyllolepids and arthrodires.

The antiarch *Bothriolepis* is perhaps the best known of the placoderms, being represented by over 100 species from all over the world, from both marine and freshwater environments. Significantly, *Bothriolepis* is characterised by elongated, arm-like pectoral fins that have been interpreted as having enabled it to pull itself out of its aquatic habitat and onto land. It is further characterised by lung-like organs that have been interpreted as having enabled it to breathe on land.

Spiny sharks or acanthodians. Spiny sharks are another extinct, Silurian–Permian, group, thought to have evolved independently from the jawless fish. They are characterised by a cartilaginous skeleton covered by a thin film of bone. They are exemplified by the climatiiforms, the ischnacanthiforms and the acanthodiforms, the last-named characterised by gill-rakers interpreted as adaptations to filter-feeding.

Sharks and allied forms, or chondrichthyans. The sharks and allied forms are an extant, Devonian–Recent, group, thought to have evolved from the armoured fish. They are characterised by a cartilaginous skeleton, interpreted as indicating an advanced adaptation, to aid buoyancy. They are exemplified by stem- and crown-group 'true' sharks, such as *Cladoselachus*, *Hybodus*, *Otodus* and *Carcharocles*; by rabbit-fish and allied forms, or holocephalans (Carboniferous–Recent), such as *Ischyodus* (similar to the modern *Chimaera*); and by rays and skates, or batoids (Jurassic–Recent), such as *Spathobatis*. Both marine and freshwater sharks are known, the latter including the wobbegong. Secondarily freshwater rays are also known, for example the stingray *Heliobatis* from the celebrated lacustrine Eocene Green River shales of Wyoming.

Interestingly, as demonstrated by Steno, the so-called 'tonguestones' (*glossopetrae*) or 'St Paul's Tongues' of folklore are in fact the fossil teeth of *Carcharocles* (see Preface). It is thought that at least the latter name derived from the abundance of *Carcharocles* on Malta, where legend had it that St Paul was once shipwrecked, and bitten by a snake, whereupon he turned the tongues of all the snakes on the island to stone, so that they would no longer be a danger. It then came to be believed that 'tonguestones' especially from Malta, could ward off the effects of snakebite or other poison, such that they came to be carried as talismans or lucky charms, and indeed traded all around Europe. In medieval banquets, mounted specimens would be displayed hung off tree-like red corals, and removed by guests to dip into their goblets of wine as an antidote to any poison that might have been put in by a rival.

Bony fish or osteichthyans: ray-finned fish or actinopterygians. The ray-finned fish are an extant, Devonian–Recent, group, thought to have evolved from the spiny sharks. They are characterised by bony skeletons, by bony rays to strengthen the fins, and by swim-bladders modified from the air sacs of their primitive ancestors. They are exemplified by evidently primitive fossil forms such as the palaeoniscoid *Palaeoniscum*, characterised by residual rigid armour, and by asymmetrical or heterocercal tails; by advanced or 'neopterygian' fossil forms such as the teleosts (Triassic–Recent), including the herring-like *Pholidophorus* and the salmon-like *Enchodus*, characterised instead by flexible scales, and by symmetrical or homocercal tails enhancing the efficiency of their swimming; and by a great diversity of living forms. Both marine and freshwater ray-finned fish are known. Indeed, some, especially some members of the salmon family, are, uniquely, able to survive both in the sea and in rivers, which they ascend to spawn. As another aside, when I was a youngster fishing in mid-Wales, I used not infrequently to see individuals of the migratory sea-trout or sewin *Salmo trutta* evidently disoriented by the density change from sea water to fresh water – my friends and I used to call them 'drunkies'.

Lobe-finned fish or sarcopterygians. The lobe-finned fish are an extant, Devonian–Recent group. They are thought to have evolved independently from the spiny sharks. They are also thought to have evolved into the tetrapods, exemplified by *Ichthyostega*.

They are characterised by lungs modified from the swim-bladders of their ancestors, and, importantly, by independently operated pectoral and pelvic fins capable of being used for locomotion as well as simply stabilisation. The pectoral and pelvic fins represent the evolutionary precursors of the arms and legs of the early tetrapods.

They are exemplified by the coelacanths or actinistids, and allied forms, including the osteolepiforms; and by the lungfish or dipnoans, which some authors treat as a separate sub-group.

The coelacanths are exemplified by the fossil *Diplocercides* from the Devonian, *Caridosuctor* ('prawn-eater') from the Carboniferous Bear Gulch limestone of Montana, and the more familiar-looking *Mawsonia* from the Cretaceous; and, famously, by the 'living fossil' marine *Latimeria* from the Comoro Islands off the east coast of Africa in the Indian Ocean and from Indonesia (Weinberg, 1999). Interestingly, they are also represented by swimming traces from the Triassic (Simon *et al.*, 2003).

The lungfish are exemplified by the fossil *Diabolepis* and *Dipterus* from the Devonian; and by the 'living fossil' freshwater *Neoceratodus* from Australia, *Lepidosiren* from South America and *Pteroperus* from Africa.

Conodonts. Conodonts are phosphatic tooth-like structures derived from an interpreted extinct jawless fish (Briggs, 1992; Donoghue *et al.*, 2000); 'an eel-like creature which appears to show several chordate synapomorphies: a head with eyes; a notochord; and myotomes' (Benton, 2005). Indeed, they somewhat resemble the so-called vomerine bones in the roofs of the mouths of some living fish. Conodonts occur generally as individual elements, occasionally as associations or apparatuses of elements, and extremely rarely in association with the remains of the parent animal – the first example of which was discovered in the Early Carboniferous 'Granton shrimp bed' of the area around Edinburgh in Scotland (Fig. 3.77) (Briggs *et al.*, 1983. Exceptionally well-preserved conodont parent animals from the Late Ordovician Soom shale of South Africa possess lobate

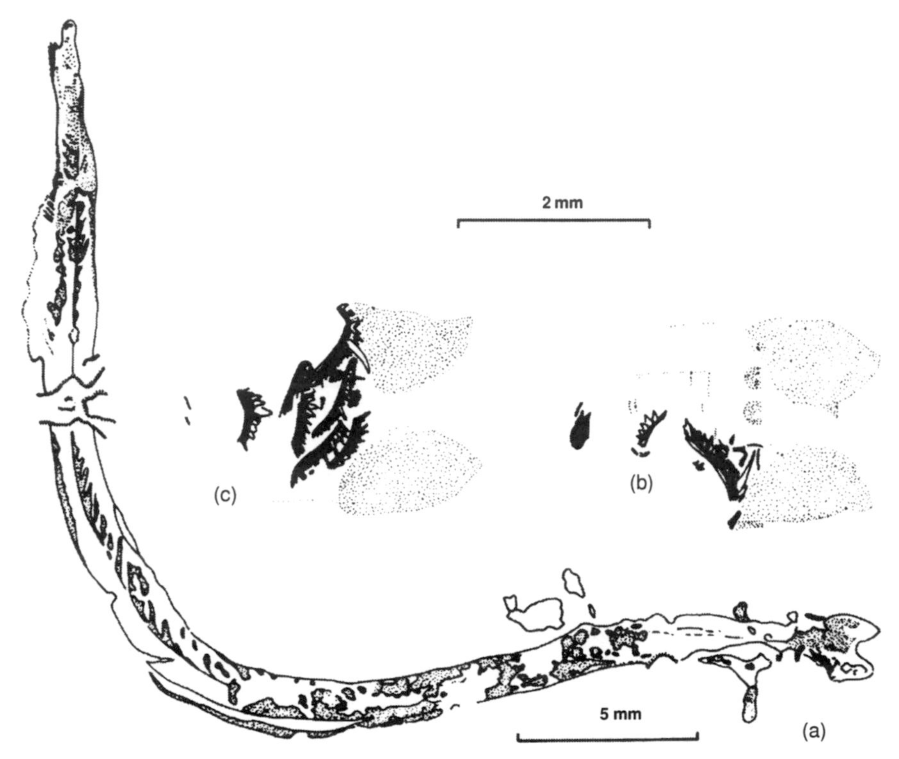

Fig. 3.77. Morphology of the conodont parent animal. (a) Body (*Clydagnathus cf. cavus-formis*); (b–c) details of head (b – part; c – counter-part.) (From Jones, 1996; reproduced with permission from Briggs *et al.*, 1983.)

structures located to the anterior of the apparatus that appear to represent sclerotic cartilages surrounding eyes, as also observed in the Silurian anapsid *Jamoytius* (Aldridge and Theron, 1993; see also Aldridge *et al.*, 1993). Conodonts are classified into form genera on the basis of details of morphology.

Ichthyoliths. Ichthyoliths are phosphatic fish teeth or dermal denticles (Fig. 3.78). They are very variable in morphology. They are classified into form genera on the basis of morphometry or biometry.

Otoliths. Otoliths are small aragonitic bones found in the inner ears of some fish (Fig. 3.79).

Palaeobiology

Fossil fish are interpreted, essentially on the basis of analogy with their living counterparts, as having been aquatic. They are further interpreted, on the basis of associated fossils and sedimentary facies, as having occupied a range of freshwater and marine environments, and on the basis of functional morphology, as having occupied a range of pelagic and benthic environments, and as having practised a variety of feeding strategies. Thusly interpreted freshwater fish include, for example, the bowfins, suckers, catfish and perch of the Oligocene lacustrine deposits of Florissant in Colorado, in the shadow of Pike's Peak, the sight that inspired the American national anthem (Meyer, 2003).

Fossil conodonts are interpreted, essentially on the basis of associated fossils and sedimentary facies, as having been deep marine. Most are also interpreted as having been nektonic, some as nektobenthic or benthic.

Fossil ichthyoliths are interpreted, essentially on the basis of analogy with their living parent animals as having been marine, some probably nektonic, others nektobenthic or benthic. Abundances, as in the Palaeocene–Eocene, are thought to reflect high palaeo-productivity.

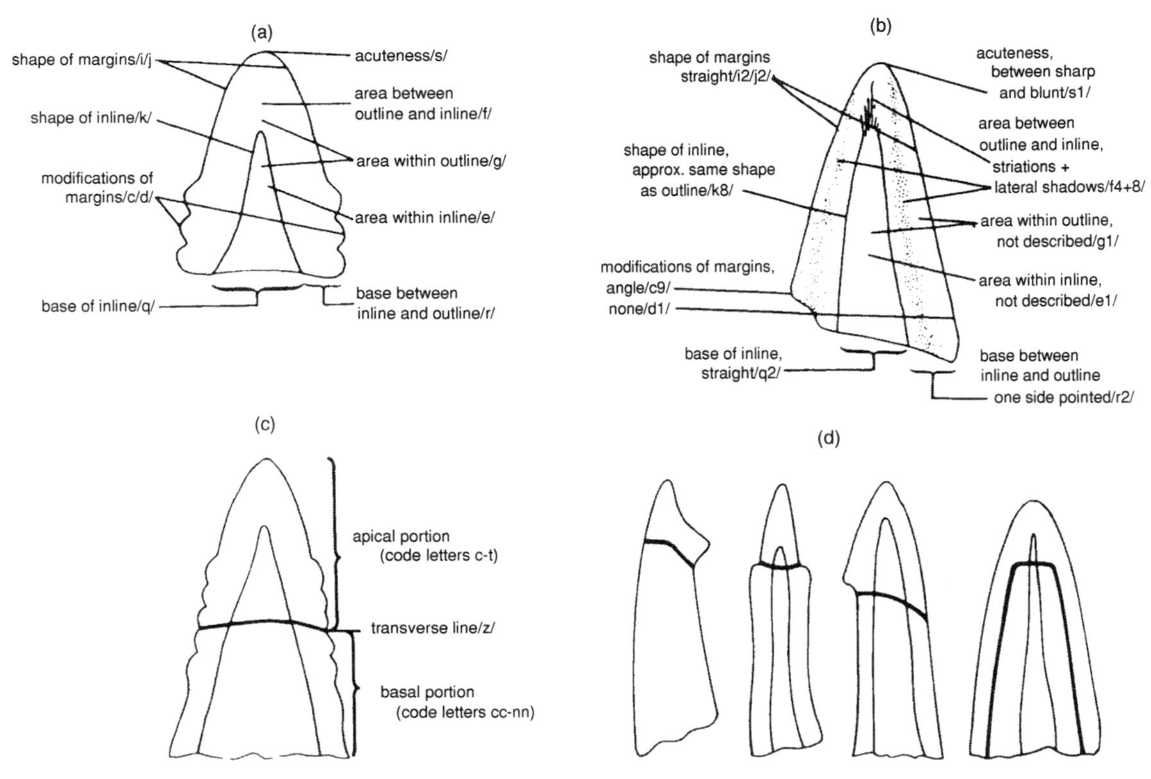

Fig. 3.78. Morphology and descriptive terminology of ichthyoliths. (a) Triangular form; (b) 'triangle with base angle', showing coding of character states; (c–d) location and variation of transverse line. (From Jones, 1996; reproduced with permission from Doyle and Riedel, in Bolli *et al.*, 1985.)

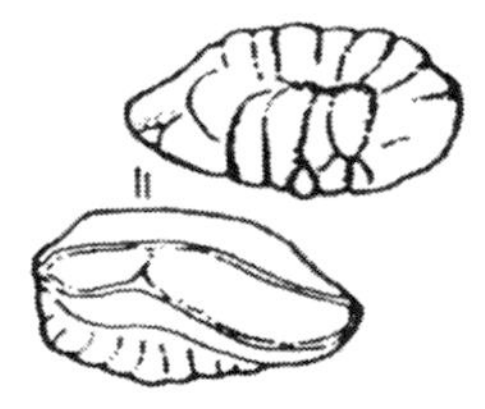

Fig. 3.79. Morphology of an otolith. (From Jones, 1996.)

Fossil otoliths are interpreted, again essentially on the basis of analogy with their living parent animals, as having been aquatic (gadid, congrid and ophidiid otoliths as marine; cichlid otoliths as freshwater).

Palaeobiogeography

Many fossil fish macrofossils and fish-derived microfossils appear to have had restricted or endemic biogeographic distributions, rendering them of some use in the characterisation of palaeobiogeographic provinces, and in turn in the constraint of plate tectonic reconstructions, for example in the Palaeozoic, in the Mesozoic and in the Cenozoic (see also Subsection 5.5.8).

In the Palaeozoic, the occurrence of endemic fish faunas, dominated by galeaspids, in the Silurian and Early Devonian of south China indicates the isolation of the South China micro-plate at that time (Lucas, 2001). In contrast, the occurrence of essentially cosmopolitan fish (and, as noted above, tetrapod) faunas in the Permian indicates the complete assembly of Pangaea by this time.

In the Mesozoic, the occurrence of the coelacanth *Mawsonia* in at least approximately coeval Early Cretaceous, Albian sediments in the Araripe basin in

Brazil in South America and in the Benue trough in Niger in Africa indicates that at this time the areas were still more-or-less contiguous, and that the rifting of the proto-South Atlantic had not proceeded particularly far (Martill, 1993).

In the Cenozoic, the occurrence of serrasalmine fish (pacus and piranhas) with Orinoco basin affinities in the Early Miocene of the Castillo formation of Falcon state in north-western Venezuela indicates that the palaeo-Orinoco discharged into the Caribbean rather than into the Atlantic at this time (Dahdul, in Sanchez-Villagra and Clack, 2004). The occurrence of sea catfish with Pacific affinities in the Early Miocene Castillo, and Middle or Late Miocene Urumaco formations of Falcon state in north-western Venezuela and in the Pliocene Cubagua of Margarita Island in Nueva Esparta State in north-eastern Venezuela indicates that the Caribbean and Pacific were connected – through the Strait of Panama – at these times (Aguilera and Rodrigues de Aguilera, in Sanchez-Villagra and Clack, 2004).

Palaeoclimatology

Conodonts and otoliths are of use in palaeoclimatic as well as in palaeobiogeographic interpretation. The use of conodonts in palaeotemperature interpretation in the Ordovician in the Iapetus Ocean is discussed by Armstrong and Owen, in Crame and Owen (2002). The use of otoliths in palaeotemperature interpretation in the Palaeogene of the Gulf Coast of the USA is discussed by Ivany *et al.*, in Prothero *et al.* (2003).

Biostratigraphy

The first fish, the jawless agnathans, evolved in the Early Cambrian. The group diversified through the Ordovician (Turner *et al.*, in Webby *et al.*, 2004), Silurian and Devonian, when the placoderms, acanthodians, chondrichthyans and osteichthyans appeared, before sustaining significant losses in the Late Devonian mass extinction, when the placoderms disappeared, and in the End-Permian mass extinction, when the acanthodians disappeared. The overall pattern throughout the Mesozoic and Cenozoic has essentially been one of ever-increasing diversification, with little or no evident loss of diversity associated with any of the main mass extinction events of the time (Forey, in Culver and Rawson, 2000). There are some 24 000 living species of teleosts alone.

Conodonts first appeared in the Late Cambrian. They diversified in the Ordovician (Albanesi and Bergstrom, in Webby *et al.*, 2004), before sustaining significant losses in the End-Ordovician mass extinction. They recovered from this event, though (Armstrong, in Hart, 1996), and diversified again in the Silurian and Devonian, when ramiform and platform types took over from conical types. They declined in the Permian, and disappeared in the Triassic.

Ichthyoliths have hitherto only been described from the Late Cretaceous to Cenozoic.

The locally rapid rate of evolutionary turnover exhibited by the fish macrofossils and fish-derived microfossils renders them of considerable use in biostratigraphy, at least in appropriate, aquatic, environments. Their usefulness is enhanced by their essentially unrestricted ecological distribution, and facies independence (see above). Locally, the usefulness of the conodonts has been further enhanced through the use of graphic correlation (Klapper *et al.*, in Mann and Lane, 1995; Kleffner, in Mann and Lane, 1995; Sweet, in Mann and Lane, 1995; Belka *et al.*, 1997). High-resolution biostratigraphic zonation schemes based on the conodonts have been established, that have essentially global applicability, across a range of palaeobiogeographic provinces (Fig. 3.80; see also Chapter 6). Local schemes are available for, for example, the Culm facies of the Devonian and Carboniferous of the British Isles, and also for the Carboniferous Limestone (Riley, 1993; Somerville, in Strogen *et al.*, 1996). Biostratigraphic zonation schemes based on ichthyoliths and otoliths have also been established, that have at least local to regional applicability.

Fish macrofossils are of use in, for example, the Silurian and Devonian of China (Lucas, 2001). Here, seven biochrons have been recognised on the basis of their successive faunas: an Early Silurian *Dayongaspis* Biochron; an early Middle Silurian *Hanyangapsis* Biochron; a late Middle Silurian *Sinogaleapsis* Biochron; an unnamed Late Silurian Biochron; an Early Devonian *Yunnanolepis* Biochron; a Middle Devonian *Bothriolepis* Biochron; and a Late Devonian

EPOCH	AGE	DCB ~m.y.	STANDARD CONODONT ZONE (PELAGIC BIOFACIES)		DEFINITIVE FIRST OCCURRENCE
EARLY CARB.			*sulcata*		*Siphonodella sulcata*
LATE DEVONIAN	FAMENNIAN	0	*praesulcata*	Late	*Protognathodus kockeli*
				Middle	
				Early	*Si. praesulcata*
		–1	*expansa*	Late	*Bispathodus ultimus*
				Middle	*Bispathodus aculeatus*
		–2		Early	*Pa. gracilis expansa*
			postera	Late	*Pa. gracilis manca*
		–3		Early	*Pa. perlobata postera*
			trachytera	Late	*Pseudo. granulosus*
		–4		Early	*Pa. rugosa trachytera*
			marginifera	Latest	*S. velifer velifer*
		–5		Late	*Pa. marginifera utahensis*
				Early	*Pa. m. marginifera*
		–6	*rhomboidea*	Late	
				Early	*Pa. rhomboidea*
			crepida	Latest	*Pa. glabra pectinata*
		–7		Late	*Pa. glabra prima*
				Middle	*Palmatolepis termini*
		–8		Early	*Palmatolepis crepida*
			triangularis	Late	*Pa. minuta minuta*
		–9		Middle	*Pa. delicatula platys*
				Early	*Pa. triangularis*
	FRASNIAN	–10	*linguiformis*		*Pa. linguiformis*
			rhenana	Late	*Pa. rhenana rhenana*
		–11		Early	*Pa. rhenana nasuta*
			jamieae		*Palmatolepis jamieae*
		–12	*hassi*	Late	*Ag. triangularis*
				Early	*Palmatolepis hassi*
		–13	*punctata*		*Palmatolepis punctata*
			transitans		*Palmatolepis transitans*
		–14	*falsiovalis*	Late	*Mesotaxis asymmetrica*
				Early	*Mesotaxis falsiovalis*
MIDDLE DEV.		–15	*disparilis*		*Klapperina disparilis*

Fig. 3.80. Stratigraphic zonation of the Devonian by means of conodonts. (From Ziegler and Sandberg, 1990.)

Remigolepis Biochron. Fish macrofossils are also of use in the non-marine latest Silurian–Devonian, Downtonian–Breconian of the Old Red Sandstone of Wales and the Welsh Borderlands (White, 1950; Ball *et al.*, 1961).

In my working experience, fish macrofossils and fish-derived microfossils have proved of some use in stratigraphic subdivision and correlation, and/or in palaeoenvironmental interpretation, in the following areas.

Palaeozoic

The Palaeozoic of the Middle East. Here, the regional Early Ordovician, Tremadoc O20 MFS of Sharland *et al.* (2001), stratotypified in the Mabrouk member of the Andam formation of central Oman, is marked by conodonts. The Early Devonian, Pragian D10 MFS, stratotypified in the Qasr limestone member of the Jauf formation of Saudi Arabia, is also marked by conodonts, and by fish. The Middle Devonian, Emsian D20 MFS, stratotypified in the Hammamiyat limestone member of the Jauf formation, is marked by the conodonts *Icriodus cf. huddlei celtibericus* and *I. cf. huddlei curvicauda*, and by fish. The Late Devonian, Famennian, Strunian D30 MFS, stratotypified in the Koprulu formation of Zap in Hakkari province in south-east Turkey, is marked by the fish *Chirodipterus* and *Danobius*. The Permian Khuff formation of south-eastern Oman is marked by a Guadalupian conodont assemblage comprising *Hindeodus excavatus, Merrilina praedivergens, Merrilina* sp. and *Sweetina* sp. (Angiolini *et al.*, 1998).

The Devonian–Carboniferous of north Africa. Here, conodonts mark the Emsian–Givetian of the Ougarta Hills in Algeria (Le Fevre, 1971), the Eifelian of the eastern Anti-Atlas in Morocco (Belka *et al.*, 1997), and the Moscovian–Gzelian of Egypt (Izart *et al.*, 1998). The measured ranges of over 50 taxa in the Jebel Ou Driss and six other sections in the Eifelian of the eastern Anti-Atlas in Morocco have been assembled into a chronostratigraphic framework by the graphic correlation method (Belka *et al.*, 1997). The local composite standard stratigraphic scheme thus developed allows subdivision of the Eifelian into over 60 composite standard units, a degree of resolution significantly greater than that attainable using conventional biostratigraphy. The framework appears to exhibit a linear fit to time, enabling calibration of conventional zones. Marker species are demonstrated to have exhibited essentially isochronous appearances, irrespective of shelf palaeobathymetry and facies.

The Devonian–Early Permian of the former Soviet Union. Here, for example, in the south Urals, the Famennian, middle *Siphonodella praesulcata* Conodont Zone, is marked by the disappearance of typically Late Devonian chondrichthyans, and the Tournaisian, *Polygnathus communis carina* Conodont Zone by the appearance of typically Early Carboniferous ones (Ivanov, in Strogen *et al.*, 1996). In the East European platform, fish assemblages have some biostratigraphy in the Tournaisian–Visean insofar as they can be used to mark specific 'horizons', and calibrated against the conodont zonation (Lebedev, in Strogen *et al.*, 1996). Incidentally, they also have palaeobiological significance here, insofar as they can be used to characterise sedimentary facies. Thus, freshwater and brackish water facies can be characterised by stenobiontic forms such as *Diplodoselache, Ageleodus, Pycnoctenion* and dipnoans, and by eurybiontic forms such as osteolepids and actinopterygians. Transitional, mixohaline (estuarine and lagoonal) facies can be characterised by stenobiontic forms such as *Deltodus, Streblodus, Copodus* and petalodontids, and by eurybiontic forms. Nearshore marine (intertidal) facies can be characterised by eurybiontic forms. Offshore marine facies can be characterised by stenobiontic forms such as *Lissodus* and actinistians, and by eurybiontic forms. (Stenobiontic forms are those that are intolerant, and eurybiontic forms tolerant, of changes in salinity, temperature, hydrodynamics and other environmental parameters).

Mesozoic

The Triassic–Cretaceous of the former Soviet Union. Conodonts are characteristic of certain levels in the Triassic of the Mangyshlak peninsula (Gaetani *et al.*, in Crasquin-Soleau and Barrier, 1998b). Fish are characteristic of certain levels in the Triassic of Ciscaucasia and the Urals (Minikh and Minikh, 1997). Selachians form markers for the Late Cretaceous of the Urals and west Siberia (Amon and de Wever, in Roure, 1994; Amon *et al.*, 1997).

The Cretaceous of Brazil. Ichthyoliths and otoliths are characteristic of parts of the Cretaceous of the Santos basin (Miller *et al.*, 2002).

Cenozoic

The Cenozoic of the North Sea basin. Here, otoliths form the basis of a regional biostratigraphic zonation (Gaemers, 1978). They are also of use in palaeoenvironmental interpretation, in the Oligocene 'molasse' of the Penzberg syncline in south Germany (Reichenbacher *et al.*, 2004).

The Cenozoic of Brazil. Here, ichthyoliths and otoliths are characteristic of parts of the Cenozoic of the Santos basin (Miller *et al.*, 2002).

The Oligocene–Miocene of central and eastern Paratethys. Here, in the west Carpathians in central Paratethys, all of the regional central Paratethyan stages are marked by particular otolith assemblages or zones (Brzobohaty, in Thon, 1983). The Kiscellian and Egerian are marked by marine assemblages, the Eggenburgian and Ottnangian by brackish assemblages, the Karpatian and Badenian by marine assemblages, and the Sarmatian by brackish assemblages. In Georgia in eastern Paratethys, certain horizons within the Oligocene–Early Miocene Maykopian regional stage are characterised by fish (Laliyev, 1964).

3.7.2 *Amphibians*

Biology, morphology and classification

Amphibians are an extant, Carboniferous–Recent, group of tetrapod vertebrates diagnosed by the laying of eggs without shells. They are thought to have evolved from lobe-finned fish, and to have evolved into reptiles.

Biology

The biology and ecology of living amphibians is well known, and more or less familiar.

Like fish, amphibians are characterised by advanced respiratory, digestive, circulatory, excretory, sexual reproductive, and central and peripheral nervous systems. Also like fish, they are 'cold-blooded' or ectothermic.

Unlike fish, amphibians are characterised by, among other features, lungs, and a urinary bladder, representing adaptations to life lived essentially on land rather than in water.

Note, nonetheless, that amphibians are ultimately dependent on water for their survival. Most living amphibians still lay eggs in water, that metamorphose by way of a tadpole developmental stage into land-living adult individuals, and their eggs are protected only by a delicate membrane, and cannot tolerate desiccation. Moreover, all living amphibians have smooth skins that are highly permeable to water, necessitating the secretion of mucus to prevent excessive and potentially lethal evaporative loss (secretion of toxins from separate, granular, glands, is another characteristic feature of amphibians). Incidentally, amphibian skins are also highly permeable to gases, and fulfil an important role in so-called cutaneous respiration.

In terms of ecology, amphibians live in a range of aquatic and – moist – terrestrial environments. As they are 'cold-blooded', most live in tropical and temperate regions. However, some live in high latitudes or at high altitudes.

Morphology

In terms of hard-part morphology, amphibians are characterised by an internal skeleton comprising a skull, backbone, ribcage and limbs, made of bone (Fig. 3.81). They are further characterised by interlocking of the vertebrae of the backbone, by attachment of a pelvic girdle to the vertebral column, and by bony limbs with ankle joints (representing adaptations to life on land, and without water to support the skeleton).

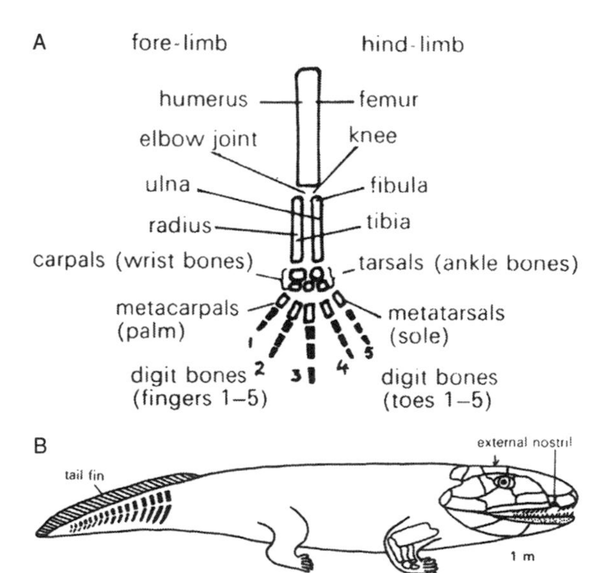

Fig. 3.81. Morphology of tetrapods and amphibians. (A) Morphology of tetrapod limb bones. (B) Morphology of an ichthyostegid amphibian. (Reproduced with permission from Black, 1989 *Elements of Palaeontology*, Cambridge University Press.)

Classification

The classification of amphibians is based on a combination of morphology and phylogeny inferred both from empirical observations on the fossil record, and on cladistic analysis of molecular data.

Two main groups have been recognised: the batrachomorphs and the reptiliomorphs. Four further sub-groups of batrachomorphs have been recognised: the nectrideans; the microsaurs; the temnospondyls; and the frogs, toads, salamanders, newts and allied forms, collectively known as the lissamphibians. Two further sub-groups of reptiliomorphs have also been recognised: the anthracosaurs and the seymouriamorphs.

Batrachomorphs: nectrideans. The nectrideans, which range from Carboniferous to Permian, are characterised by newt-like bodies. They are interpreted as having been aquatic, and as having been piscivorous, that is, feeding on (small) fish.

Microsaurs. The microsaurs (Carboniferous–Early Permian) are characterised by lizard-like bodies and strong skulls with short teeth. They are interpreted as having been both terrestrial and aquatic, and as having been insectivorous or carnivorous, feeding on insects, spiders and millipedes.

Temnospondyls. The temnospondyls (Carboniferous–Early Cretaceous) are characterised by crocodile-like bodies and broad, low, round-snouted skulls. Most are interpreted as having lived in or near fresh water, feeding on fish.

Temnospondyls are further characterised by a similar tooth structure to the extant frogs, toads, salamanders, newts and allied forms, and are interpreted as ancestral to them (an interpretation supported by the observation that some evidently had tadpole developmental stages).

Frogs, toads, salamanders, newts and allied forms (lissamphibians). The frogs, toads, salamanders, newts and allied forms (lissamphibians or 'smooth amphibians') range from Triassic to ?Recent (Holman, 2003).

Living frogs and toads (anurans or 'tail-less' amphibians) are characterised by broad skulls and by jaws lined with small teeth, and typically feed by protrusion of the tongue to capture insect prey. They are also characterised by long hindlimbs or legs adapted to hopping, jumping or climbing trees on land as well as swimming in water, and by hips strengthened to withstand the impact of landing. Fossil frogs almost identical in form to living ones are known from the Late Cretaceous of South Africa (MacRae, 1999).

Living salamanders and newts (caudatans or urodelans, or 'tailed' amphibians) are characterised by long bodies, adapted to swimming, and are active predators in aquatic as well as terrestrial environments. Incidentally, in some salamanders, respiration is entirely through the skin, or cutaneous, as the lungs have been secondarily lost, and the associated hyobranchial apparatuses modified to facilitate protrusion of the tongue.

Allied forms include the caecilians or gymnophionans, which are characterised by reduced limbs adapted to life in soil and leaf litter in terrestrial environments.

Reptiliomorphs: anthracosaurs. The anthracosaurs (Carboniferous–Permian) are characterised by narrow skulls. They are interpreted as having been predatory, both in water and on land.

Seymouriamorphs. The seymouriamorphs (Late Carboniferous–Permian) are characterised by small, high skulls, and long limbs. They are also interpreted as having been predatory, although primarily on land, and probably on microsaurs and other small tetrapods. They are exemplified by *Seymouria*.

Palaeobiology

Fossil amphibians are interpreted, essentially on the basis of analogy with their living counterparts, as having occupied a range of aquatic and – moist – terrestrial environments in low to moderate latitudes, and as having practised a variety of feeding strategies.

Some, lysophorids, from the Permian of Kansas, are interpreted, on the basis of empirical observation of burrows associated with desiccation surfaces, as having holed up during seasonal or more prolonged droughts in terrestrial environments (Hembree *et al.*,

2004). Similar behaviours are exhibited by living amphibians in the south-eastern USA (and also, incidentally, by lungfish).

Palaeobiogeography

Many fossil amphibians appear to have had restricted or endemic biogeographic distributions, rendering them of some use in the characterisation of palaeobiogeographic provinces, and in turn in the constraint of plate tectonic reconstructions, in the Palaeozoic, in the Mesozoic and in the Cenozoic.

In the Palaeozoic, Carboniferous and Permian, amphibians were apparently restricted to essentially equatorial palaeo-latitudes in Euramerica or Laurasia (Panchen, in Hallam, 1973).

Palaeoclimatology

Importantly, amphibians are of considerable use in palaeoclimatic as well as in palaeobiogeographic interpretation. Palaeoclimatic interpretation of the late Cenozoic of the East European platform based in part on amphibians is given by Ratnikov (1996); palaeoclimatic interpretation of the Miocene of central Europe, also based in part on amphibians, by Bohme (2003) (see also Sub-section 5.5.8).

Biostratigraphy

Amphibians evolved in the Carboniferous, and diversified through the Carboniferous and Permian. They then sustained severe losses in the End-Permian mass extinction, when the nectrideans, anthracosaurs and seymouriamorphs all disappeared (the microsaurs became extinct slightly earlier). They staged some form of recovery from this event, though, and underwent renewed radiation in the Triassic, when the frogs, toads, salamanders, newts and allied forms appeared. They were essentially unaffected by the End-Jurassic and End-Cretaceous mass extinctions (Fara, 2004). There are some 5000 living species of amphibians. Many are threatened or endangered species.

The slow rate of evolutionary turnover exhibited by the amphibians renders them of limited use in biostratigraphy. In my working experience, they have not proved of particular use.

3.7.3 Reptiles and birds

Biology, morphology and classification

Reptiles and birds constitute an extant, Carboniferous–Recent, group of tetrapod vertebrates diagnosed by the laying of eggs with shells (Benton, 1991; Benton and Spencer, 1995). Reptiles are thought to have evolved from amphibians; birds, from reptiles (Dingus and Rowe, 1997).

Biology

The biology and ecology of living reptiles and birds is well known, and familiar.

They are characterised by advanced respiratory, digestive, circulatory, excretory, sexual reproductive, and central and peripheral nervous systems.

Unlike amphibians, reptiles and birds are not dependent on water for their survival. They lay eggs on land rather than in water, and their eggs are protected by a semipermeable membrane and shell that prevents desiccation. Moreover, the scaly skins of the reptiles and the feathered skins of the birds prevent excessive evaporative loss of water. Reptile and bird skins are not involved in respiration, which, rather, is exclusively by means of efficient lungs. Circulation is by means of an efficient heart, enabling the generation of high blood pressure and in turn high levels of activity (the ability among living birds and mammals to breathe while engaged in such activity being achieved by the action of so-called ventilatory muscles). Importantly, living reptiles are 'cold-blooded' or ectothermic; birds 'warm-blooded' or endothermic.

In terms of ecology, reptiles live in a range of terrestrial and aquatic, freshwater, brackish and marine environments. As they are 'cold-blooded', most live in tropical and temperate regions. However, some live in high latitudes or at high altitudes, and, famously, resist freezing, and the damage it would cause to the structure of their cells, by synthesising natural 'antifreeze' substances.

Birds live in a range of terrestrial environments. As they are 'warm-blooded', they live all over the world, although the distributions of some are evidently subject to temperature control, for example, those migratory species that fly south for the winter in the northern hemisphere.

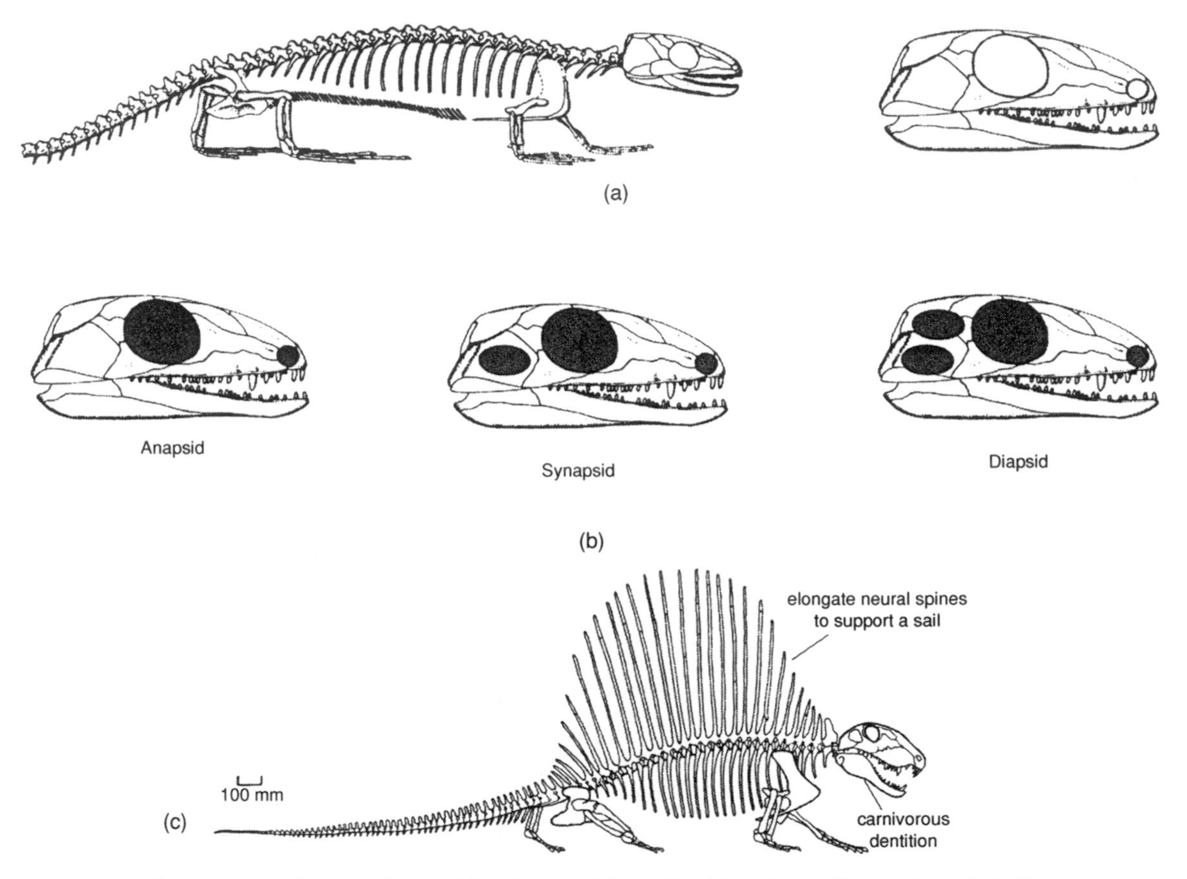

Fig. 3.82. Morphology of anaspid and synapsid reptiles. (a) Basic reptilian body and skull plan, exemplified by *Hylonomus* (anaspid), Carboniferous; (b) anapsid, synapsid and diapsid reptilian skulls; (c) *Dimetrodon* (synapsid). (Modified after Benton and Harper, 1997.)

Morphology

In terms of hard-part morphology, reptiles and birds are characterised by an internal skeleton comprising a skull, backbone, ribcage and limbs, made of bone (Figs. 3.82–3.84).

Birds are further characterised by forelimbs modified into wings and adapted for flight; by 'wishbones', representing the sites of attachment of the muscles used in flight; and by lightweight but strong, hollow but internally supported, bones generally.

Classification

The classification of reptiles and birds is based on a combination of morphology and phylogeny inferred both from empirical observations on the fossil record, and on cladistic analysis of molecular data.

Three main groups have been recognised, namely: the anapsids, the synapsids and the diapsids.

Anapsids. The anapsids are an extinct, Carboniferous–Triassic group of reptiles characterised by the absence of a single opening in the sides of their skulls (other than the eyes and nose). They include the tortoises and turtles.

The ancestral anapsid, *Hylonomus*, from the Carboniferous, is superficially similar to the amphibians of the time, especially the microsaurs, and was probably similarly insectivorous or carnivorous. However, it also exhibits certain important differences, such

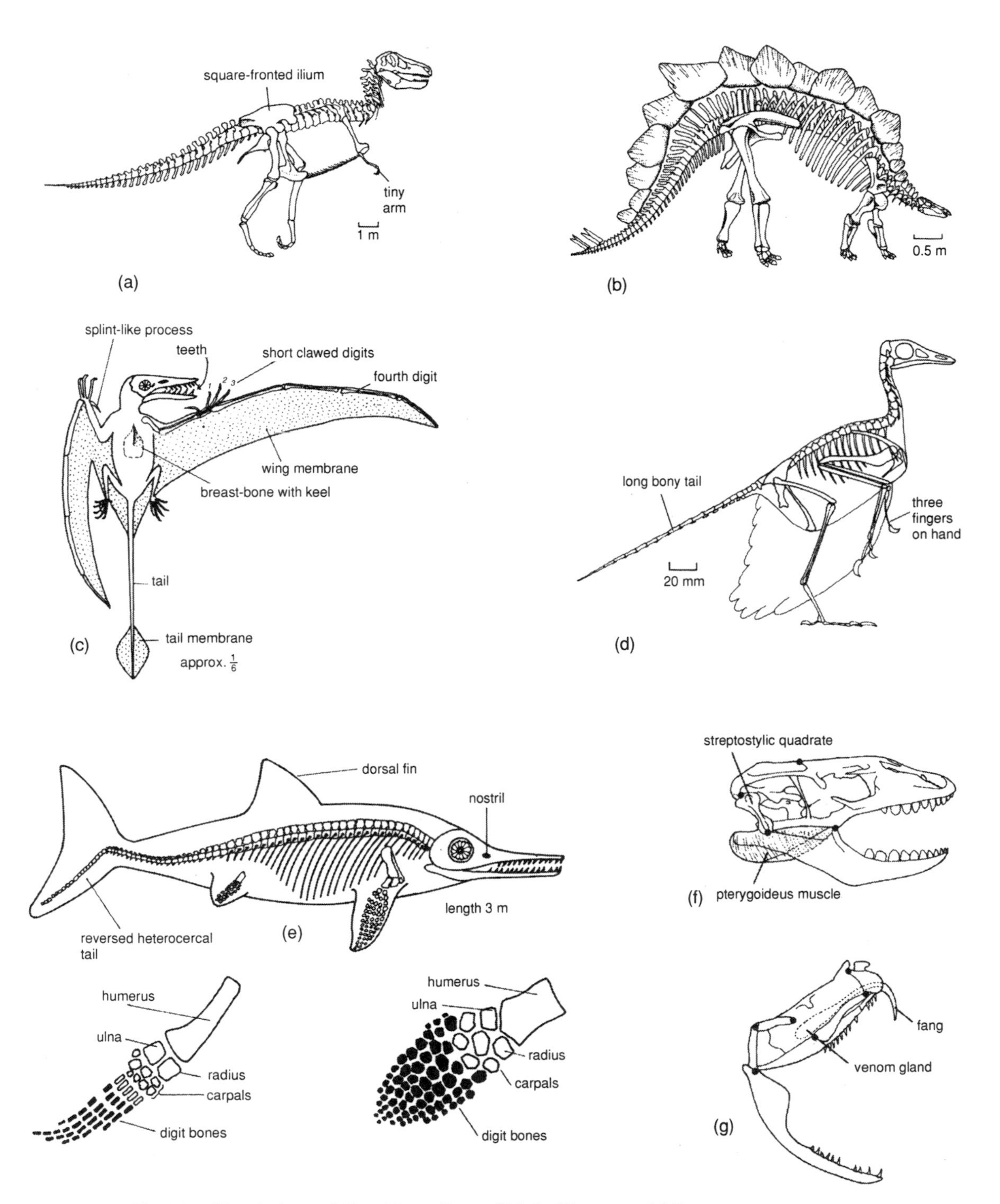

Fig. 3.83. Morphology of diapsid reptiles and birds. Dinosaurs: (a) *Tyrannosaurus* (theropod), Cretaceous; (b) *Stegosaurus* (ornithiscian), Cretaceous. Flying reptiles and birds: (c) *Rhamphorhyncus* (flying reptile); (d) *Archaeopteryx* (bird), Late Jurassic. Marine reptiles, and lizards and snakes: (e) ichthyosaur, and details of plesiosaur and ichthyosaur forelimbs (marine reptiles); (f) modern lizard skull; (g) modern snake skull. (Compiled from various sources.)

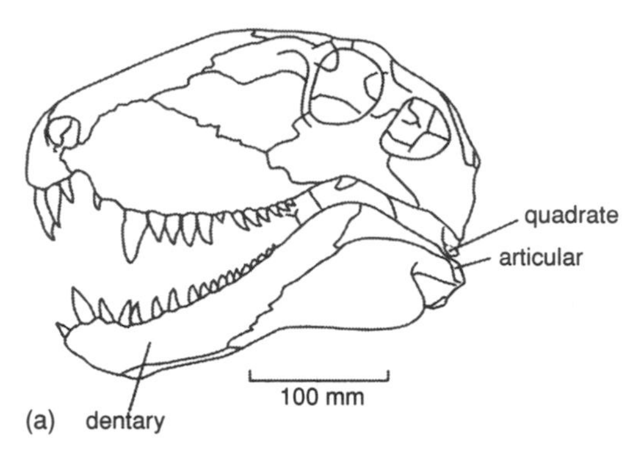

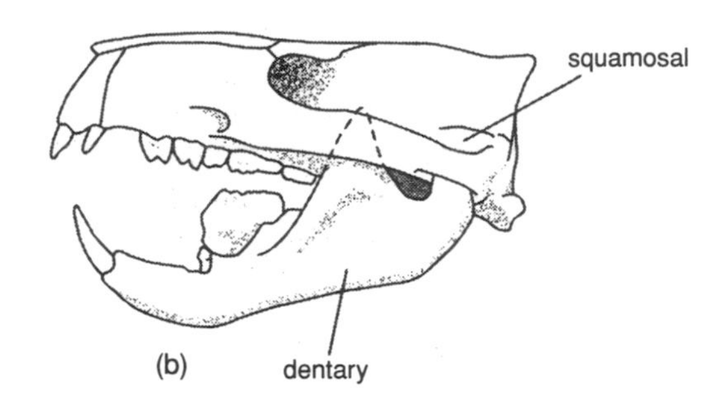

Fig. 3.84. Comparative morphology of mammal-like reptile and mammal skulls. (a) Mammal-like reptile; (b) mammal. Note the differences in lower jaw dentition and articulation. (Modified after Benton and Harper, 1997.)

as a high skull, evidence for additional jaw musculature, and an astragalus bone in the ankle, although there is as yet no evidence in *Hylonomus* of the most important feature that differentiates the reptiles from the amphibians, that is, the laying of eggs.

More advanced anapsids, from the Permian and Triassic, are characterised by broader teeth, interpreted as adaptations to a diet of tough plants as well as insects. The oldest turtles, from the Triassic, are also characterised by teeth, unlike modern forms.

Synapsids. The synapsids are an extant, Late Carboniferous–Recent, group characterised by the presence of a single opening in their skulls (excluding the eyes and nose). They in turn include the pelycosaurs and the 'mammal-like' therapsids.

Pelycosaurs (Late Carboniferous–Early Permian) are typically characterised by small to medium size, strong skulls and carnivorous dentition. They are exemplified by the Early Permian *Dimetrodon*, characterised by modified, carnivorous dentition, and by elongated spines arising from the vertebrae, which have been interpreted as having supported a sail-like structure used in the regulation of body temperature.

Therapsids (Late Permian–Recent) are characterised by differentiated dentition. They are exemplified by the Late Permian carnivorous gorgonopsians and herbivorous dicynodonts, most of which disappeared during the End-Permian mass extinction (King, 1990). They are also exemplified by the Triassic and younger cynodonts, characterised by small canals in the skull marking the former positions of nerve endings attached to sensory whiskers, and, by inference, by body hair used in the regulation of body temperature. Several lines of cynodonts developed during the Triassic, most of which disappeared during the End-Triassic mass extinction. During the Late Triassic, one of these lines gave rise to the earliest true mammals, which survived the extinction event. The transition was effected/accompanied by a shift in the position of the former articular–quadrate jaw joint into the middle ear (where it is represented by the hammer and anvil bones, used to transmit sound from the eardrum to the brain).

Diapsids. The diapsids are an extant, Late Triassic–Recent, group characterised by the presence of two openings in the sides of their skulls, the edges of which provide attachment sites for jaw muscles. They in turn include the archosaurs, the ichthyosaurs, the plesiosaurs, and the lepidosaurs. The ancestral diapsids were small- to medium-sized carnivores. Some more advanced forms, including the ancestral archosaur *Erythrosuchus*, were large carnivores, while others developed different, and in some cases extremely specialised, life styles and feeding strategies.

Archosaurs (Late Triassic–Recent) in turn include crocodilians, pterosaurs, dinosaurs and birds.

Crocodilians (Late Triassic–Recent) are characterised by an armoured exterior of bony plates. Early forms are interpreted as having been essentially terrestrial in habit, walking on all fours. Later forms appear to have adopted a more-or-less aquatic mode of life, with their limbs modified into paddles, and with powerful tails. The recognisably morphotypically modern crocodiles, alligators and gavials first appeared in the Late Cretaceous.

Pterosaurs (Late Triassic–Cretaceous) were characterised by lightweight and streamlined bodies, and by long narrow wing-like structures supported by arm and finger bones, especially a specially modified and elongated fourth-finger bone, and hence have been interpreted as having been capable of powered flight as well as gliding (Wellnhofer, 1991; Buffetaut and Mazin, 2003). They were also characterised by hatchet-shaped skulls, and have been interpreted as having fed by fishing or by hawking for insects. Some grew to a considerable size, including *Pteranodon*, with a wingspan of 5–8 m, and *Quetzalocoatlus*, with a wingspan of 11–15 m.

Dinosaurs (Late Triassic–Cretaceous) were the most important of the diapsids, at least in terms of their abundance/dominance and diversity, and their size (Carpenter and Currie, 1990; Weishampel *et al.*, 1990; Farlow and Bret-Surman, 1997; Lucas, 1997; Gardom and Milner, 2001; Martin, 2001; S. Parker, 2003; Fastovsky and Weishampel, 2005). (This listing includes a number of popular titles: perhaps unsurprisingly, dinosaurs have captured the popular imagination more than any other fossil group.) Dinosaurs include the saurischians, or 'reptile-hipped' dinosaurs, characterised by pubis and ischium bones pointing in opposite directions; and the ornithischians, or 'bird-hipped' dinosaurs, characterised by pubis and ischium bones pointing in the same direction. Saurischian dinosaurs in turn include the sauropods, interpreted as herbivorous, and the theropods, interpreted as carnivorous (Gishlick, in Briggs and Crowther, 2001). The sauropods are exemplified by the familiar *Brachiosaurus*, that grew to a staggering 23 m. They are also represented by tracks or trace fossils in the Portezuelo formation of Cuyana basin of west-central Argentina in southern South America that take their record back into the Late Triassic (Marsicano and Barredo, 2004). The theropods are exemplified by the truly awesome *Tyrannosaurus*, at 14 m the largest land predator – or scavenger – that ever lived (Fiffer, 2000; Carpenter and Smith, in Tanke and Carpenter, 2001). Ornithischian dinosaurs are all interpreted as herbivorous: many are armoured as protection against predators. They are exemplified by *Stegosaurus*, with distinctive staggered rows of bony plates on its back, which have also been interpreted as having played a role in the regulation of body temperature, and by *Triceratops*, with a three-horned head. The cranial morphology of the ceratiopsids, including frills as well as horns, is interpreted as having fulfilled a behavioural function, possibly in courtship (Sampson, in Tanke and Carpenter, 2001). Regarding the contentious issue of whether or not dinosaurs were 'warm-blooded', the current consensus is that comparatively small, active carnivorous dinosaurs probably were; but that large, passive herbivores were not, not least because such large animals with such low surface area to volume ratios would have been well able to maintain a stable core temperature without it (Bakker, 1975; Horner, in Briggs and Crowther, 2001). Some dinosaurs were able to live in polar latitudes too cold for 'cold-blooded' reptiles such as turtles and crocodiles (Buffetaut, 2004). Regarding the issue of whether or not dinosaurs cared for their young, the consensus, based essentially on finds of communal nest sites in Patagonia and in Mongolia, is that at least some did (Carpenter *et al.*, 1996; Carpenter, 1999; Chiappe and Dingus, 2001; Grellet-Tinner and Chiappe, in Currie *et al.*, 2004; Varricchio and Jackson, in Currie *et al.*, 2004).

Birds (Late Jurassic–Recent), as noted above, are characterised by feathers, wings, 'wishbones', and hollow but internally supported bones generally (Sibley and Ahlquist, 1990; Feduccia, 1996; Chiappe and Witmar, 2002; Currie *et al.*, 2004). The ancestral bird, *Archaeopteryx*, from the Late Jurassic lithographic limestone of Solenhofen in Bavaria in southern Germany, possesses a distinctly bird-like brain and brain-case, as indicated by recent computerised tomography (CT) scans (Alonso *et al.*, 2004). It also possesses a number of other bird-like characters such as feathers and wings, and a beak, together with reptile-like characters such as clawed hands and teeth (Wellnhofer, in Currie *et al.*, 2004). Its overall skeletal structure is remarkably reminiscent of that of a theropod dinosaur, clearly implying an evolutionary relationship. More advanced birds, from the Cretaceous, also still retain evidence of their dinosaur ancestry in the form of teeth. Advanced, recognisably modern birds appeared in the Palaeogene; the perching birds or passerines, so characteristic of the modern avifauna, in the Neogene.

Ichthyosaurs (Triassic–Cretaceous) were characterised by streamlined fish- or dolphin-shaped bodies, with fin- or flipper-like limbs fore and aft, and by long, thin snouts lined with sharp teeth, and are interpreted as marine, as having swum by sinuous, side-to-side movements of their bodies, and as having been carnivorous, feeding on fish, ammonites and belemnites (Callaway and Nichols, 1997). Their superficial similarity to familiar fish and marine mammals is an excellent example of convergent evolution: they themselves almost certainly evolved from land-living diapsid reptiles.

Plesiosaurs (Jurassic–Cretaceous) were characterised by larger limbs, longer limbs and, other than in the case of the pliosaurs, smaller heads, than the ichthyosaurs. They are interpreted as having been marine, as having swum by up-and-down, seabird-like, movements of their limbs, and as having been carnivorous, catching prey by darting, snake-like movements of their necks (Callaway and Nichols, 1997). Cranial, mandibular and dental morphologies indicate that *Peloneustes* preyed on fish, *Liopleurodon* on large, hard-boned vertebrates, and *Simolestes* on invertebrates (Noe, 2004).

Lepidosaurs (Late Triassic–Recent) are characterised by elongated bodies and by reduced limbs. They include the lizards and snakes, which are, and by inference were, terrestrial and aquatic, freshwater and marine.

The ancestral lizards, or sphenodontids, from the Late Triassic, are represented today only by the 'living fossil' *Sphenodon* or tuatara of New Zealand (tuatara is a Maori word, meaning 'spines on the back'). Advanced, recognisably modern lizards appeared in the Middle Jurassic. Most living forms are carnivorous; some herbivorous, such as the marine iguanas of the Galapagos, that feed on seaweed.

Snakes evolved in the Cretaceous. They are carnivorous, and kill or immobilise their prey by poisoning. Some are characterised by the ability to disarticulate the jaw, enabling large prey to be swallowed whole (Holman, 2000).

Palaeobiology

Fossil reptiles and birds are interpreted, essentially on the bases of analogy with their living counterparts, and functional morphology (Lockley, in Briggs and Crowther, 2001; Taylor, in Briggs and Crowther, 2001; Unwin, in Briggs and Crowther, 2001), as having occupied a range of terrestrial and aquatic, freshwater, brackish and marine environments, and as having practised a variety of feeding strategies.

Fossil reptiles are further interpreted as having been typically essentially restricted to low and moderate latitudes.

Palaeobiogeography

Many fossil reptiles appear to have had restricted or endemic biogeographic distributions, rendering them of some use in the characterisation of palaeobiogeographic provinces, and in turn in the constraint of plate tectonic reconstructions, in the Palaeozoic, in the Mesozoic, and in the Cenozoic.

In the Palaeozoic, Carboniferous and Permian, reptiles were apparently restricted to essentially equatorial palaeo-latitudes in Euramerica or Laurasia. In the Permian, Sakmarian–Kungurian, the occurrence of essentially identical reptiles in the Whitehill formation of the Ecca group of the Karoo supergroup of South Africa and the Irati formation of eastern South America indicates that at this time the two areas were contiguous (MacRae, 1999).

In the Mesozoic, the occurrence of identical suites of sauropod dinosaurs in coeval earliest Cretaceous, Berriasian–Hauterivian sediments in eastern Asia and Laurasia indicates that the two areas were contiguous at this time, some 20 Ma earlier than had previously been suggested (Barrett *et al.*, 2002). The occurrence of the crocodile *Araripesuchus* in at least approximately coeval Early Cretaceous, Albian sediments in the Araripe basin in Brazil in South America and in the Benue trough in Niger in Africa indicates that these areas were still probably more or less contiguous at this time, later than had previously been suggested (Martill, 1993).

In the Cenozoic, in the Middle Eocene, Lutetian (MP11), the occurrence of identical crocodilians in Purga di Bolca, Geiseltal and Messel indicates a connection between Italy and mainland Europe at this time (Kotsakis *et al.*, 2004). In contrast, the occurrence of *Crocodylus bambolii* in the *Oreopithecus–Maremmia* mammal fauna of the Monte Bamboli site in Tuscany, in Sardinia, and in Africa, in the Late Miocene, Vallesian–Turolian indicates a connection between these areas at this time. Interestingly, the *Hoplitomeryx–Microtia* mammal fauna of Apulia–Abruzzi is specifically distinct in terms of mammals, suggesting that it represents an altogether separate province.

The occurrence of a gharial with Orinoco basin affinities in the Early Miocene of the Castillo formation of Falcon state in north-western Venezuela indicates that the palaeo-Orinoco discharged into the Caribbean rather than into the Atlantic at this time (Brochu and Rincon, in Sanchez-Villagra and Clack, 2004).

Palaeoclimatology

Importantly, like amphibians, reptiles are of considerable use in palaeoclimatic as well as in palaeobiogeographic interpretation. Palaeoclimatic interpretation of the late Cenozoic of the east European platform based in part on reptiles is given by Ratnikov (1996); palaeoclimatic interpretation of the Miocene of central Europe, also based in part on reptiles, by Bohme (2003) (see also Sub-section 5.5.8).

Biostratigraphy

Reptiles evolved in the Carboniferous. They diversified through the Carboniferous and Permian, but then sustained severe losses in the End-Permian mass extinction, when many synapsids disappeared (the pelycosaurs became extinct slightly earlier). They recovered from this event, though, and diversified again in the Triassic, when the diapsids, ichthyosaurs and lepidosaurs appeared, but sustained significant losses in the End-Triassic mass extinction, when the anapsids and many early archosaurs disappeared. They recovered from this event, too, and diversified yet again in the Jurassic, when the birds and the plesiosaurs appeared, but sustained severe losses in the End-Cretaceous mass extinction, when the pterosaurs, dinosaurs, ichthyosaurs and plesiosaurs became extinct. They recovered from this event, too. There are some 6000 living species of reptiles, and 9000 of birds.

The moderate rate of evolutionary turnover exhibited by the reptiles and birds renders them of some use in biostratigraphy, at least in appropriate environments.

They are of use in, for example, the Permo-Triassic Karoo supergroup of South Africa (Fig. 3.85) (MacRae, 1999). Here, successive Zones are defined on the basis of *Aulocephalodon*, *Dicynodon*, *Lystrosaurus* and *Kannemeyeria*.

They are also of biostratigraphic use or palaeobiological importance in the Mesozoic, Triassic–Cretaceous of China (Lucas, 2001) and elsewhere (Schultz, in Koutsoukos, 2005). In China, the continental Triassic fauna is marked by the classical Pangaean *Lystrosaurus–Kannemeyeria–Shansiodon* succession, although some endemic dicynodonts are also known. Jurassic and Cretaceous faunas are important as regards elucidation of evolution, especially of the prosauropod dinosaurs, stegosaurs, ceratopsians, hadrosaurs and, importantly, birds. The feathered dinosaurs or primitive birds *Sinornis, Cathayornis, Liaoningornis* and *Confuciornis* were all first described from the Chaomidianzi, Yixian and Jiufotang formations of Liaoning in north-east China, variously interpreted as Late Jurassic, that is, around the same age as the beds that yielded *Archaeopteryx*, or, more likely, Early Cretaceous, that is, slightly younger (Pang Qiqing and Whatley, in Whatley and Maybury, 1990; Russell, in Currie *et al.*, 2004).

In Britain, ichthyosaurs and plesiosaurs are locally abundant, as in the Early Jurassic, lower Lias around Lyme Regis, also in Dorset, where many exceptional specimens now housed in museums were found by the amateur collector Mary Anning. Dinosaurs are also locally abundant, as in the marginal to non-marine 'Wealden' facies of the Early Cretaceous of the Weald and Wessex basins: indeed, they are locally sufficiently abundant as to lend their name to the containing rocks, as in the case of the *Hypsiliphodon* bed of the Isle of Wight. Famously, an *Iguanodon*, from the 'Wealden' of Sussex, was the first dinosaur ever to be described, by Gideon Mantell, although the actual name 'dinosaur' was not coined until later, by Richard Owen.

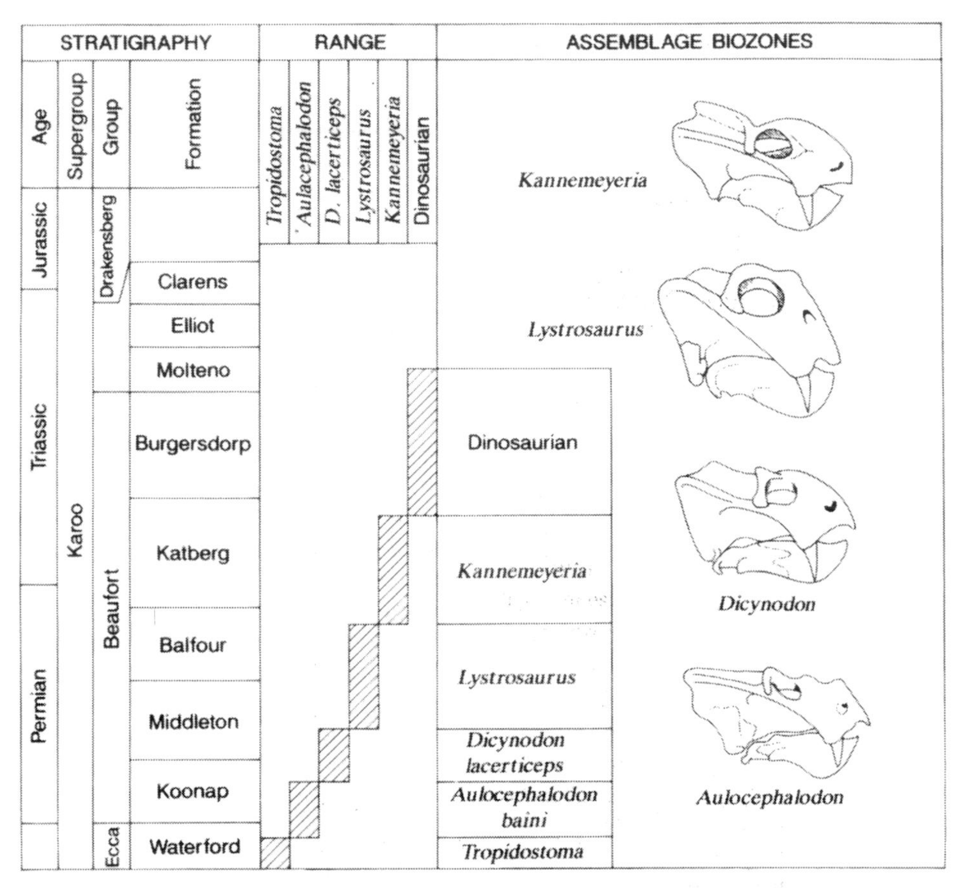

Fig. 3.85. Stratigraphic zonation of the Permian and Triassic of Gondwana by means of reptiles. (Reproduced with permission from Doyle, 1996 *Understanding Fossils*, © John Wiley & Sons Ltd.)

In my working experience, reptiles have proved of stratigraphic use only in *the Late Cretaceous of the Sokoto embayment of Nigeria*. Here, the Maastrichtian is zoned by a variety of vertebrates, including mosasaurs (Kogbe and Mehes, 1986).

Birds have proved of palaeobiological importance in *the Miocene of Paratethys and the Middle East*. Here, the occurrence of fossil ostriches in the independently dated early Late Miocene, late Sarmatian of Eldar in the Kura–Gabirry interfluve area of Azerbaijan allows the interpretation of the palaeoenvironment as terrestrial and arid (unpublished observations). The occurrence of the bones and eggs of the fossil ostrich *Struthio* in the Late Miocene Baynunah formation of Abu Dhabi allows a similar interpretation (Whybrow and Hill, 1999; unpublished observations). It is thought that the potential exists to discriminate between Late Miocene and younger sediments on the basis of the stable isotopic signal of the eggshell material. Late Miocene material is characterised by a signal consistent with a diet of both C3 (woodland) and C4 (grassland) plants. In contrast, younger material is expected to be characterised by a signal consistent with a diet dominated by C4 plants (on account of a drier climate).

3.7.4 Mammals

Biology, morphology and classification

Mammals are an extant, Late Triassic–Recent, group of tetrapod vertebrates diagnosed by the presence of mammary glands (Szalay *et al.*, 1993; Stewart and

Seymour, 1996; Kemp, 2004; Wallace, 2004). They are thought to have evolved from 'mammal-like' reptiles (Lucas and Hunt, in Fraser and Sues, 1994; Kielan-Jaworowska *et al.*, 2004). The earliest, Late Triassic and Jurassic forms, are comparatively poorly known, but are thought to have been small, nocturnal insectivores.

Biology

The biology and ecology of living mammals is well known. They are characterised by advanced respiratory, digestive, circulatory, excretory, sexual reproductive, and central and peripheral nervous systems. Respiration is by means of efficient lungs, even in aquatic forms, which therefore have to come up to the surface to breathe air. Circulation is by means of an efficient heart, enabling the generation of high blood pressure and high levels of activity (the ability to breathe while engaged in such activity being achieved by the action of ventilatory muscles). Mammals are 'warm-blooded' or endothermic.

In terms of ecology, mammals live in a wide range of terrestrial and aquatic, freshwater, brackish and marine environments, in all latitudes. Some forms, namely the bats or chiropterans, have even mastered powered flight.

A total of 23 essentially functional morphological groupings can be used to classify living and fossil mammals, based on: ecological distribution, whether aquatic, terrestrial, arboreal or subterranean; feeding strategy, whether carnivorous, omnivorous, herbivorous or frugivorous; and whether a foregut or hindgut fermenter; and size, whether small, medium or large (Rodriguez, 2004).

Morphology

In terms of hard-part morphology, mammals are characterised by an internal skeleton comprising a skull, backbone, ribcage and limbs, made of bone (Figs. 3.86–3.87).

Classification

The classification of mammals is based on a combination of morphology and phylogeny inferred in part from empirical observations on the fossil record, and in part from cladistic analysis of molecular data (Novacek *et al.*, in Benton, 1988; Murphy *et al.*, 2001). Three main groups have been recognised, namely: the monotremes, the marsupials, and the placental mammals.

Monotremes. Monotremes are a primitive, extant, Cretaceous–Recent, group of mammals characterised, like their cynodont ancestors, by the laying of eggs, and, like their mammalian relatives, by the feeding of young on milk. They are exemplified by the more-or-less familiar living duck-billed platypus and spiny anteater or echidna of Australia.

Marsupials. Marsupials are an extant, Late Cretaceous–Recent, group characterised by giving birth to live, albeit entirely dependent, young, which are reared in a specially adapted pouch on the mother's underside until weaned (Szalay, 1995). They are exemplified by the Eocene *Thylacosmilus* from South America, the marsupial equivalent of the carnivorous placental sabre-toothed cats of Europe and North America, and by the Pleistocene *Diprotodon* from Australia, the marsupial equivalent of the herbivorous placental hippopotamus of Africa. They are also exemplified by the living carnivorous Tasmanian devils, and herbivorous kangaroos, wallabies and wombats from Australia. They are thought to have dispersed from the Americas to Australia in the Oligocene.

Placental mammals. Placental mammals are an extant, Late Cretaceous–Recent, group characterised by giving birth to live young, but in a less dependent form, on account of prior development during gestation within the womb.

Edentates (Palaeocene–Recent) are characterised, as their name suggests, by an essential lack of dentition. They are exemplified by the Plio-Pleistocene giant armadillo *Glyptodon*; and by the even more spectacular giant ground-sloth *Mylodon*, which was sufficiently tall when righted to browse leaves from trees while remaining on the ground. They are also exemplified by the living armadillos, sloths and anteaters of South America, and by the pangolins or scaly anteaters of Africa and Asia. Palaeontological and molecular evidence indicates that the edentates had a entirely separate ancestry from the other mammals.

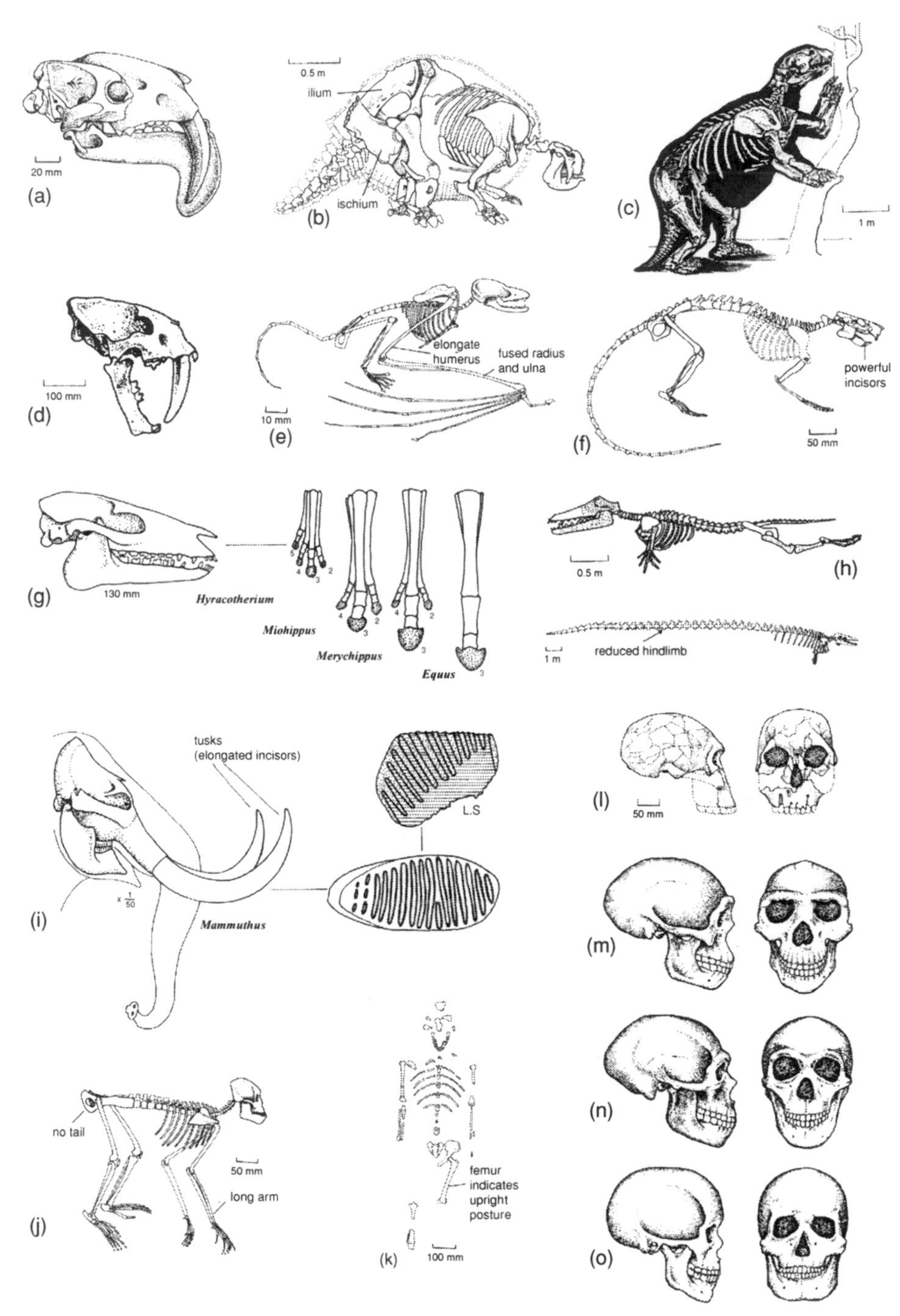

Fig. 3.86. Morphology of mammals: general morphology. (a) *Thylacosmilus* (marsupial), South America; (b) *Glyptodon* (edentate), Pleistocene, Argentina; (c) *Mylodon* (edentate), Pleistocene, Argentina; (d) *Smilodon* (carnivore, sabre-toothed cat), Pleistocene; (e) *Icaronycteris* (chiropteran, bat), Eocene; (f) *Paramys* (rodent), Eocene; (g) *Hyracotherium, Miohippus, Merychippus* and *Equus* (artiodactyls, horses), Early Eocene, Oligocene, Miocene, and Pleistocene–Recent respectively; (h) *Ambulocetus* and *Basilosaurus* (cetaceans, whales), Early Eocene and Late Eocene respectively; (i) *Mammuthus* (proboscidean, mammoth), Pleistocene (L. S. , longitudinal section). Primates: (j) *Proconsul* (ape), Miocene; (k) 'Lucy',

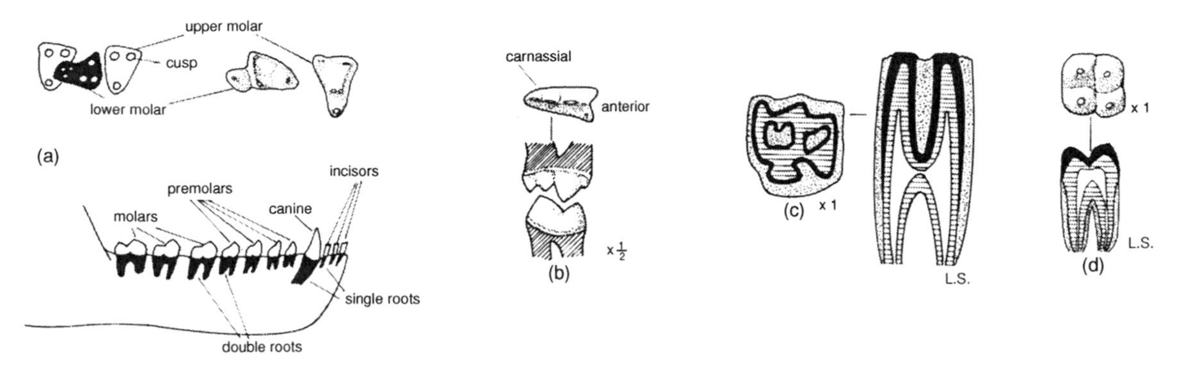

Fig. 3.87. Morphology of mammals: mammal teeth. (a) Insectivore; (b) carnivore; (c) herbivore; (d) omnivore. L.S., longitudinal section. (Modified after Black, 1989.)

Insectivores (Late Cretaceous–Recent) are characterised by long snouts, teeth with W-shaped cusps, and, typically, small size. They are exemplified by the fossil and living hedgehogs, moles and shrews.

Carnivores (Palaeocene–Recent) are characterised by sharp cheek-teeth, or carnassials, used for tearing flesh (Turner and Anton, 1997; Agusti and Anton, 2002). They are exemplified by the fossil sabre-toothed cats such as *Smilodon*, interpreted as having preyed on large, thick-skinned herbivores by tearing chunks of flesh from their bodies! They are also exemplified by the living cats, dogs, bears, hyenas, raccoons, weasels, seals, sea lions and walruses. Incidentally, the seals, sea lions and walruses are thought to have evolved from a raccoon- or weasel-like carnivore that took to an aquatic life, feeding on fish. They first appeared in the Oligocene.

Bats or *chiropterans* (Palaeocene–Recent) are characterised by greatly elongated upper and lower arm bones and fingers, forming supports for flight membranes (wings). Most living species hawk for insects at night, and roost in caves or trees during daylight. They have overcome the inherent problem of optical imaging in the dark through the development of especially light-sensitive eyes or, especially in more advanced forms, echo-location.

Rodents (Palaeocene–Recent) are characterised by deep-rooted and continuously growing incisor teeth (Luckett and Hartenberger, 1985). Their teeth are used for gnawing seeds, nuts and even wood, and are adapted accordingly (Grimes *et al.*, 2004). Rodents are exemplified by the fossil and living mice, rats, squirrels, porcupines, cavies and beavers. Fossil porcupines, cavies and beavers only appeared in the Miocene. Middle Miocene, Clarendonian beavers from Oregon in the north-western USA are associated with gleyed entisols of riparian forest aspect (Retallack, 2004).

Rabbits and *hares*, or *lagomorphs* (Palaeocene–Recent), are characterised by elongated legs, used in running and jumping. They are grazers.

Even-toed ungulates or *perissodactyls* (Eocene–Recent) are characterised by an even number of toes on each foot, i.e., either two or four (Prothero and Schoch, 1989; Prothero and Schoch, 2002; Prothero, 2005). They are exemplified by the fossil and living pigs, hippopotamuses, camels, cattle, deer, giraffe and antelope. Pigs and hippopotamuses appeared in the Oligocene. Camels appeared at about the same time, although they appear to have evolved independently, from an oreodont ancestor. Interestingly, they first appeared in North America, and only later dispersed to their present-day heartlands in the

Caption for fig. 3.86 (*cont.*) *Australopithecus afarensis* (primitive hominid), Pliocene, East Africa; (l) *Homo habilis*, (m) *H. erectus*, (n) *H. neanderthalensis* and (o) *H. sapiens* (advanced hominids), Pliocene, Early Pleistocene, Late Pleistocene, and Late Pleistocene–Holocene, respectively. (Compiled from various sources.)

deserts of north Africa, the Middle East and central Asia, and became adapted to the extreme aridity there. Middle Miocene, Clarendonian camels from Oregon are associated with alfisols of grassy woodland or wooded grassland aspect (Retallack, 2004). Cattle, deer, giraffe and antelope appeared in the Late Miocene. They are characterised by horns or antlers, perhaps the most remarkable being those of the Pleistocene giant Irish deer, *Megaloceras*. Most even-toed ungulates are ruminants, that is, they 'chew the cud', or swallow their food, partially digest it in a fore-stomach, bring it up, and swallow it again, enabling them to extract the maximum amount of nourishment from poor sources such as grasses.

Odd-toed ungulates or *artiodactyls* (Eocene–Recent) are characterised by an odd number of toes on each foot, i.e., one, three or five (MacFadden, in Hecht *et al.*, 1988; Prothero and Schoch, 1989; MacFadden, 1992; Prothero and Schoch, 2002; Prothero, 2005). They are exemplified by the fossil and living horses, rhinoceroses and tapirs. Most are ruminants. Horses, or equids, appeared in the Eocene. As early as the late nineteenth century, a model for equine evolution had been published, by Marsh, based on his observations of the succession in North America, that envisaged a progressive evolutionary trend toward increase in size, and by inference running ability, from the dog-sized horses of the Eocene, exemplified by *Hyracotherium (Eohippus)*, to those of the present day. In fact, the actual pattern is rather more complex, with several evolutionary radiations, dispersals from North America to the Old World, and lineages, now recognised, rather than simply one (with small, medium and large species coexisting, at least in the Miocene). Nonetheless, the observation of an overall increase in size through time still holds true. Accompanying trends are toward reduction in the number of toes on the fore and hind feet, from four and three respectively to one; and from short-crowned teeth adapted to browsing to long-crowned teeth adapted to grazing. Middle Miocene, Clarendonian horses from Oregon are associated with alfisols of grassy woodland or wooded grassland aspect (Retallack, 2004). Rhinoceroses also appeared in the Eocene. Like horses, they, too, exhibit an overall increase in size through time. Early forms were of modest size, and not dissimilar to the horses of the time. Some later forms grew gigantic, however, including the Miocene *Indricotherium*, which, standing at 5.5 m tall at the shoulder and weighing in at 15 tonnes, was the largest land mammal that ever lived! Tapirs probably also appeared in the Eocene, but have never been conspicuously common, and at the present time are restricted to parts of central and South America and south-east Asia.

Whales, dolphins and *allied taxa*, or *cetaceans* (Eocene–Recent), are characterised by their aquatic life style (Prothero and Schoch, 2002). Whales evolved in the Eocene, probably from terrestrial carnivores. The oldest form, *Ambulocetus*, is characterised by the large head, sharp teeth and short body of its ancestors, and indeed also by vestigial limbs. Younger toothed forms, such as *Basilosaurus (Zeuglodon)*, are characterised by proportionately smaller heads and elongated and streamlined bodies well adapted to swimming in pursuit of prey, presumed to have been fish. The baleen whales evolved after the toothed forms, and were and are able to attain considerable size on account of their ability to filter enormous amounts of food from the surface waters of the seas. Remarkably well-preserved toothed whales, with their bones still articulated or at least closely associated, have been observed in the Eocene Fayum and Mokattam groups of Wadi al-Hitan ('*Zeuglodon* Valley') and elsewhere in the Greater Fayum area of Egypt (Dolson *et al.*, 2002). Baleen whales with baleen preserved have been described from the Miocene–Pliocene Pisco formation of Peru.

Elephants or *proboscideans* (Eocene–Recent) are characterised by tusks, either in the upper jaw, as in modern species, or in the lower jaw, or both, as in the ancient gomphotheres (Guthrie, 1990; Haynes, 1991; Lister and Bahn, 2000; Agusti and Anton, 2002; Prothero and Schoch, 2002; Stone, 2002; Reumer *et al.*, 2003). Ancient, Middle Miocene, Barstovian to Late Miocene, Hemphillian gomphotheres from the great plains region of the USA have been interpreted on the basis of the carbon isotopic composition of their tusk and molar enamel as having subsisted primarily by browsing on C3 or woodland plants (Fox and Fisher, 2004). The sub-fossil, Pleistocene woolly mammoth (*Mammuthus primigenius*) from the 'mammoth steppe' of northern Eurasia and contiguous

North America has been interpreted on the basis of direct observations of preserved stomach contents as having subsisted primarily by grazing on grassland plants such as grass, sedge, club-moss and the Arctic sagebrush *Artemisia frigida*, and secondarily by browsing on woodland plants such as the dwarf birch *Betula nana*, larch, willow and alder (Lister and Bahn, 2000). Interestingly, the mammoth's trunk was characterised by two opposable projections at the tip that may have functioned like a finger and thumb and enabled it to pick individual grass stems, flowers or buds – a surprisingly delicate operation for such a large animal.

Primates (Late Cretaceous–Recent) include primitive forms characterised by mobile shoulder joints, grasping hands and feet and other features, representing adaptations to an arboreal or tree-dwelling habit (Conroy, 1990; Retallack, 1991; Jablonski, 1993; Begun *et al.*, 1997; Agusti *et al.*, 1999; Andrews and Banham, 1999; Fleagle, 1999; de Bonis *et al.*, 2001; Plavcan *et al.*, 2002; Hartwig, 2003; Anapol *et al.*, 2004). These primitive forms are exemplified by the fossil plesiadapids, and by the living lemurs, lorises and tarsiers, which are fructivorous or fruit-eating. Primates also include advanced or 'anthropoid' (literally, 'human-like') forms further characterised by good binocular vision, at least relatively large brains, extended care of the young, and socialisation, representing adaptations – or pre-adaptations – to a ground-dwelling habit. These advanced forms are exemplified by the fossil and living monkeys, and apes (and humans). Advanced primates exhibit a trend through time away from herbivory and toward omnivory, that appears to represent an evolutionary response to changing environmental and climatic conditions and to changing vegetation (see Sub-section 5.5.8).

Monkeys appeared in the Oligocene, and diverged into two groups, namely the Old World monkeys, or cercopithecid catarrhines, and the New World monkeys or platyrrhines. The former are characterised by narrow, projecting noses, and by reduced tails, or by no tails at all; the latter by flat noses and long, prehensile tails.

Apes, or hominoids, first appeared sometime around the beginning of the Miocene, in Africa, and are considered to have arisen from Old World monkeys. They are characterised by the absence of a tail. Apes include primitive forms such as the Miocene *Proconsul*, characterised by comparatively large brain-cases, and also include advanced forms, such as the fossil and living gibbons, orang-utans, gorillas, chimpanzees and humans, characterised by large brain-cases, or, to use the jargon, enhanced encephalisation. Palaeontological and molecular evidence indicates that gibbons and orang-utans are the most primitive of the living groups, and gorillas, chimpanzees and humans the most advanced. Gibbons appeared in the Early Miocene; orang-utans in the Middle Miocene; gorillas in the Late Miocene; chimpanzees in the latest Miocene; and humans in the Early Pliocene.

Humans, or hominids, are characterised by the – proportionately – largest brains and highest intelligence of all of the primates (Sperber, 1990; Diamond, 1991; Jones *et al.*, 1992; Leakey and Lewin, 1992; Lewin, 1993; Tattersall, 1993; Cavalli-Sforza *et al.*, 1994; Leakey, 1994; Thomas, 1994; Leakey and Lewin, 1995; Tattersall, 1995; Tudge, 1995; Vrba *et al.*, 1995; Johanson and Edgar, 1996; Shreeve, 1996; Stringer and McKie, 1996; Conroy, 1997; Fagan, 1998; Andrews and Banham, 1999; Klein, 1999; Lewin, 1999; Bar-Yosef and Pilbeam, 2000; Berger, 2000; Cavalli-Sforza, 2000; Haviland, 2000; Sykes, 2001; Agusti and Anton, 2002; Olson, 2002; Ulfstrand, 2002; Wells, 2002; Boyd and Silk; 2003; Coppens, 2003; Leakey and Harris, 2003; Oppenheimer, 2003; Calvin, 2004; Dunbar, 2004; Jobling *et al.*, 2004; Kingdon, 2004; Lewin and Foley, 2004; Regal, 2004; Mai *et al.*, 2005; see also Subsection 5.5.8).

The oldest widely accepted representative of the human lineage is *Ardipithecus ramidus* from the Early Pliocene of Ethiopia, radiometrically dated to 4.39 Ma. (*Kenyanthropus platyops* from the Early Pliocene of Kenya, *Orrorin tugenensis*, from the Late Miocene, 6 Ma, of Kenya, and *Sahelanthropus tchadensis* from the Late Miocene, 6–7 Ma, of Chad are also candidates; and *Pierolapithecus catalunicus*, from the Middle Miocene, 12.5–13 Ma of Spain, is close to the last common ancestor of great apes and humans.) Unfortunately, *Ardipithecus ramidus* is represented by only fragmentary skeletal remains, that provide scant clues as to its mode of life.

Australopithecus appears to have evolved also in east Africa approximately 4 Ma. *Australopithecus anamensis* is represented by fragmentary skeletal remains, including a knee joint from Kenya that provides the earliest evidence for bipedalism. The oldest skeletal remains of *Australopithecus afarensis* are from Ethiopia, and dated to 3.39 Ma. Note, though, that this species is thought to have been responsible for trackways in Tanzania, dated to 3.7–3.8 Ma, and also providing evidence for bipedalism. *Australopithecus afarensis* grew to a height of 1.1–1.2 m, and had a brain capacity of 415 cm^3. *Australopithecus* diversified into a number of robust and gracile species in east – and South – Africa by the Late Pliocene. The gracile australopithecine *Australopithecus africanus* grew to a height of 1.3 m, and had a brain capacity of 480 cm^3. In contrast, the robust australopithecine *Australopithecus (Paranthopus)* grew to a height of 1.75 m, but had a brain capacity of only 520–550 cm^3.

The evolution of our own genus, *Homo* – from a gracile australopithecine ancestor – in Africa in the Late Pliocene has been interpreted as having been in response to climatic and environmental forcing. This selection pressure would have favoured the socialisation and intelligence necessary for efficient food-gathering, and hence encephalisation and evolution. *Homo habilis*, from the latest Pliocene to earliest Pleistocene, 2.4–1.5 Ma, grew to a height of 1.3 m, and had a brain capacity of 630–800 cm^3 (600 cm^3, incidentally, representing the somewhat arbitrary criterion – the so-called 'cerebral Rubicon' – used to distinguish *Homo* from *Australopithecus*). There is evidence that it was the first creature ever to use tools. The diet of *H. habilis* appears to have differed from that of its ancestors in including nuts and seeds of arid scrubland habitats. The presently comparatively poorly known *Homo rudolfensis* appears similar to *H. habilis*, and is arguably synonymous with it. *Homo erectus* is interpreted as having evolved from *H. habilis* approximately 1.8 Ma, and as having become extinct some time after 0.4 Ma. It grew to a height of 1.6 m, and had a brain capacity of 850–1100 cm^3. There is evidence that it used weapons, for hunting, as well as tools. *Homo ergaster* is similar to *H. erectus*, and is arguably synonymous with it, although some authors have maintained it as a distinct entity (also known as 'African *H. erectus*', to distinguish it from 'Asian *H. erectus*'). *Homo heidelbergensis* ('archaic' *H. sapiens*) is interpreted as having evolved from *H. erectus* approximately 600 000 years ago, and as having become extinct approximately 150 000 years ago. It grew to a height of 1.8–1.9 m, and had a brain capacity of 1300 cm^3. There is evidence that it, too, used weapons as well as tools. *Homo neanderthalensis* is interpreted as having evolved from *H. heidelbergensis* approximately 250 000–300 000 years ago, and as having become extinct approximately 27 500 years ago. It was evidently comparatively short and stocky, but had a brain capacity averaging 1400 cm^3. It appears to have evolved an advanced societal structure.

Our own species, *Homo sapiens*, is interpreted as having evolved independently from *H. heidelbergensis* ['archaic' *H. sapiens*], in Africa, approximately 150 000 years ago (recent re-dating places the oldest known specimens somewhat closer to 200 000 years ago). We have a brain capacity averaging 1360 cm^3. We have supposedly evolved still more advanced societies and cultures.

The recently discovered dwarf species *Homo floresiensis* is interpreted as having evolved from *H. erectus/ergaster* some time before 38 000 years ago, and as having become extinct 18 000 years ago. It grew to a height of only 1 m, and had a brain capacity of only 380 cm^3 – less than the 600 cm^3 currently used to distinguish *Homo* from *Australopithecus*!

Palaeobiology

Fossil mammals are interpreted, essentially on the bases of analogy with their living counterparts, and functional morphology (Pollard, in Briggs and Crowther, 1990; Jernvall and Selanne, 1999; Fortelius and Solounias, 2000; Ungar and Williamson, 2000; Purnell, in Briggs and Crowther, 2001; van Valkenburgh, in Briggs and Crowther, 2001), as having occupied a range of terrestrial and aquatic, freshwater, brackish and marine environments, and as having practised a variety of feeding strategies. Mammals that graze on abrasive grasses, typically in arid environments, are characterised by high-crowned or hypsodont teeth; those that browse on non-abrasive vegetation, typically in humid environments, by low-crowned or brachydont teeth.

Palaeobiogeography

Many fossil mammals appear to have had restricted or endemic biogeographic distributions, rendering them of some use in the characterisation of palaeobiogeographic provinces, and in turn in the constraint of plate tectonic reconstructions, in the Mesozoic and in the Cenozoic.

In the Cenozoic, Palaeocene–Eocene mammal distributions point toward the palaeogeographic and palaeoclimatic evolution of the North Atlantic (see Box 5.2). Oligocene–Holocene mammal distributions point towards the palaeogeographic and palaeoclimatic evolution of the Old World (see Subsection 5.5.8).

The occurrence of a megatheriid or mylodontid sloth with Orinoco basin affinities in the Early Miocene of Cerro La Cruz in Lara state in northwestern Venezuela indicates that the palaeo-Orinoco discharged into the Caribbean rather than into the Atlantic at this time (Sanchez-Villagra *et al.*, in Sanchez-Villagra and Clack, 2004).

Palaeoclimatology

The exacting ecological requirements and tolerances of many mammal species, and their rapid response to changing environmental and climatic conditions, render them especially useful in palaeoenvironmental interpretation, palaeoclimatology and climatostratigraphy, and in archaeology (see also Section 7.6).

Biostratigraphy

The first mammals appeared in the Late Triassic. They then diversified during the Mesozoic, but sustained significant losses in the End-Cretaceous mass extinction. They recovered from this event, though, and diversified dramatically through the Cenozoic, filling the niches left vacant by the extinctions of other groups. The overall pattern throughout the Cenozoic has essentially been one of ever-increasing diversification, although some losses were sustained during the End-Eocene and Pleistocene mass extinctions. There are some 4000 living species of mammals.

The rapid rate of evolutionary turnover exhibited by the mammals renders them of considerable use in biostratigraphy, at least in appropriate, typically non-marine, environments.

High-resolution biostratigraphic zonation schemes based on the terrestrial mammals have been established, that have at least local to regional applicability. Calibration is typically by means of magnetostratigraphy.

For example, in Europe, the Cenozoic is divided into a total of 47 zones on the basis of mammals (Fig. 3.88) (Mein, 1975; Lindsay *et al.*, 1989; de Bruijn *et al.*, 1992; Hardenbol *et al.*, in de Graciansky *et al.*, 1998; see also Sub-section 5.5.8). It is also divided into 15 'European land mammal ages' or ELMAs (see Box 5.2). Importantly, the Quaternary is divided into zones on the basis of voles and allied forms, providing a high-resolution temporal framework for assessing climatic and environmental change (see also Section 7.6). Many Quaternary species range through to the Recent, such that their present-day distributions can be used to infer the past climate.

In North America, the Late Cretaceous is divided into five 'Pre-Aquilan Faunas' and four 'North American land mammal ages' or NALMAs, namely, the Aquilan, Judithian, 'Edmontonian' and Lancian (Cifelli *et al.*, in Woodburne, 2004). The Cenozoic is divided into a total of 19 NALMAs, namely the Palaeocene Puercan, Torrejonian, Tiffanian and Clarkforkian, the Eocene Wasatchian, Bridgerian, Uintan, Duchesnean and Chadronian, the Oligocene–earliest Miocene Orellan, Whitneyan and Arikareean, the Miocene–earliest Pliocene Hemingfordian, Barstovian, Clarendonian and Hemphillian, the Pliocene–earliest Pleistocene Blancan, and the Pleistocene Irvingtonian and Rancholabrean (Woodburne and Swisher, in Berggren *et al.*, 1995; Woodburne, 2004; see also Box 5.2).

In South America, the Cenozoic is divided into 21 'South American land mammal ages' or SALMAs (Flynn and Swisher, in Berggren *et al.*, 1995; Flynn *et al.*, 2003).

In China, the Cenozoic is divided into 17 land mammal ages (Flynn *et al.*, in Berggren *et al.*, 1995; Lucas, 2001). Mammals are of palaeobiological importance as well as biostratigraphic use here, too. Chinese faunas are especially important as regards the elucidation of the evolution of the earliest mammals of the Mesozoic, and of the eutherians or

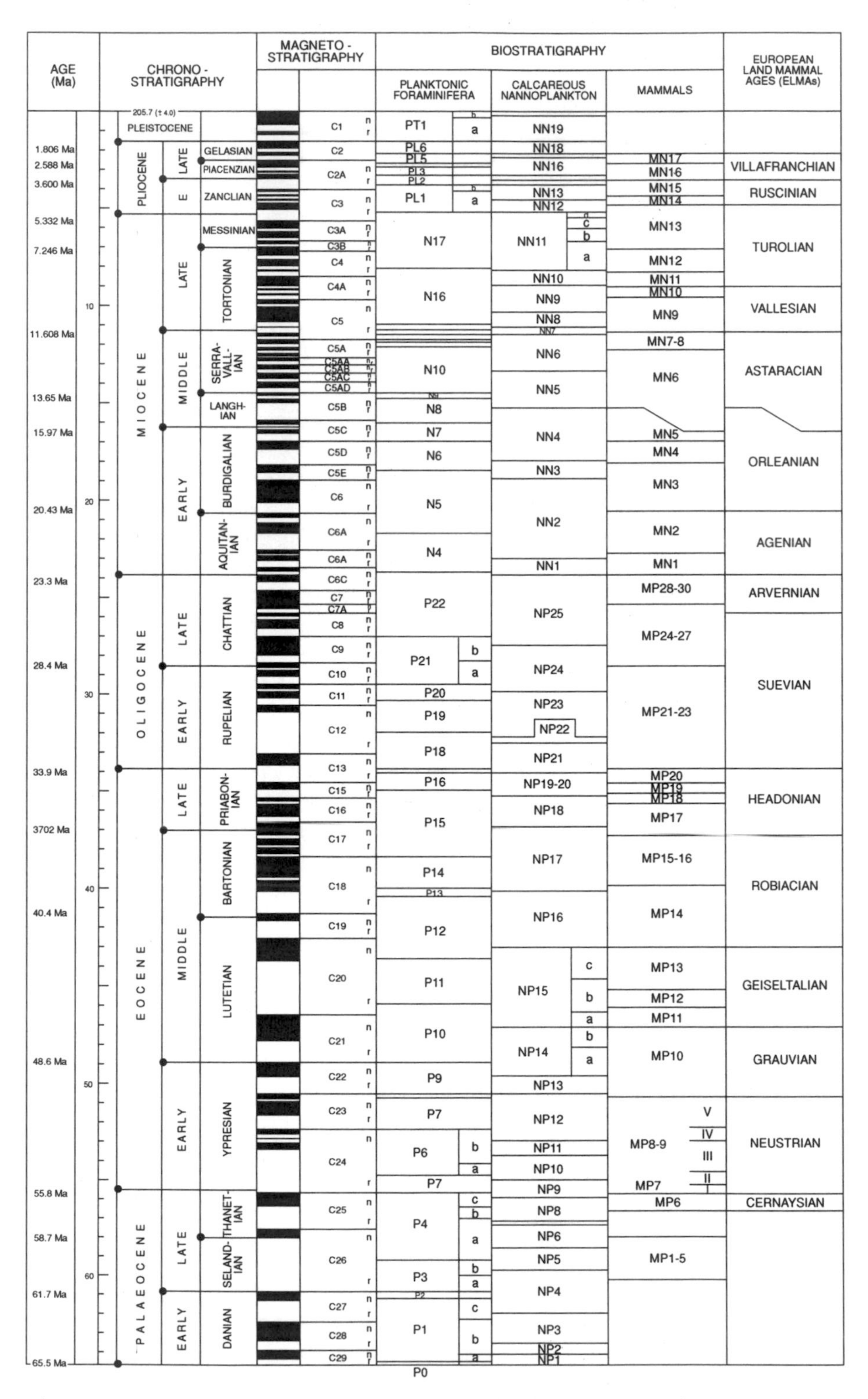

Fig. 3.88. Stratigraphic zonation of the Cenozoic of Europe by means of mammals.

placental mammals of the Cenozoic, including the carnivores, rodents and lagomorphs (Lucas, 2001).

In South Africa, the Neogene–Pleistogene is divided into six land mammal ages (MacRae, 1999). These are: the Namibian (18–14 Ma), Langebaanian (5.0–4.5 Ma), Makapanian (3.0–1.6 Ma), Cornelian (1.6–0.5 Ma), Florisian (200 000–12 000 years ago) and the Recent (12 000 years ago to present). Incidentally, mammals are of palaeobiological importance as well as biostratigraphic use here, and, to paraphrase Darwin's famous phrase, shed much light on human origins.

In my working experience, mammals have proved of some use in stratigraphic subdivision and correlation, and/or in palaeoenvironmental interpretation, in the following areas.

The Oligocene–Holocene of Paratethys. Here, for example, the occurrence of the fossil horse *Hipparion* cf. *gracile* (?*H. brachypus/primigenium*) enables an early Late Miocene, late Sarmatian or Vallesian to early Turolian, Mein Zone MN9–MN11 zone age to be assigned to sediments at the locality of Eldar in the Kura–Gabirry interfluve area of Azerbaijan (unpublished observations). The associated occurrences not only of fossil hyenas, pigs, giraffe and antelope but also of fossil seals and whales enables the environment to be interpreted as close to the palaeo-shoreline. This age assignment and environmental interpretation constrain the history of uplift in the Caucasus. Importantly, this, in turn, has implications for the provenance, and thus the reservoir quality, of the sediments in the Black Sea and Caspian.

The Miocene of the Middle East. Here, the occurrence of the fossil horse *Hipparion gracile* (?*H. brachypus/primigenium*) enables a Late Miocene, Mein Zone MN9–MN11 age to be assigned to the Lahbari member of the Agha Jari formation in the Zagros Mountains in south-west Iran (R. Bernor, personal communication; unpublished observations). This age assignment helps constrain the uplift history of the Zagros fold-belt. Importantly, this, in turn has implications for the maturation of the petroleum source-rocks in the associated foredeep. Incidentally, the occurrence of *Hipparion* also indicates a tentative correlation between the Lahbari of south-west Iran and both the Maragheh bone beds of north-east Iran, independently dated by radiometry (Bernor, Solounias *et al.*, in Bernor *et al.*, 1996), and the lower Bakhtiari of adjacent Iraq, independently dated by magnetostratigraphy (Thomas *et al.*, 1981).

The Miocene–Pleistocene of the Himalayan molasse basin. See Barry *et al.* (1982), Pilbeam *et al.*, in Bernor *et al.*, (1996), Nanda (2000), Corvinus and Rimal (2001) and Raza *et al.* (2002). The Miocene–Pleistocene Siwalik series of Pakistan is divided into 11 zones on the basis of mammals. The lower Siwalik Kamlial and Chinji formations are divided into five zones (Mein Zone MN4–MN8); the middle Siwalik Nagri and Dhok Pathan, and upper Siwalik, into six (Mein Zones MN9–MN14). The Late Miocene–Pleistocene middle–upper Siwalik of India and Nepal is divided into four zones on the basis of mammals. These are: the *Hipparion sensu lato* Interval Zone (Late Miocene, 9.5–7.4 Ma), the *Selenoportax lydekkeri* Interval Zone (Late Miocene, 7.4–5.3 Ma), the *Hexaprotodon sivalensis* Interval Zone (Pliocene, 5.3–2.9 Ma), and the *Elephas planifrons* Interval Zone (latest Pliocene–earliest Pleistocene, 2.9–1.5 Ma).

3.8 Trace fossils

Biology, morphology and classification

Trace fossils are those that provide indirect rather than direct physical indications of past life. They are most commonly manifested as trails or tracks, or as burrows or borings (Pemberton *et al.*, in Briggs and Crowther, 1990; Maples and West, 1992; Donovan, 1994; Lockley and Hunt, 1995; Bromley, 1996; Tresise and Sarjeant, 1997; Lockley and Meyer, 2000; Braddy, in Briggs and Crowther, 2001; Taylor *et al.*, 2003; McIlroy, 2004; Pemberton and MacEachern, in Koutsoukos, 2005).

The relationships of trace fossils to the trace-making organisms is only known in a minority of instances, such that form taxonomy rather than formal taxonomy is used in classification. Note, though, that proven or probable foraminiferan trace fossils have been described by Kaminski *et al.* (1988) and Plewes *et al.* (1993); '*Chondrites*-like' thyasirid bivalve traces by Pervesler and Zuschin (2002); trilobite – and possibly also non-trilobite arthropod – traces by a number of authors; insect traces by Hasiotis

(2003, 2004); callianassid crustacean traces by a number of authors, including Curran and Martin (2003); coelacanth and other fish swimming traces by Simon *et al.* (2003) and Minter and Braddy (2004); amphibian traces by Braddy *et al.* (2003); dinosaur traces by a number of authors, including Gillette and Lockley (1989), Thulborn (1990), Lockley (1991), Manning, in McIlroy (2004) and Marisco and Barredo (2004); dinosaur swimming traces by Ezquerra *et al.* (2004); and bird and mammal traces by McCrea and Sarjeant, in Tanke and Carpenter (2001). Hominid tracks have also been described (see p. 214).

Four main sub-groups of trace fossils have been recognised on the basis of the behavioural activity interpreted as having been responsible, namely: dwelling, escape, locomotion and feeding traces. In fact, in detail, eight sub-groups have been recognised, namely, dwelling traces s.s. or domichnia, farming traces or agrichnia, spreite traces or fodinichnia, escape traces s.s. or fugichnia, resting traces or cubichnia, crawling traces or repichnia, grazing traces or pascichnia and predation traces or praedichnia. Dwelling traces *sensu stricto* include *Diplocraterion, Ophiomorpha, Skolithos* and *Trypanites*. Farming traces include *Belorhaphe, Palaeodictyon* and *Spiroraphe*. Spreite traces include *Chondrites, Gyrophyllites, Phycodes* and *Rosselia*. Escape traces s.s. include nested funnels and U-in-U spreites. Resting traces include *Asteriacites, Lockeia* and *Rusophycus*. Crawling traces include *Aulichnites, Cruziana, Diplichnites* and *Scolicia*. Grazing traces include *Helminthoida, Lophoctenium, Nereites* and *Spirophycus*. Predation traces include *Centrichnus* and *Oichnus*. Interestingly, some observed feeding traces conform closely to computer simulations of optimal foraging strategies in food-poor environments (Hammer, 1998).

Palaeobiology

At least five associations of trace fossils, or ichnofacies, have been recognised on the basis of their relationships to the sedimentary environment, namely, the *Psilonichnus, Skolithos, Cruziana, Zoophycos* and *Nereites* ichnofacies: additional hardground *Trypanites*, woodground *Teredolites*, and firmground *Glossifungites* ichnofacies have also been locally recognised (Fig. 3.89) (Seilacher, in Hallam, 1967; Frey and Pemberton, 1985). A range of shallow marine sub-environments is differentiable on the basis of these ichnofacies. Some marginal marine sub-environments such as delta plains and estuaries are also locally differentiable (Bann and Fielding, in McIlroy, 2004; Bann *et al.*, in McIlroy, 2004; Mangano and Buatois, in McIlroy, 2004; McIlroy, in McIlroy, 2004). Non-marine, freshwater sub-environments are comparatively less well known in terms of their ichnofacies, although they are becoming better known (Buatois and Mangano, in McIlroy, 2004; Melchor, in McIlroy, 2004). Terrestrial sub-environments, including palaeosols, are also becoming better known in terms of their ichnofacies (Hasiotis, 2003; Tschinkel, 2003; Genise, in McIlroy, 2004; Genise *et al.*, in McIlroy, 2004; Hasiotis, 2004). Dinosaur tracks are already sufficiently well known and understood that inferences not only as to gait but also as to migratory behaviour have been drawn from them (Lockley, in Briggs and Crowther, 2001; Manning, in McIlroy, 2004).

Ichnofacies analysis is of considerable use in palaeoenvironmental interpretation, and in sequence stratigraphic interpretation (Pemberton *et al.*, in McIlroy, 2004). Thus, a shoaling-upward sequence from below storm wave-base to above the high tide mark has been inferred on the basis of the succession from *Zoophycos* ichnofacies through *Cruziana* and *Skolithos* ichnofacies to *Psilonichnus* ichnofacies, in the Cretaceous of the western interior seaway of the USA. (Interestingly, the *Zoophycos* ichnofacies is dominated by grazing and foraging ichnospecies, the *Cruziana* ichnofacies by deposit-feeding ichnospecies, and the *Skolithos* ichnofacies by filter-feeding ichnospecies.) Key omission surfaces such as sequence boundaries, amalgamated or composite sequence boundaries/transgressive surfaces ('transgressive surfaces of erosion'), and high-energy maximum flooding surfaces have been inferred on the basis of the substrate-controlled firmground *Glossifungites*, hardground *Trypanites* and woodground *Teredolites* ichnofacies (see also Box 6.1).

Storm versus fair-weather sub-environments in shallow marine environments, and turbiditic versus hemipelagic sub-environments in deep marine submarine fan environments, may also be differentiable on the basis of trace fossils (see also p. 227).

The relationship between ichnofacies and microfacies in marine environments is discussed by Jones,

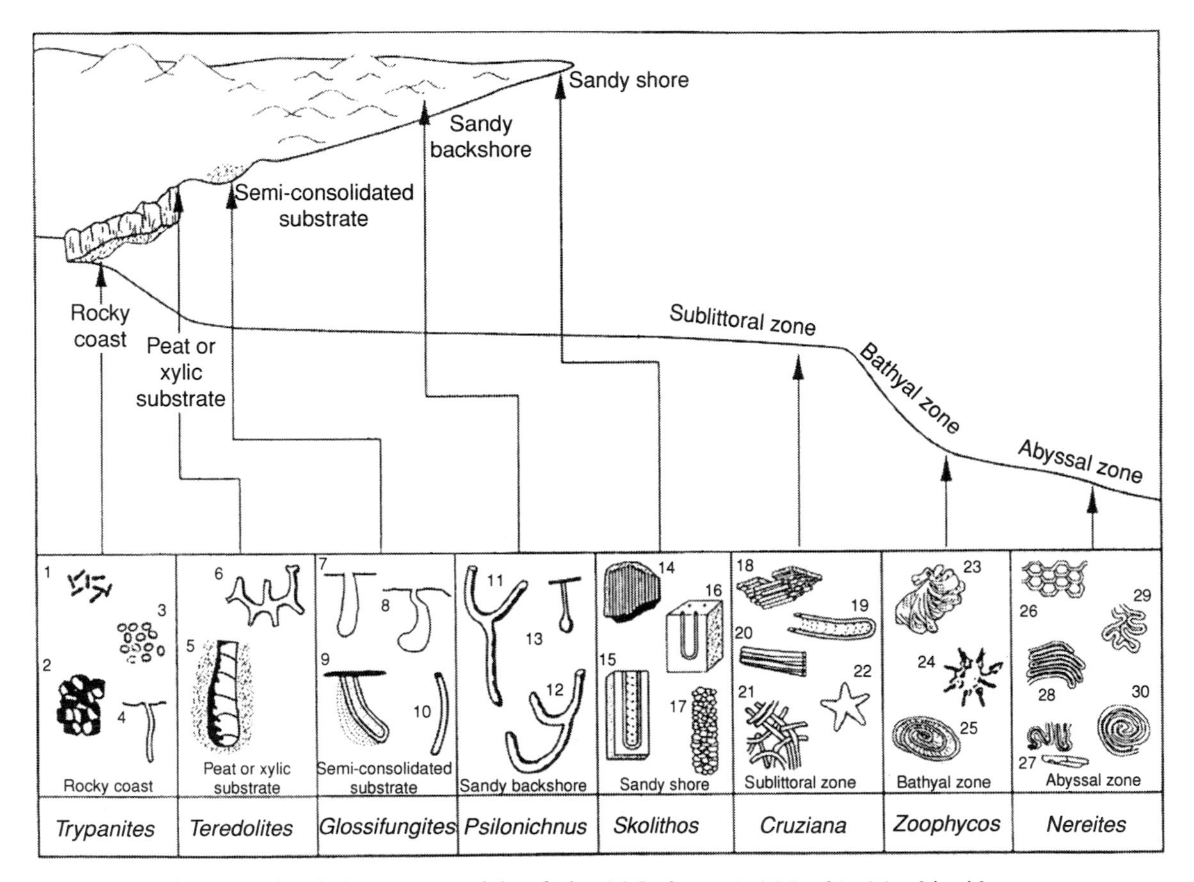

Fig. 3.89. Ichnofacies. *Trypanites* ichnofacies: (1) *Caulostrepsis*; (2) *Entobia*; (3) echinoid borings; (4) *Trypanites*. *Teredolites* ichnofacies: (5) *Teredolites*; (6) *Thalassinoides*. *Glossifungites* ichnofacies: (7–8) *Gastrochaenolites*; (9) *Diplocraterion*; (10) *Skolithos*. *Psilonichnus* ichnofacies: (11–12) *Psilonichnus*; (13) *Macanopsis*. *Skolithos* ichnofacies: (14) *Skolithos*; (15) *Diplocraterion*; (16) *Arenicolites*; (17) *Ophiomorpha*. *Cruziana* ichnofacies: (18) *Phycodes*; (19) *Rhizocorallium*; (20) *Teichichnus*; (21) *Planolites*; (22) *Asteriacites*. *Zoophycos* ichnofacies: (23) *Zoophycos*; (24) *Lorenzinia*; (25) *Zoophycos*. *Nereites* ichnofacies: (26) *Palaeodictyon*; (27) *Taphrhelminthopsis*; (28) *Helminthoida*; (29) *Cosmoraphe*; (30) *Spiroraphe*. Note that the range of *Skolithos* has been extended further landward by recent work on Early Devonian red beds in Podolia, Ukraine, that record the transition from marine to non-marine environments (Uchman *et al*., 2004). (Modified after Frey and Pemberton, 1985.)

in Jones and Simmons (1999) and Jones *et al*., in Jones and Simmons (1999).

Biostratigraphy

Trace fossils first appeared in the Late Precambrian, and have ranged through to the present day. The complete suite of interpreted behavioural types became developed by the Ordovician. Deep-water forms only became common in the Ordovician, probably through migration from shallow water (Crimes and Fedonkin, 1994). Terrestrial forms only became common in the Jurassic, coincident with the rise of the insects. Previously, such environments would not have been colonised and would not have had an ichnological signature, only a sedimentological one. Palaeoenvironmental interpretations should take observations such as this into account.

In my working experience, trace fossils have proved of limited use in biostratigraphy. Note, though, that the first appearance of the trace fossil *Phycodes pedum* has now been accepted as marking the base of the Cambrian (see Chapter 6). Note also that ichnospecies of *Cruziana* are of some biostratigraphic use in the Palaeozoic of Gondwana (McIlroy, in McIlroy, 2004), and that many ichnospecies of *Dictyodora* also appear to have restricted ranges in the Palaeozoic (Uchman, in McIlroy, 2004).

Trace fossils have proved of considerable use in palaeoenvironmental interpretation, for example, in the field in the Silurian of the Welsh basin (Crimes and Crossley, 1980), in the Cretaceous and Cenozoic of the Serrania del Interior in the eastern Venezuelan basin (Macsotay and Chachati, 1995; Vivas *et al.*, 1988), in the Palaeogene of the Spanish Pyrenees (Crimes, 1973) and the Polish Carpathians (Ksiazkiewicz, 1977), and in the Neogene of Trinidad (Jones *et al.*, in Jones and Simmons, 1999). Importantly, trace fossils are recognisable in core, and on image logs, as well as in surface outcrops. Dung beetle traces in outcrops and cores from the Early Pliocene 'productive series' of Azerbaijan have demonstrated the intermittent development of land surfaces within the predominantly lacustrine setting of the reservoir.

4 • Palaeobiology

Palaeobiology deals with the documentation, analysis and interpretation of the relationships of fossils to evolving earth and life processes and environments, and their application to the elucidation thereof. This chapter deals with applications of fossils in the interpretation of earth and life processes (excluding evolution, covered in Chapter 5), and environments. Readers interested in further details of the principles and practice of palaeobiology are referred to the works of Briggs and Crowther (1990), Dodd and Stanton (1990), Bosence and Allison (1995), Brenchley and Harper (1998), Erwin and Wing (2000), Gastaldo and DiMichele (2000), Briggs and Crowther (2001) and Cohen (2003).

Life strategy

There are a number of theoretical biological models of population dynamics, one of the most elegant and influential of which involves only two variables, *r* and *K*, at either end of a continuum (Pianka, 1970). Of these, *r* refers to the theoretically initially limitless rate of increase of the population of a species, assuming no competition, for example, on occupying a new environmental niche; *K* to the ultimately limiting, carrying capacity of that niche. Whether a species or a higher-level taxon selects in favour of a strategy that tends to maximise *K*, through the ability to compete, or *r*, through the ability to reproduce at a high rate, will depend on a number of factors, including population density etc. *K*-strategy selection typically involves slow development to reproductive maturity, at large size, sexual reproduction, small numbers of progeny, typically in more than one issue or confinement (iteroparity), long generation time, and a long lifespan; *r*-strategy selection, rapid development to reproductive maturity, at small size, asexual reproduction in some groups, large numbers of progeny, typically in one issue (semelparity), short generation time and a short lifespan.

In general, there appears to be a tendency towards conservative, or specialist, *K*-strategy selection in stable environments, and radical, or generalist opportunistic, *r*-strategy selection in unstable or otherwise stressed environments. Interestingly, environments that are both stable and stressed favour selections involving a combination of components of both strategies. For example, deserts favour plants characterised by a *K*-type tolerance of long periods of drought, and an *r*-type ability to react by reproducing rapidly whenever it rains.

4.1 Palaeoecology and palaeoenvironmental interpretation

The ecological distributions of the principal fossil groups discussed in Chapter 3 are summarised in Fig. 4.1. As intimated in Chapter 3, palaeoenvironmental interpretation based on the various groups is essentially based on analogy with their living counterparts (see appropriate sections in Chapter 3 and below), functional morphology (see below), and associated fossils and sedimentary facies. Importantly, the potential palaeoenvironmental usefulness of fossils can be impaired by natural factors (see Section 2.2), and by artificial factors such as sample acquisition and processing.

4.1.1 Palaeoenvironmental interpretation on the basis of analogy

The palaeoenvironmental interpretation of fossils on the basis of analogy with their living counterparts relies on Lyell's 'principle of uniformitarianism' (see below). In other words, fossil representatives of extant species are assumed to have occupied the same types of environment as their living counterparts and eco-homoeomorphs (see also Section 4.3). And extinct species are assumed to have occupied the same type of environment as their living congeners – or other relatives – and interpreted eco-homoeomorphs, as typically in the cases of the Palaeogene and Mesozoic. Note that, for a variety of reasons, these assumptions are not necessarily universally valid.

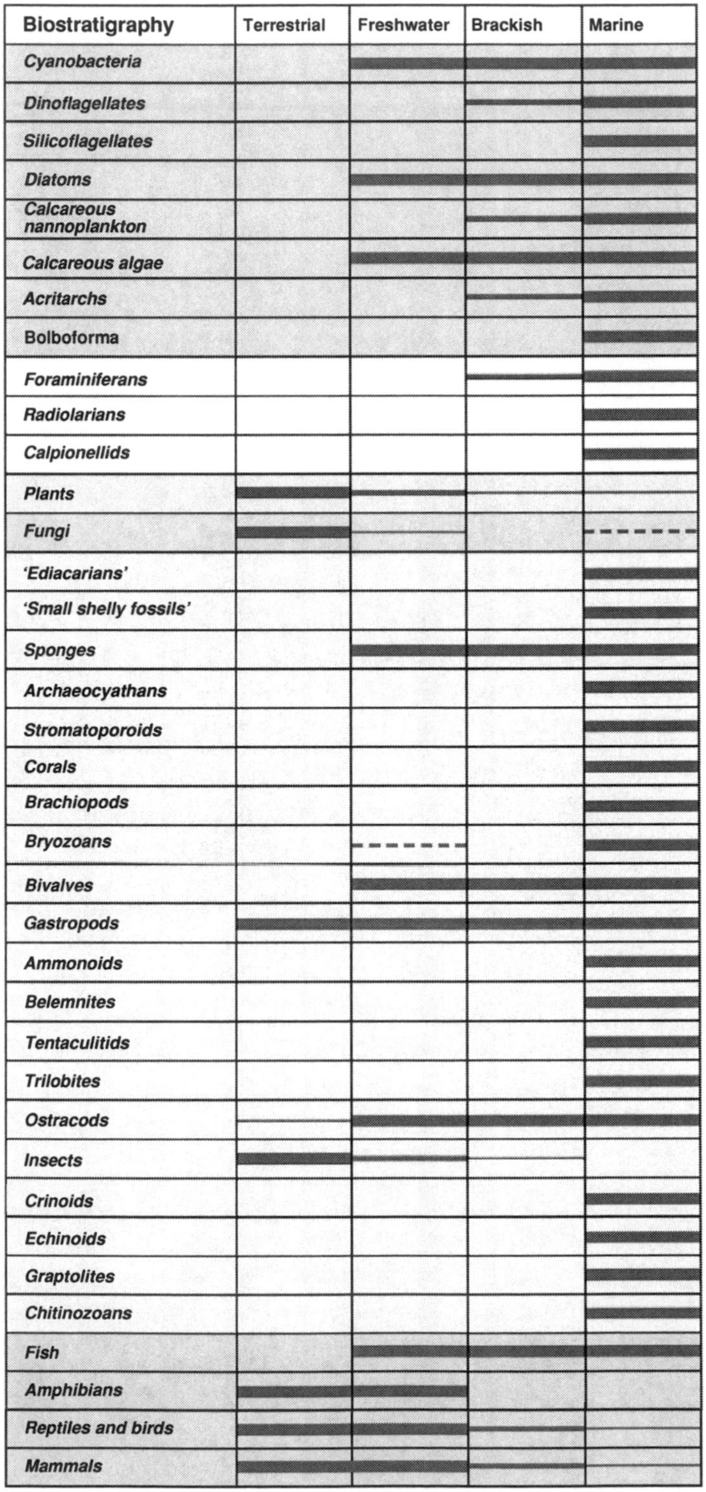

Biostratigraphy	Terrestrial	Freshwater	Brackish	Marine
Cyanobacteria				
Dinoflagellates				
Silicoflagellates				
Diatoms				
Calcareous nannoplankton				
Calcareous algae				
Acritarchs				
Bolboforma				
Foraminiferans				
Radiolarians				
Calpionellids				
Plants				
Fungi				
'Ediacarians'				
'Small shelly fossils'				
Sponges				
Archaeocyathans				
Stromatoporoids				
Corals				
Brachiopods				
Bryozoans				
Bivalves				
Gastropods				
Ammonoids				
Belemnites				
Tentaculitids				
Trilobites				
Ostracods				
Insects				
Crinoids				
Echinoids				
Graptolites				
Chitinozoans				
Fish				
Amphibians				
Reptiles and birds				
Mammals				

Fig. 4.1. Palaeoecological distribution of selected fossil groups.

The principle of uniformitarianism

The principle of uniformitarianism posits that processes operating at the present time, including, in this context, those that control environmental and other conditions and thus the environmental distributions of species, have also been operative in the past. Thus, understanding the present, and, in this context, the present environmental distributions of species, enables understanding of the past. Put simply, if not simplistically, 'understanding the present is the key to understanding the past'.

This is not to say that conditions, or even species, were the same in the past as they are at the present. Indeed, empirical observations clearly indicate that they were not. However, the very differences in species are due to the operation of the process of evolution over time, albeit at what appears through the narrow window of historical as against geological time to be at a slow rate. Thus, the observations actually validate rather than invalidate the principle, as some have argued.

Neither is it to say that rates of change of conditions or of the operations of processes have been uniform through time. Again, empirical observations clearly indicate not. For example, the otherwise apparently steady progress of (macro)evolution has been periodically interrupted by catastrophic mass extinction events over geological time. Importantly, though, uniformitarianism and catastrophism are not mutually exclusive, as some have argued.

4.1.2 Palaeoenvironmental interpretation on the basis of functional morphology

The palaeoenvironmental interpretation of fossils on the basis of functional morphology relies on the principle that function informs form, and that the analysis of form informs as to function (Thompson, 1917; La Barbeira, in Briggs and Crowther, 1990; Lugar, in Briggs and Crowther, 1990; Selden, in Briggs and Crowther, 1990; Thomason, 1995; Plotnick and Baumiller, in Erwin and Wing, 2000; Jacobs, in Briggs and Crowther, 2001; Speck and Rowe, in Briggs and Crowther, 2001; Prothero, 2004; see also sections on foraminiferans, plants, 'small shelly fossils', archaeocyathans, corals, brachiopods, bivalves, gastropods, trilobites, crinoids, graptolites and vertebrates in Chapter 3).

Note that the functional morphological principle is not necessarily universally applicable. For example, some morphological features fulfil no obvious function, but rather are 'accidents of history' that have not been eliminated by natural selection because they have no actual adverse effect on the organism. Others appear to represent preadaptions, that is, adaptations to one function or environment, retained on the change to another.

Life position and feeding strategy

Nonetheless, the functional morphological principle works well in the interpretation of the relationship of the fossil to the sedimentary environment, that is to say, not whether terrestrial or aquatic, which is usually immediately apparent from the geological context, sedimentology or ichnology, but rather whether, if marine, pelagic or benthic; and if benthic, epifaunal or infaunal; and if epifaunal, motile or sessile (Goldring, in Bosence and Allison, 1995). The principle also works well in the interpretation of locomotion in marine reptiles such as ichthyosaurs and plesiosaurs (Taylor, in Briggs and Crowther, 2001), terrestrial reptiles such as dinosaurs (Lockley, in Briggs and Crowther, 2001), and aerial reptiles such as pterosaurs (Unwin, in Briggs and Crowther, 2001).

The functional morphological principle also works well in the interpretation of feeding strategies, for example, from morphology in the case of foraminiferans (Jones and Charnock, 1985), mouth-part morphology in the case of insects (Labandeira, 1997), and tooth morphology and/or wear in the case of vertebrates (Pollard, in Briggs and Crowther, 1990; Jernvall and Selanne, 1999; Fortelius and Solounias, 2000; Ungar and Williamson, 2000; Purnell, in Briggs and Crowther, 2001; van Valkenburgh, in Briggs and Crowther, 2001).

Importantly, interpretation of the feeding strategies of individual fossil species in turn enables reconstruction of the trophic structure of the entire fossil community – loosely, the 'food pyramid' (Crame, in Briggs and Crowther, 1990). Interestingly, there appears to be bathymetric variation in the trophic structure of modern marine communities, with suspension-feeders tending to predominate in shallow waters, and detritus-feeders in deep waters (Vinogradov, 1969). However, there are also patches of the deep sea that are dominated by suspension-feeders, presumably on account of the lack of availability of organic detritus (other than phytodetritus). Deep-sea feeding strategies are also adapted to the avoidance of wastage.

4.2 Discrimination of non-marine and marine environments

4.2.1 *Non-marine environments*

Modern, and, by inference, ancient, terrestrial environments are characterised by plants, fungi, gastropods, insects, amphibians, reptiles and birds, and mammals. Freshwater environments are characterised by cyanobacteria, pennate diatoms, non-calcareous and calcareous chlorophyte algae, calcareous charophyte algae, sponges, bivalves, gastropods, cypridacean and allied ostracods, insects, fish, amphibians, reptiles and birds, and mammals. Freshwater sub-environments are differentiable on the basis of the distributions and/or shell chemistry of ostracods (Jones, 1996).

4.2.2 *Marine environments*

Brackish and marine environments are characterised by cyanobacteria, dinoflagellates, silicoflagellates, centric diatoms, calcareous nannoplankton, calcareous rhodophyte and chlorophyte algae, acritarchs, *Bolboforma*, foraminiferans, radiolarians, calpionellids, 'ediacarians', 'small shelly fossils', sponges, archaeocyathans, stromatoporoids, corals, brachiopods, bryozoans, bivalves, gastropods, ammonoids, belemnites, tentaculitids, trilobites, cytheracean and allied ostracods, crinoids, echinoids, graptolites, chitinozoans, fish, reptiles and birds, and mammals. Marine sub-environments such as depth or bathymetric zones and/or biogeographic provinces can be differentiated on the basis of the distributions of the various groups, or of sub-groups within the groups (Jones, 1996). Distance from shoreline, and hence indirectly depth, can also be inferred from the ratio between marine dinoflagellates to terrestrially derived spores and pollen in palynological preparations (Jones, 1996). Distance from shoreline, depth and palaeo-productivity related to nutrient input associated with river discharge or with upwelling can be inferred from the ratio of peridinioid to gonyaulocoid dinoflagellates (Harris and Tocher, 2003). Note, though, that the amount of information that can be gleaned about the original, autochthonous assemblages from the observed, allochthonous assemblages varies in inverse proportion to the amount of modification by transportation and other taphonomic processes.

4.3 Bathymetry and palaeobathymetry

As intimated above, marine bathymetric zones can be differentiated on the basis of the distributions of the various fossil groups, perhaps most notably the benthic foraminiferans and other microfossils (see appropriate sections in Chapter 3; see also Jones, 1996 and Olson and Leckie, 2003), but also macrofossils and trace fossils (see appropriate sections in Chapter 3; see also Farrow, in Briggs and Crowther, 1990 and Orr, in Briggs and Crowther, 2001), including, in shallow marine carbonate environments, calcareous algae and corals (see appropriate sections in Chapter 3; see also Jones, 1996 and Perrin *et al.*, in Bosence and Allison, 1995). The bathymetric distributions of modern benthic foraminiferans are comparatively well documented in the marine biological and ecological, biological oceanographic and micropalaeontological literature (Jones, 1994). Unfortunately, the various sampling, processing, staining (for 'live' specimens), picking, identification and counting procedures used do not necessarily all conform to the same standards, and therefore the resulting bathymetric distribution data-sets are not necessarily all directly comparable. Nonetheless, modern benthic foraminiferal bathymetric distributions appear typically to allow the differentiation of at least marginal, shallow and deep marine bathymetric zones in most basins. Shallow marine environments are often capable of further differentiation into inner, middle and outer shelf (or neritic); deep marine environments into upper, middle and lower slope (or bathyal) and abyssal.

Palaeobathymetric interpretation of extant species

The palaeobathymetric distributions and zonations of ancient, but extant, benthic foraminiferal species have been established largely on the basis of direct analogy with their living counterparts or ecohomoeomorphs, as typically in the cases of the Pleistogene and Neogene, and atypically in the cases of the Palaeogene and Mesozoic. Potential problems arise in attempting to use modern benthic foraminiferal bathymetric distribution data as a proxy for palaeobathymetric interpretation in those cases where distributions are evidently determined by variables other than depth, and vary from place to place. Note in this context that the upper depth limits of several species in the Gulf of Mexico have been elegantly demonstrated to be either elevated or depressed within the area of influence of the Mississippi delta, on account of the so-called 'delta effect'. Note also that the upper depth limits of some species in the North Atlantic can be either elevated or depressed depending on the latitude, on account of temperature rather than depth control.

Palaeobathymetric interpretation of extinct species

The palaeobathymetric distributions and zonations of ancient, and extinct, benthic foraminiferal species have been established partly on the basis of indirect analogy with their living relatives or other interpreted eco-homoeomorphs, as typically in the cases of the Palaeogene and Mesozoic, and partly on the basis of functional morphology and/or associated fossils and sedimentary facies, as typically in the case of the Palaeozoic. Potential problems arise in attempting to use modern distribution data as a proxy for palaeobathymetric interpretation in those cases where distributions varied from time to time. Note in this context that nodosariid and epistominid foraminiferans appear to have occupied shallow marine environments in Mesozoic, but occupy deep marine environments at the present time.

The palaeobathymetric distributions of extinct, Palaeogene benthic foraminiferans in the Mackenzie delta in Arctic Canada have been established, partly on the basis of calibration against seismic and sedimentological data, by McNeil, in Dixon *et al.* (1985). The palaeobathymetric distributions of Cenozoic and post-salt Cretaceous benthic foraminiferans in the Atlantic basins of Brazil have been established by Koutsoukos (1985) and Koutsoukos *et al.* (1991).

4.3.1 Marginal marine environments

Modern marginal marine, paralic to peri-deltaic, environments are characterised by *Ammoastuta, Ammobaculites, Ammomarginulina, Ammotium, Arenoparrella, Haplophragmoides, Jadammina, Miliammina, Reophax, Textularia, Trochammina* and *Trochamminita* (arenaceous foraminiferans), *Quinqueloculina* (miliolide), *Bolivina, Bulimina* and *Buliminella* (buliminides) and *Ammonia, Asterorotalia, Discorbis, Elphidium, Nonion* and *Protelphidium* (rotaliides). Tidally dominated delta-top – including mangrove swamp – sub-environments are characterised by *Miliammina* and *Trochammina* assemblages; delta-front environments by *Buliminella* assemblages; proximal prodelta sub-environments by *Eggerella* assemblages; and distal prodelta sub-environments by *Glomospira* and *Alveovalvulina–Cyclammina* assemblages, in the case of the Orinoco delta in the eastern Venezuelan basin (Jones, in Ali *et al.*, 1998; Jones *et al.*, in Jones and Simmons, 1999). (Fluvially dominated delta-top environments are characterised by the testate amoeban *Centropyxis*.) Channel, levee and interdistributary bay sub-environments are differentiable on the basis of cross-plots of arenaceous foraminiferal 'morphogroups' in the case of the Mississippi delta (Jones, 1996). Marginal marine sub-environments are also differentiable on the basis of the shell chemistry of foraminiferans (Reinhardt *et al.*, 2003).

Ancient marginal marine environments are characterised by taxonomically similar faunas. *Trochammina* and *Miliammina* are characteristic of paludal or marsh, and *Trochammina* and *Ammobaculites* of estuarine, sub-environments in the Late Cretaceous of the western interior seaway of the USA (Tibert *et al.*, in Olson and Leckie, 2003; Tibert and Leckie, 2004). Ostracods are also of use in palaeoenvironmental interpretation here.

4.3.2 Shallow marine environments

Shallow marine clastic environments

Modern shallow marine clastic environments are characterised by, for example, *Clavulina* and *Textularia* (arenaceous foraminiferans), *Quinqueloculina* (miliolide), *Dentalina* and *Lenticulina* (nodosariides), *Bolivina, Bulimina, Cassidulina* and *Uvigerina* (buliminides), *Hoeglundina* (robertinide), *Ammonia, Cancris, Chilostomella, Cibicides, Discorbis, Elphidium, Eponides, Gyroidina, Hanzawaia, Melonis, Nonion, Nonionella, Pullenia, Rosalina* and *Siphonina* (rotaliides) and by generally rare planktonic foraminiferans (globigerinides) (Jones, 1994, 1996). Inner shelf sub-environments are characterised by *Ammonia* and *Hanzawaia* assemblages, middle to outer shelf sub-environments by *Uvigerina* and *Bolivina floridana* assemblages, in the case of the eastern Venezuelan basin.

Ancient shallow marine clastic environments are characterised by taxonomically similar faunas.

Shallow marine carbonate environments

Modern shallow marine carbonate environments are characterised by, for example, the LBFs *Alveolinella, Amphisorus, Archaias, Borelis, Cyclorbiculina, Marginopora, Peneroplis* and *Sorites* (miliolides), *Amphistegina, Baculogypsina, Calcarina, Heterocyclina, Heterostegina, Operculina* and *Operculinella* (rotaliides) and by calcareous algae and corals (Jones, 1994, 1996). Shallow marine, euphotic, inner to middle platform, and back-reef to reef, sub-environments are characterised by miliolide LBFs with green or red algal, dinoflagellate, B-1 diatom or, exceptionally, B-3 diatom photosymbionts; typically deeper marine, sub-euphotic, outer platform, and fore-reef, sub-environments by rotaliide LBFs with B-2 or B-3 diatom photosymbionts; and deep marine, sub-photic, basinal and off-reef, sub-environments by smaller benthic and planktonic foraminiferans (Jones, 1996). (Note in this context that the distributions of LBFs are determined by water clarity rather than depth, since it is this that determines the amount of light available for their photosymbionts, and therefore that it is possible to find typically deeper marine LBFs in shallower water, when clarity is reduced, for example, by turbidity or turbulence: Taylor, in Capriulo, 1990; Hohenegger, 2004.) Wilson's standard sub-environments are differentiable on the basis of cross-plots of foraminiferal 'morphogroups'. Photosynthesis is, and hence photic zone sub-environments are, indicated by the shell chemistry of LBFs, specifically the boron isotopic composition (Honisch *et al.*, 2003).

Ancient shallow marine carbonate environments are characterised by taxonomically similar faunas (Jones, 1996).

4.3.3 ***Deep marine environments***

Modern deep marine environments are characterised by diverse arenaceous foraminiferans, *Spirosigmoilina* (miliolide), *Dentalina, Lenticulina* and *Nodosaria* (nodosariides), *Bolivina, Francesita, Globobulimina, Sphaeroidina* and *Stilostomella* (buliminides), *Hoeglundina* (robertinide), *Alabaminoides, Anomalinoides, Chiolostomella, Cibicidoides, Gyroidina, Ioanella, Oridorsalis, Osangularia, Osangulariella, Planulina* and *Pullenia* (rotaliides) and typically by abundant and diverse planktonic foraminiferans (globigerinides) (Jones, 1994, 1996). Upper to middle slope sub-environments are characterised by buliminides; lower slope environments by rotaliides; and abyssal environments, especially those below the calcite compensation depth, by arenaceous foraminiferans. Oxygen minimum zone, submarine fan, and hydrothermal vent and cold (hydrocarbon) seep sub-environments may also be differentiable (see below).

Ancient deep marine clastic environments are characterised by taxonomically similar faunas (but see also Box 3.1).

Oxygen minimum zones

Modern oxygen-poor environments, such as the upper slope oxygen minimum zone (OMZ), are characterised by infaunal buliminides (Jones, 1996). Ancient oxygen-poor environments are characterised by taxonomically similar faunas. Importantly, oxygen-poor and organic-rich environments are sites of petroleum source-rock deposition and/or preservation (see Section 7.1).

Submarine fans

Modern and, by inference, ancient, submarine fan environments are characterised by particular foraminiferal faunas or assemblages (Jones, 1996; Jones, in Jones and Simmons, 1999; Jones, 2001; Jones *et al.*, in Powell and Riding, in press). In essence, they are characterised by an autochthonous deep marine benthic component; an allochthonous, but contemporaneous, shallow marine benthic and planktonic, and non-marine, component; and, occasionally, an allochthonous, and non-contemporaneous, component. The autochthonous deep marine component is characteristically dominated by 'deep-water arenaceous foraminiferans' (DWAFs) (see Box 3.1). Instances of both horizontal variation and vertical variation in foraminiferal assemblage composition within fan complexes have been reported in the literature (horizontal variation is related to vertical variation through Walther's law). *Horizontal variation* is manifested by a transition from essentially allochthonous faunas associated with fan channel deposits, to autochthonous faunas associated with levee/overbank deposits. In the case of the modern Mississippi fan in the Gulf of Mexico, channels are characterised by the impoverished development, and overbanks by the richer development, of an autochthonous bathyal benthic assemblage. Allochthonous, but contemporaneous, shelfal benthic components are present in both settings.

Allochthonous, and non-contemporaneous, components are also present in both settings, though more obviously so in sand-prone channels, suggesting that reworking is associated with flows essentially confined to the channels. Modern and ancient channel deposits of the Monterey and other fans off California in the north-east Pacific, the Baltimore canyon in the north-west Atlantic and the Amazon fan in the western equatorial Atlantic are also characterised by contemporaneous and non-contemporaneous allochthonous assemblages. Ancient, Eocene channel axis deposits of the Ainsa fan in the Spanish Pyrenees are characterised by contemporaneous and non-contemporaneous allochthonous benthic assemblages; channel off-axis deposits by impoverished low-diversity autochthonous deep-water assemblages; and levee/overbank deposits by moderate-diversity autochthonous assemblages (Jones, in Powell and Riding, in press). Interestingly, channel off-axis deposits of the Ainsa fan are also characterised by increased incidences of epifaunal suspension-feeders ('Morphogroup' A of Jones and Charnock, 1985), interpreted as indicating an ecological preference for this site rather than a taphonomic artefact – that is to say, habitation rather than transportation. *Vertical variation* is manifested by a transition from essentially allochthonous assemblages associated with turbidites to interpreted autochthonous assemblages associated with hemipelagites. According to this interpretation, the observed variation through individual turbidite packages, from assemblages characterised by epifaunal suspension-feeders ('Morphogroup' A of Jones and Charnock, 1985) to assemblages characterised by epifaunal and infaunal detritus-feeders ('Morphogroups' B and C), reflects the greater ability of the former group to colonise the sea-floor in between turbidity flows, and to exploit the source of food in the turbid suspension (Jones, in Jones and Simmons, 1999). According to an alternative interpretation, the assemblages associated with hemipelagites, as well as those associated with turbidites, are allochthonous, and the observed vertical variation attributable to waning flow, and associated hydrodynamic sorting particles of differing settling velocity. Note in this context, though, that turbidites from the Monterey fan are characterised by hydrodynamic sorting of foraminiferans, but hemipelagites are not.

Modern and, by inference, ancient submarine fan environments are also characterised by particular types of palynological assemblage (Hoorn, in Flood *et al.*, 1997). Turbiditic sub-environments on the Crati fan in the Ionian Sea are characterised by filter-feeding bivalve morphotypes (Colella and di Geronimo, 1987). Submarine fan sub-environments are differentiable on the basis of trace fossils. Instances of both horizontal variation and vertical variation in trace fossil assemblage composition within fan complexes have been reported. Horizontal variation is manifested by a transition from low-diversity assemblages dominated by interpreted opportunistic shallow-water taxa such as *Ophiomorpha* and *Thalassinoides* in inner fan and canyon sub-environments, through moderate- to high-diversity assemblages characterised by significant proportions of shallow-water taxa in middle fan sub-environments, to high-diversity assemblages characterised by abundant deep-water taxa and only rare shallow-water taxa in fan-fringe and outer fan sub-environments. Vertical variation is manifested by a transition from low-diversity assemblages dominated by interpreted opportunistic taxa associated with turbidites, to high-diversity assemblages associated with hemipelagites, together with pervasive bioturbation and complex tiering. The rate and pattern of recolonisation appears to vary according not only to physical effects, but also to biological effects, including feeding and reproductive behaviours, such that there is no standard succession. Many of the opportunistic species observed exploiting disturbed environments in modern settings are soft-bodied, and have no body-fossil counterparts in ancient settings. Examples include the polychaetes *Streblospio benedicti, Hobsonia florida, Polydora ligni* and *Capitella capitata*, which are all deposit-feeders with short reproductive cycles embodying a planktonic larval phase.

Hydrothermal vents and cold (hydrocarbon) seeps

Modern and ancient hydrothermal vent and cold (hydrocarbon) seep faunas differ from all others on account of their primary producers being chemosynthetic bacteria rather than photosynthetic plants, and thus their ultimate dependence on a geochemical rather than a biochemical energy source. They are essentially only found where geochemical energy sources are available on the sea-floor: hydrogen sulphide in the case of hydrothermal vents, and methane in the case of cold seeps. Incidentally, a third type of fauna reliant on chemosynthetic bacteria has recently been discovered, associated with 'whale falls' or decaying whale carcasses on the floor of the sea.

Hydrothermal vents support benthic meio- and macro-biotas that ultimately depend for their survival on chemo-autotrophic bacteria. These biotas comprise over 400 species of organisms, over 80% of which are endemic to hydrothermal vent sites, including one new class, three new orders and 22 new families (Tunnicliffe, 1992). The dominant groups are soft-bodied vestamentifarian worms such as *Riftia*, and bivalves such as the giant vesicomyid clam *Calyptogena* and the 'vent-mussel' *Bathymodiolus*. *Riftia* and *Calyptogena* host chemo-autotrophic bacterial symbionts. *Bathymodiolus* hosts methane-oxidising bacterial symbionts. Modern hydrothermal vents and associated biotas are known from the mid-Atlantic, the eastern and western Pacific and the Indian Ocean (Tunnicliffe *et al.*, 1998). Ancient hydrothermal vents and associated biotas are known from the Cambrian through to the Tertiary (Little, in Briggs and Crowther, 2001).

Cold seeps support benthic meio- and macro-biotas that ultimately depend for their survival on chemo-autotrophic bacteria that metabolise the seeping hydrocarbon. These biotas comprise over 200 species of organisms, the vast majority of which are endemic to cold seep sites, and the remainder of which are only known elsewhere from hydrothermal vent sites (Sibuet and Olu, 1998). Carbon isotope analyses of the biotas from a cold seep site in the North Sea have revealed that at least the bivalve *Thyasira sarsi* is actually metabolising seeping hydrocarbons here, through symbiotic chemo-autotrophic bacteria (Dando *et al.*, 1994). Modern cold seeps and associated biotas are known from the North Sea, the North Atlantic and Gulf of Mexico, and the North Pacific (see also below). Ancient cold seeps and associated biotas are known from the Palaeozoic of Greenland (von Bitter *et al.*, 1992), the Mesozoic of Greenland, France, Arctic Canada and Argentina (Beauchamp and Savard, 1992; Gaillard *et al.*, 1992; Kelly *et al.*, in Harper *et al.*, 2000; Gomez-Perez, 2003), and the Cenozoic of Washington State in the USA, Peru, Vietnam, the Caribbean and Italy (Campbell, 1992; Wefer *et al.*, 1994; Jones, 1996; Gill *et al.*, 2003; Barbieri and Panieri, 2004).

Box 4.1 Benthic foraminiferans associated with hydrocarbon seeps

Introduction

In recent years there have been several studies on benthic foraminiferans associated with modern and ancient hydrocarbon seeps (Jones, in Jenkins, 1993; Akimoto *et al.*, 1994; Wefer *et al.*, 1994; Coles *et al.*, 1996; Jones, 1996; Kitazato, 1996; Sen Gupta *et al.*, 1997; Rathburn *et al.*, 2000; Bernhard *et al.*, 2001; Hill *et al.*, 2003; Barbieri and Panieri, 2004). The ultimate objectives of these studies have been to evaluate the potential of benthic foraminiferans as indicators of modern and ancient hydrocarbon seepage, and/or to determine the timing and duration of seepage. Respective applications are in providing sense checks on geochemical indications of hydrocarbon seepage, and in determining the timing of hydrocarbon charge in relation to that of trap formation. This section summarises the results of some of these studies.

Materials and methods

Studies have been undertaken on benthic foraminiferans associated with modern seeps in the North Sea (Jones, in Jenkins, 1993; Jones, 1996), the North Atlantic (Coles *et al.*, 1996; Jones, 1996), the Gulf of Mexico (Jones, 1996; Sen Gupta *et al.*, 1997), and the North Pacific (Akimoto *et al.*, in Tsuchi, 1994; Kitazato, 1996; Rathburn *et al.*, 2000; Bernhard *et al.*, 2001; Hill *et al.*, 2003). Studies have also been undertaken on benthic foraminiferans associated with ancient seeps in the Miocene (Jones, 1996; Barbieri and Panieri, 2004) and Pleistocene (Wefer *et al.*, 1994). The results of these studies are not all directly comparable, as they did not involve directly comparable analytical and interpretive procedures. Nonetheless, the following observations and comparisons can be made.

Results and discussion

Similarity index matrix

Samples from the North Sea and from the Gulf of Mexico were compared with one another using specific-level similarity indices (Jones, in Jenkins, 1993; Jones, 1996). In the case of the North Sea, seep and control samples proved distinct at the 75% similarity level (Jones, in Jenkins, 1993; Jones, 1996). In the case of the Gulf of Mexico, fresh seep and control samples proved distinct, and biodegraded seep and control samples indistinct, at the 37.5% similarity level (Jones, 1996). (The threshold for the Gulf of Mexico was reduced in order to reflect and compensate for the generally greater dissimilarity between deep-water samples – a function of their generally greater diversity.)

Abundance, diversity, dominance and equitability

In the North Sea, seep and control sample proved distinct in terms of abundance, diversity and dominance (Jones, in Jenkins, 1993; Jones, 1996). Abundance and diversity proved lower, and dominance higher, in seep as compared to control samples. In the Gulf of Mexico, seep and control samples proved more or less distinct in terms of abundance, diversity and dominance (Jones, 1996). Abundance and diversity proved lower in most seep as compared to control samples, the one exception being a seep sample from a mud volcano (the anomalously high abundance diversity of this sample arguably attributable to reworking from the shallow subsurface). 'Fisher diversity' and 'unit diversity', that is, diversity as a function of abundance, proved lower in seep as compared to control samples. Dominance proved higher in seep as compared to control samples. Equitability, that is, the evenness of distribution, proved lower in seep as compared to control samples. Interestingly, in the Porcupine basin in the north-east Atlantic, seep samples were characterised by early abundance but higher diversity than control samples (Coles *et al.*, 1996). One possible explanation is that the hard substrate provided by the carbonates associated with the seeps provides niches for a wider variety of species (Coles *et al.*, 1996). Notably, attached species are more abundant and diverse in the seep than in the control samples.

Taxonomic composition (ordinal level)

In the North Sea, the Gulf of Mexico and the Porcupine basin in the north-east Atlantic, seep samples were generally characterised by higher absolute and relative abundances of rotaliides and lower absolute and relative abundances of buliminides than control samples (Jones, in Jenkins, 1993; Coles *et al.*, 1996; Jones, 1996).

Taxonomic composition (generic and specific level)

In the North Sea, *Uvigerina peregrina* and *Cassidulina laevigata carinata sensu lato.* (buliminides), and *Hyalinea balthica* and *Elphidium ex gr. clavatum* (rotaliides) were tentatively positively correlated with seepage; and *Bulimina marginata* and *Trifarina angulosa* (buliminides) were tentatively negatively correlated with seepage (Jones, in Jenkins, 1993; Jones, 1996). In the Porcupine basin in the north-east Atlantic, *Sigmoilopsis schlumbergeri* (miliolide), *Lenticulina* spp. (nodosariides), *Cassidulina laevigata carinata sensu lato* and *C. obtusa sensu lato* (buliminides) and *Cibicides lobatulus, Cibicidoides pachyderma, Discoanomalina coronata, Elphidium ex gr. clavatum, Gyroidina orbicularis, Hyalinea balthica* and *Melonis affinis* (rotaliides) were positively correlated with seepage; and *Bulimina marginata, Trifarina angulosa, Uvigerina bradyana* and *U. mediterranea* (buliminides) were negatively correlated with seepage (Coles *et al.*, 1996). In the Gulf of Mexico, *Lenticulina* spp. (nodosariides) and *Cibicidoides pachyderma* (rotaliide) were positively correlated with seepage; and *Vaginulinopsis subaculeatus* (nodosariide), *Bolivina albatrossi* (buliminide) and *Alabaminoides exiguus* and *Chilostomella oolina* (rotaliides) were negatively correlated with seepage (Jones, 1996). In parts of the Gulf of Mexico, *Bolivina albatrossi, B. ordinaria, Cassidulina laevigata carinata sensu lato* and *Trifarina bradyi* (buliminides) and *Gavelinopsis translucens* and *Osangularia rugosa* (rotaliides) appear positively correlated with hydrogen sulphide and methane seepage, and also incidentally, with the sulphide-oxidising bacterium *Beggiatoa* (Sen Gupta *et al.*, 1997). It has been hypothesised that these species are either microaerophiles or facultative anaerobes (see above). In Sagami Bay off Japan, *Bulimina striata* and *Rutherfordoides cornuta* (buliminides) appear positively correlated with hydrogen sulphide and methane seepage respectively, and with *Calyptogena* (Akimoto *et al.*, in Tsuchi, 1994). It has been hypothesised that at least the former species either hosts sulphide-oxidising bacterial symbionts, or is an anaerobe, able to respire in the absence of oxygen (Kitazato, 1996). The former hypothesis was preferred. Anaerobic respiration would release acid as a by-product, which would tend to dissolve calcareous tests. In California, *Bolivina, Globobulimina* and *Uvigerina* (buliminides) and *Chilostomella* and *Nonionella* (rotaliides) appear positively correlated with seepage, and also, incidentally, with the giant vesicomyid clam *Calyptogena* (Rathburn *et al.*, 2000). Carbon isotope analysis indicates that at least *Bolivina aubargentea* and *Uvigerina peregrina* (buliminides) are actually metabolising the seeping methane. In Monterey Bay, *Ammodiscus incertus, Psammosphaera* sp., *Reophax* sp. 1, *Spiroplectammina biformis, Tolypammina* sp., *Trochammina* sp. and *Verneuilinella* sp. (arenaceous foraminiferans), *Adelosina* sp. and *Triloculinella* sp. (miliolides), *Lagenosolenia* sp. (nodosariide), *Bolivina pacifica, B. spissa, Cassidulina delicata, Fursenkoina rotundata, Globobulimina* spp., *Rutherfordoides cornuta* and *Trifarina angulosa* (buliminides) and *Alabaminoides exiguus, Astrononion* sp., *Cibicides* sp., *Cibicidoides wuellerstorfi, Epistominella pacifica, E. smithi* and *Gyroidina altiformis* (rotaliides) appear positively correlated with seepage, and with (?*Calyptogena*) clam fields and flats (Bernhard *et al.*, 2001; author's unpublished observations on their data). *Phthanotrochus arcanus* (allogromiide), *Miliammina* sp., *Reophax gracilis, Reophax* sp. 2 and *Textularia* sp. (arenaceous foraminiferans), *Bolivina* sp., *Bulimina mexicana, Buliminella tenuata, Loxostomum pseudobeyrichi, Praeglobobulimina spinescens, Uvigerina peregrina* and *Uvigerina* sp. (buliminides), and *Chilostomella oolina, Nonion* sp. and *Nonionella globosa* (rotaliides) appear negatively correlated with seepage (author's unpublished observations).

Proportions of epifaunal and infaunal 'morphotypes'

In the North Sea, the Gulf of Mexico, and Monterey in California, seep samples were characterised by generally higher incidences of epifaunal 'morphotypes', and correspondingly lower incidences of infaunal 'morphotypes', than control samples (Jones, in Jenkins, 1993; Jones, 1996; Bernhard *et al.*, 2001; author's unpublished observations on their data). Similar trends have also been observed in studies of the meio- and macro-fauna of the North Sea (Dando *et al.*, 1991). Whether the disproportionately low incidences of infaunal 'morphotypes' in seep samples are caused directly by the seeping hydrocarbons or indirectly by associated adsorbed toxins, or the products of bacterial methanotrophism or sulphur oxidation, or hypo- or hyper-saline waters, is as yet unclear. However, the results of elemental analyses of seep samples from the Barra fan area of the north-east Atlantic did not reveal any abnormally high concentrations of heavy metals shown to have deleterious effects on foraminiferal populations (Jones, 1996).

(Modified after Jones, in Jenkins (1993), Coles *et al.* (1996) and Jones (1996).

4.4 Biogeography and palaeobiogeography

Many fossil groups appear to have had restricted or endemic biogeographic distributions, rendering them of use in the characterisation of palaeobiogeographic provinces or realms, and in turn in the constraint of plate tectonic and terrane reconstructions (Adams and Ager, 1967; Middlemiss and Rawson, 1971; Hallam, 1973; Hughes, 1973; Gray and Boucot, 1979; Fortey and Cocks, in Briggs and Crowther, 1990; Newton, in Briggs and Crowther, 1990; Hallam, 1994; Meyerhoff *et al.*, 1996; Humphries and Parenti, 1999; Golonka and Ford, 2000; Lieberman, 2000; Hill, in Briggs and Crowther, 2001; Shi, in Briggs and Crowther, 2001; Crame and Owen, 2002; Wagreich, 2002; Ford and Golonka, 2003; Golonka *et al.*, 2003; Huggett, 2004). Importantly, fossils provide at least some measure of longitude as well as latitude (not provided by palaeomagnetism).

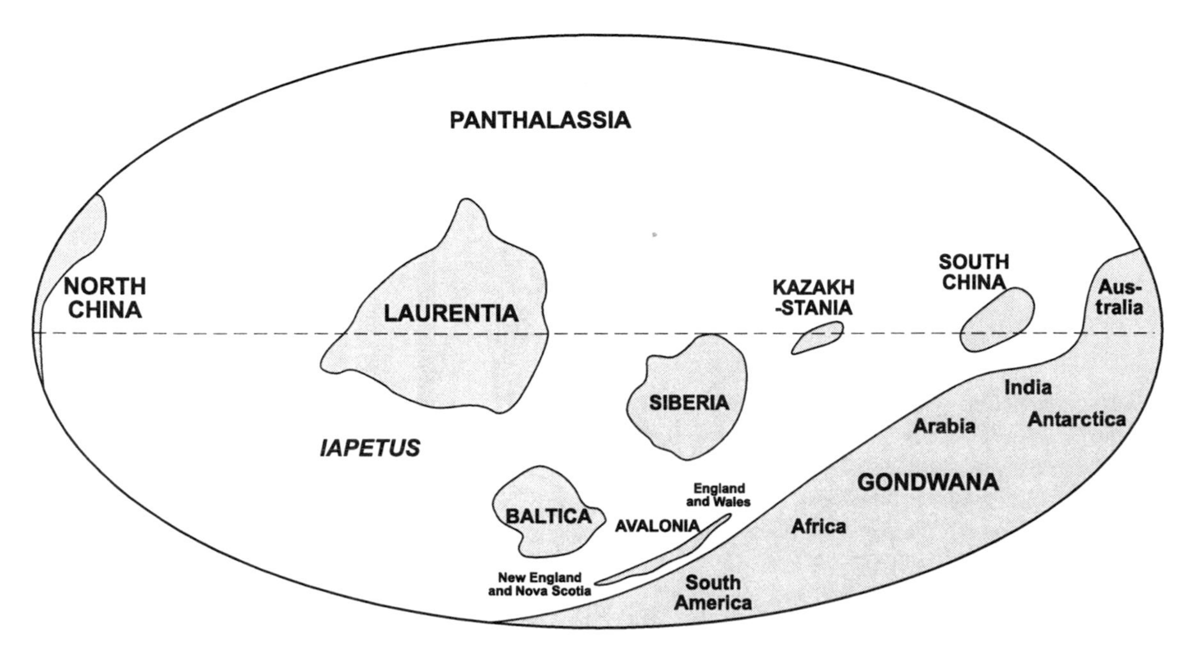

Fig. 4.2. Cambrian palaeogeography and palaeobiogeography.

Some of the more palaeobiogeographically important fossil groups are the dinoflagellates, diatoms, calcareous nannoplankton, calcareous algae, acritarchs, foraminiferans, radiolarians, plants, sponges, stromatoporoids, corals, brachiopods, bryozoans, bivalves, gastropods, ammonoids, belemnites, trilobites, ostracods, crinoids, echinoids, graptolites, fish, amphibians, reptiles and mammals (see appropriate sections in Chapter 3).

4.4.1 *Palaeozoic*

Many fossil groups appear to have had restricted or endemic biogeographic distributions in the Palaeozoic (McKerrow and Scotese, 1990; Cocks, 2001; Servais *et al.*, 2003; Torsvik and Cocks, 2004). Acritarch, brachiopod, trilobite, ostracod and graptolite distributions point toward the development of three marine palaeobiogeographic provinces in the Cambrian, namely the Laurentia–Baltica–Avalonia, Siberia and Gondwana provinces; and four provinces in the Ordovician, namely, the Laurentia, Baltica–Avalonia, low-latitude Gondwana and high-latitude Gondwana provinces (Figs. 4.2–4.3). In the Ordovician, Baltica–Avalonia, of which England and Wales were a part, was separated from Laurentia, of which Scotland was a part, by the Iapetus Ocean, a substantial body of deep water that constituted an insuperable barrier to the dispersal of shallow-water benthic organisms. Baltica–Avalonia was also separated from Gondwana by the Rheic Ocean. Towards the end of the Ordovician, provincialism generally decreased, and cosmopolitanism increased, as Baltica–Avalonia, Laurentia and Gondwana drifted towards one another. Cosmopolitanism remained the norm through the Silurian (Fig. 4.4). Provincialism became re-established in the Devonian, and the Malvino-Kaffric province, centred on South America, the Falklands or Malvinas, and South Africa, became established for the first time, within Gondwana (Fig. 4.5). Further marine provinces became established by the Permian. Plant and vertebrate distributions point toward the development of four generally separate terrestrial palaeobiogeographic provinces in the Devonian–Permian, namely the Angara, Euramerica, Cathaysia and Gondwana provinces (Figs. 4.5–4.7). The plant macrofossils and microfossils of the Permian of Oman are similar to those of South China, supporting the so-called 'Pangaea B' plate reconstruction (Berthelin *et al.*, 2003).

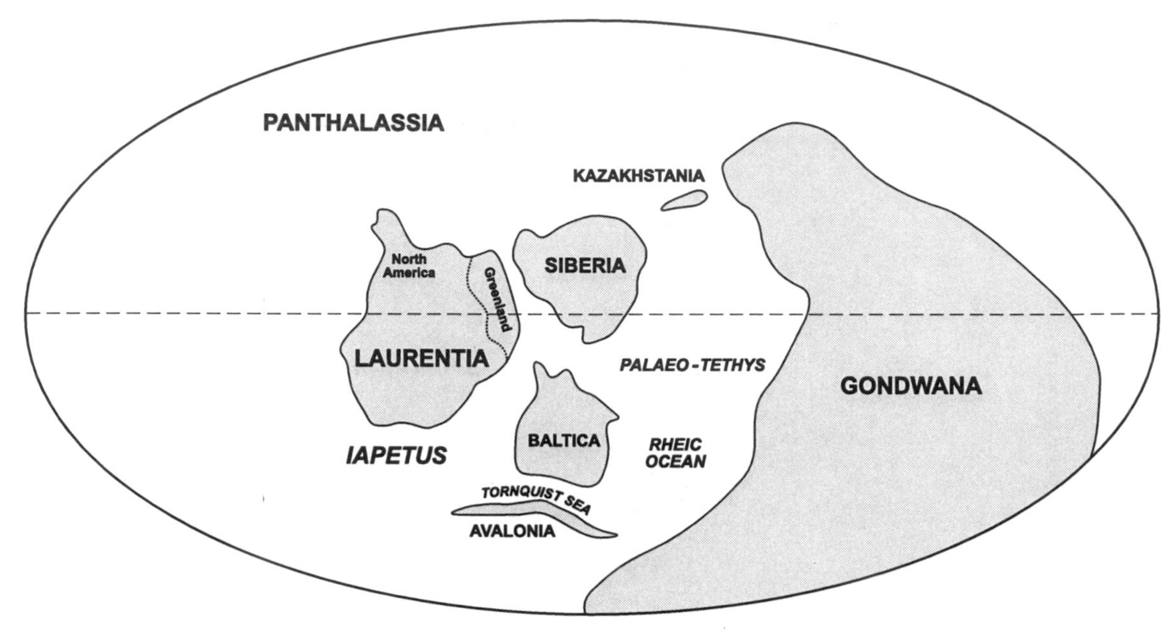

Fig. 4.3. Ordovician palaeogeography and palaeobiogeography.

In terms of terrane reconstruction, Early Carboniferous radiolarians from the Shan-Thai terrane of north-west Thailand are constituted for the most part of species known from European Palaeo-Tethys, within 30° of the palaeo-equator (Quinglai Feng *et al.*, 2004). This indicates that by this time the Shan-Thai terrane had already rifted off from Gondwana, then situated in higher latitudes to the south. The

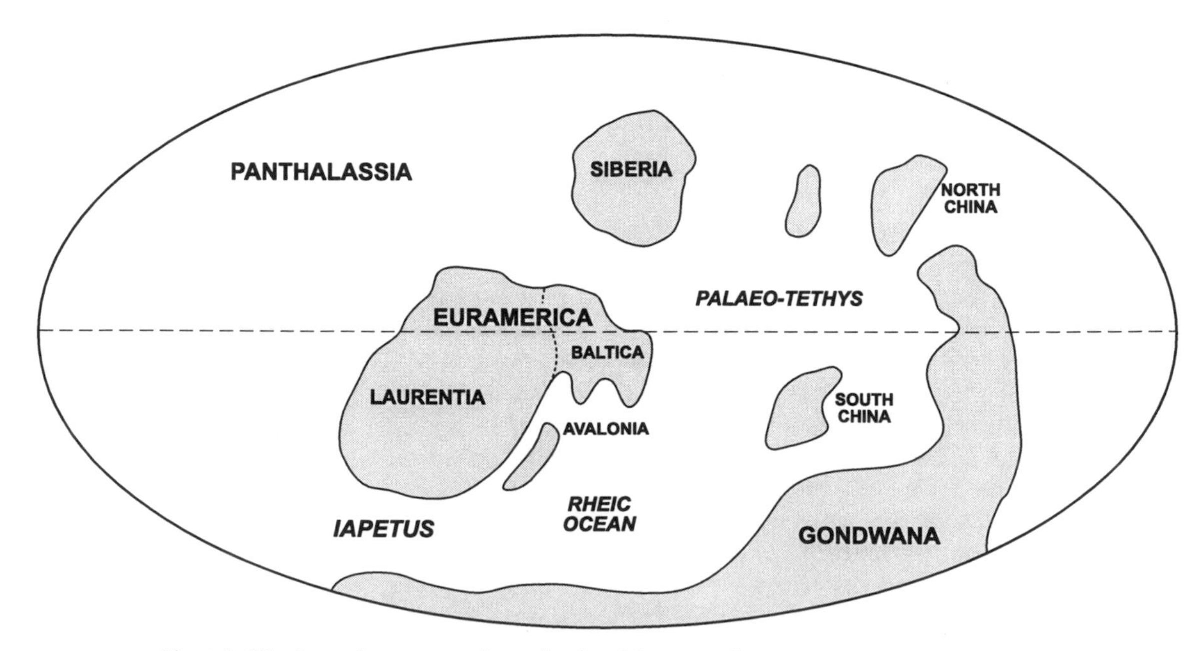

Fig. 4.4. Silurian palaeogeography and palaeobiogeography.

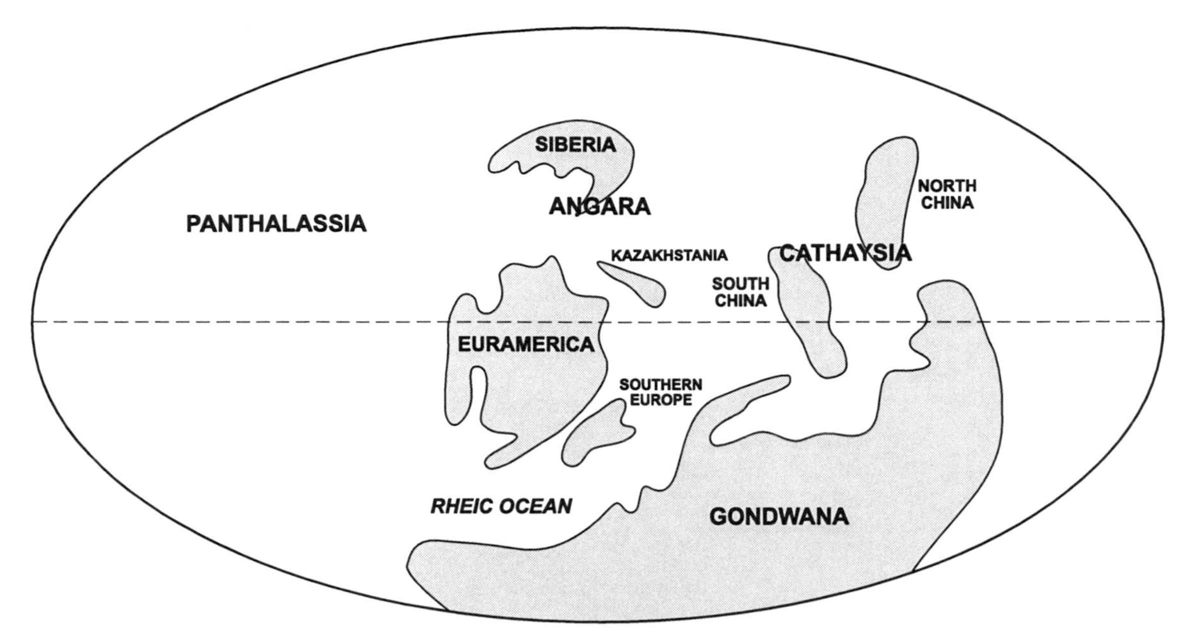

Fig. 4.5. Devonian palaeogeography and palaeobiogeography.

presence of the calcareous alga *Palaeoaplysina* in the Late Carboniferous–Early Permian of the Klamath Mountains of California indicates a palaeobiogeographic link with Idaho, British Columbia, Yukon, and Ellesmere and Axel Heiberg Islands elsewhere in North America, and further indicates that the Klamath Mountains do not form part of a far-travelled terrane, as has been hypothesised (Watkins

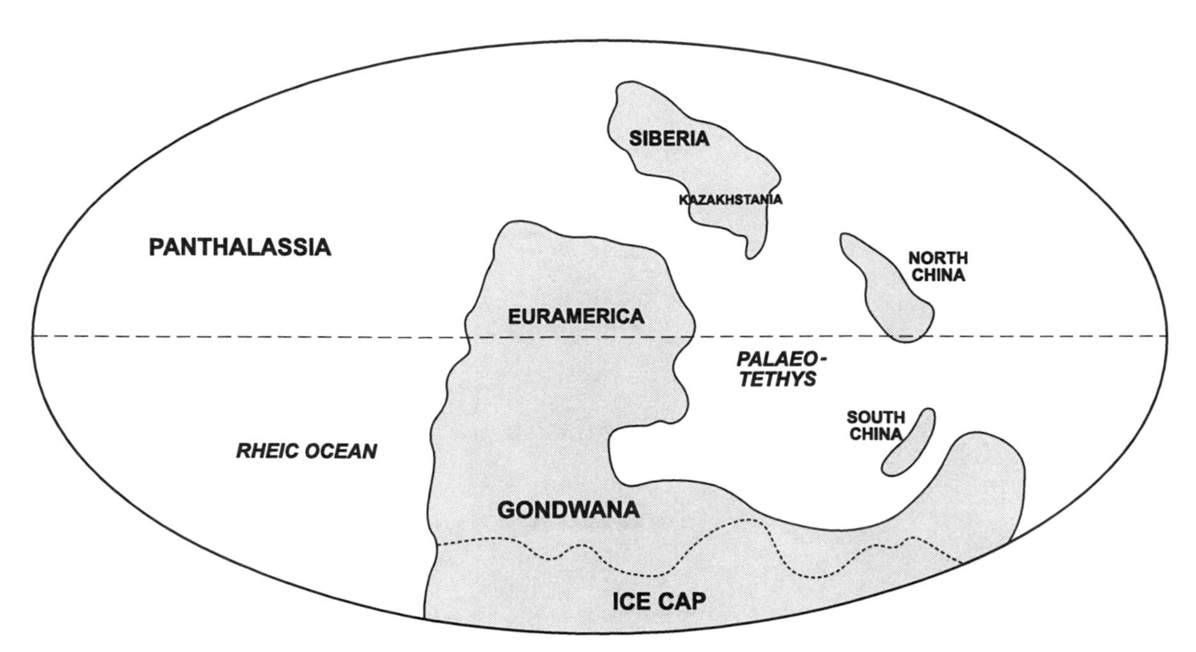

Fig. 4.6. Carboniferous palaeogeography and palaeobiogeography.

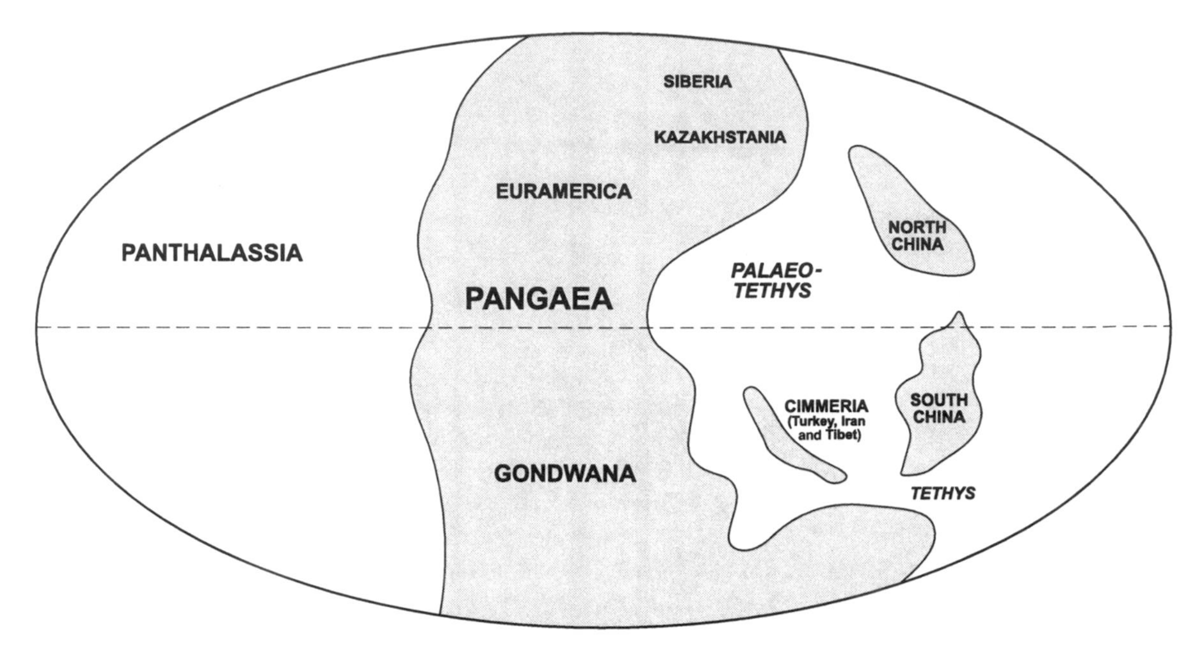

Fig. 4.7. Permian palaeogeography and palaeobiogeography.

and Wilson, 1989). *Palaeoaplysina* is also found in Svalbard, Bjornoya, Timan–Pechora and the Urals.

4.4.2 Mesozoic

Many fossil groups appear to have had restricted or endemic biogeographic distributions in the Mesozoic (Figs. 4.8–4.10) (Westermann, 2000). Foraminiferal, radiolarian, bivalve, ammonoid, belemnite and ostracod distributions point toward the development of three marine palaeobiogeographic provinces in the Triassic–Cretaceous, namely the Boreal, Tethyan and Austral provinces. A number

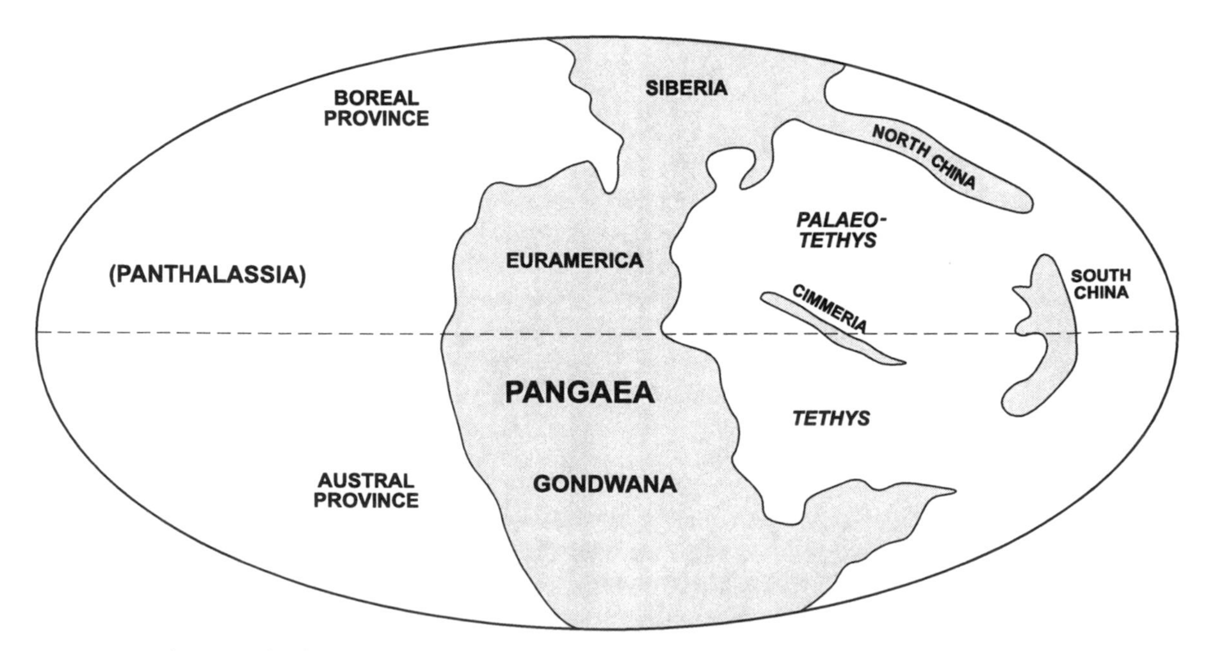

Fig. 4.8. Triassic palaeogeography and palaeobiogeography.

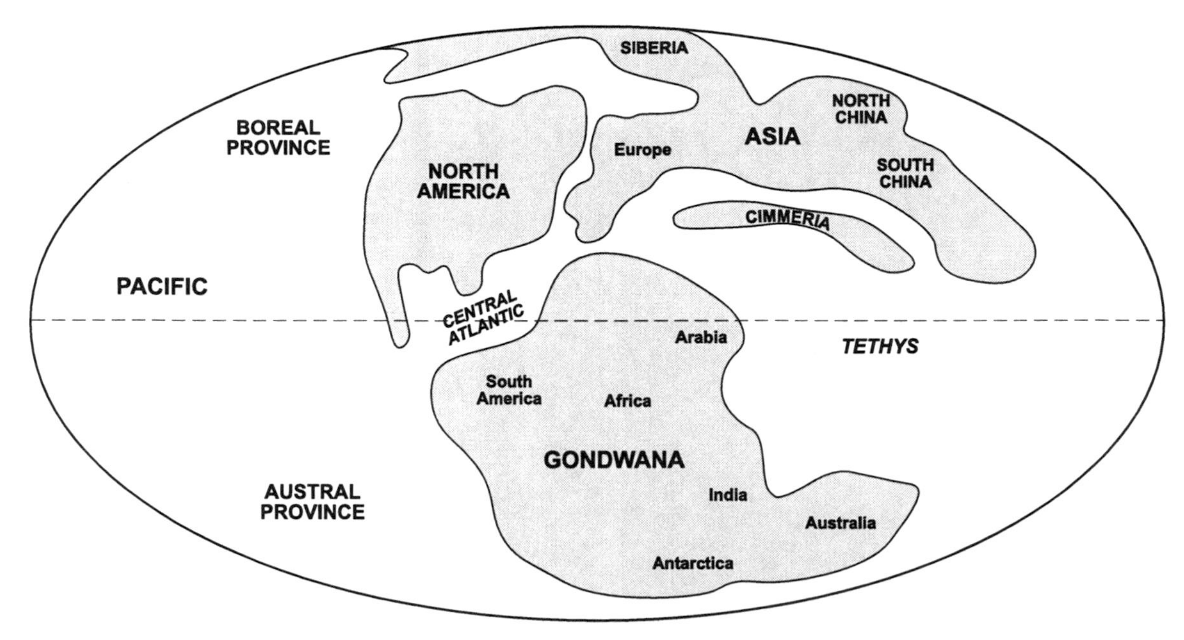

Fig. 4.9. Jurassic palaeogeography and palaeobiogeography.

of sub-provinces are also recognisable within each of these three provinces, for example, the Arctic, Boreal–Atlantic and Boreal–Pacific sub-provinces in the Boreal province. The boundaries between the various provinces and sub-provinces changed through time in response to tectonic, climatic and tectono- or glacioeustatic sea-level changes, and the consequent creation and destruction of dispersal routes. Nonetheless, as a general rule, the low-latitude Tethyan province is

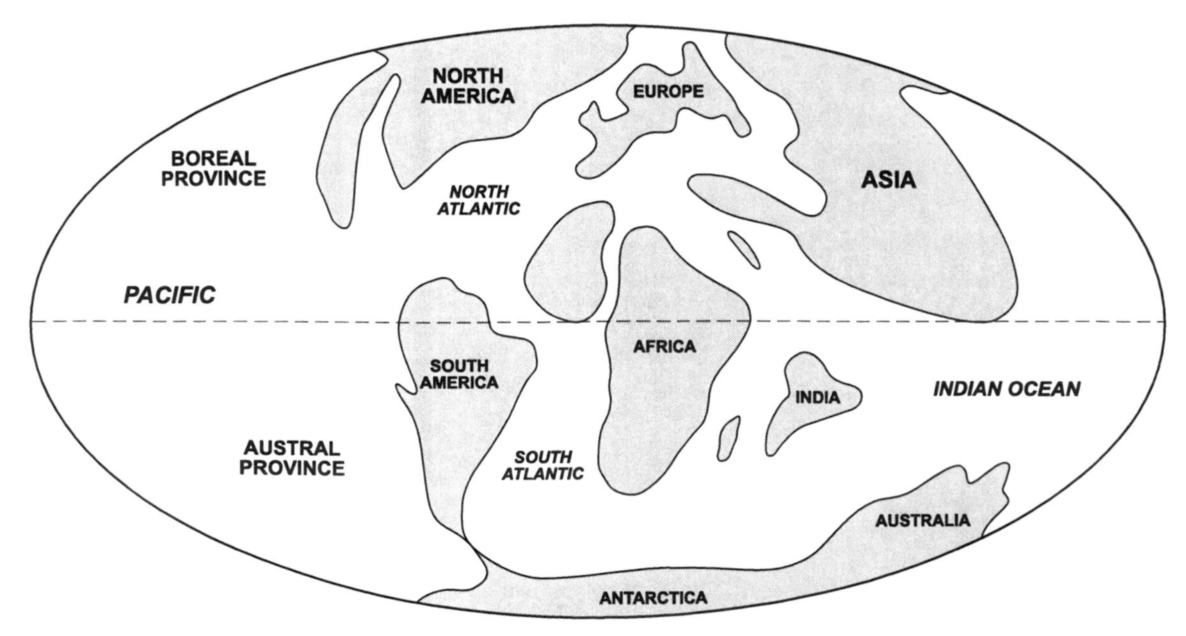

Fig. 4.10. Cretaceous palaeogeography and palaeobiogeography.

distinguishable by the presence, and the high-latitude Boreal and Austral provinces by the absence, of algal–coral–rudist reefs (the Tethyan province is also distinguished by the low abundance and diversity of foraminiferans, sponges, bivalves, ammonoids, crinoids and echinoids). During most of the Cretaceous, the western interior of the North American continent was covered by an epi-continental sea that did not interchange with the open oceans, because of its distance from them, and which therefore came to be characterised by only a few cosmopolitan taxa, and many more endemic ones. Plant and vertebrate distributions point toward the development of a single terrestrial palaeobiogeographic province throughout much of the Mesozoic, namely the Gondwana province. The plant pollen *Classopollis*, the fish *Mawsonia* and identical suites of reptiles are found throughout Africa and South America in the Triassic to Early Cretaceous. However, in the 'Middle' Cretaceous, provincialism became established, as the South Atlantic began to rift open between Africa and South America, and to form a barrier to the dispersal of terrestrial organisms. Also at this time, a number of sub-provinces became recognisable on the basis of plant and pollen distributions. Thus, during the Cretaceous, the configuration of the continents changed from prevailing Mesozoic pattern of two supercontinents (Laurasia and Gondwana) straddling an equatorial ocean (Tethys), to one with several continents separated by oceans, like that of today.

In terms of terrane reconstruction, palaeontological data indicate both north-west–south-east and south-east–north-west transportation of terranes in Mexico, along 'megashears'. Jurassic and Cretaceous foraminiferans and radiolarians from the Vizcaino Peninsula in Baja California Sur are of boreal aspect, and similar to those of the California Coast Ranges, Alaska and Japan to the north-west, indicating north-west–south-east transportation, along the 'Walper megashear' (Pessagno *et al.*, in Mann, 1999). In contrast, Middle Jurassic ammonites from Oaxaca and Guerrero states are of generally austral aspect, and are similar to those of the Andes, specifically Argentina, to the south-east, indicating south-east–north-west transportation, along the 'Cserna megashear' (Cantu-Chapa, in Bartolini *et al.*, 2001). Note, though, that *Stephanoceras*, also recorded in Oaxaca, is of undifferentiated eastern Pacific aspect. The presence of *Wagnericeras* and *Reineckia*, also of eastern Pacific aspect, recorded in south-eastern and eastern Mexico and in the Gulf of Mexico, indicates the opening of a marine connection from the eastern Pacific to the Gulf of Mexico in the Middle Jurassic (the 'Balsas portal' or 'Hispanic corridor' of authors, or the 'Intra-Pangean seaway' of Itturalde-Vinent, in Bartolini *et al.*, 2003).

4.4.3 Cenozoic

Many fossil groups appear to have had restricted or endemic biogeographic distributions in the Cenozoic (see also Sub-sections 5.5.7 and 5.5.8).

Vertebrate distributions point toward the palaeogeographic evolution of northern South America in the Oligo-Miocene (Diaz de Gamero, 1996; Sanchez-Villagra and Clack, 2004). The Oligocene to Middle or Late Miocene serrasalmine fish (including pacu and piranha), reptile (including gharial) and mammal (including megatheriid or mylodontid sloth) faunas of western Venezuela are of Orinoco basin palaeobiogeographic affinity, indicating that the palaeo-Orinoco discharged into the Caribbean, in western Venezuela, at this time (and only began to discharge into the Atlantic, in eastern Venezuela, following uplift of the Merida Andes in the Late(st) Miocene). Vertebrate and other distributions point toward the palaeogeographic evolution of the Caribbean in the Mio-Pliocene (Perfit and Williams, in Woods, 1989). The Miocene marine fish faunas of Venezuela are of Pacific affinity, and indicate a marine connection from the Pacific to the Caribbean at this time, through the Strait of Panama (Aguilera and Rodrigues de Aguilera, in Sanchez-Villagra and Clack, 2004). Foraminiferal evidence indicates that the Strait of Panama shoaled in the Pliocene, and implies that the Isthmus of Panama emerged (Duque Caro, 1990; McDougall, 1996). The evident severance of the marine connection through the Strait of Panama led to the separate evolution of Pacific and Caribbean bivalves and gastropods (Vermeij, 1978). The emergence of the Isthmus of Panama led to the interchange of North and South American terrestrial faunas and floras (Woodburne and Swisher, in Berggren *et al.*, 1995). Incidentally, it has also been interpreted as resulting initially in an intensification of the Gulf Stream, and global warming; and ultimately in an intensification of equatorial Atlantic summer storms (hurricanes), increased precipitation, global cooling and glaciation (Lear *et al.*, 2003).

Bivalve and gastropod distributions point toward the palaeogeographic evolution of the North Pacific in the Late Miocene–Pleistocene (Durham and McNeil, in Hopkins, 1967; Vermeij, 1991; Reid, 1996; Marincovich and Gladenkov, 1999; Gladenkov *et al.*, 2002). The dispersal of the bivalve *Astarte* from the Arctic into the North Pacific in the Late Miocene indicates that the Bering Strait was open at this time. The separate evolution of North Atlantic and North Pacific gastropods in the Pleistocene indicates that the Bering Strait was closed at this time.

4.5 Palaeoclimatology

Many fossil groups appear to have had restricted bathymetric and/or biogeographic, and hence temperature, distributions, rendering them of use in palaeotemperature and palaeoclimatic interpretation (alongside abiotic, lithologic indicators, such as reefal carbonates and coals, restricted to tropical latitudes and, by inference, palaeo-latitudes) (Spicer, in Briggs and Crowther, 1990; Crowley and North, 1991; Frakes *et al.*, 1992; Vrba *et al.*, 1995; Parrish, 1998; Bradley, 1999; Gerhard *et al.*, 2001). The exacting ecological requirements and tolerances of many dinoflagellate, diatom, foraminiferal, radiolarian, plant, ostracod, insect and mammal species, and their rapid response to changing environmental and climatic conditions, render them useful not only in palaeoclimatology, but also in climatostratigraphy, and in environmental archaeology (see also Section 7.6). Transfer functions based on their distributions are also useful (see Section 4.6).

Some fossil groups, most notably foraminiferans, brachiopods, belemnites, fish and mammals, are useful in the direct measurement rather than indirect interpretation of palaeotemperature or palaeoclimate, based on the oxygen isotopic record of changing temperature or ice volume that they preserve in their shells, bones or teeth. The theory is that the heavier isotope of oxygen, ^{18}O, is proportionately commoner in the atmosphere and oceans, and in the teeth, bones or shells of terrestrial and marine organisms, during glaciations, on account of differential evaporation and sequestration into ice build-ups of the lighter isotope, ^{16}O (Emiliani, 1955; Ericson and Wollin, 1956; Shackleton *et al.*, 1990). High-resolution palaeoclimate curves have been constructed on the basis of the marine oxygen isotope record that also serve, when calibrated against biostratigraphy, magnetostratigraphy or absolute chronostratigraphy, as the basis of a workable 'marine isotope stage' (MIS) or 'oxygen isotope stage' (OIS) climatostratigraphy, at least for the Tertiary and Quaternary (see Sub-section 6.3.1). Long-term climatic maxima are recorded in the Palaeocene–Middle Eocene, Early–early Middle Miocene and Early Pliocene, and long-term climatic minima in the Late Eocene–Oligocene, and late Middle–Late Miocene; medium- to short-term climatic minima in the glacials and stadials of the Late Pliocene to Pleistocene, and medium- to short-term climatic maxima in the interglacials and interstadials of the Late Pliocene to Pleistocene; and in the Holocene. According to power spectrum analyses, 50% of the variance in the palaeoclimate data is attributable to a 100-ka cycle, 25% to a 40-ka cycle associated with variation in the obliquity of the Earth's axis and 10% to a 20-ka cycle associated with variation in the orientation or precession in the Earth's axis (Hays *et al.*, 1976). These astrophysical phenomena had already previously been argued to have important consequences for insolation and climate (Milankovitch, 1938).

4.5.1 *Palaeozoic*

Foraminiferans, plants, corals, brachiopods, trilobites and fish are important in the palaeoclimatic interpretation of the Palaeozoic. Plant macrofossils and microfossils, and brachiopods, have recently been used in multidisciplinary studies of climate change in the Permian of Oman following the glaciation of Gondwana and the opening of the Neotethyan Ocean (Angiolini, in Brunton *et al.*, 2001; Angiolini *et al.*, 2003a, b; Berthelin *et al.*, 2003; Stephenson *et al.*, 2003a). Brachiopods from the early stages of the deglaciation are cold-adapted, while those from the later stages of the deglaciation are warm-adapted (Angiolini, in Brunton *et al.*, 2001; Angiolini *et al.*, 2003a, b). Plants from the later stages of the deglaciation are of essentially tropical – rain-forest – aspect (Berthelin *et al.*, 2003).

4.5.2 *Mesozoic*

Dinoflagellates, silicoflagellates, diatoms, calcareous nannofossils, calcareous algae, foraminiferans, radiolarians, plants, corals, brachiopods, bivalves, gastropods, belemnites, ostracods, insects, echinoids, fish, amphibians and reptiles are important in the palaeoclimatic interpretation of the Mesozoic.

Plants have recently been used in a multidisciplinary study of the Triassic Fremouw formation of Antarctica, deposited at a palaeo-latitude of 70–75°S (Cuneo *et al.*, 2003). Here, forest density, forest cover per hectare and mean separation of trees have all been inferred using various quantitative methods, and the overall nature of the forest community established by integrating this evidence with taphonomic and sedimentological information. The forest appears to have grown along river banks and in proximal flood-plain environments. The growing season, inferred from tree-ring analysis, appears to have been surprisingly long, and even frost-sensitive plants such as the cycad *Antarcticycas* were evidently able to survive the winter. These observations point toward a general climatic amelioration, as does the replacement of the *Glossopteris* flora of the Permian by the seasonally deciduous *Dicroidium* flora – and *Lystrosaurus* fauna – of the Triassic, although this may have been local rather than regional or global, and associated with the removal of an earlier orographic barrier to humid winds (a volcanic arc). Whatever the cause of the climatic amelioration, the actual empirical observations are difficult to reconcile with the extreme temperature ranges of previous models.

Belemnites are of particular use in palaeotemperature interpretation in the marine Jurassic–Cretaceous (Grocke *et al.*, 2003; McArthur *et al.*, 2004; Rosales *et al.*, 2004; Wierbowski, 2004). This is on account of their enhanced resistance to chemical change during diagenesis, and their consequent ability to preserve intact geochemical signals such as oxygen – and strontium – isotope ratios: indeed, the 'Pee Dee' belemnite standard is that against which such measurements are calibrated. The belemnite record from the Late Jurassic to Early Cretaceous indicates cooling in the middle Volgian followed by warming in the late Volgian to Berriasian–Ryazanian. The record from the Hauterivian indicates warming from 11 °C to 15 °C in the *Regale* Zone, followed by cooling to 11 °C in the *Inversum* Zone. The record from the Barremian indicates warming to 20 °C, followed by cooling to 14 °C – at the time of the onset of volcanism on the Ontong–Java plateau, in the *Elegans* Zone – and subsequent warming to 16 °C. Interestingly, although reflecting the same trends, the oxygen isotope ratio recorded by *Hibolites* is consistently more positive than that of *Acroteuthis*, possibly on account of vital or bathymetric effects (differing depth habitats). Importantly, in *Acroteuthis* and in *Aulacoteuthis*, although not in *Hibolites*, the oxygen isotope ratio is positively correlated with the content of the trace elements sodium, strontium and magnesium, implying that the latter can also be used as a proxy of palaeotemperature.

The planktonic foraminiferal record from the Late Cretaceous indicates sea-surface temperatures (SSTs) of as much as 30 °C or even more in the tropics in the Cenomanian–Turonian, while the benthic foraminiferal record indicates slightly lower sea-bottom temperatures, although not as low as those of today (Skelton, 2003). The high SSTs would probably have been associated with raised salinities in arid areas, especially in the central Atlantic and in Tethys. The high-density brines thus formed in these equatorial areas would have sunk and could have created a deep-sea circulation pattern driven by saline water, and hence quite different from that of today, driven by cool water. The evaporation in equatorial areas would also have brought about the transfer of significant amounts of water vapour to the atmosphere, and the transfer of heat towards the poles. This may in turn have brought about an increase in storm activity. Indeed, increased storm activity is indicated by increased wavelengths of storm-induced hummocky cross-stratification coincident with the temperature maximum in the Cenomanian–Turonian.

4.5.3 *Cenozoic*

Dinoflagellates, silicoflagellates, diatoms, calcareous nannofossils, calcareous algae, foraminiferans, radiolarians, plants, corals, brachiopods, bivalves, gastropods, ostracods, insects, echinoids, fish, amphibians, reptiles and mammals are important in the palaeoclimatic interpretation of the Cenozoic (see also Sub-sections 5.5.7 and 5.5.8).

4.6 Quantitative and other interpretive techniques in palaeobiology

Statistical databases are of use in documenting the history of life as represented in the fossil record (Benton, in Harper, 1999), and in describing and quantifying biotic change over time (Johnson and McCormick, in Harper, 1999; Markwick and Lupia, in Crame and Owen, 2002). Statistical techniques such as multivariate analysis, factor analysis and principal component analysis are of use in palaeoenvironmental interpretation. For example, hierarchical cluster analysis has been used in the differentiation of restricted and fully marine environments in the Miocene, Eggenburgian of Austria (Mandic and Steininger, 2003). General computer-assisted aspects of palaeoenvironmental interpretations and applications to petroleum exploration are described by Robinson and Kohl (1978) (Cenozoic, Gulf of Mexico) and Lesslar (1987) (Cenozoic, Sarawak).

4.6.1 Palaeobathymetry

The ratio of planktonic to benthic foraminiferans (P: B) can be used as a measure of bathymetry (Jones, 1996). However, the trend toward a higher ratio in deeper water is non-universal and commonly locally reversed, such that values are not necessarily unique to any particular bathymetry or bathymetric zone. For example, the ratio is anomalously low in deep waters off the mouths of major rivers, due to low planktonic productivity in the freshwater plume; on the upper slope, due to low predation and hence high benthic productivity in the oxygen minimum zone; and also on the abyssal plain, due to preferential dissolution of planktonics below the calcite compensation depth. The trend toward a higher ratio is also non-linear, such that the degree of bathymetric differentiation or resolution that it allows is variable.

Cross-plots of benthic foraminiferal 'morphogroups' allow the differentiation of bathymetric zones and/or sedimentary sub-environments (Jones, 1996). Abundance, diversity, dominance and equitability are also useful guides to bathymetry, environment and/or systems tract. Abundance, that is, the number of specimens per unit volume or weight of sample, and diversity, that is, most simply expressed, the number of species per sample, are typically highest in deep water and/or in maximum flooding surfaces. Note, though, that depth is only one of a number of variables controlling abundance and diversity. Note also that abundance and diversity are not always co-variant. High abundance associated with low diversity often indicates some form of environmental stress, such as temperature, salinity, oxicity or toxicity. High dominance, most simply expressed as relative abundance of dominant species, and low equitability, that is, evenness of distribution, also often indicate environmental stress. The biosequence stratigraphic utility of 'SHE' diversity analysis is discussed by Wakefield, in Olson and Leckie (2003).

In the oil industry, palaeobathymetric interpretation is becoming automated through the use of various proprietary data entry, manipulation and display software packages containing links to bathymetric distribution data-sets or 'look-up' tables. The interpretation is typically displayed in the form of a palaeobathymetric curve, capable of being imported into the work-station environment for integration with well log and seismic data. Excursions in the palaeobathymetric curve can be used alongside the well log and seismic data to identify facies dislocations and/or stratigraphic discontinuities. Running displays of correlation coefficients between samples can also be used to identify facies dislocations and/or stratigraphic discontinuities (Olson *et al.*, in Olson and Leckie, 2003).

Transfer functions based on the environmental distributions and tolerances of diatoms as well as foraminiferans have been used in sea-level reconstructions (see also Section 7.6).

4.6.2 Palaeobiogeography and palaeoclimatology

Transfer functions are also of use in palaeobiogeographic and palaeoclimatological interpretation. For example, transfer functions based on the environmental distributions and tolerances of terrestrial and marine plant macrofossils and microfossils (palynomorphs) have been used in establishing terrestrial and marine climate, hardiness, moisture needs and salinity tolerances. Aspects of plant leaf physiognomy and the percentage of entire-margined species have been used as proxies for mean annual temperature and mean annual range of temperature. Cladistic techniques have been used in the palaeobiogeographic interpretation of the Early Ordovician of Gondwana (Turvey, in Crame and Owen, 2002; see also Humphries and Parenti, 1999). GIS databases of fossil brachiopod and bivalve distributions have been used in the palaeobiogeographic interpretation of the Middle–Late Devonian (Rode and Liebermann, 2004). GIS databases of Recent non-avian tetrapod distributions have been used in the palaeoclimatic interpretation of the Middle Eocene Messel Lagerstatte in Germany (Markwick, in Crame and Owen, 2002).

5 • Key biological events in earth history

This chapter deals with the generalities and specifics of the evolutionary and extinction events and trends that have controlled past and present, and will control future, biodiversity on Earth.

Readers interested in further details of a general nature are referred to Nisbet (1991), van Andel (1994), Moore *et al.* (1996), Smith and Szathmary (1999), Stanley (1999), Wicander and Monroe (2000), Groombridge and Jenkins (2002), Conway Morris (2003), Southwood (2003), Dawkins (2004), Erwin, in Moya and Font (2004), Prothero and Dott (2004) and Cowen (in press).

5.1 Evolution and extinction

The history of life on Earth, insofar as it can be captured in words, can be said to have been one of general evolutionary diversification intermittently interrupted by mass extinction and recovery (Benton, 1995; Rosenzweig, 1995; Foote, in Erwin and Wing, 2000; Miller, in Erwin and Wing, 2000; Benton, in Briggs and Crowther, 2001; Groombridge and Jenkins, 2002). Certain of the difficulties in description arise from difficulties in quantifying diversity, in part in turn arising from uncertainties as to the representativeness of the fossil record, and from biases and inconsistencies in sampling and interpreting same (see Box 5.1). There are particular problems associated with taxonomic bias and inconsistency, that is, the definition of taxa across different fossil groups. The two main models that have been proposed to date for the evolution of diversity are the equilibrium model, which assumes – logistic – diversification only up to and not beyond a critical point, and the expansion model, which assumes continuous – either additive or exponential – diversification (Benton, in Briggs and Crowther, 2001).

5.1.1 *Evolution*

This is not the appropriate forum for a detailed discussion of the complex and contentious subject of evolution, that is, the process whereby species and higher-level taxa evolve – or are transformed – into and replaced by others. Suffice it to say here that empirical observations of the fossil record should leave little doubt, even in the mind of the most ardent creationist, that evolution has manifestly evidently been taking place throughout geological time. The difficulties that sceptics have in understanding and accepting evolution perhaps arise in part from the fact that the process cannot actually be observed, other than under artificial laboratory conditions, as it occurs over generations (Lazarus, in Briggs and Crowther, 2001).

The means or mode of evolution is 'natural selection' (Wallace, 1858; Darwin, 1859). Put simply, if not oversimplistically, natural selection acts through some form of pressure, such as environmental pressure, on a naturally varying population. Over generations, this pressure selects for those individuals that possess characters that are both beneficial and heritable, and against those that do not; and ultimately results in a new set of characters, and a new species. The currently most widely accepted version of evolutionary theory, 'Neodarwinism', has replaced Darwin's original concept of 'blending' inheritance with Mendel's concept of 'particulate' or 'genetic' inheritance. (Thankfully, the so-called 'Social Darwinism' of the late nineteenth century and related 'eugenics' of the early twentieth, with their connotations of class and racial superiority, are now recognised as among the worst ideas on the dark and lamentable catalogue of human history.) It is now believed that the process is facilitated by the reproductive isolation of parts of the population (at least on the case of so-called allopatric speciation). Such isolation is most commonly brought about by a geographical barrier, such as a rising mountain range, or a propagating rift, or a rising sea level; or by a climatic change, and associated changes to distribution fronts.

The timing or tempo of evolution remains disputed even within the scientific community. To me, though, the two main models that have been proposed, the progressive 'phyletic gradualism' model attributable to Darwin, although not named as such by him, and the effectively – within the limits of temporal resolution – instantaneous 'punctuated equilibrium' model of Eldredge and Gould, in Schopf (1972) are not mutually exclusive, as some have argued. Either can be pre-eminent at a given place at a given time; the other at the same place but at a different time, or at the same time but at a different place.

Readers interested in details of the scientific as against the religious view of evolutionary biology, or in the products as against the processes thereof, are referred to the recent works of Skelton (1993), Osborne and Benton (1996), Futuyma (1998), Page and Holmes (1998), Patterson (1999), Stearns and Hoekstra (2000), Carroll *et al.* (2001), Dixon *et al.* (2001), Levinton (2001), Nielsen (2001), Pigliucci (2002), Rothschild and Lister (2003) and Moya and Font (2004), or to the numerous popularised accounts, notably those provided by Richard Dawkins or the late Stephen Jay Gould. Those interested in the specifics of individual (macro-)evolutionary diversification events are referred to the appropriate sections below.

Evolutionary events

Key evolutionary events in the history of life include: the origin of life (prokaryotes); the evolution of complex life (eukaryotes); the evolution of multicellularity; the evolution of mineralised skeletons, and the Cambrian evolutionary diversification (the 'Cambrian explosion'); the evolution of reefs; the Ordovician evolutionary diversification; the evolution of vertebrates; the evolution of life on land; the evolution of trees and forests; the evolution of flight; the Mesozoic evolutionary diversification; the evolution of flowering plants; the evolution of grasses and of grassland animals; and the evolution of humans. Each of these events is dealt with, in turn, in its stratigraphic context, below.

Evolutionary biotas

Evolutionary biotas are sets of higher taxa that have similar histories of diversification, dominance and decline over geological time (Fig. 5.1) (Sepkoski, 1981). The principal three, as recognised on the basis of factor analysis, are the Cambrian, the Palaeozoic, and the Mesozoic–Cenozoic – or modern – evolutionary biotas. The succession of these evolutionary biotas records long-term change in the nature and diversity of life on Earth.

'Ecological evolutionary units' are somewhat different in nature, as they represent reorganisation

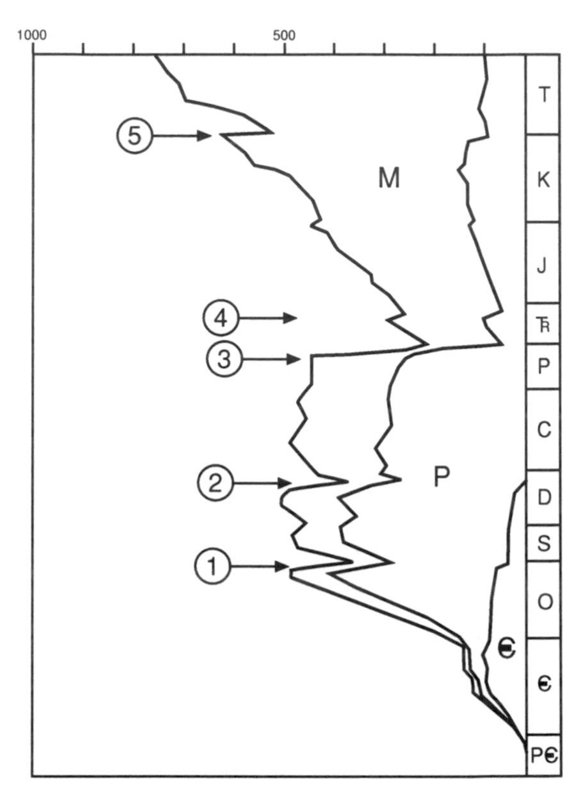

Fig. 5.1. Evolutionarybiotas and mass extinctions. (Modified after Sepkoski, 1981.) C, Cambrian evolutionary fauna; P, Palaeozoic evolutionary faura; M, Mesozoic evolutionary fauna. Units on horizontal axis are millions of years; units on vertical axis are numbers of families ①–⑤ denote mass extinction events.

on the timescale of a few million years rather than several tens or hundreds of millions of years. They appear to represent short-term adjustments to mass extinction events (see below).

The Cambrian evolutionary biota

The Cambrian evolutionary biota includes 'small shelly fossils', inarticulate brachiopods, trilobites, eocrinoids and other higher taxa. It appeared in the Precambrian, Vendian and ranges through to the Recent. However, it was most important in the Cambrian, representing the principal constituent of the evolutionary diversification or explosion of that time, and has been in decline thereafter, the decline accentuated by the End-Ordovician and Late Devonian mass extinctions.

The Palaeozoic evolutionary biota

The Palaeozoic evolutionary biota comprises articulate brachiopods, anthozoans, cephalopods, crinoids, graptolites and other higher taxa. It appeared in the Cambrian and ranges through to the Recent, although it was most important in the Ordovician–Devonian, representing the principal constituent of the evolutionary diversification of that time, and has been in decline thereafter, the decline accentuated by the rise of the Mesozoic–Cenozoic biota in the Carboniferous and Permian, and by the End-Permian mass extinction. The Palaeozoic fauna of the terrestrial realm includes the primitive labyrinthodonts, anapsids and synapsids that dominated the Devonian–Permian (Benton, 1985). The Palaeozoic flora of the terrestrial realm includes the early vascular plants that dominated the Silurian–Devonian and the pteridophytes that dominated the Carboniferous–Permian (Niklas *et al.*, 1983).

The Mesozoic–Cenozoic or modern evolutionary biota

The Mesozoic–Cenozoic or modern evolutionary biota comprises bivalves, gastropods, echinoids, vertebrates and other higher taxa. It appeared in the Palaeozoic and ranges through to the present day, and is most important in the Post-Palaeozoic, representing the principal constituent of the evolutionary diversification of that time. The Mesozoic–Cenozoic fauna of the terrestrial realm includes the dinosaurs and pterosaurs that dominated the Mesozoic and died out at the time of the End-Cretaceous mass extinction, and the reptiles, birds and mammals that have dominated the Cenozoic (Benton, 1985). The Mesozoic–Cenozoic flora of the terrestrial realm includes the gymnosperms or seed-plants which dominated the Jurassic and the angiosperms or flowering plants which have dominated the Cretaceous–Recent (Niklas *et al.*, 1983).

5.1.2 *Extinction*

Background extinction, like evolution, has evidently been taking place throughout geological time, although mean rates are similarly difficult to quantify (Lawton and May, 1995).

Interestingly, through time, there appear to have been a number of mass extinction events, during which extinction rates were significantly elevated above the background level. The effects of these events were severe, and they were experienced over short time-frames and over wide geographical areas. Indeed, they were apparently essentially global and non-selective (but see also below).

Readers interested in further details of extinction are referred to, for example, the recent work of MacLeod, in Rothschild and Lister (2003). Those interested in the generalities of mass extinction events are referred to Sharpton and Ward (1990), Glen (1994), Jablonski, in Lawton and May (1995), Hart (1996), Hallam and Wignall (1997), Courtillot (1999), Ward (2000), Koeberl and MacLeod (2002) and Taylor (2004). Those interested in the specifics of individual mass extinction events are referred to the appropriate sections below.

Mass extinction events

Major mass extinction events include the Late Precambrian, Early Cambrian, Late Cambrian, End-Ordovician, Late Devonian, End-Permian, End-Triassic, Late Cretaceous (Cenomanian–Turonian), End-Cretaceous, End-Palaeocene, End-Eocene, Pleistocene and Holocene events, which are discussed in detail, in context, below. (Comparatively minor mass extinction events include the Early Jurassic, Middle Jurassic, End-Jurassic, Early Cretaceous, Oligocene, Middle Miocene and Pliocene events, which are not discussed).

The End-Ordovician, Late Devonian, End-Permian, End-Triassic and End-Cretaceous mass extinction events constitute the 'Big Five' of Raup and Sepkoski (1982) and other authors (Fig. 5.1). The 'Big Five' appear to differ from other mass extinction events, and indeed from background extinctions, only in degree, and not in kind (MacLeod, in Rothschild and Lister, 2003). All types of extinction can therefore be interpreted as the effects of the same types of underlying cause, 'mediated through the waxing and waning of . . . ecological hierarchies' (MacLeod, in Rothschild and Lister, 2003). Interestingly in this context, there is a statistically significant relationship between the intensity of extinction and both sea-level fall, from the Cambrian to the Permian, and flood-basalt eruption, from the Permian to Recent. Also interestingly, there has been a statistically significant decrease in the intensity of extinction from at least the Permian to Recent. This mirrors a concomitant increase not only in the delivery of nutrients to shallow marine habitats, as measured by various geochemical proxies, but also in the number of recycler niches and in the length of food chains – which would appear therefore to have had some sort of buffering effect.

Interestingly, the taxonomic and ecological impacts of the 'Big Five' mass extinction events were apparently at least to an extent decoupled (Droser *et al.*, 2000; McGhee *et al.*, 2004). The End-Ordovician mass extinction was extremely significant in terms of taxonomic impact, but not in terms of ecological impact. Conversely, the End-Cretaceous mass extinction was comparatively insignificant, leastwise the least significant of the 'Big Five', in terms of taxonomic impact, but the second most significant after the End-Permian mass extinction in terms of ecological impact. It seems that the component species and structure of an ecosystem are at least as important as diversity in maintaining its integrity, and that the loss of those species with the highest ecological value can lead to ecological crises.

Effects of mass extinction events

Selectivity. There is evidence that the effects of some mass extinction events were selective (McKinney, in Briggs and Crowther, 2001). Large species appear to have been disproportionately susceptible to mass extinction, as do species or communities occupying restricted geographical ranges or specialised ecological niches, and those exhibiting low or unstable population densities (see Box 5.1). Low-diversity communities also appear to have been disproportionately susceptible, leading to the suggestion that high diversity might somehow buffer the effects of the events.

Recovery. Kauffman and Harries have proposed a general model for recovery following mass extinction events, in which the aftermath is characterised by an initial survival interval and a later recovery interval proper (Kauffman and Harries, in Hart, 1996; Bottjer, in Briggs and Crowther, 2001). Interestingly, there are certain similarities between the recovery following mass extinction events and the succession following non-lethal ecological disturbance (Sole *et al.*, 2002). The survival interval is characterised initially by blooms of 'disaster' species and of 'opportunistic' species, and later by increases of preadapted survivors and ecological generalists, and, importantly, by the evolution and/or radiation of crisis-adapted 'progenitor taxa'. The recovery interval proper is characterised by further evolutionary diversification, initially within surviving and newly established crisis progenitor lineages, and later within new lineages. It is also characterised by the reappearance of so-called 'Lazarus taxa' from refugia.

Recovery time appears approximately proportional to the percentage loss sustained (see Box 5.1). Interestingly in this context, although it took 10 Ma for diversity to stabilise after the End-Cretaceous mass extinction event, the initial recovery was actually to a higher level, indicating some kind of rebound effect (of the sort that financial analysts are wont to term a 'dead cat bounce'). Note in this context that recovery time appears to vary across trophic groups, with primary producers (re)appearing first, followed by secondary consumers, that is, herbivores and carnivores. Detritivores appear able to survive even drastic declines in primary production. Note also that recovery time appears to vary across entire ecosystems. For example, recovery following the End-Permian mass extinction was effectively geologically instantaneous in terrestrial but not in marine ecosystems.

Causes of mass extinction events

Intrinsic, Earth-bound causes that have been invoked as responsible for mass extinction events include volcanism, sea-level change and associated anoxia, and climate (Hallam, in Briggs and Crowther, 1990). Importantly, there appears to be some interrelatedness between at least volcanism, sea-level change and climate. Sea-floor spreading affects sea level and hence weathering, consumption of the greenhouse gas carbon dioxide, and greenhouse-driven climate. Interestingly, biotic forcing could also be capable of affecting climate, as suggested by Lovelock's 'Gaia' hypothesis. For example, excessively high phytoplankton productivity could result in sufficient abstraction of carbon dioxide from the atmosphere to cause global cooling. Similarly, low phytoplankton productivity could result in sufficient build-up of carbon dioxide in the atmosphere to cause global warming.

Extrinsic, extra-terrestrial causes include bolide impacts and associated effects such as darkness and cooling ('impact winter'), greenhouse warming, acid rain etc. (Jablonski, in Briggs and Crowther, 1990). Evidence for extra-terrestrial forcing comes from cratering, iridium anomalies, shocked quartz, microtektites (melt-glass) etc.

Periodicity of mass extinction events

Raup and Sepkoski (1984) have proposed that there is a 26 Ma periodicity associated with mass extinction events (see also Sepkoski, in Briggs and Crowther, 1990). The proposed periodicity has been elegantly explained by extrinsic, extra-terrestrial causes, according to the 'Nemesis' and 'Planet X' hypotheses. The former hypothesis involves the existence of a dim binary companion star to the sun, Nemesis, and the latter, a tenth planet in the solar system, 'Planet X', both with eccentric orbits, which would cause passage through the Oort cloud and a comet shower every 26–28 Ma. Note, though, that the existence of Nemesis has never been proven, and that simulations suggest that if it did exist it would probably be unstable and easily thrown off course. Note also that the existence of 'Planet X' has never been proven either, and that even if it did exist it might not have sufficient mass to cause a comet shower. The proposed periodicity associated with mass extinction events has also been explained by intrinsic causes, such as dynamic instability in the mantle – and associated increased mantle plume activity – on a periodicity of approximately 30 Ma (coincident with that of magnetic-field reversal).

Many authors have questioned whether there is actually any periodicity associated with mass extinction events. For example, Whatley, in Whatley and Maybury (1990), states: 'The present author's data for the extinction of Mesozoic Ostracoda . . . clearly indicates . . . intervals between . . . Jurassic extinction peaks . . . very close to Raup and Sepkoski's 26 MY interval . . .' However, he also adds that 'In the Cretaceous, the interval is less regular', and that 'no . . . clear support is forthcoming for the Cainozoic data'. He concludes that 'No doubt the data could be "massaged" in order to improve . . . fit to existing dogma. The author, however, prefers to leave this task to a professional data manipulator.' MacLeod, in Rothschild and Lister (2003) has also eloquently argued that there is no periodic component to mass extinction events. His Monte Carlo simulations of Phanerozoic extinction data – and comparison of the simulation results with the Fourier transformation of an extinction intensity time series – appear to demonstrate that the proposed 26 Ma periodicity is not statistically significant.

Box 5.1 Foraminiferal diversity trends through time

The fossil record of the foraminiferans is sufficiently good, and sufficiently well documented, to allow detailed observations to be made on the evolution and control of the diversity of the group through time, some of which may be of more general applicability (author's unpublished observations).

Foraminiferans

Foraminiferans are a class of sarcodine protists characterised by granulo-reticulose protoplasm. Most known modern forms are also characterised by agglutinated arenaceous or secreted calcareous shells or tests (the soft-bodied allogromiids excepted). Ancient – preservable – forms have an extensive and comparatively well-documented fossil record, extending from latest Precambrian–earliest Cambrian – in the case of *Platysolenites* – to Recent. Foraminifera constitute part of the Mesozoic–Cenozoic or modern evolutionary biota of Sepkoski (1981). Little cladistic work has been done on the group.

The principal suprageneric classification schemes are those of Haynes (1981) and Loeblich and Tappan (1987) (see also Tappan and Loeblich, 1988). Haynes's scheme recognises 99 families; Loeblich and Tappan's 288 (in part because Loeblich and Tappan tended to apply a familial – and in some cases even higher – level weighting to what Haynes regarded as generic- and lower-level classificational criteria such as details of wall structure).

Diversity trends through time

Familial-level diversity data from a database based on Haynes's scheme has been plotted against time, enabling observations on trends through time

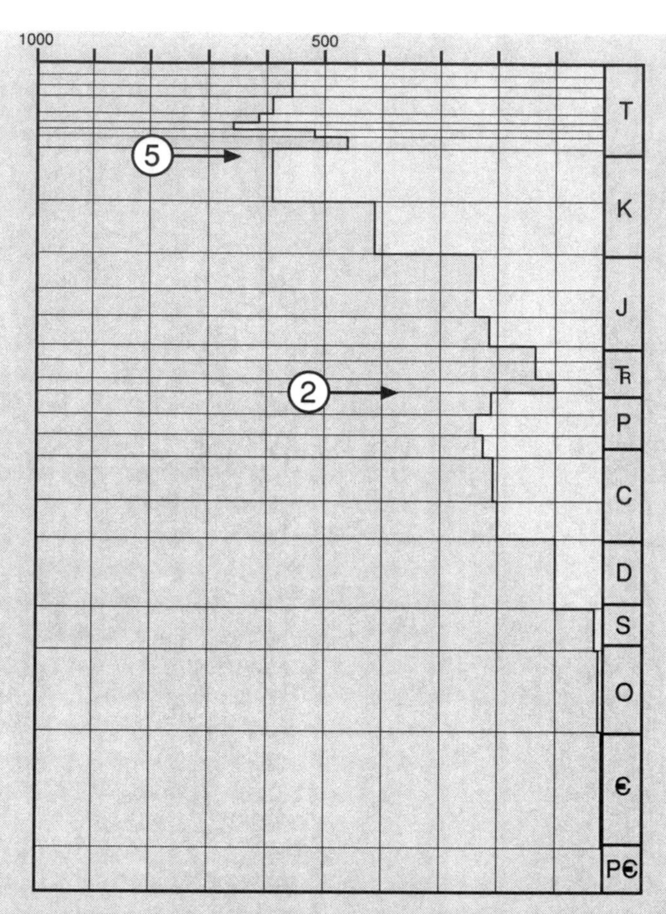

Fig. 5.2. Foraminiferal diversity trends through time. Units on horizontal axis are millions of years; units on vertical axis are numbers of families. ②, End-Permian mass extinction event; ⑤ K/T mass extinction event.

(Fig. 5.2). Diversity data from Loeblich and Tappan's scheme has also been plotted for comparative purposes. 'Taxonomic bias' is evident only in the case of the apparent End-Carboniferous event in the Loeblich and Tappan data (when a large number of monogeneric or oligogeneric 'families' died out), and other than in this case can be effectively eliminated. Other potential sources of bias remain, including preservational bias (that to do with the representativeness of the fossil record), sampling bias (palaeontologist interest) and interpretive bias (lack of normalisation of studied time intervals) (Jones, in Agusti *et al.*, 1999; Smith, in Briggs and Crowther, 2001).

Palaeozoic

Empirical observations indicate that foraminiferans underwent a slow expansion of diversity in the Early Palaeozoic, followed by a rapid expansion in the Carboniferous, especially of the specialist shallow marine larger benthic fusulinides. Then 64% of all families were wiped out at the time of the End-Permian mass extinction (one of the 'Big Five' of Raup and Sepkoski, 1982), all of them fusulinides. Recovery of pre-extinction diversity took until the Jurassic.

Mesozoic

There was another period of slow expansion in the Jurassic, followed by rapid expansion in the Cretaceous, especially of the specialist shallow marine larger benthic rotaliides and the planktonic globigerinides. Then 19% of all families were wiped out at the time of the End-Cretaceous mass extinction (another of the 'Big Five'), all of them rotaliides, which lost a total of 43% of all of their families, and globigerinides, which lost a total of 86% of all of their families. Many infaunal taxa underwent extinctions or at least 'Lazarus' extinctions over the boundary section, interpreted as indicating a collapse in primary productivity (Peryt *et al.*, in Bubik and Kaminski, 2004). Recovery of pre-extinction diversity took until the Eocene.

Cenozoic

According to data based on Haynes's scheme, diversity attained an all-time maximum in the Eocene, which, coincidentally or otherwise, was characterised by a climatic optimum and sea-level high-stand, and has been declining over the time interval from the Oligocene to the present, which has been characterised by generally deteriorating climate and falling sea level. In detail, 13% of families have become extinct between the Eocene and the present time, all of them specialist larger benthic rotaliids and globigerinides. (According to data based on Loeblich and Tappan's scheme, diversity attained a maximum in the Holocene. This is interpreted as representing preservational bias, as reflecting the poor preservational potential of some of the features Loeblich and Tappan used to distinguish some of their families.)

Observations on foraminiferal diversity trends

The rapid expansion of the specialist shallow marine larger benthic fusulinides in the Carboniferous may have been on account of the occupation of niches vacated by stromatoporoids and tabulate corals during the Late Devonian mass extinction. That of the specialist shallow marine larger benthic rotaliides and planktonic globigerinides in the Cretaceous may have been on account of the occupation of niches newly created by the sea-level rise of the time. (Note also in this context that the diversity of the globigerinides has been independently interpreted as exhibiting cyclicity paralleling coeval cyclic changes in the environment, such as sea-level change, perturbed by intermittent extinction events: Prokoph *et al.*, 2004.)

It appears that specialist groups that have just undergone rapid expansion suffer disproportionately in mass extinction events. This can be interpreted as indicating either that rapid expansion is unsustainable and an unsuccessful life strategy in the longer term, or that it is the specialisation itself that is ultimately unsuccessful. Interestingly in this context, many if not all modern specialist LBF appear to be essentially extreme *K*-strategists as opposed to *r*-strategists (Pianka, 1970), characterised by a facultative or obligate symbiotic relationship with photosynthetic algae, a degree of dependence rendering them extremely vulnerable to mass extinction during any change of conditions or relationships (Brasier, in Bosence and Allison, 1995). Analogy with their living counterparts indicates that many, if not all, fossil LBFs also harboured symbionts (as does carbon isotope evidence).

Observations of general applicability

The observation that recovery time following a mass extinction is proportional to the percentage loss sustained may apply more generally (see also Erwin, 2001). Note, though, that the recovery from the End-Triassic mass extinction appears to have been unusually slow, possibly on account of the low diversity going in to the event, in turn inherited from the End-Permian mass extinction.

The observation that specialist groups suffer disproportionately – 'selectivity by trait' – may also apply more generally (see also Erwin, 1998; McKinney, in Briggs and Crowther, 2001). It may even apply to entire communities of specialist organisms, such as reef communities. This may be on account of limited spatial distribution (Jablonski and Raup, 1995; Jablonski, in Rothschild and Lister, 2003) or population density (Brown, 1995).

5.2 Proterozoic

5.2.1 *The origin of life (prokaryotes)*

The most widely accepted hypothesis is that life originated spontaneously through the natural biochemical synthesis of complex organic compounds under ambient conditions (Miller, 1953; Knoll, in Briggs and Crowther, 1990; Woese and Wachterhauser, in Briggs and Crowther, 1990; Miller, in Schopf, 1992; Schopf, 1993; Bengtson, 1994; Brack, 1999; Dyson,

1999; Schopf, 1999; Fry, 2000; Lazcano, in Briggs and Crowther, 2001; Fenchel, 2002; Knoll, 2003; Raven and Skene, in Rothschild and Lister, 2003). Importantly, a series of elegant laboratory experiments conducted by Stanley Miller in the 1950s demonstrated that at least amino acids could indeed be spontaneously generated under simulated Precambrian, Hadean oceanic and atmospheric conditions, although the precise mechanism by which these proto-proteins aggregated into actual life forms remains unclear.

Interesting alternative hypotheses are that life originated in hydrothermal vent systems or similarly extreme environments (Nisbet and Sleep, in Rothschild and Lister, 2003), or through the introduction of organic compounds in carbonaceous chondrites of extra-terrestrial origin (Horneck, in Rothschild and Lister, 2003).

Isotopic signals arguably indicating the existence of life have been recorded from as long ago as 3800 Ma (Mojzsis *et al.*, 1996). However, incontrovertible fossils first appeared in the rock record, in the form of prokaryotic bacteria, in rocks dating to 3200 Ma (Schopf, in Schopf, 1992; Rasmussen, 2000). At least some primitive prokaryotes apparently subsisted by photosynthesising. Importantly, this process in turn produced free oxygen, which ultimately made the surface of the planet habitable by other groups of organisms for the first time (Lenton, in Rothschild and Lister, 2003; Nisbet and Sleep, in Rothschild and Lister, 2003). Incidentally, oxygenation also resulted in the disappearance of the previously widespread 'banded ironstone formations' (BIFs), formed only under reducing conditions.

5.2.2 *The evolution of complex life (eukaryotes)*

The most widely accepted hypothesis is that eukaryotes evolved through the development of a symbiotic relationship between two or more different types of prokaryotes, or by the incidental – and non-lethal – ingestion of one or more, by another (Sagan, 1967; Khakhina, 1992; Dyer and Oban, 1994; Moreira and Lopez-Garcia, 1998; Hartman and Fererov, 2002; Knoll, 2003; Lake *et al.*, in Rothschild and Lister, 2003; but see also Martin and Muller, 1998). Eukaryotes differ from prokaryotes in their possession of membrane-bounded nuclei and organelles, and in their capability to reproduce sexually, and correspondingly greater capacity to mutate and to evolve.

Eukaryotes first appeared in the rock record, in the form of sphaeromorph acritarchs, in rocks dating to the Palaeoproterozoic, 1600–1800 Ma (Javaux *et al.*, 2004). Molecular evidence indicates that they might have actually evolved as long ago as 2800 Ma.

5.2.3 *The evolution of multicellularity*

Multicellular organisms or metazoans are interpreted as having evolved through the aggregation of multiple eukaryotic unicells, and their organisation into different organs and tissue types with different functions (Glaessner, 1984; Fedonkin, in Briggs and Crowther, 1990; McMenamin and McMenamin, 1990; Lipps and Signor, 1992; Runnegar, in Schopf, 1992; Fortey *et al.*, 1996; Knoll and Carroll, 1999; Smith, 1999; Erwin, in Briggs and Crowther, 2001; Knoll, 2003). (Interestingly, multicellularity also appears to have arisen independently in a number of evolutionary lineages.)

Cladistic analysis incorporating molecular data clearly indicates a divergence at least as long ago as 1000 Ma (Wray *et al.*, 1996; Erwin, in Briggs and Crowther, 2001).

Direct fossil evidence for the existence of metazoans from as long ago as this is generally lacking, although annulated filamentous body fossils of at least arguable metazoan origin have been recorded from rocks as old as 700–800 Ma in China, and trace fossils of at least arguable metazoan origin from rocks as old as 1000 Ma in India and in the Grand Canyon (Erwin, in Briggs and Crowther, 2001). In contrast, there is abundant fossil evidence for the existence of a diverse array of metazoans in the so-called 'Precambrian surge' of the Vendian, from approximately 650 Ma, most notably in the form of the 'ediacarians' and certain 'small shelly fossils', but also notably in the form of the recently discovered enigmatic early metazoans of the Doushantou formation of China, dated to 570–580 Ma (Erwin, in Briggs and Crowther, 2001). The oldest 'ediacarians' thus far known are from the Ediacaran, immediately above Varangerian tillites dated to 620–650 Ma (Fedonkin, in Briggs and Crowther, 1990). Incidentally, it has been hypothesised that certain metazoans could have sought refuge at deep marine hydrothermal vent sites during the Varangerian glaciation, and only emerged to recolonise shallow marine environments in the Cambrian, contributing to an apparent rather than real evolutionary diversification (see Steiner *et al.*, 2001; see also below).

The abundance and diversity of the 'ediacarian' metazoans already in the Vendian is arguably, albeit indirectly, indicative of a still older origin, in the Riphean.

Another indication that this was the case comes from the observation that stromatolites began to decline in the Riphean, from around 1000 Ma, which some authors have suggested was due to an early rise of 'ediacarians', and associated excessive grazing activity (Walter and Heys, 1985).

Note, though, that some other authors have hypothesised that the rise of the 'ediacarians' and decline of the stromatolites at this time was brought about by environmental change associated with a series of glaciations, ultimately resulting in a so-called 'snowball Earth' in the Sturtian (Harland and Rudwick, 1964; Walker, 2003). Incidentally in this context, recent calculations have shown that weathering of the Laurentian flood basalts extruded in the Riphean could have resulted in sufficient consumption of greenhouse carbon dioxide to cause global cooling and initiate a 'snowball' glaciation.

5.2.4 The Late Precambrian mass extinction

There was a mass extinction at or near the end of the Vendian, at approximately 650 Ma, although it is fair to say that the precise timing and indeed the severity of the event remain somewhat poorly constrained (McMenamin, in Briggs and Crowther, 1990).

The principal groups to have been affected were the stromatolites, the acritarchs and the 'ediacarians'. A number of ichnogenera also became extinct.

In the case of the stromatolites, the supposed mass extinction may simply have been the last in a long line of extinction events dating back to the Riphean (Conway Morris, in Briggs and Crowther, 1990; McMenamin, in Briggs and Crowther, 1990). Various causal mechanisms have been proposed for this longer-term decline in stromatolites, including excessive grazing by early 'ediacarians', and environmental change associated with a series of glaciations (see above).

In the case of the acritarchs, there appears to have been a genuine intra-Vendian extinction event and turnover, with taxa such as *Trachysphaeridium laufeldi* and *Kildinella lophostriata* disappearing, and taxa such as the chlorococcocean? *Bavlinella foveolata* and the vendotaenids appearing (Vidal and Knoll, 1982; Vidal and Moczydlowska Vidal, 1997). The recovery of the acritarchs from this event appears to have taken until the Cambrian.

In the case of the 'ediacarans', some authors have interpreted an intra-Vendian extinction event and turnover, coincident in time with that affecting the acritarchs, citing as evidence the apparent absence of taxa such as *Charnia* and *Charniodiscus*, from the Kotlin Horizon of the Russian Platform (Fedonkin, 1987). However, other authors have argued that the extinction of the 'ediacarians' was at the end of the Vendian (Bowring and Erwin, 1998). Moreover, still other authors have argued that the apparent disappearance of the 'ediacarians' at the end of the Vendian is essentially an artefact of poor preservation, and that in fact the group actually ranged through to the Cambrian, an argument seemingly supported by the recent discovery of 'ediacarian'-like forms in Cambrian Lagerstatten.

5.3 Palaeozoic

Note that there was an additional, minor, evolutionary diversification event in the Devonian, possibly initiated by the creation of new niches during the transgression of the time. Brachiopods, ammonoids, trilobites and crinoids all evolved or diversified (House, 2002).

There was also a minor extinction event in the Silurian, Wenlock of the east European platform in Baltica, resulting in the stepwise extinction of graptolites in the *Lundgreni*, *Testis* and *Flemingii–Dubius* Zones (Porebska *et al.*, 2004). Extinction, survival and recovery phases have been documented. Only *Pristiograptus dubius* survived the combined extinction events, of which the most severe was in the *Testis* Zone. It was joined in the survival phase by the opportunistic *Gothograptus nassa* and *Pristiograptus parvus*, and in the recovery phase by a number of new species. Phytoplankton abundance and diversity patterns, total organic carbon (TOC) fluctuations and strong positive carbon and oxygen isotope anomalies indicate that the productivity regime was unstable during the extinctions in the *Testis* and *Flemingii–Dubius* Zones. There appears to be a relationship between the extinctions and eutrophication and anoxia associated with the prolonged global sea-level high-stand of the time, and/or with enhanced Rheic Ocean upwelling.

5.3.1 *The Cambrian evolutionary diversification*

There appears to have been a major evolutionary diversification in the earliest Cambrian, Nemakit-Daldynian, the so-called 'Cambrian evolutionary diversification' or 'Cambrian explosion' (Conway Morris, in Briggs and Crowther, 1990; Runnegar and Bengtson, in Briggs and Crowther, 1990; Runnegar, in Schopf, 1992; Brasier, in Walliser, 1995; Valentine *et al.*, 1999; Conway Morris, in Briggs and Crowther, 2001; Zhuravlev and Riding, 2001; Kowalewski and Kelly, 2002; Kelly *et al.*, 2003; A. Parker, 2003).

Note, though, that much of the observed diversification is of organisms with mineralised skeletons with enhanced preservation potential, and as such may represent an apparent – preservational – rather than a real phenomenon. Incidentally, the process of mineralisation appears to have been facilitated by a change in the chemistry of the ocean, forcing organisms to ingest and excrete increased quantities of minerals, and/or an increase in the amount of oxygen in the atmosphere, enabling them to precipitate minerals more easily (Cook and Shergold, 1984).

Interestingly, the mineralisation has been interpreted as an evolutionary response to selection pressure exerted by increased predation, which has in turn been interpreted by some authors as associated with the evolution of the eye (A. Parker, 2003). Importantly in this regard, Cambrian biotas appear to have included, for the first time, representatives of all the known trophic types, including primary producers (phytoplanktonic acritarchs), secondary consumers (zooplanktonic filter-feeding arthropods), and tertiary and higher-level consumers (benthic filter-feeding hyolithids, benthic priapulids, nektobenthic trilobites and nektonic anomalocarids) (Butterfield, in Briggs and Crowther, 2001). In contrast, Precambrian populations of the 'small shelly fossil' *Cloudina* from China exhibit only occasional instances of boring predation (Bengtson, in Bengtson, 1994).

The (co-)evolution of predator–prey systems, whereby prey and predator orgamisms evolve ever-better defensive and offensive strategies respectively, became a characteristic feature not only of the Cambrian, but also of the Phanerozoic as a whole (Kelley *et al.*, 2003).

5.3.2 *The evolution of reefs*

Reefs are more or less rigorously defined as resistant organic frameworks forming raised relief on the sea-floor (Rosen, in Briggs and Crowther, 1990; Scrutton, in Briggs and Crowther, 1990; Wood, 1999; Wood, in Briggs and Crowther, 2001). (Interestingly, the alternative usage of the word is as a danger to shipping.) At the present time, reefs are essentially restricted to shallow waters in low latitudes, as indeed they appear to have been throughout their geological history (Kiessling *et al.*, 1999; Kiessling *et al.*, in Insalaco *et al.*, 2000; Kiessling, 2001; Stanley, 2001; Kiessling *et al.*, 2002; Kuznetsov, in Zempolich and Cook, 2002). They are important reservoirs of biodiversity.

Reefs range from the Early Cambrian to the present time. Different types of organisms have contributed to the construction of reefs over time, as plate configurations and climates have changed, and as evolution and extinction have proceeded (Rigby, 1971; Heckel, 1974; James, in Scholle *et al.*, 1983). There have even been times, following mass extinction events, when there have been no reef-building organisms, and consequently no reefs.

In the Palaeozoic, Early Cambrian reefs were constructed chiefly by stromatolites and archaeocyathans. The Middle–Late Cambrian, following on from the Early Cambrian mass extinction, was a time during which there was essentially no reef construction. Ordovician–Devonian reefs were constructed chiefly by sponges, stromatoporoids and rugose and tabulate corals. The latest Devonian to Early Carboniferous, following the Late Devonian mass extinction, was another time during which there were essentially no reefs, but rather mud-dominated reef-like mounds known as 'Waulsortian mounds' – of probable microbial origin – that did not form rigid frameworks (Wendt, 1993; Monty *et al.*, 1995; Wendt *et al.*, 1997). Late Carboniferous–Permian reefs were constructed chiefly by calcareous algae.

In the Mesozoic, Triassic reefs were constructed chiefly by calcareous algae and scleractinian corals; Jurassic reefs by corals; and Cretaceous reefs by scleractinian corals and rudist bivalves. The Early Triassic, following the End-Permian mass extinction, was a time during which there were essentially no reefs, but rather microbial mounds. The Early Jurassic, following the End-Triassic mass extinction, was another such time.

In the Cenozoic, reefs were constructed chiefly by scleractinian corals – as indeed they are at the present time. The Palaeocene, following the End-Cretaceous mass extinction, was a time during which there was essentially no reef construction.

5.3.3 The Early Cambrian mass extinction

There was a mass extinction at the end of the Early Cambrian, sometimes known as the 'Botomian–Toyonian crisis' (Brasier, in Walliser, 1995). The principal group to have been affected was the archaeocyathans, the result being that archaeocyathan reefs, which had previously been widespread, for example in east Siberia, foundered (Jones, in Simmons, in press). The recovery from the Early Cambrian mass extinction event is discussed by Zhuravlev, in Hart (1996).

5.3.4 The Late Cambrian mass extinction

There was another mass extinction, in fact in detail consisting of a series of as many as five distinct extinction events, in the Late Cambrian (Brasier, in Walliser, 1995). The principal groups to have been affected were the inarticulate brachiopods, and, in North America and around the world, the trilobites.

5.3.5 The Ordovician evolutionary diversification

There was a major evolutionary diversification in the Ordovician, possibly initiated by the competition for the niches evacuated by the earlier extinction event (see above). Rugose and tabulate corals, articulate brachiopods, stenolaemate bryozoans, molluscs, trilobites, echinoderms and graptolites all evolved or diversified at this time (Crame and Owen, 2002; Webby *et al.*, 2004).

5.3.6 The evolution of vertebrates

Vertebrates, in the form of marine fish, are interpreted as having evolved, probably from a primitive lancelet-like chordate such as *Pikaia*, in the Cambrian (Ostrom, in Schopf, 1992; Gee, 1996; Long, in Briggs and Crowther, 2001; Smith and Sansom, in Briggs and Crowther, 2001; see also Section 3.7). However, they did not diversify until the Silurian, by which time they had become able to penetrate non-marine environments such as rivers and lakes. Fish are interpreted as having evolved into amphibians in the Devonian. Reptiles and birds are interpreted as having evolved, from amphibians, in the Carboniferous; mammals, from mammal-like reptiles, in the Triassic.

5.3.7 The End-Ordovician mass extinction

There was a significant mass extinction at the end of the Ordovician, resulting in the extinction of somewhere between 70% and 85% of all species, and 22% to 33% of all families (Brenchley, in Briggs and Crowther, 1990; Brenchley *et al.*, 1995; Brenchley, in Briggs and Crowther, 2001; Groombridge and Jenkins, 2002; Brenchley, in Webby *et al.*, 2004). In fact, in detail, there appear to have been two distinct events, separated in time by some 0.5–2 Ma, the first in the *Normalograptus extraordinarius* Graptolite Zone, at the beginning of the Hirnantian, and the second in the *Normalograptus persculptus* Zone, within the Hirnantian. The first event accelerated an already evident decline in some planktonic groups, such as acritarchs, chitinozoans and graptolites, causing a dramatic decline in many shallow-marine, benthic groups, and leaving only a low-diversity fauna, dominated everywhere by the newly evolved and interpreted opportunistic brachiopod *Hirnantia*. The second event eliminated many elements of the *Hirnantia* fauna, and many remaining elements of the pre-extinction fauna.

The principal groups to have been affected were the acritarchs, corals, brachiopods, stenolaemate bryozoans, nautiloids, ostracods, trilobites, echinoderms, graptolites and chitinozoans (and also trace fossils: Twitchett and Barras, in McIlroy, 2004). Entire reef communities were also affected. Interestingly, however, the structure of the ecosystem as a whole was but little affected.

The recovery from the End-Ordovician mass extinction is discussed by Armstrong, by Berry and by Kaljo, all in Hart (1996).

The cause that has been most often invoked for the End-Ordovician mass extinction is global cooling – of up to 8 °C – associated with glaciation, rapidly followed by global warming, accompanied by transgression and oceanic anoxia, associated with deglaciation (Eyles, 1993; Brenchley *et al.*, 1994; Herrmann *et al.*, 2004a, b; Pope and Harris, 2004).

A 'superplume' event has also been implicated (Barnes, in Webby *et al.*, 2004).

There is abundant chemical evidence for glaciation and deglaciation in the form of carbon and oxygen isotope excursions, worldwide, and physical evidence in the form of erosional glacial valleys and syn- and post-glacial deposits, throughout Gondwana. Global cooling associated with glaciation, and accompanying glacioeustatic sea-level fall and shallow-marine habitat destruction, could certainly account for certain of the observed effects on the benthic fauna at the time of the first extinction event. Accompanying changes in oceanic circulation, including advection of toxic, hypertrophic bottom waters to the surface, could account for the observed effects on the plankton (Berry *et al.*, in Stanley *et al.*, 1995). Succeeding global warming associated with deglaciation, and accompanying transgression and – S-State – oceanic anoxia, could account for the remaining observed effects on the pre-extinction benthic faunas, and the Hirnantia fauna, at the time of the second extinction event (Armstrong, in Hart, 1996).

5.3.8 *The evolution of life on land*

The first recognisable land plants, in the form of mosses, club-mosses, ferns and allied forms, appeared in the rock record in the Silurian (Edwards and Burgess, in Briggs and Crowther, 1990; Richardson, in Schopf, 1992; Gordon and Olson, 1995; Thomas and Cleal, 2000; Edwards, in Briggs and Crowther, 2001; Gensel and Edwards, 2001; Kenrick and Davis, 2004). These early plants – probably together with fungi – generated their own soils, providing fertile new ground for further colonisation, and also stabilising the substrate, and slowing down erosion rates (Wright, in Briggs and Crowther, 1990). Seed-plants are interpreted as having evolved, from club-mosses, ferns and allied forms, in the Devonian; flowering plants, from seed-plants, in the Cretaceous.

The first land invertebrate animals to appear, at least as long ago as the Devonian, were arthropods, whose flexible exoskeletons were well preadapted to life in the miniature forests of the time (Selden, in Briggs and Crowther, 1990; Gordon and Olson, 1995; Trewin and McNamara, 1995; Ash, in Briggs and Crowther, 2001; Selden, in Briggs and Crowther, 2001). Indeed, there is some trace fossil evidence to suggest that they might have appeared in the Ordovician (Johnson *et al.*, 1994; Lockley and Meyer, 2000) or Silurian (Brooks *et al.*, 2003). Land gastropods appeared in the Carboniferous.

The first land vertebrate animals to appear – to exploit the growing plant and invertebrate food sources – were amphibians, interpreted as having evolved, from an air-breathing fish ancestor, in the Devonian (Milner, in Briggs and Crowther, 1990; Ostrom, in Schopf, 1992; Gordon and Olson, 1995; Sues, 2000; Ash, in Briggs and Crowther, 2001; Coates, in Briggs and Crowther, 2001). The oldest known land vertebrate body fossils are from the Late Devonian, while the oldest known land vertebrate trace fossils are from the Middle? Devonian (Stossel, 1995; Lockley and Meyer, 2000). Reptiles and birds are interpreted as having evolved, from amphibians, in the Carboniferous (Ostrom, in Schopf, 1992; King, 1996; Chiappe, in Briggs and Crowther, 2001). Complete terrestrialisation may be said to have occurred at this time, when the amniotic egg evolved, and the former tie to the aquatic environment was finally severed (Ostrom, in Schopf, 1992). Mammals are interpreted as having evolved, from mammal-like reptiles, in the Triassic (Ostrom, in Schopf, 1992; Hopson, in Briggs and Crowther, 2001).

5.3.9 *The Late Devonian mass extinction*

There was a significant mass extinction in the Late Devonian, resulting in the extinction of somewhere between 70% and 80% of all species, and 20% of all families (McGhee, in Briggs and Crowther, 1990; McGhee, 1996; McGhee, in Briggs and Crowther, 2001; Groombridge and Jenkins, 2002). In fact, in detail, there appear to have been five distinct extinction events, separated in time by as much as some 0.8 Ma (Walliser, in Walliser, 1995). The first was the so-called lower Kellwasser event, at 364.7 Ma, within the Late *Rhenana* Conodont Zone of the Frasnian; the second, third and fourth, the upper Kellwasser events at 364.2 Ma, 364.1 Ma and 364.0 Ma, within the *Linguifoirmis* Zone of the Frasnian; and the fifth, the Homoctenid event, at 363.9 Ma, within the Early *Triangularis* Zone of the Famennian.

The principal groups to have been affected were the cyanophytes, phytoplankton, calcareous algae, plants, stromatoporoids, rugose and tabulate corals, brachiopods, molluscs, ostracods, trilobites, echinoids, crinoids, fish and amphibians. Tentaculitid molluscs entirely disappeared. Entire reef ecosystems were affected, as substantial groups of reef-building organisms disappeared, at least temporarily. In the Timan–Pechora basin in the north-eastern part of the Russian platform, solitary rugose corals were the first group to disappear, in the Frasnian, followed by dendroid, massive and encrusting stromatoporoids and colonial rugose corals, later in the Frasnian, and finally, cyanophytes such as *Renalcis*, and red calcareous algae, in the Famennian (author's unpublished observations). Stromatoporoids and rugose and tabulate corals failed to recover their pre-extinction diversity at any time in their subsequent evolutionary history. Brachiopods arguably never really recovered either, becoming displaced from their formerly pre-eminent position among the filter-feeding benthos by the bivalve molluscs. Recovery of reef ecosystems took until the Late Carboniferous. The recovery from the Late Devonian mass extinction is further discussed by Cejchan and Hladil, by House, by Kossovaya and by diMichele and Phillips, all in Hart (1996).

The causes that have been invoked for the Late Devonian mass extinction are intrinsic, and associated with global cooling and associated sea-level fall, or global warming and associated sea-level rise and oceanic anoxia; or extrinisic, and associated with impacts of extra-terrestrial bodies; or some combination thereof (Daizhao Chen and Tucker, 2003).

Global cooling could account for all the observed effects on marine and terrestrial faunas and floras not (pre)adapted to cold conditions, including the apparently increasing restriction of their range towards the tropics. However, there is no chemical or physical evidence for glaciation in the Late Devonian.

Global warming, to temperatures indicated by isotope evidence to have been as high as 34 °C, could have killed off a wide range of shallow marine carbonate-producing organisms (Ormiston and Oglesby, in Huc, 1995), as, incidentally, high-temperature El Niño–southern oscillation (ENSO) events do at the present time (Glynn, in Insalaco *et al.*, 2000). Associated sea-level rise and oceanic anoxia could account for the observed effects of the lower Kellwasser event and the first of the upper Kellwasser events, which were apparently principally on shallow marine benthic faunas not (pre)adapted to low-oxygen conditions (Godderis and Joachimski, 2004). Oceanic anoxia and overturn of the type invoked in the case of the End-Ordovician mass extinction could account for the observed effects of the second and third of the upper Kellwasser events and of the Homoctenid event, which were apparently principally on marine planktonic faunas and floras. However, oceanic anoxia could not account for any of the observed effects on terrestrial faunas and floras.

The impacts of extra-terrestrial bodies, especially in combination with – ?consequent – global cooling, could account for all the observed effects of all the events. Perhaps significantly, there is reliable evidence of impact in Belgium at the time of the Homoctenid event, in the form of microtektite, or melt-glass. There is also less reliable evidence of impacts in Austria and China at the times of the Kellwasser events, in the form of microspherules, interpreted as fragmentary microtektites, or of iridium anomalies (possibly the result of biological concentration). Intriguingly, though, there is also evidence of impacts immediately pre- and post-dating the mass extinction event(s), that apparently had no obvious effect on the biota, in the *Punctata* Conodont Zone of the Frasnian, and the *Crepida* Conodont Zone of the Famennian, respectively.

5.3.10 The evolution of forests

Evidence for forests first appeared in the rock record in the Late Devonian (Scheckler, in Briggs and Crowther, 2001). However, evidence for the extensive development of forests did not appear until the Carboniferous, with the diversification of the gymnosperms (Thomas and Cleal, 1993; White, 1996; DiMichele, in Briggs and Crowther, 2001; Kenrick and Davis, 2004). Vertical tiering of forest habitats evidently also first appeared in the Carboniferous, enabling the evolutionary diversification not only of plants, but also of invertebrates and vertebrates, in essentially entirely new niches. Recognisably modern rain-forests first appeared in the Cretaceous, coincident with the evolution of the angiosperms.

5.3.11 The evolution of flight

The first flying animals to appear – to exploit the growing plant food source – were winged insects, interpreted as having evolved from wingless insects, in the Carboniferous (Wootton, in Briggs and Crowther, 1990; Labandeira, in Briggs and Crowther, 2001). Flight has since arisen independently in a number of lineages of vertebrates (Padian, in Briggs and Crowther, 1990; Shipman, 1998; Chiappe, in Briggs and Crowther, 2001; Padian *et al.*, in Tanke and Carpenter, 2001; Paul, 2002; Unwin, in Buffetaut and Mazin, 2003). Reptiles capable of powered flight, as opposed to gliding, appeared, in the form of pterosaurs, in the Triassic; birds in the Jurassic; and flying mammals, in the form of bats, in the Palaeocene. The full range of flying or gliding vertebrates today includes flying fish, frogs, lizards and snakes, birds, and marsupial and placental mammals.

5.3.12 The End-Permian mass extinction

There was a significant mass extinction, the most significant of all the mass extinction events, at the end of the Permian, resulting in the extinction of somewhere between 70% and 95% of all species, and 50% to 60% of all families (Erwin, in Briggs and Crowther, 1990; Erwin, 1993; Wignall, in Briggs and Crowther, 2001; Groombridge and Jenkins, 2002; Benton, 2003). In fact, in detail, there appear to have been at least two distinct extinction events, separated in time by as much as 8–10 Ma, the actual figure varying from one timescale to another. The first event was at the beginning of the Late Permian, Wuchiapingian or Wujiapingian stage, at approximately 260 Ma, and particularly affected marine faunas in low latitudes, most particularly the fusuline foraminiferans (Ross and Ross, 1995). The second, more severe, event, or series of events, took place towards the end of the Late Permian, Changhsingian or Changxingian stage, at 251.1 Ma, and affected almost the entire range of marine and terrestrial faunas and floras worldwide (Wignall, in Briggs and Crowther, 2001).

The principal groups affected were the phytoplankton, plants, foraminiferans, radiolarians, sponges, rugose and tabulate corals, articulate brachiopods, stenolaemate bryozoans, goniatite ammonoids, trilobites, insects, echinoderms and tetrapods (and also trace fossils: Twitchett and Barras, in McIlroy, 2004). In the marine benthic realm, the rugose and tabulate corals disappeared entirely. The brachiopods and crinoids were also badly affected, and never recovered their pre-extinction diversity. The bryozoans underwent significant turnover, as did the bivalves and echinoids, with formerly insignificant groups rising to replace those lost. In the pelagic realm, the zooplanktonic radiolarians experienced almost complete extinction, possibly on account of the effects on their phytoplanktonic food sources (carbon isotope evidence indicating a collapse of primary productivity in the oceans at the end of the Permian). On land, a wide range of plant groups were adversely affected, and the former *Glossopteris* floras of Gondwana, and the equivalent *Cordaites* forests of the northern hemisphere, were killed off, although fungi were evidently able to flourish, possibly on decaying plant matter (Steiner *et al.*, 2003). Animals were also adversely affected, and a number of early insect groups were killed off, as were a number of early tetrapod groups, including the herbivorous pareiasaurs, the carnivorous gorgonopsids and the omnivorous millerettids, although the opportunistic dicynodont *Lystrosaurus* was evidently able to thrive (Zawiskie, in Padian, 1986). The most important net effects of the End-Permian mass extinction were, in the marine benthic realm, firstly, the effective elimination of the former sessile epifauna of reef-forming sponges, corals, brachiopods, bryozoans and crinoids, and its replacement by a motile epifauna and infauna, albeit initially of only low diversity and high dominance, of bivalves and gastropods; and secondly, the resultant effective elimination of reefs – other than those formed by microbial activity – which were only able to become re-established in the Middle Triassic (Wignall, in Briggs and Crowther, 2001; Pruss and Bottjer, 2003).

The recovery from the End-Permian mass extinction is discussed by Schubert and Bottjer (1995), Erwin and Hua-Zhang, in Hart (1996), Bottjer, in Briggs and Crowther (2001) and Twitchett (2004). In the marine benthic realm, there was a succession from faunas dominated by *Lingula, Claraia*, microgastropods, crinoids and stromatolites over the recovery interval.

The causes that have been most often invoked for the End-Permian mass extinction are global climatic change, sea-level change, anoxia, volcanic eruption, the dissociation of methane from deep-sea hydrates, or some combination thereof (Wignall, in Briggs and Crowther, 2001; de Wit *et al.*, 2002; Kidder and

Worsley, 2004). Impacts of extra-terrestrial bodies have also been implicated, and indeed fullerenes of possible, although not proven, extra-terrestrial origin have been reported from the event horizon (Metcalfe *et al.*, 2001).

Palaeopedological or palaeosol evidence from the Karoo basin of South Africa, from eastern Australia and Antarctica indicates that there was a significant global climatic change at the end of the Permian, from a predominantly humid regime characterised on land by the extensive development of flood-plains, to a much more arid regime characterised by calcretes, even at high palaeo-latitudes, with low palaeol-atitudes probably at least bordering on uninhabitable (Wignall, in Briggs and Crowther, 2001).

There is also evidence for a major sea-level fall at the time of the first extinction event at the beginning of the Wuchiapingian, and a minor fall, or a series of minor falls, followed by a major rise, at least arguably associated with global warming, at the time of the second extinction event at the end of the Changhsingian. However, there is no palaeontological evidence for such sea-level change in the shallow marine facies of Nanpanjiang basin in South China, or, arguably, in the Sichaun basin in China, or in Japan (Lehrmann *et al.*, 2003). Here, the event horizon is marked by a calcimicrobial framestone of enigmatic origin, possibly related to an 'anomalous oceanic event'.

There is also evidence for 'superanoxia' affecting the deep oceans over the entire Late Permian, and not only the deep oceans but also the shallow seas at the time of the major transgression at the end of the Changhsingian (Knoll *et al.*, 1996; Isokazi, 1997). Interestingly, though, sedimentological evidence indicates a sea-level fall and rise, and a change in marine chemistry, but no anoxia in the Abadeh section in Iran (Heydari *et al.*, 2003).

The Siberian Trap volcanic eruption, dated to at least close to the end of the Changhsingian, at approximately 250 Ma, would almost certainly have resulted in the release of significant quantities of ash into the atmosphere, which could have blocked out sunlight, prevented photosynthesis, and caused a collapse in primary productivity on land and at sea, thus accounting for at least some of the observed effects of the mass extinction (Campbell *et al.*, 1992).

According to Kidder and Worsley (2004), an increase in the 'greenhouse gases' carbon dioxide (associated with the Siberian Trap volcanism) and methane (associated with dissociation of methane hydrates in the deep sea) in the atmosphere at the end of the Permian caused global warming, sea-level rise and oceanic anoxia, through the much-discussed 'greenhouse effect'. According to Kidder and Worsley's interpretation, in the oceans, this warming caused transgression through melting of polar ice, a weakened pole-to-equator thermal gradient, weakened wind-driven upwelling, sluggish circulation, and ultimately widespread anoxia and mass extinction. Also according to Kidder and Worsley's interpretation, on land, the effects of the warming included the extinction of the extensive coal-forests of the Permian and the communities they supported, trapped in low latitudes by the peculiar disposition of Pangaea. Importantly, the effects on land in turn fed back into and reinforced not only the atmospheric 'greenhouse effect', through the release of carbon dioxide no longer consumed in the coal-forests, but also the oceanic transgression and anoxia, through the release of water no longer sequestered. (No doubt the ongoing extinctions in the oceans also resulted in a release of carbon dioxide no longer consumed by the phytoplankton.)

Whatever the cause, and the details do still remain to be worked out – Kidder and Worsley's interpreted chain of events being only partly supported by current quantitative palaeoclimatic models – the resulting mass extinction was on a scale never encountered before or since, probably because of the dramatic effects on the primary producers forming the basis of the trophic structure in terrestrial and marine communities.

5.4 Mesozoic

Note that there were additional, minor, mass extinction events in the Early, Middle and Late Jurassic and in the Early Cretaceous. The Early and Late Jurassic events involved extinctions of benthic brachiopods, bivalves and gastropods, and nektonic ammonites as a result of anoxia, although apparently only in Europe. Many other marine animals were unaffected, and terrestrial animals appear to have been entirely unaffected. The Middle Jurassic event is poorly constrained, but appears to have involved extinctions of cephalopods. The Early Cretaceous event is also poorly defined, but appears to have involved extinctions as a result of an oceanic anoxic event. The release of methane from deep-sea hydrates has been invoked as the cause of the Early Jurassic event.

5.4.1 *The Mesozoic evolutionary diversification*

There was a major evolutionary diversification in the Mesozoic, the so-called 'Mesozoic marine revolution', possibly initiated by the competition for the niches evacuated by the earlier extinction event (see above). Cyclostomate and gymnolaemate bryozoans, scleractinian corals, bivalves, ammonites, belemnites and echinoids all evolved or diversified at this time (Kelley and Hansen, in Briggs and Crowther, 2001). Echinoids replaced crinoids as the dominant echinoderms. On land, dinosaurs, crocodiles, pterosaurs, turtles, lizards, frogs and salamanders, and mammals, replaced labyrinthodonts, thecodonts, procolophonids, prolacertiforms, rhyncosaurs and mammal-like reptiles as the dominant tetrapods (Benton *et al.*, in Padian, 1986; Olsson and Sues, in Padian, 1986; Benton, in Briggs and Crowther, 1990). New groups apparently replaced old ones either by simply occupying the niches evacuated by them, or by out-competing them, for example, the dinosaurs out-competing the thecodonts and rhyncosaurs – advantage in this instance being bestowed by superior locomotory or other skills (possibly including the ability to self-regulate body temperature and metabolic rate).

Other causes that have been invoked for the Mesozoic evolutionary diversification include increased dispersal and endemism associated with the break-up of Gondwana, increased predation pressure, and the co-evolution of predator–prey systems (Aberhan, in Crame and Owen, 2002; Harper *et al.*, 2003). The evolutionary diversification of the planktonic foraminiferans in the Jurassic and Cretaceous has been linked to oceanic anoxic events (Hart *et al.*, in Crame and Owen, 2002; Leckie *et al.*, 2002).

5.4.2 *The End-Triassic mass extinction*

There was apparently a significant mass extinction event at the end of the Triassic, resulting in the extinction of approximately 80% of all species, and 20% of all families (Benton, 1986; Benton, in Briggs and Crowther, 1990; Benton, in Fraser and Sues, 1994; Groombridge and Jenkins, 2002; Tanner *et al.*, 2003). Note, though, that the actual severity of the extinction event has been questioned (Fraser and Sues, in Fraser and Sues, 1994; Hallam, 2002; Tanner *et al.*, 2003). In detail, there were apparently at least two distinct extinction events (Benton, 1986). There was one event in the Carnian at the beginning of the Late Triassic, and one in the Rhaetian at the end, separated in time by as much as 28 Ma (the actual figure varying from one timescale to another).

The principal groups affected were the plants, sponges, corals, brachiopods, stenolaemate bryozoans, bivalves, gastropods, insects, echinoderms, amphibians and reptiles (and also, at least locally, trace fossils: Twitchett and Barras, in McIlroy, 2004). Interestingly, the groups affected by the first, Carnian event were both marine and terrestrial, whereas those affected by the second, Rhaetian event were mostly marine. At sea, the ceratite ammonoids and conodonts disappeared entirely over the course of the Late Triassic, as did the strophomenid brachiopods, conulariids, nothosaurs and placodonts, and the bivalves lost an estimated 92% of all of their species, and 42% of all of their genera. On land, the formerly dominant tetrapods, that is, the labyrinthodonts, the mammal-like reptiles, the thecodonts, the procolophonids, the prolacertiforms and the rhyncosaurs, were essentially wiped out, and new groups, including the dinosaurs, crocodiles, pterosaurs, turtles, lizards, frogs and salamanders, and mammals, came to rise to prominence. It has been suggested that the turnover was gradual, with new groups replacing old ones either by simply occupying the niches vacated by them in a series of extinctions, for example, the crocodiles replacing the phytosaurs, or the lizards the procolophonids and prolacertiforms, or by out-competing them, for example, the dinosaurs out-competing the mammal-like reptiles, rhyncosaurs and thecodonts. It has also been suggested that the observed turnover was at least in part caused by environmental change rather than extinction (Cuny, 1995). One important effect of the End-Triassic mass extinction(s) was, in the marine benthic realm, the effective elimination of reefs, which were only able to become re-established towards the end of the Early Jurassic (Hallam, in Hart, 1996).

The causes that have been most often invoked for the End-Triassic mass extinction(s) are global climatic change, sea-level change, volcanic eruption, and the dissociation of methane from deep-sea hydrates. Impacts of extra-terrestrial bodies have also been implicated, and indeed there is evidence for impact in the form of the Manicougan impact crater in Canada, dated to 206–213 Ma, although there is no evidence for associated shocked quartz or an iridium anomaly actually at the mass extinction event horizon.

There is abundant evidence of climatic change at the time of the extinction event(s), associated with the onset of rifting of Pangaea and the subsequent drifting of Gondwana and Laurasia away from the tropical belt, although the relationship, if any, between the same and the extinction event(s) remains unclear. It may be stated, though, that there was apparently a change to a more arid climate over the period of the extinctions, as evidenced on land by the replacement of the old, Gondwanan, *Dicroidium* flora by a new, essentially cosmopolitan, conifer–benettitalean flora, and that this would have stressed at least land animal populations, and could account for the observed extinctions of the same.

There is also evidence of associated sea-level change, although again the relationship to the extinction event(s) remains unclear. It may be stated, though, that sea level over the period of the extinction event(s) was apparently generally extremely low. This would have resulted in habitat loss in the marine realm. It would also have resulted in the removal of barriers between populations in the terrestrial realm, and hence reduction in the rate of speciation, if not actual extinction.

There is evidence, too, of flood-basalt volcanic eruption in the central Atlantic magmatic province at and immediately after the end of the Triassic, although it is not known exactly what the environmental and other effects of this would have been.

There is evidence of dissociation of methane from deep-sea hydrates, or some other perturbation to the carbon cycle, in the form of a significant carbon isotope excursion at the end of the Triassic. Early Jurassic stromatolitic bioherms in the Neuquen basin in Argentina have been interpreted as related to methane seepage (Gomez-Peres, 2003).

5.4.3 *The evolution of flowering plants*

Flowering plants first appeared in the Early Cretaceous (Collinson, in Briggs and Crowther, 1990; Hughes, 1994; Taylor and Hickey, 1995; Friis *et al.*, in Briggs and Crowther, 2001; Friis, 2003; Skelton, 2003). Recognisably modern rain-forests also appeared at this time (Richards, 1996; Morley, 2000; Kenrick and Davis, 2004).

5.4.4 *The Late Cretaceous (Cenomanian–Turonian) mass extinction*

There was a mass extinction in the Late Cretaceous, at the Cenomanian–Turonian boundary, at the end of the *Rotalipora cushmani* Zone (Harries and Little, 1999; Skelton, 2003). The principal groups affected were the dinoflagellates, calcareous nannoplankton, foraminiferans, sponges, caprinid rudists, cephalopods, echinoids, bony fish and ichthyosaurs. The recovery from the Cenomanian–Turonian mass extinction is discussed by Tewari *et al.*, by Tur, by Hart and by Fitzpatrick, all in Hart (1996), and by Xiaoqiao Wan *et al.* (2003). In Tibet, there is evidence of the existence of a 'survival' interval in the *Whiteinella archaeocretacea* Zone at the base of the Turonian and a 'rapid recovery' interval thereafter, the latter characterised by the development of a diverse community of planktonic foraminiferans including both shallow and deep water-column dwellers (Xiaoqiao Wan *et al.*, 2003). The extinction and survival intervals are characterised by dysaerobic facies. The causes that have been most often invoked for the Cenomanian–Turonian mass extinction are global climatic change, sea-level change, anoxia, volcanic eruption, or some combination thereof. Impacts of extra-terrestrial bodies have also been implicated. There is good evidence of global warming, sea-level rise, and an oceanic anoxic event (OAE2 of Jenkyns, 1980). There is also good evidence of oceanic plateau-forming volcanic eruption in the Ontong-Java plateau, Caribbean/Colombian plateau, Madagascar Rise and Broken Ridge, totalling some 1 000 000 km^3 (Kerr, 1998). Associated emission of large quantities of carbon dioxide would have led to greenhouse warming. It would also have led to enhanced productivity in surface waters, which would in turn have exacerbated the anoxia in bottom waters.

5.4.5 *The End-Cretaceous mass extinction*

There was a significant mass extinction event at the Cretaceous–Tertiary (K–T) boundary, resulting in the extinction of somewhere between 40% and 76% of all species, and 14% of all families (Alvarez *et al.*, 1980; Halstead, in Briggs and Crowther, 1990; Surlyk, in Briggs and Crowther, 1990; Archibald, 1996; MacLeod and Keller, 1996; Officer and Page, 1996; Ryder *et al.*, 1996; Alvarez, 1997; MacLeod *et al.*, 1997; Frankel, 1999; Norris, in Briggs and Crowther, 2001; Wolfe and Russell, in Briggs and Crowther, 2001; Groombridge and Jenkins, 2002; Skelton, 2003).

The principal groups affected were, in the seas, the foraminiferans, diatoms, cyclostomate and some gymnolaemate bryozoans, inoceramid and rudist bivalves, ammonites, belemnites, and marine reptiles/mosasaurs and plesiosaurs; and on land, the plants, dinosaurs and pterosaurs. (Note in this context, though, that the inoceramids, rudists and dinosaurs all appear to have been declining before the event, although this may simply be due to an artefact of sampling.) Interestingly, the effects of the event appear to have been selective across taxonomic groups, with some scarcely affected at all, such as the brachiopods and echinoids; and some, such as ferns, even flourishing, as they do today in areas of disturbance. The effects also appear to have been selective across size ranges within groups, with large taxa tending to die or be killed off, and small taxa to survive.

The recovery from the End-Cretaceous mass extinction has been discussed by Koutsoukos, by Bugrova, by Speijer and van der Zwaan and by Kelley and Hansen, all in Hart (1996), and by Stilwell (2003). The initial recovery from the event, in the form of renewed diversification, took place in the time range 1–10 ka in the case of one of the most intensively studied groups, the planktonic foraminiferans, and 40–400 ka in the case of another, the calcareous nannoplankton. The final recovery of pre-extinction diversity took anything up to 15 Ma for the marine invertebrate biota as a whole, and up to 20 Ma for certain groups, such as the bivalves and gastropods. The recovery of reef ecosystems appears to have taken place relatively rapidly, in 5–10 Ma, possibly on account of the rapid emergence of new groups of organisms in the niche formerly occupied by the rudists. The recovery of terrestrial ecosystems also appears to have taken place relatively rapidly, possibly in this instance on account of the rapid emergence, and, incidentally, size increase, of the mammals in the niche formerly occupied by the dinosaurs. Thus, the ultimate long-term effect of the mass extinction event was the redirection of the interaction between different groups of organisms.

The cause that has been most often invoked for the End-Cretaceous mass extinction is the catastrophic impact of an extra-terrestrial bolide, possibly combined with flood volcanism (Morgan *et al.*, 1997). As is now well established, there is abundant evidence for an impact, in the form of a crater at Chicxulub on the Yucatan Peninsula in Mexico, and in the form of iridium anomalies, microtektites, spherules and shocked minerals elsewhere. Interestingly, evidence of earlier impacts has recently come to light at Mishor Rotem in Israel (Keller *et al.*, 2003). Here, spherules have been recorded in three red marl layers (RL1–3) within the late Maastrichtian, Zone CF1, and spherules and an iridium anomaly in a fourth red layer (RL4) at the Maastrichtian/Danian (K/T) boundary.

Apart from its immediate effects (wildfires on land, tsunamis at sea), such an impact could also ultimately have resulted in the release of sufficient particulate matter into the atmosphere to block out the sunlight for months or years, causing global cooling, or a so-called 'impact winter'. The ultimate result could have been a prevention of photosynthesis and a collapse of primary production on land and in the sea. Carbon isotope evidence indeed indicates a major perturbation in the carbon cycle in the ocean, such as would be expected from a collapse of primary phytoplankton – and secondary zooplankton – production. The resulting, essentially lifeless ocean has been dubbed the 'Strangelove Ocean', in reference to remarks on the effects of nuclear weapons in Stanley Kubrick's 1964 film *Dr Strangelove*.

The longer-term after-effects of the impact could have included global warming associated with the 'greenhouse effect', perhaps accounting for the evidence from leaf physiognomy of warming following the 'impact winter' in the western interior of the USA (Wolfe, in Iwatsuki and Raven, 1997).

It is possible, though, that climate change proceeded entirely independently of any extra-terrestrial event.

5.5 Cenozoic

Note that there were additional evolutionary or diversification events through the course of the Cenozoic (Crame and Rosen, in Crame and Owen, 2002). The evolution and diversification of land plants is discussed by Collinson, in Briggs and Crowther (2001), that of land animals by Alroy *et al.*, in Erwin and Wing (2000) and Janis, in Briggs and Crowther (2001), that of primates by Rose, in Briggs and Crowther (2001).

Note also that there were minor mass extinction events in the Oligocene, Miocene and Pliocene (Sundelius, in Briggs and Crowther, 1990). The Oligocene event involved extinctions of mammals, although apparently only in North America. The Miocene event involved extinctions of plankton. The Pliocene event also involved extinctions of plankton, and of benthic bivalves and gastropods, although apparently essentially only in the tropics. Note, though, that there is also a distinct turnover of mammals between the Chapadmalalan and Barrancalobian land mammal ages of the Pampean region of Argentina in South America, dated to approximately 3.3 Ma (Vizcaino *et al.*, 2004). Approximately 37% of all genera and 53% of all species became extinct at this time. Palaeoecological analysis indicates that the Chapadmalalan fauna is balanced from the point of view of trophic structure, whereas the Barrancalobian, although less well known, appears to contain disproportionately few carnivores, especially in the medium to large size category. Perhaps significantly, there is evidence in the succession of an impact event.

5.5.1 *The End-Palaeocene mass extinction*

There was a mass extinction at the end of the Palaeocene. The cause that has been most often invoked is global warming, possibly in turn triggered by the impact of a comet (Kent *et al.*, 2003).

5.5.2 *The End-Eocene mass extinction*

There was another mass extinction at the end of the Eocene (Prothero, 1994; Prothero and Emry, 1996; Prothero *et al.*, 2003). The principal groups affected were the plankton and open-water bony fish in the sea, and mammals on land, at least in Europe. The recovery from the End-Eocene mass extinction is discussed by Zhuravlev, in Hart (1996). The cause that has been most often invoked is global cooling. Impacts of extra-terrestrial bodies have also been implicated (Poag, 1997; Edwards and Powars, 2003). An impact crater has been found at Chesapeake Bay, in the eastern USA.

5.5.3 *The evolution of grasses and grassland animals*

Grasses first appeared in the Eocene, and became abundant in the Miocene, coincident with a change to a more arid climate (Thomasson and Voorhies, in Briggs and Crowther, 1990; Sues, 2000; Cerling, in Briggs and Crowther, 2001; Beuning and Wooller, 2002; Stromberg and Feranec, 2004). Grassland animals also became abundant in the Miocene.

5.5.4 *The evolution of humans*

'Anatomically modern' human beings evolved in Africa approximately 150 000 years ago, and dispersed 'Out of Africa' approximately 85 000 years ago, to colonise the remainder of the world (Susman, in Briggs and Crowther, 1990; Wood, in Briggs and Crowther, 2001; see also Sub-section 5.5.8). They, or rather we, are characterised by enhanced encephalisation, arguably in consequence of selection pressure associated with climatic and environmental forcing. However, neither our ability to manufacture tools, nor our ability to communicate, is unique in the animal kingdom, as was once thought. What – we think – sets us aside from all the other animals that have ever lived, including our immediate ancestors, is our consciousness: that is, our awareness of ourselves; our overview of ourselves in space and time; and our individual and collective ability to process information on our world past and present to plan for the future, and, for good or ill, to influence our own destiny.

5.5.5 *The Pleistocene mass extinction*

There was yet another mass extinction in the Pleistocene, albeit accounting for the extinction of only 1% of all species (Martin and Wright, 1967; Martin and Klein, 1984; Lundelius, in Briggs and Crowther, 1990; Stuart, 1991; Beck, 1996; Webb and Rancy, in Jackson *et al.*, 1996; Alroy, in MacPhee and Sues, 1999; Alroy, 2001; Roy, in Briggs and Crowther, 2001; Ficcarelli *et al.*, 2003).

The principal groups affected were large land mammals, that is, those with mean adult body weights in excess of – and in many cases considerably in excess of – 44 kg, such as mammoths, mastodons and woolly rhinoceroses. The effects were also selective across geographic regions, with 73% of large mammal genera becoming extinct in North America, and 79% in South America, but only 29% in Europe, and 14% in Africa. The effects were such as to change forever the structures of communities in the affected areas.

The causes that have been most often invoked are global warming and environmental change associated with the end of the Pleistocene ice age, human hunting activity, or some combination thereof.

The change from a glacial to an interglacial climate could certainly have stressed animal populations, not least on account of the accompanying effects on the structure and distribution of plant communities, and could account for the observed End-Pleistocene mass extinction. For example, according to the so-called 'co-evolutionary disequilibrium' model, the herbivores that had evolved to feed on the mosaic of vegetation that characterised the glacial periods would not have been able to survive the change to the climatically zoned vegetation that characterised the interglacial periods, in consequence of which they could have died out, as could the carnivores and scavengers that depended on them. The problem with this model is that it does not adequately account for the apparent absence of mass extinctions associated with changes from glacial to interglacial climates earlier in the Pleistocene. The principal biotic response to glaciation and deglaciation appears to have been a simple shift in geographic distribution (Coope, in Lawton and May, 1995; Budd *et al.*, in Jackson *et al.*, 1996; FAUNMAP Working Group, 1996).

Human hunting activity could also have stressed animal populations, especially those taking a long time to attain reproductive maturity, or producing only small numbers of offspring, and could also account for the observed End-Pleistocene mass extinction. Circumstantial evidence in support of the so-called 'overkill hypothesis' comes from the coincidence in time of the observed mass extinction at 10 500–11 500 years ago with the first appearance of anatomically modern humans, known as the Clovis people, in North America. One problem with this hypothesis is that it does not in itself account for the apparent absence of a mass extinction associated with the first appearance of anatomically modern humans in Eurasia 35 000 years ago. Another is distinguishing any anthropogenic effects from coeval climatic effects.

5.5.6 *The Holocene mass extinction*

One mass extinction that is often ignored is that apparently ongoing at the present day (Leakey and Lewin, 1995; Rosenzweig, 1995; Quammen, 1996; Groombridge and Jenkins, 2002). Ongoing extinction rates are difficult to quantify (May *et al.*, in Lawton and May, 1995). However, for birds, it is thought that 116 species (1% of the total number) have become extinct since reasonably reliable records began to be kept in the seventeenth century, and that a further 1029 species (11%) are threatened or endangered, and could disappear within the next century. In the case of mammals, 59 species (1%) have become extinct since records began, and a further 505 species (11%) – or, according to more pessimistic estimates, a further 1130 species (24%) – are threatened or endangered. If these rates are scaled appropriately, they equal those found in ancient mass extinctions. The most likely cause of the ongoing mass extinction is natural or artificial – anthropogenic – climatic and associated environmental change, or human hunting activity or habitat destruction, or some combination thereof. Note in this context that anthropogenic global warming – or at least an offsetting of global cooling – may have been ongoing for thousands of years, ever since forests were first cleared for farmland, and significant quantities of the greenhouse gas carbon dioxide were released into the atmosphere (Ruddiman, 2004). Note also, though, that the local mid-Holocene extinction of the black-flanked rock wallaby (*Petrogale lateralis*) at the Tunnel Cave archaeological site in south-western Australia is now attributed not to human agency, as was once the case, but rather to natural climatic and vegetational change, specifically a post-glacial increase in rainfall and an associated encroachment of closed habitat at the expense of open grassland and grazing areas (Dortch, 2004).

5.5.7 *New evidence for land mammal dispersal across the northern North Atlantic in the Early Eocene*

In the northern hemisphere, there is a well known and close similarity between Early Eocene, Wasatchian North American land mammal age (NALMA) and equivalent Neustrian European land mammal age (ELMA) land animals, including mammals, which have been intensively studied (McKenna, 1971, 1975, 1983; McKenna, in Bott *et al.*, 1983; Hooker, in Culver and Rawson, 2000), and also amphibians, reptiles and birds (Milner *et al.*, in Culver and Rawson, 2000) and, arguably, freshwater fish (Forey, in Culver and Rawson, 2000); together with land plants (Collinson, in Culver and Rawson, 2000) (Fig. 5.3). Land mammal groups appearing in both North America and Europe for the first time at the beginning of the Eocene include even-toed ungulates, odd-toed ungulates and primates (Hooker, in Culver and Rawson, 2000). Most of the recorded representatives of these groups are herbivores, frugivores or insectivores; only something of the order of 10% are carnivores (Hooker, in Culver and Rawson, 2000).

The similarity between the Early Eocene land animals and plants of North America and Europe has been interpreted as indicating dispersal across the North Atlantic, by way of a Greenland–Iceland–Faeroes–Scotland or 'Thulean' land bridge, a Norwegian–Greenland Sea or 'de Geer' land bridge, or a North America–Eurasia or 'Bering' land bridge (Fig. 5.4).

The generic- and specific-level commonality among land mammals is greatest in the 'early' Wasatchian (Wa0–Wa4) and equivalent 'early' Neustrian (PEI–'early' PEIII), implying effectively unimpeded interchange at this time (Fig. 5.3) (Hooker, in Culver and Rawson, 2000). According to the most recent calibrations of the NALMAs and ELMAs against the geomagnetic timescale, this was in 'late' Chron C24r between 55 and 53.5 Ma (Woodburne, 2004). Commonality among land

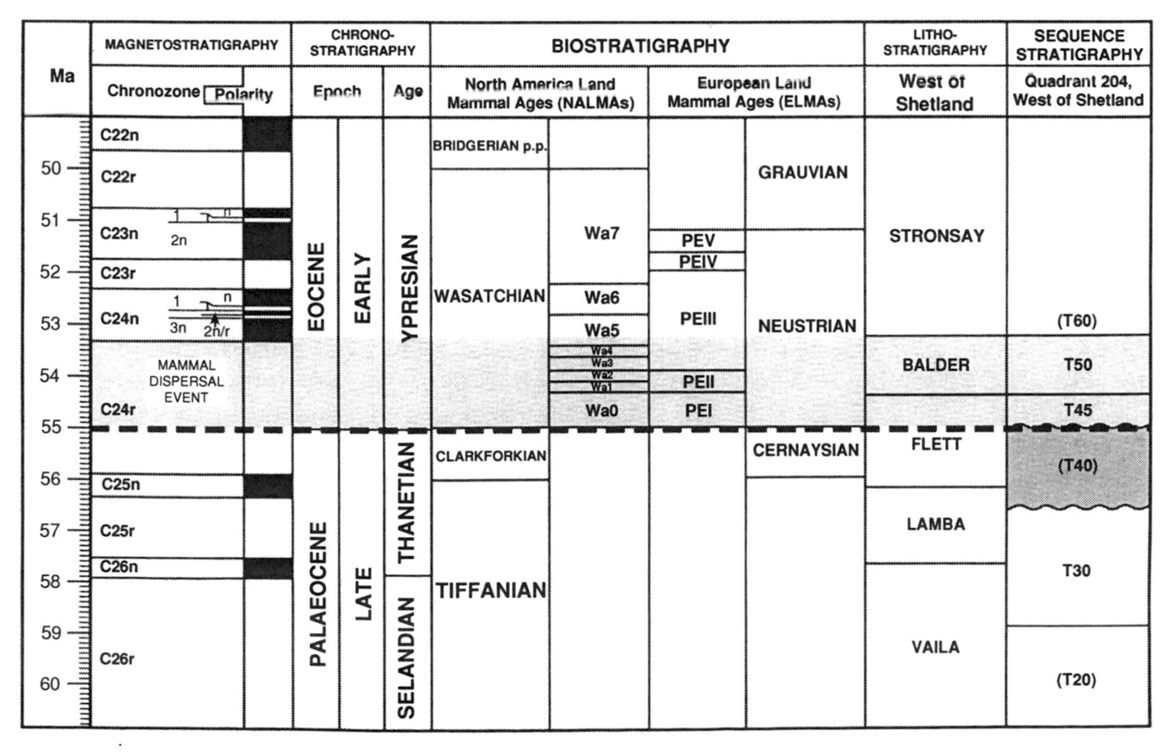

Fig. 5.3. Eocene stratigraphy of Europe and North America.

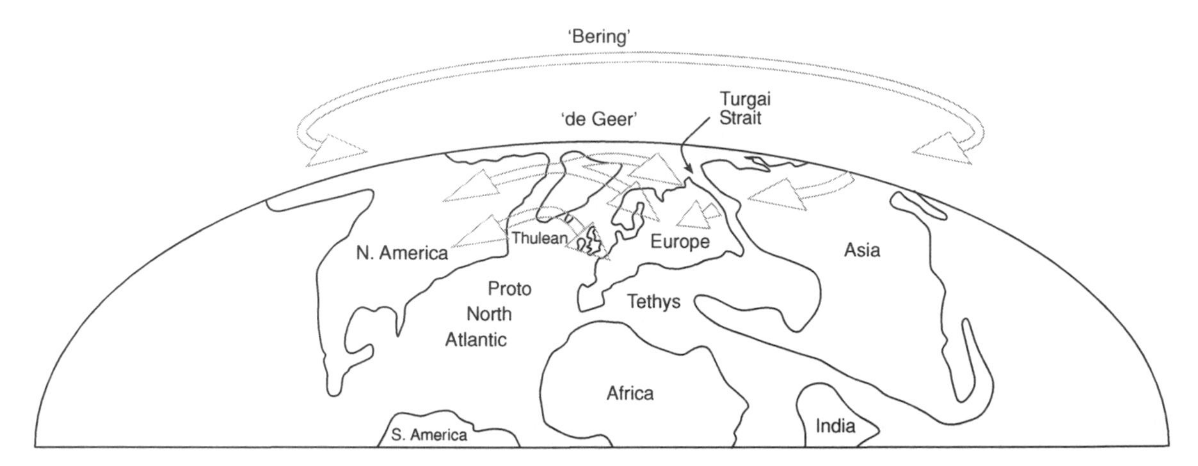

Fig. 5.4. Eocene palaeogeography of the North Atlantic, showing proposed land bridges and mammal dispersal routes.

mammals is markedly lower in preceding and succeeding stages, implying the existence of a barrier to dispersal at these times; whereas, interestingly, commonality among land plants persists from the Early into the Middle Eocene (Collinson, in Culver and Rawson, 2000), implying that the barrier was not effective in preventing the dispersal of land plants, possibly because of the role played by winds and currents (or birds and insects).

The time interval over which the North American and European land mammal faunas are most similar and over which the dispersal across the North Atlantic is inferred to have taken place is thus comparatively short (55 to 53.5 Ma, so <2 Ma).

In contrast, the time interval over which at least the Greenland–Iceland–Faeroes part of the proposed Greenland–Iceland–Faeroes–Scotland or 'Thulean' land bridge is inferred to have been in existence is comparatively long (55 to 33.5 Ma, so >20 Ma). The latter calculation is based in part on data that show that the subaerial basalts of Greenland and the Faeroes, contiguous prior to rifting in Chron C24n, are of Chron C24r age (*c*.55 Ma), and in part on data that show that the Iceland–Faeroe Ridge was subaerially emergent and lateritised in the Late Eocene (*c*.33.5 Ma). The most likely explanation for this discrepancy is that there were intermittent barriers or filters along the proposed land mammal dispersal route.

Here, the evidence for and against the various land bridges (land mammal dispersal routes) that have been proposed is discussed; and new geological and supportive palaeontological evidence for the formation and flooding of the Greenland–Iceland–Faeroes–Scotland or 'Thulean' land bridge is presented, based on data from the Faeroe–Shetland Basin.

Land bridges and land mammal dispersal

The Greenland–Iceland–Faeroes–Scotland or 'Thulean' land bridge

McKenna, in Bott *et al.* (1983) and McKenna (1983), among others, proposed that this was the most likely land bridge across the North Atlantic in the Early Eocene (Fig. 5.4). This model invokes linkage of the subaerial basalts of Greenland and the Faeroes – part of the Brito-Arctic igneous province formed in response to mantle hot-spot activity – prior to spreading in Chron C24n. However, there are problems with this model as it stands (Roberts *et al.*, in press). Firstly, the Faeroe–Shetland basin would have been a deep marine basin during the dispersal interval, as deep-water sands derived from Scotland to the east and volcanics extruded from the Faeroes to the west would have only partially filled it by this time, leaving a seaway several tens of kilometres wide and several hundreds of metres deep, presenting a significant barrier to land mammal dispersal. Secondly, as noted above, the brevity of the dispersal interval is inconsistent with the longevity of subaerial exposure indicated by the Early Eocene ages of the basalts in Greenland and the Faeroes and the Late Eocene age of the laterites on the Iceland–Faeroe ridge.

The Norwegian–Greenland Sea or 'de Geer' land bridge

McKenna, in Bott *et al.* (1983) and McKenna (1983), among others, proposed that there was a land bridge connecting Greenland to the European land mass by way of Svalbard or the Outer Voring plateau (Fig. 5.4). However, there is a problem with this model, too, in that the land connection would only have extended as far as the Hammerfest–Nordkapp basin or Inner Voring plateau (Roberts *et al.*, in press). According to Nagy *et al.*, in Hass and Kaminski (1997), the Hammerfest–Nordkapp basin would have been several tens of kilometres wide and several hundreds of metres deep during the dispersal interval, presenting a significant barrier to land mammal dispersal.

The North America–Eurasia or 'Bering' land bridge

Simpson (1946), among others, proposed that there was a land bridge connecting North America to Asia, where, incidentally, many of the land mammals in question have been interpreted as having originated, and ultimately to Europe, at the beginning of the Eocene (Fig. 5.4). However, there is a problem with this model, too, in that many authors have interpreted Asia and Europe as having been separated by a broad seaway known as the Obik Sea or Turgai Strait until at least the middle if not the end of the Eocene (Jones, in Agusti *et al.*, 1999; Roberts *et al.*, in press). Note, though, that, in contrast, Marincovich *et al.*, in Grantz *et al.* (1990) have argued that the Obik Sea was 'not fully and consistently open during the Paleogene', and Khokhlova and Oreshkina (1999) that 'the geography of the [marine] connection between the Middle Volga . . . and West Siberian basins . . . and the North European basins . . . is . . . not clear'. Note also that Iakovleva *et al.* (2001) have recently intimated that it would have been ineffective as a barrier to dispersal during periods of sea-level low-stand, as between 57.1 and 56.4 Ma and again between 55.3 and 54.5 Ma. Certainly, it did not appear to have acted as a barrier to the dispersal of the Asiatic crocodilians *Asiatosuchus* and *Pristichampus* into Europe in the Early to Middle Eocene (Kotsakis *et al.*, 2004). It could have acted as a filter rather than a barrier to dispersal, though, allowing the dispersal of aquatic reptiles but not terrestrial mammals.

Discussion: new evidence for the Greenland–Iceland–Faeroes–Scotland or 'Thulean' land bridge

New three-dimensional seismic images integrated with well-log data provide a new insight into how this seaway in the Faeroe–Shetland basin was – albeit briefly – closed in the Early Eocene to provide a land bridge across the North Atlantic and a land mammal dispersal route between North America and Europe (Roberts *et al.*, in press).

A three-dimensional seismic image of a time-slice through Early Eocene Sequence T45/50 (Fig. 5.3) in the Faeroe–Shetland basin clearly shows dendritic drainage on an interpreted subaerial delta-top and also channelisation on an interpreted submarine delta-front (Roberts *et al.*, in press). The delta progrades to the north-east, onlaps the subaerial Upper Series basalts of the Faeroes, and closes the former seaway of the Faeroe–Shetland basin. The role, if any, of eustasy in this remains poorly understood, although conceivably any one or more of a number of proposed eustatic sea-level falls around the Palaeocene/Eocene boundary could have contributed.

Well-log, sedimentological and biostratigraphic data pertaining to 204/24-1A in the Foinaven field in the Faeroe–Shetland basin (Fig. 5.3) confirm a general shallowing upward from deep marine environments in the Late Palaeocene to marginal to non-marine environments in the Early Eocene. The Late Palaeocene is characterised by submarine fan sandstones; the Early Eocene by peri-deltaic sandstones interbedded with coals (and tuffs). Biostratigraphic data presented by Mudge and Bujak (2001) further indicate that Early Eocene sequence T45/50, comprising the upper part of the Flett formation and the Balder formation (Fig. 5.3), is characterised in 204/24-1A by exclusively non-marine palynomorph assemblages containing abundant *Taxodiaceaepollenites* (cypress and/or swamp-cypress pollen), and is confirmed as having been deposited in a subaerial delta-top environment. (Recently acquired proprietary biostratigraphic data indicate that Late Palaeocene sequence T40, comprising the lower part of the Flett formation and the upper part of the Lamba formation (Fig. 5.3), is absent.)

This is significant because stratigraphic correlation indicates that the time interval represented by the upper part of the Flett formation and the Balder formation, which, as indicated above, at the 204/24-1A well locality was characterised by non-marine deposition, corresponds to the time of the land mammal dispersal event, 55–53.5 Ma (Fig. 5.3). There is a seismostratigraphic correlation between the upper part of the Flett formation and the Balder formation of the area west of Shetland and the Sele and Balder formations of the North Sea, and a biostratigraphic correlation between the Sele and Balder formations of the North Sea and the Woolwich and Reading and Harwich formations of the London basin in south-east England, which have been magnetostratigraphically dated to 'late' Chron C24r, and which, moreover, contain 'early' Neustrian (PEI–'early' PEIII) land mammals (Fig. 5.3) (Hooker and Millbank, 2001). The biostratigraphic correlation between the Sele and Balder formations and the Woolwich and Reading and Harwich formations is on the basis of the abundant occurrence of the – paratropical – dinoflagellate *Apectodinium* in the lower part of the Sele and in the Woolwich and Reading (Powell *et al.*, in Knox *et al.*, 1996; Bujak and Brinkhuis, in Aubry *et al.*, 1998), and on the occurrence of the diatom *Fenestrella antiqua* (*Coscinodiscus* sp. 1) in the upper part of the Sele and Balder and in the Oldhaven member of the Harwich (Jones, in Jones and Simmons, 1999).

The Greenland–Iceland–Faeroes–Scotland land connection was, as noted above, short-lived, probably lasting for <2 Ma between 55 and 53.5 Ma. It was then severed by marine transgression associated with thermal subsidence at the time of sea-floor spreading in Chron C24n time (Fig. 5.3). This transgression is marked in wells in the Faeroe–Shetland basin by the return of marine palynomorph assemblages in the Stronsay formation (Mudge and Bujak, 2001). (Incidentally, it appears that while the land bridge was in place, it presented at least a partial barrier to the interchange of marine organisms between the Norwegian Sea to the north and the proto-Atlantic to the south: Kaminski and Austin, 1999.)

The land mammal dispersal associated with the land connection may have been facilitated by an essentially coincident and evidently also short-lived climatic optimum, before and after which land mammals might have found it less easy to access the high-latitude area where the land bridge is interpreted to have been (Norris and Rohl, 2000). This climatic optimum has recently been interpreted as having been generated by a runaway 'greenhouse effect' caused by decomposition of methane hydrates in the deep sea (Dickens *et al.*, 1997). It has also been interpreted as having been generated by a comet impact (Kent *et al.*, 2003). Direct supportive palaeontological evidence of a climatic optimum derives from the observations that the structure, composition and diversity of land mammal faunas from North America and Europe from the time of dispersal closely resemble those from the present-day tropical rain-forest, with coeval land plants also of essentially tropical aspect (Collinson, in Culver and Rawson, 2000). Indirect evidence derives from the observation that tooth enamel from mammals, shell material from freshwater bivalve molluscs and carbonate from soils from the 'early' Wasatchian (Wa0) of the Bighorn and adjacent basins in Wyoming in North America and from age-equivalent sections in the San Juan basin in New Mexico to the south and from Ellesmere Island in Arctic Canada to the north, and from the 'early' Neustrian (PEI) of the Paris basin in Europe, all record strong negative carbon isotope or positive oxygen isotope excursions, confirming a significant warming event at this time (Hooker, in Culver and Rawson, 2000). Incidentally, the carbon isotope excursion associated with this warming event has recently been recommended as the global standard for the Palaeocene/Eocene boundary, essentially because it is globally recognisable, that is, not only within the terrestrial but also within the marine record (Fig. 5.3). In the marine realm it is coincident with a major benthic foraminiferal extinction event, and/or a major turnover in 'morphogroups' *sensu* Jones and Charnock, 1985 (Jones, in Jones and Simmons, 1999).

One might speculate that on account of the essentially tropical climate obtaining at the time, there may well have been abundant vegetation available in the mires of the delta-top of the Balder formation of Scotland to nourish the land mammals after their journey across the recently volcanic landscape of the Greenland–Faeroes ridge!

Modified after Roberts *et al.* (in press).

5.5.8 *Aspects of the palaeogeography and palaeoclimate of the Oligocene–Holocene of the Old World, and consequences for land mammal evolution and dispersal*

Old World palaeogeography and palaeoclimate over the Oligocene–Holocene interval were affected by plate movements and mountain-building (tectonism) and associated effects on atmospheric and oceanic circulation (for example, the convergence of India and Eurasia and the formation of the Himalayas; and the divergence of South America and Australasia from Antarctica and the formation of a circum-Antarctic seaway, and hence a source of cold, dense water with which to power the psychrosphere), and by cyclic fluctuations in sea level (glacioeustasy). Palaeoclimate fluctuated cyclically, generally, though never irreversibly, deteriorating until the Pleistocene, which was characterised by widespread glaciation.

Before the closure of the Tethyan seaway in the Middle Miocene, marine faunas, the best documented being invertebrates such as larger benthic and planktonic foraminiferans, corals and molluscs, were able to disperse between the Indo-Pacific and Mediterranean; and afterwards, terrestrial faunas, the best documented being vertebrates such as mammals, were able to disperse between Afro-Arabia and Eurasia (Jones, in Agusti *et al.*, 1999).

All available evidence pertinent to the palaeogeographic and palaeoclimatic evolution of the Old World over the Oligocene–Holocene is collated below, and the consequences for land mammal evolution and dispersal are described, in the form of a series of time-slice maps. Complementary and some supplementary information is given by Rogl, in Agusti *et al.* (1999), Rosenbaum *et al.* (2002), Bohme (2003) and Meulenkamp and Sissingh (2003).

The time-slices selected for palaeogeographical mapping are: Early Oligocene (Rupelian), Late Oligocene–earliest Miocene (Chattian–Aquitanian), late Early–early Middle Miocene (Burdigalian–Langhian), Middle Miocene (early–middle Serravallian), late Middle–early Late Miocene (late Serravallian–Tortonian), latest Miocene (Messinian), Early Pliocene (Zanclian), Late Pliocene (Piacenzian) and Pleistocene–Holocene (an Eocene time-slice is also included for the purposes of comparison: Fig. 5.5). Importantly, the mean duration of these time-slices, dictated by biostratigraphic resolution, is much shorter than the measured frequency of climate change over the same time interval, such that both climate and climatically induced sea-level could have changed within each time-slice, and each time-slice is best regarded as a composite.

The time-slice palaeogeographies are constrained in part by larger benthic foraminiferal palaeobiogeographic distribution data (see Jones, in Agusti *et al.*, 1999; see also Box 3.2). Palaeoclimate is constrained in part by LBF diversity data (see Jones, in Agusti *et al.*, 1999; see also Box 3.2), and by coral diversity data (Rosen, in Agusti *et al.*, 1999); and, on land, by land plant – palaeobotanical and palynological – data (Suc *et al.*, in Agusti *et al.*, 1999), and by land animal data (Reumer, in Agusti *et al.*, 1999).

Early Oligocene (Rupelian)

Palaeogeography

LBF commonality data indicates that at this time the Tethyan seaway extended from the Indo-Pacific in the east, through the area of the modern Mid-East Gulf and the Mediterranean basin, and into the Aquitaine Basin and Atlantic in the west (Fig. 5.6). The Gulf of Aden was open, the Red Sea unopened, Africa and Arabia contiguous.

Palaeoclimate

LBF diversity data indicates a range of SSTs from a low of 15 °C in Australasia, thought to have been located almost entirely in extra-tropical palaeo-latitudes, to a high of 25 °C in the northern Mediterranean. Note, though, that LBF diversity is sensitive not only to temperature but also other factors such as facies, evolution, dispersal/migration, duration of studied interval and/or size of studied area, sampling artefact and taxonomic artefact (see Box 3.2).

Palynological and palaeobotanical evidence points toward a warm temperate to tropical humid climate. There was evidently extensive development of boreo-tropical elements in Europe, including

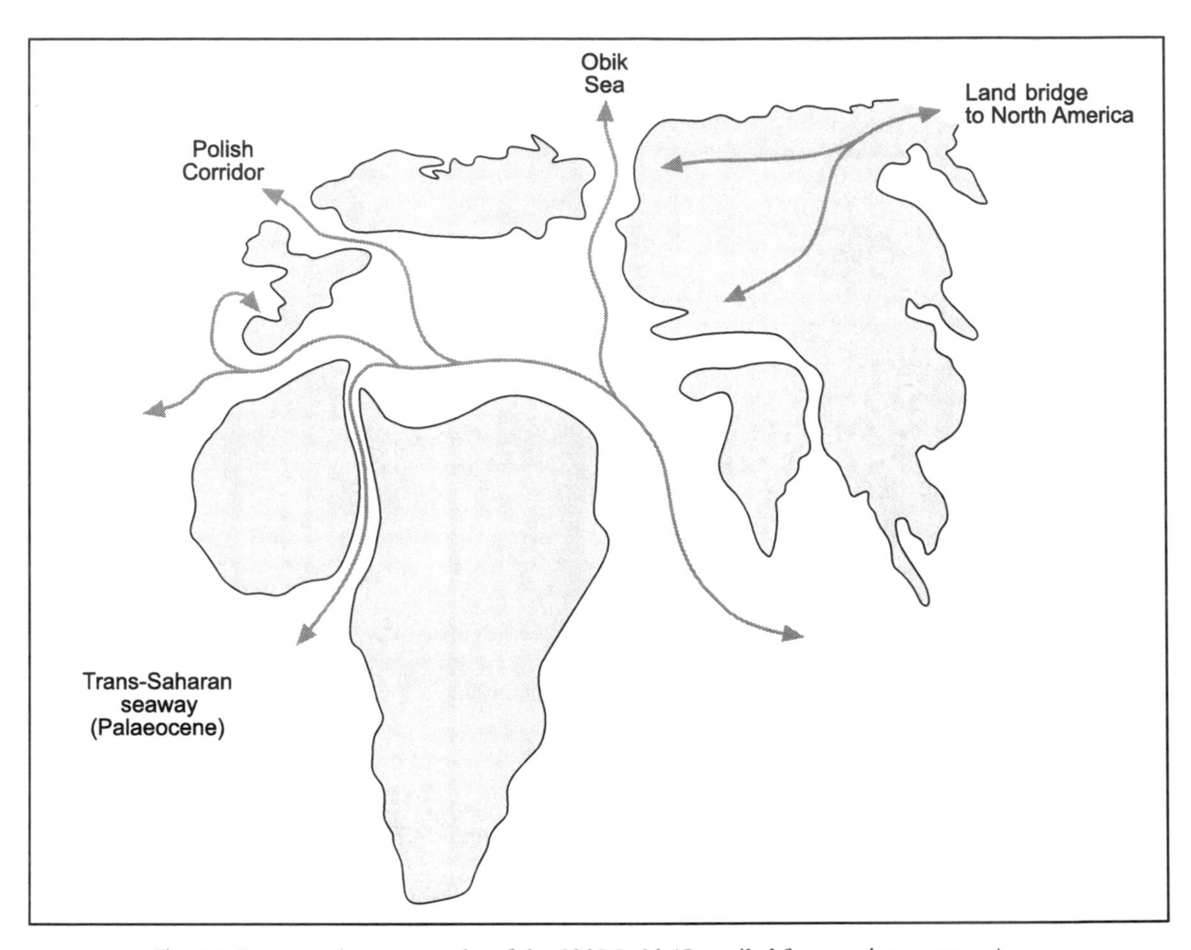

Fig. 5.5. Eocene palaeogeography of the Old World. (Compiled from various sources.)

mixed mesophytic forest, marsh and aquatic vegetation, in turn including mangroves, in Germany (Barthelt, 1989; Mai, 1998; Morley, 2000; Utescher *et al.*, 2000); although there were also separate northern, central and southern European provinces (Akhmetiev, 2000). There was also evidently extensive development of tropical rain-forest in equatorial Africa, the Indian subcontinent, parts of south-east Asia and northernmost Australasia (Morley, 2000).

Land mammal evolution and dispersal
Camels evolved in the Oligocene, in North America, and later dispersed to their present-day heartlands in the deserts of north Africa, the Middle East and central Asia, and became adapted to the extreme aridity there. Indeed, the dispersals of various mammals from the Americas into Eurasia, by way of land bridges across the Bering Sea and Obik Sea/Turgai Strait appear to have taken place at this time (see Box 5.1). This event is sometimes referred to as the 'grande coupure'.

Late Oligocene–earliest Miocene (Chattian–Aquitanian)

Palaeogeography
LBF commonality data indicate that at this time the Tethyan seaway from the Indo-Pacific to the Mediterranean was probably still more often open, at times of relative high-stand of sea level, than closed, at times of relative low-stand (Fig. 5.7). However, there may also have been an at least intermittent land bridge between Italy, Sicily and Africa (Rosenbaum *et al.*, 2002). By the end of this time interval, the Red Sea had rifted open.

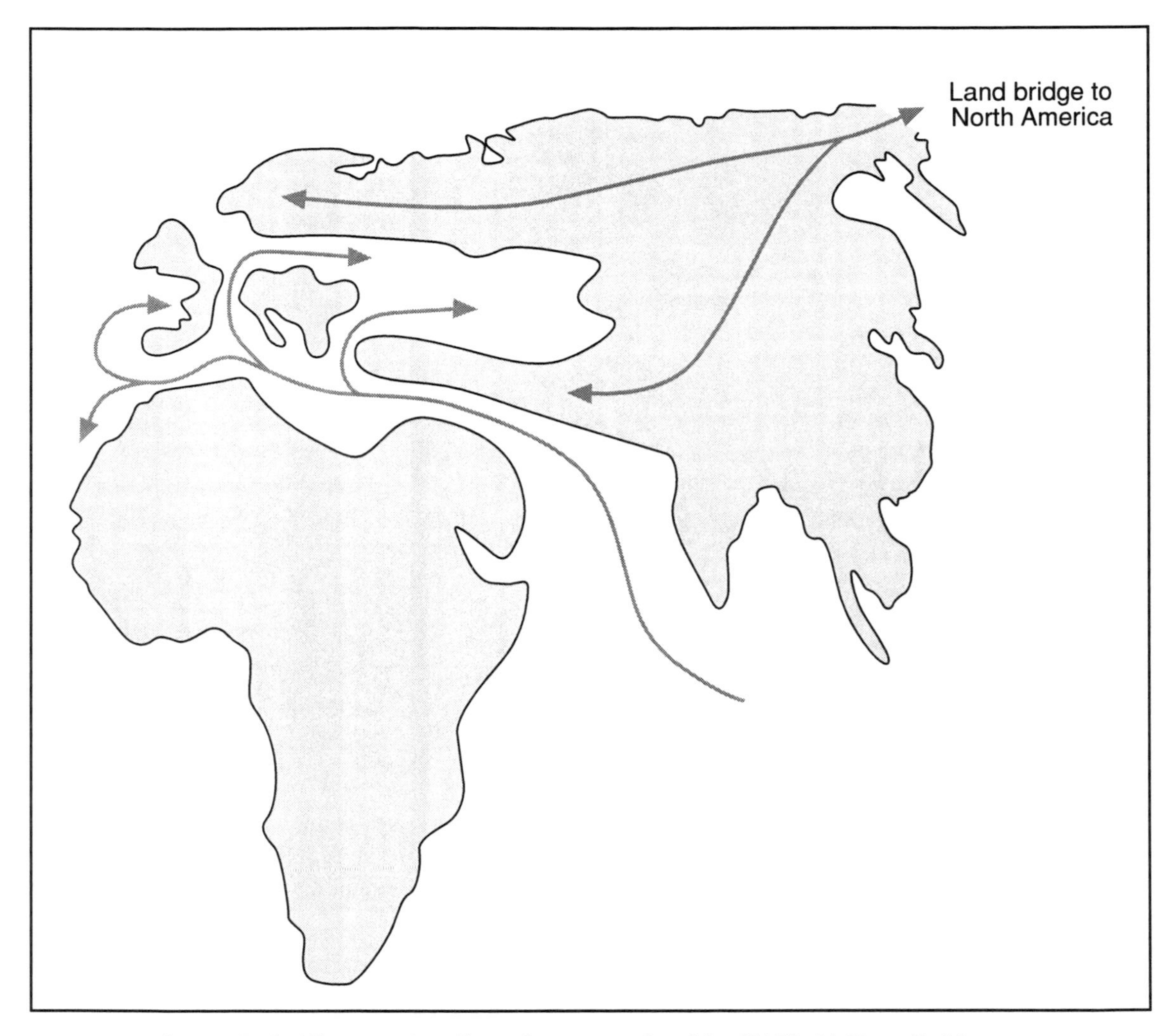

Fig. 5.6. Early Oligocene, Rupelian palaeogeography of the Old World. (Compiled from various sources.)

Planktonic foraminiferans confirm a connection between Tethys and Paratethys (Bicchi *et al.*, 2003).

Foraminiferal and ostracod data indicate that there are three palaeobiogeographic units in the Late Oligocene to Early Miocene of central Europe (Gebhardt, 2003). These are: (1) a northern unit, (2) the upper Rhine sub-province (URSP), comprising the Mainz basin, northern upper Rhine graben and Hanau basin/Wetterau, and (3) the western Paratethys. Progressive isolation of the upper Rhine sub-province through time is indicated by progressive reduction in similarity between it and the other sub-provinces from the basal Miocene onwards. (Similar observations can be made on the basis of molluscs and fish.)

Palaeoclimate

LBF diversity data indicate a range of SSTs from a low of 15 °C in Australasia and east Africa to a high of 25 °C in south-east Asia in the Chattian, and a low of 15 °C in central Europe, thought to have been located at approximately 40 °N, that is, close to the limit of tolerance of modern LBFs, to a high of 25 °C in south-east Asia in the Aquitanian. Note, though, that diversity is sensitive not only to temperature but also to other factors.

Fig. 5.7. Late Oligocene–Early Miocene, Chattian–Aquitanian palaeogeography of the Old World. (Compiled from various sources.)

Palaeobotanical and palynological evidence points toward a warm temperate to tropical, (sub-) humid climate, although with significant provincial differences (Akhmetiev, 2000). There was extensive development of boreo-tropical or Palaeotropic elements in Europe (Kovar-Eder *et al.*, 1998; Utescher *et al.*, 2000). There was multi-tiered forest characterised by canopy, sub-canopy, liana screen and herbaceous ground cover, similar to that of the south-eastern USA wetlands, in the Weisselster basin in Germany in the Chattian (Gastaldo *et al.*, 1998); and acid-tolerant upland riparian forest, swamp-forest and aquatic vegetation in northern Bohemia in the Aquitanian (Kvacek, 1998). There was sub-tropical to tropical so-called mosaic vegetation, with dry, open landscapes adjacent to humid forests, in the circum-Mediterranean (Suc *et al.*, in Agusti *et al.*, 1999). There was tropical rain-forest in equatorial Africa, the Indian subcontinent, parts of south-east Asia and northernmost Australasia (Morley, 2000). There was ever-wet and cool, but frost-free, southern beech or *Nothofagus* forest in New Zealand (Pole, 2003).

Palaeobotanical and palaeopedological (palaeosol) evidence points toward a mosaic vegetation of rain-forest, dry peripheral forest (possibly precursor Afromontane forest), riparian woodland and 'miambo', and dry deciduous woodland and seasonally waterlogged 'dambo' or 'kiewo' in Africa.

Oxygen, strontium and neodymium isotopic evidence from shark teeth indicates a cooling of as much as 4 °C between 22 and 17 Ma in parts of Paratethys, a local effect of the Alpine orogeny (Vennemann and Hegner, 1998).

Land mammal evolution and dispersal

Apes evolved from Old World monkeys, in Africa, sometime around the end of the Oligocene, and diversified through the Miocene (see also Sub-section 3.6.4). Primitive forms, characterised by comparatively small brain-cases, include *Proconsul*, found in association with the 'dambo' or 'kiewo'. *Proconsul* is a quadruped, interpreted as having moved about on all fours, either in trees or on the ground (seeking sanctuary in trees when threatened). It is further interpreted as having been a frugivore, that is, as having subsisted largely on fruit. (Incidentally, it takes its name from 'Consul', a chimpanzee who entertained visitors to London Zoo in the 1930s by performing tricks like riding a bicycle and smoking a pipe.)

The dispersals of various mammals from the Americas into Eurasia and Africa, again by way of the Bering land bridge, appear to have taken place during the Agenian land mammal age, Mein Zones MN1–MN2.

Late Early–early Middle Miocene (Burdigalian–Langhian)

Palaeogeography

LBF commonality data indicate that the Tethyan seaway was closed in the Burdigalian, and open in the Langhian (Figs. 5.8–5.9) (Jones, 2001; Jones *et al.*, in press).

Planktonic foraminiferans confirm a connection between Tethys and Paratethys (Bicchi *et al.*, 2003), as do bivalve and gastropod molluscs (Harzhauser *et al.*, 2003). The relative abundances of cool and warm water species indicate a warming with respect to the Aquitanian.

Palaeoclimate

LBF diversity data indicate a range of SSTs from a low of 15 °C in the southern Mediterranean to a high of 25 °C in south-east Asia. Note again, though, that diversity is sensitive not only to temperature but also to other factors.

Palaeobotanical and palynological evidence again points toward a warm temperate to tropical, (sub-) humid climate. There was extensive development of warm, humid vegetation, locally including mangroves, indicating average temperatures of >20 °C (Plaziat, in Bosence and Allison, 1995), in central Europe (Kovar-Eder *et al.*, 1998; Kvacek, 1998; Utescher *et al.*, 2000). There was continuing development of sub-tropical to tropical mosaic vegetation in the circum-Mediterranean (Suc *et al.*, in Agusti *et al.*, 1999), and of tropical rain-forest in equatorial, eastern and southern Africa, the Indian subcontinent, much of south-east Asia and northern Australasia (MacRae, 1999; Morley, 2000). There was seasonally dry vegetation, characterised by eucalypts and palms, subject to bushfires, in New Zealand (Pole, 2003).

Palaeobotanical and palaeopedological evidence again points toward a mosaic vegetation in Africa.

Independent isotopic evidence indicates that there was a climatic optimum at this time, associated with increased atmospheric carbon dioxide concentration (Pagani *et al.*, 1999). The evolution of various LBFs at this time is interpreted as related to niche diversification during the sea-level high-stand associated with this climatic optimum. As well as diverse LBFs, other sub-tropical to tropical climatic indicators such as coral reefs, indicating averaging temperatures of >16 °C, were present as far north as central Europe, which was situated in a palaeo-latitude of 45 °N (Rosen, in Agusti *et al.*, 1999). Palaeotemperatures in the Styrian basin in Austria have been revealed by analysis of the isotopic composition of brachiopod and bivalve shells (Bojar *et al.*, 2004).

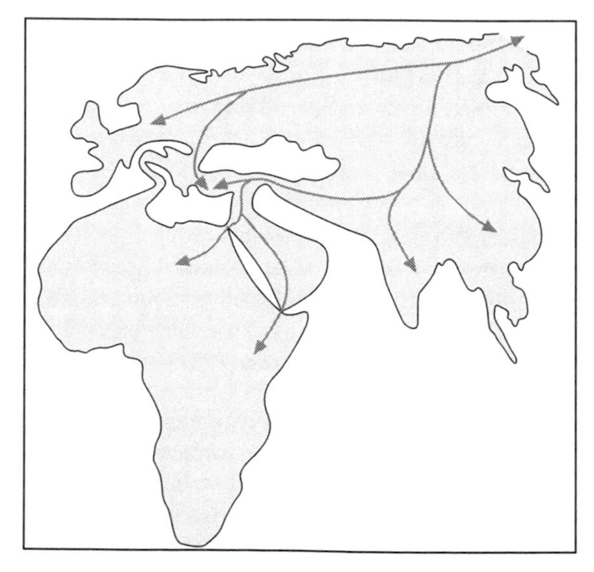

Fig. 5.8. Early Miocene, Burdigalian palaeogeography of the Old World. (Compiled from various sources.)

Fig. 5.9. Middle Miocene, Langhian palaeogeography of the Old World. (Compiled from various sources.)

Land mammal evolution and dispersal

Further dispersals of mammals from the Americas into Eurasia and Africa appear to have taken place during the Orleanian, MN3–MN5, again by way of the Bering land bridge.

Dispersals of various mammals between Africa and Eurasia appear to have taken place at this time, by way of the so-called *Gomphotherium* land bridge, as did the dispersals of mammals within Eurasia. In Eurasia, pliopithecid hominoids of this age are recorded in China to the east and throughout central and western Europe to the west.

Incidentally, dispersals of freshwater fish, and of the crocodilian *Tomistoma*, between Africa and Eurasia, also appear to have taken place at this time (Otero and Gayet, 2001; Kotsakis *et al.*, 2004).

Middle Miocene (early–middle Serravallian)

Palaeogeography

LBF commonality data indicate that at this time the Tethyan seaway was probably more often closed than open.

Planktonic foraminiferal and other data indicates that it was closed in the early Serravallian, and open in the middle Serravallian (Figs. 5.10–5.11) (Jones, in Agusti *et al.*, 1999; Rogl, in Agusti *et al.*, 1999).

Planktonic foraminiferans also indicate an at least intermittent connection between Tethys and Paratethys (Bicchi *et al.*, 2003).

Note, though, that circulation in the Mid-East Gulf and Red Sea and in parts of central Paratethys was at least intermittently extremely restricted, resulting in the formation of hypersaline waters that could have constituted a barrier to the dispersal of marine invertebrates.

Palaeoclimate

LBF diversity data indicate SSTs of $<20\,^{\circ}\mathrm{C}$ throughout the region, that is, a significant cooling with respect to the preceding time interval. Note yet again, though, that diversity is sensitive not only to temperature but also to other factors.

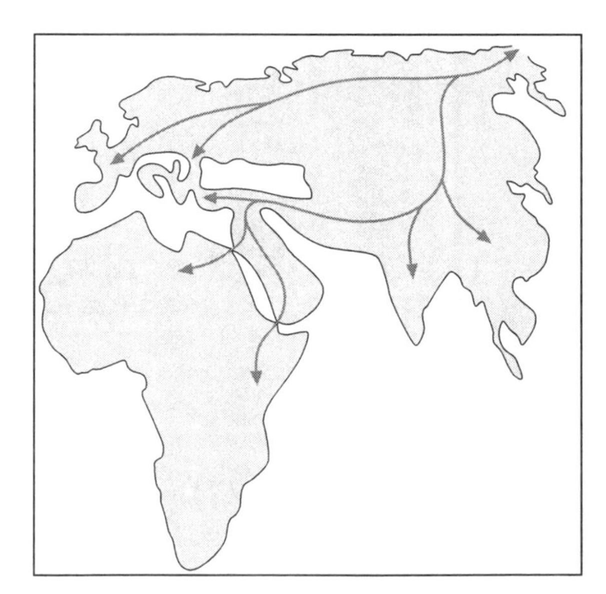

Fig. 5.10. Middle Miocene, early Serravallian palaeogeography of the Old World. (Compiled from various sources.)

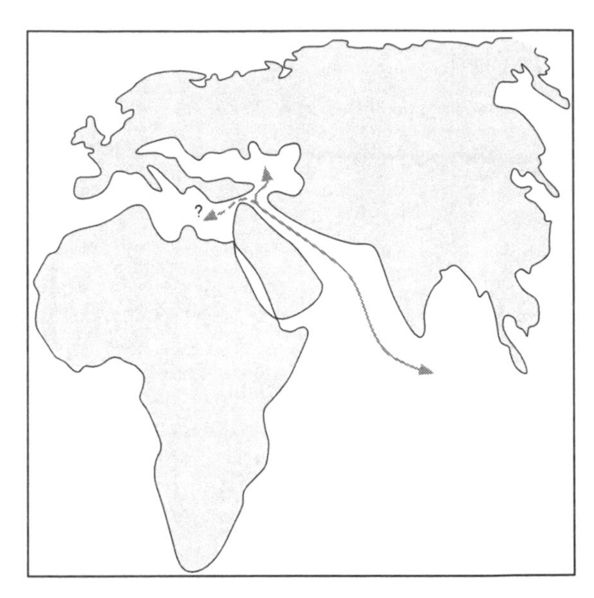

Fig. 5.11. Middle Miocene, middle Serravallian palaeogeography of the Old World. (Compiled from various sources.)

The relative abundances of cool and warm water species of planktonic foraminiferans also indicate a cooling (Bicchi *et al.*, 2003).

Palaeobotanical and palynological evidence points also toward a cooling and/or drying. There was development not only of Palaeotropic but also of Arctotertiary vegetation in central Europe, and of 'altitudinal' and non-arboreal elements in the circum-Mediterranean (Suc *et al.*, in Agusti *et al.*, 1999). There was comparatively dry vegetation, characterised by casuarinaceans, in New Zealand, from 14–15 Ma to possibly as late as 10 Ma (Pole, 2003).

Palaeobotanical and palaeopedological evidence points toward a mosaic vegetation of Afromontane forest, lowland riparian woodland, grassy woodland, wooded grassland and seasonally waterlogged grassland or 'dambo' in Kenya in east Africa.

Land mammal evolution and dispersal

The evolution of numerous mammals – including the hominoid *Ramapithecus* – at this time is regarded by some authors as related to selection pressure associated with the change alluded to above to a cooler and/or more arid climate, which caused its arboreal ancestor to move out of its contracting forest habitat and onto an expanding savannah (Whybrow and Andrews, in Culver and Rawson, 2000; Janis, in Rothschild and Lister, 2003). *Pierolapithecus catalanicus*, from the Middle Miocene, Astaracian, MN7–MN8, 12.5–13 Ma of Pierola in Catalunya, exhibits some primitive monkey-like and some advanced ape-like morphological traits, and has been interpreted as close to the last common ancestor of great apes and humans (Moya-Solà *et al.*, 2004).

The dispersals of numerous mammals – including not only *Ramapithecus* but also *Dryopithecus* and *Sivapithecus* – from Africa into Eurasia appear to have taken place at this time (Astaracian, MN6–MN8). In Eurasia, all three groups are recorded in the Chinji, Nagri and Dhok Pathan stages of the Siwalik series in India and Pakistan; *Ramapithecus* also in China to the east and in Turkey and Greece and along the banks of the Danube to the west; and *Dryopithecus* also in China, in Turkey, in the Caucasus, in Slovakia and Poland, along the banks of the Danube, Eber and Rhone rivers in Hungary, Austria and Germany, and in France and Spain.

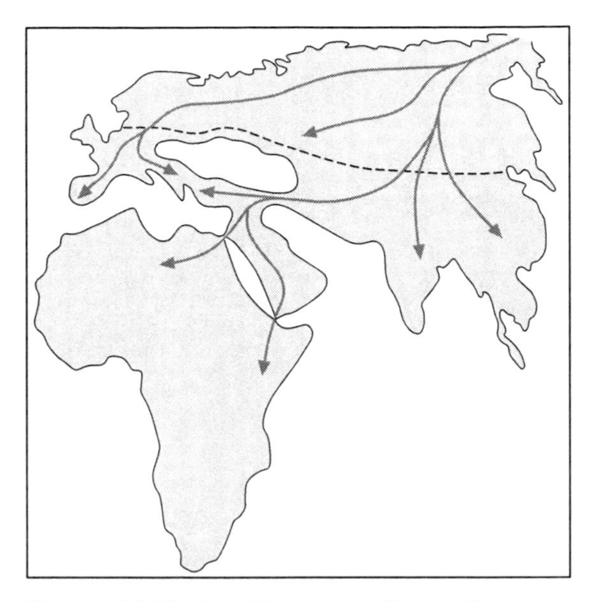

Fig. 5.12. Middle–Late Miocene, late Serravallian–Tortonian palaeogeography of the Old World. Dashed line indicates approximate northern limit of Pikermian province. (Compiled from various sources.)

Late Middle–early Late Miocene (late Serravallian–Tortonian)

Palaeogeography

By now, the Tethyan seaway was probably completely and permanently closed (Fig. 5.12).

Palaeoclimate

The permanent closure of Tethys, together with uplift in the Himalayas, Karakoram and Hindu Kush, apparently led to the development of a distinctly cooler, drier, more seasonal – monsoonal – climate.

Palaeobotanical, palynological and palaeopedological evidence points toward a generally cool, dry climate. There was evidently extensive development of Arctotertiary vegetation in central Europe, and of 'altitudinal' and non-arboreal elements in the circum-Mediterranean (Suc *et al.*, in Agusti *et al.*, 1999). There was a range of habitats featuring tropical forest, monsoon forest, riparian woodland, stream margin scrub, seasonally dry swamp, seasonally waterlogged grassland, and grassland, in Pakistan. There was dry vegetation, characterised by chenopods and asteraceans, in New Zealand (Pole, 2003).

Palaeobotanical, palynological and palaeopedological evidence also indicates a significant vegetational change, from a dominantly C3 to a dominantly C4 system, over the Late Miocene time interval, regionally (Quade and Cerling, 1995; Hoorn *et al.*, 2000; Ghosh *et al.*, 2004; Sanyal *et al.*, 2004). On the Potwar plateau in Pakistan, moist monsoon forest gave way in turn to dry monsoon forest and to grassland over this time interval. In central Nepal, sub-tropical to temperate broad-leaved forest gave way to woodland and grassland. Increases in discharge related to intensification of the monsoonal system resulted in increased incidences of the aquatic taxon *Potamogeton*. Localised development of lakes on the overbanks resulted in increased incidences of the alga *Spirogyra*, and of pteridophyte or fern spores.

Note, though, that the floras of the so-called Pikermian province (France, Greece (Pikermi, Samos), Turkey, Bulgaria, Romania, Moldavia, Ukraine, parts of Russia, Iran (Maragheh) and parts of China) have recently been reinterpreted on palaeobotanical, palynological and palaeopedological evidence as having been of sclerophyllous evergreen woodland (C3) aspect, similar to that of the modern Mediterranean, rather than savannah (C4) aspect – and the faunas that eventually migrated from it into Africa as preadapted rather than adapted to the savannah. The reinterpretation is further supported by the masticatory morphology of the mammals, patterns of tooth micro-wear, and tooth carbonate $\delta^{13}C$ data (a function of dietary $\delta^{13}C$ intake), all of which indicate not only grazing but also browsing and mixed feeding.

Microvertebrate evidence from the Chindigarh area, in the form of the replacement of survival-oriented cricetids by reproduction-oriented murid rodents, points toward inter-annual seasonal variation, and intensification of the monsoon (Patnaik, 2003). Non-grazers were replaced by grazers over the same time interval.

Land mammal evolution and dispersal

A variety of ruminants, adapted to extract the maximum amount of nourishment from poor sources such as grasses, evolved in the Late Miocene, a time of comparatively cold, arid climate, and associated widespread development of grassland. Horse teeth evolved from short-crowned, and adapted to browsing, to long-crowned, and adapted to grazing.

Vallesian, MN9–MN10. The effectively instantaneous dispersal of the three-toed horse '*Hipparion*' from the Americas into Eurasia and Africa, by way of the Bering land bridge, appears to have taken place in the Vallesian. In Eurasia, '*Hipparion*' is recorded in China, Mongolia and Kazakhstan, in the Nagri and Dhok Pathan stages of the Siwalik series in India and Pakistan, in the United Arab Emirates, Iran, Crete, Greece, Turkey and the Caucasus, in the Ukraine, and in Bulgaria, Hungary, Austria, Germany, France and Spain; in Africa, it is recorded in Tunisia, Algeria, Morocco and Kenya. The '*Hipparion*' event has been independently linked to the coincident glacioeustatic sea-level low-stand at approximately 11.1 Ma. The dietary regimes of two populations of *Hippotherium primigenium* from southern Germany, one from palaeo-Rhine deposits of Eppelsheim, and the other from lacustrine deposits of Howenegg, have been established using the 'meso-wear' and 'extended meso-wear' methods (Kaiser, 2003). The Eppelsheim population has been interpreted on the basis of analogy with the common waterbuck, which exhibits similar patterns of wear, as having fed essentially by grazing and foraging in reed beds and fringing woodlands. The Howenegg population has been interpreted on the basis of analogy with the Sumatran rhinoceros as having fed by browsing in sub-tropical mesophytic forests.

The dispersals of numerous mammals both between Africa and Eurasia and within Eurasia also appear to have taken place in the Vallesian. *Anapithecus* made its first appearance in Austria and Hungary, and *Ankarapithecus* made its first appearance in Turkey, during the 'early' Vallesian, MN9; and *Graecopithecus* or *Oranopithecus* made its first appearance in Greece during the 'late' Vallesian, MN10. Significantly, if sampling and preservational effects can be excluded, most hominoids disappeared from western Eurasia toward the end of the Vallesian, although some persisted until the Turolian in southerly refuges such as the Tusco-Sardinian and Pikermian provinces. This has been interpreted as a direct or indirect response to significant climatic and vegetational change, perhaps 'variously delayed and moderated by geographic and biotic filters' (Fortelius *et al.*, in Bernor *et al.*, 1996).

Incidentally, dispersals of crocodilians between Africa and the Tusco-Sardinian province of Europe also appear to have taken place at this time (Kotsakis *et al.*, 2004).

'Early'–'Middle' Turolian, MN11– MN12. Further dispersals of arid steppe mammals (certain rodents, hyenas and antelope) within Eurasia appear to have taken place during the 'early' to 'middle' Turolian, MN11–MN12. The cercopithecid (Old World monkey) *Mesopithecus* made its first appearance in Turkey, Greece and Hungary (Pikermian province) at this time. The dispersals of arid steppe mammals within Eurasia at this time has been interpreted as evidence of further cooling and/or drying. Warmer, more humid conditions, evidenced by the presence of woodland-dwelling murids, were only locally developed. Development of woodland in the Turolian of Afghanistan is indicated by the carbon isotopic signal in the tooth enamel of the bovid *Tragoportax* (Zazzo *et al.*, 2002). Associated development of grassland is indicated by the micro-wear pattern on the teeth of artiodactyls (Merceron *et al.*, 2004). Micro-wear patterns, as revealed by cross-plots of numbers of pits versus numbers of scratches, are similar to those exhibited by Przewalski's horse, *Equus przewalskii*.

Latest Miocene (Messinian)

Palaeogeography

At this time, due to climatic change (cooling/drying), base levels in the Mediterranean and Paratethyan basins fell to such an extent that they became isolated not only from one another but also from the rest of the world's oceans, and, ultimately, during the so-called 'Messinian salinity crisis', desiccated (Fig. 5.13) (Warny *et al.*, 2003).

A multidisciplinary study of the Messinian of the Velona basin in central Italy has recently been undertaken, with a view to establishing its palaeoenvironmental and palaeogeographic context and evolution (Ghetti *et al.*, 2002). This involved detailed geochemical analyses on biogenic carbonates, sedimentological analyses and palaeontological analyses (charophytes, plant pollen, molluscs, ostracods and mammals). The succession is interpreted as representing a brackish lacustrine or paludal environment rich in aquatic vegetation, surrounded by lowland swamp-forests dominated by swamp-cypresses (Taxodiaceae), and uplands dominated by warm-temperate deciduous forests. The lake or swamp was not connected to the Mediterranean Sea, the previous Tusco-Sardinian palaeobioprovince already disrupted. The molluscs, ostracods and mammals are of central European palaeobiogeographic affinity. The ostracods show no affinity with those of Paratethys.

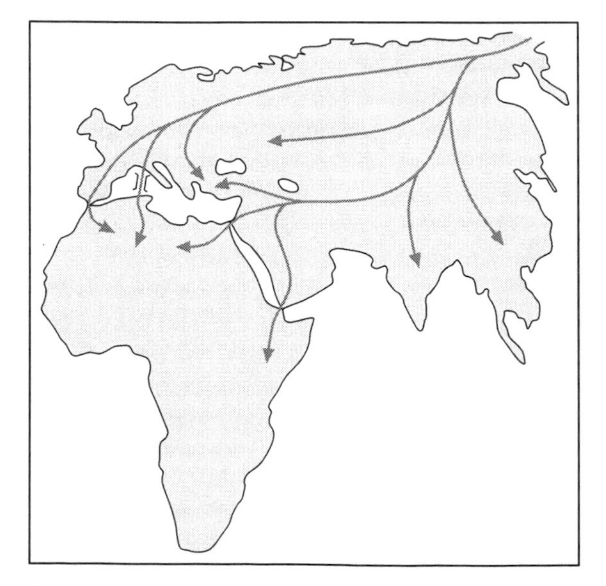

Fig. 5.13. Late Miocene, Messinian palaeogeography of the Old World. (Compiled from various sources.)

Palaeoclimate

Palaeobotanical and palynological evidence again points toward a cool, dry climate. There was continuing extensive development of Arctotertiary vegetation in central Europe, and of non-arboreal elements, lacking mangroves, in the circum-Mediterranean (Suc *et al.*, in Agusti *et al.*, 1999). There was also development, for the first time, of semi-desert elements in the circum-Mediterranean, and of an arid-adapted – Saharan – flora in Africa.

Land mammal evolution and dispersal

The candidate oldest representatives of the human lineage, *Sahelanthropus tchadensis* and *Orrorin tugenensis*, appear to have evolved at this time. *Sahelanthropus tchadensis* is known from the Late Miocene, 6–7 Ma, of Toros–Menalla in the Djurab desert of northern Chad in central Africa (Brunet *et al.*, 2002); *Orrorin tugenensis* from the Late Miocene, 6 Ma, of Kenya (Senut *et al.*, 2001). *Sahelanthropus tchadensis* appears to exhibit a combination of primitive and derived characters, and thus to indicate that the divergence between the hominids and their ape ancestors took place earlier than indicated by molecular data. The fauna and flora found in association with it indicate desert, savannah, gallery forest environments in the vicinity, together with lakes (Vignaud *et al.*, 2002).

The dispersals of numerous mammals between the Americas and Eurasia, by way of the Bering land bridge, appear to have taken place during the 'late' Turolian, MN13. The continued dispersal of arid steppe mammals within Eurasia has been taken to imply continuing climatic deterioration. The dispersals of numerous mammals between Africa and Eurasia, by way of a land bridge at the foot of the Iberian peninsula, an island chain across the Mediterranean, or, ultimately, the dried-up seabed of the Mediterranean, also appear to have taken place at this time. The cercopithecids *Dolichopithecus* and *Macaca* made their first appearances in Europe at this time.

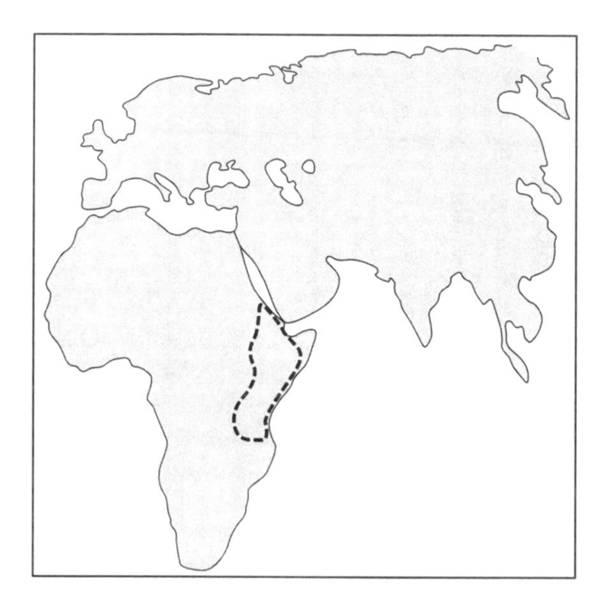

Fig. 5.14. Early Pliocene, Zanclian palaeogeography of the Old World. Dashed line indicates known range of *Ardipithecus* and early *Australopithecus*. (Compiled from various sources.)

Early Pliocene (Zanclian)

Palaeogeography

At this time, the Mediterranean and Paratethyan basins re-opened and re-filled from the Atlantic (Fig. 5.14). Non-marine, 'molasse-type' sedimentation then came to characterise the Caspian borderlands (Jones, 1996; Jones and Simmons, 1996) and much of the Middle East (Jones and Racey, in Simmons, 1994; Jones, 1996). Eurasia thus assumed something approaching its present-day geography.

Palaeoclimate

Palaeobotanical and palynological evidence points toward a temporary amelioration of climate in the earliest Pliocene, from approximately 5.5 to 4.5 Ma (Poore and Sloan, 1996; Thompson and Fleming, 1996; Haywood *et al.*, 2000, 2002). At this time, warm, at least seasonally humid conditions, evidenced by warm-temperate to sub-tropical elements, prevailed in the north-west Mediterranean, and warm, arid conditions, evidenced by open vegetational and grassland elements, in the south-west Mediterranean (Suc *et al.*, in Agusti *et al.*, 1999). Tropical rain-forest developed over much of Africa, with some sub-tropical elements and some savannnah in the north and south in the Langebaanian (MacRae, 1999).

Palaeobotanical and palynological evidence also points toward a deterioration of climate, from approximately 4.5 Ma. In the Shanxi plateau of central China, an increase in *Picea* (spruce) and *Abies* (fir) indicates cooling commencing at 4.4 Ma, possibly in response to a change in the intensity of the east Asian monsoon (Li *et al.*, 2004).

Microvertebrate evidence from the Chindigarh area points toward the development of a range of pond, pond bank, wooded grassland, sandy plain, bushland and temperate montane habitats (Patnaik, 2003). The presence of gerbils and lizards in some deposits points toward the local development of semi-arid or arid conditions.

Land mammal evolution and dispersal

The oldest widely accepted representative of the human lineage, *Ardipithecus ramidus*, appears to have evolved in east Africa at this time. The species is known from the Early Pliocene Aramis member of the Sagantole formation of the middle Awash valley of the Afar rift in Ethiopia, radiometrically dated to 4.39 Ma, and from at least approximately coeval deposits in the Baringo region of Kenya (White *et al.*, 1994; Renne *et al.*, 1999; Wood, in Andrews and Banham, 1999; Wood, in Briggs and Crowther, 2001). Unfortunately, it is represented by only fragmentary skeletal remains – teeth, jaws and an arm – that provide scant clues as to its mode of life.

The evolution of *Ardipithecus* appears to have been in response to selection pressure associated with environmental change (Janis, in Rothschild and Lister, 2003; Potts, in Rothschild and Lister, 2003). As intimated above, the climate became cooler and drier at this time, and at the same time the east African rift valley opened up. The combined effects were the opening up of tracts of grassland or savannah, and the partitioning of the rain-forest. Our ancestors climbed down from the trees and onto the grasslands, while arboreal as against terrestrial apes continued to thrive in rain-forest refuges.

Australopithecus appears to have evolved in east Africa at approximately 4 Ma (Wood, in Briggs and Crowther, 2001). (The recently discovered gracile australopithecine *Kenyanthropus* also appears to have evolved at this time.) *Australopithecus anamensis* is represented by fragmentary skeletal remains, including a knee joint from the shores of Lake Turkana in northern Kenya dated to approximately 4 Ma, the so-called 'carrying angle' of which provides the earliest evidence for bipedalism (Leakey, 1995). The oldest skeletal remains of *Australopithecus afarensis* are from an unnamed formation overlying the Sagantole (see above) in Ethiopia, dated to 3.39 Ma (Renne *et al.*, 1999). Note, though, that this species is thought to have been responsible for the famous trackways at Laetoli in Tanzania, dated to 3.7–3.8 Ma, and also providing evidence for bipedalism. *Australopithecus afarensis* is perhaps best known through the substantially complete skeleton of 'Lucy', discovered in Ethiopia in the 1970s, and dated to 3.2 Ma (Johanson and Edey, 1981; Johanson and Edgar, 1996). ('Lucy' was named after the Beatles song 'Lucy in the sky with diamonds', which was playing on the radio at the time.) 'Lucy' has a pelvis and hindlimb structure of human-like aspect, but a brain of ape-like aspect. Her pelvis is short and horizontal, rather than long and vertical, as in arboreal apes, and her toes are unsuited to grasping; but her brain has a capacity of 415 cm^3, which, for her height of 1–1.2 m, recalls the ratio in modern chimpanzees. As intimated above, *Australopithecus afarensis* and *A. anamensis* are both interpreted as having been capable of bipedalism, that is, walking through grasslands on their hindlimbs, with their forelimbs free, scrutinising the expanding horizon for opportunity and for danger. (Importantly, their upright stances would also have helped keep their braincases cool, the temperature of the air at head height being significantly lower than that at ground level.) However, there are also indications, for example from the lengths of their arms in proportion to those of their legs, that they would have been equally at home climbing trees in remaining woodlands and forests.

There is no evidence that *Australopithecus* dispersed out of Africa in the Early Pliocene.

The dispersals of numerous small mammals both between Africa and Eurasia and within Eurasia appear to have taken place during the Ruscinian, MN14–MN15 (Steininger *et al.*, in Stanley and Wezel, 1985).

Late Pliocene (Piacenzian)

Palaeogeography and palaeoclimate

At least initially, the Late Pliocene, like the early part of the Early Pliocene, was essentially a period of climatic amelioration, with temperatures for the most part higher than today, as evidenced by the widespread occurrence of 'cold-blooded' amphibians and reptiles (Ratnikov, 1996). Palaeobotanical and palynological evidence indicates that warm, humid forests expanded and cool, arid grasslands contracted at this time (Dupont and Leroy, in Wrenn *et al.*, 1999).

However, the Late Pliocene was also a period of high-frequency climatic oscillation and, by its end, of significant climatic deterioration, with widespread evidence of high-latitude glaciation from approximately 3.2 to 2.8 Ma (Thompson and Fleming, 1996). Associated desiccation is indicated by the increased incidence of wind-transported terrigenous material, or loess. In Africa, the effects of cooling and drying were exaggerated by the development of a rain-shadow in the lee of the rising Ethiopian–Kenyan dome (Partridge *et al.*, in Vrba *et al.*, 1995; Morley *et al.*, in Morley, 1999). Palaeobotanical, palynological and palaeopedological evidence indicates that, in both Africa and Eurasia, arid grasslands expanded and humid forests contracted in response to cooling or drying after the time of the onset of glaciation (Dupont and Leroy, in Wrenn *et al.*, 1999; Suc *et al.*, in Agusti *et al.*, 1999; Utescher *et al.*, 2000). In Africa, there was extensive development of arid savannah in the north and south in the Makapanian, in the period 3.0–1.6 Ma (MacRae, 1999). In the Shanxi plateau of central China, arid elements such as *Artemisia*, Chenopodiaceae and *Ephedra* became conspicuously more common at about 2.5 Ma (Li *et al.*, 2004).

Microvertebrate evidence from the Chindigarh area, in the form of a dramatic diversification of murids, points toward a further intensification of the monsoon system at 2.5 Ma (Patnaik, 2003).

Land mammal evolution and dispersal

Australopithecus diversified into a number of species in east – and south – Africa by the Late Pliocene (Fig. 5.15). Gracile australopithecines then appear to have given rise to our genus, *Homo* (see below). (*Gorilla* and *Pan* – gorillas and chimpanzees – probably evolved independently from separate stocks in the forests of West Africa.) The gracile australopithecine *Australopithecus africanus* is perhaps best known through the skull of the 'Taung child', discovered in the Buxton limeworks near Taung in

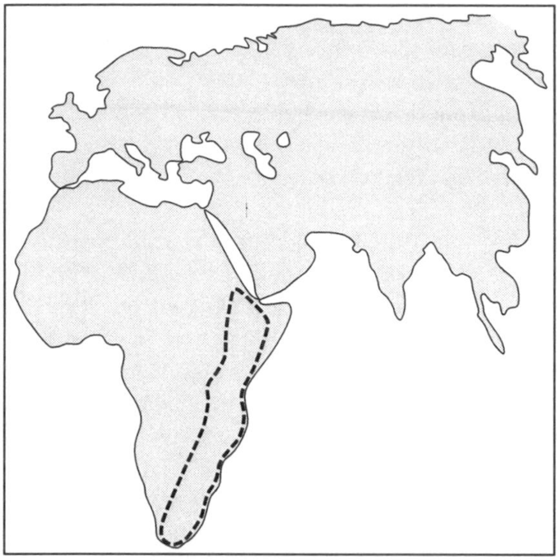

Fig. 5.15. Late Pliocene, Piacenzian palaeogeography of the Old World. Dashed line indicates known range of late *Australopithecus* and early *Homo (H. habilis)*. (Compiled from various sources.)

the Cape Province of South Africa, and dated to approximately 3 Ma (Tobias, 1984; Johanson and Edgar, 1996; MacRae, 1999). The skull of the 'Taung child' is exquisitely preserved, and still bears brain impressions in its interior. However, it does show some damage, consistent with an attack by a large raptor such as an African crowned eagle (Berger and Clarke, 1995) – an event vividly reconstructed by Turner and Anton (2004). Adult individuals of *Australopithecus africanus*, represented, for example, by the synonymous *Plesianthropus transvaalensis* or 'Mrs Ples', from the Sterkfontein limeworks near Krugersdorp, grew to a height of 1.3 m – apparently with disproportionately short legs and long arms – and had a brain capacity of 480 cm^3 (Berger, 2000). In contrast, the robust australopithecine *Australopithecus (Paranthopus)* grew to a height of 1.75 m, but had a brain capacity of only 520–550 cm^3. The oldest representative of the extinct robust australopithecine lineage, *Australopithecus (Paranthropus) aethiopicus*, has been dated to approximately 2.5 Ma, in unnamed deposits in Baringo in Kenya (Wood, in Andrews and Banham, 1999).

The evolution of our own genus, *Homo*, in Africa in the Late Pliocene has once again been interpreted as having been in response to climatic and environmental forcing (Brain, 1981; Susman, in Briggs and Crowther, 1990; Tobias, in Schopf, 1992; de Menocal, 1995; de Menocal, in Vrba *et al.*, 1995; Janis, in Rothschild and Lister, 2003; Potts, in Rothschild and Lister, 2003). This selection pressure would have favoured the socialisation and intelligence necessary for efficient food-gathering, and hence encephalisation and evolution. Interestingly, though, the bipedalism exhibited by *Homo* is regarded by Retallack (1991) as a preadaptation rather than an adaptation to mosaic grasslands, and hence independent of climatic change.

Homo habilis, from the latest Pliocene to earliest Pleistocene, 2.4–1.5 Ma, of Africa, is represented mainly by a number of more-or-less complete skulls, but also by approximately 300 skull and skeletal fragments interpreted as representing a single individual from the Olduvai Gorge in Tanzania – Olduvai Hominid (OH) 62 (Johanson and Edgar, 1996). It apparently grew to a height of 1.3 m, and had a brain capacity of 630–800 cm^3 (600 cm^3 representing the somewhat arbitrary criterion used to distinguish *Homo* from *Australopithecus*). There is abundant circumstantial evidence, in the form of association with worked stones, that *Homo habilis* was the first creature ever to use tools (incidentally, its name translates as 'handy man'). Its diet appears to have differed from that of its ancestors in including nuts and seeds of arid scrubland habitats (Jolly, 1970).

There is no evidence that *Homo* dispersed out of Africa in the Late Pliocene.

The dispersals of numerous large browsing mammals from Africa and the Americas and into Eurasia appear to have taken place during the 'early' to 'middle' Villafranchian, MN16–MN18 (Coryndon and Savage, 1973; Lindsay *et al.*, 1980; Steininger *et al.*, in Stanley and Wezel, 1985).

Pleistocene–Holocene

Palaeogeography and palaeoclimate (Figs. 5.16–5.17 refer)

The Pleistocene was a time of significant climatic deterioration marked by extensive – although intermittent – glaciation in high latitudes and at high altitudes in low latitudes (Fig. 5.16). Glacial stages alternated with interglacials, during which

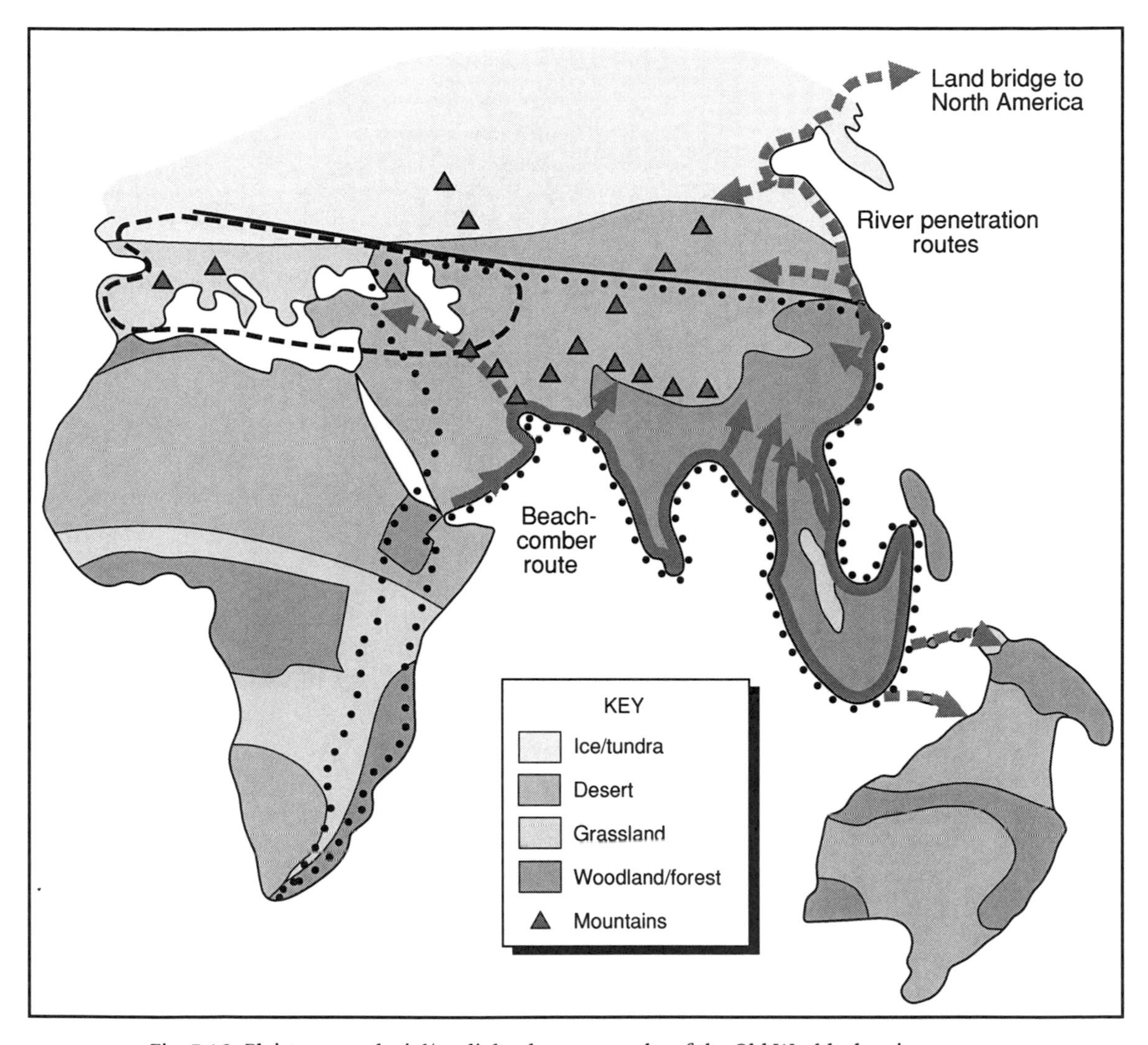

Fig. 5.16. Pleistocene, glacial/stadial palaeogeography of the Old World, showing reconstructed vegetation at the time of a representative glacial/stadial. Also shown are the distributions of the early *Homo* species, *H. erectus, H. heidelbergensis* and *H. neanderthalensis*. Dotted line indicates known range of *H. erectus*; solid line indicates approximate northern limit of *H. heidelbergensis*; dashed line indicates known range of *H. neanderthalensis*. Solid arrows indicate possible dispersal routes used by early *Homo* species, and also by *H. sapiens*, in its colonisation of the world from 85 000 years ago onwards. Dashed arrows indicate dispersal routes probably used only by *Homo sapiens*. (Compiled from various sources.)

the climate was as warm as or warmer than that at the present day (Fig. 5.17). During glacials, vegetation belts in the northern hemisphere were displaced southward by up to as much as 1500–2000 km in the former Soviet Union (Grichuk, in Velichko, 1984; Zubakov and Borzenkova, 1990). Biogeographic provinces in the oceans were displaced southward by similar amounts in the north-east Atlantic (Jones and Whittaker, in press; see also Section 7.6), presumably on account of the associated southward

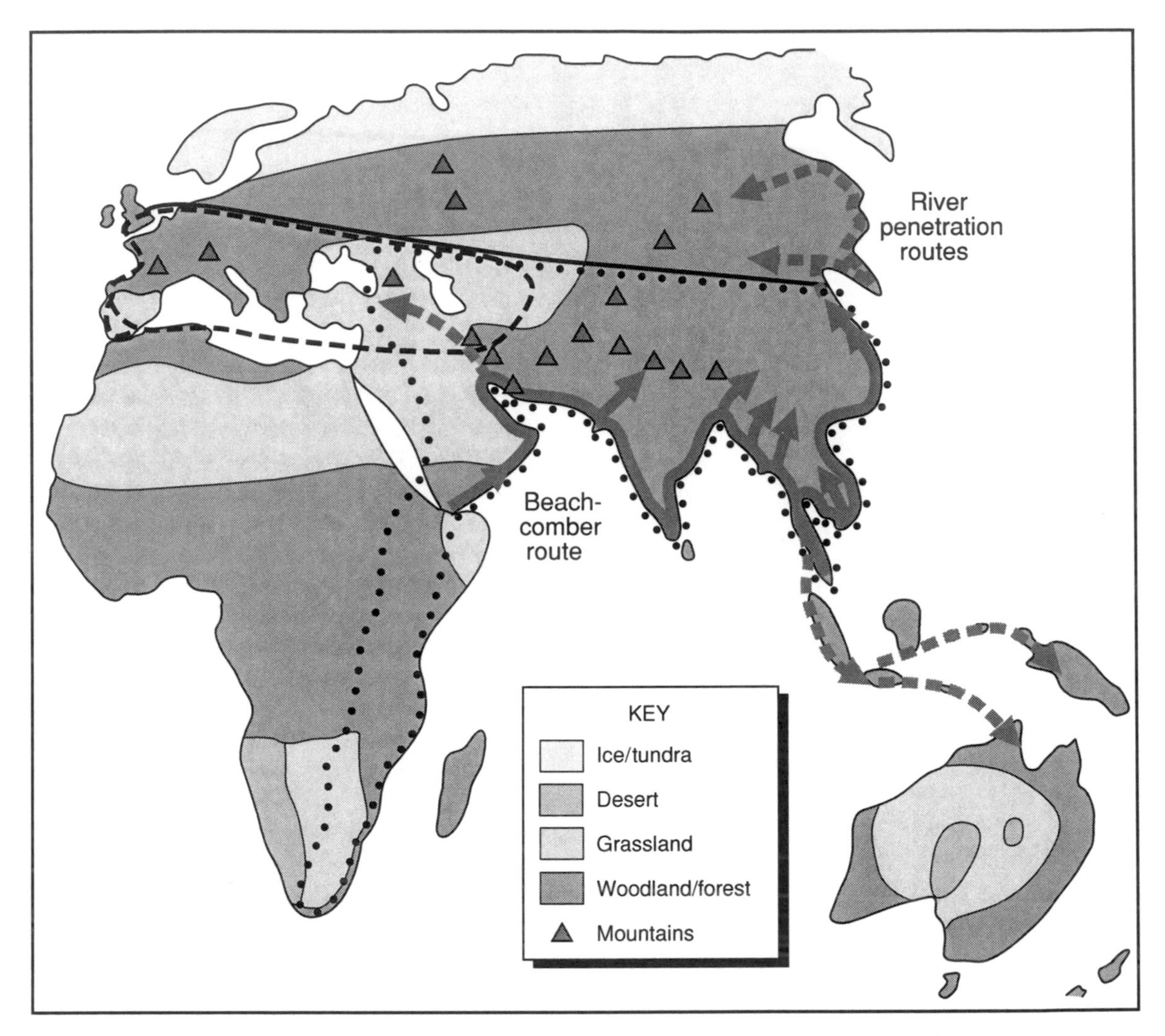

Fig. 5.17. Pleistocene, interglacial/interstadial palaeogeography of the Old World, showing reconstructed vegetation at the time of a representative interglacial/ interstadial. Also shown are the distributions of the early *Homo* species, *H. erectus*, *H. heidelbergensis* and *H. neanderthalensis*. Dotted line indicates known range of *H. erectus*; solid line indicates approximate northern limit of *H. heidelbergensis*; dashed line indicates known range of *H. neanderthalensis*. Solid arrows indicate possible dispersal routes used by early *Homo* species, and also by *H. sapiens*, in its colonisation of the world from 85 000 years ago onwards. Dashed arrows indicate dispersal routes probably used only by *Homo sapiens*. (Compiled from various sources.)

displacement of the North Atlantic drift current or 'gulf stream', which, incidentally, also had a marked effect on the maritime provinces of north-west Europe.

As intimated above, palaeobotanical, palynological and palaeopedological evidence indicates that grasslands expanded and forests contracted during glacials. (Note, though, that recent evidence from molecular isotope stratigraphy of long-chain *n*-alkanes suggests that there was no particular aridity during the last glaciation in 'Sundaland' in south-east Asia: Jianfang Hu *et al.*, 2003.) In Africa as a whole, there was extensive development of tropical forest in the Cornelian, 1.6–0.5 Ma and Florisian, 200 000–12 000 years ago only in – western – equatorial regions, with significant development of savannah elements in the north and south, and dry desert in the extreme north and south (MacRae, 1999). In west Africa, Afromontane forest expanded at the expense of lowland rain-forest other than in the warmer, wetter, more monsoonal interglacials (Dupont *et al.*, 1998).

Land mammal evolution and dispersal

The evolution and dispersal of numerous mammals of modern aspect, and the extinctions of those of archaic aspect, appear to have taken place during the 'late' Villafranchian (Azzaroli, 1983).

Homo diversified into a number of species in the Pleistocene (Wood, in Briggs and Crowther, 2001). The interpretation of their evolutionary relationships is still controversial, and the argument remains unresolved as to how a number of nominal species – beyond those considered here – fit in. Nonetheless, the consensus seems to be that *Homo erectus/ergaster* evolved from *H. habilis*, and evolved into *H. heidelbergensis*, or 'archaic *H. sapiens*' (and *H. floresianus*). Certainly, this interpreted lineage seems to have been characterised by the most rapid proportional brain growth, although there is still something of a 'muddle in the middle' in between *Homo erectus/ergaster* and *H. sapiens* (Elton *et al.*, 2001; Oppenheimer, 2003). DNA evidence indicates that *Homo neanderthalensis* is not part of the lineage leading to *H. sapiens* (Krings *et al.*, 1997; Jones, 2001).

Homo erectus/ergaster is interpreted as having evolved from *H. habilis* in Africa approximately 1.8 Ma, and as having become extinct some time after 250 000 years ago (Boaz and Ciochon, 2004). It is perhaps best known through the virtually complete skeleton of the 'Nariokotome boy' or 'Turkana boy', discovered at Nariokotome on the west side of Lake Turkana in Kenya, and dated to approximately 1.6 Ma (Johanson and Edgar, 1996), but also through the remains of 'Java man' from Trinil on Java in Indonesia (Curtis *et al.*, 2001) and 'Peking man' from Zhoukoudian or Dragon Bone Hill 40 km south of Beijing in China (Institute of Vertebrate Palaeontology and Anthropology, Chinese Academy of Sciences, 1980). *Homo erectus/ergaster* grew to a height of 1.6 m, and had a brain capacity of 850–1100 cm^3. The large, robust skull of the species is interpreted as having fulfilled one or more of a number of possible functions, including: providing sites of attachment for stronger chewing muscles; encasing a larger brain; cooling the brain; and protecting the brain, spinal cord and eyes (Boaz and Ciochon, 2004). Interestingly, there is some palaeopathological evidence to support the interpretation that it fulfilled a primarily protective function: skulls have been found with depressed fractures that had healed rather than proved fatal. Protection may have been against attack from other individuals, which is hypothesised as having been the preferred method of conflict resolution! There is some evidence that *Homo erectus/ergaster* used weapons, for hunting, as well as – Oldowan – tools; that it had mastered the use of fire, for cooking; and that it occupied semi-permanent settlements, although it may also have migrated seasonally and/or over the longer term (?to track game). It may even have evolved a basic, tribe-like societal structure, and a means of communication through either speech or sign language.

Homo erectus/ergaster is interpreted as having dispersed out of Africa and as far as the Far East (China, and, by way of the then-emergent Sunda shelf, Java) by 1.8?–1.2 Ma, perhaps in response to further climatic and environmental forcing. By this time, due to climatic change, the formerly impenetrable forests of Africa no longer presented a barrier to dispersal, having given way to treeless grassland and desert: significantly, the diet of *H. erectus/ergaster* appears to have differed from that of its ancestors in including meat from scavenging and hunting animals of grassland habitats. In Eurasia, the former shelf seas of the south no longer presented a barrier to dispersal either, having retreated and given way to coastal plains. However, the intermittently glaciated highlands and periglacial lowlands of the north – the so-called 'cold wall' – did. Moreover, the 'Wallace line', delineating the boundary between the essentially continental land mass of Eurasia to the west and the essentially oceanic archipelago of south-east Asia and Australasia to the east, also still presented a barrier to dispersal to the east of Java. Incidentally, the event that emplaced the lower Lahar in the Sangiran formation of central Java at approximately 1.9 Ma transformed previously marine environments into estuarine and paludal ones, immediately prior to the immigration of *Homo erectus* (Bettis *et al.*, 2004).

Homo heidelbergensis, or 'archaic' *H. sapiens*, is interpreted as having evolved from *H. erectus/ergaster* approximately 600 000 years ago, and as having become extinct approximately 150 000 years ago. It is perhaps best known through the skulls from Broken Hill in Zambia, Bodo d'Ar in Ethiopia, Arago in France, Steinheim in Germany, and Petralona in Greece, the lower jaw from Mauer, near Heidelberg, in Germany, and the recently discovered tooth and leg bone from a raised beach at Boxgrove on the Sussex coastal plain on the south coast of England (Johanson and Edgar, 1996; Stringer, in Roberts and Parfitt, 1999). Although it is impossible to be certain from the somewhat fragmentary fossil record, *Homo heidelbergensis* apparently grew to a height of 1.8–1.9 m, and had a brain capacity of 1300 cm^3. Circumstantial evidence in the form of associated finds from the Boxgrove site indicates that it used weapons, for hunting, as well as – Acheulian – tools. *Homo heidelbergensis* remains, and/or Acheulian tools, are found throughout Africa and Eurasia.

Homo neanderthalensis is interpreted as having evolved from *H. heidelbergensis* approximately 250 000–300 000 years ago, and as having become extinct approximately 27 500 years ago (Stringer and Gamble, 1993; Trinkaus and Shipman, 1993; Shreeve, 1996; Aiello, in Briggs and Crowther, 2001; Oppenheimer, 2003; Finlayson, 2004). Interestingly, *H. neanderthalensis* evidently coexisted with *H. sapiens* between 35 000 and 27 500 years ago in southern France and the Iberian Peninsula, which at the time would have formed refuges from the glacial ice further north (Oppenheimer, 2003). *Homo neanderthalensis* was evidently comparatively short and stocky, on account of adaptation to the generally cold climate obtaining at the time – an adaptation also exhibited by modern Inuit peoples – but had a brain capacity averaging 1400 cm^3. It appears to have evolved an advanced societal structure, characterised by communal hunting, the preparation and wearing of clothes of animal skins, care of the elderly and infirm, and, arguably, such respect for the dead, including burial, as to suggest some form of religious belief system. Incidentally, burial of the dead led to a significant improvement of fossilisation potential, by denying scavengers their former freedoms. *Homo neanderthalensis* remains, and/or Mousterian tools, are found throughout the area bounded by southern England, Spain, the Levant and Kazakhstan.

'Anatomically modern' *Homo sapiens* is interpreted as having evolved independently from *H. heidelbergensis*, or 'archaic' *H. sapiens*, in Africa, approximately 150 000 years ago (Tobias, in Schopf, 1992; Stringer and McKie, 1996; Stringer, in Culver and Rawson, 2000; Wood, in Briggs and Crowther, 2001). We have a brain capacity averaging 1360 cm^3 – marginally smaller than that of *Homo neanderthalensis*! We have supposedly evolved still more advanced societies and cultures, the latter manifested in the adornment of the body in life and after death, and in the making of images, from at least 35 000 years ago (Bronowski, 1973; Gombrich, 1995; Mellars, in Cunliffe, 1998; Mithen, 1998; Lewis-Williams, 2002).

According to the most widely accepted hypothesis, supported by molecular biological – mitochondrial DNA – evidence, *H. sapiens* dispersed 'out of Africa' and into Asia, by inference, through the Bab el Mandeb or 'gate of grief' in the south-western Arabian Peninsula, approximately 85 000 years ago (Sykes, 2001; Mithen, 2003; Oppenheimer, 2003). By 75 000 years ago, groups of *H. sapiens* had reached the then-emergent 'Sundaland' in South-East Asia, by way of the coastal so-called 'beachcomber route'; and by 65 000 years ago, eastern Indonesia, New Guinea and Australia, by 'island-hopping' (Oppenheimer, 1998; Stringer, 2000; Walter *et al.*, 2000; Xiangjun Sun *et al.*, 2000; Oppenheimer, 2003). By 50 000 years ago, they had reached Europe, by way of the overland so-called 'fertile crescent' or 'Caucasus' routes; and by 40 000 years ago, the Indian subcontinent and central Asia, also, obviously, by overland, or by 'river penetration', routes (Oppenheimer, 2003). By 30 000 years ago, they had reached Siberia, in north-east Asia, and by 25 000 years ago, Alaska, in northernmost North America, by inference, by way of the then-emergent 'Beringia' (Guidon and Delebrias, 1986). Alternatively, they may have entered the Americas by way of a Pacific crossing, or, according to the 'Solutrean' hypothesis, by way of an Atlantic crossing from Spain (Jett, 2004). By 12 500 years ago, they had reached southernmost South America (Haynes, 2002). The first permanent – as opposed to seasonal, nomadic – settlements date to approximately 11 000 years ago, coincident with the Younger

Dryas stadial and the enforced widespread implementation of agriculture (Harris, in Rothschild and Lister, 2003; Watkins, 2004).

The recently discovered dwarf species *H. floresiensis* is interpreted as having evolved from *H. erectus/ ergaster*, in the Far East, some time before 38 000 years ago, and as having become extinct 18 000 years ago (Brown *et al.*, 2004; Morwood *et al.*, 2004). It grew to a height of only 1 m, and had a brain capacity of only 380 cm^3, although it apparently manufactured stone tools with which to hunt Komodo dragons, giant rats and the pygmy elephant *Stegodon*.

6 • Biostratigraphy and sequence stratigraphy

Biostratigraphy deals with the documentation, analysis and interpretation of the ordered succession of fossils, their relationships to evolving earth and life history, and their application to the elucidation thereof (Emery and Myers, 1996; Jones, 1996; Doyle and Bennett, 1998; Backman, in Briggs and Crowther, 2001; Bowring and Martin, in Briggs and Crowther, 2001; Marshall, in Briggs and Crowther, 2001; Rawson, in Briggs and Crowther, 2001; Coe, 2003; Harries, 2003; Gradstein *et al.*, in Gradstein *et al.*, 2005; McGowran, in press). Together with sequence stratigraphy, it is one of the principal bases for chronostratigraphic subdivision and correlation of lithological units, thus providing a spatio-temporal context for their interpretation, and is a fundamental building-block of earth science. This chapter deals with applications of biostratigraphy and sequence stratigraphy in the interpretation of earth – and life – history.

6.1 Biostratigraphy

The stratigraphic distributions of the principal fossil groups discussed in Chapter 3 are summarised in Fig. 6.1. The ranges over which they are stratigraphically useful are shown by broad bands. It is evident that most are only useful over certain time intervals, and then only in the appropriate facies. Note also that the potential biostratigraphic usefulness of fossils can be impaired by natural factors, such as *post-mortem* transportation and diagenetic effects (see Section 2.2), and reworking. The biostratigraphic usefulness of fossils can also be impaired by artificial factors, such as sample acquisition and processing, and subjectivity in specific identification.

Some of the more stratigraphically useful fossil groups are the dinoflagellates, diatoms, calcareous nannoplankton, acritarchs, *Bolboforma*, foraminiferans, radiolarians, calpionellids, plants, archaeocyathans, ammonoids, belemnites, tentaculitids, trilobites, graptolites, chitinozoans, conodonts and mammals (see appropriate sections in Chapter 3; see also Sub-section 6.1.1). Other groups are only locally important.

Significantly, the more stratigraphically useful fossil groups share two common characteristics: firstly, relatively rapid rates of evolutionary turnover, and hence restricted stratigraphic distributions, and/or essentially isochronous first and last appearances; and secondly, essentially unrestricted ecological distributions (for example, throughout the marine realm, and across a range of biogeographic provinces, in the case of many planktonic or nektonic forms). The most useful groups for practical purposes are also, typically, abundant, well preserved, and easy to identify. These are referred to as 'marker fossils' or 'index fossils'. Conversely, the less stratigraphically useful groups characteristically exhibit relatively slow rates of evolutionary turnover, and/or diachronous or time-transgressive first and last appearances. Alternatively, they may exhibit restricted ecological distributions (for example, to individual bathymetric zones, in the case of many benthic forms). Note, though, that the very ecological restriction exhibited by these groups renders them palaeobiologically useful 'facies fossils' (see Chapter 4).

6.1.1 Biostratigraphic zonation or biozonation

The fundamental unit in biostratigraphy is the biozone (Fig. 6.2) (Bassett, in Briggs and Crowther, 1990). The main types of biozone are defined on a combination of: evolutionary or first appearance datums or FADs of marker species; extinction or last appearance datums or LADs; total, partial or concurrent ranges; and abundances or acmes. In the oil industry, the main type of zone is defined on LADs (colloquially known as 'tops'), because the precise positions of FADs ('bases') are often impossible to locate accurately using ditch-cuttings samples, owing to the problem of down-hole contamination or 'caving'. Integration of biostratigraphy and seismic sequence stratigraphy (see below) enables the discrimination

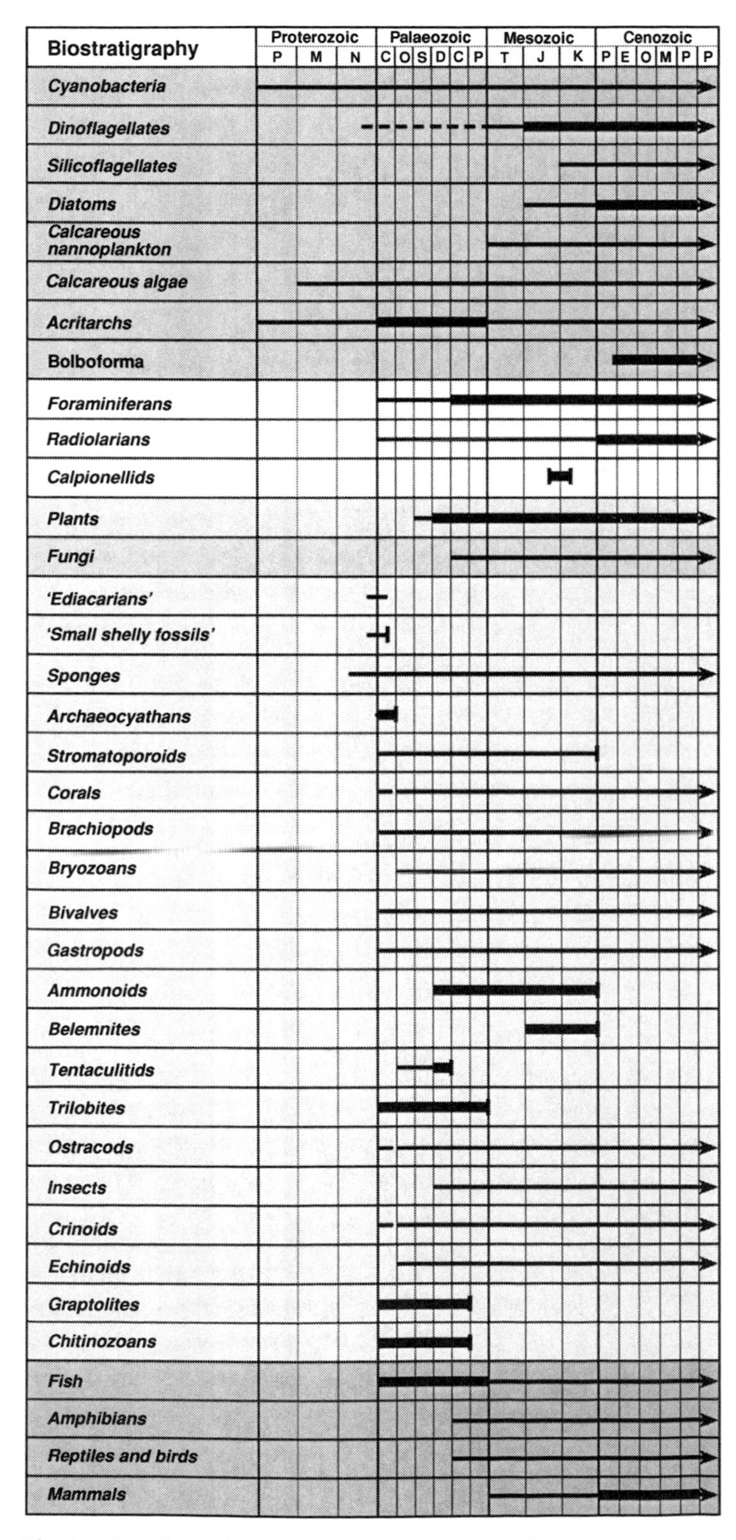

Fig. 6.1. Stratigraphic distribution of selected fossil groups.

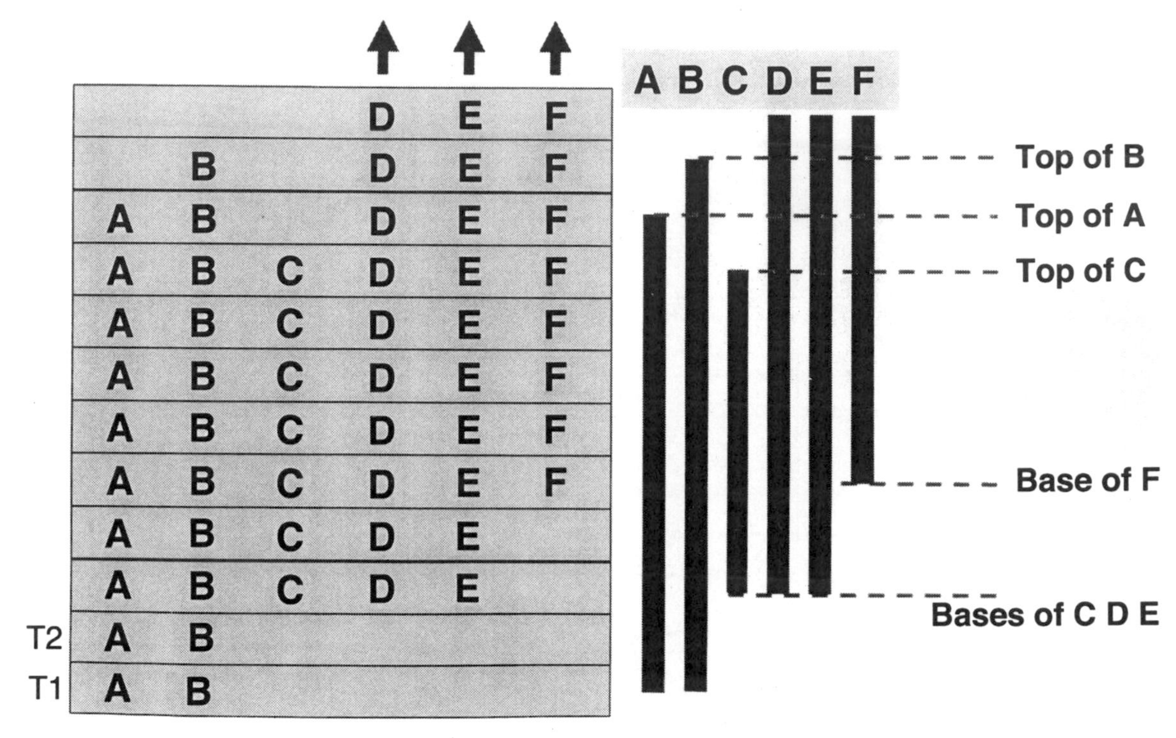

Fig. 6.2. The basis of biostratigraphic zonation. T1, T2 etc. are successive time-slices represented by rock units (I make no formal distinction between chronostratigraphy and lithostratigraphy). A–F are fossils occurring in the various time-slices. Their overall ranges, between their first appearances or bases and their last appearances or tops, are indicated on the right. Biostratigraphic zones or biozones have been defined between the bases of A and B, and C, D and E; between the bases of C, D and E and F; between the base of F and the top of C; between the top of C and the top of A; and between the top of A and the top of B.

of real extinctions from apparent ones – or 'ecologically depressed tops' – and prevents miscorrelations.

Biostratigraphic zonation or biozonation schemes have been established for part of the Proterozoic, and for the whole of the Phanerozoic, that is, the Palaeozoic, Mesozoic and Cenozoic (see below). Biozonation schemes based on acritarchs have been established for part of the Proterozoic. Widely applicable schemes based on acritarchs, foraminiferans, plants, archaeocyathans, ammonoids, tentaculitids, trilobites, graptolites, chitinozoans and conodonts have been established for the Palaeozoic. Schemes based on dinoflagellates, calcareous nannoplankton, foraminiferans, radiolarians, calpionellids, plants, ammonoids and belemnites have been established for the Mesozoic. Schemes based on dinoflagellates, diatoms, calcareous nannoplankton, *Bolboforma*, foraminiferans, plants and mammals have been established for the Tertiary and Quaternary, or Cenozoic. The global standard calcareous nannoplankton and planktonic foraminiferal biozonation schemes for the later Tertiary and Quaternary are applicable in most offshore marine environments worldwide, and have been calibrated against the absolute chronostratigraphic time-scale by so-called 'astronomical tuning' (Hinnov, in Gradstein *et al.*, 2005).

6.1.2 Correlation

The 'law of strata identified by organised fossils', established by William Smith, states that particular ages of rock can be correlated by means of stratigraphic index fossils, irrespective of facies (Fig. 6.3).

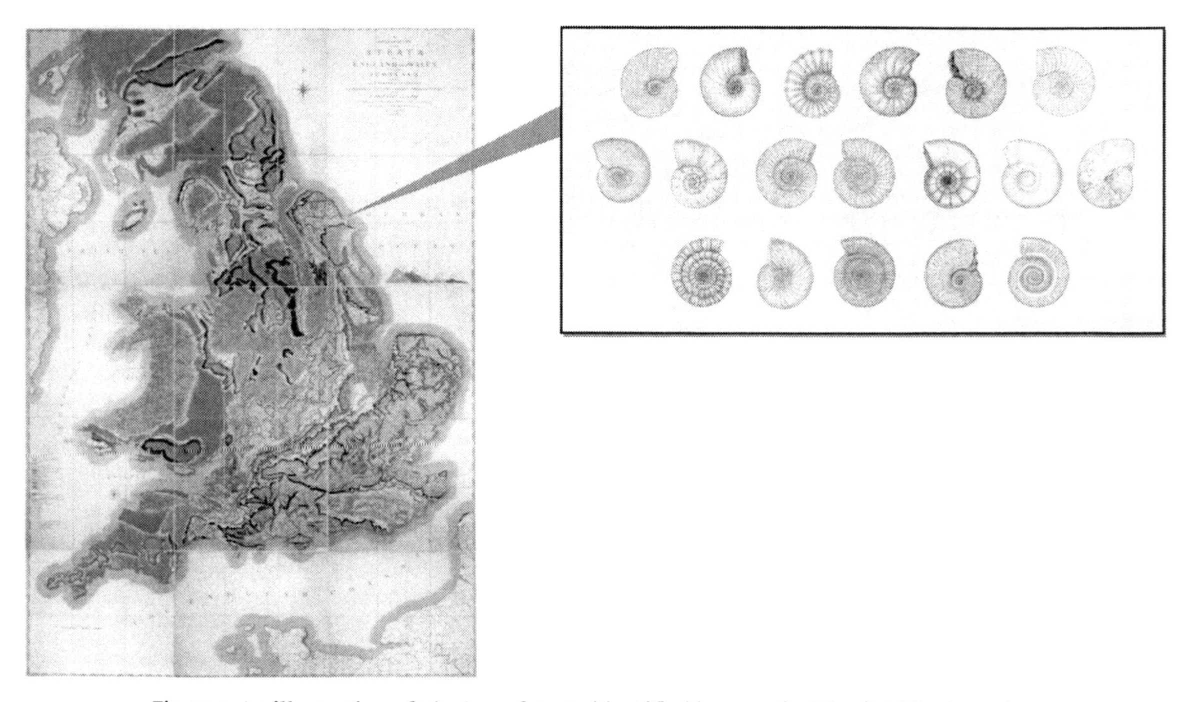

Fig. 6.3. An illustration of 'the law of strata identified by organised fossils'. The Jurassic strata of England, shown here on Smith's geological map cropping out from Yorkshire in the north-east to Dorset in the south-west, have been correlated by means of ammonites (inset).

6.1.3 *Resolution*

The mean duration or resolution of the biozones in many of the published global standard biozonation schemes, such as the ammonite schemes for the Mesozoic, or Martini's calcareous nannoplankton scheme for the Cenozoic, is typically of the order of 1 Ma. The resolution of the proprietary BP calcareous nannoplankton biozonation scheme for the Neogene–Pleistogene of the Gulf of Mexico is of the order of 0.1 Ma, 100 ka, or 100 000 years. Temporal resolution is essentially a function of evolution and extinction rates, rock accumulation rate and sample spacing.

6.2 Biostratigraphic technologies

A number of quantitative techniques have been applied in the field of palaeontology (Armstrong, in Harper, 1999). In my working experience in the oil industry, graphic correlation, and ranking and scaling, have proved of particular use (see below). Biometric or morphometric techniques have also proved of use here and elsewhere, for example, in the establishment of LBF lineages in the Cenozoic of the Tethyan province. Other statistical techniques such as hierarchical cluster analysis have also proved of use, for example, in the establishment of a mollusc-based stratigraphy for the Eggenburgian of Austria.

6.2.1 *Graphic correlation*

Graphic correlation involves correlation of an outcrop or well section against a global composite standard section by treating the two as coordinate axes and plotting the fossil events common to both as a series of points (Fig. 6.4) (Aurisano *et al.*, in Mann and Lane, 1995; Carney and Pierce, in Mann and Lane, 1995; see also Fig. 7.12). The line of correlation (LOC), fitted between the points so as to honour palaeontological and other geological data, provides information as to sediment accumulation rates and stratigraphic breaks that is of use in sequence stratigraphy and in petroleum systems analysis. Rates are indicated by, and indeed can be calculated from, the

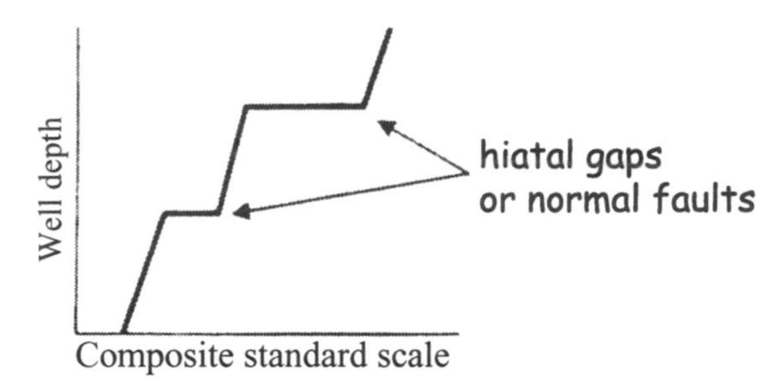

Fig. 6.4. The basis of graphic correlation. Fossil events in the well – or outcrop – section are plotted on the vertical axis, and those in the composite standard on the horizontal axis, and a line of correlation is fitted between them. Plateaux in the line of correlation (LOC) indicate hiatuses (or normal faults).

gradient of the LOC; breaks, and/or consensed sections, from plateaux. Projecting composite standard units through the LOC into the section provides a precise and accurate indication of which units are represented and which unrepresented in the section, and of the degree of expansion or condensation.

Examples of applications of graphic correlation technology in the oil industry have been published by, among others, Neal *et al.*, in Mann and Lane (1995) (Palaeogene, North Sea), Martin and Fletcher, in Mann and Lane (1995) (Plio-Pleistocene, Gulf of Mexico) and Groves and Brenckle (1997) (Permo-Carboniferous, Tarim basin, China) (see also Fig. 7.12). In the oil industry, graphic correlation has become fully automated, and capable of being imported into the workstation environment for integration with well-log and seismic data (Wescott *et al.*, 1998; Jones *et al.*, in Jones and Simmons, 1999). Alternative LOCs are fitted using pathfinding or slotting algorithms (Gary *et al.*, in Powell and Riding, in press). The 'best fit' is selected by appropriate weighting of defining events. Confidence limits can be assigned.

6.2.2 Ranking and scaling

Ranking provides a probabilistic sequence of biostratigraphic events in sections determined by manual scoring or computer-driven matrix permutation. Scaling provides a measure of the inter-event distance determined by cross-over frequency (partly a function of sample spacing). Examples of applications of ranking and scaling technology in the oil industry have been published by, among others, Gradstein *et al.*, 1988 (Cenozoic, North Sea) (see also Fig. 7.10).

6.3 Allied disciplines

6.3.1 Chemostratigraphy

Carbon isotope stratigraphy

There have been fluctuations in $\delta^{13}C$ through geological time, in response to various biological and physico-chemical controls. A standard $\delta^{13}C$ curve has been constructed on the basis of measured values in pelagic carbonates in Tethys. Measured values from elsewhere have been calibrated against this standard in order to establish a local carbon isotope stratigraphy. In my working experience in the petroleum industry, the carbon isotope technique has proved useful in establishing a local stratigraphy for the platform carbonates of the Early Cretaceous Kharaib and Shuaiba formations of Oman and the United Arab Emirates in the Middle East (Vahrenkamp, 1996). In this case, stratigraphic resolution is of the order of 1 Ma, comparable with that attainable using conventional biostratigraphic techniques. Importantly, this is sufficient to resolve the timing of the formation of the intra-shelf Bab basin. The carbon isotope technique is only applicable to sediments essentially unaltered by diagenesis.

Oxygen isotope stratigraphy

As intimated in Section 4.5, high-resolution palaeoclimate curves based on the oxygen isotopic record of changing temperature or ice volume that some fossil groups preserve in their shells, bones or teeth serve, when calibrated against biostratigraphy, magnetostratigraphy or absolute chronostratigraphy, as the basis of a workable marine isotope stage (MIS) or oxygen isotope stage (OIS) climatostratigraphy, at least for the Tertiary and Quaternary. A total of 63 numbered stages are recognised over the Quaternary (Weaver, in Hailwood and Kidd, 1993). Odd-numbered stages represent interglacials; even-numbered ones, glacials.

Strontium isotope stratigraphy

The ratio between strontium-87 and strontium-86 in the marine realm has fluctuated through geological time, in response to various physico-chemical controls. A standard curve of the ratio has been constructed on the basis of biostratigraphically constrained measured values worldwide (McArthur and Howarth, in Gradstein *et al.*, 2005). Measured values from specific localities can be calibrated against this standard in order to establish a local strontium isotope stratigraphy. In my working experience in the petroleum industry, the strontium isotope technique has proved useful in establishing a local stratigraphy for the platform carbonates of the Oligo-Miocene Darai limestone of Papua New Guinea (Eisenberg *et al.*, in Buchanan, 1996). Here, stratigraphic resolution is of the order of 1 Ma (note, though, that the resolution varies according to stratigraphic age, being highest in the Oligo-Miocene). The strontium isotope technique is only applicable in marine environments of normal salinity, and is not applicable in either hypo- or hypersaline environments. Analysis can be performed on macrofossils, microfossils, marine mudstones or marine cements. Epifaunal rather than infaunal species should be selected on account of chemical differences between marine and interstitial pore waters.

In isolated basins, the strontium isotope ratio diverges from that of the open ocean, and the degree of divergence can be used as a measure of the degree of isolation, as in the case of the eastern Mediterranean in the Cenozoic (Flecker and Ellam, 1997). In this case, the isolation inferred from the strontium isotope ratio is significantly earlier than that inferred from faunal evidence. This is interpreted to be because the fauna tracks salinity rather than isolation, and because salinity is partly controlled by evaporation, which is independent of isolation.

Trace element stratigraphy

The abundance or ratio of certain trace elements can be related to provenance, and used for provenance-based stratigraphic correlation. Data is acquired using mass spectrometry, and processed using various statistical techniques. The technique is useful in biostratigraphically barren strata, such as the continental red beds of the Triassic of the North Sea (Goldsmith *et al.*, in Evans *et al.*, 2003).

6.3.2 Cyclostratigraphy

Various cyclostratigraphic techniques and tools are available, for example, spectral analysis of gamma log signatures, which can be used to identify changes in rock accumulation rate through time (Perlmutter and de Azambuja Filho, in Koutsoukos, 2005).

6.3.3 Heavy minerals

The abundance or ratio of certain heavy minerals in sandstones can be related to provenance, and used for provenance-based stratigraphic correlation. The technique is useful in biostratigraphically barren strata, such as the continental red beds of the Devonian west of Shetland (A. Morton, unpublished observations). Provenance-sensitive mineral pairs and apatite grain roundness have been used as the basis for geosteering in the Devonian reservoir in Clair field (Morton *et al.*, in Carr *et al.*, 2003).

6.3.4 Magnetostratigraphy

Magnetic polarity has reversed repeatedly through geological time, on various time-frames, as evidenced by bands of normal and reversed polarity oceanic crust on either side of the active spreading centre at the mid-oceanic ridge (Coe, 2003; Ogg and Smith, in Gradstein *et al.*, 2005). A global standard geomagnetic timescale has been constructed using independently dated magnetic reversals. Continuous measured – outcrop or oriented core – sections can be calibrated against this standard in order to establish a local magnetostratigraphy. Resolution is typically high, but varies according to stratigraphic age, and is low in the Cretaceous 'long normal' or 'quiet zone'. The technique is useful in biostratigraphically barren strata, such as the continental Triassic Otter sandstone of Devon.

In my working experience in the petroleum industry, magnetostratigraphy has proved of use in the following areas.

The Palaeocene–Eocene of the area west of the Shetland Islands. (Unpublished observations.)

The Oligocene–Pliocene of Paratethys. (See Jones and Simmons (1996), Jones and Simmons, in Robinson (1997), Molostovskoy (1997) and Molostovskoy and Guzhikov (1999). Here, specifically in the Pliocene of the south Caspian, the resolution of the magnetostratigraphy is sufficient to detect short-lived unconformities generated by structuration, with important implications for the timing of growth of prospective structures in relation to petroleum charge.

The Miocene–Pliocene of the Middle East. Here, magnetostratigraphy – and vertebrate palaeontology – have been used to date the thick 'molasse' sequence of the Zagros Mountains in south-west Iran, with important consequences for the timing of maturation of underlying Mesozoic and Palaeogene source-rocks (author's unpublished observations).

6.3.5 Radiometric dating

Various radiometric dating techniques are available, for example, argon–argon dating, and 'sensitive high resolution ion micro-probe' (SHRIMP) dating of zircons (Coe, 2003). Note that potassium–argon dating of authigenic glauconites is inherently inaccurate because of degassing.

In my working experience in the petroleum industry, radiometric dating has proved of use in constraining the age of the 'productive series' reservoir in Azerbaijan, which is bracketed by datable volcanic ash horizons. It has also proved of use in determining the provenance of Mesozoic and Cenozoic sediments surrounding the Caucasus, with important implications for reservoir quality. Sediments of local Caucasus provenance and regional Russian platform provenance are characterised by zircon suites of different ages.

6.3.6 Quaternary dating methods

A number of methods are employed in the dating of the Quaternary or Pleistogene, that is, Pleistocene plus Holocene (see also Section 7.6). These include not only conventional biostratigraphy and oxygen isotope stratigraphy (see above), but also accelerator mass spectrometry (AMS), amino-acid racemisation, cosmogenic chlorine-36 rock exposure, dendrochronology, electron spin resonance (ESR), magnetic susceptibility, molecular stratigraphy, optically stimulated luminescence (OSL), radiocarbon dating, thermoluminescence (TL) and uranium-series dating.

In my working experience, only conventional palynostratigraphy, AMS and OSL have proved of particular use, in dating and correlating superficial sediments offshore Azerbaijan in the course of site investigation work (see also Section 7.4). Conventional palynostratigraphy provides a form of climatostratigraphy that can be indirectly calibrated against the deep-sea oxygen isotope record (Woillard, 1978; Tzedakis *et al.*, 1997). AMS is essentially a form of radiocarbon dating that requires a smaller sample, and is therefore less likely to be affected by contamination. It provides precise and accurate age determinations to 40 000 years before present. OSL is a form of dating of mineral grains that provides age determinations to at least 130 000 years before present, that is, the last interglacial.

6.4 Stratigraphic timescales

The principal published timescales integrating biostratigraphic, magnetostratigraphic, isotope and absolute chronostratigraphic data and covering substantial periods of time are those of Harland *et al.* (1982), Haq *et al.* (1987), Harland *et al.* (1990), Berggren *et al.* (1995), de Graciansky *et al.* (1998) and Gradstein *et al.* (2005). The Haq *et al.* and de Graciansky *et al.* timescales are widely used, and indeed have become unofficial standards, in the oil industry, undoubtedly because they also incorporate sequence stratigraphic data (see Section 6.5). Unfortunately, neither deals in adequate detail with the Palaeozoic. Fortunately, though, comprehensive and up-to-date stratigraphic information on the entire stratigraphic column, including the Palaeozoic, is available through the International Commission on Stratigraphy (2005). Up-to-date information regarding 'global stratotype sections and points' (GSSPs) is also available through the Commission's website.

Global stratotype sections and points (GSSPs)

Palaeozoic

The position of the GSSP for the Precambrian–Cambrian (P–Є) boundary has been fixed at the first appearance of the trace fossil *Phycodes pedum* near the base of the Fortune Head section in south-eastern Newfoundland on the Avalon plate, dated to approximately 542 Ma (Fig. 6.5). Unfortunately, the first appearance of the nominate taxon for this zone, the trace fossil *Phycodes pedum*, has been demonstrated to be diachronous when calibrated against the locally established isotopic stratigraphic framework for the boundary interval. The GSSP for the Cambrian–Ordovician boundary has been fixed at the first appearance of the conodont *Iapetognathus*

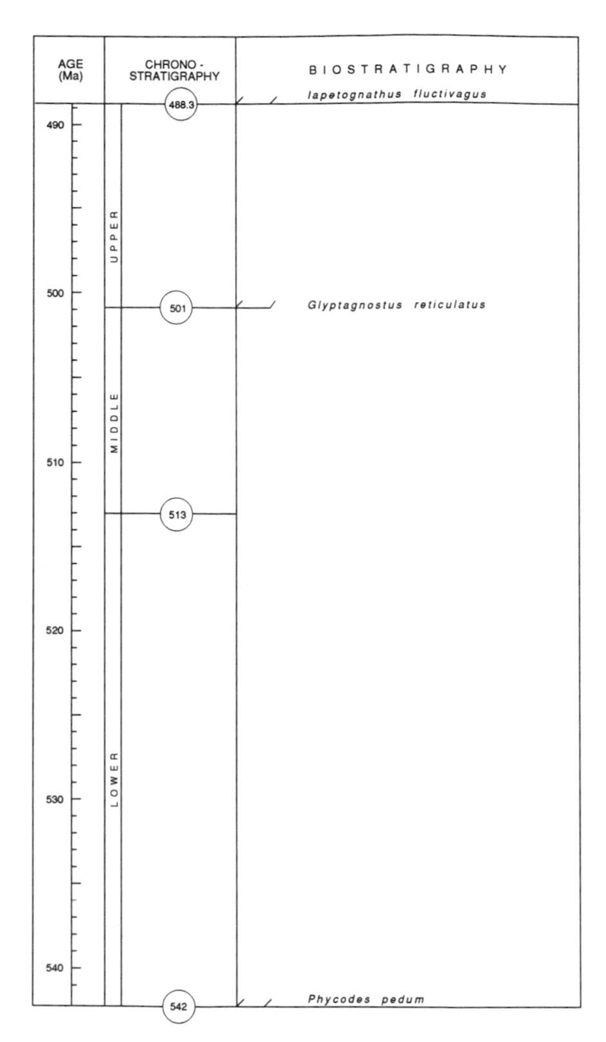

Fig. 6.5. Cambrian timescale. Also showing important trace fossil, trilobite and conodont events. Circled ages refer to GSSPs. (From International Commission on Stratigraphy, 2005.)

fluctivagus in Bed 23 of the Green Point section in western Newfoundland, dated to 488.3 Ma (Fig. 6.6). The GSSP for the Ordovician–Silurian boundary has been fixed at the first appearance of the graptolite *Parakidograptus acuminatus* near the base of the Birkhill shale in the Dob's Linn section, near Moffat in the Southern Uplands of Scotland, dated to 443.7 Ma (Fig. 6.7). The GSSP for the Silurian–Devonian boundary has been fixed at the first appearance of the graptolite *Monograptus uniformis* in Bed 20 of the Klonk section, near Prague in the Czech Republic, dated to 416 Ma (Fig. 6.8). The GSSP for the Devonian–Carboniferous boundary has been fixed at the first appearance of the conodont *Siphonodella sulcata* in Bed 89 of the La Serre section near Cabrières in the Montagne Noir region of southern France, dated to 359.2 Ma (Fig. 6.9). Incidentally, geochemical work has recently been undertaken on brachiopods from the boundary stratotype section. It is evident that the strontium isotopic composition of unaltered shells is a potentially powerful tool in precisely and accurately delineating the boundary (and also, incidentally, in inter-regional correlation). Systematic variations in values seem to represent responses to changes in continental weathering patterns and riverine fluxes in turn associated with a glacial event in the late Middle *Praesulcata* Sub-Zone. Variations in oxygen and carbon isotopic compositions of shells from the boundary section also appear to represent responses to this event. The GSSP for the Carboniferous–Permian boundary has been fixed at the first appearance of the conodont *Streptognathus isolatus* in Bed 19 of the Aidaralash River section near Aktobe in the southern Urals region of northern Kazakhstan, dated to 299 Ma (Fig. 6.10).

Mesozoic

The GSSP for the Permian–Triassic boundary has been fixed at the first appearance of the conodont *Hindeodus parvus* in Bed 27C in Meishan section D in Changxing County, Zhejiang Province, China, dated to 251 Ma (Fig. 6.11). The GSSP for the Triassic–Jurassic boundary, dated to 199.6 Ma, has not yet been selected (Fig. 6.12). The GSSP for the Jurassic–Cretaceous boundary, dated to 145.5 Ma, has not yet been selected either (Fig. 6.13).

Cenozoic

The GSSP for the Cretaceous–Tertiary boundary has been fixed at the iridium anomaly in the Boundary Clay in the El Kef section in Tunisia, dated to 65.5 Ma (Fig. 6.14). The GSSP for the Palaeocene–Eocene boundary has been fixed at the base of the carbon isotope excursion in the Dababiya section near Luxor in Egypt, dated to 55.8 Ma. The GSSP for the Eocene–Oligocene boundary has been fixed at the last appearance of the planktonic foraminifer *Hantkenina* in the marl bed at the base of the exposed section in the Massigano section near Ancona in Italy, dated

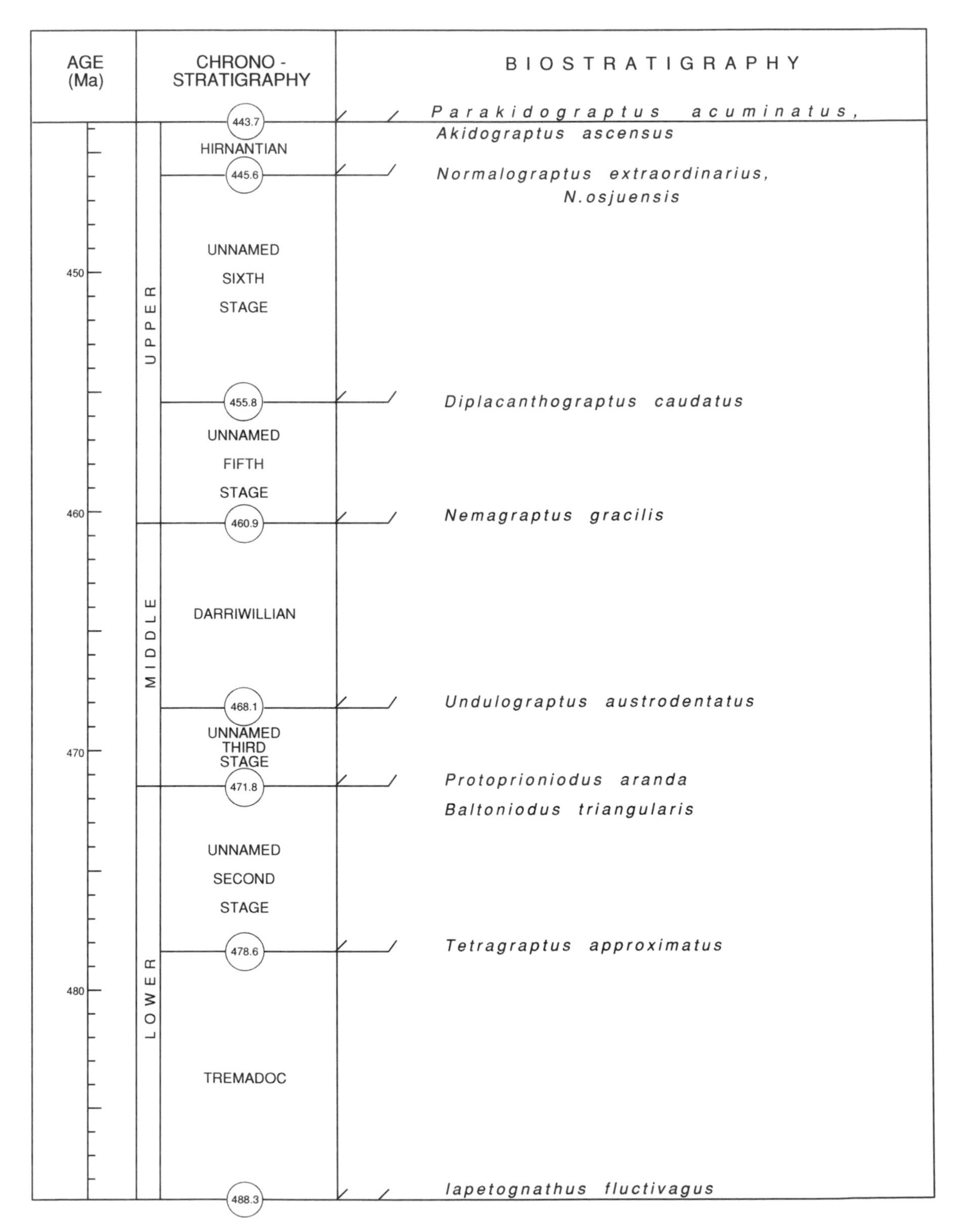

Fig. 6.6. Ordovician timescale. Also showing important conodont and graptolite events. Circled ages refer to GSSPs. (From International Commission on Stratigraphy, 2005.)

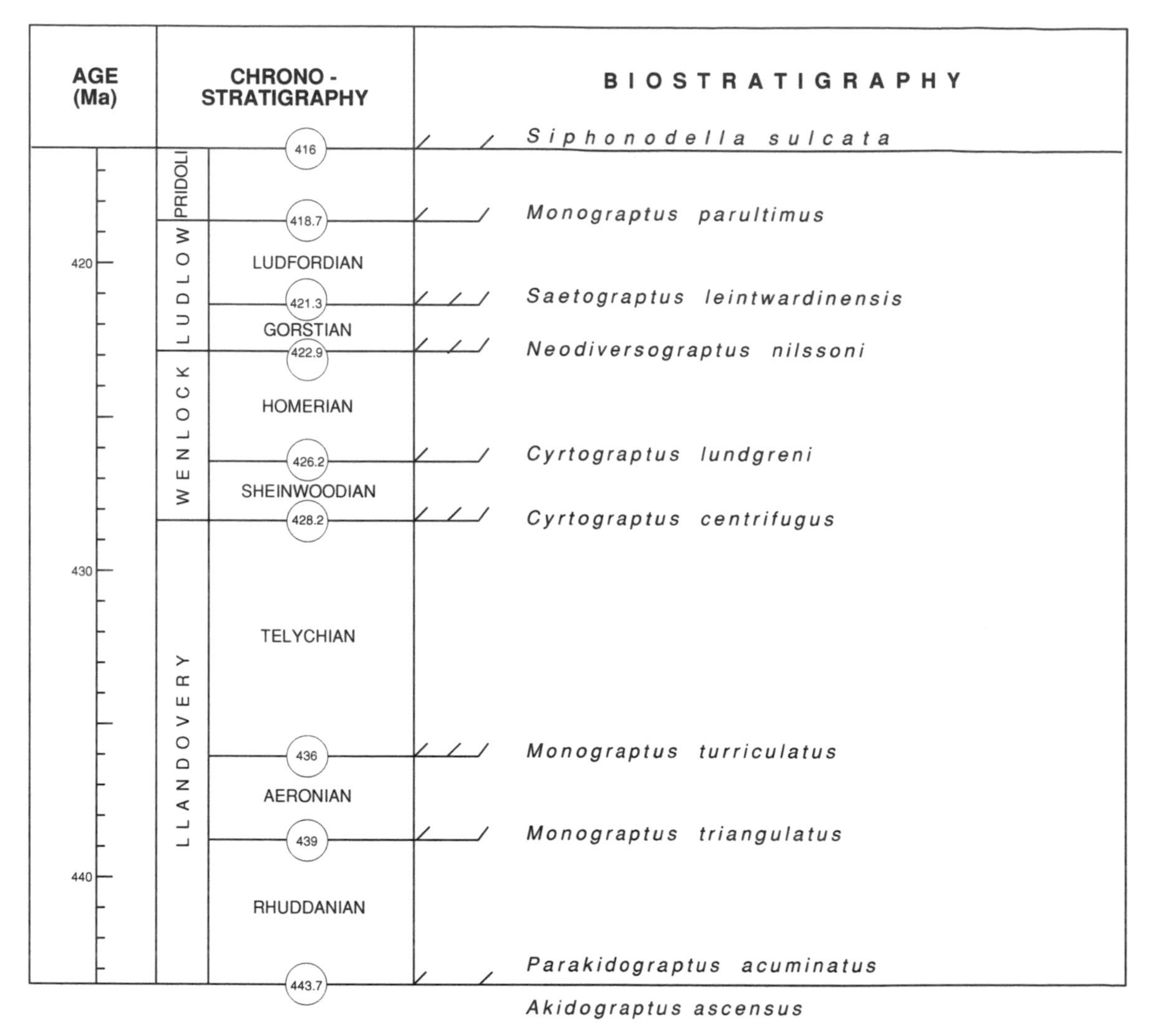

Fig. 6.7. Silurian timescale. Also showing important graptolite events. Circled ages refer to GSSPs. (From International Commission on Stratigraphy, 2005.)

to 33.9 Ma. The GSSP for the Oligocene–Miocene boundary has been fixed at the base of magnetic polarity chronozone C6Cn.2n, 35 m from the top of the Lemme–Carrosio section in Carrosio village near Genoa in Italy, dated to 23.03 Ma. The best biostratigraphic proxies for the GSSP are the first appearance of the planktonic foraminifer *Paragloborotalia kugleri* and the last appearance of the calcareous nannofossil *Reticulofenestra bisecta*. The GSSP for the Miocene–Pliocene boundary has been fixed at the top of magnetic polarity chronozone C3r, at the base of carbonate cycle 1 of the Trubi formation in the Eraclea Minoa section in Sicily, dated to 5.332 Ma. The best biostratigraphic proxies for the GSSP are the first appearance of the calcareous nannofossil *Ceratolithus acutus* and the last appearance of the calcareous nannofossil *Triquetrorhabdulus rugosus*. The GSSP for the Pliocene–Pleistocene boundary has been fixed just above the top of magnetic polarity chronozone C2n (Olduvai), at the top of sapropel layer 'e' in the Vrica section in Calabria, Italy, dated to 1.806 Ma. The best biostratigraphic proxy for the GSSP is the first appearance of sinistrally coiled representatives

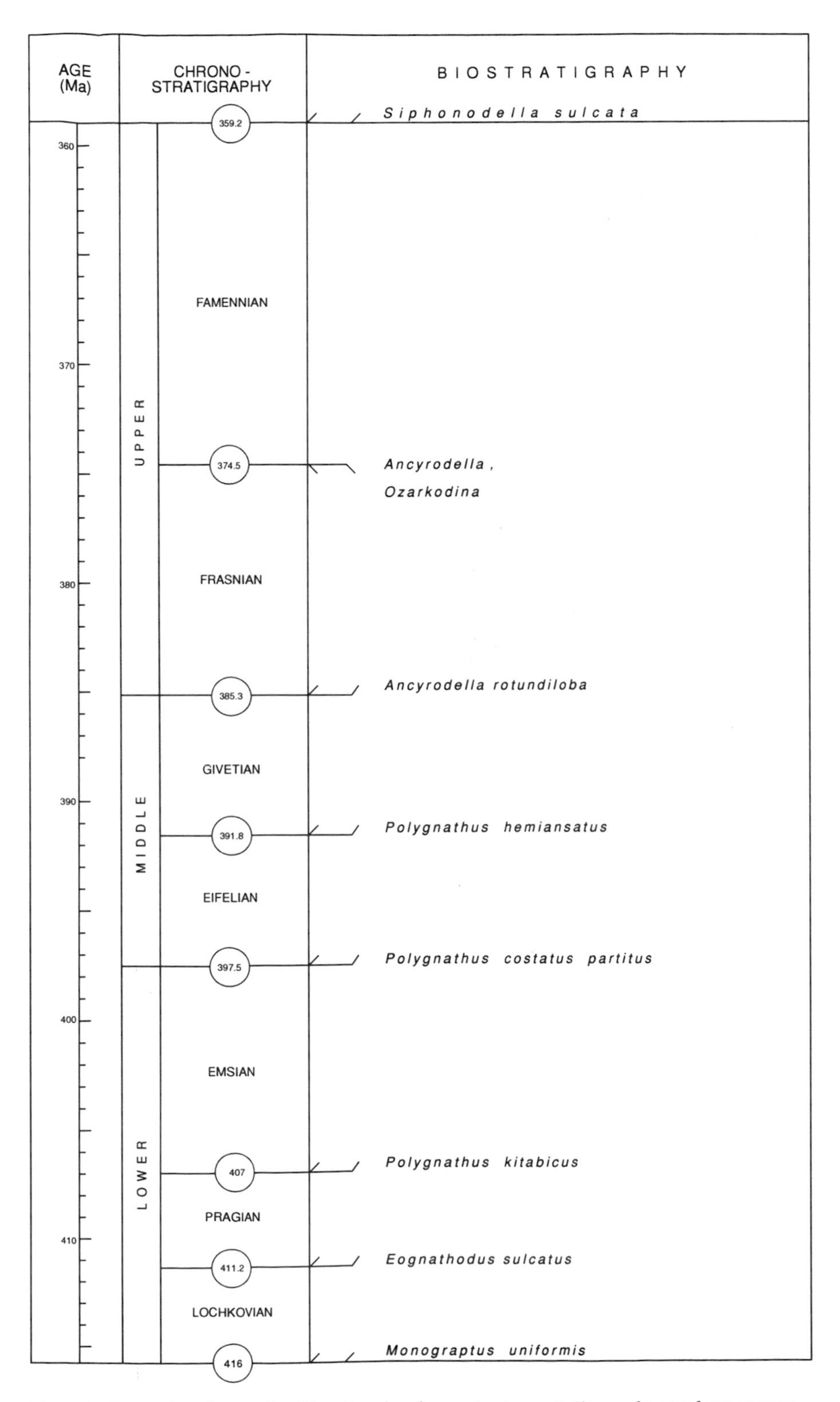

Fig. 6.8. Devonian timescale. Also showing important graptolite and conodont events. Circled ages refer to GSSPs. (From International Commission on Stratigraphy, 2005.)

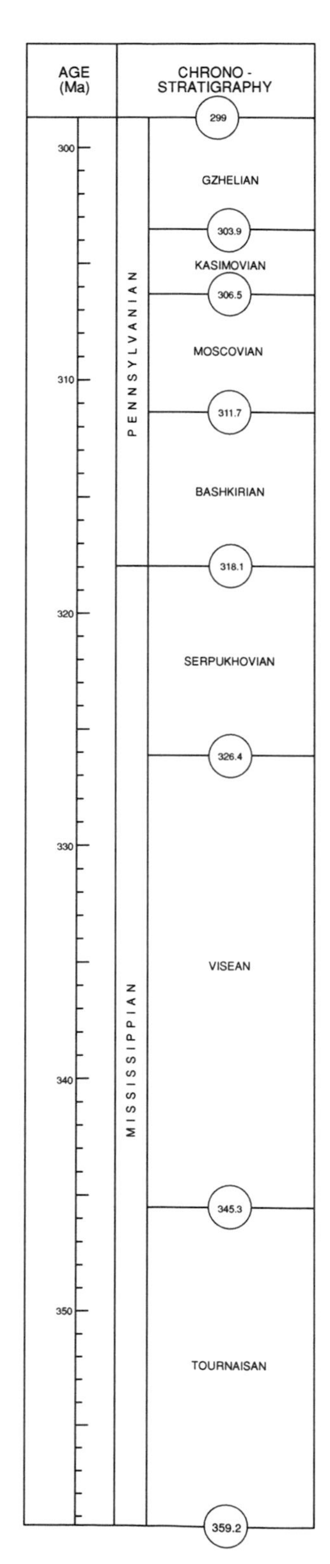

Fig. 6.9. Carboniferous timescale. Circled ages refer to GSSPs. (From International Commission on Stratigraphy, 2005.)

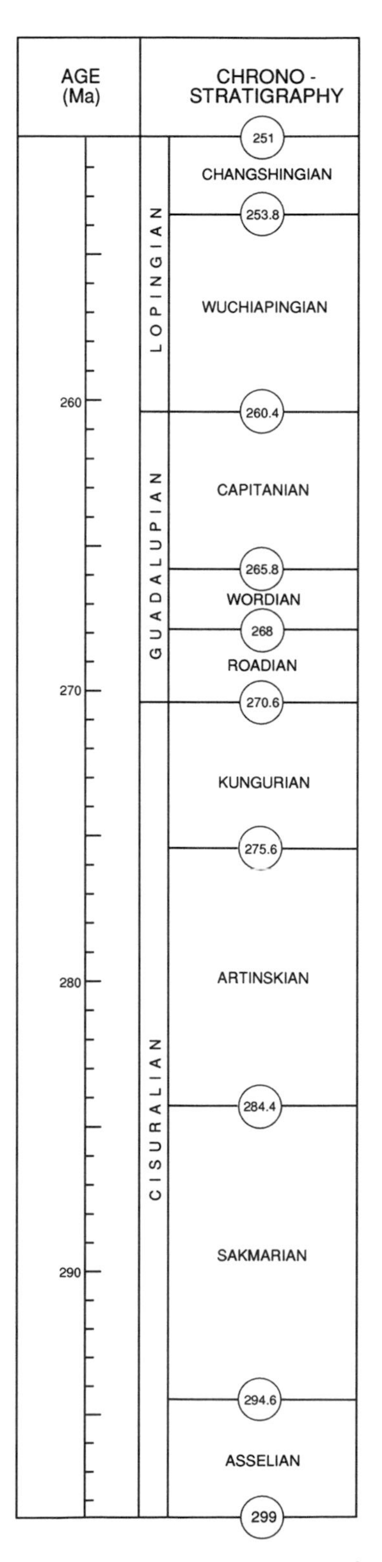

Fig. 6.10. Permian timescale. Circled ages refer to GSSPs. (From International Commission on Stratigraphy, 2005.)

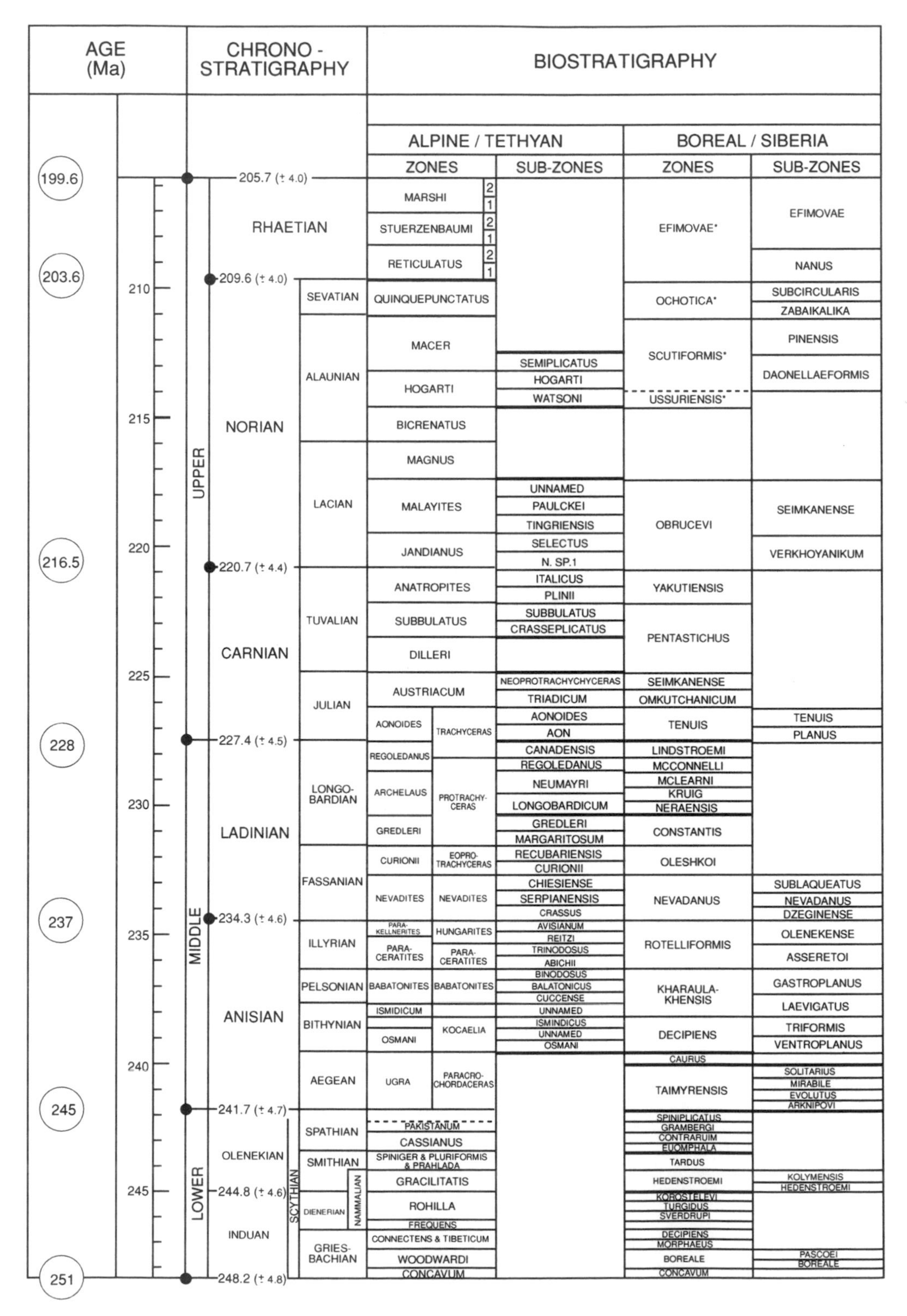

Fig. 6.11. Triassic timescale. Circled ages refer to GSSPs. From International Commission on Stratigraphy (2005.) (Tethyan and Boreal ammonite zones) Chronostratigraphy and biostratigraphy modified after Gradstein *et al.*, in Berggren et al., (1995).

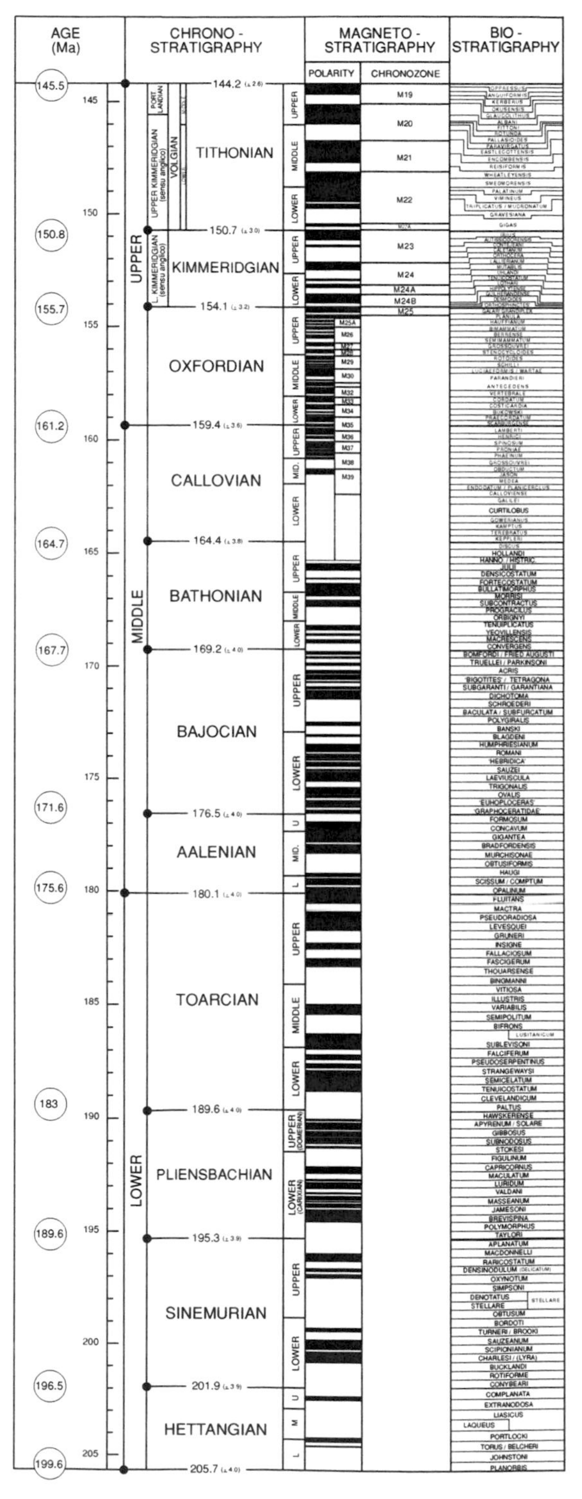

Fig. 6.12. Jurassic timescale. Circled ages refer to GSSPs. Age (Ma) from International Commission on Stratigraphy (2005). Chronostratigraphy, magnetostratigraphy and biostratigraphy (composite ammonite zones) modified after Gradstein *et al*.. in Berggren *et al*. (1995).

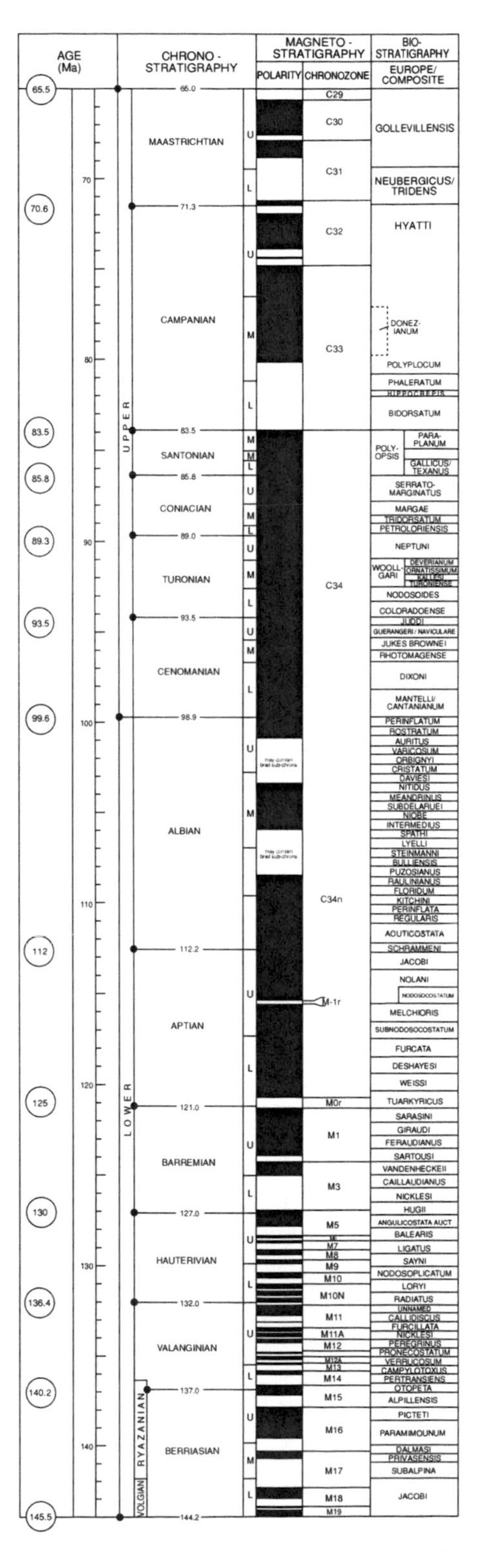

Fig. 6.13. Cretaceous timescale. Circled ages refer to GSSPs. Age (Ma) from International Commission on Stratigraphy (2005). Chronostratigraphy, magnetostratigraphy and biostratigraphy (composite ammonite zones for lower cretaceous, European ammonite zones for upper cretaceous) modified after Gradstein *et al*., in Berggren *et al*. (1995).

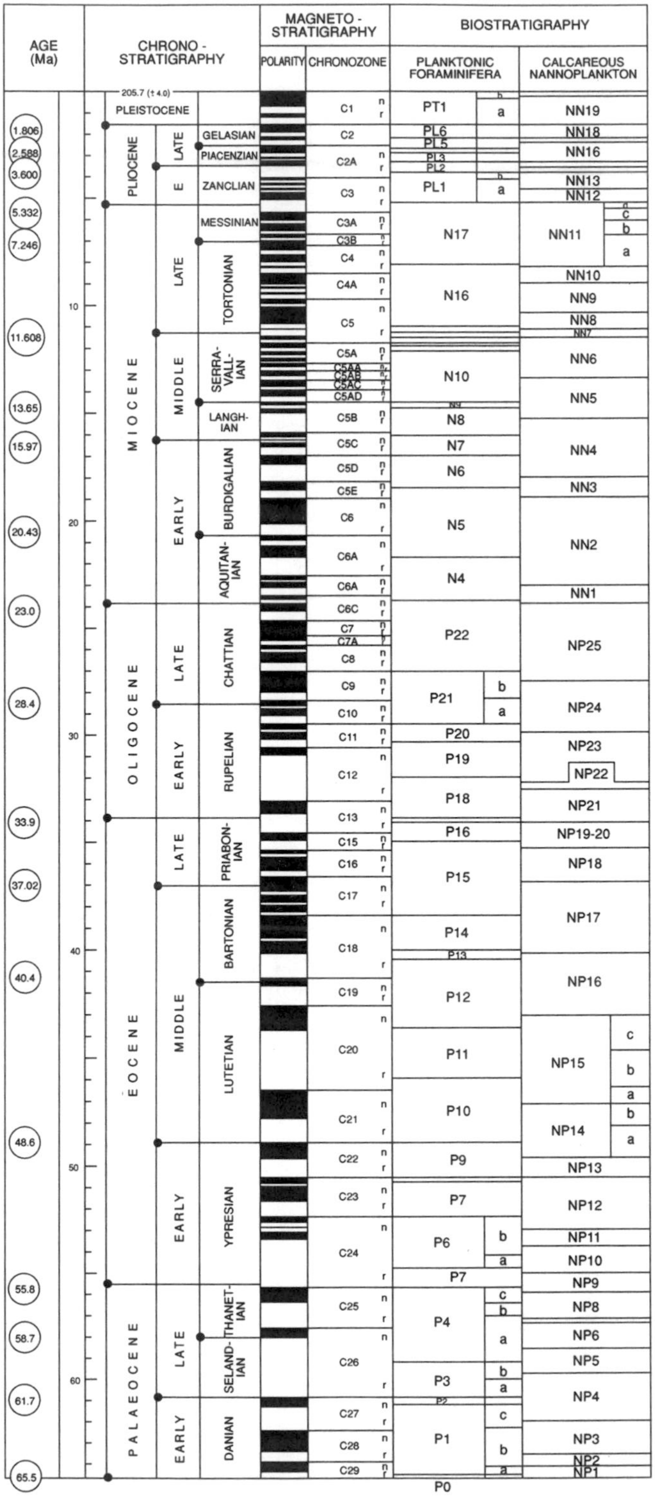

AGE (Ma)

CHRONO-STRATIGRAPHY

MAGNETO-STRATIGRAPHY: POLARITY, CHRONOZONE

BIOSTRATIGRAPHY: PLANKTONIC FORAMINIFERA, CALCAREOUS NANNOPLANKTON

205.7 (±4.0)

PLEISTOCENE

PLIOCENE: LATE – GELASIAN, PIACENZIAN; E – ZANCLIAN

MIOCENE: LATE – MESSINIAN, TORTONIAN; MIDDLE – SERRAVALLIAN, LANGHIAN; EARLY – BURDIGALIAN, AQUITANIAN

OLIGOCENE: LATE – CHATTIAN; EARLY – RUPELIAN

EOCENE: LATE – PRIABONIAN; MIDDLE – BARTONIAN, LUTETIAN; EARLY – YPRESIAN

PALAEOCENE: LATE – THANETIAN, SELANDIAN; EARLY – DANIAN

Circled ages: 1.806, 2.588, 3.600, 5.332, 7.246, 11.608, 13.65, 15.97, 20.43, 23.0, 28.4, 33.9, 37.02, 40.4, 48.6, 55.8, 58.7, 61.7, 65.5

Age scale: 10, 20, 30, 40, 50, 60

Chronozones: C1, C2, C2A, C3, C3A, C3B, C4, C4A, C5, C5A, C5AA, C5AB, C5AC, C5AD, C5B, C5C, C5D, C5E, C6, C6A, C6A, C6C, C7, C7A, C8, C9, C10, C11, C12, C13, C15, C16, C17, C18, C19, C20, C21, C22, C23, C24, C25, C26, C27, C28, C29 (n / r)

Planktonic foraminifera: PT1 (a, b), PL6, PL5, PL3, PL2, PL1 (a, b), N17, N16, N10, N9, N8, N7, N6, N5, N4, P22, P21 (a, b), P20, P19, P18, P16, P15, P14, P13, P12, P11, P10, P9, P7, P6 (a, b), P7, P4 (a, b, c), P3 (a, b), P2, P1 (a, b, c), P0

Calcareous nannoplankton: NN19, NN18, NN16, NN13, NN12, NN11 (a, b, c, d), NN10, NN9, NN8, NN7, NN6, NN5, NN4, NN3, NN2, NN1, NP25, NP24, NP23, NP22, NP21, NP19-20, NP18, NP17, NP16, NP15 (a, b, c), NP14 (a, b), NP13, NP12, NP11, NP10, NP9, NP8, NP6, NP5, NP4, NP3, NP2, NP1

Fig. 6.14. Cenozoic timescale. Circled ages refer to GSSPs. Age (Ma) from International Commission on Stratigraphy (2005). Chronostratigraphy, magnetostratigraphy and Biostratigraphy (global standard planktonic foraminiferal and calcareous nannoplankton zones) modified after Berggren *et al.*, in Berggren *et al.* (1995).

of the planktonic foraminifer *Neogloboquadrina pachyderma*. The GSSP for the Pleistocene–Holocene boundary, dated to 11.5 ka, has not yet been officially selected.

6.5 Sequence stratigraphy

Sequence stratigraphy attempts to subdivide the rock record into genetically related – commonly unconformity-bounded – rock units or sequences (Emery and Myers, 1996; Jones, 1996; Miall, 1997; Holland, in Erwin and Wing, 2000; Holland, in Briggs and Crowther, 2001; Sharland *et al.*, 2001; Coe, 2003). The methodology lends itself readily to the interpretation of seismic sections, where unconformity surfaces are identifiable by erosional truncation below and transgressive onlap above. The unconformities that form the basis of sequence stratigraphy are generated by base-level fall, in turn driven by glaciation or glacioeustasy, and/or by structuration or tectonism (see also Sub-section 6.5.1). Global coastal onlap and ?glacioeustatic sea-level charts have been constructed for the period Carboniferous to Recent. Sea-level curves have also been constructed for the Cambrian, Ordovician, Silurian and Devonian.

Definitions

Sequences

A Vailian sequence is defined as a 'stratigraphic unit composed of a relatively conformable succession of genetically related strata bounded at its top and base by unconformities or their correlative conformities' (Vail *et al.*, 1977). As intimated above, Vailian sequence boundaries (SBs) (bounding unconformities) are readily recognisable on seismic stratigraphic criteria (truncation below, and onlap above). They are also recognisable on palaeontological, sedimentological and petrophysical criteria (basinward shifts in facies, and associated lithological breaks on wireline logs).

A Gallowayan genetic stratigraphic sequence is defined as 'a package of sediments recording a significant episode of basin-margin outbuilding and basin-filling' (Galloway, 1989). Gallowayan SBs (Maximum flooding surfaces or MFSs: see also below) are also readily recognisable on seismic stratigraphic criteria (landward shifts in bathymetrically ordered facies belts below, and downlap and basinward shift in facies belts above), and on palaeontological, sedimentological and petrophysical criteria (landward shifts in facies, indications of maximum bathymetry, and associated lithological breaks on wireline logs).

The spatio-temporal distribution of sequences is dictated by 'accommodation space', which is defined as 'the space... available for potential sediment accumulation'. Accommodation space is in turn dictated by the complex interplay of a number of dynamically variable processes, including *(glacio) eustasy, tectonism, sediment compaction* and *sediment supply. Eustasy*, that is, rising or falling absolute sea level, can be either creative and destructive of accommodation space. *Tectonism*, that is – independent – subsidence and falling, or uplift and rising, absolute land level, can also be either creative and destructive of accommodation space. *Sediment compaction* is always creative. *Sediment supply* is always destructive. At any given point in a basin, if the rate of sediment supply exceeds that of the creation of accommodation space by the processes described above, there will be a basinward shift in bathymetrically ordered facies belts. This process is termed *progradation*. Conversely, if the rate of sediment supply exceeds that of the creation of accommodation space, there will be a landward shift in facies belts. This process is termed *retrogradation*. If the rates of sediment supply and the creation of accommodation space are in equilibrium, facies belts will remain in place, and build vertically. This process is termed *aggradation*.

Systems tracts

Sequences can be internally subdivided into so-called 'systems tracts' (STs). An ST is defined as 'a linkage of contemporaneous depositional systems' or bathymetrically ordered facies belts (for example, continental–paralic–shelfal–bathyal–abyssal). The principal types are the *low-stand systems tract (LST)*, the *shelf-margin systems tract* or *shelf-margin wedge (SMST* or *SMW)*, the *transgressive systems tract (TST)* and the *high-stand systems tract (HST)*. The LST can be further subdivided into a lower, *low-stand fan (LSF)* and an upper, *low-stand wedge* or *low-stand prograding complex (LSW* or *LPC)*. (Note, though, that marine geologists interpret LSFs and LSWs as attached rather than detached systems: Nelson and Damuth, 2003). The LST and TST are separated by the *transgressive surface (TS)*. The TST and HST are separated by the *maximum flooding surface (MFS). Condensed sections (CSs)* commonly occur at the TS and MFS.

The LST is defined as 'the earliest systems tract in a depositional sequence . . . if it lies directly on a Type 1 sequence boundary', and the related SMST or SMW as 'the earliest systems tract in a depositional sequence associated with a Type 2 sequence boundary'; the TST as 'the middle systems tract of both Type 1 and Type 2 sequences'; and the HST as 'the late systems tract in either a Type 1 or a Type 2 sequence'. Note that a Type 1 SB is generated by a sea-level fall to a position basinward, and a Type 2 SB by a sea-level fall to a position landward, of the shelf edge or 'offlap break' of the preceding sequence. The TS is defined as 'the first significant marine flooding surface . . . within a sequence'. The MFS is defined as 'the surface corresponding to the time of maximum flooding'. CSs are defined as 'thin marine stratigraphic units consisting of pelagic to hemipelagic sediments characterised by very low sedimentation rates . . . most areally extensive at the time of maximum regional transgression'. They commonly contain 'abundant and diverse planktonic and benthic microfossil assemblages', which enable easy identification, and also, importantly, calibration against global standard biostratigraphic zonation schemes.

6.5.1 General and clastic sequence stratigraphy

Haq *et al.* (1987) have generated a conceptual model of the development of sequences in relation to cycles of sea-level change (Fig. 6.15). This model, or modifications of it, can be used to predict the spatio-temporal distribution of lithofacies, even in tectonically active areas such as the compressional Californian Borderland or the extensional Gulf of Mexico growth-fault province, or in entirely non-marine settings.

Vail, Haq and others have argued that at least the so-called 'third-order' cyclicity is, essentially, (glacio)eustatically – rather than tectonically – mediated, and that the (glacio)eustatic signal, while it may be strengthened or weakened by the tectonic signal, is always stronger than it. This argument has been criticised by a number of authors, on the grounds that it fails to explain the discrepancy between the frequencies of the modelled 'third-order' cycles (of the order of a million years) and the observed incontrovertibly glacioeustatically mediated cycles (of the order of tens to hundreds of thousands of years) of the 'ice-house' world of the Pleistogene; and that it also fails to explain the observed cyclicity in the 'greenhouse' world of the Mesozoic (when obviously there would have been no glaciation or glacioeustasy). This criticism has in turn been countered by arguments that 'fourth- to sixth-order' cycles are in fact resolvable, although only in areas of high sedimentation rate, such as the Plio-Pleistocene of the Gulf of Mexico, the Niger delta or the south Caspian; and that 'third-order' cycles are mediated not only by (glacio)eustasy but also by tectonism, in the form of – episodic–local and regional stress release.

6.5.2 Carbonate sequence stratigraphy

Sarg, in Wilgus *et al.* (1988) and Handford and Loucks, in Loucks and Sarg (1993) have generated conceptual models of the development of carbonate sequences in relation to cycles of sea-level change (Fig. 6.16). Handford and Loucks's models are comprehensive, and take into account criticisms that earlier models of carbonate sequences relied on analogy with clastic sequences, which, in view of the more complex controls on carbonate sequences, is probably only partly appropriate. In particular, they take into account observations that modern and ancient carbonate systems, unlike clastic systems, deposit comparatively insignificant volumes of sediment during LSTs, when they are subject to chemical as against physical erosion, and much more significant volumes during HSTs.

6.5.3 Mixed sequence stratigraphy

McLaughlin *et al.* (2004) have generated models of the development of mixed clastic–carbonate sequences in the Ordovician of Kentucky and Ohio. They have also argued that the primary control on sequence architecture is (glacio)eustatic, with only a secondary tectonic component associated with the Taconian orogeny.

6.5.4 Seismic facies analysis

As intimated above, seismic reflection terminations form the basis for sequence stratigraphic analysis.

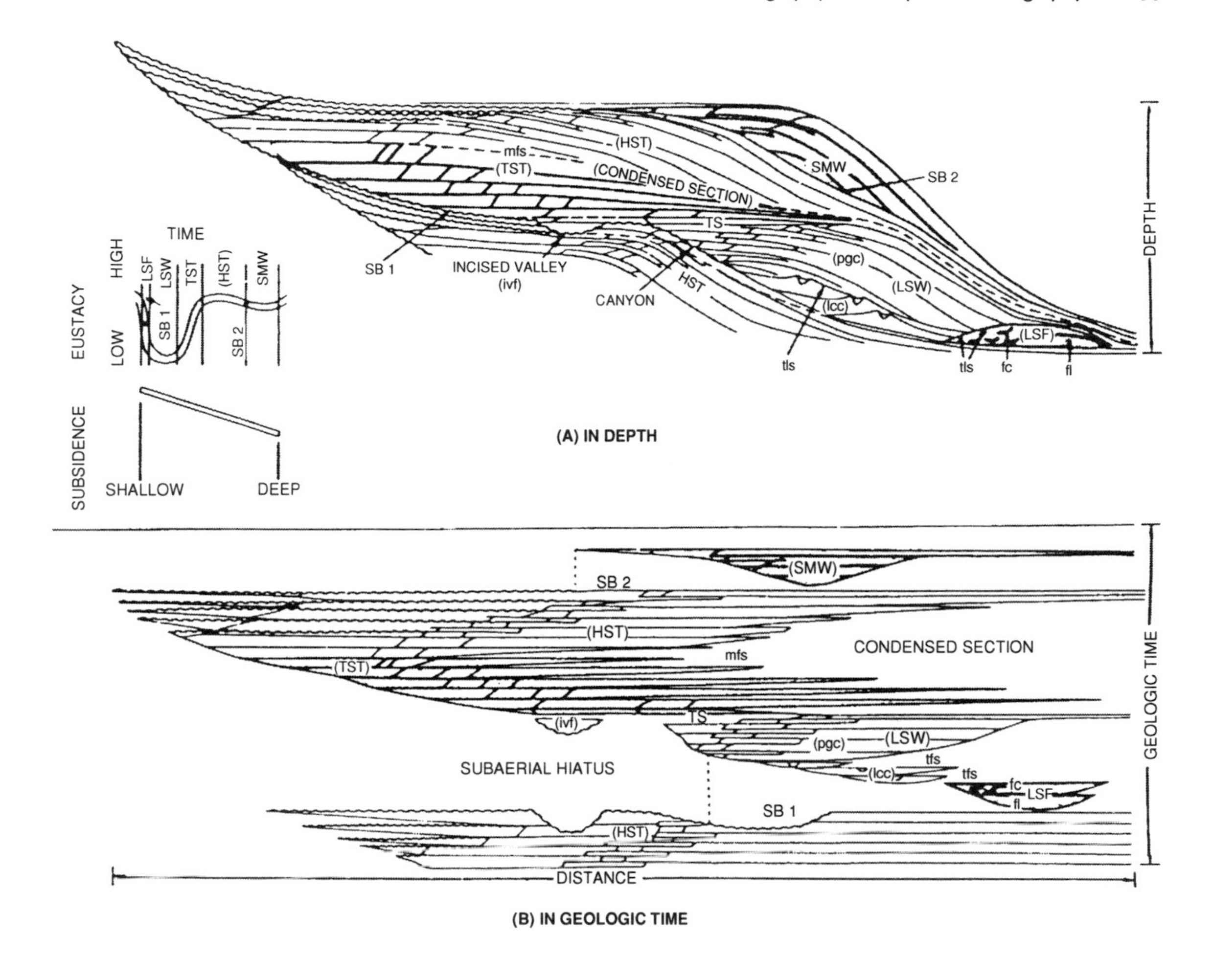

LEGEND

SURFACES

(SB) SEQUENCE BOUNDARIES
(SB 1) = TYPE 1
(SB 2) = TYPE 2

(DLS) DOWNLAP SURFACES
(mfs) = maximum flooding surface
(tfs) = top fan surface
(tls) = top leveed channel surface

(TS) TRANSGRESSIVE SURFACE
(First flooding surface above maximum regression)

Z FACIES BOUNDARIES
(Proximal at left: Distal at right)

SYSTEMS TRACTS

HST = HIGH-STAND SYSTEMS TRACT
TST = TRANSGRESSIVE SYSTEMS TRACT
LSW = LOW-STAND WEDGE SYSTEMS TRACT

ivf = incised valley fill
pgc = prograding complex
lcc = leveed channel complex

LSF = LOW-STAND FAN SYSTEMS TRACT

fc = fan channels
fl = fan lobers

SMW = SHELF-MARGIN WEDGE SYSTEMS TRACT

Fig. 6.15. Siliciclastic sequence stratigraphic model. (From Jones, 1996; in turn after Haq *et al.*, 1987.)

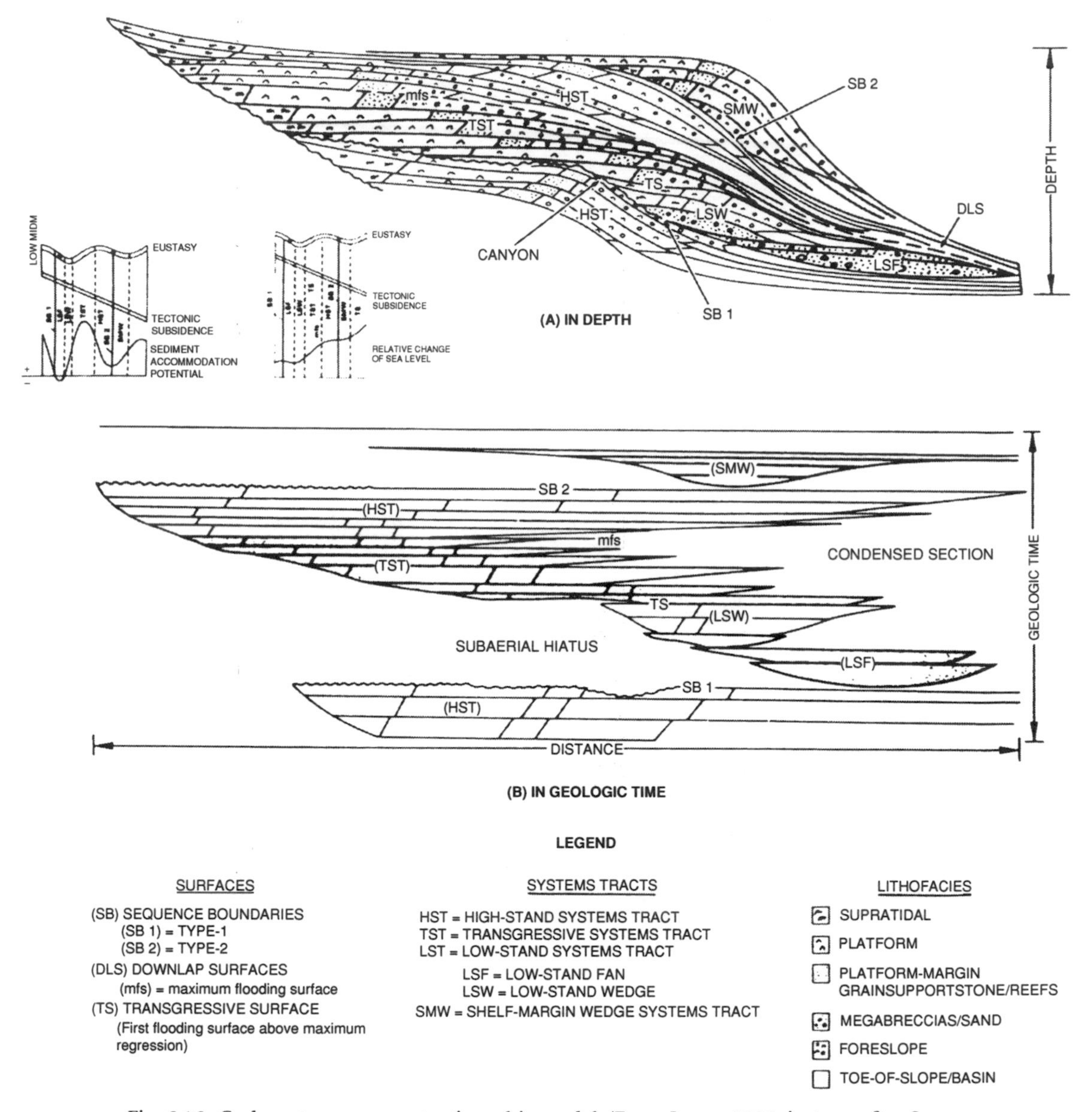

Fig. 6.16. Carbonate sequence stratigraphic model. (From Jones, 1996; in turn after Sarg, in Wilgus *et al.*, 1988.)

For example, onlap defines Vailian SBs; downlap, Gallowayan SBs or MFSs. Reflection patterns form the basis for seismic facies analysis. To simplify, reflection geometry (e.g., whether clinoform etc.) provides an indication of the depositional process. Reflection amplitude provides an indication of the product of the depositional process, that is, lithology, and also porosity and pore fluid content. Time-slices through three-dimensional seismic volumes allow the identification of individual sedimentary – and reservoir – bodies, such as deltas and submarine fans (author's unpublished observations). Other seismic attributes such as spectral decomposition and coherency can also be used in palaeoenvironmental interpretation.

Box 6.1 Palaeontological characterisation of systems tracts

The micropalaeontological characterisation of STs in marine clastic and carbonate sequences is summarised in Figs. 6.17 and 6.18 respectively. The micropalaeontological characterisation by means of freshwater ostracods of the STs of the non-marine sequences of the pre-salt Cretaceous of west Africa is described by Bate, in Cameron *et al.* (1999). Here, lake level is measurable by means of ostracod diversity. Incidentally, it is highest in the Barremian, coincident with the widespread development of lacustrine source-rock.

The palaeontological characterisation of CSs, which commonly occur at TSs and MFSs, is typically by means of measures of abundance and diversity not only of microfossils (Olson and Thompson, in Koutsoukos, 2005) but also of macrofossils (Sharland *et al.*, 2001; Holz and Simoes, in Koutsoukos, 2005).

The ichnological characterisation of key omission surfaces such as SBs, amalgamated or composite SBs/TSs ('transgressive surfaces of erosion'), and high-energy MFSs, is by means of substrate-controlled firmground *Glossifungites*, hardground *Trypanites* and woodground *Teredolites* Ichnofacies (Pemberton, in McIlroy, 2004).

(Vailian) sequence boundaries

Sequence boundaries (SBs) in clastic sequences are typically characterised by the abrupt juxtaposition of younger, more proximal fossil assemblages, sometimes containing reworked specimens, on older, more distal ones ('faunal discontinuity events') (Olson and Thompson, in Koutsoukos, 2005). Note, though, that the contrast in assemblages above and below the SB varies along depositional dip, typically being more pronounced in proximal, up-dip settings, where a biostratigraphically resolvable unconformity may also be in evidence, and less so in distal, down-dip settings, where the relationship may be closer to that of a correlative conformity. Note also that in the most proximal, up-dip settings, the assemblage above the SB is actually more distal

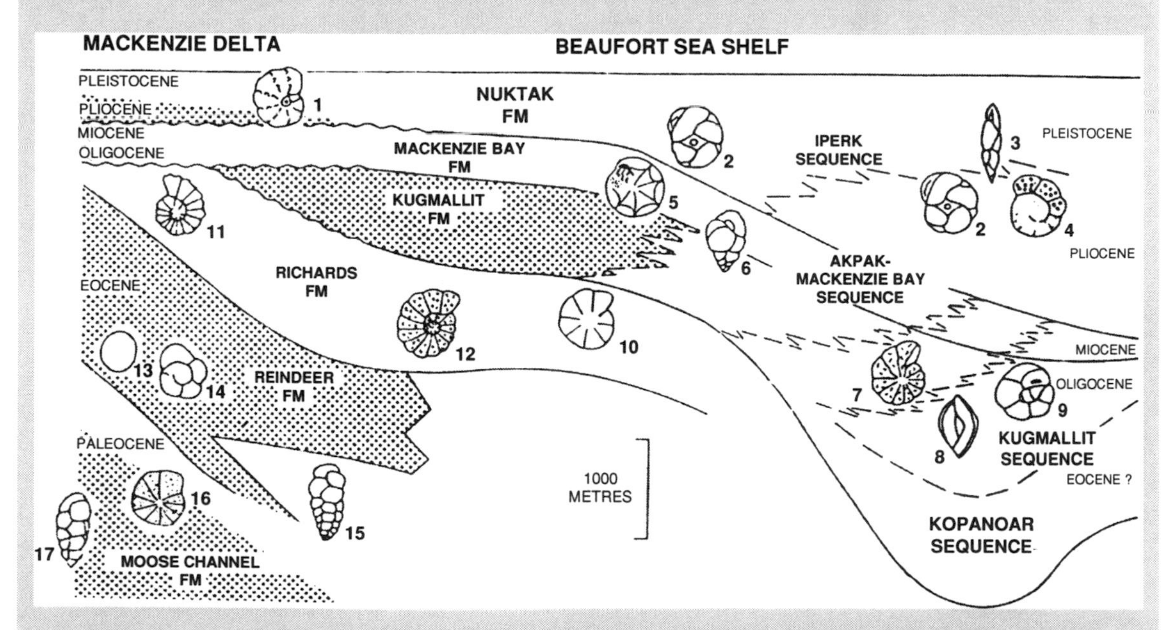

Fig. 6.17. Foraminiferal characterisation of systems tracts in clastic sequences, Cenozoic, Mackenzie delta, Canada. (From Jones, 1996; reproduced with permission from McNeil, in Dixon *et al.*, 1985.)

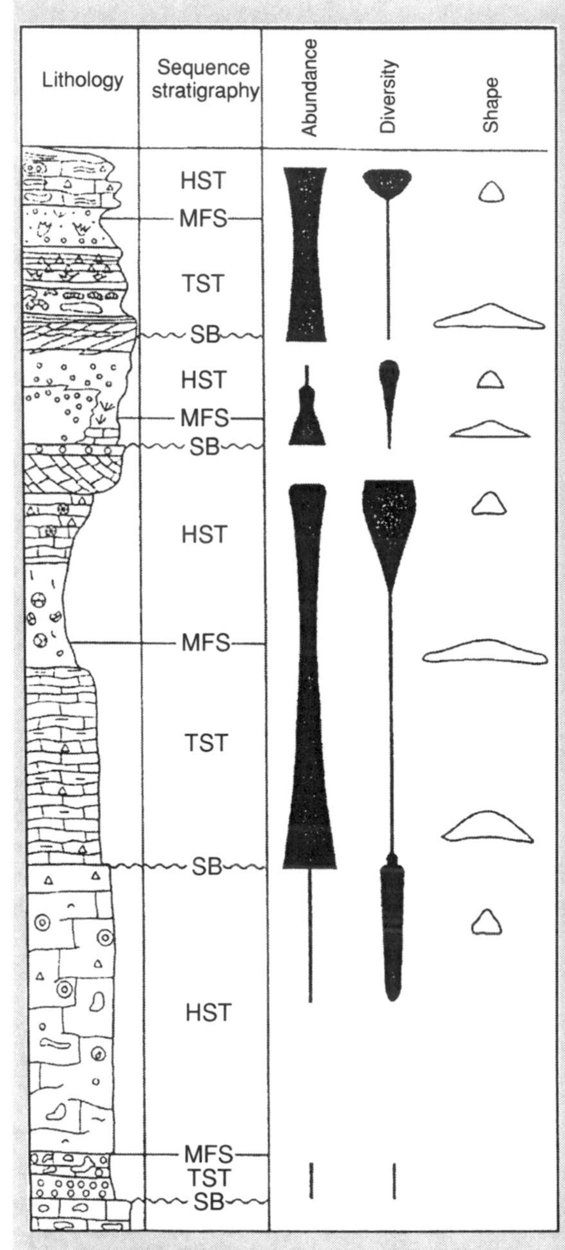

Fig. 6.18. Foraminiferal characterisation of systems tracts in carbonate sequences, 'Middle' Cretaceous, Middle East. HST, high-stand systems tract; MFS, maximum flooding surface; SB, sequence boundary; TST, transgressive systems tract. (From Jones *et al.*, in Bubik and Kaminski, 2004; in turn after Simmons *et al.*, in Hart *et al.*, 2000.)

than that below, as it is not directly associated with the SB but rather with the succeeding TS.

Low-stand systems tracts

Low-stand systems tracts (LSTs) in clastic sequences are characterised by erosion in up-dip settings, and by the deposition of submarine fans and associated sediments in down-dip settings. Fossil assemblages in the fans are characterised by an autochthonous deep marine benthic component, an allochthonous, but contemporaneous, shallow marine benthic and planktonic, and non-marine, component, and, occasionally, an allochthonous, and non-contemporaneous component (Olson and Thompson, in Koutsoukos, 2005; see section on 'Submarine fans' in Chapter 4). The deep-water component is of shallower aspect than that of the underlying sequence, although the difference is not necessarily always resolvable (because bathymetric resolution is comparatively poor – of the order of hundreds of metres – in deep water).

LSTs in the Neogene of south-east Asia are typically characterised micropalaeontologically by a dominance of deep-water arenaceous foraminiferans or DWAFs (see section on 'DWAFs' in Chapter 3), as are those in the Palaeogene of the Beaufort Sea in Arctic Canada (Fig. 6.17) (McNeil, in Dixon *et al.*, 1985), and those in the post-rift Cretaceous sequences of the Pletmos and Bredasdorp basins of South Africa (Brown *et al.*, 1995).

LSTs in the Neogene of the tropics, including south-east Asia, are characterised palynologically by terrestrially derived spores; those in the Plio-Pleistogene of the Gulf of Mexico by terrestrially derived pollen of cool/dry aspect; and those in the late Cenozoic of the Niger delta by charred grass pollen and cuticle, indicating widespread development of arid savannah prone to – ?seasonal – bush fire. In the Pleistocene glacial/interglacial cycles of the Amazon fan LSTs are characterised by terrigenous Andean and tropical lowland elements, organic debris, and reworked Palaeozoic, Cretaceous and Tertiary elements. Ferns and fungal spores are well represented. Perhaps counter-intuitively, lowland savannah elements are not (although undifferentiated grasses are). In the latest Jurassic to earliest Cretaceous of Papua New Guinea LSTs are characterised by high palynofloral abundance, but also by high dominance and low diversity, reflecting colonisation of 'stressed' ecological niches by opportunistic species. The percentage of terrestrially derived palynomorphs is in the range 25–95%. Values are higher at proximal and lower at distal sites. Plant cuticle and fresh-water algal material is locally common. In the latest Jurassic to earliest Cretaceous of the Vocontian basin of south-east France LSTs are characterised by plant fragments or phytoclasts.

Low-stand systems tracts in the earliest Cretaceous Barrow group sequences of the north-west shelf of Australia are characterised by zero to moderate microfaunal abundance and diversity, and by low to high palynofloral diversity, leastwise on the palaeo-prodelta. Arenaceous and calcareous benthic foraminiferans occur in approximately equal proportions. The percentage of terrestrially derived palynomorphs is generally >90% at proximal sites, and 60–90% at distal sites.

Mixed sequences

The LSTs in the Early Jurassic mixed clastic–carbonate sequences of Quercy in south-west France are characterised micropalaeontologically by low abundances and diversities of benthic foraminiferans, and by upwardly increasing proportions of uncoiling morphotypes of *Lenticulina*.

Transgressive surfaces

Transgressive surfaces (TSs) in clastic sequences are characterised by condensed sections, and typically by abundant and diverse fossil assemblages. They are further characterised by the juxtaposition of more distal assemblages over more proximal ones, irrespective of location along depositional dip.

Transgressive surfaces atop low-stand fans in the post-rift Cretaceous sequences of the Pletmos and Bredasdorp basins of South Africa are characterised micropalaeontologically by an acme of the foraminiferan *Trochammina* (Brown *et al.*, 1995), a probable phytodetritivore.

In the Early Cretaceous of the Bredasdorp basin of South Africa TSs – and TSTs – are characterised by palynomorph abundance maxima and diversity minima (Valicenti *et al.*, 1991). This is attributable to the blooming of opportunistic species in newly created niches (Valicenti, pers. comm.).

Transgressive systems tracts

Transgressive systems tracts (TSTs) in clastic sequences are characterised by overall retrogradation, and by fossil and trace fossil assemblages indicating an upward deepening (Olson and Thompson, in Koutsoukos, 2005).

In the Neogene of south-east Asia TSTs are characterised micropalaeontologically by 'clear-water' calcareous benthic foraminiferans, including larger benthic foraminiferans or LBFs thought to be adapted to oligotrophic conditions (see Box 3.2). Those in the post-rift Cretaceous sequences of the Pletmos and Bredasdorp basins of South Africa are characterised by upward increases in abundance of deep-water foraminiferans (Brown *et al.*, 1995). Those in the Late Cretaceous of the Basco-Cantabrian basin in Spain are characterised by the foraminiferan *Tritaxia* (Grafe, 1999).

Transgressive systems tracts in the Neogene of the tropics are characterised palynologically by terrestrially derived palmae (palm) pollen; those in the Plio-Pleistogene of the Gulf of Mexico by pollen of warm/dry aspect; and those of the Early Carboniferous of Britain (abandonment surfaces) by miospores. Those in the latest Jurassic to earliest Cretaceous of the Vocontian basin of south-east France are characterised by an upward decrease in the abundance and angularity terrestrially derived plant fragments, and an upward increase in the abundance and diversity of marine dinoflagellates.

Transgressive systems tracts in the earliest Cretaceous Barrow group sequences of the north-west shelf of Australia are characterised by zero microfaunal abundance and diversity, and low to moderate palynofloral diversity.

Carbonate sequences

Transgressive systems tracts in carbonate sequences in the Early Cretaceous of the Middle East and central Asia are characterised micropalaeontologically by upward increases in incidence of orbitolinid LBFs and planktonic foraminiferans. In the Middle East, they are also typically characterised by flattened orbitolinids (Fig. 6.18) (Simmons *et al.*, in Hart *et al.*, 2000; Jones *et al.*, in Bubik and Kaminski, 2004). Note, though, that flattening of orbitolinid LBFs probably represents an adaptation to the maximisation of exposure to light to enable interpreted hosted algal symbionts to photosynthesise, and as such is determined not so much by water depth as by water clarity (Pittet *et al.*, 2002).

Mixed sequences

Transgressive systems tracts in the Early Jurassic mixed clastic–carbonate sequences of Quercy in south-west France are characterised by moderate abundances and diversities of benthic foraminiferans, and by an upward increase in the proportion of uncoiling morphotypes of *Lenticulina*.

Maximum flooding surfaces

Maximum flooding surfaces (MFSs) are typically characterised by condensed sections, and by abundant and diverse fossil assemblages, including, in the Middle East, a wide range of macrofossils as well as microfossils (Sharland *et al.*, 2001). Note, though, that high fossil abundance and diversity is not in itself diagnostic of MFSs (it can also characterise debris flows, for example). MFSs are diagnosed by fossil assemblages indicating maximum flooding: for example, maximum bathymetry; and, at least in down-dip settings, the maximum abundance of plankton (enabling calibration against global standard biozonation schemes established in the open ocean).

Maximum flooding surfaces in the Neogene of the tropics are characterised palynologically by coastal mangrove pollen. Those in the Miocene of central Tunisia are characterised by marine dinoflagellate abundance and diversity peaks, as are those in the Oligocene–Miocene of Denmark (Dybkjaer, 2004), the Early Cretaceous of the Bredasdorp basin of South Africa, and the latest Jurassic to earliest Cretaceous of the Vocontian basin of south-east France. At least in the case of the Early Cretaceous of South Africa, this is attributable to environmental stabilisation and increased intra-specific competition (V. H. Valicenti, personal communication). In the Early Carboniferous of Britain (marine bands) MFSs are characterised by acritarchs.

Maximum flooding surfaces in the earliest Cretaceous Barrow group sequences of the north-west shelf of Australia are characterised by zero to moderate microfaunal abundance and diversity, and by low to moderate palynofloral diversity. Arenaceous and calcareous benthic foraminiferans occur in approximately equal proportions. Maximum flooding surfaces in the post-rift Cretaceous sequences of the Pletmos and Bredasdorp basins of South Africa are characterised by microfaunal and palynofloral and diversity peaks, and by an acme of the deep-water foraminiferan *Verneuilina* (Brown *et al.*, 1995).

Carbonate sequences

Maximum flooding surfaces in carbonate sequences in the Early Cretaceous of the Middle East and central Asia are characterised micropalaeontologically by peaks in abundance of planktonic foraminiferans.

Mixed sequences

Maximum flooding surfaces in the Early Jurassic mixed clastic–carbonate sequences of Quercy in south-west France are characterised by peak abundances and diversities of benthic foraminiferans, and of uncoiling morphotypes of *Lenticulina*.

High-stand systems tracts

High-stand systems tracts (HSTs) in clastic sequences are characterised by overall progradation, and by fossil and trace fossil assemblages indicating an upward shoaling (Olson and Thompson, in Koutsoukos, 2005).

High-stand systems tracts in the Neogene of south-east Asia are characterised micropalaeontologically by 'turbid-water' calcareous benthic foraminiferans, including buliminides, thought to be adapted to eutrophic conditions. Those in the post-rift Cretaceous sequences of the Pletmos and Bredasdorp basins of South Africa are characterised by upward decreases in deep-water foraminiferans such as

Verneuilina and radiolarians such as *Dictyomitra*, and increases in shallow-water foraminiferans such as *Conorotalites* (Brown *et al.*, 1995). Those in the Late Cretaceous of the Basco-Cantabrian basin in Spain are characterised by the foraminiferans *Praebulimina*, *Arenobulimina* and *Frondicularia* (Grafe, 1995).

Early HSTs in the Neogene of the tropics are characterised palynologically by terrestrially derived Rubiaceae and Euphorbiaceae (open forest and swamp plant) pollen; late HSTs by Gramineae (grass) or *Casuarina* (a littoral plant) pollen. Early HSTs in the Plio-Pleistogene of the Gulf of Mexico are characterised by pollen of warm/wet aspect; late HSTs by pollen of cool/wet aspect. In the Pleistocene glacial/interglacial cycles of the Amazon fan HSTs are characterised by low concentrations of terrigenous palynomorphs and a predominance of marine microplankton. Those in the Miocene of central Tunisia are characterised by upward decreases in the abundance of and diversity of marine dinoflagellates; and those in the latest Jurassic to earliest Cretaceous of the Vocontian basin of south-east France are characterised by an upward increase in the abundance and angularity of terrestrially derived plant fragments, and an upward decrease in the abundance and diversity of marine dinoflagellates. High-stand systems tracts in the latest Jurassic to earliest Cretaceous of Papua New Guinea are characterised by high palynofloral abundance, but also by high dominance and low diversity. The percentage of terrestrially derived palynomorphs is in the range 8–25%. Values are higher at proximal than at distal sites. They tend to increase upwards both through the HST as a whole and through individual constituent coarsening-upward or progradational parasequences.

High-stand systems tracts in the earliest Cretaceous Barrow group sequences of the north-west shelf of Australia are characterised by zero to moderate microfaunal abundance and diversity, and by low to high palynofloral diversity. Arenaceous and calcareous benthic foraminiferans occur in approximately equal proportions. The percentage of terrestrially derived palynomorphs is generally 60–90% at both the proximal site and the distal site.

Carbonate sequences

Early HSTs in carbonate sequences in the Early Cretaceous of the Middle East and central Asia are characterised by algal (*Lithocodium*) reef facies; late HSTs by aggradational to progradational rudist–coral reef facies, orbitolinid–valvulinid–echinoderm wackestone and miliolid–dasycladacean (*Hensonella*) wackestone–packstone back-reef facies. In the Middle East, they are also typically characterised by conical orbitolinids (Fig. 6.18) (Simmons *et al.*, in Hart *et al.*, 2000; Jones *et al.*, in Bubik and Kaminski, 2004).

Mixed sequences

High-stand systems tracts in the Early Jurassic mixed clastic–carbonate sequences of Quercy in south-west France are characterised by moderate abundances and diversities of benthic foraminiferans, and by an upward decrease in the proportion of uncoiling morphotypes of *Lenticulina*.

Modified after Jones *et al.*, in Jenkins (1993) and Jones (1996).

7 • Case histories of applications of palaeontology

This chapter deals with how our knowledge of fossils and earth and life history is applied in industry and elsewhere.

7.1 Petroleum geology

Readers interested in further details are referred to Jones (1996), Laudon (1996), Selley (1998) and Gluyas and Swarbrick (2003). Those interested in the environmental impact of petroleum exploration and reservoir exploitation are referred to Evans (1997); those interested in impact mitigation, to his discussion of BP's Wytch Farm oilfield underlying Poole Harbour, Dorset, UK, and under development using shore-based 'extended reach' directional drilling.

7.1.1 Principles and practice of petroleum geology

Play components

There are essentially three components to a successful play, namely, a working petroleum source-rock and system, to provide the petroleum charge; a reservoir, to store, and ultimately to flow, same; and a cap-rock (seal) and trap, to prevent it from escaping. Each of these play components is discussed in more detail below. A discussion of stratigraphic control on play component distribution follows.

Petroleum source-rocks and systems

Petroleum source-rocks. Petroleum source-rocks are those from which petroleum is derived. They typically consist of abundant organic matter or kerogen, that can be of either terrestrial land-plant or humic, or marine algal or sapropelic, origin. This organic matter is transformed, or matured, into oil and gas by heating on burial. The product will depend on the nature of the source-rock, and on the transformation ratio, or degree of maturation, in turn determined by burial history and by basin type. Typically, oil is generated from so-called 'type I' or 'type II' source-rocks, whose organic matter is of marine algal origin; and gas either from these same source rocks subjected to more maturation, or from 'type III" source-rocks, whose organic matter is of terrestrial land-plant origin. Maturation and migration can be modelled using various publicly available and proprietary one-, two- and three-dimensional basin-modelling software packages.

Petroleum systems. A petroleum system has been variously defined as 'a dynamic petroleum generating and concentrating system functioning in geological space and time', or 'a pod of active source rock and the resulting oil and gas accumulations'. As recently noted by Katz and Mello, in Mello and Katz (2000), although the term has been defined, it nonetheless still appears to mean different, albeit thematically related, things to different individuals and organisations. As interpreted by Katz and Mello, in Mello and Katz (2000), the petroleum system concept is 'a means of formalising the relationship between the geologic elements in time and space that are required for the development of a commercial petroleum accumulation'.

Petroleum systems analysis is concerned with the presence of these elements, that is, source-, reservoir- and cap-rock and trap, and with their effectiveness (see also section on 'Play fairway analysis' below), and may also be said to be concerned with how much petroleum a basin has generated, where it has migrated, and where it is trapped, and in what quantity and phase. The relationships between the elements are conventionally presented in the form of a series of structural cross-sections showing not only the fill of the basin but also the fluid flow therein, and, importantly, the progression of the petroleum migration front through time, and hence the limits of the system. An understanding of the limits of petroleum systems within basins is critical to an evaluation of their potential prospectivity, with exploration risk increasing in proportion to the distance from an established petroleum system. For example, in the North Sea basin, where migration is almost exclusively vertical, the limit of prospectivity is essentially controlled by the limit of the mature Kimmeridge clay formation source-rock. An understanding of the certainty with which limits can be proscribed is also important. For example, in the Campos basin, offshore Brazil, the limit of the petroleum system and of prospectivity was recently significantly extended when advances in drilling technology enabled exploration in deep water for the first time.

Petroleum systems are conventionally named after the source-rock rather than the reservoir-rock, since one source may charge only one or more than one reservoir. There may be only one petroleum system in a basin, or there may be more than one, in which case they may be stratigraphically or geographically discrete, or they may be overlapping. Identification of individual petroleum systems in basins where more than one is operative requires typing of reservoired petroleum to the source-rock. This is nowadays relatively easily achievable using a range of geochemical techniques including, for example, biomarker analysis.

There are certain empirical observations with regard to the distribution of oil and gas reserves within a basin that constitute general rules (although there are also specific exceptions, especially when more than one petroleum system is operative). As a general rule, oil gravity decreases with depth. In other words, heavy oil is found at shallow depths, light oil and condensate at intermediate depths, and gas at depth. This is a function of the combined effects of source-rock maturity and of biodegradation, or rather the lack thereof at depth, where temperature conditions are inimical to the bacteria responsible. Also as a general rule, oil gravity decreases towards the centre of the basin. In other words, heavy oil is found at the basin margin, light oil and condensate in an intermediate position, and gas in the basin centre. This has been interpreted as indicative of displacement of oil by gas along a so-called 'fill–spill chain' (Gussow, 1954). 'Gussow's principle' relies on the existence of laterally extensive permeable carrier beds, in other words, low impedance to fluid flow. Even in such a system, though, charge can be limited, and those structures most remote from the source kitchen in the centre of the basin run the risk of remaining uncharged.

Box 7.1 Palaeontological inputs into petroleum systems analysis

Biostratigraphy and chronostratigraphy

One key input in petroleum systems analysis is absolute age data, which is used to establish the ages of the source-rock and overburden, and, critically, to model the timing of migration as against that of trap formation. Chronostratigraphic data is in turn an output of routine palaeontological or biostratigraphic analysis, or of graphic correlation. Resolution is typically best – better than 1 Ma – in marine environments (see Chapter 6). However, unconventional biostratigraphic and/or non-biostratigraphic techniques can also provide precise and accurate ages in non-marine environments, for example vertebrate palaeontology and magnetostratigraphy in the 'molasse' sequences of Transcaucasia and Iran.

Interestingly, some sort of stratigraphy can also be established on the basis of 'biomarkers'. For example, C_{11}–C_{19} paraffins derived from the cyanophyte *Gloeocapsomorpha prisca* indicate an essentially Ordovician age. Nordiacholestane, a diatom derivative, indicates a Jurassic or younger age (or, in abundance, Cretaceous or younger). Oleanane, a higher land-plant derivative, indicates a Cretaceous or younger age (or, in abundance, Cenozoic).

Palaoenvironmental interpretation

Another key input is palaeoenvironmental interpretation, covering palaeobathymetry, palaeobiogeography, source-rock depositional modelling and palynofacies.

Palaeobathymetry

Another output of routine palaeontological analysis, palaeobathymetric interpretation – essentially based on proxy living benthic foraminiferal distribution data – enables the characterisation of a range of depth environments (see Chapter 4). Resolution is typically sufficiently good to enable recognition of the following depth environments: non-marine (obviously, on the basis of freshwater algae, diatoms and ostracods, terrestrially derived spores and pollen, or land-plants and animals, rather than foraminiferans); marginal marine; shallow marine, inner shelf or neritic (0–50 m); middle shelf (50–100 m); outer shelf (100–200 m); deep marine, late slope or bathyal (200–1000 m); middle slope (1000–1500 m); early slope (1500–2000 m); and abyssal plain (>2000 m). (Note that the resolution is of the order of 10s of metres in shallow marine environments, but only of the order of 100s or even 1000s of metres in deep marine environments.) Importantly, high-resolution palaeobathymetric curves can be constructed and imported into workstation environments through presently proprietary but hopefully ultimately publicly available software packages containing links to bathymetric distribution data-sets or 'look-up' tables (author's unpublished observations).

Additionally, essentially depth-independent dysoxic and anoxic environments can be characterised micropalaeontologically or palynologically. For example, oxygen minimum zones or OMZs can be characterised by infaunal bolivinid–buliminid–uvigerinid benthic foraminiferan assemblages (see also section on 'Palynofacies' below).

Palaeobiogeography

Palaeobiogeographic data can provide indications as to the palaeolatitude and palaeotemperature history of a basin. For example, the palaeo-latitude and palaeotemperature history of Kuwait has been established partly on the basis of biological and lithological indicators, such as reefal carbonates and evaporites (Sharland *et al.*, 2001). Interestingly, most of the principal source-rocks of the Middle East region, in the Middle Jurassic to 'Middle' Cretaceous, appear to have been associated with palaeo-equatorial latitudes.

Lithological prediction

Importantly, palaeontological, that is to say both biostratigraphic and palaeobiological or palaeoenvironmental data and interpretations, constrain the sequence stratigraphic model that enables predictions to be made as to lithology away from the areas of control.

Source-rock depositional modelling

Source-rock deposition – and preservation – is caused by an excess of production over consumption of organic matter.

Primary production of organic matter is controlled essentially by biological activity, by photosynthetic land-plants and marine algae. In the marine realm, it is typically highest in areas or at times of upwelling of nutrient-rich bottom waters. Present-day upwelling is indicated by phytoplankton blooms on satellite images. Past upwelling can be indicated by palaeontological evidence (Martinez, 2003). Past upwelling can be modelled using palaeoclimatological computer software (Kruijs, 1989).

Consumption of organic matter is controlled essentially by the availability of oxygen, leading to oxidation of organic matter in early diagenetic decomposition reactions. In the marine realm, it is typically lowest in areas of dysoxia or anoxia, such as the Indian Ocean, or the Black Sea; or at times of dysoxia or anoxia, such as the Cretaceous, Barremian–Aptian ('oceanic anoxic event' or OAE1 of Jenkyns, 1980), Cenomanian–Turonian (OAE2) and Senonian (OAE3) (Koutsoukos *et al.*, 1990; Kuhnt and Wiedmann, in Huc, 1995; Leckie *et al.*, 2002; Skelton, 2003; Luning *et al.*, 2004), or the late Early–early Middle Miocene, Burdigalian–Langhian (Smart and Ramsay, 1995).

Upwelling models

Wind-driven upwelling occurs as a result of prevailing offshore winds causing onshore movement of nutrient-rich bottom waters, through frictional drag – a phenomenon termed 'Ekman transport'. It is characteristic of the western seaboards of continental land masses, such as the California coast in the Middle–Late Miocene, where the Monterey formation source-rock was deposited. *Current-driven upwelling* occurs as a result of divergence of water

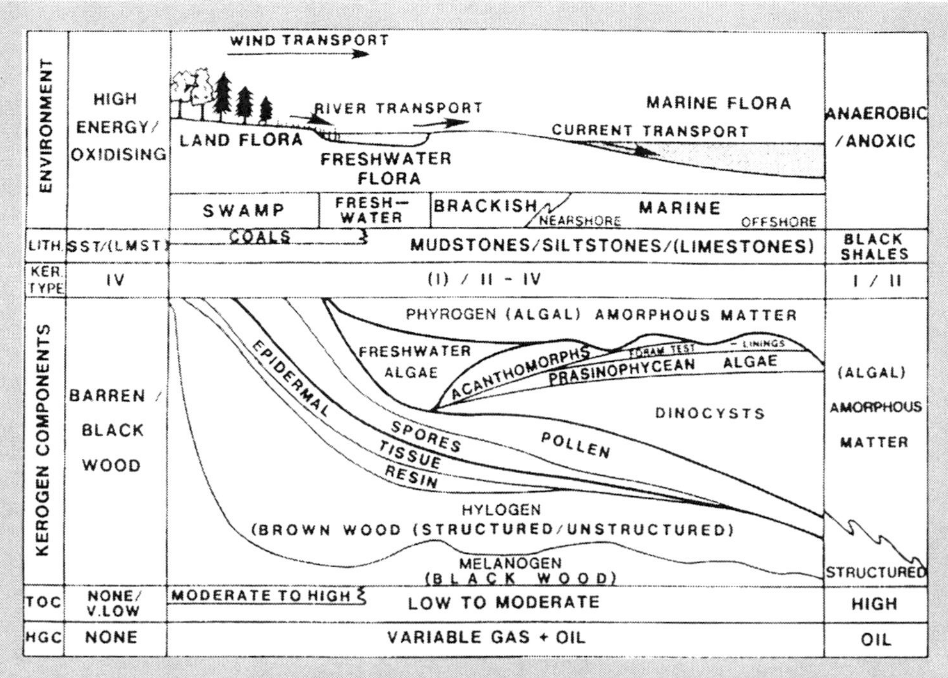

Fig. 7.1. Use of palynofacies in palaeoenvironmental interpretation. (From Jones, 1996.)

masses, for example, along the 'equatorial divergence'. *Obstruction upwelling* occurs as a result of deflection of surface currents around obstructions.

Indian Ocean model

There is a distinct and enhanced preservation of organic matter, on the upper part of the continental slope in the Indian Ocean – and indeed elsewhere. This is on account of the consumption of oxygen by zooplankton, and the absence of production of oxygen by phytoplankton (below the photic zone).

Black Sea model

There is anoxia, and enhanced preservation of organic matter, below the sill depth in the Black Sea at the present time. There was evidently similar anoxia and enhanced preservation of organic matter in, for example, the similarly restricted syn- to immediately post-rift, lacustrine to marine, sequences of the South Atlantic basins, in the Early Cretaceous (Jones, 1996; author's unpublished observations).

Palynofacies

Palynofacies analysis enables at least a degree of palaeoenvironmental interpretation, with terrestrial and marine, and proximal and distal marine, environments generally distinguishable (Fig. 7.1) (Jones, 1996; Batten and Stead, in Koutsoukos, 2005). However, it should be stressed that the extent to which the studied palynomorphs and phytoclasts have been transported is not always easy to establish.

Thermal maturity indication

An independent measure of thermal maturity is provided by the 'spore colour index' (Fig. 7.2) (Jones, 1996). It is also provided by the acritarch and chitinozoan colour indices (Marshall, in Briggs and Crowther, 1990), the foraminiferal colour index (McNeil *et al.*, 1996), the ostracod colour index (Ainsworth *et al.*, 1990), and the conodont colour index (Epstein *et al.*, 1977). The

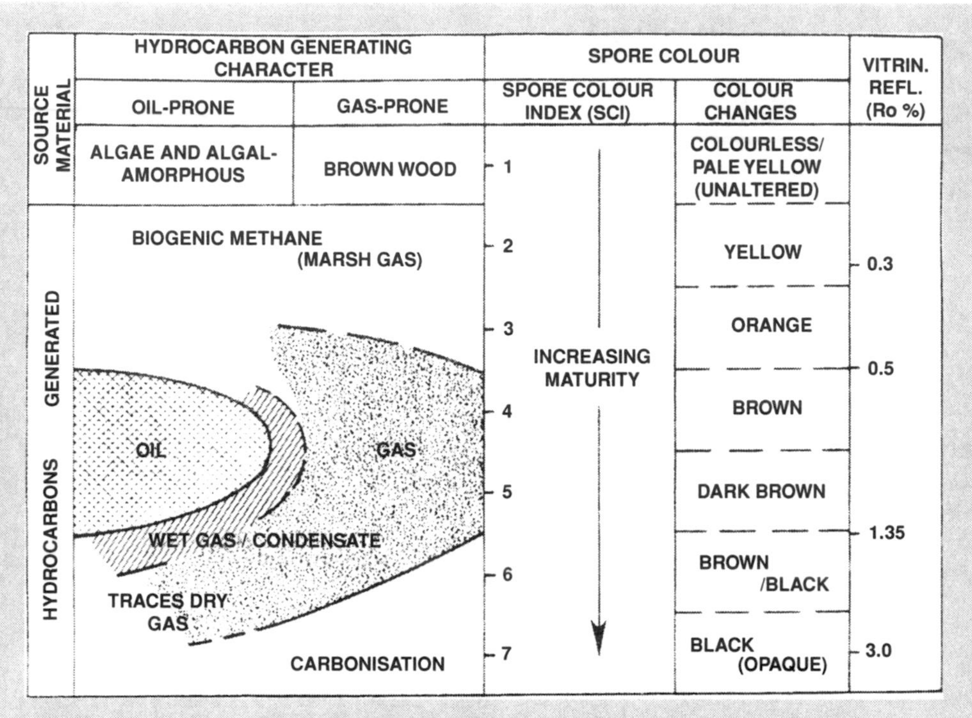

Fig. 7.2. Use of palynofacies in thermal maturity assessment. (From Jones, 1996.)

spore, foraminiferal and ostracod colour indices have been independently calibrated against vitrinite reflectance. The acritarch, chitinozoan and conodont colour indices have not, as these groups of organisms largely pre-date the land-plants that constitute the source of the vitrinite.

Reservoir-rocks

Reservoir rocks are those in which petroleum is reservoired. They require two properties, namely porosity and permeability. *Porosity* (ϕ) is the space between sedimentary particles that enables the reservoir to store pore fluid, including petroleum. It is measured by point counting of petrographic thin-sections or by petrophysical analysis of well-logs, or by experimental analysis. Values are quoted as percentages or porosity units. Obviously, only open pores constitute effective porosity, that is, that which is capable of flowing hydrocarbons. Porosity can be either primary, and related to the original depositional facies or fabric, or secondary, and related to diagenesis. Primary porosity can be either enhanced or occluded by secondary diagenetic processes. Typically, dissolution, dolomitisation and fracturing are enhancive, and compaction and cementation occlusive, processes. Importantly, an early hydrocarbon charge can cause effective cessation of occlusive diagenetic processes. *Permeability* (k) is the ability of the rock to flow petroleum. It is measured by experimental analysis. Values are measured in darcies, or, more commonly, millidarcies (md). Permeability distribution within a reservoir is commonly characterised by anisotropy, with vertical permeability (k_v) typically lower than horizontal permeability (k_h), owing to bed effects. This has important consequences for production.

Cap-rocks (seals) and traps

Cap-rocks (seals). Cap-rocks (seals) are those beneath which petroleum is trapped. They require a low permeability (see above) and a high sealing capacity, quantified using capillary entry pressure measurements. They can be of any lithology, although mudstones are the commonest, and evaporites such as anhydrite and halite the most effective.

Box 7.2 Micropalaeontological characterisation of mudstone cap-rocks

Introduction

Various types of mudstones from the Oligo-Miocene sections of approximately 30 wells from offshore west Africa have been analysed micropalaeontologically, in order to establish whether those that constitute barriers or baffles to fluid flow can be characterised.

Results

Condensed sections in the Oligo-Miocene of offshore west Africa, interpreted as representing hemipelagites, have been distinguished from expanded sections, interpreted as representing turbidites, on the basis of rock accumulation rate data derived from thickness/age relationships and/or graphic correlation (Fig. 7.3). The former have been distinguished essentially on the basis of rock accumulation rates of less than 100 m/Ma, comparable to those exhibited by – uncompacted – Recent pelagites from comparable depths in the North Atlantic (Jones, 1984). Condensed sections have also been distinguished on the basis of maximum benthic foraminiferal abundance and diversity, especially of infaunal morphotypes, and of maximum bathymetry. High abundance and

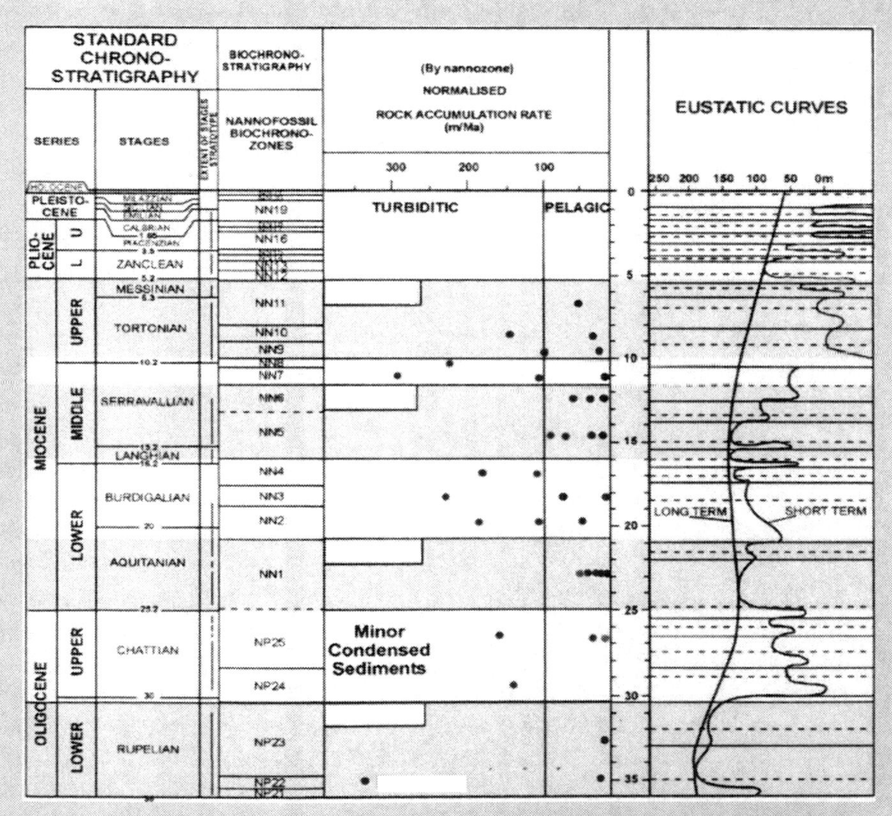

Fig. 7.3. Mudrock seal capacity, Cenozoic, Angola. Shading indicates stratigraphic position of condensed sections interpreted as regional seals in Early Oligocene and Early, Middle and Late Miocene. Condensation identified on basis of normalised rock accumulation rate by nannozone; seal capacity borne out by capillary entry pressure and other data. Condensed, pelagic sedimentation identified on basis of rates <100 m/Ma; expanded, turbiditic sedimentation on rates >100 m/Ma. Note coincidence between condensed sections/seals and ?glacioeustatically mediated second-order sea-level high-stands.

diversity is characteristic of condensed sections or MFSs. Note, however, that high abundance and diversity are not diagnostic of condensed sections. Condensed sections can be characterised by low abundance and diversity, especially if affected by dissolution of calcareous species below the lysocline. Debris flows can be characterised by high abundance and diversity, although they can still be distinguished from high abundance and diversity condensed sections on the basis of their allochthonous components, identified on the basis of bathymetric interpretation. (Proportionately) high abundance and diversity of infaunal 'morphogroup' C does appear to be diagnostic of condensed sections, reflecting colonisation of the sea-floor under the comparatively tranquil conditions obtaining in between turbidite episodes, and over a comparatively long time.

Discussion

Importantly, the micropalaeontologically characterised condensed sections in the Oligo-Miocene of offshore west Africa correlate well with regional seals identified on seismic facies, wireline log and capillary entry pressure data. Interestingly, they are also coincident in time with tectonoeustatically or glacioeustatically mediated second-order sea-level high-stands in the Early Oligocene (global standard calcareous nannoplankton zones NP21–NP23), earliest Miocene (NN1), Middle Miocene (NN5–NN6) and latest Miocene (NN9–NN11). This suggests that the methodology has essentially global applicability.

Incidentally, ineffective as well as effective seals have also been characterised micropalaeontologically, elsewhere, on the basis of the similarity of their contained benthic foraminiferal assemblages to those associated with modern seeps (see also Box 4.1). For example, the seal sequence overlying the platform carbonate reservoir target in the Miocene of the Nam Con Son basin, offshore Vietnam, is characterised by *Arenomeandrospira glomerata*, a species originally described from the Skagerrak and Kattegat, regions of shallow gas occurrence and sea-bed 'pockmarks' (Jones and Wonders, in Hart *et al*., 2000).

Based on Jones (2003b) and the author's unpublished observations.

Traps. Traps are configurations of reservoir and seal that do not allow the escape of petroleum. They are classified into three main types: structural, for example, anticlinal and diapiric, traps; stratigraphic, for example, pinch-out, reef and sub-unconformity traps, with no structural component other than regional dip; and combination traps. Structural traps are by far the most common, not least because they are typically readily imaged on seismic data. Stratigraphic traps are typically much more subtle, and require a deliberate search.

Stratigraphic control on play component distribution

Source-rocks. There is control on source-rock development within the sequence stratigraphic framework, with preferential development during TSTs (Jones, 1996; Tyson, in Hesselbo and Parkinson, 1996).

There is also extrinsic stratigraphic control on source-rock development outwith this framework, with preferential development during the 30% of Phanerozoic time represented by the Silurian, Late Devonian–Early Carboniferous, Late Carboniferous–Early Permian, Middle–Late Jurassic, 'Middle' Cretaceous and Oligo-Miocene, accounting for a disproportionate 90%+ of the world's discovered oil (Silurian – 9%; Late Devonian–Early Carboniferous – 8%; Late Carboniferous–Early Permian – 8%; Late Jurassic – 25%; 'Middle' Cretaceous – 29%; Oligocene–Holocene – 13%) (Jones, 1996). The preferential development of source-rocks at these times indicates a considerable excess of production of organic material over consumption. In the case of the Silurian, this probably reflects both high primary production of oil-producing phytoplankton during times of transgression, and low consumption, in anoxic bottom waters. In the case of the Late Devonian, at least of the Domanik formation of the Timan–Pechora basin in the north-eastern portion of the Russian platform, it probably principally reflects low consumption in a restricted, anoxic basin formed by the aggradation of carbonate platforms during times of transgression (House *et al.*, in Insalaco *et al.*, 2000). In the case of the Middle Jurassic to 'Middle' Cretaceous of the Middle East, it probably principally reflects low consumption, in restricted, anoxic intra-shelf basins (see below). In the case of the 'Middle' Cretaceous globally, it probably reflects both high production during times of transgression, and, especially in the Proto-Atlantic, low consumption, in restricted, anoxic bottom waters. The comparative leanness of the Late Cretaceous and Cenozoic globally probably reflects not only early production but also higher consumption of phytoplankton, possibly in turn enabled by the evolution of a new group of zooplankton, the planktonic foraminiferans.

Reservoir-rocks. There is control on reservoir-rock development within the sequence stratigraphic framework, with preferential development of deep marine clastic reservoirs during LSTs, and shallow marine clastic and carbonate reservoirs during HSTs (Jones, 1996). Secondary enhancement of the quality of shallow marine carbonate reservoirs takes place preferentially during LSTs in the case of dissolution, and during TSTs in the case of – brine-reflux – dolomitisation. Examples are known from the Palaeozoic of the North Caspian, the Mesozoic of Mexico, and the Mesozoic and Cenozoic of the Middle East.

There may also be some stratigraphic control on reservoir-rock development outwith this framework. The primary reservoir properties of the chalk reservoirs of the North Sea and the nummulite bank reservoirs of the circum-Mediterranean may be controlled in part by the evolution of the fossils that constitute them (see Chapter 3).

Cap-rocks (seals) and traps. There is some control on cap-rock development within the sequence stratigraphic framework, with preferential development during TSTs and MFSs (Jones, 1996; see also Box 7.2).

There may also be some control on – structural – trap development outwith this framework, with the Eocene an important time of plate reorganisation and associated structuration, as in the Caribbean, the Indian subcontinent and adjacent areas of central Asia, and Papua New Guinea, and the Miocene–Holocene an important time of structural trap development in the Zagros thrust-belt of Iran (Jones, 1996). Elsewhere in the Middle East, the Carboniferous and the 'Middle' Cretaceous were also important times of structuration (Sharland *et al.*, 2001; author's unpublished observations).

Petroleum exploration

Petroleum exploration involves a number of geological, geophysical and integrated techniques. It also involves a number of drilling and petrophysical logging technologies.

Geological techniques

Geological techniques include not only field geological mapping and sampling, undertaken principally to provide data and interpretations on the structural configuration on the surface and in the subsurface, and to provide data and interpretations on source-, reservoir- and cap-rocks, but also petroleum systems analysis and basin modelling (see above). Field geological mapping is particularly important in areas where the acquisition of geophysical data is difficult, for example, in mountainous areas, in areas of structural complexity and/or steep dips, or in areas where the surface outcrops are of limestones or evaporites.

Geophysical techniques

Geophysical techniques include seismic surveying, gravity and magnetics surveying, and other remote sensing techniques.

Seismic surveying is undertaken to provide further data and interpretations on the structural configuration in the subsurface, to identify prospects suitable for drilling, to infer lithology, porosity and pore fluid content, and to identify 'direct hydrocarbon indications' (DHIs) in reservoirs. Acquisition of seismic data on land and at sea is achieved by detonating small explosive charges at the surface and recording the time taken for the shock waves generated by them to reflect back from the various subsurface layers. Processing of seismic data leads to a product that is essentially a structural cross-section through the subsurface, albeit in time rather than depth (until converted to depth using velocity data). Seismic rock property data and other attributes are used to infer lithology, porosity and pore fluid content, and to identify DHIs in reservoirs. Three-dimensional seismic data can be used to infer the geometry of individual sedimentary bodies within reservoirs.

Gravity and magnetics surveying is undertaken to provide further data and interpretations on the structural configuration on the surface and in the subsurface, and on the structure and composition of the basement. Acquisition of gravity and magnetic data is achieved by land, marine or airborne surveying. Processing leads to a product that is essentially a map of the local gravity or magnetic field (with the effects of the regional field filtered out). The map highlights density and magnetic contrasts, such as those between the basement and basin fill, and allows interpretations to be made as to the structure and composition of the basement.

Other remote-sensing techniques include high-resolution aerial photography, satellite, radar and multispectral. Processing of remote sensing data leads to images or maps highlighting particular aspects of culture and geology, such as, for example, petroleum seeps or the alteration products thereof.

Integrated techniques

Integrated techniques include basin analysis and play fairway analysis.

Basin analysis involves the integration of geological and geophysical data and interpretations (see above), and the development of exploration models and strategies. Interestingly, empirical observations in intensively explored basins suggest some relationship between basin type and contained petroleum reserves. The richest in terms of proven reserves are the 'type IV' basins or 'continental borderland downwarps' of the Middle East. Note, though, that some of the largest reserves proven recently have been in 'type VIII' basins or 'Tertiary deltas' such as those of the deep-water Gulf of Mexico, and Brazil and west Africa in the South Atlantic, that have only recently become drillable through advances in technology.

Play fairway analysis is a methodology for mapping play component presence and effectiveness, play component presence and effectiveness risk, and composite play risk, in basin exploration. It is becoming standard practice for the mapping and associated databasing to be undertaken using GIS technology (Coburn and Yarus, 2000). Play component presence maps are essentially palaeogeographic and facies maps highlighting areas of proven or interpreted potential petroleum source-, reservoir- or cap-rock development. Play component effectiveness maps are those highlighting areas of proven or interpreted potential effective petroleum source-, reservoir- or cap-rock development, determined by burial depth etc. 'Common risk segment' (CRS) – or play component presence and effectiveness risk – maps colour-code highlighted areas by perceived risk, conventionally green for low risk, yellow for moderate risk and red for high risk. 'Composite CRS' or CCRS – or play risk – maps are made by compositing play component presence and effectiveness risk maps (and, ideally, uncertainty or data confidence maps). The only segments that can be characterised as low risk on the composite risk map are those that are characterised as low risk on all of the individual play component risk maps. Similarly, segments characterised as moderate risk on the composite risk map are those that are characterised as no higher than moderate risk on all of the individual play component risk maps. Segments that are characterised as high risk on any of the individual play component risk maps are characterised as high risk on the composite risk map.

Drilling technologies

Once a prospect has been matured, the technical and commercial risk of drilling evaluated, and the decision to drill approved, drilling can go ahead. The choice of rig will depend on a number of well-site and geological factors, including, for example, whether there is environmental legislation governing site planning or operations, and whether shallow gas pockets or overpressured intervals are anticipated from seismic data, site investigation or basin modelling, necessitating fitting or upgrading of blowout preventers (see Box 7.3). Incidentally, the sort of deep-water drilling currently being undertaken in the Gulf of Mexico and South Atlantic is close to the limits of available technology, and presents particular rig design, mobilisation and other logistical problems.

Put simply, drilling is effected by rotation of a diamond bit at the end of a drill string formed of lengths of pipe connected by collars, the weight of the apparatus providing the necessary downward force (Fig. 7.4). Periodic casing of the borehole minimises down-hole contamination or caving and generally maintains the condition of the hole, reducing the risk of the drill-pipe sticking and having to be fished out, which can be a time-consuming and costly exercise. Importantly, casing can also be used to contain overpressured zones, reducing the risk of potentially extremely dangerous leaks or blow-outs.

Specially prepared drilling mud pumped down the drill-pipe lubricates the bit, and seals the walls of the bore. It returns to the surface between the drill-pipe and the walls of the bore, bringing with it loose chippings broken off by the action of the bit. These chippings, generally known as ditch-cuttings, are collected in a trap or 'shale shaker' prior to recycling of the drilling mud. Ditch-cuttings samples are generally collected every 3 m (10 ft) for the purposes of lithological description and micropalaeontological and other analysis. In the increasingly cost-conscious oil industry, in which the main aim of drilling is to reach the reservoir target as quickly and cheaply as possible, conventional coring, which is time-consuming and costly, is generally undertaken over critical sections of the reservoir, over which it allows extraction of invaluable data on depositional and diagenetic facies, and porosity and permeability characteristics. Side-wall coring is a useful adjunct to the conventional coring programme.

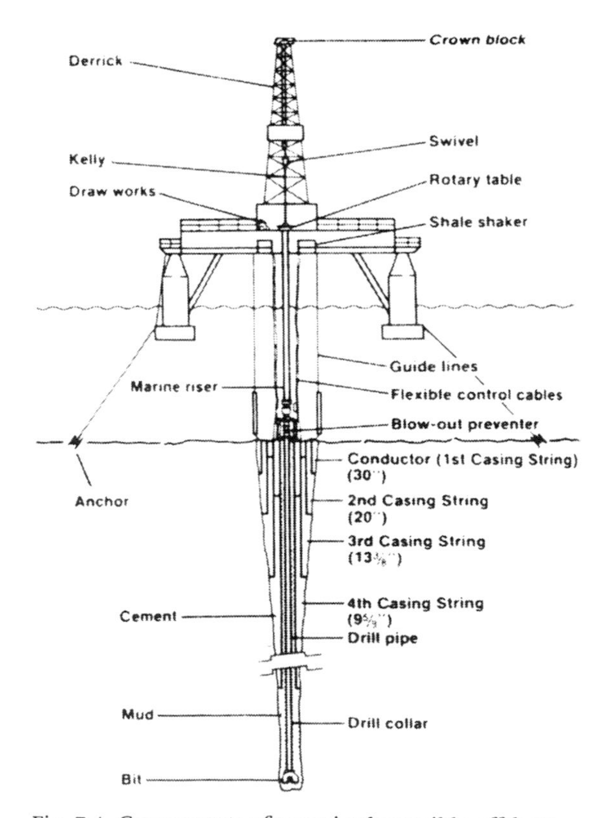

Fig. 7.4. Components of a semi-submersible offshore drilling rig. (From Jones, 1996; reproduced with permission from Copestake, in Jenkins, 1993.)

From the point of view of the industrial palaeontologist, in general, the most useful samples are conventional cores, and the least useful ditch-cuttings. This is because ditch-cuttings samples are prone to contamination not only from caved material (see above), but also on occasion from the drilling mud. For example, in my working experience, ditch-cuttings samples from the Pedernales field in the eastern Venezuelan basin became contaminated by mangrove pollen in drilling mud formulated from local river water that were indistinguishable from those in the – Pliocene – reservoir (Jones, in Jones and Simmons, 1999).

It is important to note that sample quality is to an extent also dependent on the drilling technology, and bit type, employed. 'Turbo-drilling' can have a particularly destructive effect on samples, effectively metamorphosing them and rendering them useless for analytical purposes. Coiled tubing, slim-hole and underbalance technologies have, though, recently been successfully employed in combination in the drilling of the Sajaa field in Sharjah in the United Arab Emirates, without unduly adversely affecting the quality of the samples used in micropalaeontological analysis and in 'biosteering' (see below).

Petrophysical logging technologies

A wide range of petrophysical or wireline logging technologies are available for formation evaluation.

(Note in this context that formation evaluation techniques practised by former eastern-bloc geologists differ somewhat from those practised by western geologists.) The most commonly run petrophysical or wireline logs are gamma, sonic and resistivity logs. Together, these provide an indication not only of lithology but also of porosity and pore fluid content. 'Logging-while-drilling' or LWD technologies have recently been developed that allow these logs to be run in real-time in the reservoir. Other logs that are becoming more widely used on account of their usefulness specifically in structural geological and sedimentological interpretation are dipmeter logs and image logs. Interestingly from the palaeontologist's point of view, individual fossils and trace fossils are occasionally identifiable on image logs.

Gamma logs measure radioactivity. Because in sedimentary rocks this is essentially a function of lithology, being high in fine clastics in which radioactive constituents tend to be concentrated, and low in coarse clastics and carbonates, gamma logs also provide a measure of this parameter. Note, though, that sandstones containing igneous clasts can yield anomalously high values, and non-radioactive clay minerals such as kaolinites anomalously low values. Note also, though, that the so-called natural gamma ray spectrometry tool can be used to identify individual clay minerals. Incidentally, it is also useful in source-rock evaluation, since it can distinguish uranium, commonly associated with organic matter, from other radioactive elements.

Sonic logs measure sonic transit time. Because this is essentially a function of density and porosity, sonic logs provide measures of these parameters. Density values are highest in dense carbonates, and lowest in uncompacted clastics. Porosity values are highest in porous, and lowest in tight, carbonates and clastics. Porosity can also be measured using neutron logs. Porosity and permeability can be measured using nuclear magnetic resonance logs.

Resistivity logs measure resistivity (the inverse of electrical conductivity). Because this is essentially a function of pore fluid content, resistivity logs provide a measure of this parameter (and of hydrocarbon saturation). Values are highest in rocks whose pore spaces are occupied by hydrocarbons, and lowest in rocks whose pore spaces are occupied by water or brine.

Reservoir exploitation

Reservoir exploitation involves a number of geological and geophysical techniques, reservoir modelling, well monitoring, borehole seismic and four-dimensional seismic (essentially 'time-lapse' three-dimensional, seismic illustrating the movement of reservoir fluids through time) imaging. It also involves a number of drilling and petrophysical logging technologies (see above), and testing, and completion and production technologies (including 'enhanced oil recovery' (EOR) technologies).

Testing

Testing is a process initiated on indication of hydrocarbons, whereby the bottom-hole formation is made to flow some of its contents into the drill-pipe for recovery and analysis. It provides valuable information on potential flow rate, or reservoir deliverability.

Box 7.3 Palaeontology and health, safety and environmental issues in the petroleum industry

There are five principal areas in which palaeontology impacts on health, safety and environmental (HSE) issues – or HSE issues on palaeontology – in the petroleum industry. These are: site investigation; pressure prediction; well-site operations; environmental monitoring; and environmental impact assessment. Each is discussed in turn below.

Site investigation

In my working experience in the petroleum industry, micropalaeontology has proved of use in site investigation in Azerbaijan in the south Caspian, specifically in determining the safest sites for locating drilling rigs, pipelines and other facilities. Here, it has been run on surface and shallow subsurface core samples in order to discriminate between mud volcano flow and background sediments with different engineering properties, as determined by geomechanical testing. Mud volcano flows, whose engineering properties are such that they constitute significant geohazards, have been distinguished on the basis of their reworked fossil content; background sediments on the basis of their coherent stratigraphy. Micropalaeontology and palynology, essentially in the form of climatostratigraphy tentatively tied to the marine oxygen isotope record, have also been used to date and correlate the surface and shallow subsurface core samples, together with radiocarbon dating and OSL (see Sub-section 6.3.6). High-resolution shallow seismic has been used to trace mud volcano flows identified in surface and shallow subsurface core samples outwith these areas of control, and to generate geohazard maps.

Micropalaeontology is of potential applicability elsewhere in the petroleum industry, for example in the discrimination of slope failure and background sediments with different engineering properties in areas of slope instability, and in the identification of active erosion in areas of channelisation.

It is also of proven applicability elsewhere in the civil engineering industry (see Section 7.4).

Pressure prediction

Micropalaeontology provides critical inputs into pressure prediction from petroleum systems analysis and basin modelling (see Box 7.1).

Well-site operations

There are a number of areas in which micropalaeontology impacts HSE in the field of well-site operations, including not only making real-time casing and terminal depth (TD) calls, thus avoiding pressure kicks, but also providing an independent assessment of the condition and stability of the borehole, from cavings (see also subsections on 'Operational biostratigraphy' and 'Biosteering' below).

There are also areas in which HSE impacts micropalaeontology. For example, there are particular HSE risks associated with the use of hazardous and potentially lethal chemicals such as hydrochloric acid, nitric acid and hydrofluoric acid in conventional palynology processing. However, these risks are capable of being mitigated by the appropriate training, equipment and supervision of palynology processing technicians, and the implementation of safety checks and audit processes. Importantly, the risks are capable of further mitigation through the containment of the hazardous chemicals in portable palynology processing units, developed by BP in the 1980s, which technique has since proved of use worldwide. Indeed, the risks are capable of being mitigated altogether through the use of a non-conventional, 'green, non-acid well-site processing technique' (GNAWPT) developed by BP contractors in the 1990s. This technique has also proved of use almost worldwide. It has proved of particular use in the Norwegian sector of the North Sea, where strict environmental legislation prohibits the use of the conventional, acid palynology processing technique.

Environmental monitoring

The living biota is of proven use in environmental monitoring, for example of domestic and industrial pollution (see also Section 7.5).

Interestingly, the living and fossil biota is also of proven use in the detection of modern and ancient natural hydrocarbon seeps.

Environmental impact assessment

The living biota is of proven use in environmental impact assessment (EIA) in the civil engineering industry (see also Section 7.5).

It is also of proven use in EIA in the petroleum industry. (Historically, it has been somewhat underutilised in this sector, although it is anticipated that in the future it will become increasingly important, especially in the case of rare or sensitive habitats, such as the *Lophelia* reefs of the Atlantic margin.) For example, the macro- and meiobiotas, and to a lesser extent the microbiota, have been used in baseline studies on the potential impacts of petroleum industry infrastructure projects in the oilfield area of Alaska, centred around Prudhoe Bay on the north slope and situated between the National Petroleum Reserve of Alaska to the west and the Arctic National Wildlife Refuge to the east (Truett and Johnson, 2000). Here, potential impacts have been mitigated through the implementation of a number of measures approved by the appropriate regulatory bodies (Gilders and Cronin, in Truett and Johnson, 2000). These measures include:

consolidation of facilities [into smaller areas than originally planned]; use of ice road technology to eliminate gravel roads adjacent to pipelines, and elevating . . . pipelines . . . to allow free movement of wildlife; directional drilling to reduce the number of gravel pads; [and] improved waste handling, and the elimination of . . . pits for surface storage of drilling muds and cuttings (these drilling by-products are now re-injected into confining geological formations) . . . (Gilders and Cronin, in Truett and Johnson, 2000.)

The microbiota has also been used in baseline studies on the potential impacts of infrastructure projects, in the Columbus basin off the east coast of Trinidad (author's unpublished observations).

Importantly, palaeontology provides a record of long-term global biodiversity and causal environmental and climatic change, and a means of discriminating artificial, anthropogenic from natural effects (see appropriate sections in Chapter 3, Section 4.5 and Sub-section 5.5.6; see also McKinney and Drake, 1998; Wood *et al.*, 2000; Gerhard *et al.*, 2001; Groombridge and Jenkins, 2002; Battarbee *et al.*, 2003). An Inter-Governmental Panel on Climate Change concluded in 1996 that anthropogenic greenhouse gas emissions were responsible for global warming and sea-level rise, as had long been suspected, and that action was required to limit such emissions. The case and timetable for action was set out in the Kyoto Protocol in 1997 (Grubb *et al.*, 1999).

Based on the author's unpublished observations.

7.1.2 *Applications of biostratigraphy and palaeobiology in petroleum exploration*

Chronostratigraphy and palaeoenvironmental interpretation

Palaeontology plays a key role in the absolute chronostratigraphic age-dating and palaeoenvironmental interpretation of samples acquired during field geological mapping and drilling. It constrains the timing of structural geological events and enables the establishment of a 'mechanical stratigraphy', and calibrates seismic and well data. It provides input into petroleum systems analysis, integrated basin analysis, and play fairway analysis (see above). In play fairway analysis, the principal input is in the identification of time-slices for the purposes of palaeogeographic and facies, and CRS and CCRS mapping, and in the population of palaeogeographic and facies maps with palaeoenvironmental data and interpretations (made on the basis of a predictive sequence stratigraphic model). An example of a play fairway map, for the Cenozoic of the Middle East, is given by Goff *et al.*, in Al-Husseini (1995) (see subsection on 'Middle East' below).

Operational biostratigraphy

The main areas in which palaeontology is applied in exploration drilling are monitoring of stratigraphic position while drilling, determination of casing, coring and terminal depths, and post-well analysis and auditing. Note that successful application is dependent on appropriate sampling (see comments on drilling above).

Monitoring of stratigraphic position while drilling

Monitoring while drilling exploration wells is seldom called for nowadays other than in frontier ('wildcat') areas. However, monitoring while drilling production wells has become a virtual necessity in many areas (see below).

Determination of casing, coring and terminal depths

Casing, coring and terminal depths are normally determined prior to drilling, with appropriate input from the palaeontologist. In many cases the depths are derived from two-way time to a seismic horizon, and may be inaccurate owing to uncertainty as to which is the most appropriate velocity function to use to convert time to depth. Well-site palaeontology provides an accurate determination, effectively in real-time (Lowe *et al.*, 1988). It therefore enables decisions to be made also effectively in real-time, thereby saving significant amounts of time and money. It can also have important safety implications, as when casing off overpressured intervals.

Post-well analysis and auditing

Post-well analysis generally aims to provide a stratigraphic breakdown of the well within 6 weeks of the receipt of the final sample. Analytical data can be entered into software packages for the purposes of interpretation and integration. Auditing evaluates the technical and commercial success of the drilling. One of the most important parts involves the comparison of the prognosed well stratigraphy with the actual. This comparison is generally made by means of overlaying a synthetic seismogram of the well at the appropriate location, and transferring the actual well stratigraphy onto the seismic grid. This in turn helps to refine the predictive stratigraphic model.

7.1.3 *Case histories of applications in petroleum exploration*

Central and northern North Sea

Geological setting

The following is an abridged account of the geology and petroleum geology of the central and northern North Sea basins (Fig. 7.5). Readers interested in further details are referred to Jones (1996), Glennie (1998), Gluyas and Hichens (2002) and Evans *et al.* (2003).

The North Sea basin essentially formed in response to extensional episodes associated with rifting in the North Atlantic in the Permo-Triassic and Jurassic. Regional evidence from the onshore UK also points towards an earlier – Variscan or Hercynian – compressional episode, during which back-arc rift systems formed and deformed in response to the closure of the Rheic Ocean.

The Permo-Triassic witnessed multiple phases of rifting and post-rift thermal subsidence, and was characterised by the deposition of continental clastics and evaporites. The Early–Middle Jurassic was a time of thermal doming prior to another extensional phase in the Late Jurassic, and was characterised by deposition of volcanics, volcaniclastics and marginal marine, peri-deltaic clastics, and by locally significant erosion at the 'mid-Cimmerian unconformity'. The – diachronous – Late Jurassic syn-rift sequence was characterised by deposition of deep marine shales and submarine fan sandstones in the basin centre, and of shallow marine, coastal plain sandstones at the basin margins. Significant erosion took place at or near the Jurassic/Cretaceous boundary 'late Cimmerian unconformity'. The post-rift package was characterised initially, in the Early Cretaceous, by deposition of deep marine marls and shales and submarine fan sandstones, and later, in the Late Cretaceous to Early Palaeocene, by deposition of marls in the northern North Sea and chalks in the central North Sea.

The Late Palaeocene saw the onset of sea-floor spreading in the North Atlantic, and attendant uplift of hinterland areas in Scotland and northern England. The amount of uplift has been estimated to have been as much as 300–400 m at the time of the 'Chron 25 n hiatus' or 'Forties low-stand'. Rejuvenation of sediment supply resulted in the progradation of successive sequences of deltaic and associated submarine fan clastics into the basin in the Late Palaeocene and Eocene. The morphology of the fan units was controlled by the evolving structural configuration of the basin. Inversion at the end of the Eocene resulted in the generation of the 'Pyrenean unconformity'. The succeeding Oligocene–Holocene was characterised by renewed basin fill.

Petroleum geology

Source-rocks. The most important source-rock, and also incidentally one of the most important regional cap-rocks in the central and northern North Sea, is the latest Jurassic to earliest Cretaceous Kimmeridge clay formation (Fig. 7.6). The Kimmeridge clay is modelled as having been deposited in a basin characterised by at least intermittently anoxic bottom conditions brought about by salinity stratification, and inimical to most life forms. These conditions favoured the low consumption and hence

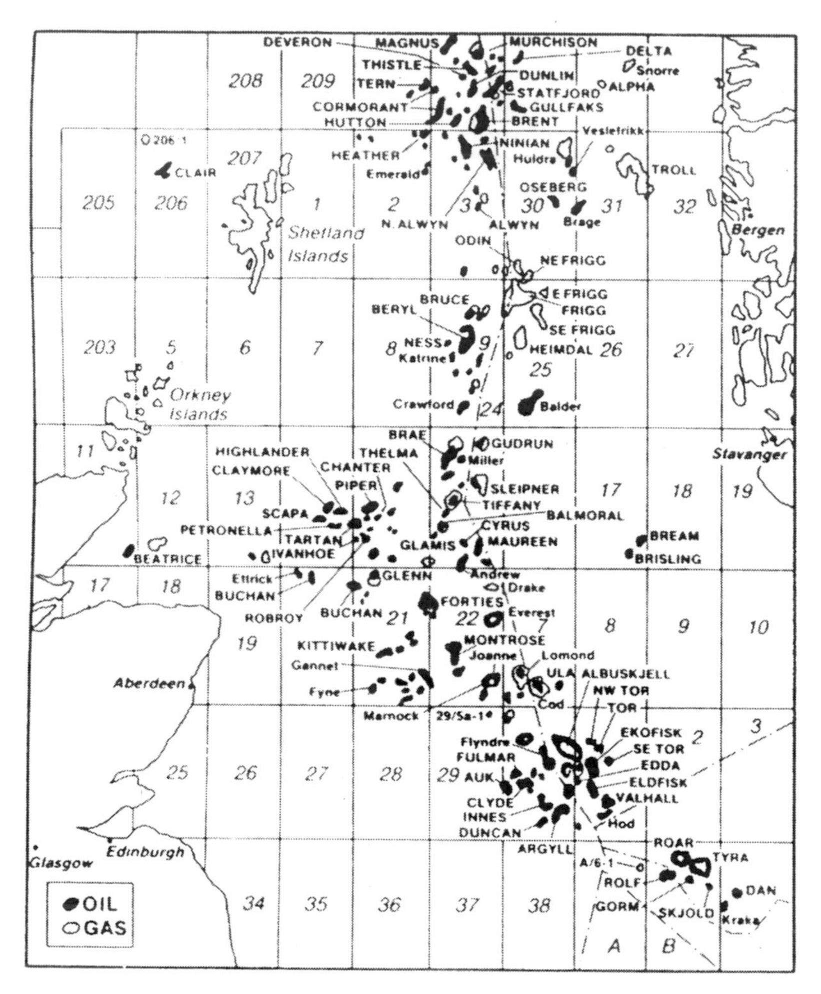

Fig. 7.5. Location map of North Sea. (From Jones, 1996; reproduced with permission from Copestake, in Jenkins, 1993.)

high preservation of organic material observed in palynological preparations.

Reservoir-rocks. Reservoir-rocks are developed at several stratigraphic horizons (Fig. 7.6). In the central North Sea, the most important reservoirs are Late Jurassic to Early Cretaceous marginal and shallow marine sandstones, Late Cretaceous to Early Palaeocene chalks, and Late Palaeocene to Middle Eocene submarine fan sandstones. In the northern North Sea, the most important reservoirs are Middle Jurassic peri-deltaic sandstones, and Late Jurassic, Late Palaeocene and Early Eocene submarine fan sandstones. The quality of the Late Cretaceous to Early Palaeocene chalk reservoirs in the central North Sea is controlled partly by primary porosity associated with the constituent calcareous dinoflagellates and nannofossils (see appropriate sections in Chapter 3).

Cap-rocks (seals) and traps. Cap-rocks are predominantly 'intraformational'. Traps are predominantly structural, those of the Brent province in the northern North Sea being classical tilted fault-blocks. However, combination and stratigraphic traps are becoming increasingly important exploration targets. Mobilisation of Permian salt has locally played an important role in trap development.

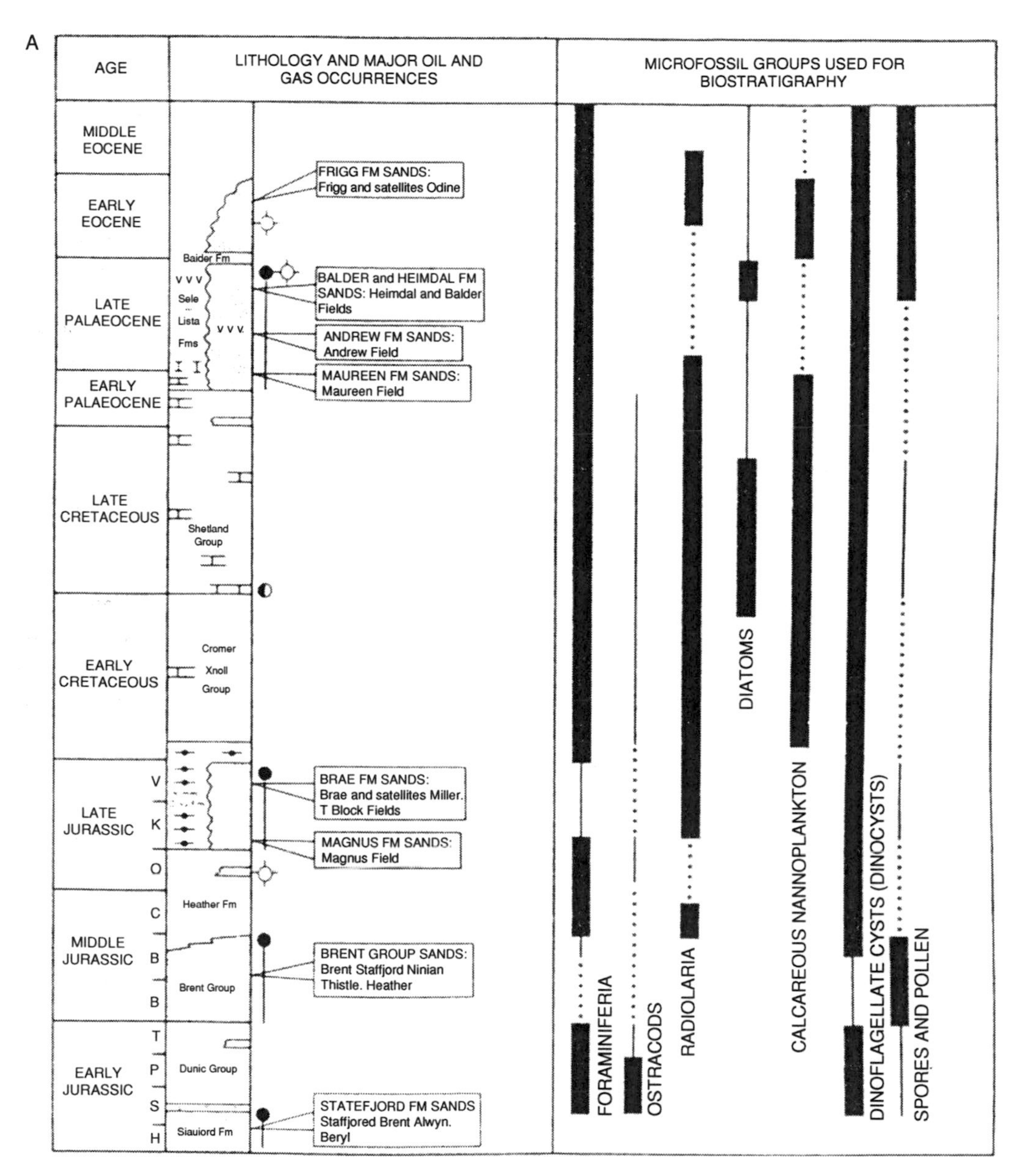

Fig. 7.6. Stratigraphic distribution of reservoir rocks in (A) the northern and (B) the central North Sea. Ranges of microfossil groups used in (reservoir) zonation are also shown. (From Jones, 1996; reproduced with permission from Copestake, in Jenkins, 1993.)

Applications of micropalaeontology: biostratigraphy

A number of biostratigraphic zonation schemes are applicable to the central and northern North Sea, the scale ranging from global through regional to reservoir, the resolution in inverse proportion to the scale (Jones, 1996). The resolution of some schemes is high enough to enable detection of intraformational unconformities, facies changes and faults, and an assessment of their implications for reservoir

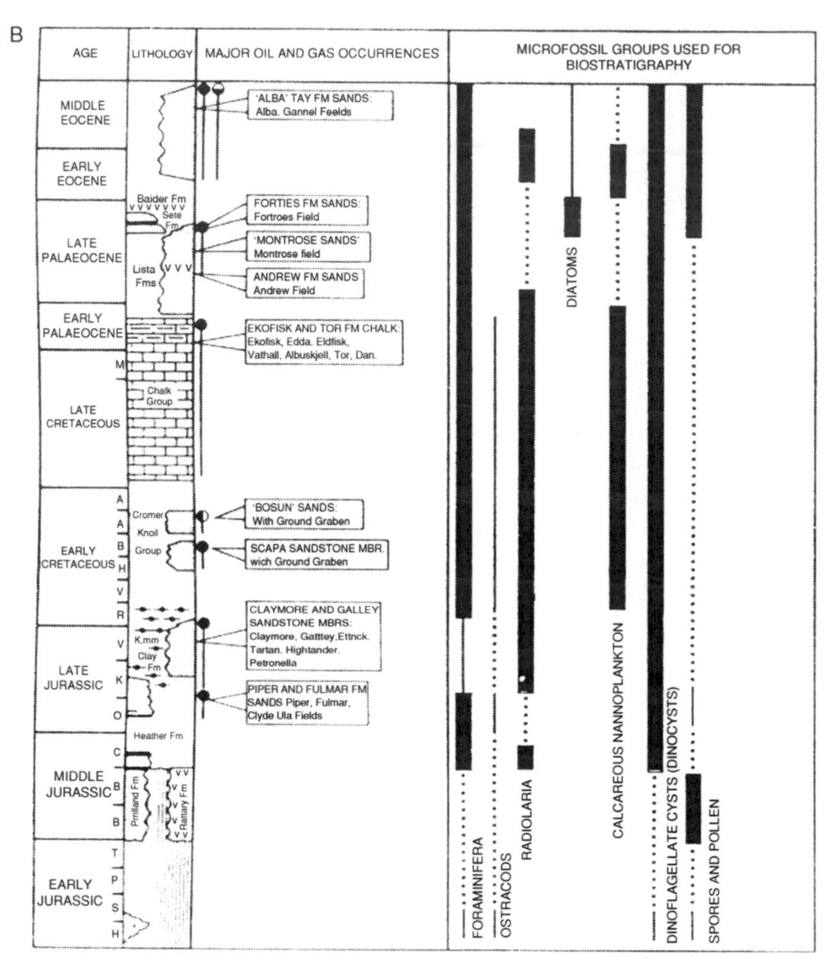

Fig. 7.6 *(cont.)*

distribution. Unfortunately, on account of their commercial sensitivity, many such schemes remain either unpublished or published only in coded or otherwise unusable form.

Mesozoic. Palynostratigraphy is the most important tool in the zonation and correlation of Jurassic–Early Cretaceous of the central and northern North Sea (Fig. 7.7) (Jones, 1996). A number of applicable onshore-based palynozonation schemes incorporating ammonite control have been established in academia and are hence available in the public domain, one recent example being that of Ainsworth *et al.*, in Underhill (1998), which also, and importantly, incorporates independent micropalaeontological zonations based on benthic foraminiferans and ostracods (see also Hesketh and Underhill, 2002). Unfortunately, though, most offshore-based schemes established in the oil industry remain in the proprietary domain, notable exceptions including that of Partington *et al.*, in Parker (1993).

Global biostratigraphic zonation schemes based on calcareous plankton, that is, calcareous nannoplankton and planktonic foraminiferans, are essentially inapplicable in the prevalent proximal or otherwise unfavourable facies of the Jurassic to Early Cretaceous. Indeed, there is no published micro- or nannopalaeontological zonation of any sort available for the Jurassic to Early Cretaceous of the offshore. However, the established onshore ranges of benthic foraminiferans and ostracod species and the

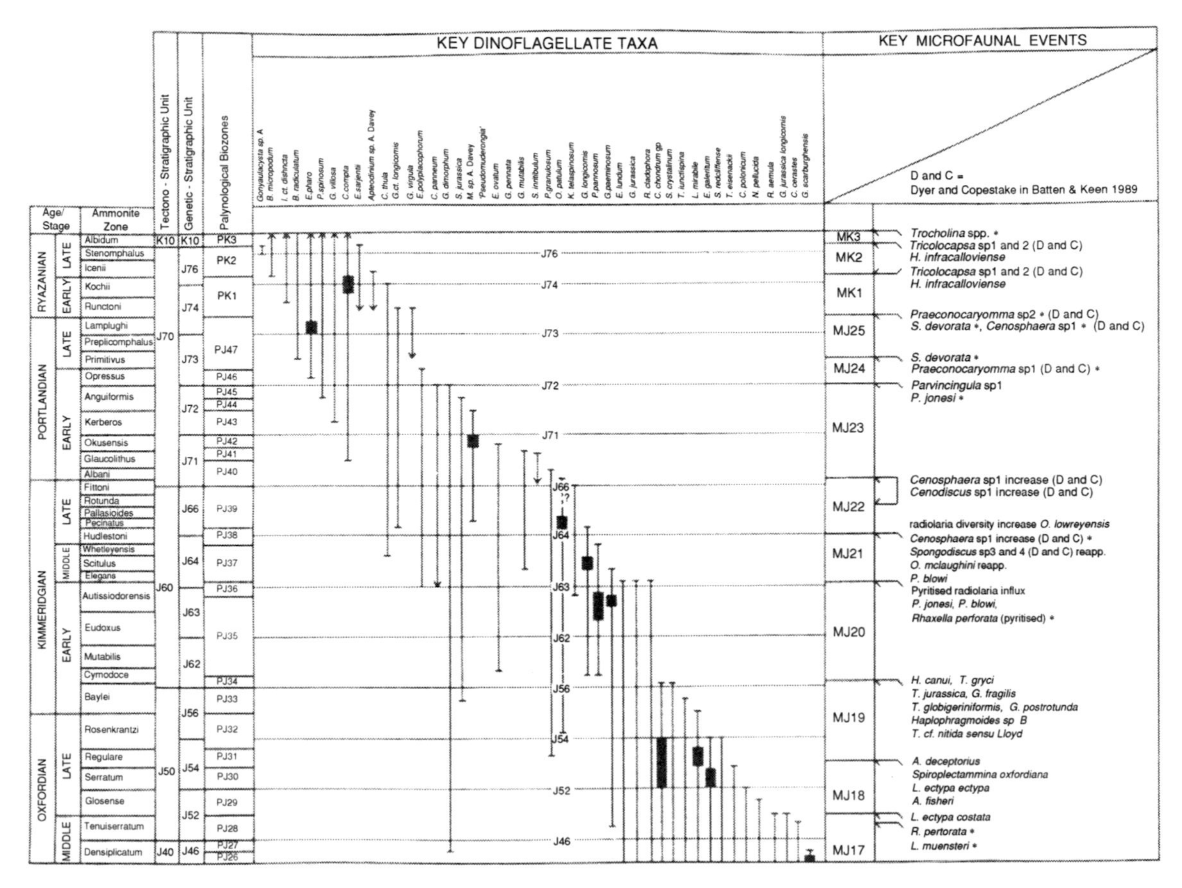

Fig. 7.7. Micropalaeontological and palynological biostratigraphic framework for the Late Jurassic of the North Sea. (From Jones, 1996; in turn after Partington, Mitchener *et al.*, in Parker, 1993.) Solid circles in 'Key microfaunal events' column denote common occurrences.

established onshore zonations of Ainsworth *et al.*, in Underhill (1998), alluded to above, can be applied with varying degrees of confidence. Moreover, offshore ranges have been established for some species of benthic foraminiferans, and for some species of radiolarians. The radiolarian datums, in particular, are of considerable chronostratigraphic and correlative value in the 'central graben', where Jurassic sediments are deeply buried, and where contained palynomorphs are often rendered unidentifiable by thermal alteration. Modified global schemes based on calcareous plankton are applicable in the calcareous facies of the Late Cretaceous (Fig. 7.8) (Jones, 1996). There are also local micropalaeontological schemes not only for the onshore but also for the offshore.

Applications of micropalaeontological and palynological biostratigraphy to problem solving in the Mesozoic – and Cenozoic – of the North Sea are discussed by, among others, Copestake, in Jenkins (1993) and Jones (1996). Copestake, in Jenkins (1993) highlighted the importance of integration of micropalaeontological and palynological data both in exploration and in reservoir exploitation, with specific reference to the Jurassic reservoir of Don field, where models established on the basis of lithostratigraphic rather than biostratigraphic correlations proved seriously flawed.

Cenozoic. Micropalaeontological and palynological stratigraphy are equally important tools in the zonation and correlation of the Cenozoic of the

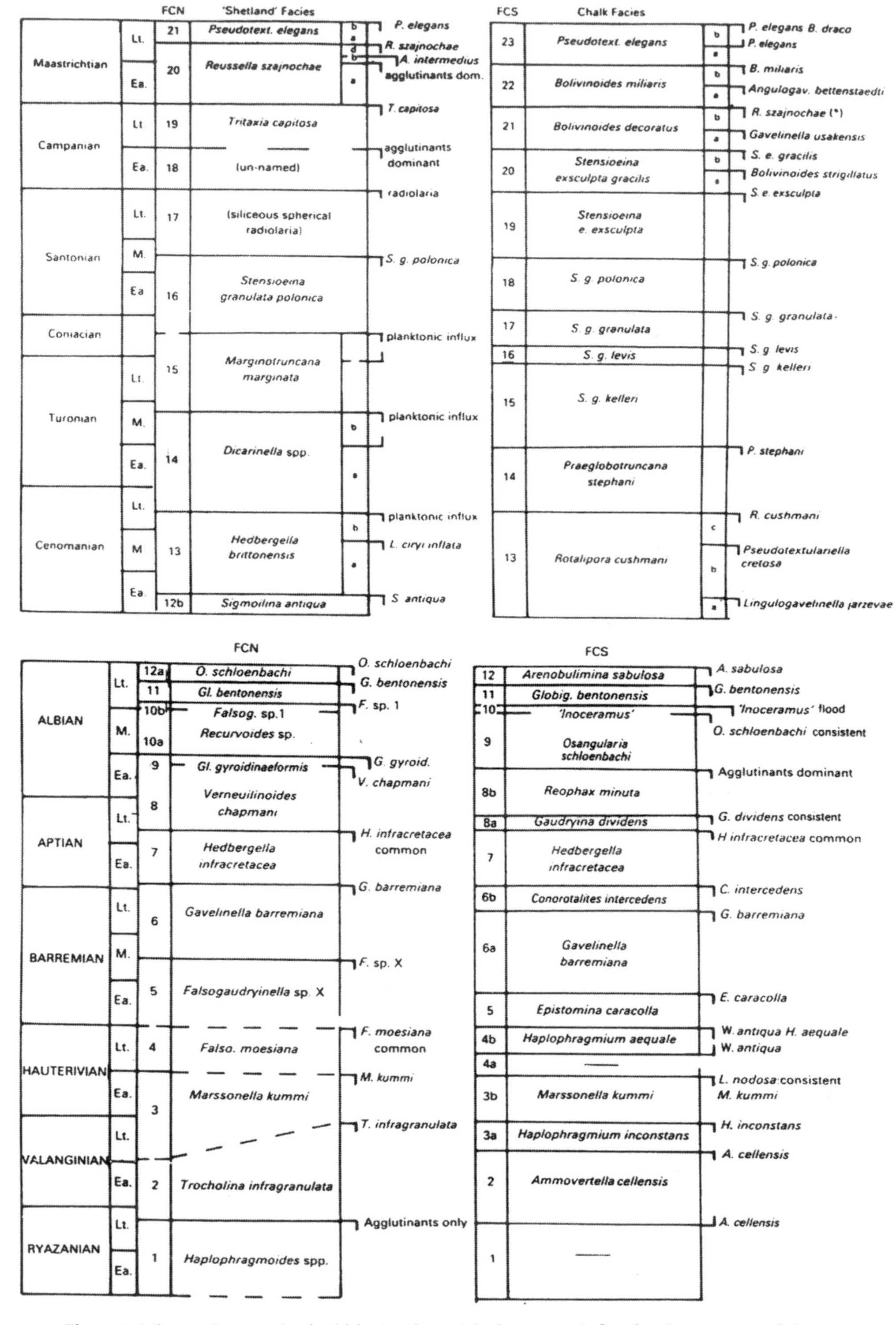

Fig. 7.8. Micropalaeontological biostratigraphic framework for the Cretaceous of the North Sea. (From Jones, 1996; reproduced with permission from King *et al.*, in Jenkins and Murray, 1989.)

NSP Zones		
Zone	Name	Assemblage
16b	*Neo. pachyderma* (S)	*N.p.*
16a	*N pachyderma* (D)	*N.p.*
15d	*Globoratalia puncticulata*	*N. atlasntica*
15c	*Neogloboquadrina atlantica*	*N. atlasntica*
15b	*N. atlantica* (D)	*N. atlasntica*
15a	*N. acostaensis*	*N. atlasntica*
14b	(*Bolboforma metzmacheri*)	*B. metz.*
14a	(*Bolboforma spiralis*)	*B. metz.*
13	(*Bolboforma clodiusi*)	
12	*Sphaeroid disjuncta*	
11	*Globorotalia praescitula*	
10	(Diatom sp. 4 King 1983)	
9c	(Diatom sp. 3 King 1983)	Diatom sp.3
9b	(unnamed)	Diatom sp.3
9a	*Globorotalia danvillensis*	Diatom sp.3
8c	*Globigerinatheka index*	*G. index*
8b	(unnamed)	*G. index*
8a	*Truncorotaloides* spp.	*G. index*
7	*Pseudohastigerina* spp.	
6	(*Cenosphaera* sp.)	
5b	*Pseudohastigerina wilcoxensis*	*S. lin.*
5a	*Subbotina* gr. *linaperta*	*S. lin.*
4	(*Cenosphaera* sp. 1)	
3	(unnamed)	
2	(*'Cenodiscus' sp.*)	
1c	*Globorotalia chapmani*	*G. pseu.*
1b	*Globorotalia pseudobulloides*	*G. pseu.*
1a	*Globoconusa daudjergensis*	*G. pseu.*

NSB Zones		
Zone	Name	Assemblage
16x	*Nonion labradoricum*	*M.p.C.g.*
15b	*Cibicides grossus*	*M.p.C.g.*
15a	*Cibicidoides pachyderma*	*M.p.C.g.*
14b	*M. pseudotepida*	*M.p.C.g.*
14a	*Cib. limbatosuturalis*	*M.p.C.g.*
13b	*Uvigerina venusta saxonica*	*U. venusta*
13a	*Uvigerina pigmea*	*U. venusta*
12c	*Uvigerina* sp. A	*U. pig.*
12b	*Elphidium antoninum*	*U. pig.*
12a	*U. semiormata saprophila*	*U. pig.*
11	*Asterigerina g. staeschei*	
10	*Uvigerina tenuipustulata*	
9	*Plectofrondicularia seminuda*	
8c	*Bolivina antiqua*	*B. antiqua*
8b	*Elphidium subnodosum*	*B. antiqua*
8a	*Asterigerina g. guerichi*	*B. antiqua*
7b	*Rotaliatina bulimoides*	
7a	*Cassidulina carapitana*	
6b	*Uvigerina germanica*	*U. ger.*
6a	*Cibicidoides truncanus*	*U. ger.*
5c	*Planulina costata*	*P. costata*
5b	*Lenticulina gutticostata*	*P. costata*
5a	*Neoeponides karsteni*	*P. costata*
4	(unnamed)	
3b	*Bulimina* sp. A	*G. hilt.*
3a	*Gaudryina hitermanni*	*G. hilt.*
2	(unnamed)	
1c	*Bulimina trigonalis*	*S. spectabilis*
1b	*Stensioeina beccariiformis*	*S. spectabilis*
1a	*Tappanina selmensis*	*S. spectabilis*

NSA Zones	
Zone	Name
12	(unnamed)
11	*Martinottiella bradyana*
10	*Spirosigmoilinella* sp. A
9	*Ammodiscus* sp. B
8	*Karreriella chilostoma*
7	*Cribrostomoides scitulus*
6b	*Karrerulina conversa*
6a	*Amm. macrospira*
5	*Spiroplectammina* aff. *spectabilis*
4b	*Reticulo. amplectens*
4a	*Textularia plummerae*
3	(unnamed)
2	*Verneuilinoides subeocaenus*
1b	*T. ruthvenmurrayi*
1a	*S. spectabilis*

Fig. 7.9. Micropalaeontological biostratigraphic framework for the Cenozoic of the North Sea. (From Jones, 1996; reproduced with permission from King *et al.*, in Jenkins and Murray, 1989.)

central and northern North Sea (Jones, 1996). Global biostratigraphic zonation schemes based on calcareous plankton are essentially inapplicable, although some modified schemes and many datums are locally applicable, especially in the Early Palaeocene Chalk group.

There are established micropalaeontological zonation schemes not only for the onshore but also for the offshore (Fig. 7.9). The principal microfossil group utilised is the benthic foraminiferans. Stratigraphically and environmentally important species of arenaceous benthic foraminiferans are illustrated by Charnock and Jones, in Hemleben *et al.* (1990) and Charnock and Jones (1997). Siliceous microfossils, namely silicoflagellates, diatoms and radiolarians, problematic microfossils, such as *Bolboforma*, and macrofossils, such as pteropods, scaphopods and fish otoliths, are also of some use. Ranking and scaling techniques have been applied to the interpretation of micropalaeontological data (Fig. 7.10) (Gradstein *et al.*, 1988).

A number of applicable onshore-based palynozonation schemes, some incorporating calcareous nannoplankton control, have been established in academia and hence are available in the public domain. Unfortunately, though, most offshore-based schemes established in the oil industry remain in the proprietary domain, notable exceptions including that of Schroder (1992) (Fig. 7.11).

Graphic correlation techniques have been applied to the interpretation of micropalaeontological, nannopalaeontological and palynological data (Fig. 7.12) (Neal *et al.*, in Mann and Lane, 1995).

Palaeoenvironmental interpretation

Mesozoic. Microfacies and palynofacies analysis are equally important tools in the palaeoenvironmental interpretation of the peri-deltaic systems of the Early–Middle Jurassic. Case histories include those of Nagy (1992) (microfacies) and Williams, in Morton (1992) (palynofacies). Nagy (1992) was able to discriminate between interdistributary bay, delta front and prodelta sub-environments of the Brent delta using microfacies analysis, specifically triangular cross-plots of epifaunal, infaunal and surficial 'morphogroups'. Nagy's 'morphogroups' are essentially the same as those recognised earlier by Jones and Charnock (1985). Williams, in Morton (1992) was able to discriminate between various peri-deltaic sub-environments of the Brent group reservoirs in the Ninian and Thistle fields using palynofacies analysis.

Cenozoic. Microfacies is the most important tool in the palaeoenvironmental interpretation of the Palaeogene. Case histories include those of Charnock and Jones, in Hemleben *et al.* (1990) and Jones (1996). We were able to demonstrate on the basis of those species present in the Late Palaeocene by reservoir in Forties field and ranging through to the Recent that the reservoir is of deep- rather than shallow-marine aspect (Fig. 7.13). Prior to this sort of micropalaeontological work being undertaken, the reservoir was erroneously interpreted as representing a delta rather than a submarine fan.

Micropalaeontological bathymetry has been integrated with structural restoration by Kjennerud and Gillmore (2003).

Integrated studies

There have been a number of integrated sequence stratigraphic studies on the North Sea: (Jones, 1996 (Mesozoic–Cenozoic); Fritsen *et al.*, 1999 (Late Cretaceous, Chalk); Evans *et al.* (2003) (Devonian–Recent)). A sequence stratigraphic scheme for the Eocene has been published by Jones *et al.*, in Evans *et al.* (2003) (Fig. 7.14).

Palaeogeographic and facies maps for selected time-slices have been published by Evans *et al.* (2003). The time-slices have been identified in part on the basis of biostratigraphy; and the palaeogeography and facies on the basis of palaeobiology. A palaeogeographic and facies map showing the distribution of the Late Jurassic Kimmeridge clay source facies is given by Fraser *et al.*, in Evans *et al.* (2003) (Fig 7.15); and CRS maps showing source presence and effectiveness have been published by Kubala *et al.*, in Evans *et al.* (2003). Palaeogeographic and facies maps showing the distribution of proven and potential Eocene submarine fan reservoir and associated facies have been published by Jones *et al.*, in Evans *et al.* (2003) (Figs. 7.16–7.23).

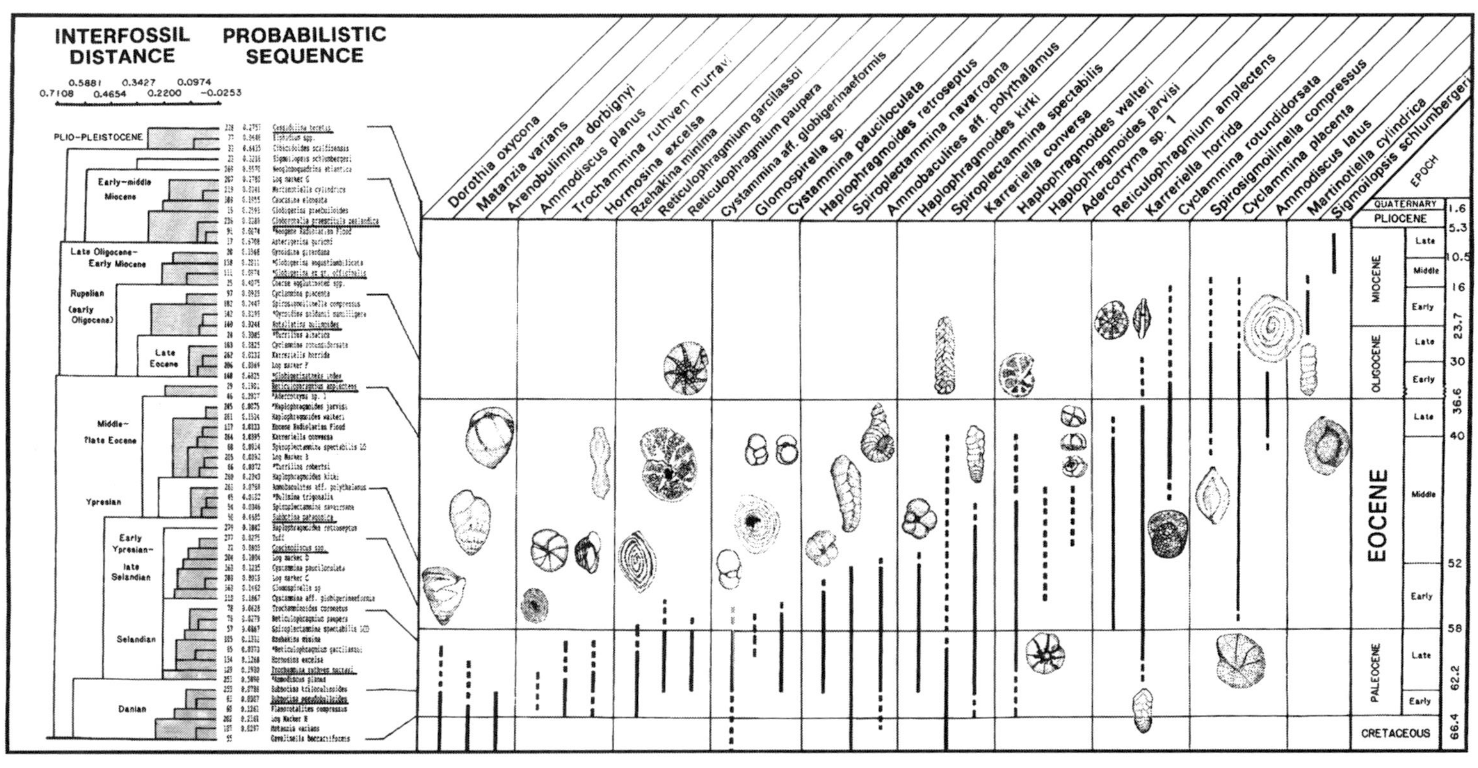

Fig. 7.10. A North Sea Cenozoic micropalaeontological zonation generated by ranking and scaling. (From Jones, 1996; reproduced with permission from Gradstein *et al.*, 1988.)

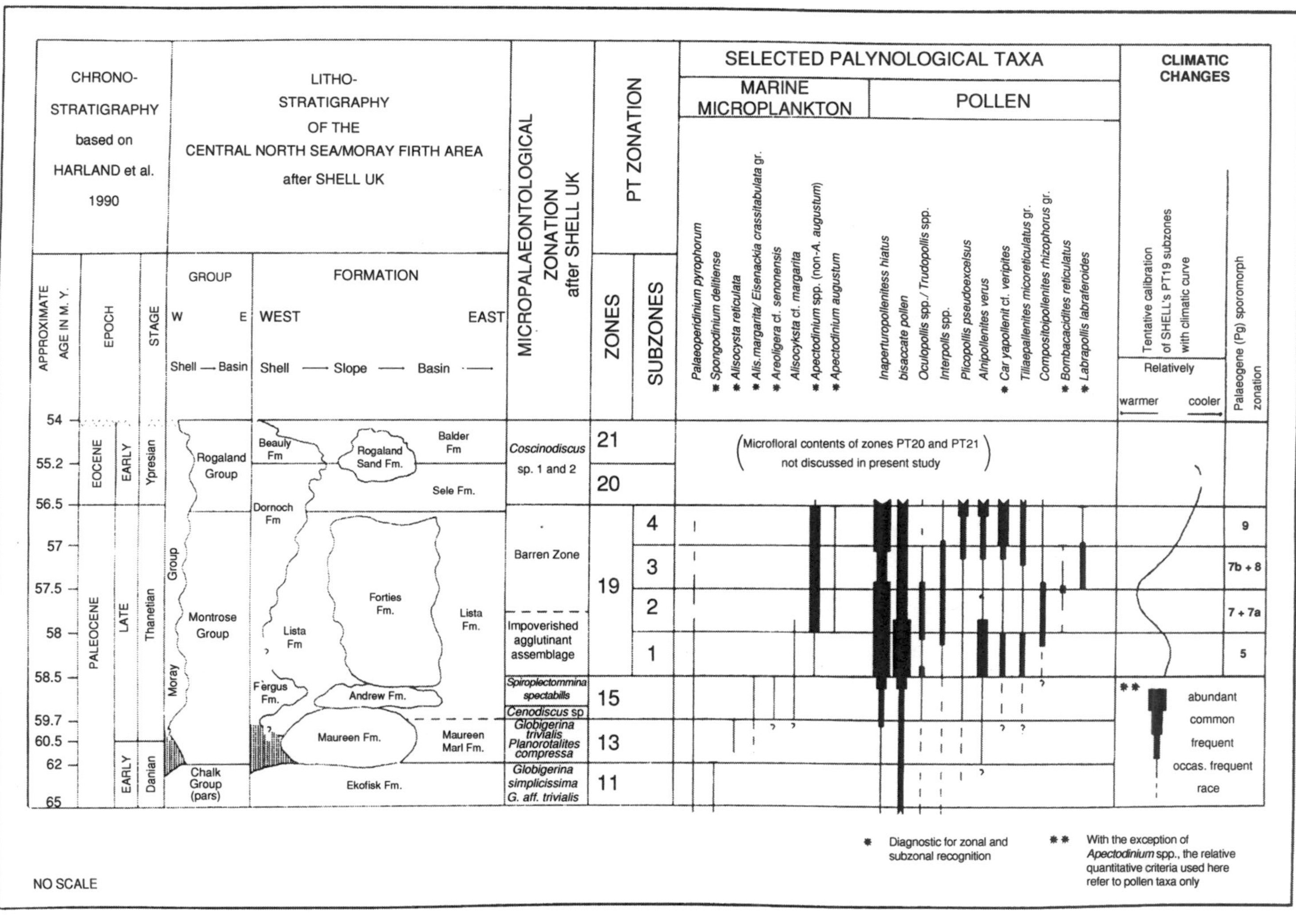

Fig. 7.11. Micropalaeontological and palynological biostratigraphic framework for the Palaeocene of the North Sea. (From Jones, 1996; reproduced with permission from Schroder, 1992.)

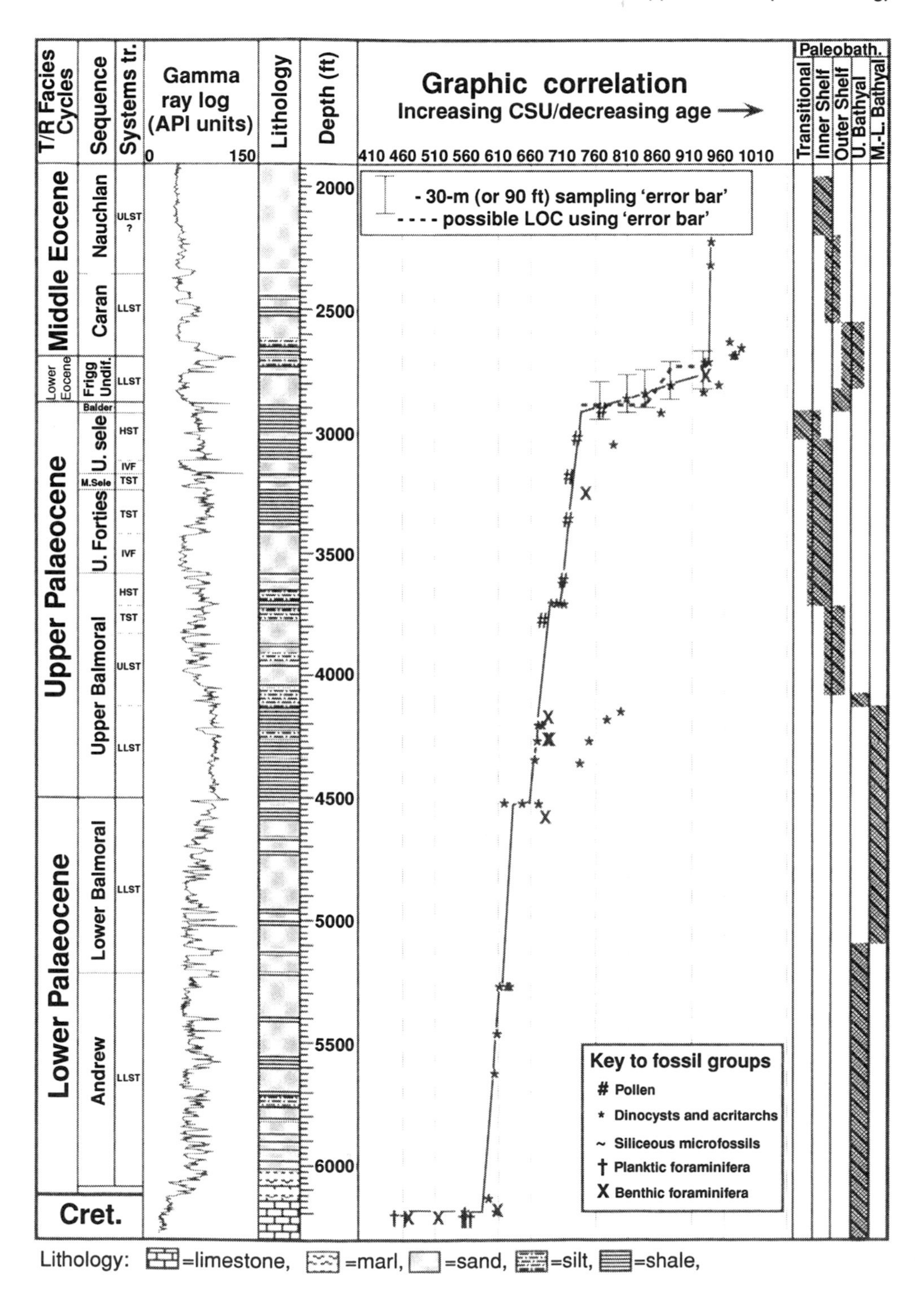

Fig. 7.12. A graphic correlation plot of a Palaeogene section in a North Sea well. CSU, Composite Standard Unit. (Reproduced with permission from Neal *et al.*, in Mann and Lane, 1995.)

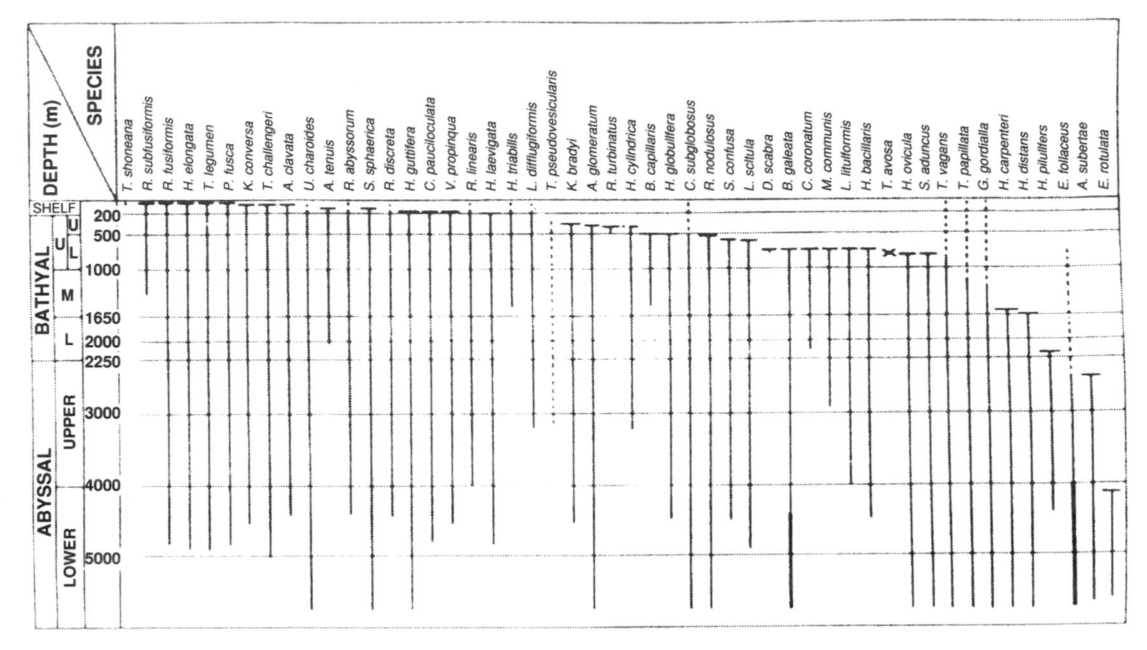

Fig. 7.13. Bathymetric ranges of arenaceous foraminiferans from the Palaeogene of the North Sea. (From Jones, 1996; in turn after Charnock and Jones, in Hemleben *et al.*, 1990.)

Timescale

Ma: 40, 45, 50

Magnetostratigraphy: 15, 16, 17, 18, 19, 20, 21, 22, 23, 24; C13, C15, C16, C17, C18, C19, C20, C21, C22, C23, C24

Epoch/age: Eocene — Late: Priabonian; 39.4; Mid: Bartonian; 42; Luietian; 49; Early: Ypresian

Global

Biostratigraphy

Planktonic foraminifera		Calcareous nannofossils	
G. cerroazulensis	P17	CP16A	NP21
	P16	CP15B	A. pseudoradians NP20
			I. recurvus NP19
G. semiinvoluta	P15	CP15A	C. oamarucnoid NP18
T. rohri	P14	CP14B	D. saipanensis NP17
O. Beckmanni	P13	CP14A	D. tani nodifer NP16
G. Lehneri	P12		
G. Subconglobata	P11	CP13C	N. alata NP15
		CP13B	
H. Aragonensis	P10	CP13A	
		CP12B	D. sublodoensis NP14
		CP12A	
A. Pentacamerata	P9	CP11	D. lodoensis NP13
G. Aragonensis	P8	CP10	T. orthostylus NP12
M. formosa formosa	P7	CP9B	D. binodosus NP11
M. Subbottinae	P6		
M. edgari		CP9A	M. contortus NP10

Sequence stratigraphy

SB: 37, 38, 39.5, 40.5, 42.5, 44, 46.5, 48.5, 49.5, 50, 50.5, 51.5, 52, 53

HFS: 36.5, 37.5, 38.8, 40, 41.2, 43, 45.5, 48, 49, 49.8, 50.3, 51, 51.8, 52.5, 53.5

ST: HS, TR/SMW, HS, TR/SMW, HS, TR, LSW, F, HS, TR/SMW, HS, TR, SMW, HS, TR, SMW, HS, TR, SMW, HS, TR/SMW, HS, TR/LSW, F, HS, TR/SMW, HS, TR/SMW, HS, TR/LSW, F, HS, TR/SMW, HS, TR/SMW, HS, TR/LSW

North Sea

Sequence stratigraphy: T100, T98, T96, T94, T92, T84, T82, T70, T60, T50, T45

Biostratigraphy: A. dictyoplokus, G. linaperta, G. index; H. porosa, B. spectabilis; D. colligerum (common); A. aubertae (very common); P. comatum (very common); B. spectabilis (very common); D. ficusoides; D. ficusoides (very common); E. ursulae; E. 'abbreviata'; D. politum, G. linaperta (very common); G. solidum, D. oebisfeldensis, G. soldadoensis; Conscinodiscus spp. and 2; A. augustum

Biostratigraphy
● Very common
◒ Common

Sequence stratigraphy
F Flooding surface
HFS High frequency sequence
HS High-stand
LSW Low-stand wedge
SB Sequence boundaries
ST System tracts
SMW Shelf margin wedge
TR Transgressive

Fig. 7.14. Sequence stratigraphic framework for the Eocene of the North Sea. (From Jones *et al.*, in Evans *et al.*, 2003.)

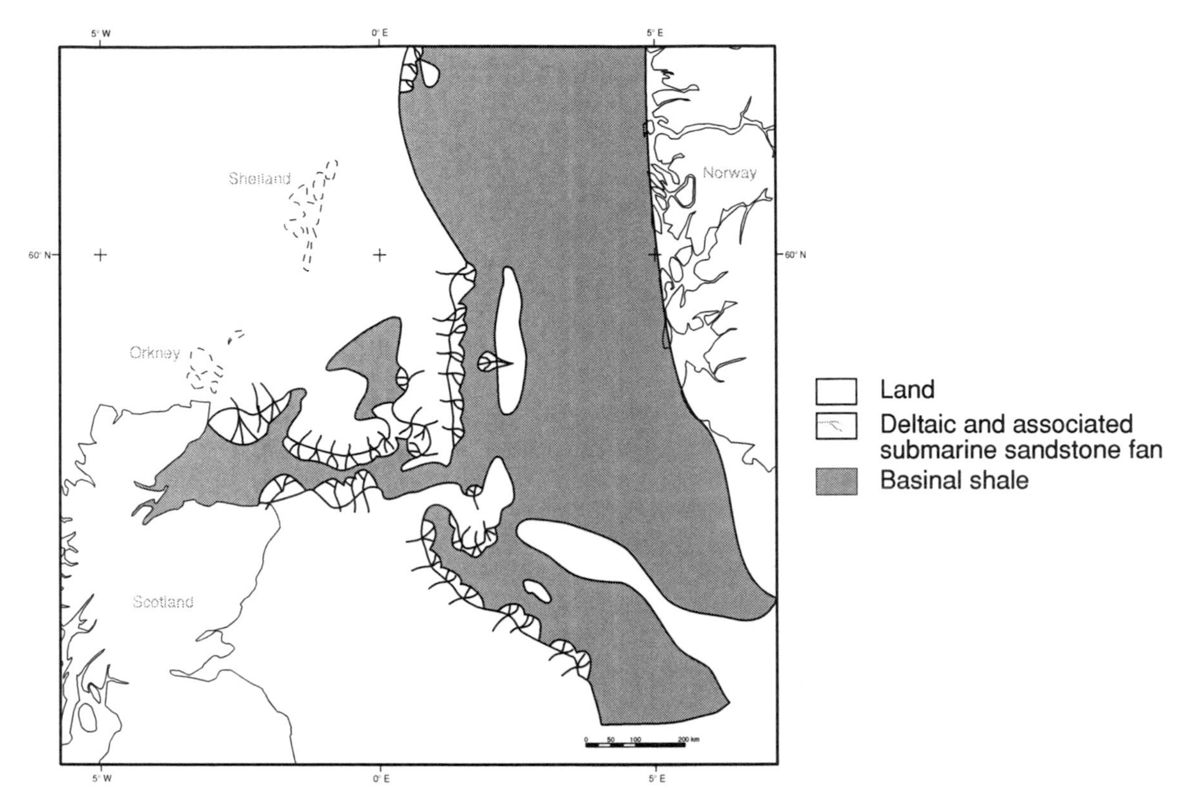

Fig. 7.15. Late Jurassic palaeogeography of the North Sea, showing maximum extent of Kimmeridge Clay source-rock. (Simplified after Fraser *et al.*, in Evans *et al.*, 2003.)

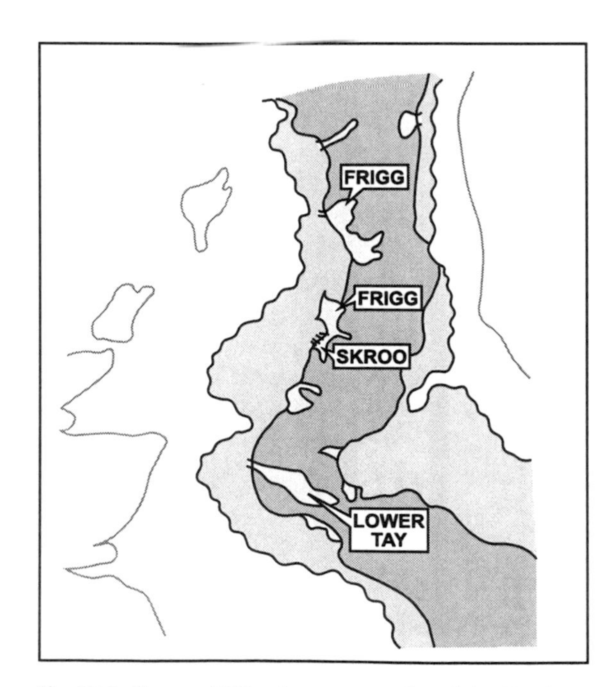

Fig. 7.16. Eocene, T60 palaeogeography of the North Sea. (Simplified after Jones *et al.*, in Evans *et al.*, 2003.)

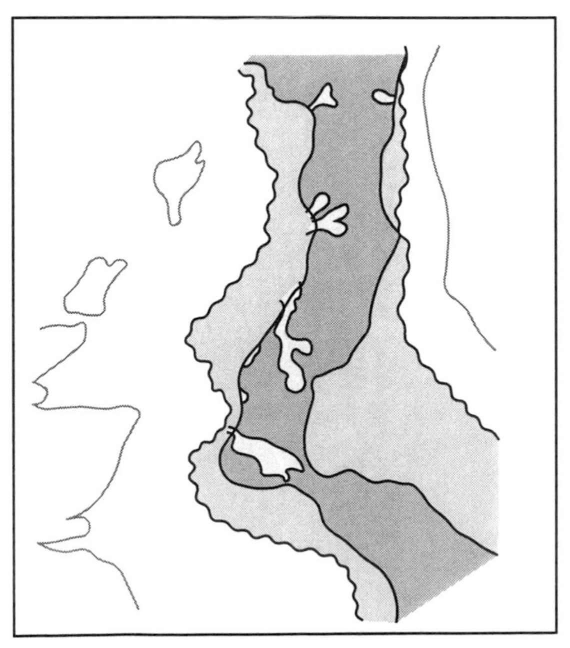

Fig. 7.17. Eocene, T70 palaeogeography of the North Sea. (Simplified after Jones *et al.*, in Evans *et al.*, 2003.)

Fig. 7.18. Eocene, T82 palaeogeography of the North Sea. (Simplified after Jones *et al.*, in Evans *et al.*, 2003.)

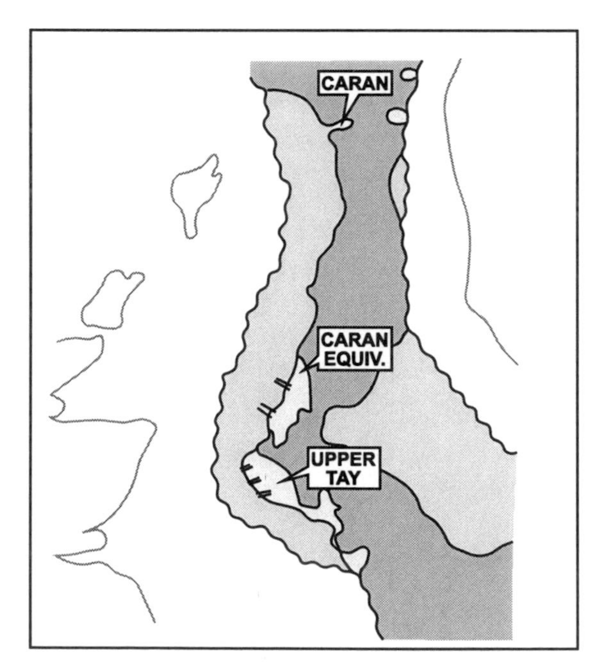

Fig. 7.19. Eocene, T84 palaeogeography of the North Sea. (Simplified after Jones *et al.*, in Evans *et al.*, 2003.)

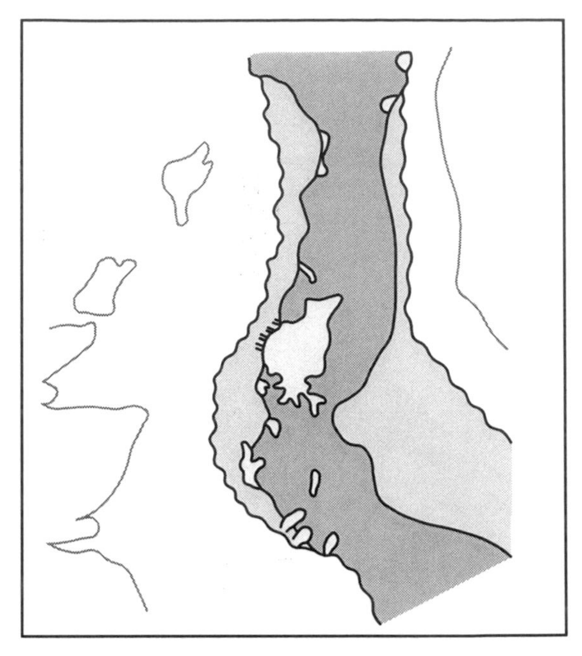

Fig. 7.20. Eocene, T92 palaeogeography of the North Sea. (Simplified after Jones *et al.*, in Evans *et al.*, 2003.)

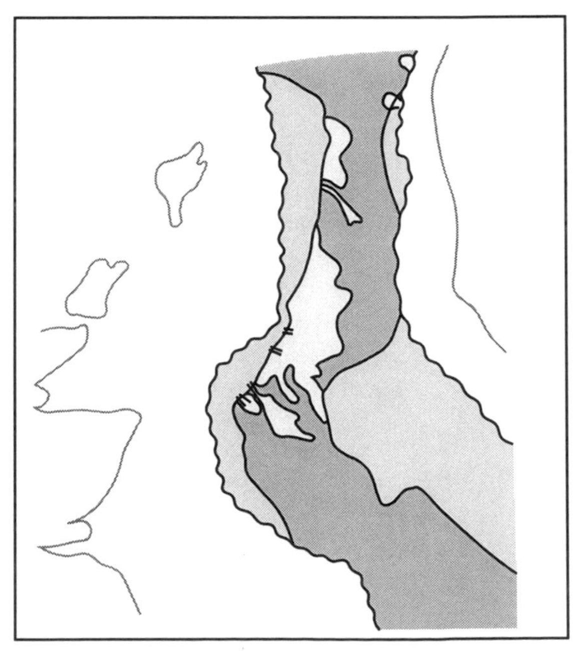

Fig. 7.21. Eocene, T94 palaeogeography of the North Sea. (Simplified after Jones *et al.*, in Evans *et al.*, 2003.)

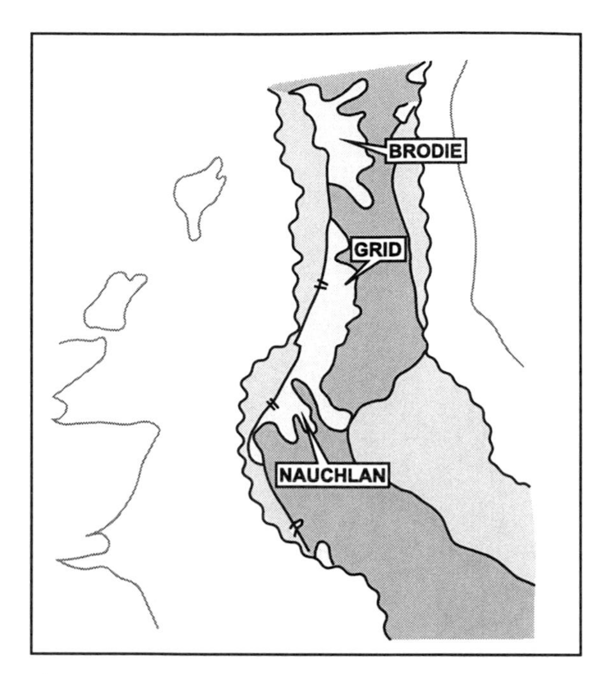

Fig. 7.22. Eocene, T96 palaeogeography of the North Sea. (Simplified after Jones *et al.*, in Evans *et al.*, 2003.)

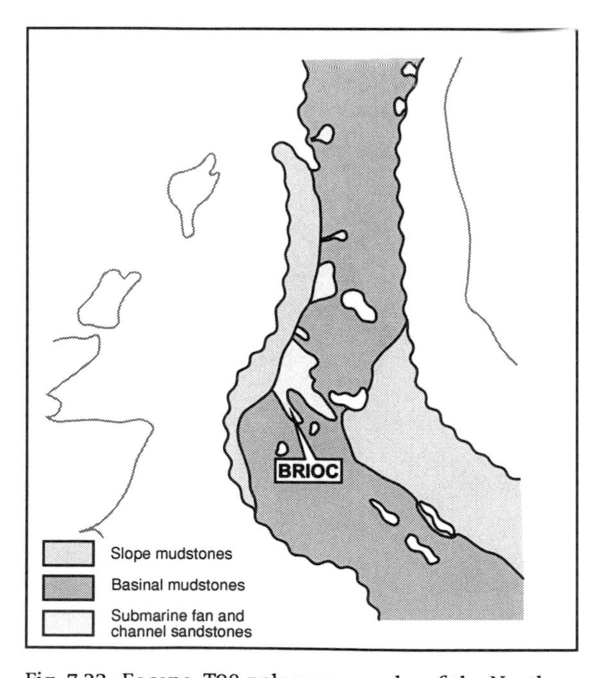

Fig. 7.23. Eocene, T98 palaeogeography of the North Sea. (Simplified after Jones *et al.*, in Evans *et al.*, 2003.)

Middle East

Geological setting

The following is an abridged account of the geology and petroleum geology of the Middle East (Fig. 7.24). Readers interested in further details are referred to Jones (1996), Alsharhan and Nairn (1997) and Sharland *et al.* (2001).

The geological history of the Middle East is long and complex (Fig. 7.25). It begins essentially with extensional? tectonism and the formation of a series of salt basins, the evidence of which is most clearly seen in the present-day south Oman salt basin, in the area around the Strait of Hormuz in the southern Gulf, and in Bandar Abbas and Fars provinces in Iran, in the Precambrian or 'Infracambrian' (Megasequence AP1 of Sharland *et al.*, 2001). These basins underwent deformation, inversion, erosion and peneplanation at the 'Infracambrian'/Cambrian boundary. The entire Arabian plate then underwent passive margin subsidence in the Cambrian to Ordovician (Megasequence AP2). A glaciation took place in the latest Ordovician, Ashgillian (Hirnantian), and a deglaciation, resulting in the widespread deposition of transgressive marine sediments, in the Early Silurian, Llandoverian, and the subsequent deposition of regressive sediments in the Late Silurian–Devonian (Megasequence AP3). The plate then underwent compressional? tectonism and uplift, ultimately resulting in extensive erosion associated with the so-called 'Hercynian unconformity' or, in Saudi Arabia, 'Pre-Unayzah unconformity', in the Early Carboniferous (Megasequence AP4). A second glaciation took place in the Late Carboniferous–Early Permian (Megasequence AP5).

Late Permian–Early Jurassic (Megasequence AP6) rifting and post-rift subsidence resulted in the break-up of the supercontinent of Gondwana, and, for the first time, extensive carbonate deposition. Middle – Late Jurassic (Megasequence AP7) rifting resulted in the break-up of Pangaea, and the formation of a series of intra-shelf basins, of which the most important were the Gotnia and Diyab basins. Carbonate deposition continued to dominate, with clastic deposition essentially confined to certain graben systems in the extreme south-west (Yemen). Extensive evaporite deposition took place at the end of the Jurassic. The Early Cretaceous, Neocomian to early Aptian, and 'Middle' Cretaceous, late Aptian to middle

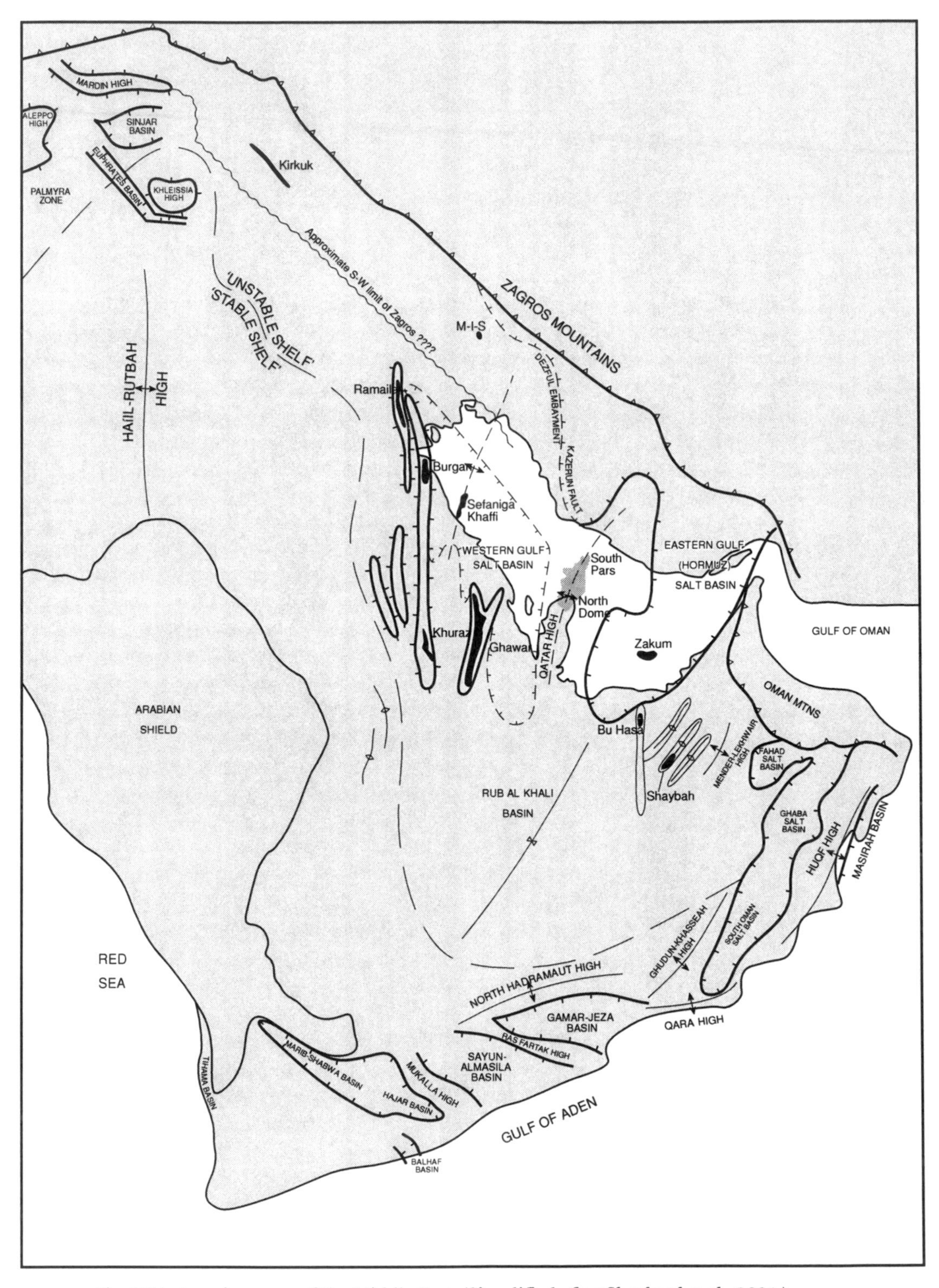

Fig. 7.24. Location map of the Middle East. (Simplified after Sharland *et al.*, 2001.)

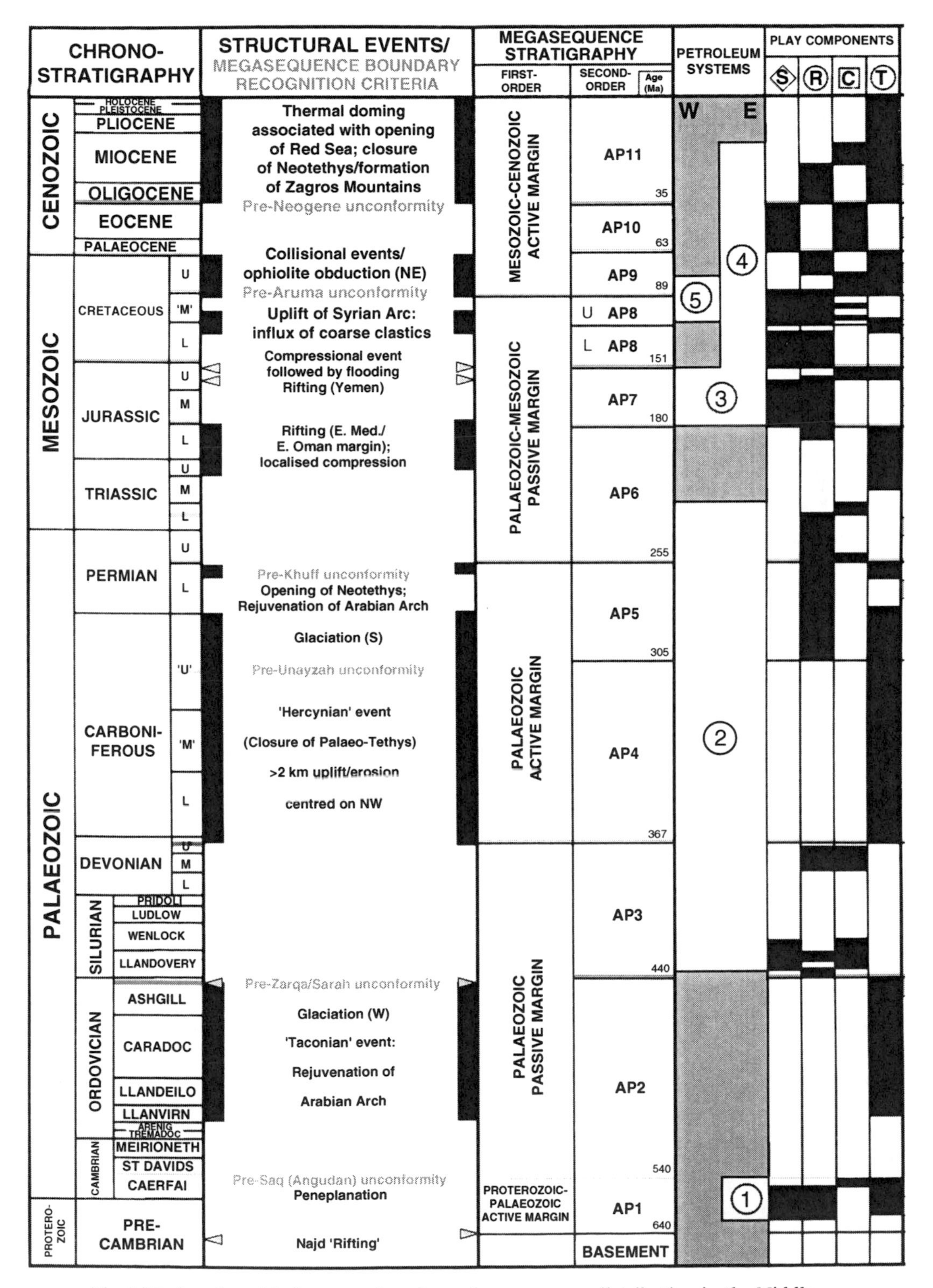

Fig. 7.25. Stratigraphic framework and petroleum resource distribution in the Middle East. Bars in 'Structural events' column indicate duration of events. Petroleum systems: (1) Infracambrian; (2) Silurian; (3) Jurassic (Sargelu/Naokelekan and Diyab basins); (4) Cretaceous–Tertiary (Garau, Kazhdumi and Pabdeh basins); (5) 'Middle' Cretaceous (Bab/Shilaif basin). S, Source-rock; R, Reservoir-rock; C, Cap-rock; T, trap.

Turonian (Megasequence AP8) was characterised by post-rift subsidence and the formation of further of intra-shelf basins, of which the most important were the Bab basin in the Early Cretaceous and the Kazhdumi and Shilaif basins in the 'Middle' Cretaceous. At this time, carbonate sedimentation in the east became balanced by clastic deposition in the west. The Late Cretaceous, late Turonian–Senonian (Megasequence AP9) saw structuration and extensive erosion associated with the so-called, in Saudi Arabia, 'Pre-Aruma unconformity', generated by ophiolite obduction in the east.

The Cenozoic saw structuration associated with the collision between the Arabian and Eurasian plates and the closure of Tethys, beginning in the Late Cretaceous and effectively ending in the Miocene, and, to a lesser extent, with the opening of the Red Sea, beginning, with rift shoulder uplift, and the shedding of coarse clastic sediments from the 'Western Arabian Highlands' to the west, in the Oligocene. The initial phases involved the formation of a deep marine foredeep known as the Pabdeh basin in what were to become the Zagros Mountains in Iran in the Palaeocene–Eocene (Megasequence AP10). The later phases witnessed the widespread deposition of a typical 'molasse' sequence of shallow marine carbonates, marginal marine evaporites and non-marine clastics in the Oligocene–Holocene (Megasequence AP11).

Petroleum geology

The Middle East is the site of 20% (103 out of 509) of the world's 'giant' oil and gas fields, that is, those containing >500 million barrels of oil – or equivalent – recoverable (Fig. 7.25). The consensus view as to why the area is so rich in giant fields is that its large and on the whole relatively simple traps were able to drain large source kitchens by way of simple migration pathways. This process was facilitated by a peculiarly fortuitous juxtaposition of source, reservoir and seal units in trap configurations, a function of the area's unique geological history. There is some stratigraphic control on play component distribution in the Middle East (see subsection on 'Stratigraphic control on play component distribution' above).

Source-rocks. Source-rocks occur at several stratigraphic horizons, the most important ones being Infracambrian, Silurian, Middle–Late Jurassic, Early Cretaceous, 'Middle' Cretaceous and Palaeocene – Eocene (Fig. 7.25). The Infracambrian source-rock is apparently of comparatively limited regional extent, only proven to be present in the south Oman salt basin. However, the Silurian, Middle–Late Jurassic, Early Cretaceous, 'Middle' Cretaceous and Palaeocene–Early Miocene source-rocks are of at least sub-regional extent (Bordenave and Burwood, in Katz, 1994; Luning *et al.*, 2000). Collectively, these prolific source-rock basins have generated a significant proportion of the world's known petroleum reserves. The Silurian petroleum system accounts for all of the oil and gas reservoired in the Palaeozoic of the Middle East, including that in North Dome/South Pars, the world's largest gas field. The Middle–Late Jurassic petroleum system accounts for all of the oil and gas reservoired in the Jurassic of the Middle East, including that in Ghawar, the world's largest oil field.

Reservoir-rocks. Reservoir-rocks occur at practically every horizon (Fig. 7.25). Important Early Permian clastic reservoirs occur in Oman and Saudi Arabia; Late Permian platform carbonate reservoirs occur across the Arabian Peninsula and Gulf and into Fars province in Iran. Jurassic platform carbonate reservoirs occur in Saudi Arabia and the United Arab Emirates. Cretaceous paralic and peri-deltaic clastic reservoirs occur in Kuwait and Iraq; Cretaceous platformal, and locally peri-reefal, carbonate reservoirs in Kuwait, Iraq, Khuzestan and Lurestan provinces in Iran, Oman and the United Arab Emirates; and Oligocene–Miocene paralic and peri-deltaic clastic reservoirs and platformal, and locally peri-reefal, reservoirs in Kuwait, Iraq and Khuzestan and Lurestan provinces in Iran. The quality of the carbonate reservoirs is controlled partly by primary depositional factors and partly by secondary diagenetic factors (fracturing also locally plays an important role in enhancing permeability, as, for example, in the Zagros Mountains of Iran and Iraq). There is some stratigraphic control on carbonate reservoir development within the sequence stratigraphic framework, with preferential deposition during HSTs, and preferential diagenetic enhancement during LSTs in the case of dissolution, and TSTs in the case of dolomitisation.

Cap-rocks (seals) and traps. Cap-rocks (seals) occur at several horizons (Fig. 7.25). The most important are Infracambrian halites and anhydrites, Early Triassic shales, Late Jurassic halites and anhydrites, 'Middle' Cretaceous shales, Late Cretaceous shales, and Middle Miocene halites and anhydrites. Locally, as in parts of the Zagros Mountains, some of the seals have been breached by surface erosion, or rendered ineffective by fracturing. Generally speaking: the Infracambrian cap-rocks delineate the top of the Infracambrian petroleum system; the Early Triassic cap-rocks the top of the Silurian petroleum system; the Late Jurassic cap-rocks the top of the Middle–Late Jurassic petroleum systems; the 'Middle' Cretaceous cap-rocks the top of the Early Cretaceous petroleum systems; the Late Cretaceous cap-rocks the top of the 'Middle' Cretaceous petroleum systems; and the Middle Miocene cap-rocks the top of the Palaeocene–Early Miocene petroleum system. Traps are predominantly structural, the style varying from inverted basement-cored anticlinal in the foreland to thrust anticlinal in the fold-belt of the Zagros Mountains of Iran and Iraq. Mobilisation of Infracambrian salt has played an important role in trap formation in Iran, the Gulf and the eastern part of the Arabian Peninsula. Combination or straight stratigraphic traps include that of the Fateh field in Dubai in the United Arab Emirates.

Applications of palaeontology: biostratigraphy

A wide range of biostratigraphically useful fossil groups have been recorded in the Proterozoic to Phanerozoic of the Middle East (see appropriate sections in Chapter 3). A range of non-biostratigraphic techniques have also been employed in dating and correlation, including carbon and strontium isotope stratigraphy (see Section 6.3).

Proterozoic. Acritarchs, 'ediacarians' and 'small shelly fossils' have proved of at least some use in biostratigraphy and palaeobiology, and/or in the characterisation of MFSs in the Proterozoic, Precambrian (Jones, 1996, 2000; Sharland *et al.*, 2001, 2004). For example, the occurrence of the 'ediacarian' *Spriggina*, with or without associated *Charnia, Dickinsonia* and unidentified medusoid forms, has enabled a correlation between the Hormuz salt formation of south-west Iran and the Rizu and Esfordi formations of north-east Iran, radiometrically dated to 595–715 Ma (Jones, 2000).

Palaeozoic. Stromatolites, palynomorphs, plants, calcareous algae, LBFs, 'small shelly fossils', archaeocyathans, corals, brachiopods, ammonoids, ostracods, trilobites, graptolites, fish, conodonts and trace fossils have proved of use in biostratigraphy and palaeobiology, and/or in the characterisation of MFSs in the Palaeozoic (Jones, 1996; Sharland *et al.*, 2001; Al-Husseini, 2004; Sharland *et al.*, 2004).

Mesozoic. Palynomorphs, calcareous algae, calcareous nannofossils, calpionellids, larger benthic and planktonic foraminiferans, ammonites, bivalves, gastropods, ostracods, crinoids and echinoids have proved of use in biostratigraphy and palaeobiology, and/or in the characterisation of MFSs in the Mesozoic (Jones, 1996; Sharland *et al.*, 2001; Al-Husseini, 2004; Sharland *et al.*, 2004). Palynozonations are applicable in the marginal marine facies of the Cretaceous, as in the Burgan formation reservoir of Kuwait (Al-Eidan *et al.*, 2001). Modified global calcareous nannoplankton and planktonic foraminiferal zonation schemes are applicable in the open marine facies of the Cretaceous (Al-Fares *et al.*, 1998). Dunnington (1955) established a Late Cretaceous planktonic foraminiferal biozonation scheme based on variations in the proportions of globotruncanid species groups. The resolution of this scheme was sufficiently high to enable it to be used not only in the prediction of the thickness of eroded Late Cretaceous sediment, but also in the prediction of the thickness of unpenetrated Late Cretaceous sediment.

The chronostratigraphic ranges of a number of key Mesozoic species of LBFs have now been established either directly or indirectly by means of correlation (Kuznetsova *et al.*, 1996; Whittaker *et al.*, 1998; Simmons *et al.*, in Hart *et al.*, 2000; Jones *et al.*, in Bubik and Kaminski, 2004), supplemented by graphic correlation (Simmons, in Simmons, 1994). Thus, for example, *Mesorbitolina lotzei* is restricted to the early Aptian, and *M. texana* to the late Aptian to middle Albian. In turn, the Early Cretaceous Shuaiba formation – an important petroleum reservoir – of the Al Huwaisah and adjacent fields in central Oman, which contains the former species throughout but not the latter, can be confidently dated to the early Aptian, while the upper part of the Shuaiba formation of the Yibal, Dhulaima and Lekhwair fields in northern Oman, which contains the latter species but not the former, can be confidently dated to late Aptian (to middle Albian). The observation that Shuaiba formation deposition persisted into the Late Aptian in northern but not in central Oman is consistent with the emerging model of a relatively deep ('Bab') intra-shelf basin in the northern area and emergence – through eustatic sea-level fall – of the formerly shallow-marine platform of the central area at this time. This is important from the petroleum geological point of view because the porosity and hence the reservoir quality of the Shuaiba formation was enhanced – through subaerial exposure and leaching – only on the margin of the basin. Stratigraphically and environmentally important species of benthic foraminiferans from the Mesozoic are illustrated by Whittaker *et al.* (1998) and Simmons *et al.* in Hart *et al.* (2000).

Cenozoic. Larger benthic and planktonic foraminiferans, calcareous algae, echinoids and vertebrates have proved of use in biostratigraphy and palaeobiology, and/or in the characterisation of MFSs, in the Cenozoic (Jones, 1996; Whybrow and Hill, 1999; Sharland *et al.*, 2001, 2004). Modified global calcareous nannoplankton and planktonic foraminiferal zonation schemes, and regional larger benthic foraminiferal zonation schemes are also applicable (Jones and Racey, in Simmons, 1994). The Cenozoic calcareous nannofossil biostratigraphy and palynostratigraphy of the Middle East is discussed by Hughes *et al.* (1991), Starkie (1994) and Hughes and Filatoff, in Al-Husseini (1995).

Palaeoenvironmental interpretation

The palaeoecological distributions of foraminiferans in the platform carbonate – reservoir – environments of the Middle East have been discussed by, among others, Banner and Simmons, in Simmons (1994) (Mesozoic) and Jones (1996) (Mesozoic–Cenozoic).

The palaeobathymetric ranges of a large number of taxa have been established by means of calibration against photosynthetic algae with known light requirements and hence water depth – or, more correctly, water clarity – preferences (Fig. 7.26) (Banner and Simmons, in Simmons, 1994; Jones, 1996; Simmons *et al.*, in Hart *et al.*, 2000; Jones *et al.*, in Bubik and Kaminski, 2004). Using this information, high-resolution palaeobathymetric interpretations have been undertaken – and indeed detailed palaeobathymetric curves constructed – whenever and wherever sufficiently closely spaced samples are available for analysis. Integrated with sedimentology, this has facilitated detailed depositional modelling and reservoir characterisation (and identification of reservoir 'sweet spots'). For example, detailed palaeobathymetric curves have been constructed for the Early Cretaceous (Barremian) Kharaib and Shuaiba formations (Thamama group) and the 'Middle' Cretaceous Nahr Umr and Natih formations (Wasia group) of parts of eastern Arabia – which are all important petroleum reservoirs (Jones *et al.*, in Bubik and Kaminski, 2004). In the case of the Kharaib and Shuaiba formations, the curves have proved useful not only in reservoir characterisation but also in correlaton between fields some tens of kilometres distant. Reservoir quality can be demonstrated to be best in palaeobathymetries of 10–30 m – marked by extensive development of porous *Bacinella/Lithocodium* boundstone facies – and thus appears to be controlled by primary depositional rather than secondary diagenetic factors.

Further subdivision of the palaeobathymetric interpretation of middle ramp sub-environments is possible on the basis of observed morphological trends in the ancient orbitolinids and comparison with analogous – but not homologous – modern counterparts (nummulitids). The inner part of the middle ramp can be characterised by conical and the outer part by flattened orbitolinids, in the same way that relatively shallow- and deep-water environments are respectively characterised by conical and flattened nummulitids at the present time, as in the Gulf of Aqaba. It is possible that the morphological trends represent adaptations to facilitate the gathering of light in order to enable hosted algal symbionts to photosynthesise to optimal efficiency (see also Pittet *et al.*, 2002).

Palaeobiogeographic aspects have been discussed by, among others, Mawson *et al.* (2000).

Integrated studies

There have been a number of integrated sequence stratigraphic studies on the Middle East: Jones (1996) (Cenozoic); van Buchem *et al.* (1996, 2002a, b),

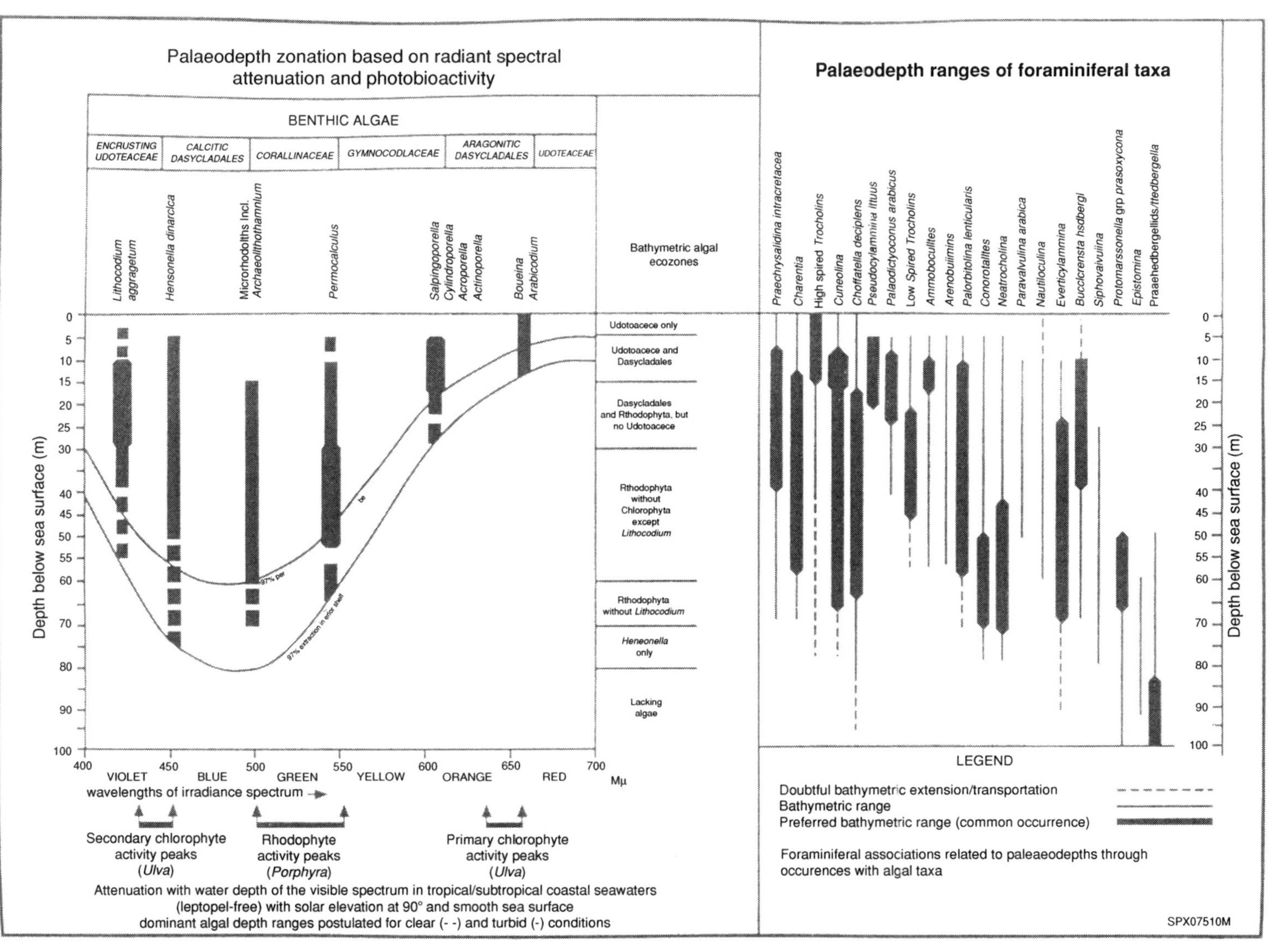

Fig. 7.26. Palaeoecological distribution of calcareous algae and foraminiferans in the Early Cretaceous of the Middle East. (From Jones, 1996; reproduced with permission from Banner and Simmons, in Simmons, 1994.)

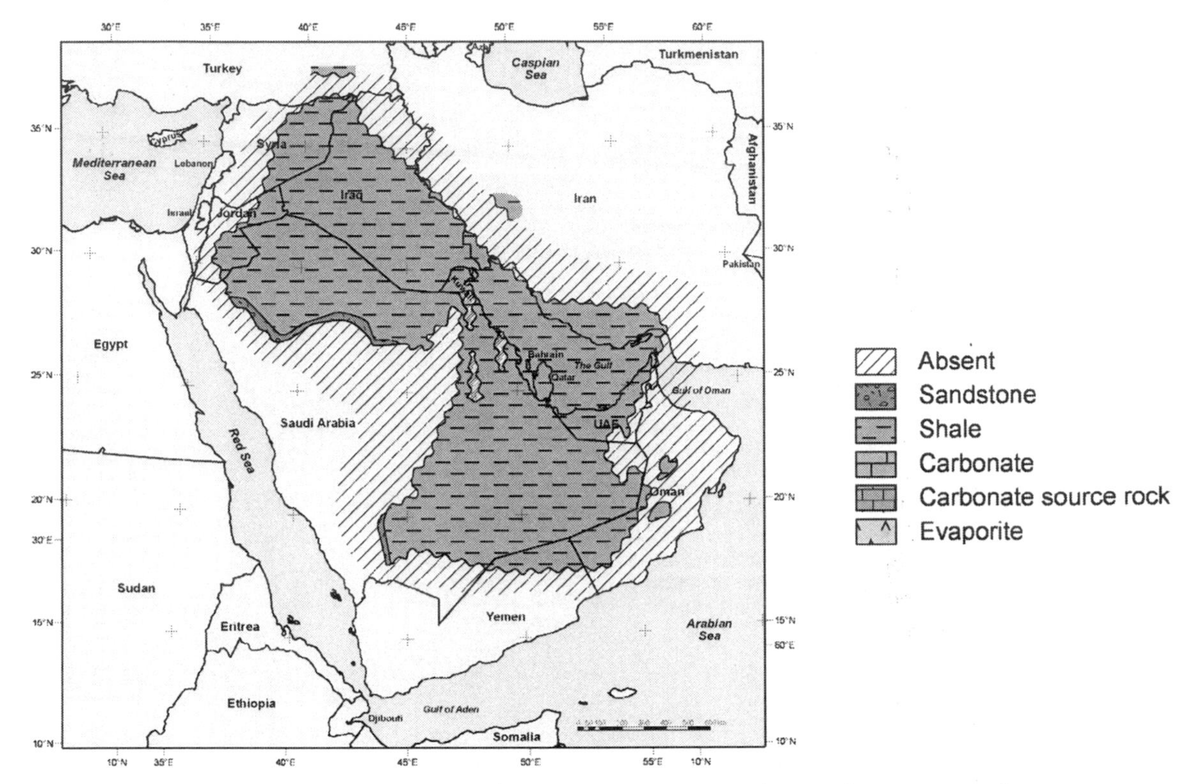

Fig. 7.27. Silurian palaeogeography of the Middle East, showing maximum extent of source facies. (Compiled from various sources.)

Al-Husseini (1997), Immenhauser *et al.* (1999) and Pittet *et al.* (2002) (Mesozoic); and Sharland *et al.* (2001, 2004) (Proterozoic–Phanerozoic).

A wide range of fossil groups has been used in the characterisation of sequence boundaries and systems tracts in the Proterozoic–Phanerozoic of the Middle East (Sharland *et al.*, 2001, 2004; see also appropriate sections in Chapter 3). (Micro) palaeontological characterisation of systems tracts is discussed by Simmons *et al.* (1992), Jones (1996), Simmons *et al.*, in Hart *et al.* (2000), Sharland *et al.*, (2001) and Jones *et al.*, in Bubik and Kaminski (2004). Transgressive systems tracts are characterised by orbitolinid and planktonic foraminiferal facies; early HSTs by essentially aggradational algal (*Lithocodium*) reef facies, and late HSTs by aggradational to progradational rudist–coral reef facies, orbitolinid–valvulinid–echinoderm and dasycladacean (*Hensonella*)–miliolid back-reef facies and, locally (Bab basin), planktonic foraminiferal fore-reef facies. Transgressive systems tracts are also characterised by upward increases in the incidences of orbitolinid foraminiferans, and the replacement of conical forms by flattened forms, and HSTs by upward decreases in the incidences of orbitolinids, and the replacement of flattened forms by conical forms. Maximum flooding surfaces are characterised by maximum abundances of planktonic foraminiferans.

Palaeogeographic and facies maps for selected time-slices have been published by Jones (1996) (Cenozoic), Ziegler (2001) (Mesozoic–Cenozoic) and Konert *et al.* (2001) (Palaeozoic). The time-slices have been identified in part on the basis of biostratigraphy; and the palaeogeography and facies on the basis of palaeobiology. Palaeogeographic and facies maps showing the distributions of the Silurian, Middle–Late Jurassic, earliest Cretaceous, 'Middle' Cretaceous and Palaeocene–Eocene source facies are given in Figs. 7.27–7.31 respectively. It is evident that the Silurian source facies extends throughout the Middle East, and indeed into north Africa (Luning *et al.*, 2000). The Middle–Late Jurassic, earliest Cretaceous and 'Middle' Cretaceous source

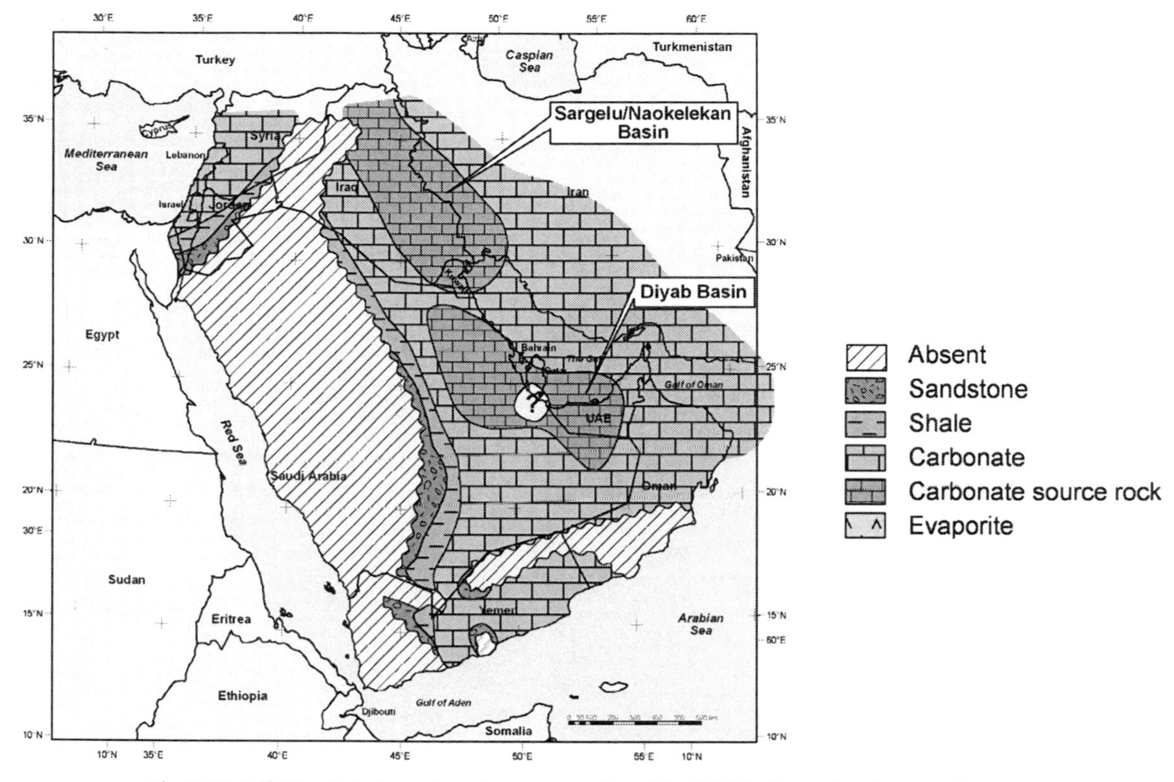

Fig. 7.28. Middle–Late Jurassic palaeogeography of the Middle East, showing maximum extent of source facies (Sargelu/Naokevekan and Diyab basins). (Compiled from various sources.)

facies are restricted to intra-shelf basins (Bordenave and Burwood, in Katz, 1994). The Palaeocene–Eocene source facies is restricted to the Zagros foredeep basin. Palaeogeographic and facies maps showing the distribution of proven Oligocene and Early Miocene paralic and peri-deltaic, and platformal and peri-reefal, reservoirs and associated facies are given in Figs. 7.32–7.33 respectively. The reservoirs are essentially restricted to the margins of the Zagros foredeep basin in Khuzestan and Lurestan provinces in Iran, and in contiguous Iraq. Note, though, that the clastic reservoirs, sourced from the 'Western Arabian highlands', emerging in response to rifting in the Red Sea, extend further west. A palaeogeographic and facies map showing the distribution of Middle Miocene evaporitic seal facies is given in Fig. 7.34. The seals are essentially restricted to the Zagros foredeep basin. A Cenozoic play fairway map is given in Fig. 7.35. The producing fields lie along the axis of the Zagros foredeep basin.

7.1.4 *Applications of biostratigraphy and palaeobiology in reservoir exploitation*

Integrated reservoir description

Palaeontology plays a key role in integrated reservoir description and in reservoir exploitation – that is, appraisal and development. Appraisal is the process whereby the extent of the reservoir and its contained reserves of hydrocarbon are determined, and its economic viability evaluated. Development is the process whereby the contained reserves of the reservoir are exploited as efficiently as possible, through the implementation of an integrated reservoir management strategy. Integrated reservoir description is the process, implemented across both appraisal and development phases, whereby general geological, geophysical and reservoir engineering, and specialist sedimentological and palaeontological data are integrated – and iterated after acquisition of new data – and a computer-simulated reservoir model is built. The objective is to model and predict reservoir

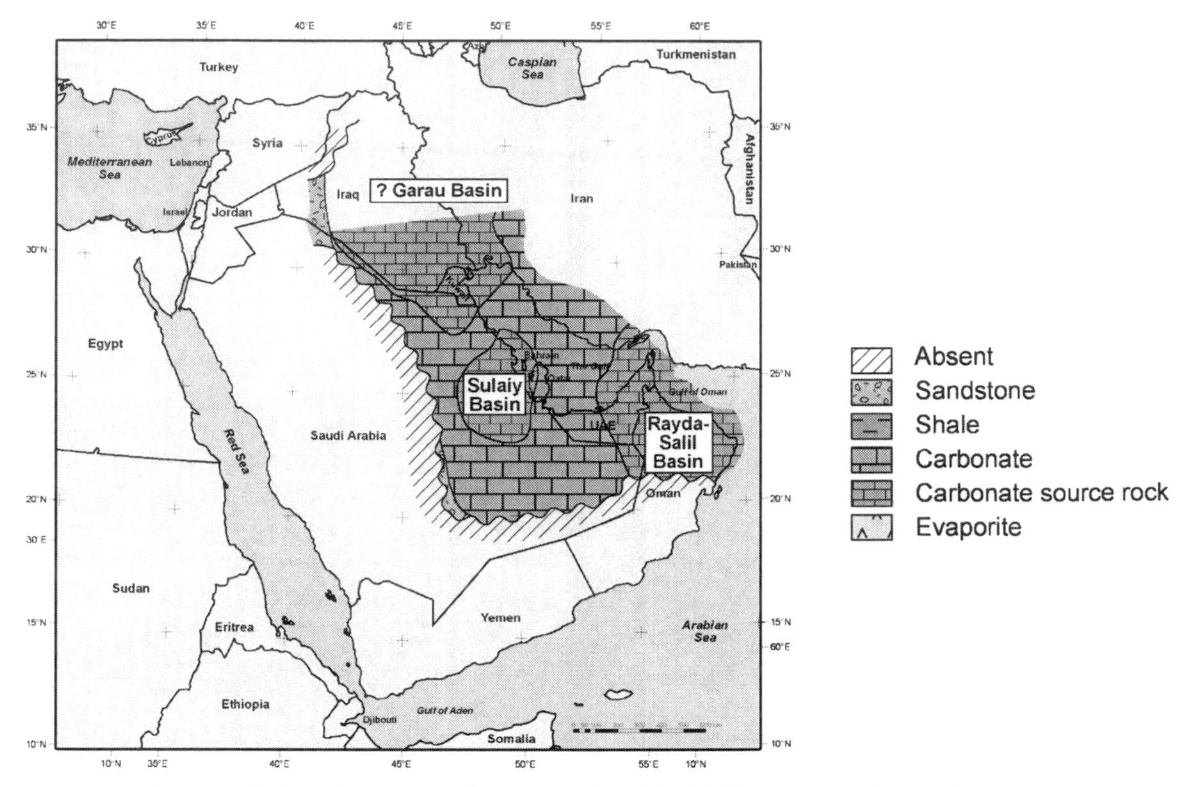

Fig. 7.29. Earliest Cretaceous palaeogeography of the Middle East, showing maximum extent of source facies (Garau, Sulaiy and Rayda/Salil basins). (Compiled from various sources.)

layering, permeability distribution and connectivity, and flow behaviour (not least in order to ensure the optimal location of production and injection wells).

Input of palaeontological data is particularly important as regards the understanding of reservoir facies, architecture, connectivity and compartmentalisation, and the identification of barriers and baffles to fluid flow (such as condensed sections). Data is acquired at a higher resolution than would be the case in routine exploration. The fossil groups used will vary depending on the reservoir age and facies. Non-biostratigraphic technologies are applicable in non-fossiliferous reservoir ages and facies (Morton *et al.*, in Carr *et al.*, 2003).

Failure to input, integrate or iterate palaeontological data can result in serious flaws in the reservoir model, and costly inefficiencies in the appraisal and development programmes.

Biosteering

An increasingly important role of palaeontology in production is in well-site operations, especially 'biosteering' (Jones *et al.*, in Koutsoukos, 2005). Biosteering involves real-time monitoring of precise stratigraphic position relative to reservoir in a – deviated – well, through the well-site application of biostratigraphic technology. It also involves, as necessary, issuing instructions to redirect the well trajectory to ensure optimal reservoir penetration, for example when encountering a sub-seismic fault or a problem with seismic depth conversion or survey data. The high resolution of the technique (its 'window'), usually established by analyses of closely spaced samples from offset wells or a pilot hole, is often of the order of only a few feet or metres. The technique is, and will remain, critical to the exploitation of many petroleum reservoirs. The value added to date through the application of biosteering in BP runs

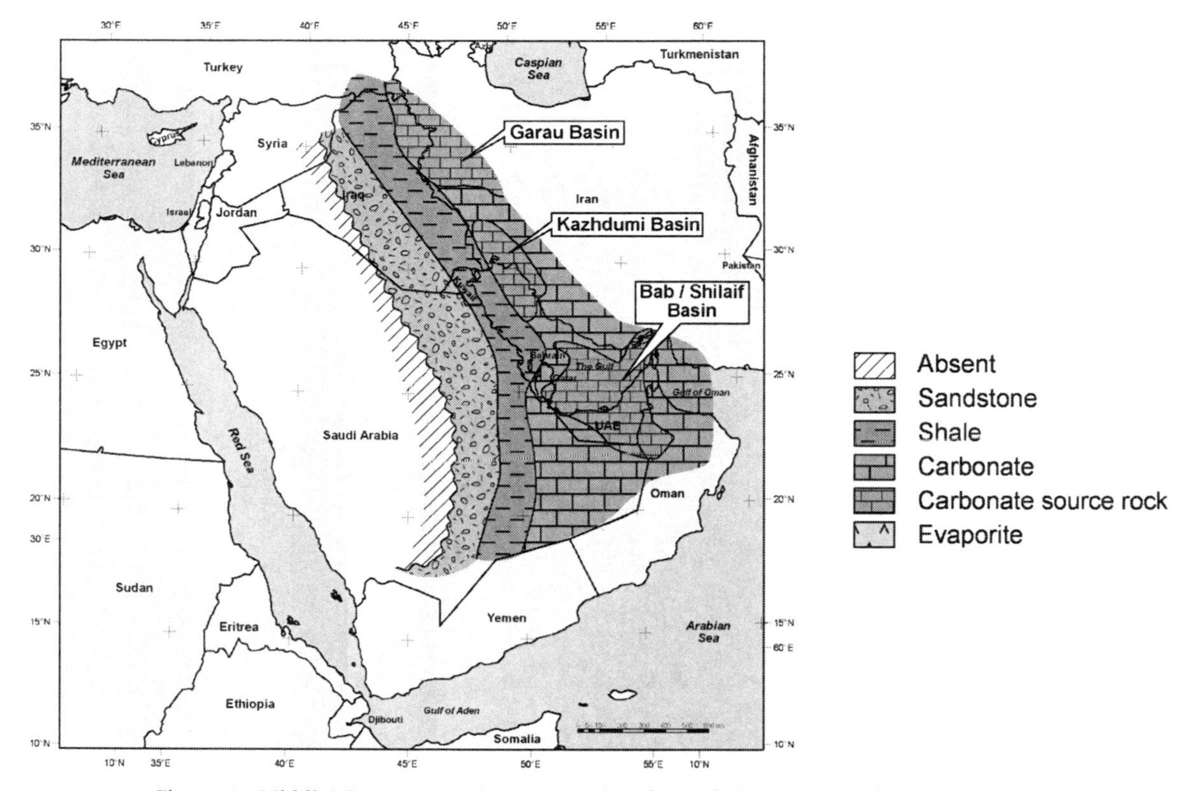

Fig. 7.30. 'Middle' Cretaceous palaeogeography of the Middle East, showing maximum extent of source facies (Garau, Kazhdumi and Bab/Shilaif basins). (Compiled from various sources.)

into hundreds of millions of dollars. It is anticipated that this figure will further increase in the future, as the technology is transferred to fields in areas only now entering into production.

7.1.5 *Case histories of applications in reservoir exploitation*

These case histories have been selected so as to include a reasonable representation of geological settings, reservoir ages and facies, and analytical techniques, with one example each from shallow marine clastic, deep marine clastic, shallow marine platform carbonate and deep marine pelagic carbonate reservoirs (Jones *et al.*, in Koutsoukos, 2005).

Cusiana field, Colombia

The Cusiana field is situated in the Llanos basin, more specifically in the frontal thrust-sheets of the Eastern Cordillera, some 150 miles north-east of Bogotá, in Colombia (Jones *et al.*, in Koutsoukos, 2005). Recoverable reserves are estimated at 1.5 billion barrels of light oil and condensate and 3.4 trillion cubic feet of gas. The – Guadalupe formation – reservoir comprises shallow marine clastics of Late Cretaceous, Santonian–Campanian age. It is characterised by significant vertical variation in quality. Production is through horizontal wells. At the time of writing, rates are in excess of 300 000 barrels per day. A palynological biostratigraphic study of cored offset wells enabled the reservoir interval to be divided into a number of zones of the order of a few tens of feet thick. Subsequent biosteering – by means of palynology – has targeted Zones GR3–GR7, which are developed within the best-quality reservoir (the 'upper Phosphate'). Incidentally, note that a BP-designed portable unit has enabled safe handling at well-site of the hazardous chemicals used in palynological processing. Note also that a BP-sponsored research programme is under way with the ultimate objective of eliminating

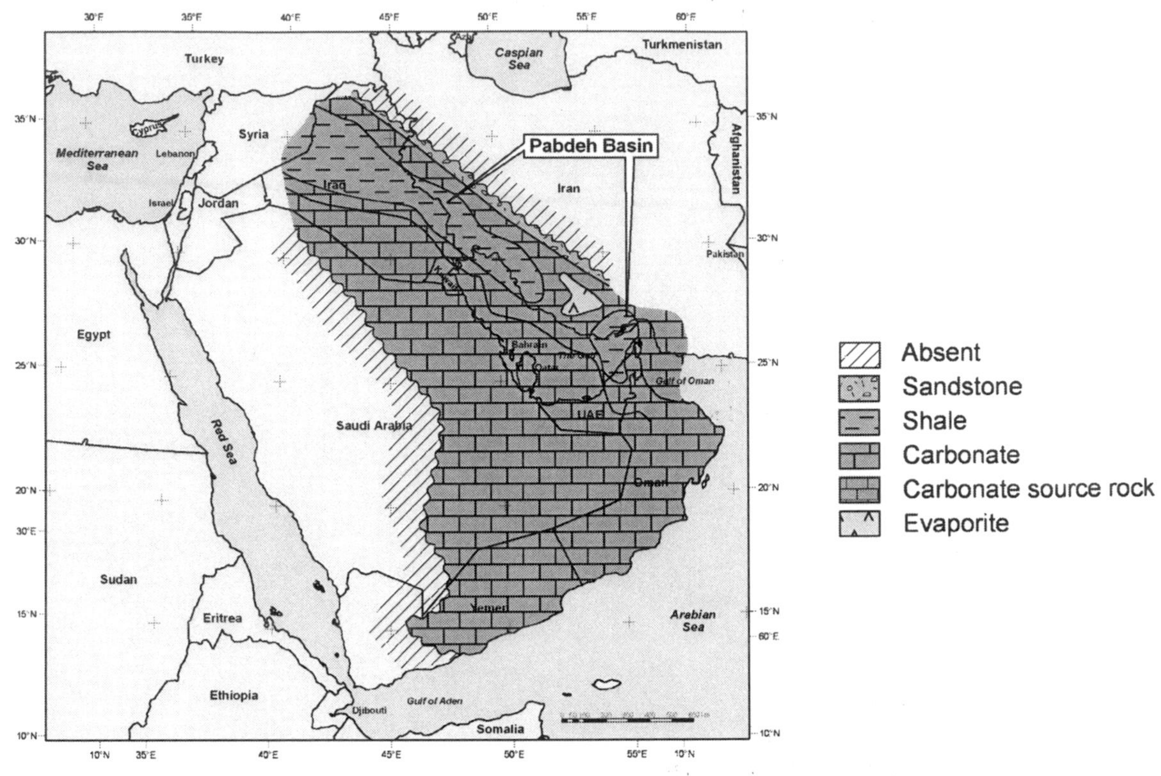

Fig. 7.31. Palaeocene–Eocene palaeogeography of the Middle East, showing maximum extent of source facies (Pabdeh Basin). (Compiled from various sources.)

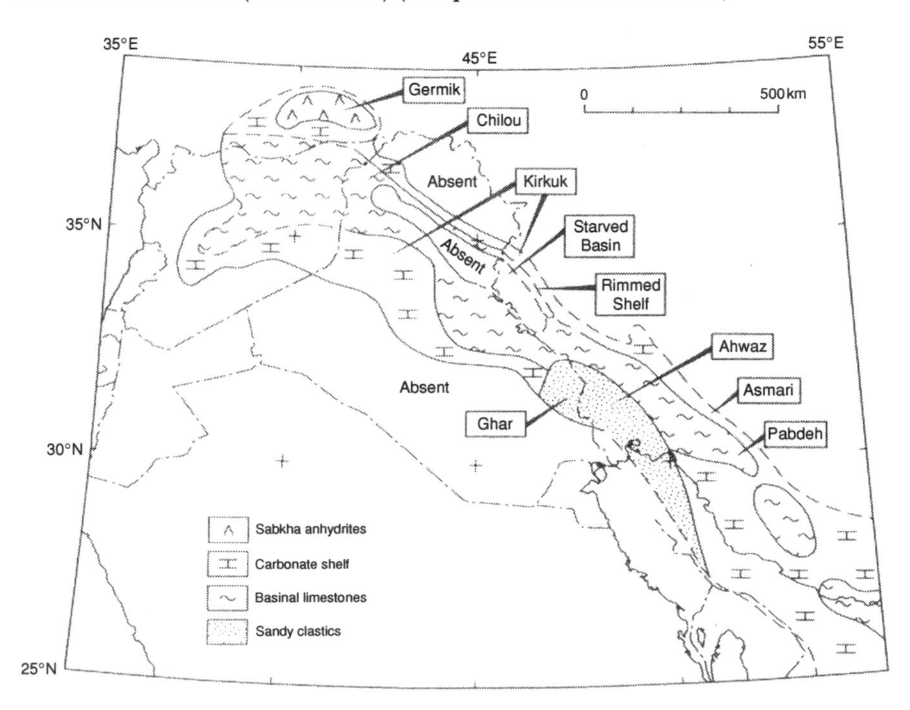

Fig. 7.32. Oligocene palaeogeography of the Middle East. (From Jones, 1996; in turn after Goff *et al.*, in Al-Husseini, 1995.)

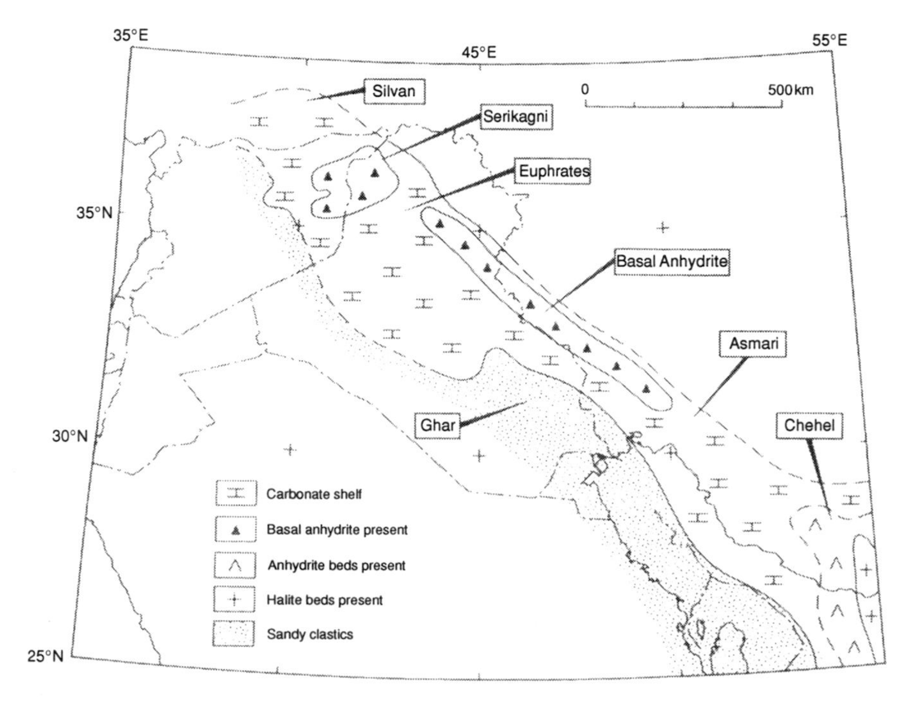

Fig. 7.33. Early Miocene palaeogeography of the Middle East. (From Jones, 1996; in turn after Goff *et al.*, in Al-Husseini, 1995.)

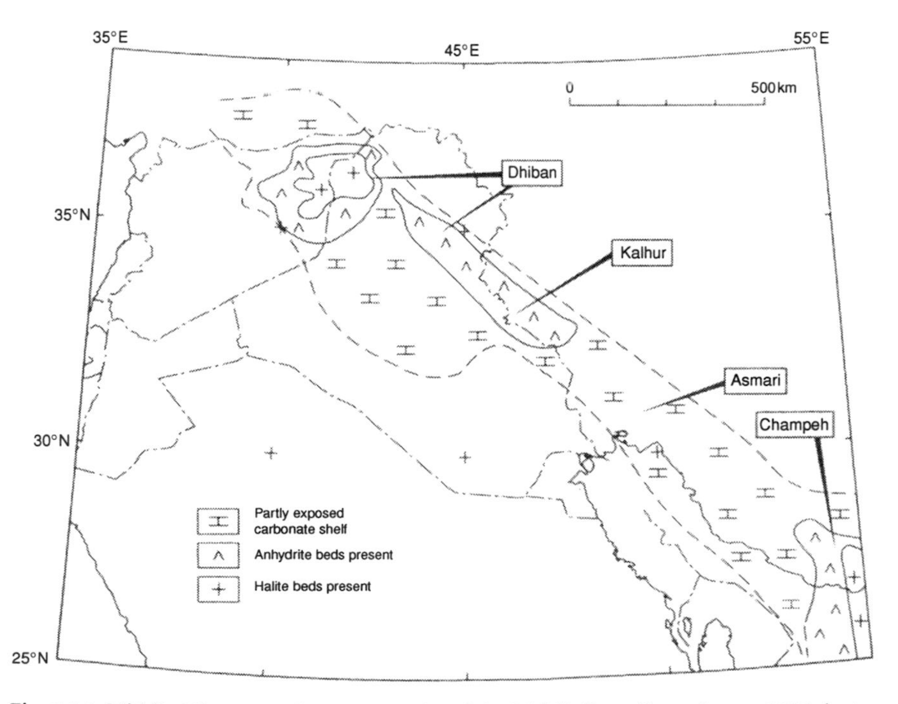

Fig. 7.34. Middle Miocene palaeogeography of the Middle East. (From Jones, 1996; in turn after Goff *et al.*, in Al-Husseini, 1995.)

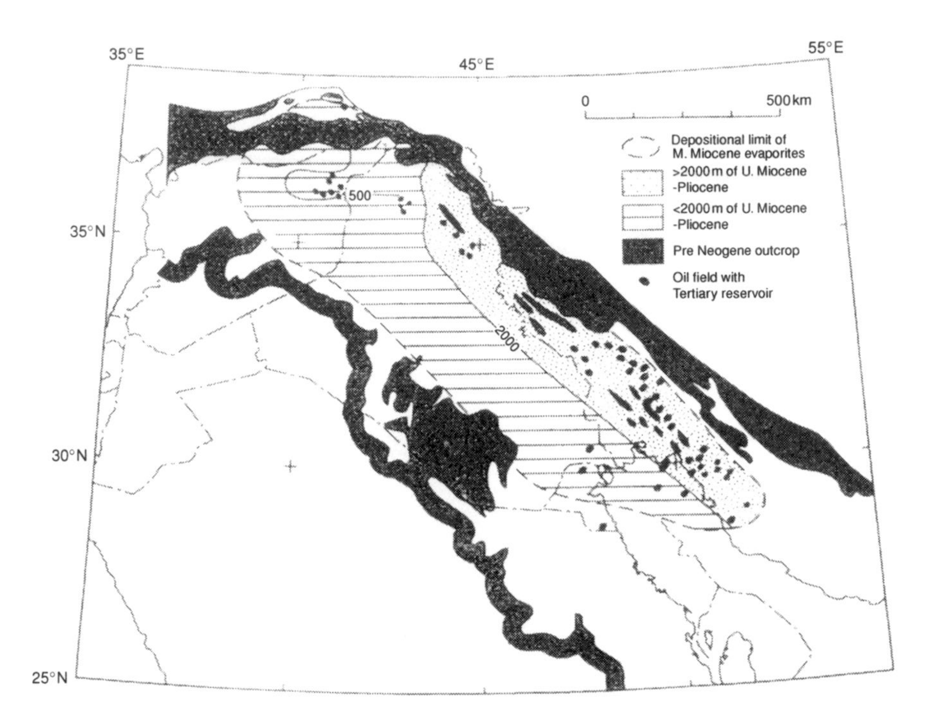

Fig. 7.35. Cenozoic play fairway map of the Middle East. (From Jones, 1996; in turn after Goff *et al.*, in Al-Husseini, 1995.)

altogether the need for the use of such chemicals. The production potential from the first biosteered horizontal well was approximately 30 000 barrels per day, as against 12 000 barrels per day from the best conventional vertical well. Moreover, one biosteered well costing US$26 million effectively does the work of three to four conventional wells costing US$15–18 million each. The three that have been drilled to date have thus resulted in significant savings in drilling costs, and value addition.

Andrew field, North Sea

The Andrew field is situated in the UK sector of the North Sea (Holmes, in Jones and Simmons, 1999; Payne *et al.*, in Jones and Simmons, 1999; Jolley *et al.*, in Carr *et al.*, 2003; Jones *et al.*, in Koutsoukos, 2005). Reserves are estimated at 118 million barrels. The – Andrew formation – reservoir comprises deep marine, submarine fan clastics of Late Palaeocene age. The pay intervals are comparatively thin. Production is through horizontal wells. At the time of writing, rates are 64 000 barrels per day. Understanding of the reservoir facies and heterogeneities and consequences for fluid flow, and optimal placement of wells with respect to fluid contacts, are critical to maximisation of oil production prior to the inevitable early gas and/or water breakthough. A micropalaeontological biostratigraphic study enabled the reservoir interval to be divided into seven zones of the order of a few tens of feet thick. Microfacies, integrated with core sedimentology, was used to identify facies and heterogeneities with each zone, thereby establishing the spatio-temporal distribution of reservoir and non-reservoir units, and potential consequences for fluid flow. Mudstones A3 and A1 were interpreted as essentially hemipelagic on the basis of contained high abundance and diversity, low dominance assemblages of arenaceous foraminiferans further characterised by comparatively high incidences of complex infaunal 'morphogroup' C (Jones and Charnock, 1985). They were therefore also interpreted as potentially of regional or field-wide extent, and constituting barriers to fluid flow. This has subsequently been confirmed by pressure data. Mudstone A2 was interpreted as interturbiditic on the basis of contained low abundance and diversity, high dominance assemblages dominated by simple epifaunal 'morphogroups' A and B. It was therefore also interpreted as of local extent, and constituting only a baffle to fluid flow. Subsequent biosteering – by means of arenaceous foraminiferal micropalaeontology and microfacies – has targeted reservoir unit B. Fortuitously, the top of this unit is effectively coincident with the gas–oil contact over the crest of the field. Realisation of this fact has allowed the biosteered well-bore to be run more medially through the reservoir than in the initial well plan, with the overlying Mudstone A3 acting as a barrier to fluid flow and hence protecting it from gas invasion. It has been estimated that the optimal well placement enabled by the biosteering has added 10 million barrels of reserves to the books.

Sajaa field, Sharjah, United Arab Emirates

The Sajaa field is situated in the frontal thrust-sheets of the Oman Mountains, some 40 km east of Sharjah town in Sharjah in the United Arab Emirates (Jones *et al.*, in Koutsoukos, 2005). Reserves are estimated at between 100–400 million barrels of condensate and 1.5–6 trillion cubic feet of gas. The – Kharaib and Shuaiba formation – reservoir comprises shallow marine, platform carbonates of Early Cretaceous, Barremian–Aptian age. Production is through horizontal wells. Rates are of the order of 60 000 barrels per day. A micropalaeontological biostratigraphic study enabled the reservoir interval to be divided into 21 zones of the order of a few tens of feet thick. Detailed palaeobathymetric interpretation of closely spaced samples has contributed significantly to the understanding of parasequence-scale reservoir facies and architecture, and to the identification of reservoir 'sweet spots'. Subsequent biosteering – by means of thin-section micropalaeontology and microfacies – has kept the well-bore within optimal parts of the reservoir over distances of several thousand feet (biosteering of multilaterals has also been possible). It has effectively replaced geosteering using coherency, which worked well in unfaulted but not in faulted sections, the approach taken on encountering a fault being to steer upwards to a known point in the stratigraphy and then back down again. This was not only time-consuming and expensive but potentially also dangerous, as it increased the risk of drilling out of the reservoir and into the seal, and causing the loss of control of the hole. Biosteering is currently being practised in conjunction with a combination of novel drilling technologies such as coiled tubing, slim-hole and underbalance (and also in conjunction with 'logging-while-drilling'). Even after the carbonate dissolution and sandblasting effects associated with these drilling technologies, sufficient cuttings are still recoverable to allow the compositing of thin-sections for micropalaeontological analysis. Note, though, that the drilling fluid has to be sufficiently buffered by the addition of alkaline substances to counter the acid dissolution (without damaging motors and seals).

Valhall field, North Sea

The Valhall field is situated in the Norwegian sector of the North Sea (Bergen and Sikora, in Jones and Simmons, 1999; Sikora *et al.*, in Jones and Simmons, 1999; Jones *et al.*, in Koutsoukos, 2005). Reserves are estimated at 705 million barrels. The – Chalk group, Tor formation – reservoir comprises deep marine, essentially pelagic carbonates, including allochthonous chalks and chalky turbidites, of Late Cretaceous, Maastrichtian age. Production is through high-angle wells, of which some 50 have already been drilled. At the time of writing, rates are 105 000 barrels per day. A micropalaeontological and nannopalaeontological biostratigraphic study enabled the reservoir interval to be divided into seven zones of the order of a few tens of feet thick. Subsequent biosteering – by means of micropalaeontology and nannopalaeontology – has targeted zones C and D, zones A and B possessing better reservoir properties in terms of porosity and permeability, but being unstable and prone to collapse under drawdown. One particularly successful well that was kept within zone D by biosteering produced 12 000 barrels per day. Other well-site applications of biostratigraphy include setting close to the base of the overburden without drilling overbalanced into the underpressured zone at the top of the reservoir, thereby causing formation damage; and terminating the well at the base of the reservoir, or 'biostopping'. Another application is identifying the origin of caved material and hence unstable zones in the tophole, impacting future well design. In terms of value added, 30% of the current production is attributed to optimal reservoir placement enabled by biosteering and associated technologies. Moreover, a cost saving of minimum of US$1 million per well – 7 days drilling, at US$150 000 per day – is achieved by being able to set casing in the correct place by means of biostratigraphy (see above).

7.2 Mineral exploration and exploitation

Geologists exploring for certain minerals know that they formed exclusively in certain settings. For example, some of the world's largest base metal deposits formed underground in a zone where mineral-laden fresh water came into contact with saline water close to the coast, the changing chemical environment causing the metals to precipitate. The ability to discern the position of the contact zone greatly increases the possibility of finding a new deposit of this type. Palaeontology provides this ability. The geologist can home in on the contact zone within the overall area of interest by plotting the localities of freshwater and marine fossils on a map.

As far as the author is aware, the only previously published accounts of the application of (micro)palaeontology to mineral exploration are those of Jones (1996) (see below) and Robbins and Burden, in Jansonius and MacGregor (1996). However, there are also apocryphal accounts of mining geologists in the Bendigo goldfields of Australia using macrofossils such as graptolites as guides to the whereabouts of gold-rich reefs.

As far as the author is aware, the only previously published account of the application of (micro)palaeontology to mineral exploitation is that of Hart, in Jenkins (1993) (see below). Hart, in Jenkins (1993) cited a number of case histories to illustrate the micropalaeontological identification of stratigraphic horizons or facies containing economic chalk (used either in the cement or paper-whitening industries). All involved Late Cretaceous chalk sections in southern England, including not only the Pitstone and East Grimstead quarry sections discussed below, but also the Quidhampton quarry section in Wiltshire. This reflects not only Hart's interest but also the peculiar problems faced in the detailed stratigraphic subdivision and surface and subsurface correlation of macroscopically essentially featureless chalky lithotypes, especially the identification of lithological and associated rock property changes associated with unconformities, faults and facies changes. Readers interested in further details of mineral exploration and exploitation, and their environmental impact, are referred to Evans (1997).

7.2.1 *Case histories of applications in mineral exploration and exploitation*

La Troya mine, Spain (mineral exploration)

Material and methods

The study was undertaken in and around the Early Cretaceous, Barremian–Aptian rudist-reef-hosted lead–zinc ore body of the La Troya mine in the Basco-Cantabrian basin in the Basque country of Spain (Lunar *et al.*, in Gibbons and Moreno, 2002). The primary objective was to collect and analyse samples from the site of the ore-body itself and from – coeval – laterally equivalent sites varying distances away. This was to test the hypothesis that micropalaeontological assemblages from sites of ore-body deposition would be characterised by environmental stress, whereas those from laterally equivalent sites would not, enabling an 'environmental stress gradient', or, in three dimensions, a 'halo', to be defined, the likes of which might serve as pointers to the locations of hitherto undiscovered ore-bodies elsewhere (Fig. 7.36). The secondary objective was to validate the established ore-body depositional model, specifically, to confirm the suspected syngenetic rather than epigenetic origin using micropalaeontological (palynological) indications of thermal maturity above and below.

Results and discussion

'Environmental stress gradient'. Unfortunately, in practice, it proved impossible to trace and sample the lateral equivalents of the ore-body, owing in part to their extreme thinness (a function of the geochemically interpreted extremely short period of time over which the ore-body formed), and in part to fault complications. The hypothesis of the 'environmental stress gradient' thus remains effectively untested either in the Basco-Cantabrian basin or elsewhere.

Depositional model. Micropalaeontological analyses of the platform carbonates underlying the ore-body indicated an Urgonian (late Barremian–early

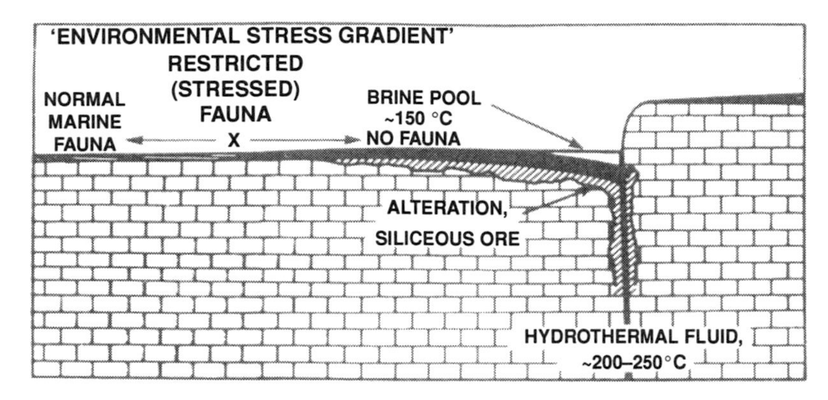

Fig. 7.36. Anticipated faunal trends at time of ore genesis, La Troya mine. Note 'environmental stress gradient'. (From Jones, 1996.)

Aptian) age. Palynostratigraphic analyses of the black pyritic shales ('margas negras') overlying it indicated a probable late Aptian age. Micropalaeontological analyses of sediments adjacent to, although, in view of the problems outlined above, not necessarily exactly laterally equivalent to, the ore-body indicated a restricted marine environment, with only rare and stress-tolerant foraminiferans and radiolarians present. Palynological analyses yielded 'spore colour indices' in the range 5–7, indicating over-maturity. No conclusions could be drawn on the basis of this evidence with regard to the syngenetic or epigenetic origin of the ore-body.

Pitstone quarry, Hertfordshire, UK (mineral exploitation)

The Pitstone quarry in Hertfordshire extracts chalk for use in the cement manufacturing industry. A balance of high- and low-calcimetry chalks provides the optimum properties. At Pitstone, calcimetry is controlled primarily by stratigraphic horizon, being low in the lower part of the lower Chalk and *Plenus* marl and high in the upper part of the lower Chalk and the lower part of the middle Chalk. A micropalaeontological study was therefore undertaken with the aim of identifying the local extent of these horizons. Closely spaced samples from a series of boreholes were analysed and the sections subdivided and correlated using a high-resolution biostratigraphic zonation scheme. The data thus obtained were presented in the form of a series of profiles projecting calcimetry (Fig. 7.37). These were utilised in formulating a pit design plan optimising the existing infrastructure and providing the most ergonomic and economic extraction of future supplies.

East Grimstead quarry, Wiltshire, UK (mineral exploitation)

The East Grimstead quarry in Wiltshire extracts chalk for use in the paper-whitener manufacturing industry. 'Bright' chalks provide the optimum properties. At East Grimstead, 'brightness' is controlled by stratigraphic horizon, being highest in the *Offaster pilula* Sub-Zone of the upper Chalk. A micropalaeontological study was therefore undertaken with the aim of identifying the local extent of this horizon. Closely spaced samples from a series of boreholes were analysed and the sections subdivided and correlated using a high-resolution biostratigraphic zonation scheme (Fig. 7.38). Results indicated that the zone of 'bright' chalks only extended for a short distance to the south of the existing quarry works.

7.3 Coal mining

Coal is formed by the compaction of land-plant and vegetable matter deposited in essentially tropical swamps and associated delta-plain and fluvial flood-plain environments (Jones, 1996). It can occur at any stratigraphic level post-dating the evolution of land-plants in the Silurian, although it is especially characteristic of the Carboniferous (Pashin and Gastaldo, 2004). Coal rank is determined by the degree of

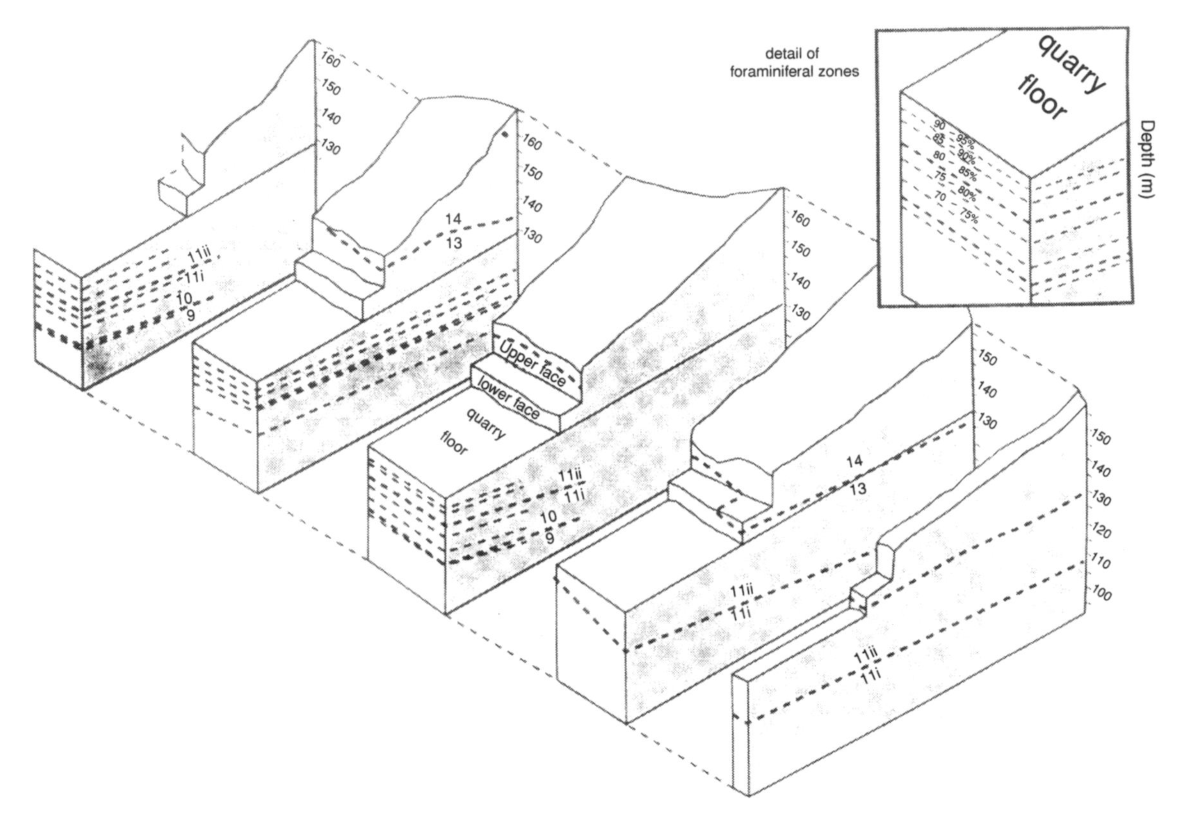

Fig. 7.37. Pitstone quarry section showing foraminiferal biozones. (From Jones, 1996; reproduced with permission from Hart, in Jenkins, 1993.)

compaction, and is lowest in brown coal and highest in anthracite. High-rank anthracite is the most sought after, as it has been compacted to the extent that it retains little of its original volatile content, and consequently produces little smoke and ash on combustion.

In Great Britain, coal is essentially restricted to the Carboniferous (Fig. 7.39). It was deposited in tropical swamp and associated environments on the northern and southern margins of an emergent mass known as St George's Land (Francis, in Craig, 1991; Kelling and Collinson, in Duff and Smith, 1992; Guion *et al.* in Woodcock and Strachan, 2002). Coal was used for cremation as long ago as the Bronze Age, and later worked by the Romans, and through into the Middle Ages. Mining operations upscaled considerably in the nineteenth century, and provided the power that drove the industrial revolution (Duff, in Duff and Smith, 1992). By the beginning of the twentieth century, Great Britain was the largest producer of coal in the world, mining 300 million tonnes annually. By the end of the twentieth century, however, it was only the eighth largest producer, mining 100 million tonnes. At the time of writing, virtually all the coal mines in the country have been closed, and in former mining communities a long-established way of life has come to an end, with dire social consequences.

In Great Britain, the marine, Tournaisian–Visean, 'Carboniferous Limestone' is zoned by means of brachiopods, corals and goniatites (see appropriate sections in Chapter 3; see also Fig. 7.39). The marine facies of the essentially Namurian, 'Millstone Grit' is zoned by means of goniatites; the non-marine facies by means of spores (Fig. 7.39). The uppermost unit, known to generations of coal miners as the 'Farewell Rock', is diachronous, and generally underlies and pre-dates, but locally overlies and post-dates, the Westphalian *Gastrioceras subcrenatum* marine band. The intermittent marine bands in the Westphalian

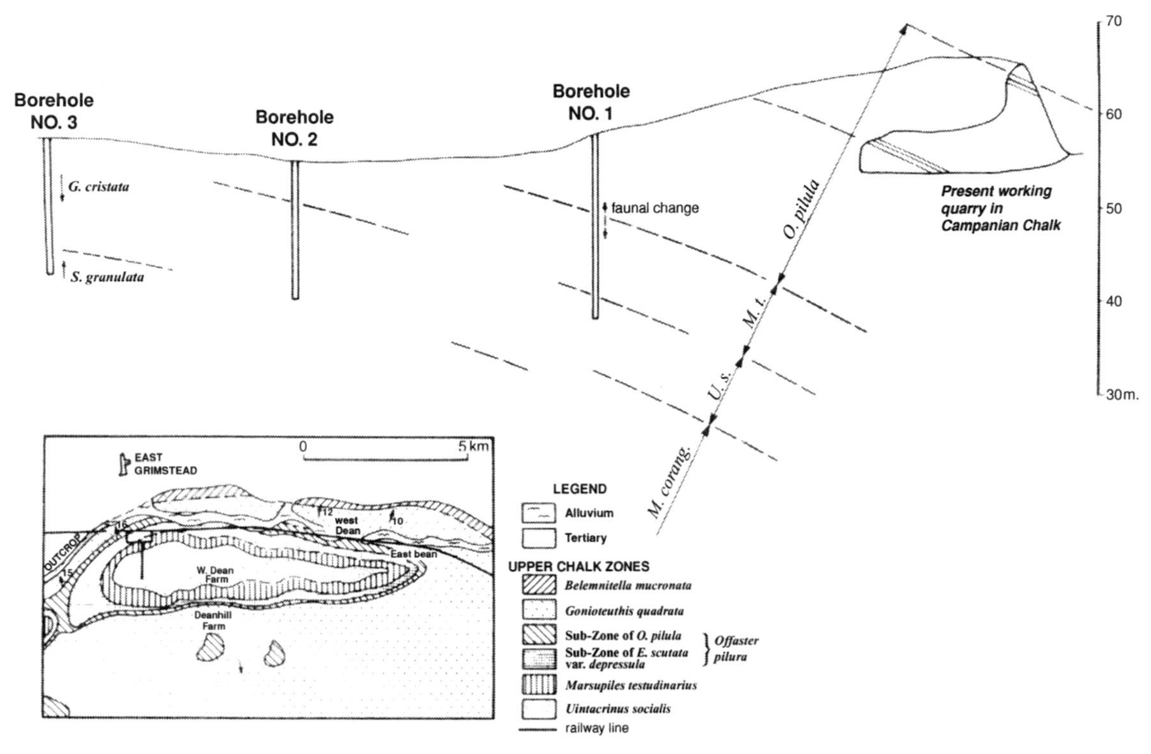

Fig. 7.38. East Grimstead quarry section showing foraminiferal and belemnite biozones. (From Jones, 1996; reproduced with permission from Hart, in Jenkins, 1993.)

'Coal Measures' are marked by goniatites; the intervening non-marine facies by plants, spores and non-marine bivalves (Figs. 7.39–7.40). In detail, successive marine bands of the Westphalian are marked by the goniatites *Gastrioceras subcrenatum*, *G. listeri*, *Anthracoceras vanderbecki*, *A. hindi*, *A. aegiranum* and *A. cambriense* (Fig. 7.39). The non-marine Westphalian is marked by, in ascending stratigraphic order, the *Densosporites annulatus*, *Radiizonates aligerens*, *Schulzospora rara*, *Dictyotriletes bireticulatus*, *Vetispora magna*, *Torispora secures* and *Thymospora obscura* spore zones (Fig. 7.39), and the *Anthracomya lenisulcata*, *Carbonicola communis*, *A. modiolaris*, *C. similis–A. pulchra*, *A. phillipsii*, *A. tenuis* and *A. prolifera* – non-marine bivalve – zones (Figs. 7.39–7.40). Spores are especially useful in the identification and correlation of Westphalian 'Coal Measure' sequences, not least as they actually constitute a significant component of the coal. Indeed, high-resolution quantitative analysis of spore assemblages can even be used in the identification and correlation of individual coal seams. Recently, spore analysis undertaken in the Pictou field in Nova Scotia revealed that the Acadia seam had not been fully worked in the Albion district, leading to the identification of an additional 22 million tonnes of coal and coal-bed methane reserves (McLean, in Jansonius and MacGregor, 1996). Spore analysis of samples from offshore oil and gas wells also revealed an extension of the Sydney field under the Gulf of St Lawrence, making it the largest coal field in eastern Canada (McLean, in Jansonius and MacGregor, 1996).

Some years ago, when speculative boreholes were sunk in central England in search of coal, the recovery of graptolites was enough to stop the drilling, and thus to save a significant amount of money – their presence indicating that the rocks being drilled were too old to contain coal (Tunnicliff *et al.*, in Palmer and Rickards, 1991). Earlier, no less a personage than William Smith attempted – albeit in vain – to stop mining for coal at what to him were also undoubtedly inappropriate stratigraphic levels (Torrens, 2003).

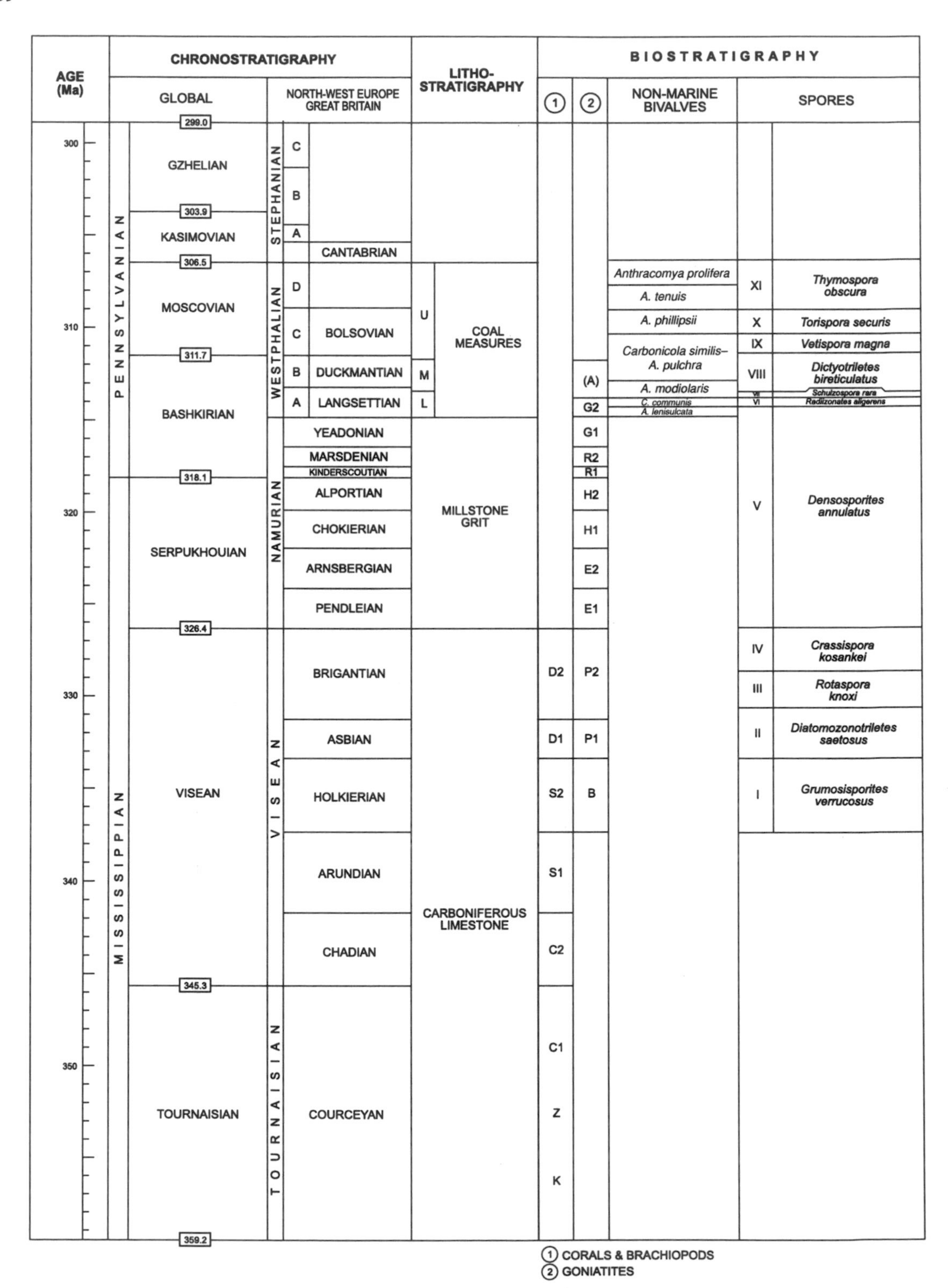

Fig. 7.39. Carboniferous stratigraphy of Great Britain. (Compiled from various sources.)

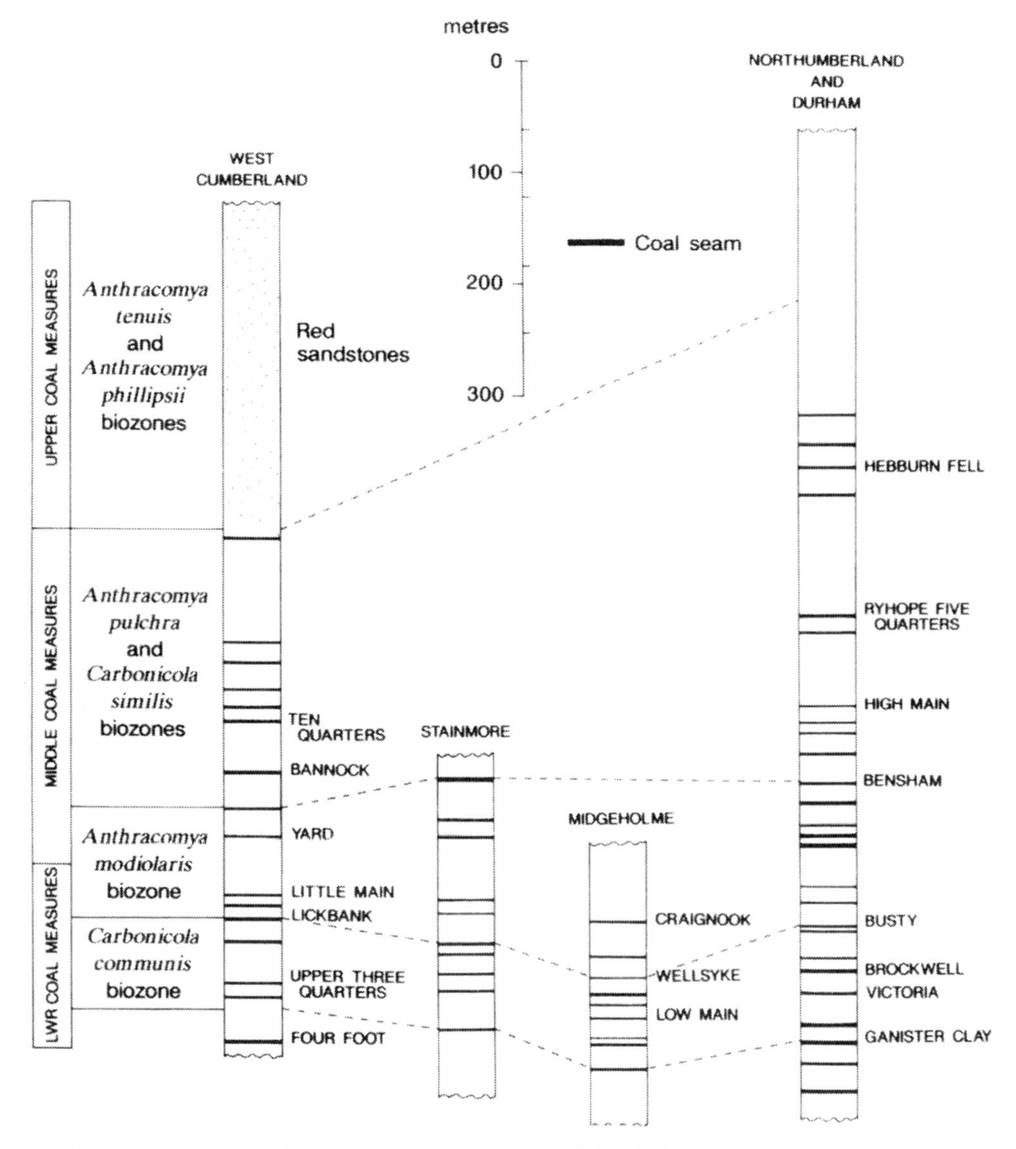

Fig. 7.40. Stratigraphic zonation and correlation of the Coal Measures of northern England by means of non-marine bivalves. (Reproduced with permission from Doyle, 1996 *Understanding fossils*, © John Wiley & Sons Ltd.)

7.4 Engineering geology

To the best of the author's knowledge, the only previously published accounts of the application of (micro)palaeontology to engineering geology are those of Hart, in Jenkins (1993) (see below), Hart, in Martin (2000) and Wright (2001) (see below). Hart, in Jenkins (1993) cited a number of case histories to illustrate the range of applications in engineering site investigations. As in the case of mineral exploitation, all involved Late Cretaceous Chalk sections in southern England. They include not only the Channel tunnel and Thames barrier site investigation sections, accounts of which are presented below, but also the Cambridge bypass section and the Blackgang Chine section on the Isle of Wight (particularly prone to landslips).

7.4.1 *Case histories of applications in engineering geology*

The living and fossil biotas are important in engineering geology in the field of site investigation (see also Box 7.3).

Channel tunnel site investigation, UK

The main application in this case was in the provision of precise and accurate stratigraphic control on the Albian Gault clay and Cenomanian lower Chalk (Hart, in Jenkins, 1993). Samples were taken with an average vertical spacing of as little as approximately 1 m from the appropriate stratigraphic section from a number of sites along a cross-channel transect. A high-resolution foraminiferal biostratigraphic zonation scheme was erected, the resolution and correlative value of which was enhanced by quantitative measures of percentages of planktonic species and of planktonic 'morphogroups' (epipelagic *Hedbergella–Whiteinella*, mesopelagic *Dicarinella–Praeglobotruncana* and bathypelagic *Rotalipora*). The resolution of the biozonation scheme was sufficient to allow recognition of local and regional onlap surfaces, as at the base of the lower Chalk, and erosional surfaces, as in the Middle Cenomanian part of the lower Chalk (Fig. 7.41). Rock property changes associated with the east–west loss through onlap of the particularly clay-rich basal Chalk have important engineering implications.

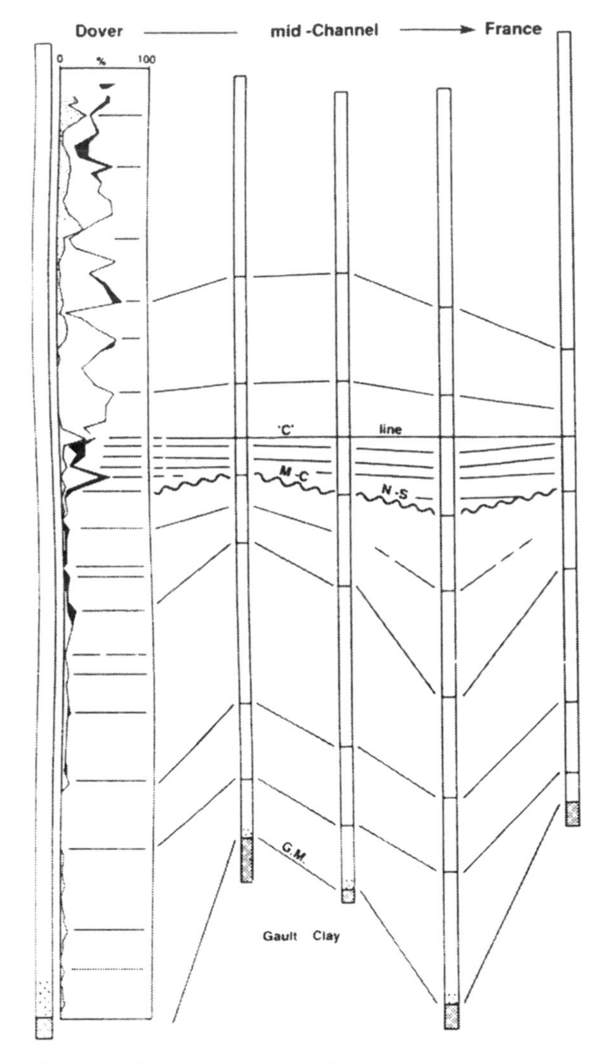

Fig. 7.41. Channel tunnel section showing foraminiferal biozones. Note onlap associated with mid-Cenomanian non-sequence (M–C N–S.) 'C', line of correlation used as reference datum; G. M., glauconite marl. (From Jones, 1996; reproduced with permission from Hart, in Jenkins, 1993.)

Thames barrier site investigation, UK

A number of boreholes were drilled to investigate the rock properties of the upper Chalk in the subsurface in order to select a suitable founding level for the 3300-tonne moveable steel gates of the Thames barrier and their supporting concrete piers (Hart, in Jenkins, 1993; Hart, in Martin, 2000). Most of the variation in engineering properties as revealed by tests was in the vertical (stratigraphic) rather than horizontal sense. There was also some overprint associated with weathering, frost-shattering and solifluction. The principal objective of the parallel micropalaeontological study was therefore to provide a detailed correlation of different stratigraphic horizons and any offset by faults. A subsidiary aim was to distinguish solifluction from *in situ* chalk using the presence of exotic components (Tertiary sand grains etc.) introduced during movement. The sample spacing was again extremely close (of the order of 1 m). Again, a high-resolution foraminiferal biozonation scheme was erected. The resolution was enhanced by separate plots, in the form of bar charts or kite diagrams, of selected superfamilies, and of the proportions of selected species or species groups within the superfamilies Cassidulinacea (*Globorotalites cushmani*,

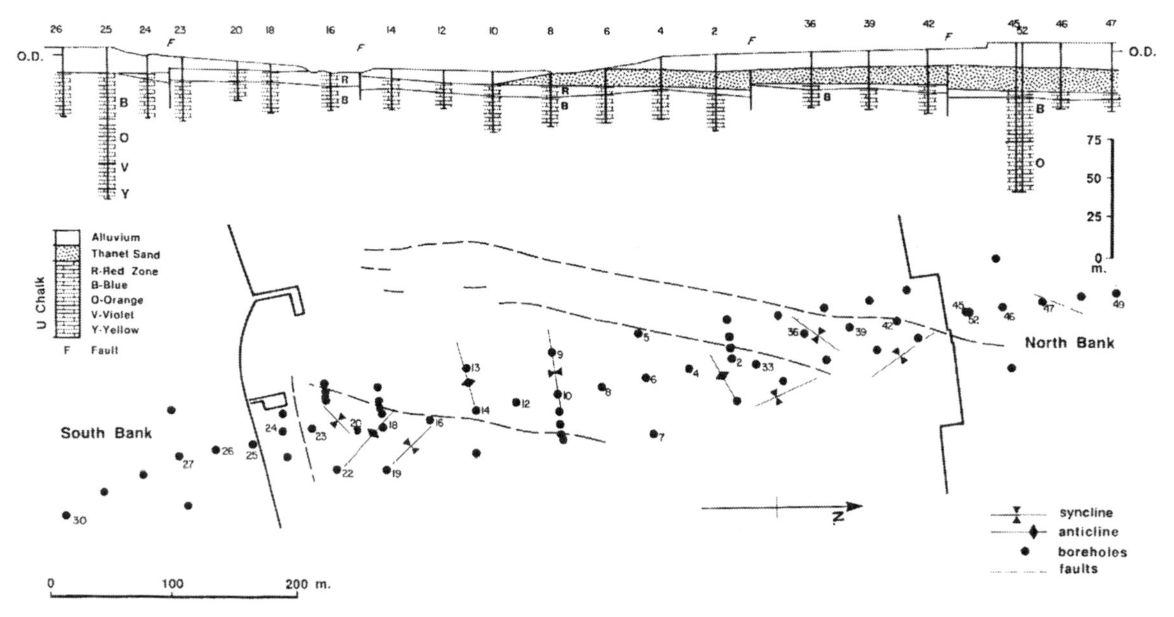

Fig. 7.42. Thames barrier site investigation section showing foraminiferal biozones. (From Jones, 1996; reproduced with permission from Hart, in Jenkins, 1993.)

Lingulogavelinella aff. *vombensis*, other) and Globigerinacea (*Globotruncana bulloides–G. marginata, G. linneiana–G. pseudolinneiana, Hedbergella*). The resolution was sufficient to allow recognition not only of a widespread erosional unconformity of varying extent at the top of the Chalk, but also of cross-cutting faults (Fig. 7.42). The engineering implications of rock property changes associated with faults at the foundation level were fully evaluated by engineering geologists prior to commencement of the construction operation.

'Project Orwell', UK

As part of the recent 'Project Orwell', completed in 1999, a 5.5-km-long tunnel was constructed to reduce flooding in and around Ipswich in Suffolk. The tunnel had to be drilled between two resistant flint bands within the Campanian *Gonioteuthis quadrata* Zone of the Chalk, since encountering either would have constituted a significant hazard. The trajectory of the tunnel was successfully kept within this narrow 'window' by means of high-resolution micropalaeontology (Wright, 2001).

7.5 Environmental science

The living biota is important in environmental science in the fields of environmental impact assessment and environmental monitoring, discussed below (see also Box 7.3). Readers interested in further details are referred to Pickering and Owen (1994), Bennett and Doyle (1997) and Ernst (2000).

7.5.1 Environmental impact assessment

According to the Department of the Environment's definition, environmental impact assessment (EIA) is 'a process by which information about the environmental effects of a project is collected, both by the developer and other sources, and taken into account by the relevant decision making body before a decision is given on whether the development should go ahead'. The living biota is of proven applicability in EIA. For example, the macro- and meiobiotas have been used in baseline studies on the potential impacts of civil engineering projects, and the mitigation of these impacts, in terrestrial environments (Morris and Emberton, in Morris and Therivel, 2001), in aquatic environments (Biggs *et al.*, in Morris and Therivel, 2001), and in marine environments (Thompson and Lee, in Morris and Therivel, 2001). The microbiota has also been used in baseline studies on the potential impacts of large-scale civil engineering projects, such as the construction of the Rance barrage in France and the Severn barrage in the UK.

7.5.2 *Environmental monitoring*

The living biota is of proven applicability in environmental monitoring. For example, the macro- and meiobiotas have been used in baseline studies on water quality (Gullan and Cranston, 2000; Morris *et al.*, in Morris and Therivel, 2001). The microbiota has been extensively used in environmental monitoring (Jones, 1996; Stoermer and Smol, 1999; Martin, 2000). In recent studies, dinoflagellates have been used to monitor anthropogenic effects, including industrial pollution (Dale, in Martin, 2000). Palynology has been used to establish the hydrostratigraphic framework at a nuclear waste disposal site at Savannah River in South Carolina, and in particular the distribution of aquifers, as these serve as preferential pathways for ground-water movement and, potentially, radioactive contaminant transport (van Pelt, in Martin, 2000). Diatoms and chrysophytes have been used in baseline studies on water quality (Dicit and Smal, in Martin, 2000). Foraminiferans have been used to monitor anthropogenic effects, including domestic and industrial anthropogenic pollution (Coccioni, in Martin, 2000; van der Zwaan, in Martin, 2000; Hayward *et al.*, 2004). They have also been used to monitor natural effects, for example historical environmental, especially salinity, change in the Everglades National Park in Florida (Ishman, in Martin, 2000), storm frequency off South Carolina (Hippensteel and Martin, in Martin, 2000) and reef vitality in Florida, Hawaii and elsewhere (Hallock, in Martin, 2000). Ostracods have also been used to monitor both anthropogenic effects (Eagar, in Martin, 2000; Rosenfeld *et al.*, in Martin, 2000) and natural effects (Schornikov, in Martin, 2000). The effects of pollution on the microbiota vary considerably. The effects of domestic pollution on foraminiferans include modifications to the structure and composition of communities, including decreases in abundance and diversity, exclusion of certain species etc. (Jones, 1996). As far as can be ascertained from the available data, the effects appear localised around the source of the pollution, and disappear almost as soon as the source of the pollution disappears. The effects of industrial, for example, toxic trace metal pollution include: modifications to the structure and composition of communities, in turn including decreases in abundance and diversity, exclusion of certain species etc.; modifications to the reproductive cycles of individual species, including increases in incidences of asexually reproducing individuals; development of test deformities; and changes in shell chemistry, including increased absorption of metals, sulphur etc., and pyritisation (Jones, 1996). The effects appear to get progressively more pronounced as the pollution continues.

7.6 Archaeology

In conventional palaeontology, fossils are typically used to provide information on the ages and environments of containing sediments.

In archaeology, the approach is subtly different, in that both sediments and fossils, including human remains and associated artefacts, are used to provide information on the stratigraphic and environmental context of human evolution and dispersal, and of ancient human life and behaviour (Straker, in Briggs and Crowther, 2001). Human remains provide information as to behaviour (Larsen, 1987), and also as to disease and/or cause of death (Aufderheide and Rodriguez-Martin, 1998). Tooth wear provides information as to diet; tooth composition, information as to the point of origination of the individual, and as to migration (Muller *et al.*, 2003). Associated animal and plant remains – including pollen grains in coprolites ('faeces facies') – provide information as to environment, climate, diet, and scavenging, hunting–gathering and farming behaviour (Davis, 1987; Piperno, 1988; Reitz and Wing, 1999; O'Connor, 2000; Pearsall, 2000). They can even provide information as to the season(s) in which sites were occupied. For example, the season in which a fish was caught can be determined by the stage of development of its ear bones, or otoliths. Fossils also provide information as to the provenance of flint, amber, pottery, building stone, and ships' ballast; and hence ancient trade links. For example, fossil foraminiferans have pinpointed the source of the clay used in the manufacture of the pottery discovered in the recent excavation of Akrotiri on the island of Thera in the Aegean (Friedrich *et al.*, in McGuire *et al.*, 2000). (Thera was part of the much larger volcanic island of Santorini until the explosive volcanic eruption of 1640 BC, which is interpreted as having been partly responsible for the decline of the Minoan civilisation on

Crete.) Incidentally, geology can also provide information as to the provenance of not only building stones, such as the standing stones of Stonehenge, but also worked metals. Assays of Bronze Age gold from all over Ireland have revealed a silver content suggestive of an origin in the Croagh Patrick area of Connemara. Archaeometallurgy indicates what type of technology was available, and when.

Note that in archaeostratigraphy, as against conventional biostratigraphy, the 'law of superposition of strata' does not always apply, as both burials and cave fills can introduce younger material below older (see also below).

Readers interested in further details are referred to Herz and Garrison (1998), Pollard (1999), Dincauze (2000), Garrison (2000) and Sobolik (2003). Those interested specifically in further details of human evolution and dispersal are referred to Sub-sections 3.6.4. and 5.5.8. Those interested in the relationship between civilisation and climate are referred to Fagan (2004).

7.6.1 Archaeostratigraphy

As intimated in the sub-section on 'Quaternary dating methods' (Sub-section 6.3.6), a number of dating methods are employed in the Quaternary, and in the construction of the archaeological or archaeostratigraphic timescale (Fig. 7.43). These methods include conventional biostratigraphy, oxygen isotope stratigraphy, accelerator mass spectometry, amino-acid racemisation, cosmogenic chlorine-36 rock exposure, dendrochronology, electron spin resonance, magnetic susceptibility, molecular stratigraphy, optically stimulated luminescence, radiocarbon dating, thermoluminescence, and uranium-series dating. Incidentally, not only dendrochronology but also temperature and climate can be derived from tree-ring analysis (Briffa *et al.*, 2004; Martinelli, 2004).

Conventional biostratigraphy

Palynology

Palynostratigraphy provides a form of climatostratigraphy that can be calibrated directly against the radiocarbon record and indirectly against the deep-sea oxygen isotope record (see Sub-section 6.3.6). In Great Britain and elsewhere in northern Europe, late glacial/older *Dryas* Zone I (Tundra) ranges from approximately 12 380 to 10 000 BC (Palaeolithic); Allerod Zone II (warmer) from 10 000 to 8800 BC (Palaeolithic); younger *Dryas* Zone III (Arctic) from 8800 to 8300 BC (Palaeolithic); post-glacial, pre-boreal Zone IV (warmer, Sub-Arctic) from 8300 to 7600 BC (Mesolithic); boreal Zones V–VI (warm, dry) from 7600 to 5500 BC (Mesolithic); Atlantic Zone VIIA (warm, wet, maritime) from 5500 to 3000 BC (Neolithic); sub-Boreal Zone VIIB (dry, continental) from 3000 to 600 BC (Bronze Age); and sub-Atlantic Zone VIII (cool, wet) from 600 BC to the present day (Iron Age to Middle Age) (Fig. 7.43). Incidentally, the vegetational and climatic trends indicated by the pollen are substantiated by other groups of organisms and by other techniques. For example, the carbon and nitrogen isotopic composition of red deer collagen from the late glacial, Bolling (Allerod) and younger *Dryas* of the northern Jura is indicative of open vegetation; and that from the post-glacial, pre-boreal and boreal, of forest (Drucker *et al.*, 2003). Also, the oxygen isotopic composition of – palaeoenvironmental water from – animal and human bone from the post-glacial of the Volga–Don steppe area of southern European Russia is indicative of a general climatic amelioration (Iacumin *et al.*, 2004).

Palynology also provides an indication of anthropogenic disturbance of the environment. For example, the increased incidence of charcoal in a bog in the Cambara do Sul region of Brazil from approximately 7400 years BP (before present) is interpreted as indicating human habitation and the widespread use of fire at this time (Behling *et al.*, 2004).

Readers interested in further details are referred to Bryant and Holloway, in Jansonius and MacGregor (1996) and Sobolik, in Jansonius and MacGregor (1996).

Vertebrate palaeontology

Vertebrate stratigraphy can also be calibrated directly against the radiocarbon record, or the magnetostratigraphic record. In Great Britain, a number of zones have been recognised on the basis of large and small mammals (Fig. 7.44) (Chaline, 1987; Currant, 1989; Currant and Jacobi, 1997; Yalden, 1999; Currant and Jacobi, 2001). Some important

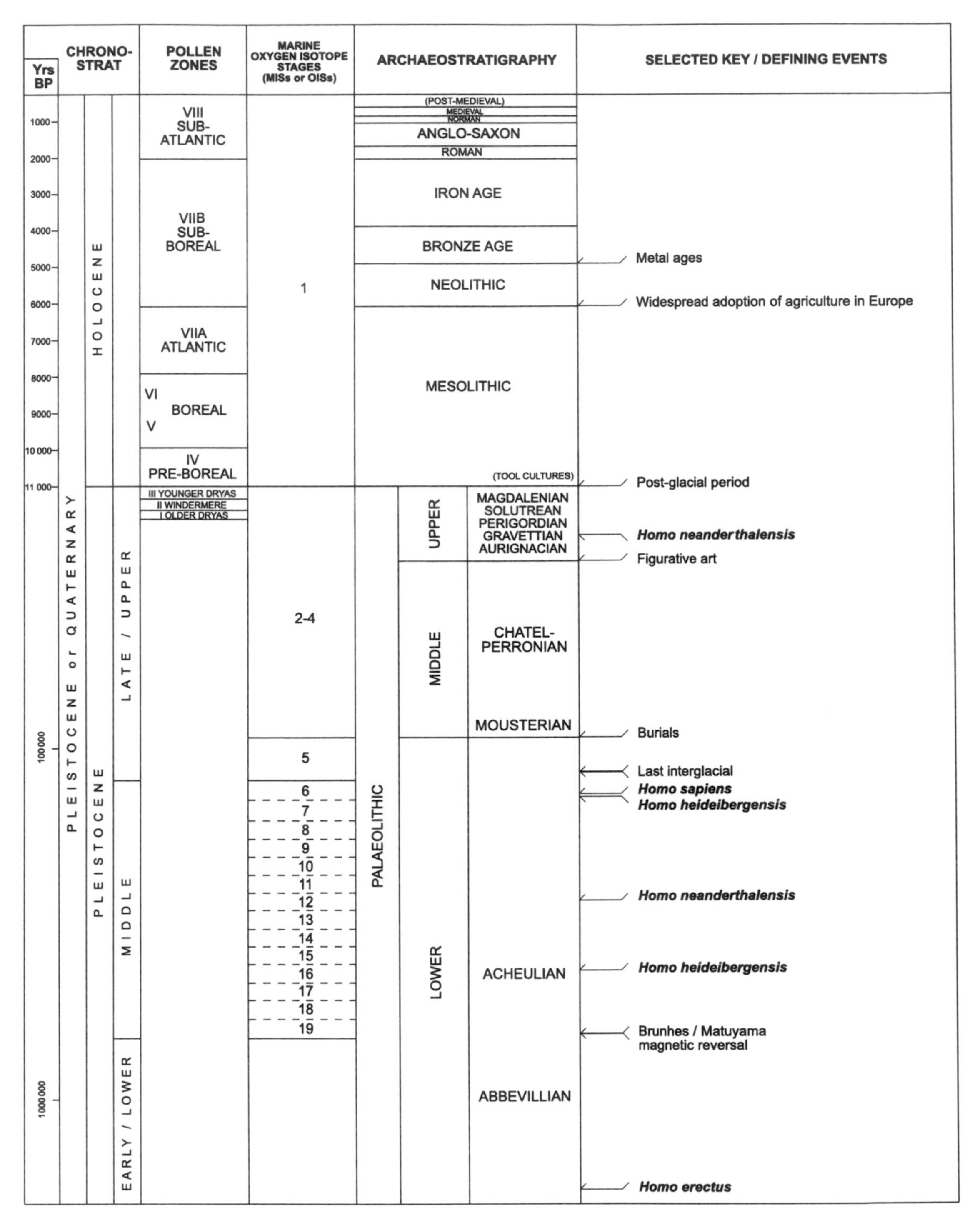

Fig. 7.43. The archaeological timescale. Note change from linear scale in Holocene to logarithmic scale in Pleistocene.

INTERGLACIAL/ OXYGEN ISOTOPE STAGE	POST-CROMERIAN	HOXNIAN	PRE-IPSWICHIAN	PRE-IPSWICHIAN	IPSWICHIAN	RECENT
	13	11	9	7	5	1
Erinaceus europhaeus Hedgehog	*	*				*
Sorex araneus Common shrew			*	*	*	*
Sorex minutus Pigmy Shrew	?	?	*	*	*	*
Neomys newtoni A water shrew	*	*				
Neomys fodiens Water Shrew			*	*		*
Crocidura sp. A white-toothed shrew				*		
Talpa europaea Mole	*	*		*		*
Macaca sylvanus *Macaque*	*		*	*		
Homo sapiens Man		*	*			*
Lepus timidus Mountain Hare	*	*	*		*	*
Sciurus whytei A squirrel	*	*				
Sciurus vulgaris Red Squirrel				*		*
Trogontherium cuvieri Giant Beaver	*	*	*			
Castor fiber Beaver	*	*	*	*		*
Muscardinus avellanarius Dormouse		*				*
Clethrionomys glareolus Bank Vole	*	*	*	*	*	*
Pliomys episcopalis An extinct vole	*	*				
Mimomys savini A water vole	*					
Arvicola cantiana A water vole		*	*	*	*	
Arvicola terrestris Water Vole					*	*
Microtus agrestis Field Vole	?	(*)	*	*	*	*
Pitymys subterraneus Pine vole	*	*	*			
Apodemus sylvaticus Wood Mouse	*	*	*	*	*	*
Canis mosbachensis A wolf	*	*				
Canis lupus Wolf			*	*	*	*
Vulpes vulpes Red Fox					*	*
Ursus deningeri / spelaeus Cave Bear	*	*	*			
Ursus arctos Brown Bear				*	*	*
Martes martes Pine Marten	*	*	*			*

Fig. 7.44. Stratigraphic subdivision of the Quaternary of Great Britain by means of mammals. (Reproduced with permission from Yalden, 1999; in turn after Currant, 1989.)

Mustela nivalis Weasel	*	*				*
Mustela erminia Stoat		*			*	*
Mustela meles Badger		*			*	*
Lutra lutra Otter	*		*	*		*
Felis silvestris Wild Cat	*		*		*	*
Equus ferus Wild Horse	*	*	*	*		
Dicerorhinus etruscus Etruscan Rhino	*	*				
Dicerorhinus kirchbergensis Merck's Rhino			*	*		
Dicerorhinus hemitoechus Narrow-nosed Rhino			*	*	*	
Hippopotamus amphibius					*	
Sus scrofa Wild Boar	*		*	*	*	*
Cervus elaphus Red Deer	*	*	*	*	*	*
Capreolus capreolus Roe Deer	*	*	*	*	*	*
Bos primigenius Aurochs			*	*	*	*
Temperature curve + –						

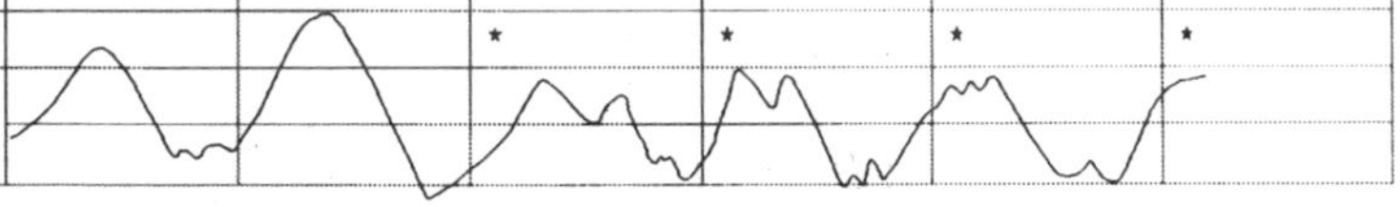

Fig. 7.44 (*cont.*)

mammals from the the British Isles are shown on Fig. 7.45. In eastern Europe, a number of zones have been recognised on the basis of progressively more evolved mammal faunas; age dates have been assigned to these zones; and correlations to western Europe and elsewhere effected (Zubakov and Borzenkova, 1990; Markova, 1999; INQUA-SEQS, 2002; Agadzhanyan, 2004; Matoshko *et al.*, 2004). Here, the late Odessa fauna is characterised by the voles *Allophaiomys pliocaenicus* and *Prolagurus ternopolitanus (P. praepannonicus)* – and also by particular pollen, molluscs and ostracods – variously dated to approximately 1.7–1.8 Ma, 1.2–1.4 Ma or 0.8–1.1 Ma. Importantly, the *Homo erectus* mandible from Dmanisi in Georgia was apparently found in association with this fauna.

Optically stimulated luminescence

Optically stimulated luminescence (OSL) has been used to date the hill-wash overlying the figure of the 'long man of Wilmington', carved into the chalk bedrock, as sixteenth century. Incidentally, molluscan stratigraphy yielded a similar age (one of the species recorded only being introduced into this country in the late medieval period). Similarly OSL has been used to date the figure of the 'white horse

Fig. 7.45 Some important Quaternary mammals. Carnivores: (a) *Crocuta crocuta spelaea* (cave hyena) right lower jaw, late Pleistocene, Kent's Cavern, Torquay, Devon, ×0.25; (b) *Ursus deningeri* (bear), left lower jaw, Cromer forest bed, Bacton, Norfolk, ×0.33. Even-toed ungulates: (c) *Bos primigenius* (aurochs), upper and lower molars, Late Pleistocene, ×0.5; (d) *Hippopotamus amphibius* (hippopotamus) molar, Pleistocene, near Bedford, ×0.5; (e) *Cervus elaphus* (red deer), antler, Late Pleistocene, Walthamstow, Essex, ×0.04; (f) *Dama clactoniana* (Clacton fallow deer), antler, Middle Pleistocene, Swanscombe, Kent, ×0.04; (g) *Megaloceros giganteus* (giant Irish deer), antler, Pleistocene, Ireland, ×0.04; (h) *Rangifer tarandus* (reindeer), antler, Late Pleistocene, Twickenham, Middlesex, ×0.04. Odd-toed ungulate: (i) *Coelodonta antiquitatis* (woolly rhinoceros) upper molar. Late Pleistocene, Kent's Cavern, Torquay, Devon, ×0.5. Proboscideans: (j) *Anancus arvernensis* (mastodon), upper molar, Red Crag (Pliocene), Suffolk, ×0.25; (k) *Mammuthus primigenius* (woolly mammoth), upper molar, Late Pleistocene, Millbank, London, ×0.25; (l) *Palaeoloxodon antiquus* (straight-tusked elephant), upper molar, Pleistocene, Greenhithe, Kent, ×0.25. (Modified after the Natural History Museum, 2000c.)

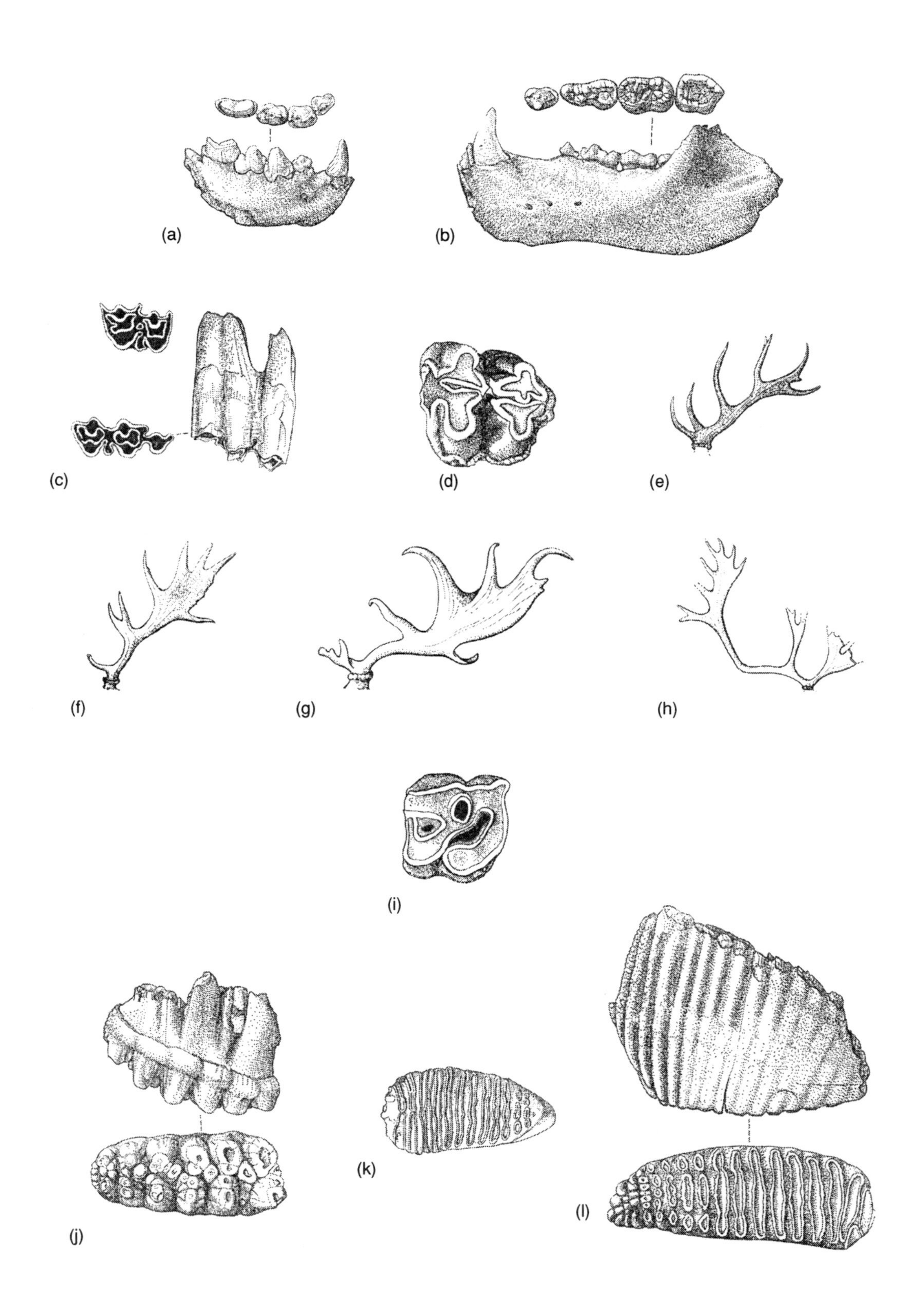
(a)
(b)
(c)
(d)
(e)
(f)
(g)
(h)
(i)
(j)
(k)
(l)

of Uffington' as Bronze Age (Renfrew and Bahn, 2000).

7.6.2 Environmental archaeology

As intimated in Chapter 3, the exacting ecological requirements and tolerances of many diatom, foraminiferan, plant, gastropod, ostracod, insect and mammal species, and their rapid response to changing environmental, and especially climatic, conditions, render them extremely useful in palaeoclimatology and climatostratigraphy, and in environmental archaeology (see also Haslett, 2002). Diatoms and foraminiferans are especially useful in sea-level reconstructions, using transfer functions (Denys and de Wolf, in Stoermer and Smol, 1999; Horton *et al.*, 1999; Juggins and Cameron, in Stoermer and Smol, 1999; Horton and Edwards, 2000; Dale and Dale, in Haslett, 2002; Gehrels, in Haslett, 2002; Horton and Edwards, in Olson and Leckie, 2003; Horton *et al.*, in Shennan and Andrews, 2004; Sejrup *et al.*, 2004).

Many species recorded in the Quaternary of the British Isles and elsewhere in northern Europe range through to the Recent, such that their present-day distributions can be used to infer the past climate. In the case of plants, detailed palaeoclimatic interpretations have been made of the well-preserved last interglacial to last glacial to post-glacial cycle of Great Britian (West, 2000). In the case of insects, even more detailed palaeoclimatic interpretations have been made (Fig. 7.46) (Coope, 1977; Elias, 1994). In the case of mammals, the occurrence of the reindeer *Rangifer tarandus* has been used to infer an Arctic climate during glacials; and that of the hippopotamus *Hippopotamus amphibius*, an Ethiopian climate during interglacials (Yalden, 1999; see also discussion on 'Boxgrove' below). An Arctic climate during glacials has also been inferred from the occurrences of the extinct but interpreted cold-adapted woolly mammoth *Mammuthus primigenius* and woolly rhinoceros *Coelodonta antiquitatis*. Note in this context that the the woolly rhinoceros *Coelodonta antiquitatis* is recorded in the Weichselian of Sourlie in western Scotland in association with plant and insect remains indicating mean annual temperatures of between −1 and +10 °C (Bos *et al.*, 2004). Note also that the small mammals from the Atelian of Cherni Yar in Astrakhan oblast in the Volga–Urals region occur in association with plant and insect remains indicating mean January temperatures of between −12 and −5 °C, mean July temperatures of between 21 and 23 °C, mean annual precipitation of 350–500 mm, and mean annual evaporation of 500–700 mm (Bidashko *et al.*, 1995). Environmental conditions were evidently similar to those obtaining on the northern steppe of the Volga–Don interfluve at the present time.

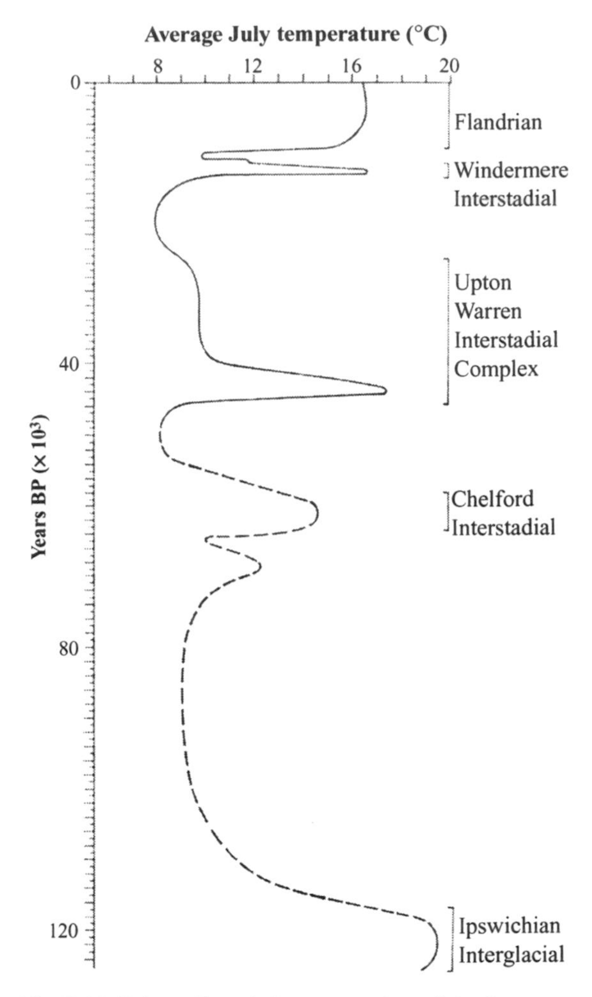

Fig. 7.46. Palaeoclimatic interpretation of the Late Pleistocene to Holocene of Great Britain by means of insects. (Reproduced with permission from Elias, 1994; in turn after Coope, 1977.)

Box 7.4 Palaeoenvironmental interpretation of the Pleistocene-Holocene of the British Isles, using proxy Recent benthic foraminiferal distribution data

Introduction

The palaeoenvironments of the Pleistocene to Sub-Recent Holocene of the British Isles have been interpreted using proxy data on the environmental – in particular the bathymetric and biogeographic – distributions of benthic foraminiferans from the Recent. The approach adopted has been qualitative and uniformitarian, that is to say, to assume that the empirically observed modern distributions are directly applicable to the interpretation of ancient environments.

Bathymetric distribution data

The bathymetric zones recognised are as follows:

Salt-marsh (locally divisible into high and low salt-marsh)
'Inner estuarine'
'Outer estuarine'
Marine.

Biogeographic distribution data

The biogeographic provinces recognised are as follows:

The Arctic/Sub-Arctic province (which includes the Arctic, the Norwegian Sea, Scandinavia and the Baltic), the southern boundary of which is approximately coincident with the 5 °C winter/10 °C summer isotherm

The (Atlantic) cool-temperate province (which includes the Iceland–Faeroe ridge, the Hatton–Rockall plateau, Rockall trough, west of Scotland, west of Ireland, the English Channel and the North Sea), the southern boundary of which is approximately coincident with the 10 °C winter/15 °C summer isotherm

The (Atlantic) warm-temperate province (which includes the Porcupine bank, the Bay of Biscay, Iberia and north-west Africa), the southern boundary of which is approximately coincident with the 15 °C winter/20 °C summer isotherm.

In the context of the Pleistocene–Holocene of the British Isles, northern cool-temperate influence has been inferred on the basis of the occurrence of the following foraminifera: *Astacolus hyalacrulus, Astrononion gallowayi, Bolivina pseudopunctata, B. skagerrakensis, Buccella tenerrima, Cassidulina norcrossi, C. reniformis, C. teretis sensu lato, Dentalina baggi, D. frobisherensis, D. ittai, Elphidiella arctica, Elphidium clavatum* (common to abundant), *E. hallandense, Esosyrinx curta, Laryngosigma hyalascidia, Nonion orbicularis, Nonionellina labradorica, Oolina apiopleura, O. borealis, Parafissurina fusuliformis, Pseudopolymorphina novangliae, Pullenia osloensis, Pyrgo williamsoni, Quinqueloculina agglutinata, Q. arctica, Robertinoides pumilum, R. suecicum, Stainforthia feylingi, S. loeblichi, Trichohyalus bartletti, Trifarina fluens* and *Triloculina trihedral*; Sub-Arctic influence on the basis of the occurrence of: *Elphidiella groenlandica, Elphidium asklundi, E. bartletti, Fissurina danica, F. serrata, Globobulimina auriculata* sspp., *Guttulina glacialis* and *Quinqueloculina stalkeri*; and Arctic influence on the basis of the occurrence of: *Cibicides grossa, Elphidiella nitida, Gordiospira arctica, Lagena flatulenta, L. parri, Laryngosigma hyalascidia* and *Oolina scalariformissulcata*, whose (essential) ranges are to the north, or which occur only in deep water to the south.

Southern cool-temperate influence has been inferred on the basis of the occurrence of: *Ammonia aberdoveyensis, A. falsobeccarii, A. flevensis, A. limnetes, A. tepida, Aubignyna perlucida, Bulimina elongata, Cornuspira selseyensis, Elphidium incertum, Fissurina haynesi, Gaudryina rudis, Lagena perlucida, Massilina secans, Nonionella* sp. A, *Oolina heronalleni, Ophthalmidium balkwilli, Quinqueloculina dimidiata, Q. intricata, Q. (?) oblonga, Rosalina* cf. *bradyi, R. milletti, Rotaliella chasteri, Siphonina georgiana, Spiropthalmidium emaciatum* and *Trifarina bradyi*; warm-temperate influence on the basis of the occurrence of: *Elphidium* cf. *advenum* and *E. earlandi*, whose (essential) ranges are to the south.

Results

Early Pleistocene: Pastonian

The Pastonian Paston member of the Cromer forest-bed formation (Norwich Crag formation) of Paston in Norfolk – the stratotype locality – is developed in non-marine to marginal marine facies, and contains no benthic foraminiferans. The slightly older Pastonian or pre-Pastonian Sidestrand member of the Norwich Crag formation of Weybourne Hope in Norfolk, formerly known as the 'Weybourne Crag', is characterised by a Sub-Arctic benthic foraminiferal assemblage.

Middle Pleistocene: Beestonian

The Beestonian Runton member of the Cromer forest-bed formation of West Runton on the coast of north-east Norfolk – the stratotype locality – and adjacent areas is developed in non-marine to marginal marine facies, and contains no benthic foraminiferans.

Cromerian, oxygen isotope stages (OISs) 17–13

The Cromerian West Runton member of the Cromer forest-bed formation of West Runton on the coast of north-east Norfolk – the stratotype locality – and

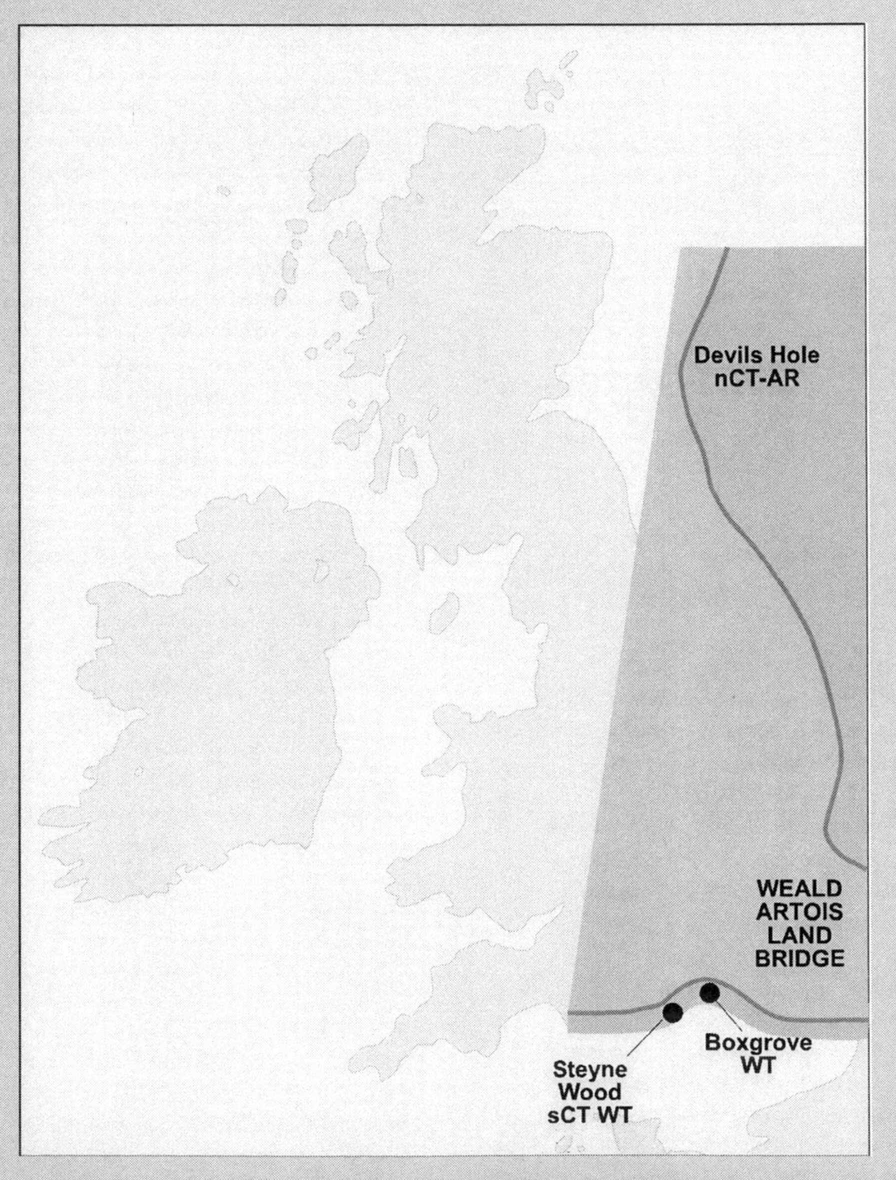

Fig. 7.47. Cromerian (oxygen isotope stage (OIS)13) palaeogeography. AR, Arctic; SA, Sub-Arctic; nCT, northern Cool-Temperate; sCT, southern Warm- Temperate; WT, Warm-Temperate.

adjacent areas is developed in non-marine to marginal marine facies, and contains no benthic foraminiferans (Fig. 7.47). The undifferentiated Cromerian of the Devil's Hole in the North Sea is characterised by a northern cool-temperate to Arctic benthic foraminiferal assemblage.

Suffolk

The Cromerian, OIS17 Cromer forest-bed formation of Pakefield in Suffolk is characterised by abundant *Elphidium asklundi*. Sub-Arctic influence could be inferred on the basis of the occurrence. However, since *Elphidium asklundi* is a marine species and the remainder of the assemblage is of brackish/estuarine, but biogeographically indeterminate, aspect, it is regarded as reworked.

Norfolk

The arguably at least in part correlative, although not as yet positively dated, 'Cromer forest-bed formation' of Norton Subcourse quarry in Norfolk also contains *Elphidium asklundi* (?again reworked), together with warm-temperate benthic foraminiferans. Note, incidentally, that the Norton Subcourse quarry section contains reworked foraminiferans of mixed, including probable Jurassic, and Early Cretaceous, rather than predominantly Late Cretaceous provenance, suggesting some sort of correlation with the comparatively flint-poor Kesgrave or Bytham sands and gravels.

Bembridge, Isle of Wight

The Cromerian, OIS13 Steyne Wood member of the Solent formation of Bembridge is characterised by a southern cool-temperate to warm-temperate assemblage.

Sussex

The Cromerian, OIS13 Slindon sand member of the West Sussex Coast formation – or Slindon formation of the West Sussex Coast group – of Boxgrove and Valdoe is characterised by a generally warm-temperate assemblage also containing rare cool-water species. It is worthy of note that the overlying Slindon silt member is exceptionally rich in vertebrate remains, including the oldest hominid remains in Britain and associated artefacts (see below).

Anglian, OIS12

The Anglian Lowestoft formation of Lowestoft in Suffolk – the stratotype locality – and adjacent areas is developed in glacial, glacio-fluvial and glacio-marine facies.

Suffolk

The Anglian Corton member of the North Sea Drift formation of Corton is characterised by a northern cool-temperate to Arctic benthic foraminiferal assemblage. The late Anglian Woolpit beds of Woolpit near Bury St Edmunds are also characterised by a northern cool-temperate to Arctic assemblage.

Norfolk

The Anglian Leet Hill member of the North Sea Drift formation of Leet Hill is barren of *in situ* benthic foraminiferans.

Hoxnian, OIS11

The Hoxnian Hoxne formation of Hoxne in Suffolk – the stratotype locality – and adjacent areas is developed in non-marine to marginal marine facies, and contains no benthic foraminiferans (Fig. 7.48).

Sussex

The Hoxnian of Earnley, Bracklesham Bay is characterised by an undifferentiated warm-temperate to Sub-Arctic benthic foraminiferal asemblage.

Inner Silver Pit, North Sea

The Hoxnian of the Inner Silver Pit is characterised by a generally Sub-Arctic to Arctic benthic foraminiferal assemblage also containing rare warm-water species.

Saalian equivalent, OISs 10–6

The Saalian equivalent, which is interpreted as including the 'Gippingian' and 'Wolstonian' of various localities in East Anglia, is developed in a range of non-marine to marine facies, and contains no benthic foraminiferans. The stratotypical Wolstonian of Wolston in the East Midlands is

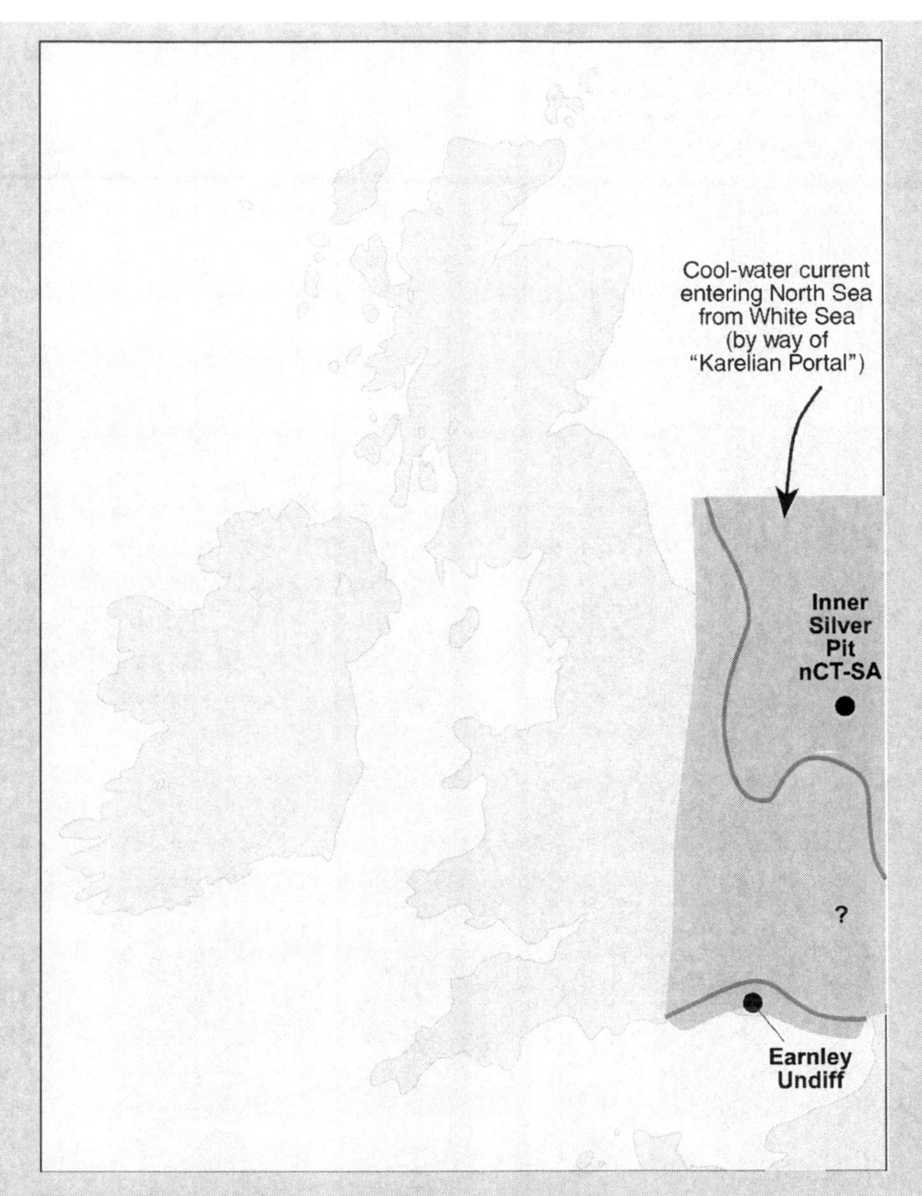

Fig. 7.48. Hoxnian (OIS11) palaeogeography. See Fig. 7.47 for key.

widely, although not universally, now regarded as correlative with the Anglian.

OIS9 (Fig. 7.49)

Sussex

The Saalian equivalent Aldingbourne member of the West Sussex Coast formation of Aldingbourne and Tangmere is characterised by a southern cool-temperate to warm-temperate benthic foraminiferal assemblage.

Essex

The Saalian equivalent of the Cudmore Grove channel, East Mersea is characterised by an undifferentiated cool-temperate to Arctic assemblage.

Norfolk

The Saalian equivalent Nar Valley clay of the Nar member of the Nar Valley formation of the East

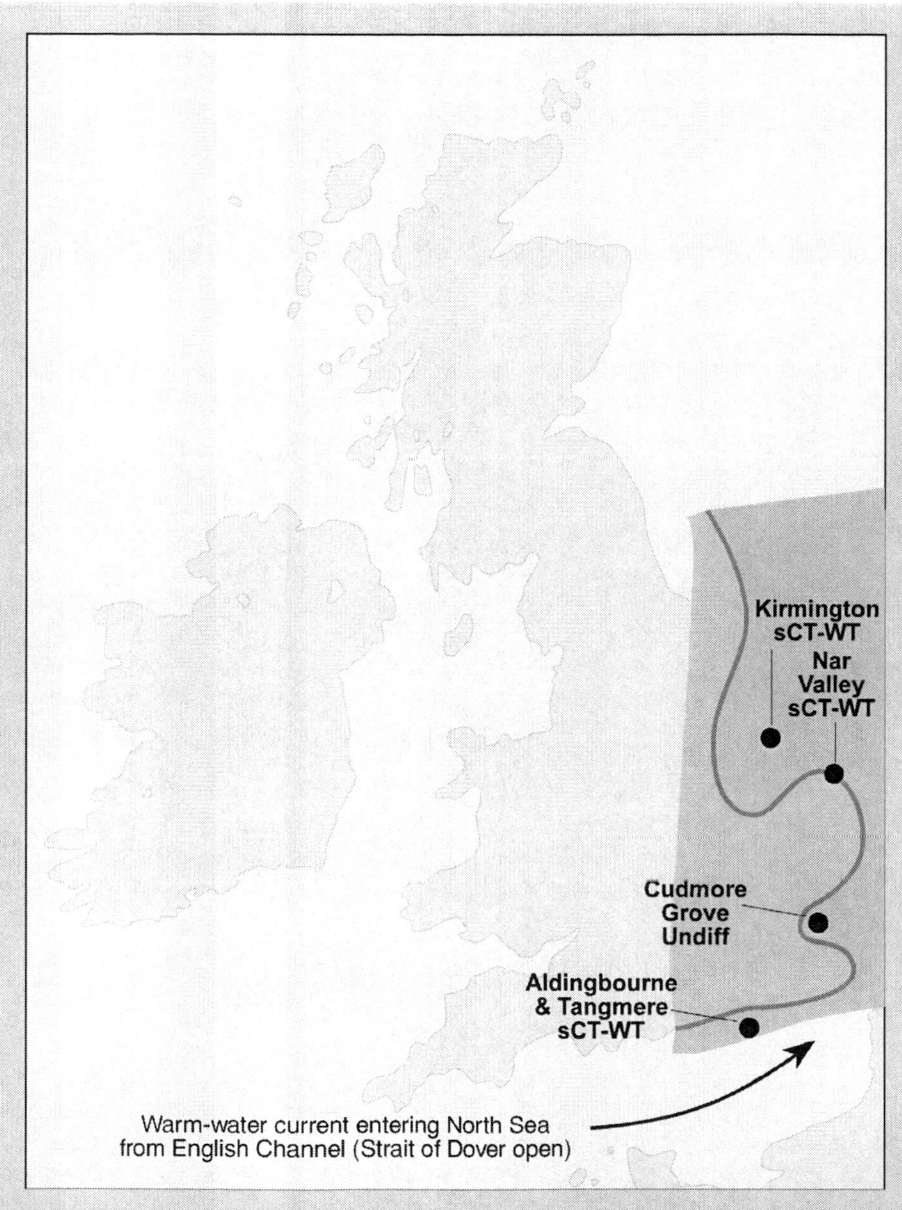

Fig. 7.49. Saalian equivalent (OIS9) palaeogeography. See Fig. 7.47 for key.

Winch No. 1 Borehole and of a trench near Tottenhill – both located close to the stratotype locality elsewhere in the Nar Valley – is characterised by a generally southern cool-temperate to warm-temperate assemblage also containing rare cool-water species.

Kirmington, Lincolnshire

The Saalian equivalent Kirmington formation of Kirmington is also characterised by a generally southern cool-temperate to warm-temperate assemblage also containing rare cool-water species.

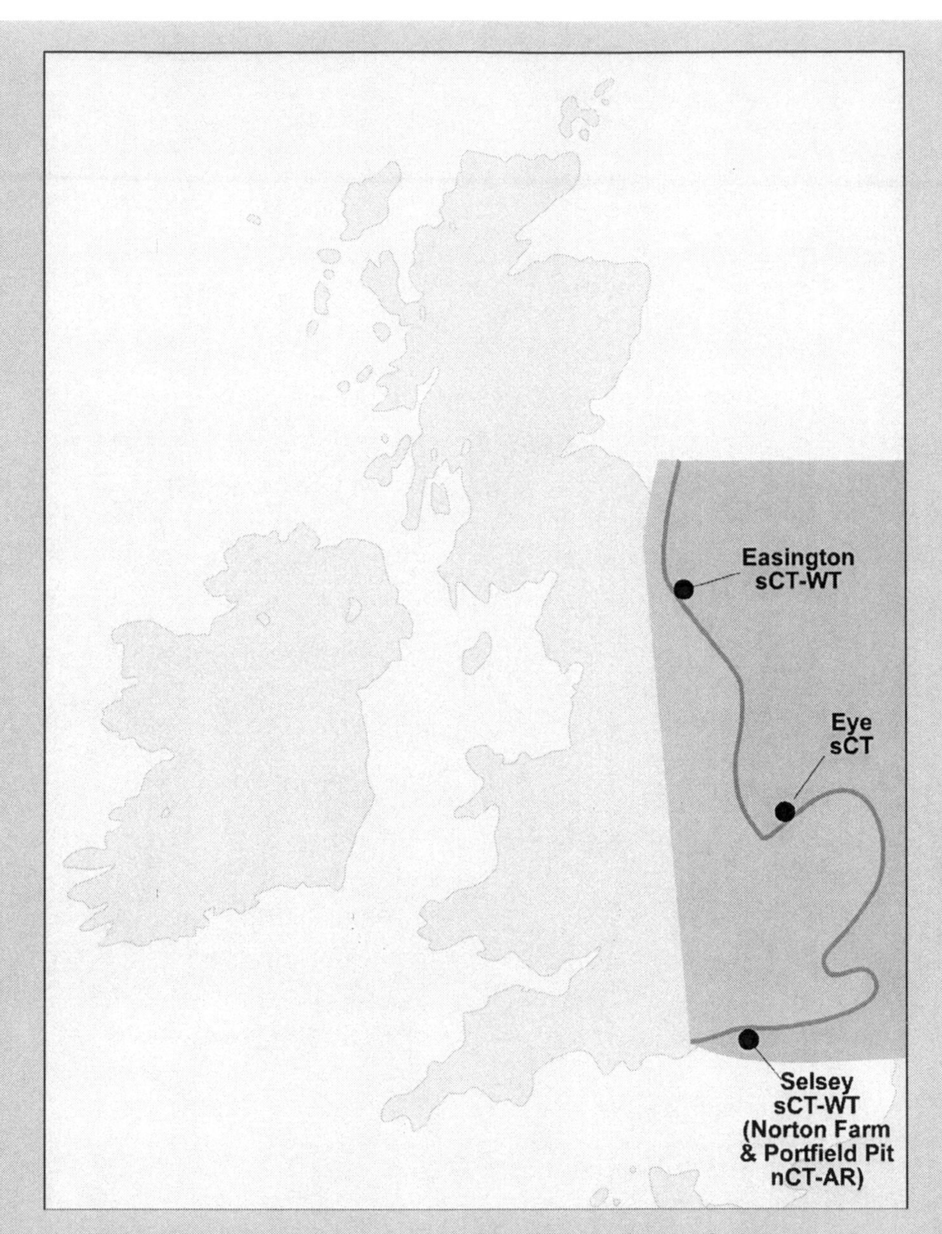

Fig. 7.50. Saalian equivalent (OIS7) palaeogeography. See Fig. 7.47 for key.

OIS7 (Fig. 7.50)

Sussex

The Saalian equivalent Lifeboat Station member of the West Sussex coast formation of the Lifeboat Station channel, Selsey is characterised by a generally southern cool-temperate to warm-temperate assemblage also containing rare cool-water species. The Saalian equivalent 'marine clay' of West Street, Selsey is characterised by a southern cool-temperate to warm-temperate assemblage. The Saalian equivalent Norton member of the West Sussex coast formation of Norton Farm and Portfield Pit is characterised by a generally northern cool-temperate or Sub-Arctic to Arctic assemblage also containing rare warm-water species. Incidentally, the fact that cool-water, northern cool-temperate to Arctic species are present in the older, shallow marine Norton sand as well as the younger, marginal marine Norton silt indicates that – ?seasonal, or long-term seasonal – climatic cooling preceded rather than accompanied glacioeustatic shallowing. Climatic cooling also appears to have preceded shallowing in the Saalian of Tancarville on the Seine estuary in France. The Saalian equivalent Black Rock member of the West Sussex coast formation of Brighton and of equivalent unnamed sediments exposed at Yeoman's Road near Worthing are characterised by a northern cool-temperate to Arctic assemblage.

Cambridgeshire

The Saalian equivalent March gravels of Northam pit, Eye is characterised by a southern cool-temperate assemblage.

Easington, Co. Durham

The Saalian equivalent Easington formation of Easington is characterised by a southern cool-temperate to warm-temperate assemblage.

Late Pleistocene: Ipswichian, OIS5

The Ipswichian Bobbitshole formation of Bobbits Hole near Ipswich in Suffolk – the stratotype locality – and adjacent areas is developed in non-marine to marginal marine facies, and contains no benthic foraminiferans (Fig. 7.51).

Cardigan Bay

The Ipswichian of Cardigan Bay is characterised by a generally warm-temperate benthic foraminiferal assemblage also containing rare cool-water species.

Somerset

The Ipswichian of the Middlezoy member of the Burtle formation of Greylake on the Somerset Levels is also characterised by a generally warm-temperate benthic containing rare cool-water species.

Devon

The Ipswichian 'blown sand' of the Thatcher Stone member of the Torbay formation of Hope's Nose and Thatcher is characterised by a warm-temperate assemblage.

Portland, Dorset

The Ipswichian of the Portland East member of the Bill of Portland formation of Portland is characterised by a southern cool-temperate to warm-temperate assemblage.

East Yorkshire

The Ipswichian Raincliff formation – formerly the 'Speeton shell bed' – of Speeton is characterised by an undifferentiated warm-temperate to Sub-Arctic assemblage.

Devensian, OISs 4–2

The Devensian Four Ashes formation in Shropshire – the stratotype locality – is developed in non-marine facies, and contains no benthic foraminiferans (Fig. 7.52).

Hebridean Shelf

The late Devensian of the Hebridean shelf is characterised by a Sub-Arctic to Arctic assemblage.

Argyll

The late Devensian of the Ardyne formation of Ardyne and Lochgilphead is characterised by an Arctic assemblage.

Sistrakeel, Co. Londonderry

The late Devensian Sistrakeel formation of Sistrakeel, Co. Londonderry is characterised by a northern cool-temperate to Arctic assemblage.

Corvish, Co. Donegal

The late Devensian of Corvish, Co. Donegal is characterised by a generally Sub-Arctic to Arctic assemblage also containing rare warm-water species.

Belderg and Glenulra, Co. Mayo

The late Devensian Belderg formation of Belderg, Co. Mayo and the Devensian of Glenulra, Co. Mayo are

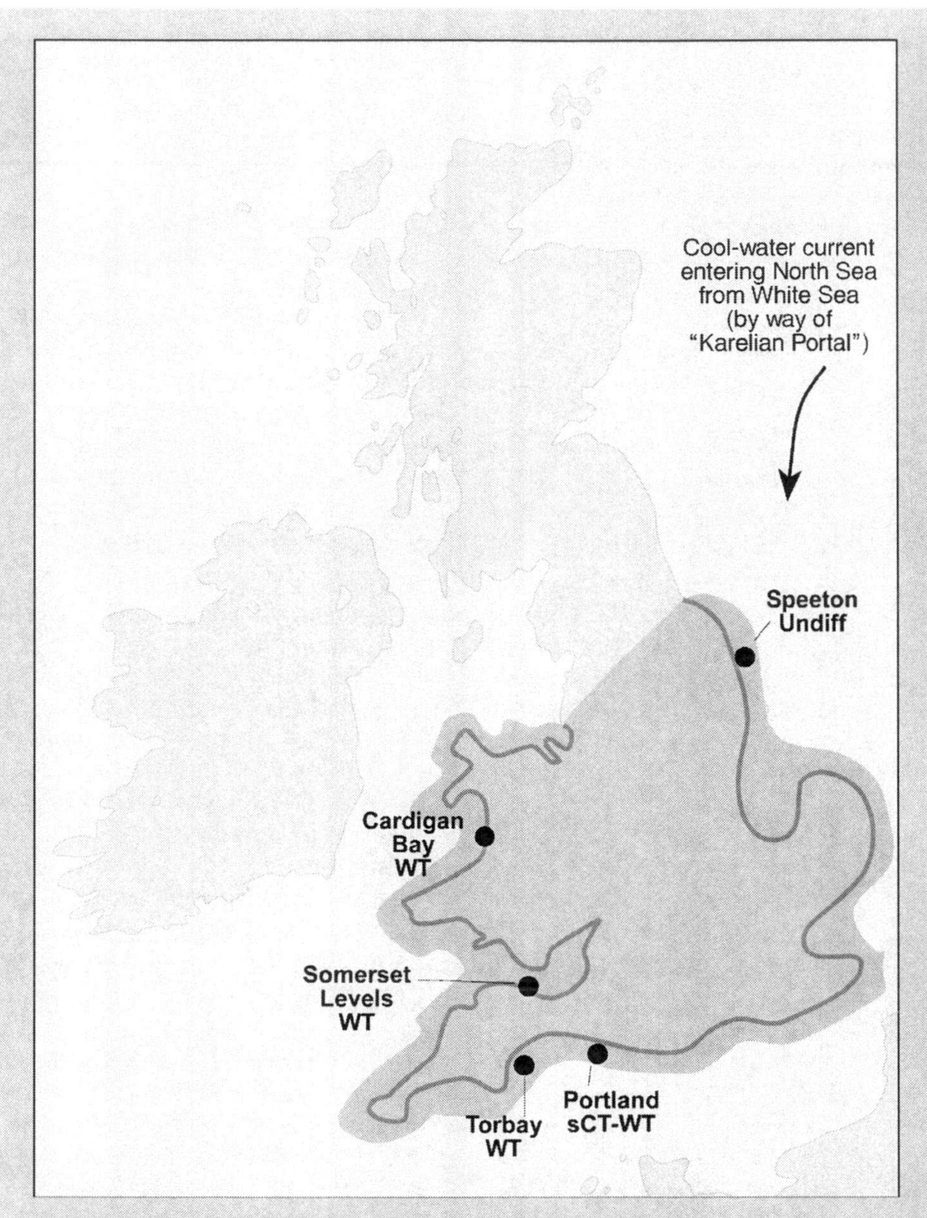

Fig. 7.51. Ipswichian (OIS5e) palaeogeography. See Fig. 7.47 for key.

also characterised by a generally Sub-Arctic to Arctic assemblage containing rare warm-water species.

Eastern Ireland

The late Devensian Cooley Farm member of the Louth formation of the margins of Dundalk Bay, Co. Louth, Derryogue member of the Mourne formation of the Mourne region, and Killard Point formation of the drumlin fields around Killard Point in Co. Down are all characterised by a northern cool-temperate to Arctic assemblage.

Isle of Man

The late Devensian Dog Mills member of the Orrisdale formation of Dog Mills on the Isle of Man is characterised by a generally Arctic assemblage also

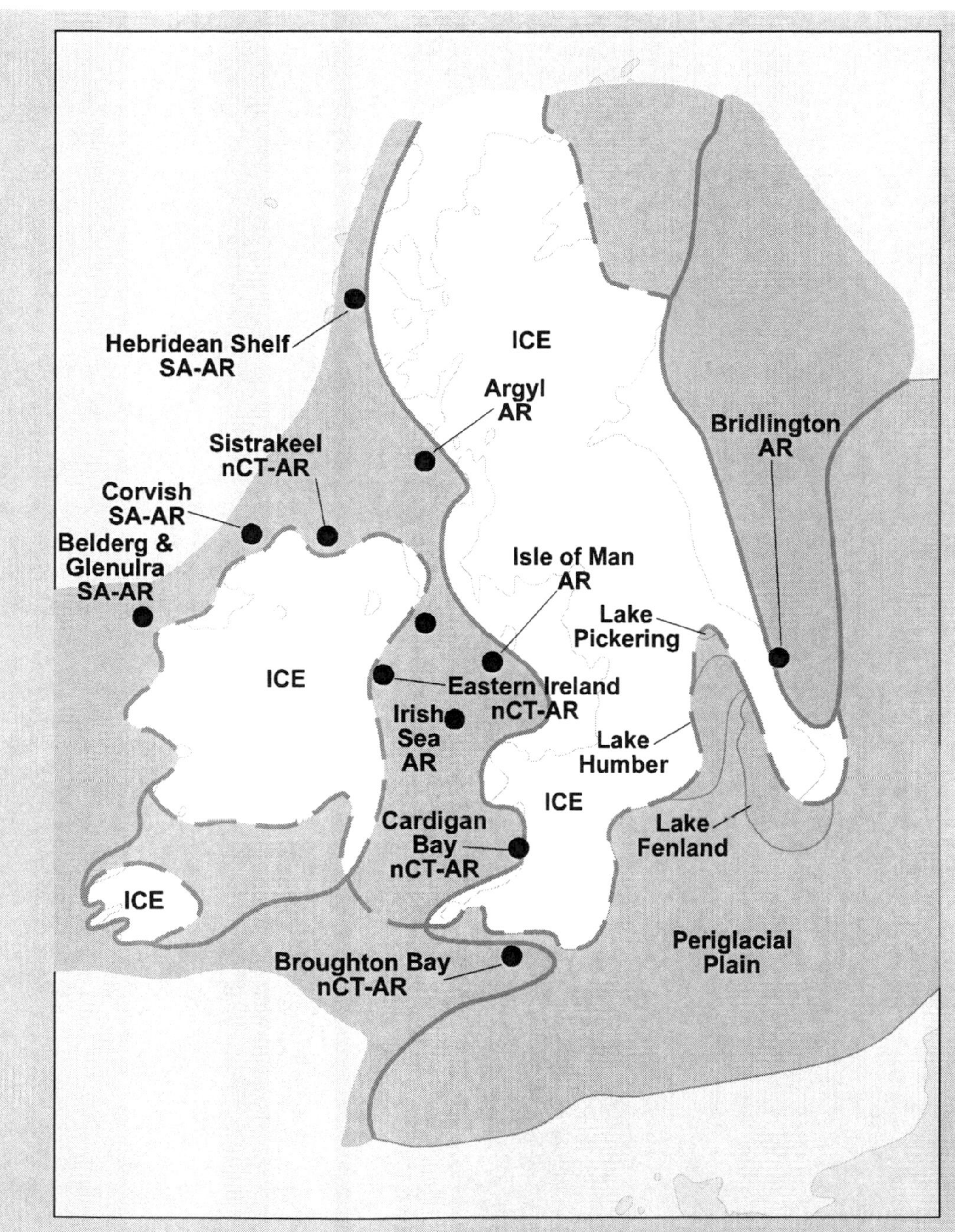

Fig. 7.52. Devensian (OIS4–2) palaeogeography, showing Late Devensian shore line. See Fig. 7.47 for key.

containing rare warm-water species, interpreted as probably reworked from the Ipswichian.

Irish Sea

The late Devensian of the ?Western Irish Sea formation of the Irish Sea is also characterised by a generally Arctic assemblage containing rare warm-water species, probably reworked from the Ipswichian.

Cardigan Bay

The late Devensian late glacial of Cardigan Bay is characterised by a generally northern cool-temperate or Sub-Arctic to Arctic assemblage also containing rare warm-water species, interpreted as probably reworked from the Ipswichian.

Glamorgan

The late Devensian of Broughton Bay is also characterised by a generally northern cool-temperate to Arctic assemblage containing rare warm-water species, probably reworked from the Ipswichian.

East Yorkshire

The late Devensian of the Bridlington member of the Holderness formation – formerly the 'Bridlington Crag' or 'basement till of Holderness' – of Dimlington near Bridlington is characterised by a generally Arctic assemblage also containing rare warm-water species, interpreted as probably reworked from the Ipswichian.

Holocene: Flandrian, OIS1

Hebridean Shelf

The Flandrian of the Hebridean shelf is characterised by a northern cool-temperate to Arctic benthic foraminiferal assemblage.

Cardigan Bay

The Flandrian of the Dovey marshes and of Cardigan Bay is characterised by a mixed assemblage containing indications of both northern cool-temperate to Arctic and southern cool-temperate to warm-temperate influence.

Devon

The Flandrian of Start Bay is characterised by a southern cool-temperate to warm-temperate assemblage.

Hampshire

The Flandrian of Fawley (Southampton Water) is also characterised by a southern cool-temperate to warm-temperate assemblage.

Suffolk

The Flandrian of the Breydon formation is also characterised by a southern cool-temperate to warm-temperate assemblage.

Northumberland

The Flandrian of the Warkworth borehole is also characterised by a southern cool-temperate to warm-temperate influenced assemblage.

Modified after Jones and Whittaker (in press).

7.6.3 Case histories in archaeology

Westbury Cave, Somerset (Palaeolithic, 600 000 years BP)

The ancient cave exposed in the quarry above the village of Westbury-sub-Mendip in Somerset is renowned for having yielded evidence, in the form of flints and cut marks on animal bone, of archaic human occupation of Britain, dating back to the early Middle Pleistocene, Cromerian interglacial, approximately 600 000 years BP, to the Early Palaeolithic (Bishop, 1975; Wymer, 1988; Barton, 1997). Incidentally, the immediately surrounding area is steeped in archaeology. There is a later Palaeolithic cave site at 'Hyaena Den', near Wookey Hole, and an important Neolithic cave site in Cheddar Gorge, where the skeleton of 'Cheddar man' was discovered. Neolithic artefacts, trackways, settlements, burial barrows and henges abound in the Somerset Levels around Glastonbury, where there are also the remains of a Bronze Age lake village.

Recent work has indicated that the flints from Westbury Cave might in fact be ecofacts – or 'incertofacts' – rather than worked artefacts, and might have been introduced naturally (Cook, in Andrews *et al.*, 1999; Schreve *et al.*, in Andrews *et al.*, 1999).

Associated animal remains are also for the most part thought to have been introduced into the cave by natural processes such as water transport, mud flow or surface collapse (Andrews and Ghaleb, in Andrews *et al.*, 1999; Andrews and Stringer, in Andrews *et al.*, 1999). Note, though, that certain of the animal remains appear to have been introduced by bears actually living in the cave, or, as droppings, by birds of prey (Andrews and Ghaleb, in Andrews *et al.*, 1999; Andrews and Stringer, in Andrews *et al.*, 1999). The remains include those of amphibians and reptiles (Andrews and Stringer, in Andrews *et al.*, 1999), small mammals (Andrews and Stringer, in Andrews *et al.*, 1999; Currant, in Andrews *et al.*, 1999), large ruminant herbivores (Gentry, in Andrews *et al.*, 1999) and large carnivores (Turner, in Andrews *et al.*, 1999).

Importantly, evidence of butchery by archaic humans has been found in the cave (Andrews and Ghaleb, in Andrews *et al.*, 1999). The evidence is in the form of cut marks, although only on a single bone of the more than 5000 examined, a metacarpal of the red deer *Cervus elaphus*.

Interestingly, the unit containing the evidence of butchery by archaic humans, Unit 19, is characterised by a Boreal fauna containing the Norway lemming *Lemmus lemmus*, the common vole *Microtus arvalis* and the ancestral narrow-headed or Siberian vole *Microtus (Stenocranius) gregaloides*, while, in contrast, most of the older units in the cave are characterised by warmer, temperate faunas (Andrews and Stringer, in Andrews *et al.*, 1999; Preece and Parfitt, in Lewis *et al.*, 2000).

It is probable that Unit 19 represents a late part of the Cromerian interglacial (Currant, in Andrews *et al.*, 1999). Note in this context that the youngest occurrence of *Microtus (Stenocranius) gregaloides* has been correlated with the Cromerian III, OIS15 in north-west Europe (Preece and Parfitt, in Lewis *et al.*, 2000). Note also, though, that the youngest occurrence of *Microtus (Stenocranius) gregaloides* has been tentatively correlated with the pre-Cromerian, OIS22, and that the oldest occurrence of *Microtus arvalis* has been tentatively correlated with the Cromerian, OIS13, at the Gran Dolina site in Atapuerca in Spain (Antonanzas and Bescos, 2002), although these tentative correlations may not be tenable (S. A. Parfitt, the Natural History Museum, personal communication).

Boxgrove, Sussex (Palaeolithic, 500 000 years BP)

The raised beach at Boxgrove on the West Sussex coastal plain has also yielded evidence of archaic human occupation of Britain dating back to the early Middle Pleistocene, or Early Palaeolithic, although apparently to a younger stratigraphic level than that represented by the Westbury site (Pitts and Roberts, 1997; Roberts, in Roberts and Parfitt, 1999; Stringer, in Roberts and Parfitt, 1999; Stringer and Trinkaus, in Roberts and Parfitt, 1999; Prior, 2003). The evidence is in the form of *Homo heidelbergensis* remains – a tibia and a tooth – and associated Acheulian flint hand-axes and other artefacts.

The human remains and associated artefacts are concentrated on an Early Palaeolithic foreshore occupation surface in the Slindon silt member of the Slindon formation of the West Sussex Coast group (Parfitt, in Roberts and Parfitt, 1999; Roberts, in Roberts and Parfitt, 1999). Note, though, that artefacts also occur in the underlying Slindon sand member of the Slindon formation, and in the overlying Eartham lower and upper gravel members of the Eartham formation.

A range of terrestrial and freshwater animal remains have been found in association with the human remains and artefacts, principally in the Slindon silt member (Parfitt, in Parfitt and Roberts, 1999; Roberts, in Roberts and Parfitt, 1999). These include terrestrial and freshwater molluscs, freshwater fish, amphibians and reptiles, birds, and mammals (Parfitt, in Parfitt and Roberts, 1999). The mammalian fauna includes insectivores, chiropterans, carnivores, proboscideans, perissodactyls, artiodactyls, rodents and lagomorphs as well as *Homo heidelbergensis* (Parfitt, in Parfitt and Roberts, 1999). Importantly, the molluscan fauna is indicative of a land surface for the most part covered with vegetation, and dotted with freshwater pools (Preece and Bates, in Roberts and Parfitt, 1999). The key environmental indicators are *Pupilla muscorum* and *Vallonia* spp., which indicate open country; *Lymnaea truncatula* and *Succinea oblonga*, which indicate some

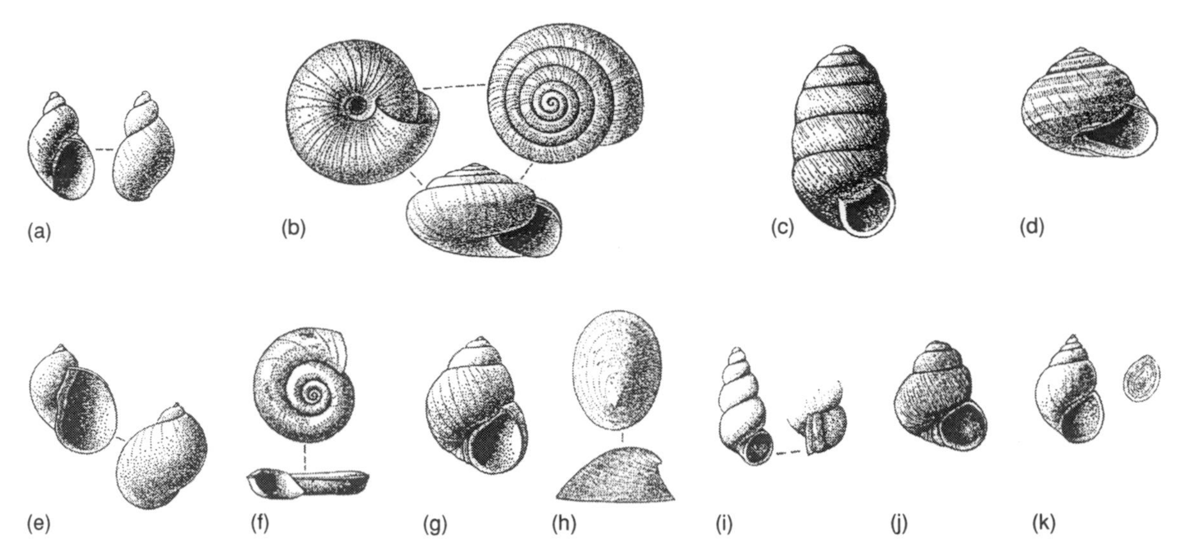

Fig. 7.53. Some important Quaternary gastropods. Terrestrial gastropods: (a) *Succinea oblonga* (the amber snail), Pleistocene, Devensian, Tottenham, ×3; (b) *Hygromia hispida* (the bristly snail), Holocene, Halling, Kent, ×3; (c) *Pupilla muscorum*, Pleistocene, Devensian, Ponders End, Middlesex, ×10; (d) *Cepaea nemoralis*, Pleistocene, ?Hoxnian, Ilford, Essex, ×1. Freshwater gastropods: (e) *Lymnaea peregra* (the pond snail), Pleistocene, ?Hoxnian, Ilford, Essex, ×1; (f) *Planorbis planorbis* (the trumpet snail), Pleistocene, West Wittering, Sussex, ×1.5; (g) *Viviparus diluvianus* (the river snail), Pleistocene, ?Hoxnian, Swanscombe, Kent, ×1; (h) *Ancylus fluviatilis* (the river limpet), Pleistocene, Clacton, Essex, ×2; (i) *Belgrandia marginata*, Pleistocene, Clacton, Essex, ×9; (j) *Valvata antiqua*, Pleistocene, ?Hoxnian, Swanscombe, Kent, ×2.5; (k) *Bithynia tentaculata* (shell and operculum), Pleistocene, ?Hoxnian, Swanscombe, Kent, ×2. (Modified after the Natural History Museum, 2000c.)

vegetation; *Aegopinella* and clausiliids, which indicate vegetation; *Spermodea lamellata* and *Acanthinula aculeata*, which indicate deeply shaded and moist hollows; and *Anisus leucostoma, Lymnaea peregra* and *Pisidium* spp., which indicate fresh water. Some of these and some other palaeoenvironmentally important terrestrial and freshwater gastropods from the the British Isles are shown in Fig. 7.53. The remains of marine as well as terrestrial and freshwater organisms have also been found at Boxgrove, principally in the Slindon sand member (Parfitt, in Roberts and Parfitt, 1999). These include foraminiferans and ostracods (Whittaker, in Roberts and Parfitt, 1999; Whittaker *et al.*, 2003; Jones and Whittaker, in press), dinoflagellates, molluscs and fish.

Clear evidence of butchery by humans of the extinct cave bear *Ursus deningeri*, the red deer *Cervus elaphus*, the extinct giant deer *Megaloceros* sp., the wild horse or tarpan *Equus ferus*, the extinct rhinoceros *Stephanorhinus hundsheimensis* and *Bison* sp. has been found at Boxgrove (Parfitt and Roberts, in Roberts and Parfitt, 1999). From impact as well as cut marks, it would appear that these animals were butchered for their marrowbone fat as well as their meat and skins. Significantly, in at least three cases, cut marks associated with human butchery clearly pre-date gnaw marks associated with non-human carnivory, indicating that humans got to the animal carcasses before hyenas and other carnivores. Whether the humans actually hunted and killed animals, or simply scavenged them, remains unknown. Note, though, that the arguable existence of a puncture wound in the scapula of a wild horse provides possible evidence of hunting. Interestingly, the technology employed in the manufacture of the flint hand-axes and other tools used in the butchery has been demonstrated by experimental archaeological techniques to have involved knapping with soft – bone and antler – as well as hard hammers (Wenban-Smith, in Roberts and Parfitt, 1999).

The age of the Slindon formation of Boxgrove has been estimated by a variety of methods, including uranium series dating, luminescence dating of brick-earth, optically stimulated luminescence, electron spin resonance and coupled ESR–uranium-series dating, amino-acid geochronology, and magnetostratigraphy, as well as conventional biostratigraphy. Conventional biostratigraphic dating was in turn undertaken employing calcareous nannoplankton and mammals. Overall, age estimations are in the range of approximately 175 000–525 000 years BP, or OISs 6–13 (Roberts and Parfitt, in Roberts and Parfitt, 1999). Those that are regarded as most reliable, namely those derived from amino-acid geochronology and conventional biostratigraphy, are in the range of approximately 350 000–525 000 years BP, or OISs 11–13. Amino-acid geochronology and calcareous nannoplankton biostratigraphy indicate approximately 350 000–425 000 years BP, or OIS11, or Hoxnian, while mammalian biostratigraphy indicates approximately 475 000–525 000 years BP, or OIS13, or Cromerian. The mammalian age estimation of approximately 525 000 years BP, or OIS13, or Cromerian, is preferred, because the amino acid and calcareous nannoplankton age estimations are difficult to reconcile with regional observations, and because the latter is based not on positive but on negative evidence (Parfitt, in Roberts and Parfitt, 1999; Roberts and Parfitt, in Roberts and Parfitt, 1999; Bates, in Wenban-Smith and Hosfield, 2001; Bridgland, in Wenban-Smith and Hosfield, 2001). Evidence of a Cromerian age is provided by the presence of the extinct shrews *Sorex runtonensis* and *S. (Drepanosprex) savini*, the vole *Pliomys episcopalis*, the cave bear *Ursus deningeri*, the rhinoceros *Stephanorhinus hundsheimensis* and the giant deer *Megaloceros dawkinsi* and *M. verticornis*. All of the aforementioned species are recorded from the Cromerian, none from the Hoxnian. Evidence of a Cromerian IV, OIS13 rather than Cromerian III, OIS15 age is provided by the presence of the narrow-headed or Siberian vole *Microtus (Stenocranius) gregalis* (Parfitt, in Roberts and Parfitt, 1999; Preece and Parfitt, in Lewis *et al.*, 2000; S. A. Parfitt, the Natural History Museum, personal communication). Incidentally, this species survives today not only in the Siberian and Alaskan Arctic but also in the steppes of central Asia (Fig. 7.54) (Yalden, 1999).

Palaeoclimatic evidence from the contained fossils supports the interpretation that the Slindon formation of Boxgrove represents the latest part of the Cromerian interglacial (Parfitt, in Roberts and Parfitt, 1999; Roberts and Parfitt, in Roberts and Parfitt, 1999; Jones and Whittaker, in press; see also above). The Eartham formation apparently represents the Anglian glacial.

Massawa, Eritrea (Palaeolithic, 125 000 years BP)

Outcrops of Abdur reef limestone on raised beaches near the port of Massawa on the Red Sea coast of Eritrea in the Horn of Africa have yielded evidence of – by inference – 'anatomically modern' human occupation of the area, dating back to the last interglacial, OIS5e, 125 000 years BP, to the Middle Palaeolithic, or, as it is known in Africa, the 'Middle Stone Age' (MSA) (Bruggemann *et al.*, 2004). The evidence is in the form of Acheulian flint bifacial hand-axes and cores, and MSA obsidian blades and flakes. The Acheulian artefacts are associated with transgressive gravels and high-stand reefal limestones, and oyster beds and shell middens. The MSA artefacts are associated with slightly later nearshore and beach facies, and both shellfish and large mammal bones. The association of both the Acheulian and MSA artefacts with shellfish is an indication of the exploitation of marine food resources as well as those to be found around the lakes and landscapes of east Africa. The exploitation of marine food resources has been interpreted as necessitated by the drying out of the inland watering holes, and the dying out of the game on the plains, during the penultimate glacial between 170 000 and 130 000 years BP (Oppenheimer, 2003). Incidentally, it may have been similar environmental stress associated with the last glacial that finally drove the successful dispersal of *Homo sapiens* out of Africa, and along the 'beachcomber route' to Eurasia (Oppenheimer, 2003; see also Subsection 5.5.8).

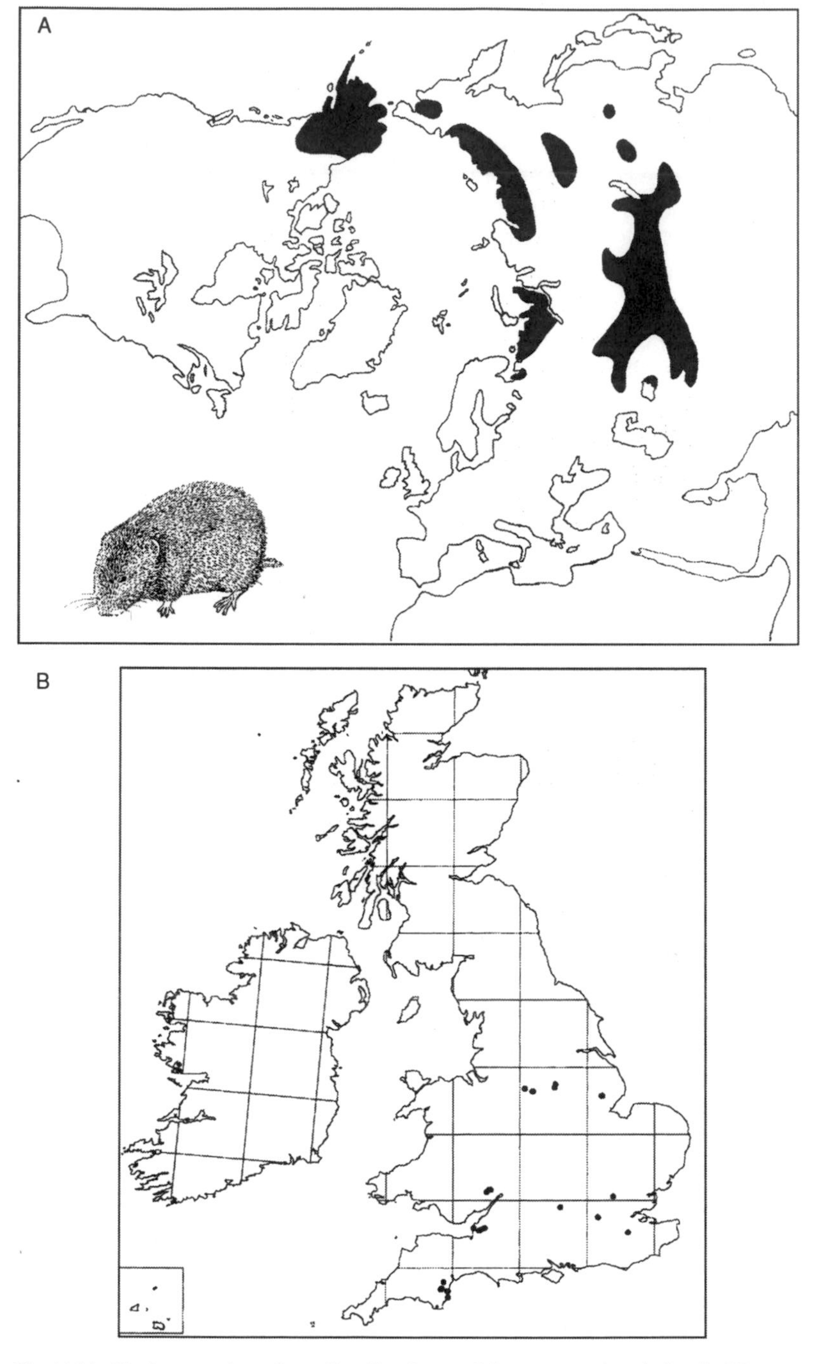

Fig. 7.54. Modern and ancient distributions of the narrow-headed vole *Microtus (Stenocranius) gregalis*. (A) Modern distribution; (B) ancient distribution in Great Britain. (Reproduced with permission from Yalden, 1999.)

Goat's Hole, Paviland, Gower (Palaeolithic, 26 000 years BP)

Famously, in 1822 or 1823, the Rev. William Buckland discovered an apparently ceremonially buried 'anatomically modern' *Homo sapiens* skeleton in Goat's Hole in Paviland in the Gower Peninsula in south Wales, dating back to the Late Palaeolithic (Swainston and Brookes, in Aldhouse-Green, 2000; Trinkaus and Holliday, in Aldhouse-Green, 2000; Prior, 2003). The skeleton came to be popularly known as the 'Red Lady', although it eventually turned out to be male. A recent archaeological survey undertaken in the late 1990s revealed that excavations in the nineteenth and early twentieth centuries had essentially removed the unit that yielded the Late Palaeolithic skeleton, leaving only older Palaeolithic units, confirmed as such by uranium-series and thermoluminescence dating (Aldhouse-Green, in Aldhouse-Green, 2000; Bowen *et al.*, in Aldhouse-Green, 2000). As an adjunct to the recent survey, excavations were undertaken in adjacent Hound's Hole and Foxhole Cave. The excavation at Hound's Hole yielded a late glacial fauna and Palaeolithic finds. The partial excavation at Foxhole Cave yielded Neolithic human remains and finds in 'modern topsoil layer 1'; a post-glacial fauna, a human tooth dated to the Mesolithic, Mesolithic artefacts, and evidence of Mesolithic occupation, in the form of a hearth and burnt bone, in 'humic scree layer 2'; and a late glacial fauna in 'soliflucted scree layer 3'. The Foxhole Cave excavation did not reach the Palaeolithic.

Uncalibrated radiocarbon evidence indicates that the Late Palaeolithic 'Red Lady' dates to 26 000 years BP, and other bones and bone artefacts from Goat's Hole to 37 800–15 250 years BP (Bowen *et al.*, in Aldhouse-Green, 2000); and lithic artefacts to the Mousterian, Aurignacian, Gravettian, Creswellian, final Late Palaeolithic and Mesolithic (Swainston, in Aldhouse-Green, 2000). The non-human bones are a mixture of carnivores, including the spotted hyena *Crocuta crocuta*, the wolf *Canis lupus*, the fox *Vulpes vulpes*, ?the Arctic fox *Alopex lagopus* and the brown bear *Ursus arctos*; and herbivores, including the woolly mammoth *Mammuthus primigenius*, the wild horse or tarpan *Equus ferus*, the woolly rhinoceros *Coelodonta antiquitatis*, the wild boar *Sus scrofa*, the reindeer *Rangifer tarandus*, the red deer *Cervus elephas*, ?the giant deer *Megaloceros giganteus*, the steppe bison *Bison priscus* and the wild sheep *Ovis aries* (Bowen *et al.*, in Aldhouse-Green, 2000; Turner, in Aldhouse-Green, 2000). The non-human carnivores are thought to have been principally responsible for the bone accumulations (Bowen *et al.*, in Aldhouse-Green, 2000). There is no evidence of butchery by ancient humans (Turner, in Aldhouse-Green, 2000). Analysis of the carbon and nitrogen isotopic composition of collagen indicates that the 'Red Lady' subsisted at certain times of year on sea fish and shellfish, implying seasonal migration to the coast, at the time located some 100 km to the west; and/or to the 'palaeo-Severn', 20 km to the south, to exploit salmon running upriver to spawn (Bowen *et al.*, in Aldhouse-Green, 2000). Interestingly, in contrast, analysis of the collagen from Mesolithic and Neolithic human remains from Foxhole Cave does not indicate any such dietary intake of sea fish or shellfish. Note, though, that the data on the Mesolithic individual is equivocal, as he or she was an infant, and arguably still breast-feeding.

Importantly, the Late Palaeolithic 'Red Lady' has yielded a mitochondrial DNA sequence corresponding to the commonest extant lineage in Europe (Bowen *et al.*, in Aldhouse-Green, 2000). This discovery does not support the so-called 'diffusion model' for the colonisation of Europe, which argues for a late, post-agricultural revolution, Neolithic replacement of the early, Palaeolithic population by migrants from the Near East (Ammerman and Cavalli-Sforza, 1994). Instead, it supports the so-called 'continuity model' (Richards *et al.*, 1996). Analysis of the body proportions of the 'Red Lady' indicates that he was slightly more warm-adapted than most modern Europeans, and slightly more cold-adapted than most modern sub-Saharan Africans, though these observations do not permit any unequivocal inferences to be drawn with regard to population movement or gene flow (Trinkaus and Holliday, in Aldhouse-Green, 2000).

'Doggerland', North Sea (Mesolithic, 9500 years BP/7500 BC)

In 1931, the fishing vessel *Colinda*, under the captaincy of the magnificently monickered Pilgrim E.

Lockwood, trawled up some peat from the bottom of the North Sea some 40 km off the coast of East Anglia that proved to contain a barbed spear-point fashioned from antler (Mithen, 2003; Prior, 2003). Later that same year, Harry Godwin of Cambridge University analysed further peat samples from the area for their pollen content, and made the even more remarkable discovery of woodland species identical to those from the north-eastern coast of England and the northern coast of mainland Europe from the period immediately following the last ice age. The whole of what is now the North Sea was evidently then part of a contiguous land mass (see also Shennan *et al.*, in Shennan and Andrews, 2000), on which ancient humans hunted for reindeer, and foraged for whatever else they could find to eat. This ancient land mass has come to be known as 'Doggerland'.

Mount Sandel, Coleraine, Co. Derry, Northern Ireland (Mesolithic, 9000 years BP/7000 BC)

Archaeological excavations in the bluffs alongside the River Bann at Mount Sandel near Coleraine in Co. Derry have yielded evidence of the oldest human occupation of Ireland, dating back to post-glacial, Littletonian, approximately 9000 years BP/7000 BC, to the Mesolithic (Mitchell and Ryan, 1997; Prior, 2003). The evidence is in the form of flint axes and small blades or points known as microliths, in securely dated stratigraphic contexts, and of associated animal and plant remains. The flint tool assemblage is comparable to that of the Mesolithic of Star Carr in Yorkshire in England, although it lacks the chisel-like so-called burin, interpreted as having been used to work bone, antler and possibly wood, all perfectly preserved here below an unusually high water table (Mitchell and Ryan, 1997; Mithen, in Cunliffe, 1998; Mithen, 2003; Prior, 2003).

Conditions in Ireland during the late glacial, Nahanagan stadial, in the period 10 600–10 000 years BP/8600–8000 BC – which preceded the post-glacial, Littletonian – were evidently sufficiently inhospitable as to have made it uninhabitable (Mitchell and Ryan, 1997). The pollen record indicates that the climate at this time was Arctic, and that the vegetation was dominated by tundra and alpine scrub with *Artemisia*. The record of the northern European beetles *Boreophilus henningianus* and *Diacheila arctica* confirms an Arctic climate. The vertebrate palaeontological record indicates that the fauna was dominated by reindeer, giant deer, wolf and brown bear (Woodman *et al.*, 1997).

Conditions during the post-glacial, Littletonian, from 10 000 years BP/8000 BC, ameliorated and made the land amenable to habitation (Mitchell and Ryan, 1997). The pollen record indicates that the climate at this time became warmer, and that in response the vegetation became dominated initially by open scrub, then by open woodland, and finally by juniper and birch woodland. The vertebrate palaeontological record indicates that at least initially the fauna was still characterised by some essentially Arctic elements, such as the lemming (Woodman *et al.*, 1997). The post-glacial return of the flora and fauna to Ireland from refugia to the south and east has been described as a 'steeplechase', complete with 'water jumps' over the English Channel and the Irish Sea, which (re-)formed 7500 years BP, in 5500 BC (Mitchell and Ryan, 1997). Interestingly, snakes appear to have refused the 'water jump' over the Irish Sea. Alternatively, as legend has it, they were banished from Ireland by the early Christian missionary St Patrick in the fifth century.

The early humans that came on to this scene – the Sandelians – evidently used their flint weapons and tools to hunt game such as wild boar, possibly with the assistance of domesticated dogs; and game birds such as capercaillie, grouse, duck and pigeon, possibly with the assistance of tamed and trained goshawk (also found at the Neolithic to Bronze Age site at Newgrange in Co. Meath). They apparently also fished for salmon, trout and eels in fresh water, and bass and flounder in salt water, and foraged for seasonal shellfish, nuts, berries, water-lily (*Nymphaea alba*) and other seeds, and roots. They are best thought of as hunter–gatherers.

Black Sea (Mesolithic/Bronze Age, 7500 years BP/5500 BC)

The marine geologists Bill Ryan and Walt Pitman and their colleagues have proposed that the reconnection between the Black Sea and Mediterranean in the Holocene took place in the form of a catastrophic flood of Mediterranean water into the Black Sea approximately 7500 years BP, in 5500 BC, and that this catastrophic flood gave rise to the

4000-year-old 'epic of Gilgamesh' and the Old Testament tale of 'Noah's flood' (Ryan *et al.*, 1997, 2003; Ryan and Pitman, 1998).

Ryan and Pitman's argument hinges in part on the apparently instantaneous appearance of euryhaline Mediterranean molluscs in the Black Sea at this time. Note, though, that alternative interpretations exist as to the nature of this event, and its significance in terms of timing. For example, some authors have interpreted it as a response to the gradual establishment of favourable conditions for colonisation in the Black Sea, in turn brought about through increasing two-way flow through the Bosphorus, a process that actually initiated approximately 12000 years BP, in 10 000 BC (Aksu *et al.*, 2002; Kaminski *et al.*, 2002; Cagatay *et al.*, 2003). Some other authors have interpreted it as a response to the gradual submergence of a sill or barrier in the Bosphorus (Erdal Kerey *et al.*, 2004).

Note that while Western Europe was still in the Mesolithic 7500 years BP, the Near East had already entered into the Bronze Age (Sherratt, in Cunliffe, 1998). Civilisation clearly spread from the east to the west.

Littleton, Co. Tipperary, Ireland (Neolithic to Medieval, 4200 BC–1200 AD)

An 8-m core from a raised bog near Littleton in Co. Tipperary has yielded evidence of human occupation and activity in this part of Ireland from the Neolithic to the present time (Mitchell and Ryan, 1997). The evidence is in the form of the pollen record of anthropogenic disturbance of the environment, specifically a decrease in the proportion of native woodland pollen, and an increase in the proportion of plantain, grass, and cultivated cereal and herb pollen, interpreted as indicating forest clearance for farming (Fig. 7.55).

The pollen record from the Neolithic indicates a generally warm, wet, maritime climate at this time, and also indicates, for the first time, subsistence farming – and, by inference, permanent settlement – from 4200 to 2000 BC. The observed decline in elm from 3900 to 2800 BC could also indicate human activity, or disease, or both. Note in this context that Stone Age – and Bronze Age – people are thought to have used elm for firewood and for fodder for their livestock, and to have pollarded it specifically for these purposes, in the process predisposing it to disease. Note also that remains of the beetle (*Scolytus scolytus*) that carries the fungus (*Ceratocystis ulmae*) that causes Dutch elm disease have been found in deposits on Hampstead Heath in London that predate the decline in elm there.

The pollen record from the Bronze Age indicates a generally dry, continental climate. It also indicates early Bronze Age farming between 2000 and 1200 BC, and evidently more extensive or efficient late Bronze Age farming between 1200 and 300 BC. The increase in efficiency in the late Bronze Age has been interpreted as associated with the use of the plough or 'ard'. Incidentally, land management practices that are now thought of as characteristically Celtic were apparently also first implemented in

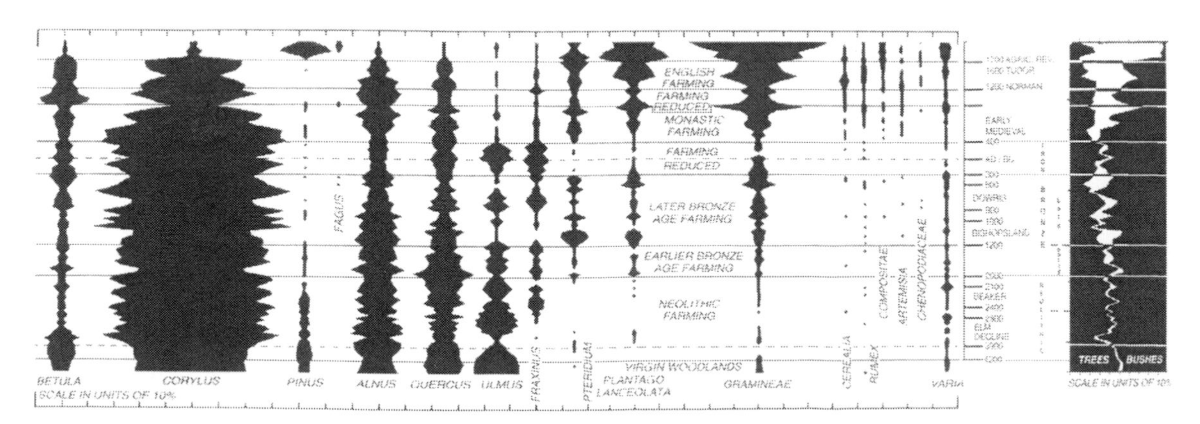

Fig. 7.55. The pollen record from Littleton Bog, Co. Tipperary, Ireland. (Reproduced with permission from Mitchell and Ryan, 1997.)

the late Bronze Age. Interestingly in this context, the Welsh word for plough or 'ard' is *aradr*.

The pollen record from the Iron Age to the present time indicates a generally cool, wet climate. That from the Iron Age indicates a reduction in farming between 300 BC and AD 400. That from the early Medieval period indicates a resumption associated with monasticism, followed by a reduction, both between AD 400 and 1200. Incidentally, that from the Anglo-Norman period to the present time indicates another resumption, from AD 1200 (and a further increase in efficiency, associated with the 'agricultural revolution', from AD 1700).

Skara Brae, Orkney (Neolithic, 3000 BC)

Archaeological excavation at Skara Brae in the Orkney Islands off the north coast of Scotland has yielded evidence of domestic and community life dating back to the late Neolithic (Sherratt, in Cunliffe, 1998; Schama, 2000; Prior, 2003). The evidence is in the form of an entire village of stone houses, which would once have been home to 50–60 people, together with stone and bone artefacts, and associated animal and plant remains. The village apparently lay perfectly preserved for thousands of years under a protective blanket of dune sand, until a storm in the mid nineteenth century revealed it once more to the world.

Conditions obtaining in the Neolithic were evidently somewhat warmer than they are today, although there were still but a few trees to soften the landscape. The Neolithic inhabitants of the island – the Orcadians – therefore built their houses out of local flagstones, which they were able to cleave from the country rock with antler picks. Like the Mesolithic Sandelians, they fished, for red bream and corkfin wrasse, and foraged, for mussels and oysters (they may also have hunted game, with dogs). Unlike the Sandelians, though, they also farmed. They farmed cattle, for meat and milk, and they grew crops such as barley and wheat, where none will grow today (the land being suitable for pastoral but not for arable purposes).

Tyrolean Alps (Neolithic/'Chalcolithic', 3000 BC)

The frozen and mummified body of an 'ice man' now named 'Otzi' was discovered in the Tyrolean Alps in 1991 (Spindler, 1994; Bortenschlager and Oeggl, 2000; Fowler, 2000). Radiocarbon dating of plant and animal remains associated with the body has yielded a consistent age of around 5300 years BP. Detailed archaeological examination of the plant and animal remains has yielded fascinating insight's into Otzi's life (and death). Remarkably, some 200 species of plants have been found in, on or around Otzi's body (Dickson *et al.*, 2003). Many are from lower altitudes than where Otzi was discovered, and thus clearly introduced by him. Some of the plant species, such as hop hornbeam and *Neckera*, also indicate a southerly provenance (as, incidentally, does the flint in Otzi's flint tool and weapons, which has been typed back to the locality of Monte Lessini in Italy by means of the siliceous microfossil assemblages that it contains). The hop hornbeam is represented by pollen in Otzi's gut (interestingly, with its cell content preserved, as it only is in late spring or early summer). Einkorn *Triticum monococcum* has also been found in Otzi's gut, together with ibex and red deer meat, representing his last meal (Rollo *et al.*, 2002). A single species of *Sphagnum* moss has also been found in Otzi's gut, and interpreted as having been ingested incidentally, possibly as part of the wrapping of the provisions (Dickson *et al.*, 2003). Another species of *Sphagnum* has been interpreted as having been used as a wound dressing. Tinder fungus has been found in a pouch together with flints and pyrite, collectively constituting all the elements of a fire-lighting kit. (Norway maple leaves have also been found, and interpreted as having been used to carry glowing embers.) Bracket fungus has also been found, and interpreted as having been used as a medicine (it has antiseptic properties). Seventeen types of wood or charcoal have been found. Otzi used wood from the wayfaring tree to make arrows; yew, to make a bow, and the haft of a copper-headed axe; ash, the handle of a dagger; and birch bark to make a bucket in which to carry a 'trail mix' of dried sloes. He also used bark, together with grass, to make a serviceable cape; and goat-skin to make underclothing and shoes.

City of London (Medieval, AD 1200?–1350)

Archaeological excavation at a site in Tudor Street in the City of London in 1978, preparatory to site redevelopment, has yielded interesting evidence of domestic and industrial activity in the Middle Ages (Boyd, in Neale and Brasier, 1981). The evidence is essentially in the form of the record of ancient organismal remains in the medieval deposits of the River Fleet, a tributary of the Thames, and of the influence on it of human activity. Incidentally, the word 'fleet' is Anglo-Saxon in origin, and refers to the ability of the river to float boats on the rising tide (Barton, 1962). The post-Anglo-Saxon history of the River Fleet has been described as 'a decline from a river to a brook, from a brook to a ditch, and from a ditch to a drain' (Barton, 1962).

The medieval deposits of the Fleet are approximately 1 m thick, between −0.4 and +0.6 m OD (ordnance datum, the UK standard for sea level), and unconformably overlie a bedrock of London clay and a layer of interpreted pre-Roman occupation deposits, and underlie a layer of post-medieval industrial waste, post-dating the period of land reclamation between AD 1350 and 1360 (Boyd, in Neale and Brasier, 1981). They have been provisionally dated to the thirteenth and fourteenth centuries on the basis of potsherds.

The medieval deposits of the Fleet altogether contain over 140 species of organisms, including diatoms, silicoflagellates, organic-walled microplankton, charophytes, angiosperms, testate amoebae, foraminiferans, sponges, thecate hydroids, polychaete worms, parasitic nematode worms (from human faeces!), freshwater bryozoans, bivalve and gastropod molluscs, cladocerans, ostracods and echinoids. The principal elements of the biota are essentially freshwater to only slightly brackish, and it is thought that the minor, marine elements are allochthonous. It is also thought that the salinity of the medieval Fleet may have been artificially low on account of the effects of the old London bridge, built in 1209 and not removed until as recently as 1832.

They also contain abundant industrial and domestic refuse such as wood chips (waste from woodworking), horn cores (waste from horn-working), hide (waste from butchery or tannery), leather shoes, meat- and domestic-animal bones, and animal and human dung! The refuse and organic waste dumped into the river evidently resulted initially in eutrophication, and the proliferation of some, opportunistic, species, including freshwater molluscs and the macrophytes *Zannicellia* and *Groenlandia* (in Bed X). However, it also resulted ultimately in hypertrophication, and the effective elimination of all species (in Bed Y), as recorded in archive records for 1343. Incidentally, the pollution became so bad that it was ordered in 1357 that 'no man shall take . . . any manner of rubbish . . . or dung, from out of his stable or elsewhere, to throw . . . the same into the rivers of Thames and Fleet. . . . And if any one should be found doing the contrary thereof, let him have the prison for his body, and other heavy punishment as well, at the discretion of the Mayor and the Aldermen'.

References

Abramovich, S., Keller, G., Stuben, D. and Berner, Z. 2003. Characterization of Late Campanian and Maastrichtian planktonic foraminiferal depth habitats and vital activities based on stable isotopes. *Palaeogeography, Palaeoclimatology, Palaeoecology*, **201**(1–2): 1–29.

Adams, A. E. and Al-Zahrani, M. S. 2000. Palaeoberesellids (dasycladaceans) from the Upper Jurassic Arab D reservoir, Saudi Arabia. *Palaeontology*, **43**(3): 591–7.

Adams, C. G. and Ager, D. V. (eds.) 1967. *Aspects of Tethyan Biogeography*, Publication No. 7. Systematics Association.

Agadzhanyan, A. K. 2003. Middle Pliocene small mammals of the Russian plain. *Stratigraphy and Geological Correlation*, **11**(6): 620–35.

Ager, D.V. 1965. The adaptation of Mesozoic brachiopods to different environments. *Palaeogeography, Palaeoclimatology, Palaeoecology*, **1**: 143–79.

Agusti, J. and Anton, M. 2002. *Mammoths, Sabretooths and Hominids*. Columbia University Press.

Agusti, J., Rook, L. and Andrews, P. (eds.) 1999. *Hominoid Evolution and Climatic Change in Eurasia*, vol. 1, *Climatic and Environmental Change in the Neogene of Europe*. Cambridge University Press.

Ahlberg, P. E. and Milner, A. R. 1994. The origin and early diversification of the tetrapods. *Nature*, **368**: 507–14.

Ahr, W. M., Harris, P. M., Morgan, W. A., Somerville, I. D. and Stanton, R. J. (eds.) 2000. *Permo-Carboniferous Carbonate Platforms and Reefs*. Society for Economic Paleontologists and Mineralogists.

Ainsworth, N. R., Burnett, R. D. and Kontrovitz, M. 1990. Ostracod colour change by thermal alteration, offshore Ireland and Western UK. *Marine and Petroleum Geology*, **7**: 288–97.

Akhmetiev, M. A. 2000. The Late Eocene and Oligocene phytogeography of western Eurasia. *Paleontological Journal*, **34**(Suppl. 1): S106–15.

Akimoto, K., Hattori, M. and Oda, M. 2002. Late Cenozoic paleobathymetry and paleogeography in the South Fossa-Magna and Enshunada regions, Japan, based on planktonic and benthic foraminifera. *Marine Geology*, **187**: 89–118.

Akinci, O., Robertson, A., Poisson, A. and Bozkurt, E. (eds.) 2003. The Isparta Angle, SW Turkey: its role in the evolution of Tethys in the eastern Mediterranean region. *Geological Journal*, **38**: 191–4.

Aksu, A. E., Hiscott, R. N., Mudie, P. J., *et al.* 2002. Persistent Holocene outflow from the Black Sea to the eastern Mediterranean contradicts Noah's Flood hypothesis. *GSA Today*, May 2002: 4–10.

Alberti, G. K. B. 1983. Unterdevonische Novakiidae (Dacryoconarida) aus dem rheinischen Schiefengebirge, aus Oberfranken und aus N-Afrika (Algerien, Marokko). *Senckenbergiana Lethaea*, **64**: 295–313.

Aldhouse-Green, S. 2000. *Paviland Cave and the 'Red Lady': A Definitive Report*. Western Academic & Specialist Press.

Aldridge, R. J. and Theron, J. N. 1993. Conodonts with preserved soft tissue from a new Ordovician *Konservat-Lagerstatte*. *Journal of Micropalaeontology*, **12**(1): 113–17.

Aldridge, R. J., Briggs, D. E. G., Smith, M. P., Clarkson, E. N. K. and Clark, N. D. L. 1993. The anatomy of conodonts. *Philosophical Transactions of the Royal Society of London B*, **19**: 279–91.

Al-Eidan, A. J., Wethington, W. B. and Davies, R. B. 2001. Upper Burgan reservoir description, northern Kuwait: impact on reservoir development. *GeoArabia*, **6**(2): 179–208.

Alem, N. 1998. Palynostratigraphy of the Lower Devonian and Lower Carboniferous from REG and TEG fields in the Timimoun Basin, west Algerian Sahara. Unpublished M.Sc. thesis, University College London.

Al-Fares, A. A., Bouman, M. and Jeans, P. 1998. A new look at the Middle to Lower Cretaceous stratigraphy, offshore Kuwait. *GeoArabia*, **3**(4): 543–60.

Al-Hajri, S. and Owens, B. (eds.) 2000. *Stratigraphic Palynology of the Palaeozoic of Saudi Arabia*, GeoArabia Special Publication No. 1. Gulf PetroLink.

Al-Hajri, S. A., Filatoff, J., Wender, L. E. and Norton, A. K. 1999. Stratigraphy and operational palynology of the Devonian system in Saudi Arabia. *GeoArabia*, **4**(1): 53–68.

Al-Husseini, M. I. (ed.) 1995. *GEO'94: Selected Middle East Papers from the Middle East Geoscience Conference, 25–27 April, 1994, Bahrain*. Gulf PetroLink.

1997. Jurassic sequence stratigraphy of the western and southern Arabian Gulf. *GeoArabia*, **2**(4): 361–82.

(ed.) 2004. *Carboniferous, Permian and Early Triassic Arabian Stratigraphy*, GeoArabia Special Publication No. 3. Gulf PetroLink.

Al-Husseini, M., Amthor, J., Grotzinger, J. and Mattner, J. 2003. Arabian plate Precambrian–Cambrian boundary interpreted in Oman's Ara Group. *GeoArabia*, **8**(4): 578–80.

Ali, W., Paul, W. and Young On, V. (eds.) 1998. *Transactions of the 3rd Geological Conference of the Geological Society of Trinidad and Tobago and the 14th Caribbean Geological Conference*. Geological Society of Trinidad and Tobago.

Allison, P. A. and Briggs, D. E. G. (eds.) 1991. *Taphonomy: Releasing the Data Locked in the Fossil Record*, Topics in Geobiology Series No. 9. Plenum Press.

Alonso, P.D., Milner, A.C., Ketcham, R.A., Cookson, M.J. and Rowe, T.B. 2004. The avian nature of the brain and inner ear of *Archaeopteryx*. *Nature*, **430**: 666–9.

Alroy, J. 2001. A multispecies overkill simulation of the end-Pleistocene megafaunal mass extinction. *Science*, **292**: 1893–6.

Alsharhan, A.S. 1987. Geology and reservoir characteristics of carbonate buildup in giant Bu Hasa oil field, Abu Dhabi, United Arab Emirates. *American Association of Petroleum Geologists Bulletin*, **71**(10): 1304–18.

1995. Facies variation, diagenesis, and exploration potential of the Cretaceous rudist-bearing carbonates of the Arabian Gulf. *American Association of Petroleum Geologists Bulletin*, **79**(4): 531–50.

Alsharhan, A.S. and Nairn, A.E.M. 1997. *Sedimentary Basins and Petroleum Geology of the Middle East*. Elsevier.

Alvarez, W. 1997. *T. rex and the Crater of Doom*. Princeton University Press.

Alvarez, L.W., Alvarez, W., Asaro, F. and Michel, H.V. 1980. Extraterrestrial cause for the Cretaceous–Tertiary extinction: experimental results and theoretical interpretation. *Science*, **208**: 1095–108.

Amini, Z.Z., Adabi, M.H., Burrett, C.F. and Quilty, P.G. 2004. Bryozoan distribution and growth form associations as a tool in environmental interpretation, Tasmania, Australia. *Sedimentary Geology*, **167**(1–2): 1–15.

Amitrov, O.V. 2000. Zoogeography of the Late Oligocene marine basins in western Eurasia using gastropods. *Paleontological Journal*, **34**(3): 244–51.

Ammerman, A.J. and Cavalli-Sforza, L.L. 1984. *The Neolithic Transition and the Genetics of Population in Europe*. Princeton University Press.

Amon, E.O., Blueford, J.R., de Wever, P. and Zhelezko, V.I. 1997. An essay on regional geology of the Upper Cretaceous deposits of southern Urals territories. *Geodiversitas*, **19**(2): 293–318.

Amthor, J., Al-Zadjali, I., Tiley, G. and El-Tonbary, M. 2000. From ice to salt: frontier oil and gas exploration of the Huqf supergroup of Oman. *Abstracts, MEGSTRAT1 Workshop*, Dubai: 3.

Amthor, J.E., Grotzinger, J.P., Schroder, S., *et al.* 2003. Extinction of *Cloudina* and *Namacalathus* at the Precambrian–Cambrian boundary in Oman. *Geology*, **31**: 431–4.

Anapol, F., German, H.Z. and Jablonski, N.G. (eds.) 2004. *Shaping Primate Evolution: Form, Function and Behaviour*. Cambridge University Press.

Anderson, D.T. (ed.) 2001. *Invertebrate Zoology*, 2nd edn. Oxford University Press.

Anderson, L.I. and Trewin, N.H. 2003. An Early Devonian arthropod fauna from the Windyfield cherts, Aberdeenshire, Scotland. *Palaeontology*, **46**(3): 467–510.

Andrews, J.E., Coletta, P., Pentecost, A., *et al.* 2004. Equilibrium and disequilibrium stable isotope effects in modern charophyte calcites: implications for palaeoenvironmental studies. *Palaeogeography, Palaeoclimatology, Palaeoecology*, **204**(1–2): 101–14.

Andrews, P. 1990. *Owls, Caves and Fossils: Predation, Preservation and Accumulation of Small Mammal Bones in Caves, with an Analysis of the Pleistocene Cave Faunas from Westbury-sub-Mendip, Somerset, UK*. Natural History Museum.

Andrews, P. and Banham, P. (eds.) 1999. *Late Cenozoic Environments and Hominid Evolution*. Geological Society.

Andrews, P., Cook, J., Currant, A. and Stringer, C. 1999. *Westbury Cave: The Natural History Museum Excavations, 1976–1984*. Western Academic & Specialist Press.

Angiolini, L., Balini, M., Garzanti, E., Nicora, A. and Tintori, A. 2003a. Gondwanan deglaciation and opening of Neotethys: the Al Khlata and Saiwan formations of interior Oman. *Palaeogeography, Palaeoclimatology, Palaeoecology*, **196**: 99–123.

Angiolini, L., Balini, M., Garzanti, E., *et al.* 2003b. Permian climatic and paleogeographic changes in northern Gondwana: the Khuff formation of interior Oman. *Palaeogeography, Palaeoclimatology, Palaeoecology*, **191**: 269–300.

Anglada, R. and Randrianasolo, A. 1985. Utilisation des foraminifères planctoniques dans la paléoocéanographie de la Tethys au Crétace. *Bulletin de la Société Géologique de France, Série 8*, **1**(5): 747–55.

Anonymous. 1991. *Proceedings of the 1st International Conference on Rudists, Beograd, 1988*, Special Publication No. 2. Serbian Geological Society.

Antonanzas, R.L. and Bescos, G.C. 2002. The Gran Dolina site (Lower to Middle Pleistocene, Atapuerca, Spain): new palaeoenvironmental data based on the distribution of small mammals. *Palaeogeography, Palaeoclimatology, Palaeoecology*, **186**(3–4): 311–34.

Aqrawi, A.A.M., Thehni, T.A., Sherwani, G.H. and Kareem, B.M.A. 1998. Mid-Cretaceous rudist-bearing carbonates of the Mishrif formation: an important reservoir sequence in the Mesopotamian basin, Iraq. *Journal of Petroleum Geology*, **21**(1): 57–82.

Archangelsky, S.W. and Wagner, R.H. 1983. *Glossopteris anatolica* sp. nov. from uppermost Permian strata in south-east Turkey. *Bulletin of the British Museum (Natural History), Geology Series*, **37**(3): 81–92.

Archibald, J.D. 1996. *Dinosaur Extinction and the End of an Era*. Columbia University Press.

Armentrout, J.M. (ed.) 1981. *Pacific Northwest Cenozoic Biostratigraphy*, Special Paper No. 184. Geological Society of America.

Athersuch, J., Horne, D.J. and Whittaker, J.R. 1989. *Marine and Brackish Water Ostracods*. E.J. Brill.

Attar, A., Fournier, J., Candilier, A.M. and Coquel, R. 1980. Etude palynologique du Devonien terminal et du Carbonifère inférieur du bassin d'Illizi (Fort-Polignac), Algérie. *Revue de l'Institut Français du Petrole*, **35**(4): 585–619.

Aubry, M.-P., Lucas, S. and Berggren, W.A. (eds.) 1998. *Late Paleocene–Early Eocene Climatic and Biotic Events in the Marine and Terrestrial Records*. Columbia University Press.

Aufderheide, A. C. and Rodriguez-Martin, C. 1998. *The Cambridge Encyclopaedia of Human Paleopathology*. Cambridge University Press.

Azzaroli, A. 1983. Quaternary mammals and the 'end-Villafranchian' dispersal event: a turning point in the history of Eurasia. *Palaeogeography, Palaeoclimatology, Palaeoecology*, **44**: 117–39.

Bakker, R.T. 1975. Dinosaur rennaissance. *Scientific American*, **232**(4): 58–78.

Baldi, T. 1986. *Mid Tertiary Stratigraphy and Paleogeographic Evolution of Hungary*. Akademiai Kiado.

Ball, H. W., Dineley, D. L. and White, E. I. 1961. The Old Red Sandstone of Brown Clee Hill and the adjacent area. *Bulletin of the British Museum (Natural History)*, **5**: 177–310.

Baraboshkin, E.Y., Alekseev, A.S. and Kopaevich, L.F. 2003. Cretaceous palaeogeography of the north-eastern Peri-Tethys. *Palaeogeography, Palaeoclimatology, Palaeoecology*, **196**: 177–208.

Barbieri, R. and Panieri, G. 2004. How are benthic foraminiferal faunas influenced by cold seeps? Evidence from the Miocene of Italy. *Palaeogeography, Palaeoclimatology, Palaeoecology*, **204**(3–4): 257–75.

Barnard, P. D. W. 1973. Mesozoic floras. *Special Papers in Palaeontology*, **12**: 175–87.

Baron-Szabo, R. C. 2000. Late Campanian–Maastrichtian corals from the United Arab Emirates–Oman border region. *Bulletin of the Natural History Museum, London, Geology*, **56**(2): 91–131.

Barrett, P. M., Hasegawa, Y., Manabe, M., Isaji, S. and Matsuoka, H. 2002. Sauropod dinosaurs from the Lower Cretaceous of eastern Asia: taxonomic and biogeographical applications. *Palaeontology*, **45**(6): 1197–217.

Barron, J.A. and Gladenkov, A.Y. 1995. Early Miocene to Pleistocene diatom stratigraphy of Leg 145. *Proceedings of the Ocean Drilling Program, Scientific Results*, **145**: 3–19.

Barry, J.C., Lindsay, E.H. and Jacobs, L.L. 1982. A biostratigraphic zonation of the middle and upper Siwaliks of the Potwar plateau of northern Pakistan. *Palaeogeography, Palaeoclimatology, Palaeoecology*, **37**: 95–139.

Bartels, C., Briggs, D. E. G. and Brassel, C. 1998. *The Fossils of the Hunsruck Slate: Marine Life in the Devonian*. Cambridge University Press.

Barthel, K.W. 1978. *Solnhofen: Ein Blick in die Erdegeschichte*. Ott.

Barthel, K.W., Swinburne, N. H. M. and Conway Morris, S. 1990. *Solnhofen: A Study in Mesozoic Palaeontology*, revised translation. Cambridge University Press.

Barthelt, D. 1989. Faziesanalyse und Untersuchung der Sedimentationmechanismen in der Unteren Brackwasser-Molasse Oberbayerns. *Münchner geowissenschaftliche Abhandlungen A*, **17**: 1–118.

Bartolini, C., Buffler, R.T. and Cantu-Chapa, A. (eds.) 2001. *The Western Gulf of Mexico Basin: Tectonics, Sedimentary Basins and Petroleum Systems*, Memoir No. 75. American Association of Petroleum Geologists.

Bartolini, C., Buffler, R.T. and Blickwede, J.F. (eds.) 2003. *The Circum-Gulf of Mexico and the Caribbean: Hydrocarbon Habitats, Basin Formation and Plate Tectonics*, Memoir No. 79. American Association of Petroleum Geologists.

Barton, N. 1962. *The Lost Rivers of London*. Leicester University Press.

1997. *Stone Age Britain*. B. T. Batsford.

Bar-Yosef, O. and Pilbeam, D. (eds.) 2000. *The Geography of Neandertals and Modern Humans in Europe and the Greater Mediterranean*. Peabody Museum of Archeology and Ethnology.

Bassett, M. G. 1984. Life strategies of Silurian brachiopods. *Special Papers in Palaeontology*, **32**: 237–63.

Bassett, M. G., King, A. H., Larwood, J. G., Parkinson, N. A. and Deisler, U. K. 2001. *A Future for Fossils*, Geological Series No. 19. National Museum of Wales.

Bate, R. H. 1972. Phosphatized ostracods with appendages from the Lower Cretaceous of Brazil. *Palaeontology*, **15**(3): 379–93.

Battarbee, R., MacKay, A., Birks, J. and Oldfield, F. 2003. *Global Change in the Holocene*. Hodder Arnold.

Batten, D.J. and Keen, M.C. (eds.) 1989. *North-West European Micropalaeontology and Palynology*. Ellis Horwood.

Baumgartner, P. O., O'Dogherty, L., Gorican, S., *et al.* 1995. Middle Jurassic to Lower Cretaceous Radiolaria of Tethys: occurrences, systematics, biochronology. *Mémoires de Geologie*, **23**: 1–1172.

Beauchamp, B. and Baud, A. 2002. Growth and demise of Permian biogenic chert along northwest Pangea: evidence for end-Permian collapse of thermohaline circulation. *Palaeogeography, Palaeoclimatology, Palaeoecology*, **184**: 37–63.

Beauchamp, B. and Savard, M. 1992. Cretaceous chemosynthetic carbonate mounds in the Canadian Arctic. *Palaios*, **7**(4): 434–50.

Beck, C. B. (ed.) 1976. *Origin and Early Evolution of Angiosperms*. Columbia University Press.

(ed.) 1988. *Origin and Evolution of Gymnosperms*. Columbia University Press.

Beck, M.W. 1996. On discerning the cause of Late Pleistocene megafaunal mass extinctions. *Paleobiology*, **22**: 91–103.

Begun, D.R., Ward, C.V. and Rose, M.D. 1997. *Function, Phylogeny and Fossils: Miocene Hominoid Evolution and Adaptations*. Kluwer.

Behling, H., DePasta Pillar, V., Orloci, L. and Bauermann, S. G. 2004. Late Quaternary *Araucaria* forest, grassland (campos), fire and climate dynamics, studied by high-resolution pollen, charcoal and multivariate analysis of the Cambara do Sul core in southern Brazil. *Palaeogeography, Palaeoclimatology, Palaeoeoclogy*, **203**: 277–97.

Behrensmeyer, A.K. and Kidwell, S.M. (eds.) 1993. *Taphonomic Approaches to Time Resolution in Fossil Assemblages*, Short Courses in Paleontology No. 6. University of Tennessee Press.

Behrensmeyer, A. K., Damuth, J. D., DiMichele, W. A., *et al.* 1992. *Terrestrial Ecosystems through Time: Evolutionary Paleobiology of Terrestrial Plants and Animals*. University of Chicago Press.

Belka, Z., Kaufmann, B. and Bultynck, P. 1997. Conodont-based quantitative biostratigraphy for the Eifelian of the eastern Anti-Atlas. *Geological Society of America Bulletin*, **109**(6): 643–51.

Bengtson, S. 1994. *Early Life on Earth*. Columbia University Press.

Bengtson, S. and Yue Zhao. 1992. Predatorial borings in Late Precambrian mineralized exoskeletons. *Science*, **257**: 367–9.

Bennett, M. R. and Doyle, P. 1997. *Environmental Geology: Geology and the Human Environment*. John Wiley.

Benoit, A. and Taugourdeau, P. 1961. Sur quelques chitinozoares de l'Ordovicien du Sahara. *Revue de l'Institut Français du Petrole*, **16**(12): 1404–21.

Benton, M. J. 1985. Patterns in the diversification of Mesozoic non-marine tetrapods and problems in historical diversity analysis. *Special Papers in Palaeontology*, **33**: 185–202.

1986. More than one event in the Late Triassic mass extinction. *Nature*, **321**: 857–61.

(ed.) 1988. *The Phylogeny and Classification of the Tetrapods*, vol. 2, *Mammals*. Clarendon Press.

1991. *The Reign of the Reptiles*. Kingfisher.

1995. Diversification and extinction in the history of life. *Science*, **268**: 52–8.

2003. *When Life Nearly Died: The Greatest Mass Extinction of All Time*. Thames & Hudson.

2005. *Vertebrate Palaeontology*, 3rd edn. Blackwell.

Benton, M. J. and Harper, D. A. T. 1997. *Basic Palaeontology*. Longman.

Benton, M. J. and Spencer, P. S. 1995. *Fossil Reptiles of Great Britain*. Kluwer.

Berger, L. R. 2000. *In the Footsteps of Eve: The Mystery of Human Origins*. National Geographic.

Berger, L. R. and Clarke, R. J. 1995. Bird of prey involvement in the collection of the Taung Child fauna. *Journal of Human Evolution*, **29**(3): 275–99.

Berggren, W. A. and Aubert, J. 1975. Paleocene benthic foraminiferal biostratigraphy, paleobiogeography and paleoecology of Atlantic–Tethyan regions: midway-type faunas. *Palaeogeography, Palaeoclimatology, Palaeoecology*, **18**: 73–192.

Berggren, W. A., Kent, D. V., Aubry, M.-P. and Hardenbol, J. (eds.) 1995. *Geochronology, Timescales and Global Stratigraphic Correlation*, Special Publication No. 54. Society of Economic Paleontologists and Mineralogists.

Bernhard, J. M. 2003. Potential symbionts in bathyal foraminifera. *Science*, **299**: 861.

Bernhard, J. M., Buck, K. R. and Barry, J. P. 2001. Monterey Bay cold-seep biota: assemblages, abundance and ultrastructure of living foraminifera. *Deep-Sea Research*, **148**: 2233–49.

Bernor, R. L., Fahlbusch, V. and Mittmann, H.-W. (eds.) 1996. *The Evolution of Western Eurasian Neogene Mammal Faunas*. Columbia University Press.

Berthelin, M., Broutin, J., Kerp, H., *et al.* 2003. The Oman Gharif mixed paleoflora: a useful tool for testing Permian Pangea reconstructions. *Palaeogeography, Palaeoclimatology, Palaeoecology*, **196**: 85–98.

Bertrand-Sarfati, J. and Monty, C. 1994. *Phanerozoic Stromatolites*, 2nd edn. Kluwer.

Bethoux, O., McBride, J. and Maul, C. 2004. Surface laser scanning of fossil insect wings. *Palaeontology*, **47**(1): 13–20.

Bettis, E. A., III, Zaim, Y., Larick, R. R., *et al.* 2004. Landscape development preceding *Homo erectus* immigration into central Java, Indonesia: the Sangiran formation lower lahar. *Palaeogeography, Palaeoclimatology, Palaeoecology*, **206**(1–2): 115–31.

Beuning, K. and Wooller, M. (eds.) 2002. Reconstruction and modelling of grass-dominated ecosystems. *Palaeogeography, Palaeoclimatology, Palaeoecology*, **177**(1–2).

Bicchi, E., Ferero, E. and Gonera, M. 2003. Palaeoclimatic interpretation based on Middle Miocene planktonic foraminifera: the Silesia basin (Paratethys) and Monferrato (Tethys) records. *Palaeogeography, Palaeoclimatology, Palaeoecology*, **197**: 265–303.

Bidashko, F. G., Proskurin, K. P. and Shatrovskii, A. G. 1995. Paleogeography of the lower Volga at the end of Atelian time based on entomological and botanical data. *Paleontological Journal*, **30**(3): 332–6.

Bignot, G. 1985. *Elements of Micropalaeontology*, English translation. Graham & Trotman.

Bishop, M. J. 1975. Earliest record of Man's presence in Britain. *Nature*, **253**: 95–7.

Black, R. M. 1989. *Elements of Palaeontology*, 2nd edn. Cambridge University Press.

Blome, C. D., Whalen, P. A. and Reed, K. M. 1996. *Siliceous Microfossils: Notes for a Short Course*, Department of Geological Sciences, Studies in Geology No. 9. University of Tennessee.

Boardman, R. S., Cheetham, A. H. and Rowell, A. J. (eds.) 1987. *Fossil Invertebrates*. Blackwell.

Boaz, N. T. and Ciochon, R. L. 2004. *Dragon Bone Hill: An Ice-Age Saga of Homo erectus*. Oxford University Press.

Bohme, M. 2003. The Miocene climatic optimum: evidence from ectothermic vertebrates of central Europe. *Palaeogeography, Palaeoclimatology, Palaeoecology*, **195**: 389–401.

Bojar, A.-V., Hiden, H., Fenninger, A. and Neubauer, F. 2004. Middle Miocene seasonal temperature changes in the Styrian basin, Austria, as recorded by the isotopic composition of pectinid and brachiopod shells. *Palaeogeography, Palaeoclimatology, Palaeoecology*, **203**(1–2): 95–106.

Bolli, H. M., Saunders, J. B. and Perch-Nielsen, K. 1985. *Plankton Stratigraphy*. Cambridge University Press.

Borgomano, J., Masse, J.-P. and Al-Maskiry, S. 2002. The lower Aptian Shuaiba carbonate outcrops in Jebel Akhdar, northern Oman: impact on static modelling for Shuaiba petroleum reservoirs. *American Association of Petroleum Geologists Bulletin*, **86**(9): 1513–29.

Boriani, M., Bonafede, M., Piccardo, G. B. and Vai, G. B. (eds.) 1989. The lithosphere in Italy: advances in earth science research. *Atti dell'Accademia Nazionale del Lincei*, **80**.

Bortenschlager, S. and Oeggl, K. (eds.) 2000. *The Iceman and his Natural Environment*. Springer-Verlag.

Bos, J. A. A., Dickson, J. H., Coope, G. R. and Jardine, W. G. 2004. Flora, fauna and climate of Scotland during the Weichselian middle pleniglacial: palynological, macrofossil and coleopteran investigations. *Palaeogeography, Palaeoclimatology, Palaeoecology*, **204**(1–2): 65–100.

Bosence, D. W. J. 1980. Sedimentary facies, production rates and facies models for Recent coralline algal gravels, Co. Galway, Ireland. *Geological Journal*, **15**(2): 91–111.

Bosence, D. W. J. and Allison, P. A. 1995. *Marine Palaeoenvironmental Analysis from Fossils*, Special Publication No. 83. Geological Society.

Bott, M. H. P., Saxov, S., Talwani, M. and Thiede, J. (eds.) 1983. *Structure and Development of the Greenland–Scotland Ridge: New Methods and Concepts*, NATO IV Marine Sciences No. 8. Plenum Press.

Bottjer, D. J., Etter, W., Hagadorn, J. W. and Tang, C. M. (eds.) 2002. *Exceptional Fossil Preservation: A Unique View on the Evolution of Marine Life*. Columbia University Press.

Boucek, B. 1964. *The Tentaculites of Bohemia: Their Morphology, Ecology, Phylogeny and Biostratigraphy*. Czechoslovak Academy of Sciences.

Boucot, A. J., Roht, D. M. and Blodgett, R. B. 1988. A marine invertebrate faunule from the Tawil sandstone (basal Devonian) of Saudi Arabia and its biogeographic–paleogeographic consequences. *New Mexico Bureau of Mines & Mineral Resources, Memoir*, **44**: 361–72.

BouDagher-Fadel, M. K., Banner, F. T. and Whittaker, J. E. 1997. *The Early Evolutionary History of Planktonic Foraminifera*, Chapman & Hall.

Bown, P. R. (ed.) 1998. *Calcareous Nannofossil Biostratigraphy*, Chapman & Hall.

Bowring, S. A. and Erwin, D. H. 1998. A new look at evolutionary rates in deep time: uniting paleontology and high-precision geochronology. *GSA Today*, **8**(9): 1–8.

Bowsher, A. 1955. *Origin and Adaptation of Platyceratid Gastropods*, Paleontological Contributions, Mollusca Art. 5. University of Kansas Press.

Boyd, R. and Silk, J. B. 2003. *How Humans Evolved*, 3rd edn. W. W. Norton.

Braccini, E., Denison, C. N., Scheevel, J. R., *et al.* 1997. A revised chrono-lithostratigraphic framework for the pre-salt (Lower Cretaceous) in Cabinda, Angola. *Bulletin des Centres Recherche Exploration–Production Elf-Aquitaine*, **21**(1): 125–51.

Brack, A. (ed.) 1999. *The Molecular Origins of Life: Assembling Pieces of the Puzzle*. Cambridge University Press.

Braddy, S. J., Morrissey, L. B. and Yates, A. M. 2003. Amphibian swimming traces from the Lower Permian of southern New Mexico. *Palaeontology*, **46**(4): 671–83.

Bradley, R. S. 1999. *Paleoclimatology: Reconstructing Climates of the Quaternary*, 2nd edn. Harcourt Academic Press.

Brain, C. K. 1981. The evolution of Man in Africa: was it a consequence of Cainozoic cooling? *Annals of the Geological Society of South Africa*, **84**: 1–19.

Brasier, M. D. 1980. *Microfossils*. George Allen & Unwin.

Brenchley, P. J. (ed.) 1984. *Fossils and Climate*. John Wiley.

Brenchley, P. J. and Harper, D. A. T. 1998. *Palaeoecology*. Chapman & Hall.

Brenchley, P. J., Marshall, J. D., Carden, G. A. F., *et al.* 1994. Bathymetric and isotopic evidence for a short-lived Late Ordovician glaciation in a greenhouse period. *Geology*, **22**: 295–8.

Brenchley, P. J., Carden, G. A. F. and Marshall, J. D. 1995. Environmental changes associated with the 'first strike' of the Late Ordovician mass extinction. *Modern Geology*, **20**: 69–82.

Briffa, K. R., Osborn, T. J. and Schweingruber, F. H. 2004. Large-scale temperature inferences from tree rings: a review. *Global and Planetary Change*, **40**(1–2): 11–26.

Briggs, D. E. G. 1992. Conodonts: a major extinct group added to the vertebrates. *Science*, **256**: 1284–6.

Briggs, D. E. G. and Crowther, P. R. 1990. *Palaeobiology*. Blackwell. (eds.) 2001. *Palaeobiology*, 2nd edn. Blackwell.

Briggs, D. E. G., Clarkson, E. N. K. and Aldridge, R. J. 1983. The conodont animal. *Lethaia*, **16**: 1–14.

Briggs, D. E. G., Erwin, D. H. and Collier, F. J. 1994. *The Fossils of the Burgess Shale*. Smithsonian Institution Press.

Brinkman, D. B., Russell, A. P., Eberth, D. A. and Peng, J. 2004. Vertebrate palaeocommunities of the lower Judith River group (Campanian) of Southeastern Alberta, Canada, as interpreted from vertebrate microfossil assemblages. *Palaeogeography, Palaeoclimatology, Palaeoecology*, **213**(3–4): 295–313.

Bromley, R. G. 1996. *Trace Fossils*, 2nd edn. Chapman & Hall.

Bronowski, J. 1973. *The Ascent of Man*. BBC Books.

Brooks, B.-G. J., Morrissey, L. B. and Braddy, S. J. 2003. Arthropod terrestrialization: new ichnological data from the Late Silurian Clam Bank formation, Newfoundland. *Palaeontological Association Newsletter*, **54**(47th Annual Meeting Abstracts): 125.

Brooks, J. M., Kennicutt, M. C., III, MacDonald, I. R., *et al.* 1989. Gulf of Mexico hydrocarbon seep communities. IV. Descriptions of known chemosynthetic communities. *Proceedings of the 21st Offshore Technology Conference*: 663–7.

Brown, J. H. 1995. *Macroecology*. University of Chicago Press.

Brown, L. F., Jr., Benson, J. M., Brink, G. J., *et al.* 1995. *Sequence Stratigraphy in Offshore South African Basins: An Atlas on Exploration for Cretaceous Lowstand Traps by Soekor (Pty) Ltd*, Studies in Geology No. 41. American Association of Petroleum Geologists.

Brown, P., Sutikna, T., Morwood, M. J., *et al.* 2004. A new small-bodied hominin from the Late Pleistocene of Flores, Indonesia. *Nature*, **431**: 1055–61.

Bruggemann, J. H., Buffler, R. T., Guillaume, M. M. M., *et al.* 2004. Stratigraphy, palaeoenviroments and model for the deposition of the Abdur Reef limestone: context for an important archaeological site from the last interglacial on the Red Sea coast of Eritrea. *Palaeogeography, Palaeoclimatology, Palaeoecology*, **203**: 179–206.

Brunet, M., Guy, F., Pilbeam, D., *et al.* 2002. A new hominid from the Upper Miocene of Chad, Central Africa. *Nature*, **418**: 145–51.

Brunton, C. H. C., Cocks, L. R. M. and Long, S. L. (eds.) 2001. *Brachiopods Past and Present*, Special Volumes Series No. 63. Systematics Association.

Bubik, M. and Kaminski, M. A. (eds.) 2004. *Proceedings of the 6th International Workshop on Agglutinated Foraminifera*, Special Publication No. 8. Grzybowski Foundation.

Buchanan, P. G. (ed.) 1996. *Petroleum Exploration, Development and Production in Papua New Guinea, Proceedings of the 3rd PNG petroleum convention, Port Moresby*. PNG Chamber of Mines and Petroleum.

Buffetaut, E. 2004. Polar dinosaurs and the question of dinosaur extinction: a brief review. *Palaeogeography, Palaeoclimatology, Palaeoecology*, **214**: 225–31.

Buffetaut, E. and Mazin, J.-M. (eds.) 2003. *Evolution and Palaeobiology of Pterosaurs*, Special Publication No. 217. Geological Society.

Bukry, D. 1974. Coccoliths as paleosalinity indicators: evidence from the Black Sea. *American Association of Petroleum Geologists Memoir*, **20**: 335–53.

1978. Cenozoic silicoflagellate and coccolith stratigraphy, southeastern Atlantic Ocean, Deep Sea Drilling Project Leg 40. *Initial Reports of the Deep Sea Drilling Project*, **40**: 635–49.

Bulman, O. M. B. 1964. Lower Palaeozoic plankton. *Quarterly Journal of the Geological Society, London*, **119**: 401–18.

Bultynck, P. (ed.) 2000. Subcommission on Devonian stratigraphy: fossil groups important for boundary definition. *Courier Forschungsinstitut Senckenberg*, **220**.

Butterfield, N. J. and Rainbird, R. H. 1998. Diverse organic-walled fossils, including 'possible dinoflagellates', from the Early Neoproterozoic of Arctic Canada. *Geology*, **26**(11): 963–6.

Buttler, C., Cherns, L. and Massa, D. 2004. Late Ordovician cool-water bryozoan mud mounds from Libya. *Palaeontological Association Newsletter*, **57**(48th Annual Meeting Abstracts): 145.

Cagatay, M. N., Gorur, N., Polonia, A., *et al.* 2003. Sea-level changes and depositional environments in the Izmit Gulf, eastern Marmara Sea, during the Late Glacial–Holocene period. *Marine Geology*, **202**(3–4): 159–73.

Cahuzac, B. and Poignant, A. 1997. Essai de biozonation de l'Oligo-Miocene dans les bassins Européens à l'aide des grands foraminifères néritiques. *Bulletin de la Société Géologique de France*, **168**(2): 155–69.

Callaway, J. M. and Nichols, E. I. (eds.) 1997. *Ancient Marine Reptiles*. Academic Press.

Callender, W. R., Staff, G. M., Powell, E. N. and MacDonald, I. R. 1990. Gulf of Mexico hydrocarbon seep communities. V. Biofacies and shell orientation of autochthonous shell beds below storm wavebase. *Palaios*, **5**: 2–14.

Calvin, W. H. 2004. *A Brief History of the Mind: From Apes to Intellect and Beyond*. Oxford University Press.

Cameron, N. R., Bate, R. H. and Clure, V. S. (eds.) 1999. *The Oil and Gas Habitats of the South Atlantic*, Special Publication No. 153. Geological Society.

Campbell, I. H., Czamanske, G. K., Fedorenko, V. A., Hill, R. I. and Stepanov, V. 1992. Synchronism of the Siberian Traps and the Permian–Triassic boundary. *Science*, **258**: 1760–3.

Campbell, K. A. 1992. Recognition of a Mio-Pliocene cold seep setting from the northeast Pacific convergent margin, Washington, USA. *Palaios*, **7**(4): 422–33.

Capriulo, G. M. (ed.) 1990. *Ecology of Marine Protozoa*. Oxford University Press.

Cariou, E. and Hantzpergue, P. (co-ords.) 1997. Biostratigraphie du Jurassique ouest-Européen et Mediterranéen. *Bulletins des Centres Recherche Exploration–Production Elf-Aquitaine, Mémoires*, **17**: 293–304.

Carlson, S. J. and Sandy, M. R. (eds.) 2001. *Brachiopods Ancient and Modern*, Paper No. 7. Paleontological Society.

Carpenter, K. 1999. *Eggs, Nests and Baby Dinosaurs*. Indiana University Press.

Carpenter, K. and Currie, P. J. (eds.) 1990. *Dinosaur Systematics: Approaches and Perspectives*. Cambridge University Press.

Carpenter, K., Hirsch, K. F. and Horner, J. R. (eds.) 1996. *Dinosaur Eggs and Babies*. Cambridge University Press.

Carr, N. G. and Whitton, B. A. (eds.) 1973. *The Biology of Blue-Green Algae*. Oxford University Press.

Carr, T. R., Mason, E. P. and Feazel, C. T. 2003. *Horizontal Wells: Focus on the Reservoir*, Methods in Exploration Series No. 14. American Association of Petroleum Geologists.

Carrillo, M., Paredes, I., Crux, J. A. and de Cabrera, S. 1995. Aptian to Maastrichtian paleobathymetric reconstruction of the eastern Venezuelan Basin. *Marine Micropaleontology*, **26**: 405–18.

Carroll, R. L. 1988. *Vertebrate Paleontology and Evolution*. Cambridge University Press.

Carroll, S. B., Grenier, J. K. and Weatherbee, S. C. 2001. *From DNA to Diversity*. Blackwell.

Casey, R. and Rawson, P. F. (eds.) 1973. *The Boreal Lower Cretaceous*. Seel House Press.

Cavalli-Sforza, L. L. 2000. *Genes, Peoples and Languages*. Penguin.

Cavalli-Sforza, L. L., Menozzi, P. and Piazza, A. 1994. *The History and Geography of Human Genes*. Princeton University Press.

Chaline, J. 1987. Arvicolid data (Arvicolidae, Rodentia) and evolutionary concepts. *Evolutionary Biology*, **21**: 227–310.

Charnock, M. A. and Jones, R. W. 1997. North Sea lituolid foraminifera with complex inner structures: taxonomy, stratigraphy and evolutionary relationships. *Annales Societatis Geologorum Poloniae*, **67**: 183–96.

Checa, A. G. and Jimenez-Jimenez, A. P. 2003. Evolutionary morphology of oblique ribs of bivalves. *Palaeontology*, **46**(4): 709–24.

Cherns, L., Wheeley, J. R. and Karis, L. 2004. Tunnelling trilobites in Middle Ordovician *Thalassinoides*. *Palaeontological Association Newsletter*, **57**(48th Annual Meeting Abstracts): 116.

Chiappe, L.M. and Dingus, L. 2001. *Walking on Eggs: Discovering the Astonishing Secrets of the World of Dinosaurs*. Little, Brown.

Chiappe, L.M. and Witmer, L.M. 2002. *Mesozoic Birds: Above the Heads of Dinosaurs*. University of California Press.

Chivas, A. R., de Deckker, P. and Shelley, G. M. G. 1985. Strontium content of ostracods indicates paleosalinity. *Nature*, **316**: 251–3.

Christensen, W. K. 1995. *Belemnitella* from the Upper Campanian and Lower Maastrichtian of Norfolk, England. *Special Papers in Palaeontology*, **51**.

Clack, J.A. 2002. *Gaining Ground: The Origin and Evolution of Tetrapods*. Indiana University Press.

Clarkson, E.N.K. 1979. The visual system of trilobites. *Palaeontology*, **22**: 1–22.

1998. *Invertebrate Palaeontology and Evolution*, 4th edn. Blackwell.

Cleal, C.J. (ed.) 1991. *Plant Fossils in Geological Investigation: The Palaeozoic*. Ellis Horwood.

Cleevely, R.J. 1983. *World Palaeontological Collections*. British Museum (Natural History).

Coates, A.G. and Jackson, J.B.C. 1987. Clonal growth, algal symbiosis, and reef formation by corals. *Paleobiology*, **13**: 363–78.

Coburn, T.C. and Yarus, J.M. (eds.) 2000. *Geographic Information Systems in Petroleum Exploration and Development*, Computer Applications in Geology No. 4. American Association of Petroleum Geologists.

Coccioni, R. and Luciani, V. 2004. Planktonic foraminifera and environmental changes across the Bonarelli event (OAE2, latest Cenomanian) in its type area: a high-resolution study from the Tethyan reference Bottaccione section (Gubbio, central Italy). *Journal of Foraminiferal Research*, **34**(2): 109–29.

Cocks, L.R.M. 2001. Ordovician and Silurian global geography. *Journal of the Geological Society, London*, **158**: 197–210.

Cocks, L. R. M. and Fortey, R. A. 1982. Faunal evidence for oceanic separations in the Palaeozoic of Britain. *Journal of the Geological Society, London*, **139**: 465–78.

Coe, A.L. (ed.) 2003. *The Sedimentary Record of Sea-Level Change*. Cambridge University Press.

Cohen, A.S. 2003. *Paleolimnology: The History and Evolution of Lake Systems*. Oxford University Press.

Colbert, E.H. and Morales, M. 1991. *Evolution of the Vertebrates*, 4th edn. Wiley-Liss.

Cole, K.M. and Sheath, R.G. 1990. *Biology of the Red Algae*. Cambridge University Press.

Colella, A. and Di Geronimo, I. 1987. Surface sediments and macrofaunas of the Crati submarine fan (Ionian Sea, Italy). *Sedimentary Geology*, **51**: 257–77.

Coles, G.P., Ainsworth, N.R., Whatley, R.C. and Jones, R.W. 1996. Foraminifera and Ostracoda from Quaternary carbonate mounds associated with gas seepage in the Porcupine Basin, offshore western Ireland. *Revista Espanola de Micropaleontologia*, **27**(2): 113–51.

Colin, J. P., de Deckker, P. and Peypouquet, J. P. (eds.) 1987. *Ostracods in Earth Sciences*. Elsevier.

Collignon, M. 1981. Faune Albo-Cenomanienne de la formation des marnes de Kazhdumi, region du Fars-Khuzestan (Iran). *Documents du Laboratoire de Géologie de la Faculté des Sciences, Lyon, Hors Série*, **6**: 251–91.

Collins, C. (ed.) 1995. *The Care and Conservation of Palaeontological Material*. Butterworth-Heinemann.

Conroy, G.C. 1990. *Primate Evolution*. W. W. Norton.

1997. *Reconstructing Human Origins*. W. W. Norton.

Conway Morris, S. 1993. Ediacaran-like fossils in Cambrian Burgess shale-type faunas of North America. *Palaeontology*, **36**: 593–635.

1998. *The Crucible of Creation*. Oxford University Press.

2003. *Life's Solution: Inevitable Humans in a Lonely Universe*. Cambridge University Press.

Conway Morris, S. and Peel, J.S. 1990. Articulated halkieriids from the Lower Cambrian of north Greenland. *Nature*, **345**: 802–5.

1995. Articulated halkieriids from the Lower Cambrian of north Greenland and their role in early protostome evolution. *Philosophical Transactions of the Royal Society of London B*, **347**: 305–58.

Cook, P.J. and Shergold, J.H. 1984. Phosphorus, phosphorites and skeletal evolution at the Precambrian–Cambrian boundary. *Nature*, **308**: 231–6.

Coope, G.R. 1977. Fossil coleopteran assemblages as sensitive indicators of climatic changes during the Devensian (last) cold stage. *Philosophical Transactions of the Royal Society of London B*, **280**: 313–40.

Cooper, R.A., Fortey, R.A. and Lindholm, K. 1991. Latitudinal and depth zonation of Early Ordovician graptolites. *Lethaia*, **24**: 199–218.

Coppens, Y. 2003. *Human Origins*. Hachette.

Copper, P. and Jin, J. (eds.) 1996. *Brachiopods, Proceedings of the 3rd International Brachiopod Congress*, Sudbury, Ontario, Canada. Balkema.

Corliss, B. H. 1985. Microhabitats of benthic foraminifera within deep-sea sediments. *Nature*, **314**: 435–8.

Corliss, B.H. and Chen, C. 1988. Morphotype patterns of Norwegian Sea deep-sea benthic foraminifera and ecologic implications. *Geology*, **16**: 716–19.

Corliss, B.H. and Fois, E. 1991. Morphotype analysis of deep-sea benthic foraminifera from the northwest Gulf of Mexico. *Palaios*, **5**: 589–605.

Corvinus, G. and Rimal, L.N. 2001. Biostratigraphy and geology of the Neogene Siwalik group of the Surai Khola and Rato Khola areas in Nepal. *Palaeogeography, Palaeoclimatology, Palaeoecology*, **165**: 251–79.

Coryndon, S. C. and Savage, R. J. G. 1973. The origin and affinities of African mammal faunas. *Special Papers in Palaeontology*, **12**: 121–35.

Costello, M.J., Emblow, C. and White, R. (eds.) 2001. *European Register of Marine Species*. Muséum National d'Histoire Naturelle.

Courtillot, V. 1999. *Evolutionary Catastrophes*, English translation. Cambridge University Press.

Cowen, R. in press. *The History of Life*, 4th edn. Blackwell.

Cozar, P. and Rodriguez, S. 2003. The palaeoecological distribution of the endothyroids (foraminifera) in the Guadiato area (SW Spain, Mississippian). *Palaeogeography, Palaeoclimatology, Palaeoecology*, **201**(1–2): 1–19.

Craig, G.Y. 1991. *The Geology of Scotland*, 3rd edn. Geological Society.

Crame, J.A. and Owen, A.W. (eds.) 2002. *Palaeobiogeography and Biodiversity Change: The Ordovician and Mesozoic–Cenozoic Radiations*, Special Publication No. 194. Geological Society.

Cramer, F.H. and Diez, M.D.C. R. 1978. Lower Paleozoic phytoplankton from North Africa and adjacent regions: general survey. *Annales des Mines et de la Géologie*, **28**: 21–34.

Crasquin-Soleau, S. and Barrier, E. (eds.) 1998a. *Peri-Tethys Memoir 3: Stratigraphy and Evolution of Peri-Tethyan Platforms*, Mémoires No. 177. Muséum National d'Histoire Naturelle.

(eds.) 1998b. *Peri-Tethys Memoir 4: Epicratonic Basins of Peri-Tethyan Platforms*, Mémoires No. 179. Muséum National d'Histoire Naturelle.

(eds.) 2004. *Peri-Tethys Memoir 5: New Data on Peri-Tethyan Sedimentary Basins*, Mémoires No. 182. Muséum National d'Histoire Naturelle.

Crasquin-Soleau, S., Broutin, J., Roger, J., *et al.* 1999. First Permian ostracode fauna from the Arabian plate (Khuff formation, Sultanate of Oman). *Micropaleontology*, **45**(2): 163–82.

Crasquin-Soleau, S., Marcoux, J., Angiolini, L. and Nicora, A. 2004. Palaeocopida (Ostracoda) across the Permian–Triassic events: new data from southwestern Taurus (Turkey). *Journal of Micropalaeontology*, **23**(1): 67–76.

Crimes, T.P. 1973. From limestones to distal turbidites: a facies and trace fossil analysis of the Zumaya flysch (Paleocene–Eocene, north Spain). *Sedimentology*, **20**: 105–31.

Crimes, T.P. and Crossley, J.D. 1980. Inter-turbidite bottom current orientation from trace fossils with an example from the Silurian flysch of Wales. *Journal of Sedimentary Petrology*, **50**: 821–30.

Crimes, T.P. and Fedonkin, M.A. 1994. Evolution and dispersal of deep-sea traces. *Palaios*, **9**: 74–83.

Croucher, R. and Woolley, A.R. 1982. *Fossils, Minerals and Rocks: Collection and Preservation*. British Museum (Natural History).

Crowley, T.J. and North, G.R. 1991. *Paleoclimatology*. Oxford University Press.

Culver, S.J. and Rawson, P.F. (eds.) 2000. *Biotic Response to Global Change*. Cambridge University Press.

Cuneo, N.R., Taylor, E.L., Taylor, T.N. and Krings, M. 2003. *In situ* fossil forest from the upper Fremouw formation (Triassic) of Antarctica: paleoenvironmental setting and paleoclimate analysis. *Palaeogeography, Palaeoclimatology, Palaeoecology*, **197**: 239–61.

Cunliffe, B. (ed.) 1998. *Prehistoric Europe: An Illustrated History*. Oxford University Press.

Cuny, G. 1995. French vertebrate faunas and the Triassic-Jurassic boundary. *Palaeogeography, Palaeoclimatology, Palaeoecology*, **119**: 343–58.

Curran, H.A. and Martin, A.J. 2003. Complex decapod burrows and ecological relationships in modern and Pleistocene intertidal carbonate environments, San Salvador Island, Bahamas. *Palaeogeography, Palaeoclimatology, Palaeoecology*, **192**: 229–46.

Currant, A. 1989. The Quaternary origins of the modern British mammal fauna. *Biological Journal of the Linnean Society*, **38**: 23–30.

Currant, A. and Jacobi, R. 1997. Vertebrate faunas from the British Late Pleistocene and the chronology of human settlement. *Quaternary Newsletter*, **82**: 1–8.

2001. A formal mammalian biostratigraphy for the Late Pleistocene of Britain. *Quaternary Science Reviews*, **20**: 1707–16.

Currie, P.J., Kopplehus, E.B., Shugar, M.A. and Wright, J.L. (eds.) 2004. *Feathered Dragons: Studies on the Transition from Dinosaurs to Birds*. Indiana University Press.

Curtis, G.H., Swisher, C.C., III and Lewin, R. 2001. *Java Man: How Two Geologists Changed the History of Human Evolution*. Little, Brown.

Cutler, A. 2003. *The Seashell on the Mountaintop: A Story of Science, Sainthood and the Humble Genius who Discovered a New History of the Earth*. Heinemann.

Dain, L.G. and Kuznetsova, K.I. 1971. Zonal subdivision of the stratotypical section of the Volgian stage based on foraminifera. *Voprosy Mikropaleontologyii*, **14**: 103–24.

Daizhao Chen and Tucker, M.E. 2003. The Frasnian–Famennian mass extinction: insights from high-resolution sequence stratigraphy and cyclostratigraphy in south China. *Palaeogeography, Palaeoclimatology, Palaeoecology*, **193**: 87–111.

Dando, P.R., Austen, M.C., Burke, R.A., Jr, *et al.* 1991. Ecology of a North Sea pockmark with an active methane seep. *Marine Ecology Progress Series*, **70**: 49–63.

Dando, P.R., Bussmann, I., Niven, S.J., *et al.* 1994. A methane seep area in the Skagerak, the habitat of the pogonophore *Siboglinum poseidoni* and the bivalve mollusc *Thyasira sarsi*. *Marine Ecology Progress Series*, **107**: 157–67.

Danielian, T., Ricordel, C. and Musavu-Moussavou, B. 2003. Oceanic anoxic events (OAEs) and plankton evolution: a case study from Mid-Cretaceous Radiolaria. *Palaeontological Association Newsletter*, **54**(47th Annual Meeting Abstracts): 128.

Darwin, C. 1859. *On the Origin of Species by Means of Natural Selection*. John Murray.

Davis, S.J.M. 1987. *The Archaeology of Animals*. Routledge.

Davison, I., Bate, R. and Reeves, C. 2004. Early opening of the South Atlantic: Berriasian rifting to Aptian salt deposition. *Extended Abstracts, 3rd Petroleum Exploration Society of Great Britain/Houston Geological Society International Conference on African Exploration and Production*.

Dawkins, R. 2004. *The Ancestor's Tale: A Pilgrimage to the Dawn of Life*. Weidenfeld & Nicolson.

De Bonis, L., Koufos, G. D. and Andrews, P. (eds.) 2001. *Hominoid Evolution and Climate Change in Europe*, vol. 2, *Phylogeny of the Neogene Hominoid Primates in Eurasia*. Cambridge University Press.

De Bruijn, H., Daams, R., Daxner-Hock, G., *et al.* 1992. Report of the RCMNS working group on fossil mammals, Reisenberg, 1990. *Newsletters on Stratigraphy*, **26**(2/3): 65–118.

De Graciansky, P.-C., Hardenbol, J., Jacquin, T. and Vail, P. R. (eds.) 1998. *Mesozoic and Cenozoic Sequence Stratigraphy of European Basins*, Special Publication No. 60. Society of Economic Paleontologists and Mineralogists.

De Menocal, P. B. 1995. Plio-Pleistocene African climate. *Science*, **270**: 53–9.

De Vargas, C., Norris, R., Zaninetti, L., Gibb, S.W. and Pawlowski, J. 1999. Molecular evidence of cryptic speciation in planktonic foraminifers and their relation to oceanic provinces. *Proceedings of the National Academy of Sciences, USA*, **96**: 2864–8.

De Verteuil, L. and Johnson, N. 2002. Upper Miocene–Pliocene subsurface palynological reference section for the Cruse, Forest and Morne l'Enfer formations, Forest Reserve field, Trinidad, West Indies. *Abstracts, Joint Meeting, American Association of Stratigraphic Palynologists–The Micropalaeontological Society–North American Micropaleontological Society, University College London*.

De Wever, P. and Popova, I. M. 1997. Cenozoic Radiolaria from the European platform: a review. *Geodiversitas*, **19**(2): 383–470.

De Wever, P. and Vishnevskaya, V. S. 1997. Mesozoic radiolarians from the European platform: a review. *Geodiversitas*, **19**(2): 319–82.

De Wever, P., Dumitrica, P., Caulet, J. P., Nigrini, C. and Cadidroit, M. 2002. *Radiolarians in the Sedimentary Record*. Gordon & Breach.

De Wit, M. J., Ghosh, J. G., de Villiers, S., *et al.* 2002. Multiple organic carbon isotope reversals across the Permo-Triassic boundary of terrestrial Gondwana sequences: clues to extinction patterns and delayed ecosystem recovery. *Journal of Geology*, **110**: 227–40.

Diamond, J. 1991. *The Rise and Fall of the Third Chimpanzee: How our Animal Heritage Affects the Way we Live*. Radius.

Diaz de Gamero, M. L. 1996. The changing course of the Orinoco River during the Neogene: a review. *Palaeogeography, Palaeoclimatology, Palaeoecology*, **123**: 385–402.

Dickens, G. R., Castillo, M. M. and Walker, J. C. G. 1997. A blast of gas in the latest Paleocene: simulating first-order effects of massive dissociation of oceanic methane hydrate. *Journal of Geology*, **25**: 259–62.

Dickson, J. H., Oeggl, K. and Hardey, L. L. 2003. The Iceman reconsidered. *Scientific American*, **288**(5): 70–80.

Dincauze, D. F. 2000. *Environmental Archaeology: Principles and Practice*. Cambridge University Press.

Dingus, L. and Rowe, T. 1997. *The Mistaken Extinction: Dinosaur Evolution and the Origin of Birds*. W. H. Freeman.

Dixon, D., Jenkins, I., Moody, R. and Zhuravlev, A. 2001. *Atlas of Evolution*. Cassell.

Dixon, J., Dietrich, J., McNeil, D. H., *et al.* 1985. *Geology, Biostratigraphy and Organic Geochemistry of Jurassic to Pleistocene Strata, Beaufort-Mackenzie Area, Northwest Canada*. Canadian Society for Petroleum Geology.

Dodd, J. R. and Stanton, R. J. 1990. *Paleoecology: Concepts and Applications*, 2nd edn. John Wiley.

Dolson, J., Barkooky, A. E., Wehr, F., *et al.* 2002. *The Eocene and Oligocene Paleo-Ecology and Paleo-Geography of Whale Valley and the Fayoum Basin: Implications for Hydrocarbon Exploration in the Nile Delta and Eco-Tourism in the Greater Fayoum Basin*, Field Trip Guidebook for Field Trip No. 7, American Association of Petroleum Geologists International Conference and Exhibition, Cairo, 2002 (Search and Discovery Article No. 10030). American Association of Petroleum Geologists.

Donoghue, P. C., Forey, P. L. and Aldridge, R. J. 2000. Conodont affinity and chordate phylogeny. *Biological Reviews*, **75**: 191–251.

Donaghue, P. and Smith, P. (eds.) 2004. *Telling the Evolutionary Time: Molecular Clocks and the Fossil Record*. Taylor & Francis.

Donovan, S. K. (ed.) 1991. *The Processes of Fossilisation*. Belhaven Press.

1994. *The Palaeobiology of Trace Fossils*. John Wiley.

1995. Pelmatozoan columnals from the Ordovician of the British Isles. Pt. 3. *Palaeontographical Society Monograph*, **149**(597): 115–93.

Dortch, J. 2004. Late Quaternary vegetation change and the extinction of black-flanked rock-wallaby (*Petrogale lateralis*) at Tunnel Cave, southwestern Australia. *Palaeogeography, Palaeoclimatology, Palaeoecology*, **211**(3–4): 185–204.

Doyle, P. 1996. *Understanding Fossils*. John Wiley.

Doyle, P. and Bennett, M. 1998. *Unlocking the Stratigraphical Record*. John Wiley.

Doyle, P. and MacDonald, D. I. M. 1993. Belemnite battlefields. *Lethaia*, **26**: 65–80.

Doyle, P. and Shakides, E. V. 2004. The Jurassic belemnite suborder Belemnotheutina. *Palaeontology*, **47**(4): 983–98.

Droser, M. L., Bottjer, D. J., Sheehan, P. M. and McGhee, G. R. 2000. Decoupling of taxonomic and ecologic severity of Phanerozoic marine mass extinctions. *Geology*, **28**: 675–8.

Drucker, D., Bocherens, H., Bridault, A. and Billiou, D. 2003. Carbon and nitrogen isotopic composition of red deer (*Cervus elaphus*) collagen as a tool for tracking palaeoenvironmental change during the Late Glacial and early Holocene in the northern Jura (France). *Palaeogeography, Palaeoclimatology, Palaeoecology*, **195**: 375–88.

Duarte-Silva, A., Palma, S., Sobrinho-Goncalves, L. and Moita, M. T. 2004. *Braarudosphaera bigelowii* in waters of the upwelling coast of Portugal. *Abstracts of the 10th International Nannoplankton Association Conference*, Lisbon: 35.

Duff, P. McL. D. and Smith, A. J. 1992. *The Geology of England and Wales*. Geological Society.

Dunbar, R. 2004. *The Human Story: A New History of Mankind's Evolution*. Faber & Faber.

Dunnington, H. V. 1955. Close zonation of Upper Cretaceous globigerinal sediments by abundance ratios of *Globotruncana* species groups. *Micropaleontology*, **1**(3): 207–19.

Dupont, L. M., Marret, F. and Winn, K. 1988. Land–sea correlation by means of terrestrial and marine palynomorphs from the equatorial East Atlantic: phasing of SE trade winds and oceanic productivity. *Palaeogeography, Palaeoclimatology, Palaeoecology*, **142**: 51–84.

Duque Caro, H. 1990. Neogene stratigraphy, paleooceanography and paleobiogeography in northwest South America and the evolution of the Panama seaway. *Palaeogeography, Palaeoclimatology, Palaeoecology*, **77**: 203–34.

Dybkjaer, K. 2004. Dinocysts stratigraphy and palynofacies studies used for refining a sequence stratigraphic model: uppermost Oligocene to Lower Miocene, Jylland, Denmark. *Review of Palaeobotany and Palynology*, **131**(3–4): 201–49.

Dyer, B. D. and Oban, R. A. 1994. *Tracing the History of Eukaryotic Cells: The Enigmatic Smile*. Columbia University Press.

Dyson, F. 1999. *Origins of Life*, 2nd edn. Cambridge University Press.

Eaton, G. L. 1996. *Seriliodinium*, a new late Cenozoic dinoflagellate from the Black Sea. *Review of Palaeobotany and Palynology*, **91**: 151–69.

Edwards, J. D. and Santogrossi, P. A. (eds.) 1990. *Divergent/Passive Margin Basins*, Memoir No. 48. American Association of Petroleum Geologists.

Edwards, L. E. and Powars, D. S. 2003. Impact damage to dinocysts from the late Eocene Chesapeake Bay event. *Palaios*, **49**(3): 275–85.

Ehrenberg, S. N., Pickard, N. A. H., Henriksen, L. B., *et al.* 2001. A depositional and sequence stratigraphic model for cold-water, spiculitic strata based on the Kapp Starostin formation (Permian) of Spitsbergen and equivalent deposits from the Barents Sea. *American Association of Petroleum Geologists Bulletin*, **85**(12): 2061–87.

Elias, S. A. 1994. *Quaternary Insects and Their Environments*. Smithsonian Institution Press.

El-Khayal, A. A. 1985. Occurrence of a characteristic Wealden fern (*Weichselia reticulata*) in the Wasia formation, central Saudi Arabia. *Scripta Geologica*, **79**: 75–88.

Elsik, W. C., Baesemann, C. B., Graham, A. K., *et al.* 1983. Annotated glossary of fungal palynomorphs. *American Association of Stratigraphic Palynologists Foundation, Contributions Series*, **11**: 1–35.

Elton, S., Bishop, L. C. and Wood, B. 2001. Comparative context of Plio-Pleistocene brain evolution. *Journal of Human Evolution*, **41**: 1–27.

Emery, D. and Myers, K. J. (eds.) 1996. *Sequence Stratigraphy*. Blackwell.

Emiliani, C. 1955. Pleistocene temperatures. *Journal of Geology*, **63**: 538–78.

Epstein, A. G., Epstein, J. B. and Harris, L. D. 1977. *Conodont Color Alteration: An Index to Organic Metamorphism*, US Geological Survey Professional Paper No. 995. US Government Printing Office.

Erdal Kerey, I., Meric, E., Tunoglu, C., *et al.* 2004. Black Sea–Marmara Sea Quaternary connections: new data from the Bosphorus, Istanbul, Turkey. *Palaeogeography, Palaeoclimatology, Palaeoecology*, **204**(3–4): 277–95.

Ericson, D. B. and Wollin, G. 1956. Correlation of six cores from the equatorial Atlantic and Caribbean. *Deep Sea Research*, **3**: 104–25.

Ernst, W. G. (ed.) 2000. *Earth Systems: Processes and Issues*. Cambridge University Press.

Erwin, D. H. 1993. *The Great Paleozoic Crisis*. Columbia University Press.

1998. The end and the beginning: recoveries from mass extinctions. *Trends in Ecology and Evolution*, **13**: 344–9.

2001. Lessons from the past: biotic recoveries from mass extinction. *Proceedings of the National Academy of Sciences, USA*, **98**: 5399–403.

Erwin, D. H. and Wing, S. L. (eds.) 2000. *Deep Time: Paleobiology's Perspective*. Paleontological Society.

Erzinclioglu, Y. Z. 2000. *Maggots, Murder and Men: Memories and Reflections of a Forensic Entomologist*. Harley Books.

Evans, A. M. 1997. *An Introduction to Economic Geology and Its Environmental Impact*. Blackwell.

Evans, D., Graham, C., Armour, A. and Bathurst, P. (eds. and co-ords.) 2003. *The Millennium Atlas: Petroleum Geology of the Central and Northern North Sea*. Geological Society/Norwegian Petroleum Society/Geological Survey of Denmark and Greenland.

Eyles, N. 1993. Earth's glacial record and its tectonic setting. *Earth-Science Reviews*, **35**(1/2): 1–248.

Ezquerra, R., Costeur, L., Galton, P. M., Doublet, S. and Perez-Lorente, F. 2004. Lower Cretaceous swimming theropod trackway from La Virgen del Campo (La Rioja, Spain). *Palaeontological Association Newsletter*, **57**(48th Annual Meeting Abstracts): 150.

Fagan, B. M. 1998. *People of the Earth: An Introduction to World Prehistory*. 9th edn. Longman.

2004. *The Long Summer: How Climate Changed Civilization*. Granta.

Fara, E. 2004. Estimating minimum global species diversity for groups with a poor fossil record: a case study of Late Jurassic–Eocene lissamphibians. *Palaeogeography, Palaeoclimatology, Palaeoecology*, **207**(1–2): 59–82.

Farlow, J. O. and Brett-Surman, M. K. (eds.) 1997. *The Complete Dinosaur*. Indiana University Press.

Fastovsky, D. E. and Weishampel, D. B. 2005. *The Evolution and Extinction of the Dinosaurs*. Cambridge University Press.

FAUNMAP Working Group 1996. Spatial response of mammals to Late Quaternary environmental fluctuations. *Science*, **272**: 1601–6.

Fedonkin, M. A. 1987. *The Non-Skeletal Fauna of the Vendian and Its Place in the Evolution of Metazoans*. Nauka. (in Russian)

Feduccia, A. 1996. *The Origin and Evolution of Birds*. Yale University Press.

Fenchel, T. 2002. *Origin & Early Evolution of Life*. Oxford University Press.

Fenerci-Masse, M., Masse, J.-P. and Chazottes, V. 2004. Quantitative analysis of rudist assemblages: a key for palaeocommunity reconstructions. the late Barremian record from SE France. *Palaeogeography, Palaeoclimatology, Palaeoecology*, **206**(1–2): 133–47.

Ficcarelli, G., Coltori, M., Moreno-Espinosa, M., *et al.* 2003. A model for the Holocene extinction of the mammal megafauna in Ecuador. *Journal of South American Earth Sciences*, **15**(8): 835–45.

Fiffer, S. 2000. *Tyrannosaurus Sue*. W. H. Freeman.

Finlayson, C. 2004. *Neandertals and Modern Humans: An Ecological and Evolutionary Perspective*. Cambridge University Press.

Fischer, J.-C., le Nindre, Y.-M., Manivit, J. and Vaslet, D. 2001. Jurassic gastropod faunas of central Saudi Arabia. *GeoArabia*, **6**(1): 63–100.

Fisher, C. G., Sageman, B. R., Asure, S. E., Acker, B. and Mahar, Z. 2003. Planktonic foraminiferal porosity analysis as a tool for paleoceanographic reconstruction, mid Cretaceous western interior sea. *Palaios*, **18**(1): 34–46.

Fleagle, J. G. 1999. *Primate Adaptation and Evolution*, 2nd edn. Academic Press.

Flecker, R. and Ellam, R. M. 1999. Distinguishing climatic and tectonic signals in the sedimentary successions of marginal basins using Sr isotopes: an example from the Messinian salinity crisis, eastern Mediterranean. *Journal of the Geological Society, London*, **156**: 847–54.

Flood, R. D., Piper, D. J. W., Klaus, A. and Peterson, L. C. (eds.) 1997. *Proceedings of the Ocean Drilling Program, Scientific Results*, **155**.

Flor, L., Pilloud, A., Crux, J. and Arelis, F. 1997. Bioestratigrafico con nanofossiles calcareos del Paleogeno en las secciones Rio Aragua y Quebrada la Galina, Oriente de Venezuela. *Memorias del I Congreso Latinamericano de Sedimentologio, Porlamar, Venezuela.*

Floyd, J. D., Williams, M. and Rushton, A. W. A. 1999. Late Ordovician (Ashgill) ostracodes from the Drummock group, Craighead inlier, Girvan district, SW Scotland. *Scottish Journal of Geology*, **35**(1): 15–24.

Flynn, J. J., Wyss, A. R., Croft, D. A. and Charrier, R. 2003. The Tinguiririca fauna, Chile: biochronology, paleoecology, biogeography, and a new earliest Oligocene South American land mammal "age". *Palaeogeography, Palaeoclimatology, Palaeoecology*, **195**: 229–59.

Follot, J. 1952. *Monographies Régionale*, 1st series, *Algerie*, No. 1, *Ahnet et Mouydir*. XIXème Congrès Geologique International.

Ford, D. and Golonka, J. 2003. Phanerozoic paleogeography, paleoenvironment and lithofacies maps of the circum-Atlantic margins. *Marine and Petroleum Geology*, **20**(3–4): 249–85.

Forey, P. L. 1997. *History of the Coelacanth Fishes*. Chapman & Hall.

Forey, P. L., Humphries, C. J., Kitching, I. L., *et al.* 1992. *Cladistics: A Practical Course in Systematics*, Systematics Association Publication No. 10. Clarendon Press.

Fortelius, M. and Solounias, N. 2000. Functional characterisation of ungulate molars using abrasion attrition wear gradient. *American Museum Novitates*, **3301**: 1–36.

Fortey, R. A. 2000. *Trilobite*. HarperCollins.

Fortey, R. A. and Cocks, L. R. M. 2003. Palaeontological evidence bearing on global Ordovician–Silurian continental reconstructions. *Earth-Science Reviews*, **61**: 245–307.

Fortey, R. A. and Owens, R. M. 1999. Feedings habits in trilobites. *Palaeontology*, **42**(3): 429–66.

Fortey, R. A., Briggs, D. E. G. and Wills, M. A. 1996. The Cambrian evolutionary "explosion": decoupling cladogenesis from morphological disparity. *Biological Journal of the Linnean Society*, **57**: 13–33.

Fourcade, E. and Michaud, F. 1987. L'Ouverture de l'Atlantique et son influence sur les peuplements des grands foraminifères des plates-formes peri-océaniques au Mesozoique. *Geodinimica Acta*, **1**(4–5): 247–62.

Fourtanier, E. and Seyve, C. 2001. Biostratigraphy of two offshore Upper Miocene drilled sections from western Africa. *Africa Geoscience Review*, **8**(1–2): 37–48.

Fowler, B. 2000. *Iceman*. MacMillan.

Fox, D. L. and Fisher, D. C. 2004. Dietary reconstruction of Miocene *Gomphotherium* (Mammalia, Proboscidea) from the Great Plains region, USA, based on the carbon isotope composition of tusk and molar enamel. *Palaeogeography, Palaeoclimatology, Palaeoecology*, **206**(3–4): 311–35.

Frakes, L. A., Francis, J. E. and Syktus, J. I. 1992. *Climate Modes of the Phanerozoic*. Cambridge University Press.

Frankel, C. 1999. *The End of the Dinosaurs: Chicxulub Crater and Mass Extinctions*, English translation. Cambridge University Press.

Fraser, N. 2000. The Early Jurassic reef gap interval: paleoecology and sclerochronology of *Lithiotis problematica*. *Abstracts and Proceedings, CalPaleo Conference*, Berkeley, CA: 17.

2001. Dissecting the "*Lithiotis*" facies: implications for the Early Jurassic "reef eclipse". *Program with Abstracts, Geological Society of America Annual Meeting, 2001*: Paper No. 156-0.

2003. Sclerochronology and stable isotopic records of "*Lithiotis*" facies bivalves: rapid growth rates not longevity. *Palaeontological Association Newsletter*, **54**(47th Annual Meeting Abstracts): 171.

Fraser, N. and Bottjer, D. 2001. The beginning of Mesozoic bivalve reefs: Early Jurassic "*Lithiotis*" facies bioherms. *Abstracts, North American Paleontological Convention*, Berkeley, CA.

Fraser, N. C. and Sues, H.-D. (eds.) 1994. *In the Shadow of the Dinosaurs: Early Mesozoic Tetrapods*. Cambridge University Press.

Fraser, R.H. and Currie, D.J. 1996. The species richness–energy hypothesis in a system where historical factors are thought to prevail: coral reefs. *American Naturalist*, **148**: 138–59.

Frey, R.W. and Pemberton, S.G. 1985. Biogenic structures in outcrops and cores. I. Introduction to ichnology. *Bulletin of Canadian Petroleum Geology*, **33**: 72–115.

Friis, E.M. 2003. Stem groups and angiosperm origins. *Palaeontological Association Newsletter*, **54**(47th Annual Meeting Abstracts): 131.

Fritsen, A., Bailey, H., Gallagher, L., *et al.* 1999. *A Joint Chalk Stratigraphic Framework*. Norwegian Petroleum Directorate.

Fry, I. 2000. *The Emergence of Life on Earth: A Historical and Scientific Overview*. Rutgers University Press.

Fugagnoli, A. 2004. Trophic regimes of benthic foraminiferal assemblages in Lower Jurassic shallow water carbonates from northeastern Italy (Calcari Grigi, Trento platform, Venetian Prealps). *Palaeogeography, Palaeoclimatology, Palaeoecology*, **205**(1–2): 111–30.

Futuyma, D. 1998. *Evolutionary Biology*, 3rd edn. Sinauer Associates.

Gaemers, P.A.M. 1978. A biozonation based on Gadidae otoliths for the north-west European younger Cenozoic, with a description of some new species and genera. *Mededlingen, Werkgroep voor Tertiare en Kwartiare Geologie*, **15**(4): 141–61.

Gaillard, C., Rio, M., Rolin, Y. and Roux, M. 1992. Fossil chemosynthetic communities related to vents or seeps in sedimentary basins: the pseudobioherms of southeastern France compared to other world examples. *Palaios*, **7**(4): 451–5.

Gallagher, S.J., Greenwood, D.R., Taylor, D., *et al.* 2003. The Pliocene climatic and environmental evolution of southeastern Australia: evidence from the marine and terrestrial realm. *Palaeogeography, Palaeoclimatology, Palaeoecology*, **193**: 349–82.

Galloway, W.E. 1989. Genetic stratigraphic sequences in basin analysis. I. architecture and genesis of flooding-surface bounded depositional units. *American Association of Petroleum Geologists Bulletin*, **73**(2): 125–42.

Gardom, T. and Milner, A. 2001. *The Natural History Museum Book of Dinosaurs*. Carlton Books.

Garrison, E.G. 2000. *Techniques in Archaeological Geology*. Springer-Verlag.

Gastaldo, R.A. and DiMichele, W.D. (eds.) 2000. *Phanerozoic Terrestrial Ecosystems: Notes for a Short Course*, Paper No. 6. Paleontological Society.

Gastaldo, R.A., Riegel, W., Puttmann, W., Linnemann, U.G. and Zetter, R. 1998. A multidiscipinary approach to reconstruct the Late Oligocene vegetation in central Europe. *Review of Palaeobotany and Palynology*, **101**(1–4): 71–94.

Gayet, M., Abisaad, P. and Belouze, A. 2003. *Les Poissons Fossiles*. Editions Desiris.

Gebhardt, H. 2003. Palaeobiogeography of Late Oligocene to Early Miocene central European Ostracoda and foraminifera: progressive isolation of the Mainz basin, northern upper Rhine graben and Hanau basin/Wetterau. *Palaeogeography, Palaeoclimatology, Palaeoecology*, **201**(3–4): 343–54.

Gee, H. 1996. *Before the Backbone: Views on the Origin of Vertebrates*. Chapman & Hall.

Gehling, J.M. 1999. Microbial mats in terminal Proterozoic siliciclastics: Ediacaran death masks. *Palaios*, **14**: 40–57.

Gensel, P.G. and Edwards, D. (eds.) 2001. *Plants Invade the Land: Evolutionary and Environmental Perspectives*. Columbia University Press.

Gerhard, L.C., Harrison, W.E. and Hanson, B.M. 2001. *Geological Perspectives of Global Climate Change*, Studies in Geology No. 47. American Association of Petroleum Geologists.

Germes, G.J.B. 1983. Implications of a sedimentary facies and depositional environmental analysis of the Nama group in South-West Africa/Namibia. *Geological Society of South Africa Special Publications*, **11**: 89–114.

Ge Sun and Dilcher, D.L. 2002. Early angiosperms from the Lower Cretaceous of Heilongjiang, China. *Review of Palaeobotany and Palynology*, **121**: 91–112.

Ghetti, P., Anadon, P., Bertini, A., *et al.* 2002. The early Messinian Velona basin (Siena, central Italy): paleoenvironmental and paleobiogeographical reconstructions. *Palaeogeography, Palaeoclimatology, Palaeoecology*, **187**: 1–33.

Ghosh, P., Padia, J.T. and Mohindra, R. 2004. Stable isotopic studies of paleosol sediment from upper Siwalik of Himachal Himalaya: evidence for high monsoonal intensity during Late Miocene? *Palaeogeography, Palaeoclimatology, Palaeoecology*, **206**(1–2): 103–14.

Gibbons, W. and Moreno, T. (eds.) 2002. *The Geology of Spain*. Geological Society.

Gili, E., Masse, J.-P. and Skelton, P.W. 1995. Rudists as gregarious sediment dwellers, not reef-builders on Cretaceous carbonate platforms. *Palaeogeography, Palaeoclimatology, Palaeoecology*, **118**: 245–67.

Gill, F.L., Little, C.T.S. and Harding, I.C. 2003. Tertiary cold seeps of the Caribbean region. *Palaeontological Association Newsletter*, **54**(47th Annual Meeting Abstracts): 133.

Gill, G.A., Santantonio, M. and Lathuiliere, B. 2004. The depth of pelagic deposits in the Tethyan Jurassic and the use of corals: an example from the Apennines. *Sedimentary Geology*, **166**(3–4): 311–34.

Gillette, D.D. and Lockley, M.G. (eds.) 1989. *Dinosaur Tracks and Traces*. Cambridge University Press.

Girouard, M. 1981. *Alfred Waterhouse and The Natural History Museum*. British Museum (Natural History).

Gladenkov, A.Y. 1998. Oligocene and Lower Miocene diatom zonation in the North Pacific. *Stratigraphy and Geological Correlation*, **6**(2): 150–63.

Gladenkov, A.Y., White, L.D., Gladenkov, Y.B. and Blueford, J.R. 2000. Cenozoic biostratigraphy of the Pogranichnyi region, eastern Sakhalin, Russia. *Palaeogeography, Palaeoclimatology, Palaeoecology*, **158**: 45–64.

Gladenkov, A.Y., Oleinik, A.E., Marincovich, L., Jr and Barnov, K.B. 2002. A refined age for the earliest opening of Bering

Strait. *Palaeogeography, Palaeoclimatology, Palaeoecology*, **183**: 321–8.

Gladenkov, Y.B. 1980. Stratigraphy of marine Paleogene and Neogene of northeast Asia (Chukotka, Kamchatka, Sakhalin). *American Association of Petroleum Geologists Bulletin*, **64**(7): 1087–93.

Gladenkov, Y.B. 1999. *Cenozoic Ecosystem of the Okhotsk Sea Region*. GEOS. (in Russian)

Glaessner, M.F. 1984. *The Dawn of Animal Life*. Cambridge University Press.

Glasspool, I.J., Hilton, J., Collinson, M.E., Wang, S.-J. and Sen, L.C. 2004. Foliar physiognomy in Cathaysian gigantopterids and the potential to track Palaeozoic climates using an extinct plant group. *Palaeogeography, Palaeoclimatology, Palaeoecology*, **205**(1–2): 69–110.

Glen, W. (ed.) 1994. *Mass-Extinction Debates: How Science Works in a Crisis*. Stanford University Press.

Glennie, K.W. 1998. *Petroleum Geology of the North Sea*, 4th edn. Blackwell.

Gluyas, J.G. and Hichens, H.M. (eds.) 2002. *United Kingdom Oil and Gas Fields: Millennium Commemorative Volume*, Memoir No. 20. Geological Society.

Gluyas, J. and Swarbrick, R. 2003. *Petroleum Geoscience*. Blackwell.

Godderis, Y. and Joachimski, M.M. 2004. Global change in the Late Devonian: modelling the Frasnian–Famennian short-term carbon isotope excursions. *Palaeogeography, Palaeoclimatology, Palaeoecology*, **202**(3–4): 309–29.

Goldring, R. 1999. *Field Palaeontology*, 2nd edn. Longman.

Golonka, J. and Ford, D. 2000. Pangean (Late Carboniferous–Middle Jurassic) paleoenvironment and lithofacies. *Palaeogeography, Palaeoclimatology, Palaeoecology*, **161**: 1–34.

Golonka, J., Bocharova, N.Y., Ford, D., *et al.* 2003. Paleogeographic reconstructions and basin development of the Arctic. *Marine and Petroleum Geology*, **20**(3–4): 211–48.

Gombrich, E. 1995. *The Story of Art*, 16th edn. Phaidon.

Gomez-Perez, I. 2003. An early Jurassic deep-water stromatolitic bioherm related to possible methane seepage (Los Molles formation, Neuquen, Argentina). *Palaeogeography, Palaeoclimatology, Palaeoecology*, **201**(1–2): 21–49.

Gonzalez de Juana, C., de Iturralde, A.J. and Picard Cadillat, X. 1980. *Geologia de Venezuela y de Sus Cuencas Petroliferas*. Foninves.

Goodman, D.K. (ed.) 1999. *Proceedings of the 9th International Palynological Congress*. American Association of Stratigraphic Palynologists Foundation.

Goolaerts, S., Dupuis, C. and Steurbaut, E. 2004. Role of palaeoenvironment in ammonite distribution: a Late Cretaceous example from Tunisia. *Palaeontological Association Newsletter*, **57**(48th Annual Meeting Abstracts): 155.

Gordon, M.S. and Olson, E.C. 1995. *Invasion of the Land: The Transition of Organisms from Aquatic to Terrestrial Life*. Columbia University Press.

Gould, S.J. 1969. An evolutionary microcosm: Pleistocene and Recent history of the land snail *P. (Poecilozonites)* in Bermuda. *Bulletin of the Museum of Comparative Zoology, Harvard University*, **138**: 407–532.

1990. *Wonderful Life: The Burgess Shale and the History of Nature*. Radius.

Gradstein, F.M., Kaminski, M.A. and Berggren, W.A. 1988. Cenozoic foraminiferal biostratigraphy of the central North Sea. *Abhandlungen der Geologischen Bundesanstalt*, **41**: 97–108.

Gradstein, F., Ogg, J. and Smith, A. 2005. *A Geologic Time Scale 2004*. Cambridge University Press.

Grafe, K.-U. 1999. Sedimentary cycles, burial history and foraminiferal indicators for systems tracts and sequence boundaries in the Cretaceous of the Basco-Cantabrian basin (northern Spain). *Neues Jahrbuch für Geologie und Paläontologie, Abhandlungen*, **212**(1–3): 85–130.

Grant, R.E. 1966. A Permian productoid brachiopod: life history. *Science*, **152**: 660–2.

Grant, S.F., Stewart, I.J. and Jones, R.W. 2000. Lower Congo basin chronostratigraphy. *GeoLuanda 2000 International Conference, 14th African Colloquium on Micropalaeontology/4th Colloquium on the Stratigraphy and Palaeogeography of the South Atlantic*, Luanda, Angola, Abstracts: 76.

Grantz, A., Johnson, L. and Sweeney, J.F. (eds.) 1990. *The Geology of North America*, vol. L, *The Arctic Ocean Region*. Geological Society of America.

Gray, J. and Boucot, A.J. (eds.) 1979. *Historical Biogeography: Plate Tectonics and the Changing Environment*. Oregon State University Press.

Green, J.C. and Leadbeater, B.S.C. (eds.) 1994. *The Haptophyte Algae*. Oxford Scientific Publications.

Green, O.R. 2001. *A Manual of Practical Laboratory and Field Techniques in Palaeobiology*. Kluwer.

Greuter, W., McNeill, J., Barrie, F.R. *et al.* (eds.), 2000. *International Code g Botanical Nomenclature, adopted by the 16th International Botanical Congress, St. Louis, M., July–August 1999*. Koeltz, Konigstein.

Griffiths, H.I. and Holmes, J.A. 2000. *Non-Marine Ostracods and Quaternary Palaeoenvironments*, Technical Guide No. 8. Quaternary Research Association.

Grimaldi, D. and Engel, M. in press. *Evolution of the Insects*. Cambridge University Press.

Grimes, S.T., Collinson, M.E., Hooker, J.J., *et al.* 2004. Distinguishing the diets of coexisting fossil theridomyid and glirid rodents using carbon isotopes. *Palaeogeography, Palaeoclimatology, Palaeoecology*, **207**(1–3): 103–19.

Grocke, D.R., Price, G.D., Ruffell, A.H., Mutterlose, J. and Baraboshkin, E. 2003. Isotopic evidence for Late Jurassic–Early Cretaceous climate change. *Palaeogeography, Palaeoclimatology, Palaeoecology*, **201**(1–2): 97–118.

Groombridge, B. and Jenkins, M.D. 2002. *World Atlas of Biodiversity*. University of California Press.

Groves, J.R. and Brenckle, P.L. 1997. Graphic correlation in frontier petroleum provinces: application to Upper Paleozoic sections in the Tarim Basin, western China. *American Association of Petroleum Geologists Bulletin*, **81**(8): 1259–66.

Grubb, M., Vroliyk, C. and Brack, D. 1999. *The Kyoto Protocol: A Guide and Assessment*. Royal Institute of International Affairs.

Grzebyk, D., Sako, Y. and Berland, B. 1998. Phylogenetic analysis of nine species of *Prorocentrum* (Dinophtceae) inferred from 18S ribosomal DNA sequences, morphological comparison and description of *Prorocentrum panamaensis sp. nov. Journal of Phycology*, **34**(6): 1055.

Guidon, N. and Delebrias, G. 1986. Carbon-14 dates point to Man in the Americas 32 000 years ago. *Nature*, **321**: 769–71.

Gullan, P.J. and Cranston, P.S. 2000. *The Insects: An Outline of Entomology*, 2nd edn. Blackwell.

Gunnell, G.F. (ed.) 2001. *Eocene Biodiversity*. Kluwer.

Gussow, W.C. 1954. Differential entrapment of oil and gas: a fundamental principle. *American Association of Petroleum Geologists Bulletin*, **38**: 816–53.

Guthrie, R.D. 1990. *Frozen Fauna of the Mammoth Steppe*. University of Chicago Press.

Haig, D.W. 2003. Palaeobathymetric zonation of foraminifera from Lower Permian shale deposits of a high-latitude southern interior sea. *Marine Micropaleontology*, **49**(4): 317–34.

Hailwood, E.A. and Kidd, R.B. (eds.) 1993. *High Resolution Stratigraphy*, Special Publication No. 70. Geological Society.

Hall, R. and Holloway, J.D. (eds.) 1998. *Biogeography and Geological Evolution of SE Asia*. Backhuys.

Hallam, A. (ed.) 1967. Depth indicators in marine sedimentary environments. *Marine Geology, Special Issue*, **5**(5/6).

(ed.) 1973. *Atlas of Palaeobiogeography*. Elsevier.

1994. *An Outline of Phanerozoic Biogeography*. Oxford University Press.

2002. How catastrophic was the end-Triassic mass extinction? *Lethaia*, **35**: 147–57.

Hallam, A. and Wignall, P. 1997. *Mass Extinctions and their Aftermath*. Oxford University Press.

Hammer, O. 1998. Computer simulation of the evolution of foraging strategies: application to the ichnological record. *Palaeontologia Electronica*, **1**(2), http://www.palaeo-electronica.org.toc.htm.

Haq, B.U. and Boersma, A. (eds.) 1978. *Introduction to Marine Micropaleontology*. Elsevier.

Haq, B.U., Hardenbol, J. and Vail, P.R. 1987. Chronology of fluctuating sea levels since the Triassic. *Science*, **235**: 1153–65.

Harland, W.B. and Rudwick, M.S. 1964. The great infra-Cambrian ice age. *Scientific American*, **211**(2): 28–36.

Harland, W.B., Cox, A.V., Llewellyn, P.G., *et al.* 1982. *A Geologic Time Scale*. Cambridge University Press.

Harland, W.B., Armstrong, R.L., Cox, A.V., *et al.* 1990. *A Geologic Time Scale 1989*. Cambridge University Press.

Harley, M.M., Morton, C.M. and Blackmore, S. (eds.) 2000. *Pollen and Spores: Morphology & Biology*. Royal Botanic Gardens, Kew/Linnean Society of London/Natural History Museum/Systematics Association.

Harper, D.A.T. (ed.) 1999. *Numerical Palaeobiology: Computer-Based Modelling and Analysis of Fossils and their Distributions*. John Wiley.

Harper, E.M. and Skelton, P.W. 1993. The Mesozoic marine revolution and epifauna: bivalves. *Scripta Geologica, Special Issue*, **2**: 127–53.

Harper, E.M., Taylor, J.D. and Crame, J.A. (eds.) 2000. *Evolutionary Biology of the Bivalvia*, Special Publication No. 177. Geological Society.

Harper, E.M., Peck, L.S. and Hendry, K. 2003. Who wants to eat a brachiopod? *Palaeontological Association Newsletter*, **54**(47th Annual Meeting Abstracts): 136.

Harries, P.J. 2003. *High-Resolution Approaches in Stratigraphic Paleontology*. Kluwer.

Harries, P.J. and Little, C.T.S. 1999. The early Toarcian (Early Jurassic) and the Cenomanian–Turonian (Late Cretaceous) mass extinctions: similarities and contrasts. *Palaeogeography, Palaeoclimatology, Palaeoecology*, **154**(1–2): 39–66.

Harris, A.J. and Tocher, B.A. 2003. Palaeoenvironmental analysis of Late Cretaceous dinoflagellate cyst assemblages using high-resolution sample correlation from the western interior basin, USA. *Marine Micropaleontology*, **48**(1–2): 127–48.

Harris, P.M. (ed.) 1983. *Carbonate Buildups: A Core Workshop*. Society of Economic Paleontologists and Mineralogists.

Harris, P.M. and Frost, S.H. 1984. Middle Cretaceous carbonate reservoirs, Fahud Field and northwestern Oman. *American Association of Petroleum Geologists Bulletin*, **68**(5): 649–58.

Harris, P.M. and Kowalik, W.S. (eds.) 1994. *Satellite Images of Carbonate Depositional Systems: Examples of Reservoir- and Exploration-Scale Geologic Facies Variation*, Methods in Exploration Series No. 11. American Association of Petroleum Geologists.

Hart, M.B. (ed.) 1987. *Micropalaeontology of Carbonate Environments*. Ellis Horwood.

(ed.) 1996. *Biotic Recovery from Mass Extinction Events*, Special Publication No. 102. Geological Society.

Hart, M.B., Kaminski, M.A. and Smart, C.W. (eds.) 2000. *Proceedings of the 5th International Workshop on Agglutinated Foraminifera*, Special Publication No. 7. Grzybowski Foundation.

Hart, M.B., Hylton, M.D., Oxford, M.J., *et al.* 2003. The search for the origin of the planktonic foraminifera. *Journal of the Geological Society, London*, **160**: 341–3.

Hartman, H. and Fererov, A. 2002. The origin of the eukaryotic cell: a genomic investigation. *Proceedings of the National Academy of Sciences, USA*, **99**: 1420–5.

Hartman, W.D., Wendt, J.W. and Wiedenmayer, F. 1990. *Sedimenta*, vol. VIII, *Living and Fossil Sponges*. Rosenstiel School of Marine and Atmospheric Sciences, University of Miami.

Hartwig, W.C. (ed.) 2003. *The Primate Fossil Record*. Cambridge University Press.

Harzhauser, M. and Kowalke, T. 2002. Late Middle Miocene (Sarmatian) potamiid-dominated gastropod assemblages of the central Paratethys as a tool for facies interpretation. *Facies*, **46**: 57–82.

Harzhauser, M. and Mandic, O. 2004. The muddy bottom of Lake Pannon: a challenge for dreissenid settlement

(Late Miocene; Bivalvia). *Palaeogeography, Palaeoclimatology, Palaeoecology*, **204**(3–4): 331–52.

Harzhauser, M. and Piller, W. E. 2001. The Sarmatian of the central Paratethys: an urgent need of a reevaluation of models. *Proceedings of the 3rd European Paleontological Congress*, Leiden, The Netherlands: 20–1.

Harzhauser, M., Piller, W. E. and Steininger, F. F. 2002. Circum-Mediterranean Oligo-Miocene biogeographic evolution: the gastropods' point of view. *Palaeogeography, Palaeoclimatology, Palaeoecology*, **183**: 103–33.

Harzhauser, M., Mandic, O. and Zuschin, M. 2003. Changes in Paratethyan marine molluscs at the early/middle Miocene transition: diversity, palaeogeography and palaeoclimate. *Acta Geologica Polonica*, **53**(4): 323–39.

Hasiotis, S. T. 2003. Complex ichnofossils of solitary and social soil organisms: understanding their evolution and roles in terrestrial paleoecosystems. *Palaeogeography, Palaeoclimatology, Palaeoecology*, **192**: 259–320.

2004. Reconnaisance of Upper Jurassic Morrison formation ichnofossils, Rocky Mountain region, USA: paleoenvironmental, stratigraphic, and paleoclimatic significance of terrestrial and freshwater ichnocoenoses. *Sedimentary Geology*, **167**(3–4): 177–268.

Haslett, S. K. 2002. *Quaternary Environmental Micropalaeontology*. Arnold.

2004. Late Neogene–Quaternary radiolarian biostratigraphy: a brief review. *Journal of Micropalaeontology*, **23**(1): 39–47.

Hass, H. C. and Kaminski, M. A. (eds.) 1997. *Contributions to the Micropaleontology and Paleoceanography of the Northern North Atlantic*, Special Publication No. 5. Grzybowski Foundation.

Haude, R., Jahnke, H. and Walliser, O. H. 1994. Scyphocrinoiden an der Wende Silur/Devon. *Aufschluss, Heidelberg*, **45**: 49–55.

Haviland, W. A. 2000. *Human Evolution and Prehistory*. Harcourt Brace.

Haynes, G. 1991. *Mammoths, Mastodons and Elephants: Biology, Behavior and the Fossil Record*. Cambridge University Press.

2002. *The Early Settlement of North America: The Clovis Era*. Cambridge University Press.

Haynes, J. R. 1981. *Foraminifera*. MacMillan.

Hays, J. D., Imbrie, J. and Shackelton, N. K. 1976. Variations in the Earth's orbit: pacemaker of the ice age. *Science*, **194**: 1121–32.

Hayward, B. W., Grenfell, H. R., Nicholson, K., *et al.* 2004. Foraminiferal record of human impact on intertidal estuarine environments in New Zealand's largest city. *Marine Micropaleontology*, **53**: 37–66.

Hayward, P. J. and Ryland, J. S. 1985. *Cyclostome Bryozoans*. E. J. Brill/ Backhuys.

Haywood, A. M., Valdes, P. J. and Sellwood, B. W. 2000. Global scale palaeoclimate reconstruction of the Middle Pliocene climate using the UKMO GCM: initial results. *Global and Planetary Change*, **25**: 239–56.

2002. Magnitude of climate variability during Middle Pliocene warmth: a palaeoclimate modelling study. *Palaeogeography, Palaeoclimatology, Palaeoecology*, **188**: 1–24.

Hecht, M. K., Wallace, B. and Prance, G. T. (eds.) 1988. *Evolutionary Biology*. Plenum Press.

Heckel, P. H. 1974. Carbonate buildups in the geological record: a review. *Society of Economic Paleontologists and Mineralogists Special Publication*, **18**: 90–154.

Hedgpeth, J. W. (ed.) 1957. *Treatise on Marine Ecology and Paleoecology*, vol. 1, *Ecology*, Memoir No. 67. Geological Society of America.

Hembree, D. I., Martin, L. D. and Hasiotis, S. T. 2004. Amphibian burrows and ephemeral ponds of the Lower Permian Speiser shale, Kansas: evidence for seasonality in the midcontinent. *Palaeogeography, Palaeoclimatology, Palaeoecology*, **203**(1–2): 127–52.

Hemleben, C., Kaminski, M. A., Kuhnt, W. and Scott, D. B. (eds.) 1990. *Paleoecology, Biostratigraphy, Paleoceanography and Taxonomy of Agglutinated Foraminifera*. Kluwer.

Henderson, P. A. 1990. *Freshwater Ostracods*. E. J. Brill.

Hennig, W. 1979. *Phylogenetic Systematics*, 2nd edn. University of Illinois Press.

Henriksen, K., Young, J. R., Bown, P. R. and Stipp, S. L. S. 2004. Coccolith biomineralisation studied with atomic force microscopy. *Palaeontology*, **47**(3): 725–44.

Herman, Y. (ed.) 1974. *Marine Geology and Oceanography of the Arctic Seas*. Springer-Verlag.

Herrera Cubilla, A. and Jackson, J. B. C. (eds.) 2000. *Proceedings of the 11th International Bryozoology Association Conference*. Smithsonian Tropical Research Institute.

Herrle, J. O. 2003. Reconstructing nutricline dynamics of mid-Cretaceous oceans: evidence from calcareous nannofossils from the Niveau Paquier black shale (SE France). *Marine Micropaleontology*, **47**(3–4): 307–21.

Herrmann, A. D., Haupt, B. J., Patzkowsky, M. E., Seidov, D. and Slingerland, R. L. 2004a. Response of Late Ordovician paleoceanography to changes in sea-level, continental drift, and atmospheric pCO_2: potential causes for long-term cooling and glaciation. *Palaeogeography, Palaeoclimatology, Palaeoecology*, **210**(2–4): 385–401.

Herrmann, A. D., Patzkowsky, M. E. and Pollard, D. 2004b. The impact of paleogeography, pCO_2, poleward ocean heat transport and sea level change on global cooling during the Late Ordovician. *Palaeogeography, Palaeoclimatology, Palaeoecology*, **206**(1–2): 59–74.

Herz, N. and Garrison, E. G. 1998. *Geological Methods for Archaeology*. Oxford University Press.

Hesketh, R. A. P. and Underhill, J. R. 2002. The biostratigraphic calibration of the Scottish and Outer Moray Firth Upper Jurassic successions: a new basis for the correlation of Late Oxfordian–Early Kimmeridgian Humber Group reservoirs in the North Sea basin. *Marine and Petroleum Geology*, **19**: 541–62.

Hess, H., Ausich, W. I., Brett, C. E., Simms, M. J. and Kindlimann, R. 1999. *Fossil Crinoids*. Cambridge University Press.

Hesselbo, S. P. and Parkinson, D. N. (eds.) 1996. *Sequence Stratigraphy in British Geology*, Special Publication No. 103. Geological Society.

Heydari, E., Hassanzadeh, J., Wade, W.J. and Ghazi, A.M. 2003. Permian–Triassic boundary interval in the Abadeh section of Iran with implications for mass extinction. I. Sedimentology. *Palaeogeography, Palaeoclimatology, Palaeoecology*, **193**: 405–23.

Hill, C.R. and El-Khayal, A.A. 1983. Late Permian plants including charophytes from the Khuff formation of Saudi Arabia. *Bulletin of the British Museum (Natural History), Geology Series*, **37**(3): 105–12.

Hill, C.R., Wagner, R.H. and El-Khayal, A.A. 1985. *Qasimia gen. nov.*, an early *Marattia*-like fern from the Permian of Saudi Arabia. *Scripta Geologica*, **79**: 1–50.

Hill, T.M., Kennett, J.P. and Spero, H.J. 2003. Foraminifera as indicators of methane-rich environments: a study of modern methane seeps in Santa Barbara channel, California. *Marine Micropaleontology*, **49**(1–2): 123–38.

Hofling, R. and Steuber, T. (eds.) 1999. *Proceedings of the 5th International Congress on Rudists, Abstracts and Field Guides*. Erlanger Geologische Abhandlungen.

Hohenegger, J. 2000. Coenoclines of larger foraminifera. *Micropaleontology*, **46**(Suppl. 1): 127–52.

2004. Depth coenoclines and environmental considerations of western Pacific larger foraminiferans. *Journal of Foraminiferal Research*, **34**(1): 9–33.

Holbourn, A., Henderson, A.S. and MacLeod, N. (eds.) 2004. *PaleoBase: Deep Sea Benthic Foraminifera*. CompuStrat (Blackwell/Natural History Museum). (CD).

Holl, C., Zonneveld, K.A.F. and Willems, H. 1998. On the ecology of calcareous dinoflagellates: the Quaternary eastern Equatorial Atlantic. *Marine Micropaleontology*, **33**(102): 1–25.

Holman, J.A. 2000. *Fossil Snakes of North America: Origin, Evolution, Distribution, Paleoecology*. Indiana University Press.

2003. *Fossil Frogs and Toads of North America*. Indiana University Press.

Holmer, L.E., Skovsted, C.B. and Williams, A. 2002. A stem group brachiopod from the Lower Cambrian: support for a *Micrina* (halkierid) ancestry. *Palaeontology*, **45**(5): 875–82.

Honisch, B., Bijma, J., Russell, A.D., *et al.* 2003. The influence of symbiont photosynthesis on the boron isotopic composition of foraminifera shells. *Marine Micropaleontology*, **49**(1–2): 87–96.

Hooker, J.J. and Millbank, C. 2001. A Cernaysian mammal from the Upnor Formation (Late Palaeocene, Herne Bay, UK) and its implications for correlation. *Proceedings of the Geologists' Association*, **112**: 331–8.

Hooper, J.N.A. and van Soest, R.W.M. (eds.) 2002. *Systema Porifera: A Guide to the Classification of Sponges*. Kluwer Academic/Plenum Press.

Hoorn, C., Ohja, T. and Quade, J. 2000. Palynological evidence for vegetation development and climatic change in the sub-Himalayan zone (Neogene, central Nepal). *Palaeogeography, Palaeoclimatology, Palaeoecology*, **163**: 133–61.

Hopkins, D.M. (ed.) 1967. *The Bering Land Bridge*. Stanford University Press.

Horne, D.J., Smith, R.J., Whittaker, J.E. and Murray, J.W. 2004. The first British record and a new species of the superfamily Terrestricytheroidea (Crustacea, Ostracoda): morphology, ontogeny, lifestyle and phylogeny. *Zoological Journal of the Linnean Society*, **142**: 253–88.

Horton, B.P. and Edwards, R.J. 2000. Quantitative palaeoenvironmental reconstruction techniques in sea-level studies. *Archaeology in the Severn Estuary*, **11**: 105–19.

Horton, B.P., Edwards, R.J. and Lloyd, J.M. 1999. A foraminiferal-based transfer function: implications for sea-level reconstruction. *Journal of Foraminiferal Research*, **29**: 117–29.

Hou Xian-Guang and Bergstrom, J. 1997. Arthropods of the Lower Cambrian Chengjiang fauna, southwest China. *Fossils and Strata*, **45**.

Hou Xian-Guang, Siveter, D.J., Williams, M., *et al.* 1996. Appendages of the arthropod *Kunmingella* from the Early Cambrian of China: its bearing on the systematic position of the Bradoriida and the fossil record of the Ostracoda. *Philosophical Transactions of the Royal Society of London B*, **351**: 1131–45.

Hou Xian-Guang, Aldridge, R.J., Bergstrom, J., *et al.* 2004. *The Cambrian Fossils of Chengjiang, China: The Flowering of Early Animal Life*. Blackwell.

House, M.R. (ed.) 1979. *The Origin of Major Invertebrate Groups*, Systematics Association Special Publication No. 12. Academic Press.

(ed.) 1993. *The Ammonoidea: Environment, Ecology and Evolutionary Change*, Systematics Association Special Volume No. 47. Clarendon Press.

2002. Strength, timing, setting and cause of mid-Palaeozoic extinctions. *Palaeogeography, Palaeoclimatology, Palaeoecology*, **181**: 5–25.

Howarth, M.K. (ed.) 1995. Palaeontology of the Qahlah and Simsima formations (Cretaceous, late Campanian–Maastrichtian) of the United Arab Emirates–Oman border region. *Bulletin of the Natural History Museum, London, Geology*, **51**(2).

Huc, A.-Y. (ed.) 1995. *Paleogeography, Paleoclimate and Source Rocks*, Studies in Geology No. 40. American Association of Petroleum Geologists.

Hudson, R.G.S. and Chatton, M. 1959. The Musandam limestone (Jurassic to Lower Cretaceous) of Oman, Arabia. *Notes et Mémoires sur le Moyen-Orient, Muséum National d'Histoire Naturelle, Paris*, **7**: 69–93.

Huggett, R.J. 2004. *Fundamentals of Biogeography*, 2nd edn. Routledge.

Hughes, G.W. 1996. A new bioevent stratigraphy of Late Jurassic Arab-D carbonates of Saudi Arabia. *GeoArabia*, **1**(3): 417–34.

1997. The Great Pearl Bank Barrier of the Arabian Gulf as a possible Shu'aiba analogue. *GeoArabia*, **2**(3): 279–304.

2000. Bioecostratigraphy of the Shu'aiba formation, Shaybah field, Saudi Arabia. *GeoArabia*, **5**(4): 545–78.

Hughes, G.W., Varol, O. and Beydoun, Z.R. 1991. Evidence for Middle Oligocene rifting of the Gulf of Aden and Late

Oligocene rifting of the southern Red Sea. *Marine and Petroleum Geology*, **8**: 354–8.

Hughes, G.W., Siddiqui, S. and Sadler, R.K. 2003. Shu'aiba rudist taphonomy using computerised tomography and image logs, Shaybah field, Saudi Arabia. *GeoArabia*, **8**(4): 585–96.

Hughes, N.F. (ed.) 1973. *Organisms and Continents through Time*, Special Paper No. 12. Palaeontological Association.

1976. *Palaeobiology of Angiosperm Origins*. Cambridge University Press.

1994. *The Enigma of Angiosperm Origins*. Cambridge University Press.

Hughes, N.F. and Moody-Stuart, J. 1966. Descriptions of schizaeceous spores taken from Early Cretaceous macrofossils. *Palaeontology*, **9**(2): 274–89.

Humphries, C.J. and Parenti, L.R. (eds.) 1999. *Cladistic Biogeography*, 2nd edn. Oxford University Press.

Hunter, V.F. 1978. Foraminiferal correlation of Tertiary mollusc horizons of the southern Caribbean area. *Geologie en Mijnbouw*, **57**: 193–203.

Iacumin, P., Nikolaev, V., Ramingi, M. and Longinelli, A. 2004. Oxygen isotope analyses of mammal bone remains from Holocene sites in European Russia: palaeoclimatic implications. *Global and Planetary Change*, **40**(1–2): 169–76.

Iakovleva, A.I. and Kulkova, I.A. 2003. Paleocene–Eocene dinoflagellate zonation of Western Siberia. *Review of Palaeobotany and Palynology*, **123**: 185–97.

Iakovleva, A.I., Brinkhuis, H. and Cavagnetto, C. 2001. Late Palaeocene–Early Eocene dinoflagellate cysts from the Turgay Strait, Kazakhstan; correlations across ancient seaways. *Palaeogeography, Palaeoclimatology, Palaeoecology*, **172**: 243–68.

Ikebe, N. and Tsuchi, R. (eds.) 1984. *Pacific Neogene Datum Planes: Contributions to Biostratigraphy and Chronology*. University of Tokyo Press.

Immenhauser, A.A., Schlager, W., Burns, S.J., *et al.* 1999. Late Aptian to late Albian sea-level fluctuations constrained by geochemical and biological evidence (Nahr Umr formation, Oman). *Journal of Sedimentary Research*, **69**: 434–46.

Ingrouille, M.J. and Eddie, W. in press. *Plants: Evolution and Diversity*. Cambridge University Press.

INQUA-SEQS 2002. *Upper Pliocene and Pleistocene of the Southern Urals Region and its Significance for Correlation of the Eastern and Western Parts of Europe*. INQUA-SEQS Conference, Ufa, Excursion Guide.

Insalaco, E., Skelton, P.W. and Palmer, T.J. 2000. *Carbonate Platform Systems: Components and Interactions*, Special Publication No. 178. Geological Society.

Institute of Vertebrate Palaeontology and Anthropology, Chinese Academy of Sciences. 1980. *Atlas of Primitive Man in China*. Science Press.

International Commission on Stratigraphy. 2005. http://www.stratigraphy.org.

Isokazi, Y. 1997. Permo-Triassic boundary superanoxia and stratified superocean: records from the lost deep sea. *Science*, **276**: 235–8.

Itaki, T. 2003. Depth-related radiolarian assemblage in the water-column and surface sediments in the Japan Sea. *Marine Micropaleontology*, **47**(3–4): 253–70.

Iwatsuki, I. and Raven, P.H. (eds.) 1997. *Evolution and Diversification of Land Plants*. Springer-Verlag.

Izart, A., Vaslet, D., Briand, C., *et al.* 1998. Stratigraphic correlations between the continental and marine Tethyan and Peri-Tethyan basins during the Late Carboniferous and the Early Permian. *Geodiversitas*, **20**(4): 521–96.

Jablonski, D. and Raup, D.M. 1995. Selectivity of end-Cretaceous marine bivalve extinctions. *Science*, **268**: 389–91.

Jablonski, N.H. (ed.) 1993. *Theropithecus: The Rise and Fall of a Primate Genus*. Cambridge University Press.

Jackson, J.B.C., Budd, A.F. and Coates, A.G. (eds.) 1996. *Evolution and Environment in Tropical America*. University of Chicago Press.

Jacobs, B.F. and Herendeen, P.S. 2004. Eocene dry climate and woodland vegetation in tropical Africa reconstructed from fossil leaves from Northern Tanzania. *Palaeogeography, Palaeoclimatology, Palaeoecology*, **213**(1–2): 115–23.

Jacobs, D.K. and Landman, N.H. 1993. *Nautilus*: a poor model for the function and behavior of ammonoids. *Lethaia*, **26**: 101–11.

Jaekel, O. 1918. Phylogenie und System der Pelmatozoen. *Paläontologische Zeitschrift*, **3**: 1–128.

Jangoux, M. (ed.) 1980. *Proceedings of the European Colloquium on Echinoderms*, Brussels, 1979. Balkema.

Jansonius, J. and MacGregor, D.C. (eds.) 1996. *Palynology: Principles and Applications*. American Association of Stratigraphic Palynologists Foundation.

Janssen, A.W. 2003. Notes on the systematics, morphology and biostratigraphy of fossil holoplanktonic Mollusca. XIII. Considerations on a subdivision of Thecosomata, with the emphasis on genus group classification of Limacinidae. *Cainozoic Research*, **2**(1–2): 163–70.

Janzen, J.-W. 2002. *Arthropods in Baltic Amber*. Ampyx.

Jardine, S. and Yapaudjian, L. 1968. Lithostratigraphie et palynologie du Devonien–Gothlandien gréseux du Bassin de Polignac (Sahara). *Revue de l'Institut Français du Petrole*, **23**(4): 439–69.

Jardine, S., de Klasz, I. and Debenay, J.-P. (eds.) 1996. Géologie de l'Afrique et de l'Atlantique Sud. *Bulletin Centres Recherches Exploration–Production Elf Aquitaine*, Mémoir, **16**.

Javaux, E.J., Marshall, C.P., Shuhai Xiao, Knoll, A.H. and Walter, M.R. 2004. Early eukaryotes in Paleoproterozoic and Mesoproterozoic oceans. *Palaeontological Association Newsletter*, **57**(48th Annual Meeting Abstracts): 123–4.

Jefferies, R.P.S. 1968. The subphylum Calcichordata, primitive fossil chordates with echinoderm affinities. *Bulletin of the British Museum (Natural History), Geology Series*, **16**: 243–339.

1975. Fossil evidence concerning the origin of the chordates. *Symposium of the Zoological Society of London*, **36**: 253–318.

1986. *The Ancestry of the Vertebrates*. British Museum (Natural History).

1997. A defence of the calcichordates. *Lethaia*, **30**: 1–10.

Jefferies, R.P.S., Brown, N.A. and Daley, P.E. 1996. The early phylogeny of chordates and echinoderms and the origin of chordate left–right asymmetry and bilateral symmetry. *Acta Zoologica*, **77**: 101–22.

Jenkins, D.G. (ed.) 1993. *Applied Micropalaeontology*. Kluwer.

Jenkins, D.G. and Murray, J.W. (eds.) 1989. *Stratigraphical Atlas of Fossil Foraminifera*, 2nd edn. Ellis Horwood.

Jenkyns, H.C. 1980. Cretaceous anoxic events: from continents to oceans. *Journal of the Geological Society, London*, **137**(2): 171–88.

Jernvall, J. and Selanne, L. 1999. Laser confocal microscopy and geographic information systems in the study of dental morphology. *Palaeontologia Electronica*, **2**(1), http://www.palaeo-electronica.org.toc.htm.

Jett, S. 2004. *Crossing Ancient Oceans: The Question of Pre-Columbian Contacts Re-Examined*. Copernicus.

Jianfang Hu, Ping'an Peng, Dianyong Fang, *et al.* 2003. No aridity in Sunda land during the last glaciation: evidence from molecular-isotopic stratigraphy of long chain *n*-alkanes. *Palaeogeography, Palaeoclimatology, Palaeoecology*, **201**(3–4): 269–81.

Jian-Wen Shen and Webb, G.E. 2004. Famennian (Upper Devonian) calcimoicrobial (*Renalcis*) reef at Miaomen, Gilin, Guangxi, south China. *Palaeogeography, Palaeoclimatology, Palaeoecology*, **204**(3–4): 373–94.

Jin Yugan, Wang Jungeung and Xu Shanhong 1991. *Palaeoecology of China*. Nanjing University Press.

Jobling, M.A., Hules, M.E. and Tyler-Smith, C. 2004. *Human Evolutionary Genetics: Origins, People and Disease*. Garland.

Johanson, D.C. and Edey, M.A. 1981. *Lucy: The Beginnings of Humankind*. Penguin.

Johanson, D.C. and Edgar, B. 1996. *From Lucy to Language*. Weidenfeld & Nicolson.

Johnson, E.W., Briggs, D.E.G., Suthren, R.J., Wright, J.L. and Tunnicliff, S.P. 1994. Non-marine arthropod traces from the subaerial Ordovician Borrowdale Volcanic group, English Lake District. *Geological Magazine*, **131**: 395–406.

Jollie, M. 1982. What are the 'Calcichordata'? And the larger question of the Chordates. *Zoological Journal of the Linnean Society*, **75**: 167–88.

Jollivet, D., Faugeres, J.-C., Griboulard, R., Desbruyeres, D. and Blanc, G. 1990. Composition and spatial organisation of a cold seep community on the south Barbados accretionary prism: tectonic, geochemical and sedimentary context. *Progress in Oceanography*, **24**: 25–45.

Jolly, C. 1970. The seed-eaters, a new model of hominid differentiation based on a baboon analogy. *Man*, **5**: 5–26.

Jones, M. 2001. *The Molecule Hunt: Archaeology and the Search for Ancient DNA*. Allen Lane/Penguin.

Jones, R.W. 1984. Late Quaternary benthic foraminifera from deep-water sites in the north-east Atlantic and Arctic Oceans. Unpublished Ph.D. thesis, University College of Wales, Aberystwyth.

1986. Distribution of 'morphogroups' of agglutinating foraminifera in the Rockall Trough: a synopsis. *Proceedings of the Royal Society of Edinburgh*, **88B**: 55–8.

1994. *The Challenger Foraminifera*. Oxford University Press.

1996. *Micropalaeontology in Petroleum Exploration*. Oxford University Press.

1997. Aspects of the Cenozoic stratigraphy of the northern Sulaiman Ranges, Pakistan. *Journal of Micropalaeontology*, **16**: 51–8.

2000. Proterozoic to Palaeozoic sequence stratigraphy of south-west Iran. *Abstracts, MEGSTRAT1 Workshop*, Dubai: 4–6.

2001. Biostratigraphic characterisation of submarine fan sub-environments, deep-water offshore Angola. *Abstracts, 6th International Workshop on Agglutinated Foraminifera*, Prague.

2003a. Micropalaeontological characterisation of submarine fan/channel sub-environments, deep-water Angola. *Abstracts, William Smith Conference ('Wrestling with Mud')*, Geological Society, London.

2003b. Micropalaeontological characterisation of mudrock seal capacity. *Abstracts, William Smith Conference ('Wrestling with Mud')*, Geological Society, London.

Jones, R.W. and Charnock, M.A. 1985. 'Morphogroups' of agglutinating foraminifera, their life positions and feeding habits and potential applicability in (paleo)ecological studies. *Revue de Paléobiologie*, **4**(2): 311–20.

Jones, R.W. and Simmons, M.D. 1996. A review of the stratigraphy of eastern Paratethys (Oligocene–Holocene). *Bulletins of the Natural History Museum, London, Geology*, **52**(1): 25–49.

(eds.) 1999. *Biostratigraphy in Production and Development Geology*, Special Publication No. 152. Geological Society.

Jones, R.W. and Whittaker, J.E. in press. Palaeoenvironmental interpretation of the Pleistocene–Holocene of the British Isles using proxy data from key Recent benthonic foraminiferal distributions. *Bulletins of The Natural History Museum*.

Jones, R.W., Simmons, M.D. and Whittaker, J.E. in press. On the stratigraphic and palaeobiogeographic significance of *Borelis melo* (Fichtel & Moll, 1798) and *B. melo curdica* (Reichel, 1937) (Foraminifera, Miliolida, Alveolinidae). *Journal of Micropalaeontology*.

Jones, S., Martin, R. and Pilbeam, D. (eds.) 1992. *The Cambridge Encyclopedia of Human Evolution*. Cambridge University Press.

Jones, T.P. and Rowe, N.P. 1999. *Fossil Plants and Spores*. Geological Society.

Kaiho, K. 1994. Benthic foraminiferal dissolved oxygen index and dissolved oxygen levels in the modern ocean. *Geology*, **22**: 719–22.

Kaiser, T.M. 2003. The dietary regimes of two contemporaneous populations of *Hippotherium primigenium* (Perissodactyla, Equidae) from the Vallesian (Upper Miocene) of southern Germany. *Palaeogeography, Palaeoclimatology, Palaeoecology*, **198**: 381–402.

Kaminski, M.A. and Austin, W.E.N. 1999. Oligocene deep-water agglutinated foraminifers at Site 985, Norwegian Basin,

Southern Norwegian Sea. *Proceedings of the Ocean Drilling Program*, **162**: 169–77.

Kaminski, M.A., Grassle, J.F. and Whitlatch, R.B. 1988. Life history and recolonisation among agglutinated foraminifera in the Panama Basin. *Abhandlungen der Geologischen Bundesanstalt*, **41**: 229–44.

Kaminski, M.A., Geroch, S. and Kaminski, D.G. (eds.) 1993. *The Origins of Applied Micropalaeontology: the school of Josef Grzybowski*, Special Publication No. 1. Grzybowski Foundation.

Kaminski, M.A., Geroch, S. and Gasinski, M.A. (eds.) 1995. *Proceedings of the 4th International Workshop on Agglutinating Foraminifera*, Krakow, Poland, 1993, Special Publication No. 3. Grzybowski Foundation.

Kaminski, M.A., Aksu, A., Box, M., *et al.* 2002. Late Glacial to Holocene benthic foraminifera in the Marmara Sea: implications for Black Sea–Mediterranean Sea connections following the last deglaciation. *Marine Geology*, **190**: 165–202.

Katsumi, U. 2003. The Permian fusulinoidean faunas of the Sibumasu and Baoshan blocks: their implications for the paleogeographic and paleoclimatic reconstruction of the Cimmerian continent. *Palaeogeography, Palaeoclimatology, Palaeoecology*, **193**: 1–24.

Katz, B.J. (ed.) 1994. *Petroleum Source Rocks*. Springer-Verlag.

Kauffman, E.G. and Johnson, C. 1988. The morphological and ecological evolution of Middle and Upper Cretaceous reef-building rudistids. *Palaios*, **3**: 194–216.

Kauffman, E.G. and Sohl, N.F. 1974. Structure and evolution of Antillean Cretaceous rudist frameworks. *Verhandlungen der Naturforschenden Gesellschaft in Basel*, **84**(1): 399–467.

Kauffman, E.G. and Walliser, O.H. (eds.) 1990. *Extinction Events in Earth History*. Springer-Verlag.

Keller, G. 2004. Low-diversity, late Maastrichtian and early Danian planktonic foraminiferal assemblages of the eastern Tethys. *Journal of Foraminiferal Research*, **34**(1): 49–73.

Keller, G. and Pardo, A. 2004. Age and paleoenvironment of the Cenomanian–Turonian global stratotype section and point at Pueblo, Colorado. *Marine Micropaleontology*, **51**(1–2): 95–128.

Keller, G., Stinnesbeck, W., Adatte, T. and Steuben, D. 2003. Multiple impacts across the Cretaceous–Tertiary boundary. *Earth-Science Reviews*, **62**: 327–63.

Keller, M.A. and Isaacs, C.M. 1985. An evaluation of temperature scales for silica diagenesis in diatomaceous sequences including a new approach based on the Miocene Monterey formation, California. *Geo-Marine Letters*, **5**: 31–5.

Kelley, P.H., Kowalewski, M. and Hansen, T.A. 2003. *Predator–Prey Interactions in the Fossil Record*. Kluwer.

Kelly, D.C. Norris, R.D. and Zachos, J.C. 2003. Deciphering the paleoceanographic significance of Early Oligocene *Braarudosphaera* Chalks in the South Atlantic. *Marine Micropaleontology*, **49**(1–2): 49–63.

Kelman, R., Feist, M. and Trewin, N.H. 2003. Charophyte algae from the Early Devonian Rhynie chert, Aberdeenshire. *Palaeontological Association Newsletter*, **54**(47th Annual Meeting Abstracts): 140.

Kemp, T.S. 2004. *The Origin and Evolution of Mammals*. Oxford University Press.

Kennedy, W.J. and Cobban, W.A. 1976. Aspects of ammonite biology, biogeography and biostratigraphy. *Special Papers in Palaeontology*, **17**.

Kenrick, P. and Davis, P. 2004. *Fossil Plants*. Natural History Museum.

Kent, D.V., Cramer, B.S., Lanci, L., *et al.* 2003. A case for a comet impact trigger for the Paleocene/Eocene thermal maximum and carbon isotope excursion. *Earth and Planetary Science Letters*, **211**: 13–26.

Kerr, A.C. 1998. Oceanic plateau formation: a cause of mass extinction and black shale deposition around the Cenomanian–Turonian boundary. *Journal of the Geological Society, London*, **155**: 619–26.

Keupp, H. 2000. *Ammoniten: Paläobiologische Erfolsspiralen*. Thorbecke.

Khakhina, L.N. 1992. *Concepts of Symbiogenesis: A Historical and Critical Account of the Research of Russian Botanists*. Yale University Press.

Khokhlova, I.E. and Oreshkina, T.V. 1999. Early Palaeogene siliceous microfossils of the Middle Volga Region: stratigraphy and palaeogeography. *Geodiversitas*, **21**(3): 429–51.

Kidder, D.L. and Worsley, T.R. 2004. Causes and consequences of extreme Permo-Triassic warming to globally equable climate and relation to the Permo-Triassic extinction and recovery. *Palaeogeography, Palaeoclimatology, Palaeoecology*, **203**(3–4): 207–37.

Kielan-Jaworowska, Z., Cifelli, R.L. and Zhe-Xi Luo 2004. *Mammals from the Age of the Dinosaurs: Origins, Evolution and Structure*. Columbia University Press.

Kiessling, W. 2001. Paleoclimatic significance of Phanerozoic reefs. *Geology*, **29**: 751–4.

Kiessling, W., Flugel, E. and Golonka, J. 1999. Paleoreef maps. *American Association of Petroleum Geologists Bulletin*, **83**(10): 1552–87.

2002. *Phanerozoic Reef Patterns*, Special Publication No. 72. Society of Economic Paleontologists and Mineralogists.

King, G. 1990. *The Dicynodonts: A Study in Paleobiology*. Chapman & Hall.

1996. *Reptiles and Herbivory*. Kluwer.

Kingdon, J. 2004. *Lowly Origin: Where, When and Why Our Ancestors First Stood Up*. Princeton University Press.

Kirk, N. 1969. Some thoughts on the ecology, mode of life, and evolution of the Graptolithina. *Proceedings of the Geological Society, London*, **1659**: 273–92.

1972. More thoughts on the automobility of the graptolites. *Quarterly Journal of the Geological Society, London*, **128**: 127–33.

Kitazato, H. 1996. Benthic foraminifera associated with cold seepages: discussion of their faunal characteristics and adaptations. *Fossils*, **60**: 48–52.

Kitching, I.J., Forey, P.L., Humphries, C.J. and Williams, D.M. 1998. *Cladistics*, 2nd edn., Systematics Association Publication No. 11. Oxford Science Publications.

Kjennerud, T. and Gillmore, G.K. 2003. Integrated Palaeogene palaeobathymetry of the northern North Sea. *Petroleum Geoscience*, **9**: 125–32.

Klein, R. 1999. *The Human Career*, 2nd edn. University of Chicago Press.

Knight, R. and Mantoura, R.F.C. 1985. Chlorophyll and carotenoid pigments in foraminifera and their symbiotic algae: analysis by high performance liquid chromatography. *Marine Ecology, Progress Series*, **23**: 241–9.

Knoll, A. H. 2003. *Life on a Young Planet*. Princeton University Press.

Knoll, A.H. and Carroll, S.B. 1999. Early animal evolution: emerging views from comparative biology and geology. *Science*, **284**: 2129–37.

Knoll, A.H., Bambach, R.K., Canfield, D.E. and Grotzinger, J.P. 1996. Comparative earth history and Late Permian mass extinction. *Science*, **273**: 452–7.

Knox, R.W.O'B., Corfield, R.M. and Dunay, R.E. (eds.) 1996. *Correlation of the Early Paleogene in Northwest Europe*, Special Publication No. 101. Geological Society.

Kobayishi, F. 1997a. Middle Permian biogeography based on fusulinacean faunas. *Cushman Foundation for Foraminiferal Research Special Publication*, **36**: 73–6.

1997b. Middle Permian fusulinacean faunas and paleogeography of exotic terranes in the circum-Pacific. *Cushman Foundation for Foraminiferal Research Special Publication*, **36**: 77–80.

Kocsis, F.A., Jr 2002. *Vertebrate Fossils: A Neophyte's Guide*. Ibis Graphics.

Koeberl, C. and MacLeod, K.G. 2002 *Catastrophic Events and Mass Extinctions: Impacts and Beyond*. Geological Society of America.

Kogbe, C.A. and Mehes, K. 1986. Micropaleontology and biostratigraphy of the coastal basins of west Africa. *Journal of African Earth Sciences*, **5**(1): 1–100.

Konert, G., Afifi, A.M., Al-Hajri, S.A. and Droste, H.J. 2001. Paleozoic stratigraphy and hydrocarbon habitat of the Arabian plate. *GeoArabia*, **6**(3): 407–42.

Kopaevich, L.F., Alekseev, A.S. and Baraboshkin, E.Y. 1999. Cretaceous sequences of the Mangyshlak peninsula (peri-Caspian area). *Geodiversitas*, **21**(3): 407–19.

Kotanski, Z., Morycowa, E., Peybernes, B. and Durand-Delga, M. 1988. Indices de l'existence d'une plate-forme carbonatée à madreporaires, algues et grands foraminifères benthiques du Malm, sur les zones internes au nord de la dorsale calcaire du Djurdjura (Algérie). *Comptes Rendus de l'Académie des Sciences, Paris, Série II*, **307**: 1809–18.

Kotsakis, T., Delfino, M. and Piras, P. 2004. Italian Cenozoic crocodilians: taxa, timing and palaeobiogeographic implications. *Palaeogeography, Palaeoclimatology, Palaeoecology*, **210**: 67–87.

Koutsoukos, E.A.M. 1985. Distribucão paleobatimetrica de foraminiferos bentonicos do Cenozoico margem continental Atlantica. *Trabahos VIII Congreso Brasiliero de Paleontologia*, Brasilia: 355–70.

(ed.) 2005. *Applied Integrated Stratigraphy in Exploration and Development Geology: New Techniques and Perspectives*. Kluwer.

Koutsoukos, E.A.M. and Hart, M.B. 1990. Radiolarians and diatoms from the mid-Cretaceous section of the Sergipe basin, northeastern Brazil: palaeoceanographic assessment. *Journal of Micropalaeontology*, **9**(1): 45–64.

Koutsoukos, E.A. and Merrick, K.A. 1985. Foraminiferal paleoenvironments from the Barremian to Maestrichtian of Trinidad, West Indies. *Transactions of the 1st Geological Conference of the Geological Society of Trinidad & Tobago*: 85–101.

Koutsoukos, E.A.M., Leary, P.N. and Hart, M.B. 1990. Latest Cenomanian–earliest Turonian low-oxygen tolerant benthonic foraminifera: a case study from the Sergipe basin (NE Brazil) and the western Anglo-Paris Basin (Southern England). *Palaeogeography, Palaeoclimatology, Palaeoecology*, **77**(2): 145–79.

Koutsoukos, E.A.M., Mello, M.R., de Azambuja Filho, N.C., Hart, M.B. and Maxwell, J.R. 1991. The upper Aptian–Albian succession of the Sergipe basin, Brazil: an integrated paleoenvironmental assessment. *American Association of Petroleum Geologists Bulletin*, **75**(3): 479–98.

Kovar-Eder, J., Meller, B. and Zetter, R. 1998. Comparative investigations on the basal fossiliferous layers at the opencast mine Oberdorf (Koflach-Voitsberg lignite deposit, Styria, Austria; Early Miocene). *Review of Palaeobotany and Palynology*, **101**(1–4): 125–47.

Kowalewski, M. and Kelly, P.H. (eds.) 2002. *The Fossil Record of Predation*, Paper No. 8. Paleontological Society.

Krings, M., Stone, A., Schmitz, R., *et al.* 1997. Neandertal DNA sequences and the origin of modern humans. *Cell*, **90**: 19–30.

Kroh, A. and Nebelsick, J.H. 2003. Echinoid assemblages as a tool for palaeoenvironmental reconstruction: an example from the Early Miocene of Egypt. *Palaeogeography, Palaeoclimatology, Palaeoecology*, **201**(1–2): 157–77.

Kruijs, E. 1989. Predicting the locations of mid-Cretaceous wind-driven upwelling and productivity: a critical evaluation. Unpublished M.S. thesis, Pennsylvania State University.

Ksiazkiewicz, M. 1977. *Trace Fossils in the Flysch of the Polish Carpathians*, Palaeontologia Polonica No. 36. Polska Akademia Nauk, Zaklad Paleobiologii/Polish Academy of Sciences, Institute of Paleobiology.

Kuznetsova, K.I., Grigelis, A.A., Adjamian, J., Jarmakani, E. and Hallaq, L. 1996. *Zonal Stratigraphy and Foraminifera of the Tethyan Jurassic (Eastern Mediterranean)*. Gordon & Breach.

Kvacek, Z. 1998. Bilina: a window on Early Miocene marshland environments. *Review of Palaeobotany and Palynology*, **101**(1–4): 111–24.

Labandeira, C.C. 1997. Insect mouthparts: ascertaining the paleobiology of insect feeding strategies. *Annual Review of Ecology and Systematics*, **28**: 153–93.

Labandeira, C. C. and Sepkoski, J. J., Jr 1993. Insect diversity and the fossil record. *Science*, **261**: 310– 5.

Laliyev, A. G. 1964. *The Maykopian Series of Georgia*. Nedra. (in Russian)

Lalli, C. M. and Gilmer, R. W. 1989. *Pelagic Snails: The Biology of Holoplanktonic Gastropod Mollusks*. Stanford University Press.

Lamb, J. 1964. The geology and paleontology of the Rio Aragua surface section, Serrania del Interior, state of Monagas, Venezuela. *Boletín Informativo Asociación Venezolano de Geología, Minería y Petroleo, Caracas*, **7**(4): 111–23.

Landing, E. 1993. *In situ* earliest Cambrian tube worms and the oldest Metazoan-constructed biostrome (Placentian series, southeastern Newfoundland). *Journal of Paleontology*, **67**: 333–42.

Landman, N. H., Tanabe, K. and Davis, R. A. (eds.) 1996. *Ammonoid Paleobiology*, Topics in Geobiology Series No. 13. Plenum Press.

Lane, P. D., Siveter, D. J. and Fortey, R. A. (eds.) 2003. Trilobites and their relatives. *Special Papers in Palaeontology*, **70**.

Langer, M. R. and Hottinger, L. 2000. Biogeography of selected larger foraminifera. *Micropaleontology*, **26**(Suppl. 1): 105–26.

Lanzoni, E. and Magloire, L. 1969. Associations palynologiques et leurs applications stratigraphiques dans le Devonien Supérieur et Carbonifère Inférieur du Grand Erg occidental (Sahara Algérien). *Revue de l'Institut Français du Petrole*, **24**(4): 441–53.

Larsen, C. S. 1997. *Bioarchaeology: Interpreting Behaviour from the Human Skeleton*. Cambridge University Press.

Latal, C., Piller, W. E. and Harzhauser, M. 2004. Palaeoenvironmental reconstructions by stable isotopes of Middle Miocene gastropods of the central Paratethys. *Palaeogeography, Palaeoclimatology, Palaeoecology*, **211**(1–2): 157–69.

Laudon, R. C. 1996. *Principles of Development Geology*. Prentice Hall.

Lawton, J. H. and May, R. M. (eds.) 1995. *Extinction Rates*. Oxford University Press.

Leakey, M. 1995. The farthest horizon. *National Geographic Magazine*, **188**: 38–51.

Leakey, M. G. and Harris, J. M. (eds.) 2003. *Lothagam: The Dawn of Humanity in Eastern Africa*. Columbia University Press.

Leakey, R. 1994. *The Origin of Humankind*. Weidenfeld & Nicolson.

Leakey, R. and Lewin, R. 1992. *Origins Reconsidered: In Search of What Makes Us Human*. Little, Brown.

1995. *The Sixth Extinction: Biodiversity and its Survival*. Phoenix.

Lear, C. H., Rosenthal, Y. and Wright, J. D. 2003. The closing of a seaway: ocean water masses and global climate change. *Earth and Planetary Science Letters*, **210**: 425–36.

Leckie, R. M., Bralower, T. J. and Cashman, R. 2002. Oceanic anoxic events and plankton evolution: biotic response to tectonic forcing during the mid-Cretaceous. *Paleoceanography*, **17**(3): 13.1–30.

Lees, A. and Buller, A. T. 1972. Modern temperate-water and warm-water shelf sediments contrasted. *Marine Geology*, **19**: 159–98.

Lees, A., Buller, A. T. and Scott, J. 1969. Marine carbonate sedimentation processes, Connemara, Ireland: an interim report. *Reading University Geological Reports*, **2**.

Le Fevre, J. 1971. Paleoecological observations on Devonian ostracodes from the Ougarta Hills (Algeria). *Bulletin du Centre Recherche de Pau (Société National des Petroles d'Aquitaine)*, **5**: 817–41.

Lehrmann, D. J., Payne, J. L., Felix, S. V., *et al.* 2003. Permian–Triassic boundary sections from shallow-marine carbonate platforms of the Nanpanjiang basin, south China: implications for oceanic conditions associated with the end-Permian extinction and its aftermath. *Palaios*, **18**: 138–52.

Lesslar, P. 1987. Computer-assisted interpretation of depositional palaeoenvironments based on foraminifera. *Geological Society of Malaysia Bulletin*, **21**: 103–19.

Lethiers, F. and Whatley, R. 1994. The use of Ostracoda to reconstruct the oxygen levels of Late Palaeozoic oceans. *Marine Micropaleontology*, **24**: 57–69.

Levi-Setti, R. 1993. *Trilobites*, 2nd edn. University of Chicago Press.

Levinton, J. S. 2001. *Genetics, Paleontology and Macroevolution*, 2nd edn. Cambridge University Press.

Lewin, R. 1993. *Principles of Human Evolution: A Core Textbook*. Blackwell.

1999. *Human Evolution: An Illustrated Introduction*. Blackwell.

Lewin, R. and Foley, R. 2004. *Principles of Human Evolution*, 2nd edn. Blackwell.

Lewis-Williams, D. 2002. *The Mind in the Cave: Consciousness and the Origins of Art*. Thames and Hudson.

Li, Y., Kershaw, S. and Mu, X. 2004. Ordovician reef systems and settings in south China before the Late Ordovician mass extinction. *Palaeogeography, Palaeoclimatology, Palaeoecology*, **205**(3–4): 235–54.

Lieberman, B. S. (ed.) 2000. *Paleobiogeography*, Topics in Geobiology Series No. 16. Kluwer Academic/Plenum Press.

Lindsay, E. H., Opdyke, N. D. and Johnson, N. M. 1980. Pliocene dispersal of the horse *Equus* and late Cenozoic mammalian dispersal events. *Nature*, **287**: 135–8.

Lindsay, E. H., Fahlbusch, V. and Mein, P. (eds.) 1989. *European Neogene Mammal Chronology*, NATO ASI Series A, Life Sciences No. 180. Plenum Press.

Lindstrom, A. and Peel, J. S. 2003. Shell repair and mode of life of *Praenatica gragaria* (Gastropoda) from the Devonian of Bohemia (Czech Republic). *Palaeontology*, **46**(3): 623–33.

Lipps, J. (ed.) 1993. *Fossil Prokaryotes and Protists*. Blackwell.

Lipps, J. H. and Signor, P. W. (eds.) 1992. *Origin and Early Evolution of the Metazoa*, Topics in Geobiology Series No. 10. Plenum Press.

Lister, A. and Bahn, P. 2000. *Mammoths: Giants of the Ice Age*. Marshall.

Lockley, M. G. 1991. *Tracking Dinosaurs: A New Look at the Ancient World*. Cambridge University Press.

Lockley, M. G. and Hunt, A. P. 1995. *Dinosaur Tracks and Other Fossil Footprints of the Western United States*. Columbia University Press.

Lockley, M. G. and Meyer, C. 2000. *Dinosaur Tracks and Other Fossil Footprints of Europe*. Columbia University Press.

Loeblich, A. R., Jr and Tappan, H. 1987. *Foraminiferal Genera and Their Classification*. Van Nostrand Reinhold.

Lomando, A. J. and Harris, P. M. (eds.) 1988. *Giant Oil and Gas Fields*. Society of Economic Paleontologists and Mineralogists.

Long, J. A. 1995. *The Rise of Fishes*. Johns Hopkins University Press.

Loucks, R. G. and Sarg, J. F. 1993. *Carbonate Sequence Stratigraphy*, Memoir No. 57. American Association of Petroleum Geologists.

Lough, J. M. 2004. A strategy to improve the contribution of coral data to high-resolution paleoclimatology. *Palaeogeography, Palaeoclimatology, Palaeoecology*, **204**(1–2): 115–43.

Lowe, S., Chitolie, J., Pearce, J. M. and Welsh, A. 1988. Application of well-site palynology to hydrocarbon exploration. *Proceedings of the Indonesian Petroleum Association 17th Annual Convention*, Jakarta: 301–3.

Loydell, D. K., Orr, P. J. and Kearns, S. 2004. Preservation of soft tissues in Silurian graptolites from Latvia. *Palaeontology*, **47**(3): 503–12.

Lucas, S. G. 1997. *Dinosaurs: The Textbook*, 2nd edn. W. C. Brown.

2001. *Chinese Fossil Vertebrates*. Columbia University Press.

Luckett, W. P. and Hartenberger, J.-L. (eds.) 1985. *Evolutionary Relationships among Rodents*. Plenum Press.

Luger, P. 2003. Paleobiogeography of late Early Cretaceous to Early Paleocene marine Ostracoda in Arabia and North to Equatorial Africa. *Palaeogeography, Palaeoclimatology, Palaeoecology*, **197**: 319–42.

Luning, S., Craig, J., Loydell, D. K., Storch, P. and Fitches, B. 2000. Lower Silurian "hot shales" in North Africa and Arabia: regional distribution and depositional model. *Earth-Science Reviews*, **49**: 121–200.

Luning, S., Kolonic, S., Belhadj, E. M., *et al.* 2004. Integrated depositional model for the Cenomanian–Turonian organic-rich strata in North Africa. *Earth-Science Reviews*, **64**: 51–117.

Lusheng Huang 1997. Calcareous nannofossil biostratigraphy in the Pearl River Mouth Basin, South China Sea, and Neogene reticulofenestrid coccolith size distribution pattern. *Marine Micropaleontology*, **32**(1–2): 31–57.

Lynam, R. L. 1994. *Vertebrate Taphonomy*. Cambridge University Press.

MacDonald, I. R., Boland, G. S., Baker, J. S., *et al.* 1989. Gulf of Mexico hydrocarbon seep communities. II. Spatial distribution of seep communities and hydrocarbons at Bush Hill. *Marine Biology*, **101**: 235–47.

MacFadden, B. J. 1992. *Fossil Horses: Systematics, Paleobiology, and Evolution of the Family Equidae*. Cambridge University Press.

MacKinnon, D. I., Lee, D. E. and Campbell, J. D. (eds.) 1991. *Brachiopods through Time*. Balkema.

(eds.) 1996. *Brachiopods through Time: Proceedings of the 2nd International Brachiopod Congress*, Dunedin, New Zealand. Balkema.

MacLeod, N. and Keller, G. 1996. *Cretaceous–Tertiary Mass Extinctions*. W. W. Norton.

MacLeod, N., Rawson, P. F., and Forey, P. L. 1997. The Cretaceous–Tertiary biotic transition. *Journal of the Geological Society*, **154**: 265–92.

MacPhee, R. D. E. and Sues, H.-D. 1999. *Extinctions in Near Time: Causes, Contexts and Consequences*. Plenum Press.

MacRae, C. 1999. *Life Etched in Stone: Fossils of South Africa*. Geological Society of South Africa.

Macsotay, O. and Chachati, B. 1995. Estructuras sedimentarias, hidrodinámicas y biogénicas del Miocene Inferior-Medio, estado Anzoategui, Venezuela. *Boletín de Geología, Ministerio de Energía y Minas, Publicación Especial*, **10**: 153–63.

Madigan, M. T., Martinko, J. M. and Parker, J. 2003. *Brock Biology of Microorganisms*, 10th edn. Prentice Hall.

Mai, D. H. 1998. Contribution to the flora of the Middle Oligocene Calau beds in Brandenburg, Germany. *Review of Palaeobotany and Palynology*, **101**(1–4): 43–70.

Mai, L. L., Owl, M. Y. and Kersting, M. P. 2005. *The Cambridge Dictionary of Human Biology and Evolution*. Cambridge University Press.

Maisey, J. G. 1996. *Discovering Fossil Fishes*. Henry Holt.

Majoran, S. 1987. Notes on mid-Cretaceous biostratigraphy of Algeria. *Journal of African Earth Sciences*, **6**(6): 781–6.

Mandic, O. and Steininger, F. F. 2003. Computer-based mollusc stratigraphy: a case study from the Eggenburgian (Lower Miocene) type region (NE Austria). *Palaeogeography, Palaeoclimatology, Palaeoecology*, **197**: 263–91.

Mann, K. O. and Lane, H. R. 1995. *Graphic Correlation*, Special Publication No. 53. Society of Economic Paleontologists and Mineralogists/Society for Sedimentary Geology.

Mann, P. (ed.) 1999. *Caribbean Basins*, Sedimentary Basins of the World Series No. 4. Elsevier.

Maples, C. G. and West, R. R. (eds.) 1992. *Trace Fossils*, Short Courses in Paleontology No. 5. Paleontological Society.

Marek, L., Parsely, R. L. and Galle, A. 1997. Functional morphology of hyoliths based on flume studies. *Vestnik Ceskeho Geologickeho Ustavu*, **72**: 351–8.

Marincovich, L., Jr and Gladenkov, A. Y. 1999. Evidence for an early opening of the Bering Strait. *Nature*, **397**: 149–51.

Markova, A. K. 1999. Early Pleistocene faunas of small mammals in eastern Europe. *Stratigraphy and Geological Correlation*, **7**(2): 179–89.

Marret, F. and Zonneveld, K. A. F. 2003. Atlas of modern organic-walled dinoflagellate cyst distribution. *Review of Palaeobotany and Palynology*, **125**(1–2): 1–200.

Marshall, J. E. A. and Hemsley, A. R. 2003. A Mid Devonian seed-megaspore from east Greenland and the origin of the seed plants. *Palaeontology*, **46**(4): 647–70.

Marsicano, C. A. and Barredo, S. P. 2004. A Triassic tetrapod footprint assemblage from southern South America: palaeobiogeographical and evolutionary implications.

Palaeogeography, Palaeoclimatology, Palaeoecology **203**(3–4): 313–35.

Martill, D. M. 1993. *Fossils of the Santana and Crato formations, Brazil*, Field Guides to Fossils No. 5. Palaeontological Association.

Martin, A. J. 2001. *Introduction to the Study of Dinosaurs*. Blackwell.

Martin, P. S. and Klein, R. G. (eds.) 1984. *Quaternary Extinctions*. University of Arizona Press.

Martin, P. S. and Wright, H. E., Jr 1967. *Pleistocene Extinction*. Yale University Press.

Martin, R. E. 1999. *Taphonomy: A Process Approach*. Cambridge University Press.

(ed.) 2000. *Environmental Micropaleontology*, Topics in Geobiology Series No. 15. Kluwer Academic/Plenum Press.

Martin, W. and Muller, M. 1998. The hydrogen hypothesis for the first eukaryote. *Nature*, **392**: 37–41.

Martinelli, N. 2004. Climate from dendrochronology: latest developments and results. *Global and Planetary Change*, **40**(1–2): 129–40.

Martinez, J. I. 2003. The paleoecology of Late Cretaceous upwelling events from the upper Magdalena basin, Colombia. *Palaios*, **18**: 305–20.

Martinez-Delclos, X., Briggs, D. E. G. and Penalver, E. 2004. Taphonomy of insects in carbonates and amber. *Palaeogeography, Palaeoclimatology, Palaeoecology*, **203**(1–2): 65–72.

Massa, D. and Moreau-Benoit, A. 1976. Essai de synthèse stratigraphique et palynologique du système Devonien en Libye occidentale. *Revue de l'Institut Français du Petrole*, **31**(2): 287–333.

Masse, J.-P. 1985. Paléobiogéographie des rudistes du domaine peri-Meditérrannéen à l'Aptian inférieur. *Bulletin de la Société Géologique de France, Série 8*, **1**(5): 715–21.

Masse, J.-P. and Gallo Maresca, M. 1997. Late Aptian Radiolitidae (rudist bivalves) from the Mediterranean and southwest Asiatic regions: taxonomic, biostratigraphic and palaeobiogeographic aspects. *Palaeogeography, Palaeoclimatology, Palaeoecology*, **128**: 101–10.

Masse, J.-P. and Skelton, P. W. (eds.) 1998. *Quatrième Congrès International sur les Rudistes*, Mémoir Spéciale No. 22. Géobios.

Masse, J.-P., Fenerci-Masse, M. and Ozer, S. 2002. Late Aptian rudist faunas from the Zonguldak region, western Black Sea, Turkey. *Cretaceous Research*, **23**(4): 523–36.

Masse, J.-P., Fenerci, M. and Pernarcic, E. 2003. Palaeobathymetric reconstruction of peritidal carbonates: Late Barremian, Urgonian, sequences of Provence (SE France). *Palaeogeography, Palaeoclimatology, Palaeoecology*, **200**(1–4): 65–81.

Masurel, H. 1989. Ostracods as palaeoenvironmental indicators in the Lower Carboniferous Yoredale series of northern England. *Journal of Micropalaeontology*, **8**(2): 156–82.

Matoshko, A. V., Gozhik, P. F. and Danukalova, G. 2004. Key late Cenozoic fluvial archives of eastern Europe: the Dniester, Dnieper, Don and Volga. *Proceedings of the Geologists' Association*, **115**: 141–73.

Matteucci, R., Basso, D. and Tomaselli, V. (eds.) 1994. *Studies on Ecology and Paleoecology of Benthic Communities*. Società Paleontologica Italiana.

Mattioli, E. and Pittet, B. 2004. Spatial and temporal distribution of calcareous nannofossils along a proximal–distal transect in the Lower Jurassic of the Umbria–Marche basin (central Italy). *Palaeogeography, Palaeoclimatology, Palaeoecology*, **205**(3–4): 295–316.

Maury, C. J. 1922. The Recent *Arcas* of the Panamic province. *Palaeontographica Americana*, **1**(4): 163–208.

1925. *A Further Contribution to the Paleontology of Trinidad (Miocene Horizons)*. Cornell University Press.

Mawson, R., Talent, J. A. and Long, J. A. (eds.) 2000. Mid-Palaeozoic biota and biogeography. *Records of the Western Australian Museum, Supplement*, **58**.

Maync, W. 1949. Notes on the Lower Cretaceous formations of Venezuela and the stratigraphical position of the *Choffatella*-bearing deposits. *Eclogae Geologicae Helvetiae*, **42**(2): 530–46.

Mayor, A. 2000. *The First Fossil Hunters: Paleontology in Greek and Roman Times*. Princeton University Press.

McArthur, J. M., Mutterlose, J., Price, G. D., *et al.* 2004. Belemnites of Valanginian, Hauterivian and Barremian age: Sr-isotope stratigraphy, composition (87Sr/86Sr, del13C, del18O, Na, Sr, Mg) and palaeo-oceanography. *Palaeogeography, Palaeoclimatology, Palaeoecology*, **202**(3–4): 253–72.

McDougall, K. 1996. Benthic foraminiferal response to the emergence of the Isthmus of Panama and coincident paleoceanographic changes. *Marine Micropaleontology*, **28**(2): 133–69.

McGhee, G. R., Jr 1996. *The Late Devonian Mass Extinction*. Columbia University Press.

McGhee, G. R., Jr, Sheehan, P. M., Bottjer, D. J. and Droser, M. L. 2004. Ecological ranking of Phanerozoic biodiversity crises: ecological and taxonomic severities are decoupled. *Palaeogeography, Palaeoclimatology, Palaeoecology*, **211**(3–4): 289–97.

McGirr, N. 1998. *Nature's Connections: An Exploration of Natural History*. Natural History Museum.

McGowran, B. in press. *Biostratigraphy: Principles and Practice*. Cambridge University Press.

McGuire, W. J., Griffiths, D. R., Hancock, P. L. and Stewart, I. S. (eds.) 2000. *The Archaeology of Geological Catastrophes*, Special Publication No. 171. Geological Society.

McIlroy, D. (ed.) 2004. *The Application of Ichnology to Palaeoenvironmental and Stratigraphic Analysis*, Special Publication No. 228. Geological Society.

McIlroy, D., Green, O. R. and Brasier, M. D. 2001. Palaeobiology and evolution of the earliest agglutinated foraminifera: *Platysolenites*, *Spirosolenites* and related forms. *Lethaia*, **34**: 13–29.

McKenna, M. C. 1971. Fossil mammals and the Eocene demise of the De Geer North Atlantic dispersal route. *Geological Society of America Abstracts with Program*, **3**: 644.

1975. Fossil mammals and .Early Eocene North Atlantic land continuity. *Annals of the Missouri Botanical Gardens*, **62**: 335–53.

1983. Holarctic landmass rearrangement, cosmic events, and Cenozoic terrestrial organisms. *Annals of the Missouri Botanical Gardens*, **70**: 459–89.

McKenzie, K.G. and Jones, P.J. (eds.) 1993. *Ostracoda in the Earth and Life Sciences, Proceedings of the 11th International Symposium on Ostracoda*, Warrnambool, Victoria, Australia, 1991. Balkema.

McKerrow, W.S. and Scotese, C.R. (eds.) 1990. *Palaeozoic Palaeogeography and Biogeography*, Memoir No. 12. Geological Society.

McKinney, F.K. and Jackson, J.B.C. 1991. *Bryozoan Evolution*. University of Chicago Press.

McKinney, F.K. and Taylor, P.D. 2001. Bryozoan species extinction and migration during the last one hundred million years. *Palaeontologia Electronica*, **4**(1), http://www.palaeo-electronica.org.toc.htm.

McKinney, M.L. and Drake, J.A. (eds.) 1998. *Biodiversity Dynamics: Turnover of Populations, Taxa, and Communities*. Columbia University Press.

McLaughlin, P.I., Brett, C.E., McLaughlin, S.L.T. and Cornell, S.R. 2004. High-resolution sequence stratigraphy of a mixed carbonate-siliciclastic, cratonic ramp (Upper Ordovician; Kentucky–Ohio, USA): insights into the relative influence of eustasy and tectonics through analysis of facies gradients. *Palaeogeography, Palaeoclimatology, Palaeoecology*, **210**(2–4): 267–94.

McMenamin, M.A.S. 1982. A case for two Late Proterozoic–earliest Cambrian faunal province loci. *Geology*, **10**: 290–2.

1986. The garden of Ediacara. *Palaois*, **1**: 178–82.

1998. *The Garden of Ediacara*. Columbia University Press.

McMenamin, M. A. S. and McMenamin, D.L.S. 1990. *The Emergence of Animals*. Columbia University Press.

McNamara, K.J. (ed.) 1990. *Evolutionary Trends*. Belhaven.

1997. *Stromatolites*, 2nd edn. Western Australia Museum.

McNeil, D.H., Issler, D.R. and Snowdon, L.R. 1996. Colour alteration, thermal maturity, and burial diagenesis in fossil foraminifers. *Geological Survey of Canada Bulletin*, **499**: 1–34.

Meier, K.J.S. and Willems, H. 2003. Calcareous dinoflagellate cysts in surface sediments from the Mediterranean Sea: distribution patterns and influence of main environmental gradients. *Marine Micropaleontology*, **48**(3–4): 321–54.

Mein, P. 1975. *Report on Activity, RCMNS Working Groups (1975–1981)*. Regional Commission on Mediterranean Neogene Stratigraphy.

Meisch, C. 2000. *Freshwater Ostracoda of Western and Central Europe*. Heidelberg: Spektrum.

Meister, C., Alzouma, K., Lang, J. and Mathey, B. 1992. Les ammonites du Niger (Afrique occidentale) et la transgression trans-Saharienne au cours du Cenomenien–Turonien. *Geobios*, **25**: 55–100.

Melchin, M.J. and DeMont, M.E. 1995. Possible propulsion modes in Graptoloidea: a new model for graptoloid locomotion. *Paleobiology*, **21**(1): 110–20.

Mello, M.R. and Katz, B.J. (eds.) 2000. *Petroleum Systems of South Atlantic Margins*, Memoir No. 73. American Association of Petroleum Geologists.

Melnikova, L.M., Siveter, D.J. and Williams, M. 1997. Cambrian Bradoriida and Phosphatocopida (Arthropoda) of the former Soviet Union. *Journal of Micropalaeontology*, **16**: 179–91.

Melnyk, D.H. and Maddocks, R.F. 1988a. Ostracode biostratigraphy of the Permo-Carboniferous of central and north-central Texas. I. Paleoenvironmental framework. *Micropaleontology*, **34**(1): 1–20.

1988b. Ostracode biostratigraphy of the Permo-Carboniferous of central and north-central Texas. II. Ostracode zonation. *Micropaleontology*, **34**(1): 21–39.

Merceron, G., Blondel, C., Brunet, M., *et al.* 2004. The Late Miocene paleoenvironment of Afghanistan as inferred from dental microwear in artiodactyls. *Palaeogeography, Palaeoclimatology, Palaeoecology*, **207**(1–2): 143–63.

Meszaros, N. 1992a. Nannoplankton zones of the Miocene deposits in the Transylvanian basin. *International Nannoplankton Association Newsletter*, **13**(2): 59–60.

1992b. Nannoplankton zones of the Paleogene deposits in the Transylvanian basin. *International Nannoplankton Association Newsletter*, **13**(2): 60–1.

Metcalfe, I., Nicoll, R.S., Mundil, R., *et al.* 2001. The Permian–Triassic boundary and mass extinction in China. *Episodes*, **24**(4): 239–44.

Meulenkamp, J.E. and Sissingh, W. 2003. Tertiary palaeogeography and tectonostratigraphic evolution of the northern and southern peri-Tethys platforms and the intermediate domains of the African–Eurasian convergent plate boundary zone. *Palaeogeography, Palaeoclimatology, Palaeoecology*, **196**: 209–28.

Meyer, D.L. 1985. Evolutionary implications of predation on Recent comatulid crinoids from the Great Barrier Reef. *Paleobiology*, **11**: 154–64.

Meyer, D.L. and Macurda, D.B., Jr 1977. Adaptive radiation of the comatulid crinoids. *Paleobiology*, **3**: 74–82.

Meyer, H.W. 2003. *The Fossils of Florissant*. Smithsonian Books.

Meyerhoff, A.A., Boucot, A.J., Meyerhoff Hull, D. and Dickins, J.M. 1996. *Phanerozoic Faunal & Floral Realms of the Earth: The Intercalary Relations of the Malvinokaffric and Gondwana Faunal Realms with the Tethyan Faunal Realm*, Memoir No. 189. Geological Society of America.

Miall, A.D. 1997. *The Geology of Stratigraphic Sequences*. Springer-Verlag.

Mickulic, D.G. (ed.) 1990. *Arthropod Paleobiology*, Short Courses in Paleontology No. 3. University of Tennessee Press.

Middlemiss, F.A. and Rawson, P. 1971. Faunal provinces in space and time. *Geological Journal, Special Issue*, **5**.

Milankovitch, M. 1938. Astronomische Mittel zur Erforschung der erdgeschichten Klimate. *Handbuch der Geophysik*, **9**: 593–698.

Mildenhall, D.C. 1990. Forensic palynology in New Zealand. *Review of Palaeobotany and Palynology*, **64**: 227–34.

Miller, C.G., Richter, M. and Do Carmo, D.A. 2002. Fish and ostracod remains from the Santos basin (Cretaceous to Recent), Brazil. *Geological Journal*, **37**: 297–316.

Miller, S.L. 1953. Production of amino acids under possible primitive Earth conditions. *Science*, **117**: 527–8.

Milsom, C. and Rigby, S. 2004. *Fossils at a Glance*. Blackwell.

Ming-Mei Liang, Bruch, A., Collinson, M., *et al.* 2003. Testing the climatic estimates from different palaeobotanical methods: an example from the Middle Miocene Shanwang flora of China. *Palaeogeography, Palaeoclimatology, Palaeoecology*, **198**: 279–301.

Minikh, M.G. and Minikh, A.V. 1997. Ichthyofaunal correlation of the Triassic deposits from the northern cis-Caspian and southern cis-Urals regions. *Geodiversitas*, **19**(2): 279–92.

Minter, N.J. and Braddy, S.J. 2004. Unravelling fish swimming trails: a review of the ichnogenus *Undichna*. *Palaeontological Association Newsletter*, **57**(48th Annual Meeting Abstracts): 165.

Mishra, V.P. 2000. Neogene megaplants from Himalaya. *Geological Survey of India Miscellaneous Publication*, **64**: 23–30.

Mitchell, F. and Ryan, M. 1997. *Reading the Irish Landscape*. Town House.

Mithen, S. 1998. *The Prehistory of the Mind*. Orion.

2003. *After the Ice: A Global Human History, 20 000–5000 BC*. Weidenfeld & Nicolson.

Mitlehner, A.G. 1994. The occurrence and preservation of diatoms in the Paleogene of the North Sea basin. Unpublished Ph.D. thesis, University College London.

Moczydlowska, M. 2004. Ediacaran phytoplankton and cyanobacteria diversity recovery after the Snowball Earth glaciation. *Palaeontological Association Newsletter*, **57**(48th Annual Meeting Abstracts): 167–8.

Mojzsis, S.J., Arrhenius, G., McKeegan, K.D., *et al.* 1996. Evidence for life on Earth before 3800 million years ago. *Nature*, **384**: 55–9.

Molostovskoy, E.A. 1997. Magnetostratigraphy of the Pliocene deposits in the Black Sea, Caspian region and adjacent areas. *Geodiversitas*, **19**(2): 471–95.

Molostovskoy, E.A. and Guzhikov, A.Y. 1999. Some peculiarities concerning the Pliocene evolution of the Pliocene Black Sea and Caspian basins. *Geodiversitas*, **21**(3): 477–89.

Monks, N. and Palmer, P. 2002. *Ammonites*. Natural History Museum.

Monty, C., Bosence, D.W.J., Bridges, P.H. and Pratt, B. (eds.) 1995. *Carbonate Mud-Mounds: Their Origin and Evolution*, Special Publication No. 23. International Association of Sedimentologists.

Moore, P.D., Chaloner, B. and Slott, P. 1996. *Global Environmental Change*. Blackwell.

Moreira, D. and Lopez-Garcia, P. 1998. Symbiosis between methanogenic Archaea and delta-Proteobacteria as the origin of eukaryotes: the syntrophic hypothesis. *Journal of Molecular Evolution*, **47**: 517–30.

Moreno-Vasquez, J. 1995. Neogene biofacies in eastern Venezuela and their calibration with seismic data. *Marine Micropaleontology*, **26**: 287–302.

Morgan, J., Warner, M., Brittan, J., *et al.* 1997. Size and morphology of the Chicxulub impact crater. *Nature*, **390**: 472–6.

Morley, C.K. (ed.) 1999. *Geoscience of Rift Systems: Evolution of East Africa*, Studies in Geology No. 44. American Association of Petroleum Geologists.

Morley, R.J. 2000. *Origin and Evolution of Tropical Rain Forests*. John Wiley.

2003. Interplate dispersal paths for megathermal angiosperms. *Perspectives in Plant Ecology, Evolution and Systematics*, **6**(1–2): 5–20.

Morris, P. and Therivel, R. (eds.) 2001. *Methods of Environmental Impact Assessment*, 2nd edn. Spon.

Morton, A.C. (ed.) 1992. *Geology of the Brent Group*, Special Publication No. 61. Geological Society.

Morwood, M.J., Soejono, R.P., Roberts, R.G., *et al.* 2004. Archaeology and age of a new hominin from Flores in eastern Indonesia. *Nature*, **431**: 1087–91.

Motoyama, I. and Maruyama, T. 1998. Neogene diatom and radiolarian biochronology for the middle to high latitudes of the northwest Pacific region: calibration to the Cande and Kent's geomagnetic polarity time scales (CK92 and CK95). *Journal of the Geological Society of Japan*, **104**: 171–83.

Moya, A. and Font, E. (eds.) 2004. *Evolution: From Molecules to Ecosystems*. Oxford University Press.

Moya-Solà, S., Kohler, M., Alba, D.M., Casanovas-Vilar, I. and Galindo, J. 2004. *Pierolapithecus catalunicus*, a new Middle Miocene great ape from Spain. *Science*, **306**: 1339–44.

Mudge, D.C. and Bujak, J.P. 2001. Biostratigraphic evidence for evolving palaeoenvironments in the lower Paleogene of the Faeroe–Shetland basin. *Marine and Petroleum Geology*, **18**: 577–90.

Mudie, P.J. and Rochon, A. 2001. Distribution of dinoflagellates in the Canadian Arctic marine region. *Journal of Quaternary Science*, **16**(7): 603–20.

Mudie, P.J., Aksu, A.E. and Yasar, D. 2001. Late Quaternary dinoflagellate cysts from the Black, Marmara and Aegean Seas: variations in assemblages, morphology and paleosalinity. *Marine Micropaleontology*, **43**: 155–78.

Muller, J. 1981. Fossil pollen records of extant angiosperms. *Botanical Review*, **47**(1): 1–142.

Muller, W., Fricke, H., Haliday, A.N., McCulloch, M.T. and Wartho, J.-A. 2003. Origin and migration of the Alpine Iceman. *Science*, **302**: 862–6.

Mullins, G.L. 2003. Aggregates of the acritarch *Dilatisphaera laevigata*: faecal pelletisation, phytoplankton bloom or defence against phagotrophy? *Journal of Micropalaeontology*, **22**: 163–7.

Murphy, W.J., Eizirik, E., O'Brien, S.J., *et al.* 2001. Resolution of the early placental mammal radiation using Bayesian phylogenetics. *Science*, **294**: 2348–51.

Nagy, J. 1992. Environmental significance of foraminiferal morphogroups in Jurassic North Sea deltas. *Palaeogeography, Palaeoclimatology, Palaeoecology*, **95**: 111–34.

Nagymarosy, A. and Voronina, A. A. 1993. Nannoplankton from the lower Maykopian beds (Early Oligocene), Soviet Union. *Miscellanea Micropaleontologica, Proceedings of the 4th International Nannoplankton Association International Conference*, Prague, **146**(2): 189–221.

Nanda, A. C. 2000. Siwalik Group: palaeontologic and magnetostratigraphic aspects. *Geological Survey of India Miscellaneous Publication*, **64**: 53–8.

Narbonne, G. 2004. Modular construction of early Ediacaran complex life forms. *Science*, **305**: 1141–4.

Nations, J. D. and Eaton, J. G. (eds.) 1991. *Stratigraphy, Depositional Environments and Sedimentary Tectonics of the Western Margin, Cretaceous Western Interior Seaway*, Special Paper No. 260. Geological Society of America.

Natural History Museum. 2001a. *British Palaeozoic Fossils*. Intercept.

2001b. *British Mesozoic Fossils*. Intercept.

2001c. *British Cenozoic Fossils*. Intercept.

Naylor, P. 2003. *Great British Marine Animals*. Sound Diving Publications.

Neale, J. W. and Brasier, M. D. 1981. *Microfossils from Recent and Fossil Shelf Seas*. Ellis Horwood.

Nelson, C. H. and Damuth, J. E. 2003. Myths of turbidite system control: insights provided by modern turbidite studies. *International Conference on Deep-Water Processes in Modern and Ancient Environments*, Barcelona, *Abstracts*: 32.

Nestor, H. and Stock, C. W. 2001. Recovery of the stromatoporoid fauna after the Late Ordovician extinction. *Bulletin, Tohoku University Museum*, **1**: 333–41.

Neuman, B. 1988. Some aspects of the life strategies of Early Palaeozoic rugose corals. *Lethaia*, **21**: 97–114.

Nielsen, C. 2001. *Animal Evolution: Interrelationships of the Living Phyla*, 2nd edn. Oxford University Press.

Niklas, K. J. 1997. *The Evolutionary Biology of Plants*. University of Chicago Press.

Niklas, K. J., Tiffney, B. H. and Knoll, A. H. 1983. Patterns of vascular land plant diversification. *Nature*, **303**: 614–16.

Nisbet, E. G. 1991. *The Living Earth: A Short History of Life and Its Home*. HarperCollins.

Nitecki, M. H. (ed.) 1979. *Mazon Creek Fossils*. Academic Press.

Noe, L. F. 2004. The evidence for and implications of an invertebrate diet in Jurassic pliosaurs (Reptilia: Sauropterygia). *Palaeontological Association Newsletter*, **57**(48th Annual Meeting Abstracts): 127.

Norris, R. D. and de Vargas, C. 2000. Evolution all at sea. *Nature*, **405**: 23–4.

Norris, R. D. and Rohl, U. 2000. Astronomically-tuned chronology for the Paleocene-Eocene transition: *GFF*, **122**: 117–18.

Nudds, J. R. 1994. *Directory of British Geological Museums*, Miscellaneous Paper No. 18. Geological Society.

Nudds, J. R. and Pettitt, C. W. (eds.) 1997. *The Value and Valuation of Natural Science Collections*. Geological Society.

O'Connor, T. 2000. *The Archaeology of Animal Bones*. Sutton.

Officer, C. and Page, J. 1996. *The Great Dinosaur Extinction Controversy*. Addison Wesley.

Oji, J. 1996. Is predation intensity reduced with increasing depth? Evidence from the west Atlantic stalked crinoid *Endoxocrinus parrae* (Gervais) and implications for the Mesozoic marine revolution. *Paleobiology*, **22**: 339–51.

Oloriz, F. and Rodriguez-Tovar, F. J. 1999. *Advancing Research on Living and Fossil Cephalopods*. Kluwer.

Olson, H. C. and Leckie, R. M. (eds.) 2003. *Micropaleontologic Proxies for Sea-Level Change and Stratigraphic Discontinuities*, Special Publication No. 75. Society of Economic Paleontologists and Mineralogists.

Olson, S. 2002. *Mapping Human History*. Bloomsbury.

Olsson, A. A. 1932. Contributions to the Tertiary stratigraphy of northern Peru. V. The Peruvian Miocene. *Bulletin of American Paleontology*, **9**(68): 1–272.

1961. *Mollusks of the Tropical Eastern Pacific, Particularly from the Southern Half of the Panamic–Pacific Faunal Province (Panama to Peru): Panamic–Pacific Pelecypoda*. Paleontological Research Institute.

Oppenheimer, S. 1998. *Eden in the East: The Drowned Continent of Southeast Asia*. Weidenfeld & Nicolson.

2003. *Out of Eden: The Peopling of the World*. Constable & Robinson.

Osborne, R. and Benton, M. 1996. *Atlas of Evolution*. Viking.

Otero, O. and Gayet, M. 2001. Palaeoichthyofaunas from the Lower Oligocene and Miocene of the Arabian plate: palaeoecological and palaeobiogeographic implications. *Palaeogeography, Palaeoclimatology, Palaeoecology*, **165**: 141–69.

Owens, B., Al-Tayyar, H., van der Eem, J. G. L. A. and Al-Hajri, S. (eds.) 1995. Palaeozoic palynostratigraphy of the Kingdom of Saudi Arabia. *Review of Palaeobotany and Palynology*, **89**: 1–150.

Owens, B., McLean, D. and Bodman, D. 2004. A revised palynozonation of British Namurian deposits and comparisons with eastern Europe. *Micropaleontology*, **50**(1): 89–103.

Padian, K. (ed.) 1986. *The Beginning of the Age of Dinosaurs: Faunal Change across the Triassic–Jurassic Boundary*. Cambridge University Press.

Pagani, M., Arthur, M. A. and Freeman, K. H. 1999. Miocene evolution of atmospheric carbon dioxide. *Paleoceanography*, **14**: 273–92.

Page, R. D. M. and Holmes, E. C. 1998. *Molecular Evolution*. Blackwell.

Palmer, A. A. 1988. Radiolarians from the Miocene Pungo River formation of Onslow Bay, North Carolina continental shelf. *Special Publications of the Cushman Foundation for Foraminiferal Research*, **24**: 163–78.

Palmer, D. and Rickards, B. (eds.) 1991. *Graptolites*. Boydell Press.

Papp, A. 1954. Die Molluskenfauna im Sarmat des Wiener Beckens. *Mitteilungen der Geologischen Gesellschaft Wien*, **45**: 1–112.

Paramonova, N. P., Ananova, E. N., Andreeva-Grigorovich, A. S., *et al.* 1979. Palaeontological characteristics of the

Sarmatian and Maeotian of the Ponto-Caspian area. *Annales Géologiques des Pays Helléniques*, **1979**(11): 961–71.

Paredes, I., Guzman, J.I., Helenes, J., *et al.* 1994. Sea-level fluctuations from the Barremian to the Paleogene, eastern Venezuela. *V Simposio Bolivariano Exploración Petrolera en las Cuencas Subandinas:* 234–236. (in Spanish).

Parker, A. 2003. *In the Blink of an Eye: The Cause of the Most Dramatic Event in the History of Life*. Free Press.

Parker, J.R. (ed.) 1993. *Petroleum Geology of North-West Europe, Proceedings of the 4th Conference*. Geological Society.

Parker, M.E., Clark, M. and Wise, S.W., Jr 1985. Calcareous nannofossils of Deep Sea Drilling Project sites 558 and 563, North Atlantic Ocean: biostratigraphy and the distribution of Oligocene braarudosphaerids. *Initial Reports of the Deep Sea Drilling Project*, **82**: 559–89.

Parker, S. 2003. *The Complete Book of Dinosaurs*. Apple.

Parrish, J.T. 1998. *Interpreting Pre-Quaternary Climate from the Geologic Record*. Columbia University Press.

Pashin, J.C. and Gastaldo, R.A. (eds.) 2004. *Sequence Stratigraphy, Paleoclimate and Tectonics of Coal-Bearing Strata*, Studies in Geology No. 51. American Association of Petroleum Geologists.

Patnaik, R. 2003. Reconstruction of upper Siwalik palaeoecology and palaeoclimatology using microfossil palaeocommunities. *Palaeogeography, Palaeoclimatology, Palaeoecology*, **197**: 133–50.

Patterson, C. 1999. *Evolution*, 2nd edn. Natural History Museum.

Paul, G.S. 2002. *Dinosaurs of the Air: The Evolution and Loss of Flight in Dinosaurs and Birds*. Johns Hopkins University Press.

Pearsall, D.B. 2000. *Paleoethnobotany*, 2nd edn. Academic Press.

Peirano, A., Morri, C., Bianchi, C.N., *et al.* 2004. The Mediterranean coral *Cladophora caespitosa*: a proxy for past climate fluctuations? *Global and Planetary Change*, **40**(1–2): 195–200.

Peleo-Alampay, A.M., Mead, G.A. and Wei Wuchang. 1999. Unusual Oligocene *Braarudosphaera*-rich layers of the South Atlantic and their palaeoceanographic implications. *Journal of Nannoplankton Research*, **21**(1): 17–26.

Perrier, V., Vannier, J., Siveter, D.J., Kriz, J. and Manda, S. 2004. Silurian midwater communities dominated by ostracods: evidence from fossil assemblages in France and Bohemia. *Palaeontological Association Newsletter*, **57**(48th Annual Meeting Abstracts): 175.

Pervesler, P. and Zuschin, M., 2002. Chemosymbiosis, fossil lucinoid bioturbations and the *Chondrites* enigma. *World Congress of Malacology*, Vienna, Austria, 5–7 April 2002, Abstracts.

Petters V. 1954. Typical foraminiferal horizons in the Lower Cretaceous of Colombia, S.A. *Contributions from the Cushman Foundation for Foraminiferal Research*, **5**(3): 128–37.

1955. Development of Upper Cretaceous foraminiferal faunas in Colombia. *Journal of Paleontology*, **29**(2): 212–25.

Philip, G.M. 1979. Carpoids: echinoderms or chordates? *Biological Review*, **54**: 439–71.

Philip, J. 1985. Sur les relations des marges Tethysiennes au Campanien et au Maastrichtien déduites de la distribution des rudistes. *Bulletin de la Société Gélogique de France, Serie 8*, **1**(5): 723–31.

1998. Biostratigraphie et paleobiogeographie des rudistes: évolution des concepts et progrès récents. *Bulletin de la Société Géologique de France*, **169**(5): 689–708.

Philip, J. and Jaillard, E. 2004. Revision of Upper Cretaceous rudists from northwestern Peru. *Journal of South American Earth Sciences*, **17**(1): 39–48.

Pianka, E.R. 1970. On *r*- and *K*-selection. *American Naturalist*, **104**: 592–7.

Piccoli, G., Sartori, S., Franchino, A., *et al.* 1991. Mathematical model of faunal spreading in benthic palaeobiogeography (applied to Cenozoic Tethyan molluscs). *Palaeogeography, Palaeoclimatology, Palaeoecology*, **86**: 139–86.

Pickering, K.T. and Owen, L.A. 1994. *An Introduction to Global Environmental Issues*. Routledge.

Pigliucci, M. 2002. *Denying Evolution: Creationism, Scientism and the Nature of Science*: Sunderland, MA: Sinauer Associates.

Pinniger, D. 2001. *Pest Management in Museums, Archives and Historic Houses*. Archetype.

Pinous, O.V., Karogodin, Y.N., Ershov, S.V. and Sahagian, D.L. 1999a. Sequence stratigraphy, facies and sea level change of the Hauterivian productive complex, Priobskoe oil field (west Siberia). *American Association of Petroleum Geologists Bulletin*, **83**(6): 972–89.

Pinous, O. V., Sahagian, D.L., Sharyugin, B.N. and Nikitenko, B.L. 1999b. High-resolution sequence stratigraphic analysis and sea-level interpretation of the Middle and Upper Jurassic strata of the Nyurolskaya depression and vicinity (south-western west Siberia, Russia). *Marine and Petroleum Geology*, **16**: 245–57.

Pinous, O.V., Levchuk, M.A. and Sahagian, D.L. 2001. Regional synthesis of the productive Neocomian complex of west Siberia: sequence stratigraphic framework. *American Association of Petroleum Geologists Bulletin*, **85**(10): 1713–30.

Piperno, D.R. 1988. *Phytolith Analysis: An Archaeological and Geological Perspective*. Academic Press.

Pittet, B., van Buchem, F.S.P., Hillgartner, H., *et al.* 2002. Ecological succession, palaeoenvironmental change, and depositional sequences of Barremian–Aptian shallow-water carbonates of northern Oman. *Sedimentology*, **49**: 555–81.

Pitts, M. and Roberts, M. 1997. *Fairweather Eden: Life in Britain Half a Million Years Ago as Revealed by the Excavations at Boxgrove*. Century.

Plavcan, J.M., Kay, R.F., Jungers, W.L. and van Schaik, C.P. (eds.) 2002. *Reconstructing Behavior in the Primate Fossil Record*. Kluwer Academic/Plenum Press.

Plewes, C.R., Palmer, T.J. and Haynes, J.R. 1993. A boring foraminiferan from the Upper Jurassic of England and northern France. *Journal of Micropalaeontology*, **12**(1): 83–9.

Poag, C.W. 1997. The Chesapeake Bay bolide impact: a convulsive event in Atlantic coastal plain evolution. *Sedimentary Geology*, **108**: 45–90.

Poinar, G., Jr and Poinar, R. 1994. *The Quest for Life in Amber*. Addison-Wesley.

1999. *The Amber Forest*. Princeton University Press.

Pole, M. 2003. New Zealand climate in the Neogene and implications for global atmospheric circulation. *Palaeogeography, Palaeoclimatology, Palaeoecology*, **193**: 269–84.

Pollard, A.M. (ed.) 1999. *Geoarchaeology*, Special Publication No. 165. Geological Society.

Ponomarenko, A.G. 1985. Fossil insects from the Tithonian "Solnhofener Plattenkalke" in the Museum of Natural History, Vienna. *Annalen des Naturhistorisches Museum, Wien*, **A87**: 135–44.

Poore, R.Z. and Sloan, L.C. (eds.) 1996. Climates and climate variability of the Pliocene. *Marine Micropaleontology, Special Volume*, **27**(1/4).

Pope, M. and Harris, M. 2004. New insights into Late Ordovician climate, oceanography and tectonics. *Palaeogeography, Palaeoclimatology, Palaeoecology*, **210**(2–4): 117–18.

Popkhadze, L.I. 1977. *Maeotian Microfauna (Foraminifera and Ostracoda) from Western Georgia*. Akademia. Nauk Gruzenskoi SSR. (in Russian)

Popov, S.V. 1995. Zoogeography of the Early Miocene basins of western Eurasia according to bivalve molluscs. *Stratigraphy and Geological Correlation*, **3**(4): 393–405.

Popova, I.M. 1993. Significance and paleoecological interpretations of Early–Middle Miocene radiolarians from south Sakhalin, Russia. *Micropaleontology, Special Publication*, **6**: 161–74.

Porebski, E., Kozlowska-Dawidziuk, A. and Masiak, M. 2004. The *lundgreni* event in the Silurian of the east European platform, Poland. *Palaeogeography, Palaeoclimatology, Palaeoecology*, **213**(3–4): 271–94.

Porter, S.M. and Knoll, A.H. 2000. Testate amoebae in the Neoproterozoic era: evidence from vase-shaped microfossils in the Chuar group, Grand Canyon. *Paleobiology*, **26**: 360–85.

Pough, F.H., Janis, C.M. and Heiser, J.B. 2002. *Vertebrate Life*, 6th edn. Prentice Hall.

Pour, M.G. 2004. New Early to Mid Ordovician trilobite faunas of Iran and their biogeographical significance. *Palaeontological Association Newsletter*, **57**(48th Annual Meeting Abstracts): 154.

Powell, A., 1986. A new species of *Bolboforma* from the Miocene of the Voring plateau, northern Norway. *Journal of Micropalaeontology*, **5**(2): 71–4.

Powell, A.J. and Riding, J. (eds.) in press. *Exploration and Development Biostratigraphy*. Micropalaeontological Society.

Prentice, J.E. and Thomas, J.M. 1960. The Carboniferous goniatites of north Devon. *Proceedings of the 3rd Conference on the Geology and Geomorphology of South-West England, Royal Geological Society of Cornwall, Abstracts*: 6–8.

Prior, F. 2003. *Britain BC*. HarperCollins.

Prokoph, A., Rampini, M.R. and El Bilali, H. 2004. Periodic components in the diversity of calcareous plankton and geological events over the past 230 Myr. *Palaeogeography, Palaeoclimatology, Palaeoecology*, **207**: 105–25.

Prothero, D.R. 1994. *The Eocene–Oligocene Transition*. Columbia University Press.

2004. *Bringing Fossils to Life: An Introduction to Paleobiology*. McGraw-Hill.

2005. *The Evolution of North American Rhinoceroses*. Cambridge University Press.

Prothero, D.R. and Dott, R.H., Jr 2004. *Evolution of the Earth*, 7th edn. McGraw-Hill.

Prothero, D.R. and Emry, R.J. (eds.) 1996. *The Terrestrial Eocene–Oligocene Transition in North America*. Cambridge University Press.

Prothero, D.R. and Schoch, R.M. (eds.) 1989. *The Evolution of Perissodactyls*. Oxford University Press.

(eds.) 1994. *Major Features of Vertebrate Evolution*. Short Courses in Paleontology No. 7. Paleontological Society.

2002. *Horns, Tusks, and Flippers: The Evolution of Hoofed Mammals and Their Relatives*. Johns Hopkins University Press.

Prothero, D.R., Ivany, L.C. and Nesbitt, E.A. (eds.) 2003. *From Greenhouse to Icehouse: The Marine Eocene–Oligocene Transition*. Columbia University Press.

Pruss, S.B. and Bottjer, D.J. 2004. Late Early Triassic microbial reefs of the western United States: a description and model for their deposition in the aftermath of the end-Permian mass extinction. *Palaeogeography, Palaeoclimatology, Palaeoecology*, **211**(1–2): 127–37.

Purves, W.K., Sadava, D., Orians, G.H. and Heller, H.C. 2004. *Life: The Science of Biology*. 7th edn. Sinauer Associates/ W.H. Freeman.

Quade, J. and Cerling, T.E. 1995. Expansion of C4 grasses in the Late Miocene of northern Pakistan: evidence from stable isotopes in paleosols. *Palaeogeography, Palaeoclimatology, Palaeoecology*, **114**: 91–116.

Quammen, D. 1996. *The Song of The Dodo: Island Biogeography in an Age of Extinctions*. Hutchinson.

Quinglai Feng, Helmcke, D., Chonglakmani, C., Invagat-Helmcke, R. and Benpei Liu. 2004. Early Carboniferous radiolarians from north-west Thailand: palaeogeographical implications. *Palaeontology*, **47**(2): 377–394.

Racey, A. 2001. A review of Eocene nummulite accumulations: structure, formation and reservoir potential. *Journal of Petroleum Geology*, **24**(1): 79–100.

Racheboeuf, P.R. and Emig, C.C. (eds.) 1986. Les Brachiopodes fossiles et actuels. *Biostratigraphie du Paleozoique*, 4.

Radionova, E.P. and Khokhlova, I.E. 2000. Was the North Atlantic connected with the Tethys via the Arctic in the Early Eocene? Evidence from the siliceous plankton. *GFF*, **122**: 133–4.

Radinsky, L.B. 1987. *The Evolution of Vertebrate Design*. University of Chicago Press.

Rahtz, P. 1993. *Glastonbury*. B. T. Batsford/English Heritage.

Raisossadat, S. N. 2003. The relationship between ammonite distributions and sea-level changes in the Sarcheshmeh and Sanganeh formations (upper Barremian-lower Albian) in the Kopet Dagh basin in north east Iran. *Palaeontological Association Newsletter*, **54**(47th Annual Meeting Abstracts): 150.

Ramsay, A. T. S. (ed.) 1977. *Oceanic Micropalaeontology*. Academic Press.

Ramsbottom, W. H. C., Calver, M. A., Eagar, R. M. C., *et al.* 1978. *Silesian (Upper Carboniferous)*, Special Report No. 81. Geological Society of London.

Rasmussen, B. 2000. Filamentous microfossils in a 3235-million-year-old volcanogenic massive sulphide deposit. *Nature*, **405**: 676–9.

Rasmussen, H. W. 1961. A monograph of the Cretaceous Crinoidea. *Kongelige Danske Videnskabernes Biologiske Skrifter*, **12**.

Rasnitsyn, A. P. and Quicke, D. L. J. 2002. *History of Insects*. Kluwer.

Rathburn, A. E., Levin, L. A., Held, Z. and Lohmann, K. C. 2000. Benthic foraminifera associated with cold methane seeps on the northern California margin: ecology and stable isotopic composition. *Marine Micropaleontology*, **38**: 247–66.

Ratnikov, V. Y. 1996. Methods of paleogeographic reconstruction based upon fossil remains of amphibians and reptiles of the late Cenozoic of the east European platform. *Paleontological Journal*, **30**(1): 75–80.

Raup, D. M. and Sepkoski, J. J., Jr. 1982. Mass extinctions in the marine fossil record. *Science*, **215**: 1501–3.

1984. Periodicity of extinctions in the geological past. *Proceedings of the National Academy of Sciences, USA*, **81**: 801–5.

Raza, S. M., Cheema, I. U., Downs, W. R., Rajpar, A. R. and Ward, S. C. 2002. Miocene stratigraphy and mammal fauna from the Sulaiman range, southwestern Himalayas, Pakistan. *Palaeogeography, Palaeoclimatology, Palaeoecology*, **186**: 185–97.

Reboulet, S., Mattioli, E., Pittet, B., *et al.* 2003. Ammonoid and nannoplankton abundance in Valanginian (Early Cretaceous) limestone–marl successions from the southeast France Basin: carbonate dilution or productivity? *Palaeogeography, Palaeoclimatology, Palaeoecology*, **201**(1–2): 113–19.

Regal, B. 2004. *Human Evolution: A Guide to the Debates*. ABC-Clio.

Rehanek, J. and Cecca, F. 1993. Calcareous dinoflagellate cysts: biostratigraphy in upper Kimmeridgian-lower Tithonian pelagic limestones of Marches Apennines (central Italy). *Revue de Micropaléontologie*, **36**(2): 143–63.

Reichenbacher, B., Uhlig, U., Kowalke, T., *et al.* 2004. Biota, palaeoenvironments and biostratigraphy of continental Oligocene deposits of the south German molasse basin (Penzberg Syncline). *Palaeontology*, **47**(3): 639–78.

Reid, D. G. 1996. *Systematics and Evolution of Littorina*. Ray Society.

Reid, R. G. B. and Brand, D. G. 1986. Sulfide-oxidising symbiosis in lucinaceans: implications for bivalve evolution. *Veliger*, **29**: 3–24.

Reinhardt, E. G., Fitton, R. J. and Schwarcz, H. P. 2003. Isotopic (Sr, O, C) indicators of salinity and taphonomy in marginal marine systems. *Journal of Foraminiferal Research*, **33**(3): 262–72.

Reitz, E. J. and Wing, E. S. 1999. *Zooarchaeology*. Cambridge University Press.

Renfrew, C. and Bahn, P. 2000. *Archaeology*, 3rd edn. Thames & Hudson.

Renne, P. R., Wolde Gabriel, G., Hart, W. K., *et al.* 1999. Chronostratigraphy of the Miocene – Pliocene Sagantole formation, middle Awash Valley, Afar rift, Ethiopia. *Geological Society of America Bulletin*, **111**(6): 869–85.

Rentner, J. and Keupp, H. (eds.) 1991. *Fossil and Recent Sponges*. Springer-Verlag.

Retallack, G. J. 1991. *Miocene Paleosols and Ape Habitats of Pakistan and Kenya*. Oxford University Press.

2004. Late Miocene climate and life on land in Oregon within a context of Neogene global change. *Palaeogeography, Palaeoclimatology, Palaeoecology*, **214**(1–2): 97–123.

Reumer, J. W. F., de Vos, D. and Mol, D. (eds.) 2003. *Advances in Mammoth Research, Proceedings of the 2nd International Mammoth Conference*, Rotterdam 16–20 May 1999. Backhuys.

Reyment, R. A. and Bengtson, P. (eds.) 1981. *Aspects of Mid-Cretaceous Regional Geology*. Academic Press.

Rice, C. M., Anderson, L. I. and Trewin, N. H. 2002. Stratigraphy and structural setting of the Lower Devonian Rhynie chert, Aberdeenshire, Scotland: an early terrestrial hot-spring system. *Journal of the Geological Society, London*, **159**: 203–14.

Richards, M., Corte-Real, H., Forster, P., *et al.* 1996. Palaeolithic and Neolithic lineages in the European mitochondrial gene pool. *American Journal of Human Genetics*, **59**: 185–203.

Richards, P. W. 1996. *The Tropical Rain Forest*, 2nd edn. Cambridge University Press.

Richardson, J. B. and MacGregor, D. C. 1986. Silurian and Devonian spore zones of the Old Red Sandstone continent and adjacent regions. *Bulletin of the Geological Survey of Canada*, **36**: 1–79.

Rickards, R. B. 1975. Palaeoecology of the Graptolithina, an extinct class of the phylum Hemichordata. *Biological Reviews*, **50**: 397–436.

Rickards, R. B., Wright, A. J. and Hamedi, M. A. 2000. Late Ordovician and Early Silurian graptolites from southern Iran. *Records of the Western Australian Museum, Supplement*, **58**: 103–22.

Ride, W. D. L., Cogger, H. G., Dupuis, C., *et al.* 1999. *International Code of Zoological Nomenclature*. International Commission on Zoological Nomenclature.

Rigby, J. K. (ed.) 1971. *Reefs through Time*. Allen Press.

Rigby, S. 1991. Feeding strategies in graptoloids. *Palaeontology*, **34**: 797–814.

Rigby, S. and Rickards, R. B. 1989. New evidence for the life habit of graptoloids from physical modelling. *Palaeobiology*, **15**: 402–13.

Riley, N. J. 1993. Dinantian (Lower Carboniferous) biostratigraphy and chronostratigraphy in the British Isles. *Journal of the Geological Society, London*, **150**: 427–46.

Riveline, J., Berger, J.-P., Feist, M., *et al.* 1996. European Mesozoic–Cenozoic charophyte biozonation. *Bulletin de la Société Géologique de France*, **167**: 453–68.

Roberts, D.G., Mitchener, B., Ewen, D., Jones, R.W. and Hooker, J.J. in press. Palaeogene land mammal dispersal across the northern North Atlantic: "the nearest run thing". *Geology*.

Roberts, M.B. and Parfitt, S.A. (eds.) 1999. *Boxgrove: A Middle Pleistocene Hominid Site at Eartham Quarry, Boxgrove, West Sussex*, Archaeological Report No. 17. English Heritage.

Robertson, A.H.F., Searle, M.P. and Ries, A.C. (eds.) 1990. *The Geology and Tectonics of the Oman Region*, Special Publication No. 49. The Geological Society.

Robinson, A.G. (ed.) 1997. *Regional and Petroleum Geology of the Black Sea and Surrounding Region*, Memoir No. 68. American Association of Petroleum Geologists.

Robinson, E. 1993. Some imperforate larger foraminifera from the Paleogene of Jamaica and the Nicaragua rise. *Journal of Foraminiferal Research*, **23**(1): 47–65.

Robinson, G.S. and Kohl, B. 1978. Computer assisted paleoecologic analyses and application to petroleum exploration. *Transactions, Gulf Coast Association of Geological Societies*, **28**: 433–47.

Rode, A.L. and Lieberman, B.S. 2004. Using GIS to unlock the interactions between biogeography, environment and evolution in Middle and Late Devonian brachiopods and bivalves. *Palaeogeography, Palaeoclimatology, Palaeoecology*, **211**(3–4): 354–9.

Rodriguez, J. 2004. Stability in Pleistocene Mediterranean mammalian communities. *Palaeogeography, Palaeoclimatology, Palaeoecology*, **207**(1–2): 1–22.

Roehl, P.O. and Choquette, P.W. (eds.) 1985. *Carbonate Petroleum Reservoirs*. Springer-Verlag.

Rögl, F. 1998. Palaeogeographic considerations for Mediterranean and Paratethys seaways. *Annalen des (K.K) Naturhistorischen (Hof) Museums Wien*, **99A**: 279–310.

Rollo, F., Ubaldi, M., Ermini, L. and Marota, I. 2002. Otzi's last meals: DNA analysis of the intestinal content of the Neolithic glacier mummy from the Alps. *Proceedings of the National Academy of Sciences, USA*, **99**(20): 12594–9.

Rosales, I., Quesada, S. and Robles, S. 2004. Paleotemperature variations of Early Jurassic seawater recorded in geochemical trends of belemnites from the Basque–Cantabrian basin, northern Spain. *Palaeogeography, Palaeoclimatology, Palaeoecology*, **203**(3–4): 253–75.

Rosen, B.R. 1988. Progress, problems and patterns in the biogeography of reef corals and other tropical marine organisms. *Helgolander Meeresuntersuchungen*, **42**: 269–301.

Rosenbaum, G., Lister, G.S. and Duboz, C. 2002. Reconstruction of the tectonic evolution of the western Mediterranean since the Oligocene. *Journal of the Virtual Explorer*, **8**: 107–26.

Rosenzweig, M.L. 1995. *Species Diversity in Space and Time*. Cambridge University Press.

Ross, C.A. and Ross, J.R.P. 1995. Foraminiferal zonation of Late Paleozoic depositional sequences. *Marine Micropaleontology*, **26**: 469–78.

Ross, D.J. and Skelton, P.W. 1993. Rudist formations of the Cretaceous: a palaeoecological, sedimentological and stratigraphical review. *Sedimentology Review*, **1**: 73–91.

Rostovtsev, K.O. (ed.) 1985. *The Jurassic Deposits of the South Part of the Transcaucasus*. Nauka

Rostovtseva, Y.V. and Soloveva, N.A. 1999. Deep-water Sarmatian and Maeotian rocks in the Black Sea coast area, Taman peninsula. *Lithology and Mineral Resources*, **34**(3): 289–91.

Roth, P.H. 1978. Cretaceous nannofossil biostratigraphy and oceanography in the northwestern Atlantic Ocean. *Initial Reports of the Deep Sea Drilling Project*, **44**: 731–59.

Rothschild, L.J. and Lister, A.M. (eds.) 2003. *Evolution on Planet Earth: The Impact of the Physical Environment*. Academic Press.

Round, F.E., Crawford, R.M. and Mann, D.G. 1990. *The Diatoms: Biology and Morphology of the Genera*. Cambridge University Press.

Roure, F. (ed.) 1994. *Peri-Tethyan Platforms*. Editions Technip.

Rowe, M.P. 2000. Inferring the retinal anatomy and visual capacities of extinct vertebrates. *Palaeontologia Electronica*, **3**(1), http://www.palaeo-electronica.org.toc.htm.

Ruddiman, W. 2004. Human influences on climate began thousands of years ago. *Abstract, Public Lecture*, Queen Mary College, University of London.

Rudwick, M.J.S. 1961. The feeding mechanism of the Permian brachiopod *Prorichthofenia*. *Palaeontology*, **3**: 450–7.

1964. The inference of structure from function in fossils. *British Journal for the Philosophy of Science*, **15**: 27–40.

Ruiz, F., Gonzale-Regalado, M.L., Munoz, J.M., *et al.* 2003. Population age structure techniques and ostracods: applications in coastal hydrodynamics and paleoenvironmental analysis. *Palaeogeography, Palaeoclimatology, Palaeoecology*, **199**: 51–69.

Rull, V. 2002. High-impact palynology in petroleum geology: applications from Venezuela (northern South America). *American Association of Petroleum Geologists Bulletin*, **86**(2): 279–300.

Runnegar, B. 1982. Oxygen requirements, biology and phylogenetic significance of the late Precambrian worm *Dickinsonia*, and the evolution of the burrowing habit. *Alcheringa*, **6**: 223–39.

1995. Vendobionta or Metazoa? Developments in understanding the Ediacara 'fauna'. *Neues Jahrbuch für Geologie und Paläontologie, Abhandlungen*, **195**: 303–18.

Rutsch, R. 1942. Larkinien (Arcidae) aus dem Jung Tertiar von Trinidad (B.W.I.). *Eclogae Geologicae Helvetiae*, **35**(2): 213–23.

Ryan, W. and Pitman, W. 1998. *Noah's Flood: The New Scientific Discoveries about the Event that Changed History*. Simon & Schuster.

Ryan, W., Pitman, W., Major, C., *et al.* 1997. An abrupt drowning of the Black Sea shelf. *Marine Geology*, **138**: 119–26.

Ryan, W., Major, C.O., Lericolais, G. and Goldstein, S.L. 2003. Catastrophic flooding of the Black Sea. *Annual Review of Earth and Planetary Science*, **31**: 525–54.

Ryder, G., Fastovsky, D. and Gartner, S. (eds.) 1996. *The Cretaceous–Tertiary Event and Other Catastrophes in Earth History*, Special Paper No. 307. Geological Society of America.

Sadooni, F.N. and Alsharhan, A.S. 2003. Stratigraphy, microfacies and petroleum potential of the Mauddud formation

(Albian-Cenomanian) in the Arabian Gulf basin. *American Association of Petroleum Geologists Bulletin*, **87**(10): 1653–80.

Sagan, L. 1967. On the origin of mitosing cells. *Journal of Theoretical Biology*, **14**: 225–74.

Sahagian, D., Pinous, O., Olferiev, A. and Zakharov, V. 1996. Eustatic curve for the Middle Jurassic–Cretaceous based on Russian platform and Siberian stratigraphy: zonal resolution. *American Association of Petroleum Geologists Bulletin*, **80**(9): 1433–58.

Salem, M.J. and Busrewil, M.T. (eds.) 1980. *The Geology of Libya*. Academic Press.

Sanchez-Villagra, M.R. and Clack, J.A. (eds.) 2004. Fossils of the Miocene Castillo formation, Venezuela: contributions on Neotropical palaeontology. *Special Papers in Palaeontology*, **71**.

Sanfilippo, A. and Nigrini, C. 1998. Code numbers for Cenozoic low latitude radiolarian biostratigraphic zones and GPTS conversion tables. *Marine Micropaleontology*, **33**: 109–56.

Sanyal, P., Bhattacharya, S.K., Kumar, R., Ghosh, S.K. and Sangode, S.J. 2004. Mio-Pliocene monsoonal record from Himalayan foreland basin (Indian Siwalik) and its relation to vegetational change. *Palaeogeography, Palaeoclimatology, Palaeoecology*, **205**(1–2): 23–41.

Saunders, J.B. 1958. Recent foraminifera of mangrove swamps and their fossil counterparts in Trinidad. *Micropaleontology*, **4**(1): 79–92.

(ed.) 1968. *Transactions of the 4th Caribbean Geological Conference*. Caribbean Printers.

Saunders, J.B. and Bolli, H.M. 1985. Trinidad's contribution to world biostratigraphy. *Transactions of the 4th Latin American Geological Congress, Trinidad & Tobago*, 1979: 781–95.

Saunders, J.B., Jung, P., Geister, J. and Biju-Duval, B. 1982. The Neogene of the south flank of the Cibao Valley, Dominican Republic: a stratigraphic study. *Transactions of the 9th Caribbean Geological Conference*, Santo Domingo 1980, **1**: 151–60.

Saunders, J.B., Bernoulli, D., Muller-Merz, E., *et al.* 1984. Stratigraphy of the late Middle Eocene to Early Oligocene Bath cliff section, Barbados, West Indies. *Micropaleontology*, **30**(4): 390–425.

Saunders, J.B., Jung, P. and Biju-Dival, B. 1986. *Neogene Paleontology in the Northern Dominican Republic*, vol. 1, *Field Surveys, Lithology, Environment and Age*. Paleontological Research Institution.

Savarese, M. 1992. Functional analysis of archaeocyathan skeletal morphology and its paleobiological implications. *Paleobiology*, **18**: 464–80.

Savazzi, E. (ed.) 1999. *Functional Morphology of the Invertebrate Skeleton*. John Wiley.

Savitskii, V.O., Boldyreva, V.P., Danchenko, R.V. and Mitrofanova, L.I. 1979. The Oligocene–Miocene deposits of southern Sakhalin (boundary basin). *Vestnik Moskovskogo Universiteta, Geologiya*, **34**(1): 81–6.

Saxena, R.K. 2000. Palynology of the Neogene sediments of northwest India. *Geological Survey of India Miscellaneous Publication*, **64**: 11–22.

Scarparo Cunha, A.A. and Koutsoukos, E.A.M. 1998. Calcareous nannofossils and planktic foraminifers in the upper Aptian of the Sergipe basin, northeastern Brazil: palaeoecological inferences. *Palaeogeography, Palaeoclimatology, Palaeoecology*, **142**: 175–84.

Schaal, S. and Ziegler, W. 1992. *Messel: An Insight into the History of Life and of the Earth*. Clarendon Press.

Schama, S. 2000. *The History of Britain*. BBC Books.

Schmidt, A.R., Schonborn, W. and Schafer, U. 2004. Diverse fossil Amoebae in German Mesozoic amber. *Palaeontology*, **47**(2): 185–98.

Scholle, P.A., Bebout, D.C. and Moore, C.H. (eds.) 1983. *Carbonate Depositional Environments*, Memoir No. 33. American Association of Petroleum Geologists.

Schopf, J.W. 1992. *Major Events in the History of Life*. Jones and Bartlett.

(ed.) 1993. *Earth's Earliest Biosphere*. Princeton University Press.

1999. *Cradle of Life*. Princeton University Press.

Schopf, T.J.M. 1972. *Models in Paleobiology*. W.H. Freeman.

Schroder, T. 1992. A palynological zonation of the North Sea basin. *Journal of Micropalaeontology*, **11**(2): 113–26.

Schrödinger, E. 1944. *What Is Life?* Cambridge University Press.

Schubert, J.K. and Bottjer, D.J. 1995. Aftermath of the Permian–Triassic mass extinction: palaeoecology of Lower Triassic carbonates in the western USA. *Palaeogeography, Palaeoclimatology, Palaeoecology*. **116**: 1–39.

Schultze, H.-P. and Trueb, L. (eds.) 1991. *Origins of the Higher Groups of Tetrapods*. Cornell University Press.

Schuster, F. and Wielandt, U. 1999. Oligocene and Early Miocene coral faunas from Iran: palaeoecology and palaeobiogeography. *International Journal of Earth Sciences*, **88**: 571–81.

Scott, R.W. 1990. *Models and Stratigraphy of Mid-Cretaceous Reef Communities, Gulf of Mexico*, Concepts in Sedimentology and Paleontology No. 2. Society of Economic Paleontologists and Mineralogists.

Scrutton, C.T. and Clarkson, E.N.K. 1990. A new scleractinian coral from the Ordovician of the Southern Uplands of Scotland. *Palaeontology*, **34**: 179–94.

Seeling, J., Colin, J.P. and Fauth, G. 2004. Global Campanian (Upper Cretaceous) ostracod palaeobiogeography. *Palaeogeography, Palaeoclimatology, Palaeoecology*, **213**(3–4): 379–98.

Seilacher, A. 1992. Vendobionta and Psammocorallia: lost constructions of Precambrian evolution. *Journal of the Geological Society, London*, **149**: 607–13.

Seilacher, A., Drozdewski, G. and Haude, R. 1968. Form and function of the stem in a pseudoplanktonic crinoid (*Seirocrinus*). *Palaeontology*, **11**(2): 275–82.

Sejrup, H.P., Birks, H.J.B., Kristensen, D.K. and Madsen, H. 2004. Benthonic foraminiferal distributions and quantitative transfer functions for the northwest European continental margin. *Marine Micropaleontology*, **53**: 197–226.

Selden, P. and Nudds, J. 2004. *Evolution of Fossil Ecosystems*. Manson.

Selley, R.C. 1998. *Elements of Petroleum Geology*, 2nd edn. Academic Press.

Semenenko, V.N. 1979. Correlation of Mio-Pliocene of the eastern Paratethys and Tethys. *Annales Géologiques des Pays Helléniques*, **1979**(3): 1101–11.

Sen Gupta, B.K. (ed.) 1999. *Modern Foraminifera*. Kluwer.

Sen Gupta, B.K., Platon, E., Bernhard, J.M. and Aharon, P. 1997. Foraminiferal colonization of hydrocarbon-seep bacterial mats and underlying sediment, Gulf of Mexico slope. *Journal of Foraminiferal Research*, **27**(4): 292–300.

Senut, B., Pickford, M., Gommery, D., *et al.* 2001. First hominid from the Miocene (Lukeino formation, Kenya). *Earth and Planetary Science Letters*, **332**: 137–44.

Sepkoski, J.J., Jr 1981. A factor analytical description of the Phanerozoic marine fossil record. *Paleobiology*, **7**: 36–53.

Serra-Kiel, J., Hottinger, L., Caus, E., *et al.* 1998. Larger foraminiferal biostratigraphy of the Tethyan Paleocene and Eocene. *Bulletin de la Société Géologique de France*, **169**(2): 281–99.

Servais, T., Javier Alvaro, J. and Blieck, A. 2003. Early Palaeozoic palaeo(bio)geographies of western Europe and North Africa. *Palaeogeography, Palaeoclimatology, Palaeoecology*, **195**(1–2): 1–228.

Setudehnia, C. 1975. The Palaeozoic sequence of Zard Kuh and Kuh-e Dinar. *Iranian Petroleum Institute Bulletin*, **60**: 16–33.

Sey, I.I. and Kalacheva, E.D. 1983. *The Mesozoic of the Soviet Arctic*. Nauka. (in Russian)

Shackleton, N.J., Berger, A. and Peltier, W.R. 1990. An alternative astronomical calibration of the Lower Pleistocene timescale based on ODP site 677. *Transactions of the Royal Society of Edinburgh (Earth Sciences)*, **81**: 251–61.

Shainyan, S.K., *et al.* 1990. Neogene stratigraphy of western Terpeniya Bay, southeastern Sakhalin. *Byulleten Moskovskogo Obshchestva Ispytatelei Prirody, Otdel Geologicheskii*, **65**(3): 36–46.

Sharland, P.R., Archer, R., Casey, D.M., *et al.* 2001. *Arabian Plate Sequence Stratigraphy*, GeoArabia Special Publication No. 2. Gulf PetroLink.

Sharland, P.R., Casey, D.B., Davies, R.B., Simmons, M.D. and Sutcliffe, O.E. 2004. Arabian plate sequence stratigraphy: revisions to SP2. *GeoArabia*, **9**(1): 199–214.

Sharpton, V. and Ward, P. (eds.) 1990. *Global Catastrophes in Earth History: An Interdisciplinary Conference on Impacts, Volcanism, and Mass Mortality*, Special Paper No. 247. Geological Society of America.

Sheldon, P.R. 1997. Parallel gradualistic evolution of Ordovician trilobites. *Nature*, **330**: 561–3.

Shennan, I. and Andrews, J. (eds.) 2000. *Holocene Land–Ocean Interaction and Environmental Change around the North Sea*, Special Publication No. 166. The Geological Society.

Shipman, P. 1998. *Taking Wing: Archaeopteryx and the Evolution of Bird Flight*. Weidenfeld & Nicolson.

Shreeve, J. 1996. *The Neanderthal Enigma*. Viking.

Shuster, C.N., Jr, Barlow, R.B. and Brockmann, H.J. (eds.) 2003. *The American Horseshoe Crab*. Harvard University Press.

Sibley, C.G. and Ahlquist, J.E. 1990. *Phylogeny and Classification of Birds*. Yale University Press.

Sibuet, M. and Olu, K. 1998. Biogeography, biodiversity and fluid dependence of deep cold-seep communities at active and passive margins. *Deep-Sea Research, II*, **45**: 517–67.

Siesser, W.G., Bralower, T.J. and de Carlo, E.H. 1992. Mid-Tertiary *Braarudosphaera*-rich sediments on the Exmouth Plateau. *Proceedings, Ocean Drilling Program, Scientific Results*, **122**: 653–663.

Simmons, M.D. (ed.) 1994. *Micropalaeontology and Hydrocarbon Exploration in the Middle East*. Chapman & Hall.

(ed.) in press. *Micropalaeontology and Hydrocarbon Exploration in the Former Soviet Union*. Chapman & Hall.

Simmons, M.D., Preobrazhensky, M.B. and Bugrova, I.J. 1992. Biostratigraphic characterisation of carbonate sequences and systems tracts: examples from the Early Cretaceous of the Middle East and Turkmenia. *International Symposium on Mesozoic and Cenozoic Sequence Stratigraphy*, Dijon, Abstracts: 290–1.

Simms, M.J. 1986. Contrasting lifestyles in Lower Jurassic crinoids: a comparison of benthic and pseudopelagic Isocrinoida. *Palaeontology*, **29**: 475–93.

Simo, J.A., Scott, R.W. and Maasse, J.-P. 1993. *Cretaceous Carbonate Platforms*, Memoir No. 56. American Association of Petroleum Geologists.

Simon, T., Hagdorn, H., Hagdorn, M.K. and Seilacher, A. 2003. Swimming trace of a coelacanth fish from the lower Keuper of south-west Germany. *Palaeontology*, **46**(5): 911–26.

Simonetta, A. and Conway Morris, S. 1991. *The Early Evolution of Metazoa and the Significance of Problematic Taxa*. Cambridge University Press.

Simpson, G.G. 1946. Tertiary land bridges. *Transactions of the New York Academy of Sciences, Series II*, **8**(8): 255–8.

Siveter, D.J. and Williams, M. 1997. Cambrian bradoriid and phosphatocopid arthropods of North America. *Special Papers in Palaeontology*, **57**.

Siveter, D.J., Williams, M., Peel, J.S. and Siveter, D.J. 1996. Bradoriida (Arthropoda) from the Early Cambrian of north Greenland. *Transactions of the Royal Society of Edinburgh (Earth Sciences)*, **56**: 113–21.

Skelton, P. (ed.) 1993. *Evolution: A Biological and Palaeontological Approach*. Addison-Wesley.

(ed.) 2003. *The Cretaceous World*. Cambridge University Press.

Skelton, P.W. and Smith, A.B. 2002. *Cladistics: A Practical Primer on CD-ROM*. Cambridge University Press.

Skelton, P.W. and Wright, V.P. 1987. A Caribbean rudist bivalve in Oman: island hopping across the Pacific in the Late Cretaceous. *Palaeontology*, **30**: 505–29.

Skelton, P.W., Gili, E., Vicens, E. and Obrador, A. 1995. The growth fabric of gregarious rudist elevators (Hippuritids) in a Santonian carbonate platform in the southern central Pyrenees. *Palaeogeography, Palaeoclimatology, Palaeoecology*, **119**: 107–26.

Skelton, P.W., Gili, E., Rosen, B.R. and Valldeperas, F.X. 1997. Corals and rudists in the Late Cretaceous: a critique of the hypothesis of competitive displacement. *Boletín de la Real Sociedad Española de Historia Natural, Sección Geológica*, **92**(1–4): 225–39.

Smart, C.W. and Ramsay, A.T.S. 1995. Benthic foraminiferal evidence for the existence of an Early Miocene oxygen-depleted oceanic water mass? *Journal of the Geological Society, London*, **152**: 735–8.

Smith, A. 1999. Dating the origins of metazoan body plans. *Evolution and Development*, **1**: 138–42.

Smith, A.B. 1994. *Systematics and the Fossil Record: Documenting Evolutionary Patterns*. Blackwell.

2004. Phylogeny and systematics of holasteriod echinoids and their migration into the deep sea. *Palaeontology*, **47**(1): 123–50.

Smith, A.B. and Peterson, K. 2002. Dating the time of origin of major clades: molecular clocks and the fossil record. *Annual Review of Earth and Planetary Sciences*, **30**: 65–89.

Smith, A.B., Simmons, M.D. and Racey, A. 1990. Cenomanian echinoids, larger foraminifera and calcareous algae from the Natih formation, central Oman Mountains. *Cretaceous Research*, **11**: 29–69.

Smith, A.H.V. and Butterworth, M.A. 1967. Miospores in the coal seams of the Carboniferous of Great Britain. *Special Papers in Palaeontology*, **1**.

Smith, F.A. and White, J.W.C. 2004. Modern calibration of phytolith carbon isotope signatures for C3/C4 paleograssland reconstruction. *Palaeogeography, Palaeoclimatology, Palaeoecology*, **207**(3–4): 277–304.

Smith, J.M. and Szathmary, E. 1999. *The Origins of Life: From the Birth of Life to the Origins of Language*. Oxford University Press.

Smith, R.J. 2000. Morphology and ontogeny of Cretaceous ostracods with preserved appendages from Brazil. *Palaeontology*, **43**(1): 63–98.

Snell, J.F. 2004. Bryozoa from the Much Wenlock limestone (Silurian) formation of the West Midlands and Welsh borderland. *Monograph of the Palaeontographical Society*, **621**.

Sobolik, K.D. 2003. *Archaeobiology*. Altamira.

Sohl, N.F. 1987. Cretaceous gastropods: contrasts between Tethys and the temperate provinces. *Journal of Paleontology*, **61**: 1085–111.

Sole, R.V., Montoya, J.M. and Erwin, D.H. 2002. Recovery after mass extinctions: evolutionary assembly in large-scale biosphere dynamics. *Philosophical Transactions of the Royal Society of London B*, **357**: 697–707.

Southwood, R. 2003. *The Story of Life*. Oxford University Press.

Spalding, M.D., Ravilious, C. and Green, E.P. 2001. *World Atlas of Coral Reefs*. University of California Press.

Sperber, G. (ed.) 1990. *From Apes to Angels: Essays in Honor of Philip V. Tobias*. Wiley-Liss.

Spindler, K. 1994. *The Man in The Ice: The Preserved Body of a Neolithic Man Reveals the Secrets of the Stone Age*. Weidenfeld & Nicolson.

Stanley, D.J. and Wezel, F.-C. (eds.) 1985. *Geological Evolution of the Mediterranean Basin*. Springer-Verlag.

Stanley, D.G., Jr (ed.) 2001. *The History and Sedimentology of Ancient Reef Systems*, Topics in Geobiology Series No. 17. Kluwer Academic/Plenum Press.

Stanley, E.A. 1992. Application of palynology to establish the provenance and travel history of illicit drugs. *Microscope*, **40**: 149–52.

Stanley, S.H. 1999. *Earth System History*. W.H. Freeman.

Stanley, S.M. 1970. *Relation of Shell Form to Life Habits in the Bivalvia (Mollusca)*, Memoir No. 125. Geological Society of America.

Stanley, S.M., Kennett, J.P., Knoll, A.H., *et al.* (eds.) 1995. *Effects of Past Global Change on Marine Life*. National Academy Press.

Starkie, S. 1994. Calcareous nannofossil biostratigraphy and depositional history of the Late Cretaceous to Early Miocene sequence of Iraq. Unpublished Ph.D. thesis, University College London.

Stearn, C.W. 1975. The stromatoporoid animal. *Lethaia*, **8**: 89–100.

1997. Biostratigraphy of the Devonian reef facies of western and Arctic Canada based on stromatoporoids. *Boletín de la Real Sociedad Española de Historia Natural, Sección Geológica*, **92**(1–4): 325–38.

Stearn, W.T. 1998. *The Natural History Museum at South Kensington*. Natural History Museum.

Stearns, S.C. and Hoekstra, R.F. 2000. *Evolution*. Oxford University Press.

Steiner, M., Wallis, E., Erdtmann, B.D., Zhao, Y. and Yang, R. 2001. Submarine hydrothermal exhalative ore layers in black shales from south China and associated fossils: insights into a Lower Cambrian facies and bio-evolution. *Palaeogeography, Palaeoclimatology, Palaeoecology*, **169**(3–4): 165–92.

Steiner, M.B., Eshet, Y., Rampino, M.R. and Schwindt, D.M. 2003. Fungal abundance spike and the Permian–Triassic boundary in the Karoo supergroup (South Africa). *Palaeogeography, Palaeoclimatology, Palaeoecology*, **194**: 405–14.

Steininger, F., Rogl, F. and Martini, E. 1976. Current Oligocene/Miocene biostratigraphic concept of the central Paratethys (middle Europe). *Newsletters on Stratigraphy*, **4**(3): 174–202.

Stephenson, M.H. and McLean, D. 1999. International correlation of Early Permian palynofloras from the Karoo sediments of Morupule, Botswana. *South African Journal of Geology*, **102**(1): 3–14.

Stephenson, M.H., Leng, M.J., Vane, C.H. and Osterloff, P.L. 2003a. Investigating Permian climate change from a multidisciplinary study of deglacial argillaceous sediments in Oman. *Abstracts, William Smith Meeting ('Wrestling with Mud')*, Geological Society, London.

Stephenson, M.H., Osterloff, P.L. and Filatoff, J. 2003b. Palynological biozonation of the Permian of Oman and Saudi Arabia: progress and challenges. *GeoArabia*, **8**(3): 467–96.

Steuber, T. 2003. Strontium isotope stratigraphy of Cretaceous hippuritid rudist bivalves: rates of morphological change and heterochronic evolution. *Palaeogeography, Palaeoclimatology, Palaeoecology*, **200**(1–4): 221–43.

Steuber, T. and Loser, H. 2000. Species richness and abundance patterns of Tethyan Cretaceous rudist bivalves (Mollusca: Hippuritacea) in the central-eastern Mediterranean and Middle East, analysed from a palaeontological database. *Palaeogeography, Palaeoclimatology, Palaeoecology*, **162**: 75–104.

Stewart, K.M. and Seymour, K.L. 1996. *Palaeoecology and Palaeoenvironments of Late Cenozoic Mammals*. University of Toronto Press.

Stewart, W.N. and Rothwell, G.W. 1993. *Palaeobotany and the Evolution of Plants*, 2nd edn. Cambridge University Press.

Stilwell, J.D. 2003. Patterns of biodiversity and faunal rebound following the K–T boundary extinction event in Austral Palaeocene molluscan faunas. *Palaeogeography, Palaeoclimatology, Palaeoecology*, **195**: 319–56.

Stoddart, D.R. and Yonge, C.M. (eds.) 1971. *Regional Variation in Indian Ocean Coral Reefs*, Symposium No. 28. Zoological Society of London.

Stoermer, E.F. and Smol, J.P. (eds.) 1999. *The Diatoms: Applications for the Environmental and Earth Sciences*. Cambridge University Press.

Stokes and Webb 1824. Description of some fossil vegetables of the Tilgate forest in Sussex. *Transactions of the Geological Society*, **2**(1): 421(bis)–424.

Stone, H.M.I. 1998. On predator deterrence by pronounced shell ornament in epifaunal bivalves. *Palaeontology*, **41**(5): 1051–68.

Stone, R. 2002. *Mammoth: The Resurrection of an Ice Age Giant*. Fourth Estate.

Stossel, I. 1995. The discovery of a new Devonian tetrapod trackway in SW Ireland. *Journal of the Geological Society, London*, **152**: 407–13.

Stringer, C. 2000. Coasting out of Africa. *Nature*, **405**: 24–7.

Stringer, C. and Gamble, C. 1993. *In Search of the Neanderthals*. Thames & Hudson.

Stringer, C. and McKie, R. 1996. *African Exodus: The Origins of Modern Humanity*. Jonathan Cape.

Strogen, P., Somerville, I.D. and Jones, G.L. 1996. *Recent Advances in Lower Carboniferous Geology*, Special Publication No. 107. The Geological Society.

Stromberg, C.A.E. 2004. Using phytolith assemblages to reconstruct the origin and spread of grass-dominated habitats in the great plains of North America during the Late Eocene to Early Miocene. *Palaeogeography, Palaeoclimatology, Palaeoecology*, **207**(3–4): 239–75.

Stromberg, C.A.E. and Feranec, R.S. 2004. The evolution of grass-dominated ecosystems during the late Cenozoic. *Palaeogeography, Palaeoclimatology, Palaeoecology*, **207**(3–4): 199–201.

Stuart, A.J. 1991. Mammalian extinctions in the Late Pleistocene of northern Eurasia and North America. *Biological Reviews*, **66**: 453–562.

Stukalina, G.A. 1988. Studies in Paleozoic crinoid columnals and stems. *Palaeontographica A*, **204**: 1–66.

Sues, H.-D. (ed.) 2000. *Evolution of Herbivory in Terrestrial Vertebrates*. Cambridge University Press.

Sutton, M.D., Briggs, D.E.G., Siveter, D.J. and Siveter, D.J. 2003. Arms with feet: an exceptionally preserved starfish from the Silurian Herefordshire Lagerstatte. *Palaeontological Association Newsletter*, **54**(47th Annual Meeting Abstracts): 153.

Sykes, B. 2001. *The Seven Daughters of Eve*. Bantam.

Szalay, F.S. 1995. *Evolutionary History of the Marsupials and an Analysis of Osteological Characters*. Cambridge University Press.

Szalay, F.S., Novacrek, M.J. and McKenna, M.C. (eds.) 1993. *Mammal Phylogeny*. Springer-Verlag.

Takahashi, O., Mayama, S. and Matsuoka, A. 2003. Host-symbiont associations of polycystine radiolaria: epifluorescence microscopic observation of living Radiolaria. *Marine Micropaleontology*, **49**(3): 187–94.

Talwani, M., Hay, W. and Ryan, W.B.F. (eds.) 1979. *Deep Drilling Results in the Atlantic Ocean: Continental Margins and Paleoenvironment*, Maurice Ewing Series Monograph No. 3. American Geophysical Union.

Tanacredi, J.T. 2001. *Limulus in the Limelight*. Kluwer Academic/Plenum Press.

Tanke, D.H. and Carpenter, K. (eds.) 2001. *Mesozoic Vertebrate Life*. Indiana University Press.

Tanner, L.H., Lucas, S.G. and Chapman, M.G. 2003. Assessing the record and causes of Late Triassic extinctions. *Earth-Science Reviews*, **65**(1–2): 103–39.

Tapanila, L. and Holmer, L.E. 2004. Endosymbiosis in corals and stromatoporoids: a new lingulid with preserved pedicle and its trace from the Ordovician and Silurian of eastern Canada. *Palaeontological Association Newsletter*, **57**(48th Annual Meeting Abstracts): 180–1.

Tappan, H. and Loeblich, A.R. Jr 1988. Foraminiferal evolution, diversification, and extinction. *Journal of Palaeontology*, **62**: 695–714.

Tattersall, I. 1993. *The Human Odyssey: Four Million Years of Human Evolution*. Prentice Hall.

1995. *The Fossil Trail: How We Know What We Think We Know about Human Evolution*. Oxford University Press.

Taylor, A., Goldring, R. and Gowland, S. 2003. Analysis and application of ichnofabrics. *Earth-Science Reviews*, **60**: 227–59.

Taylor, D.W. and Hickey, L. 1995. *Flowering Plant Origin, Evolution and Phylogeny*. Kluwer.

Taylor, J. (ed.) 1996. *Origin and Evolutionary Radiation of the Mollusca*. Oxford University Press.

Taylor, P. 2004. *Extinctions in the History of Life*. Cambridge University Press.

Taylor, P. and Allison, P. 1998. Bryozoan carbonates through time and space. *Geology*, **26**: 459–62.

Taylor, P. and Lewis, D. in press. *Fossil Invertebrates*. Natural History Museum.

Taylor, T.N. and Taylor, E.L. 1993. *The Biology and Evolution of Fossil Plants*. Prentice Hall.

Thierstein, H. and Young, J. R. 2003. *Coccolithophores: From Molecular Processes to Global Impact*. Springer-Verlag.

Thomas, B. A. and Cleal, C. J. 1993. *The Coal Measure Forests*. National Museum of Wales.

2000. *Invasion of the Land*. National Museum of Wales.

Thomas, B. A. and Spicer, R. A. 1995. *Evolution and Palaeobiology of Land Plants*, 2nd edn. Chapman & Hall.

Thomas, B. A., Cleal, C. J. and Barthel, M. 2004. Palaeobotanical applications of incident-light darkfield microscopy. *Palaeontology*, **47**(6): 1641–5.

Thomas, H. 1994. *L'Homme avant l'homme: le scénario des origines*. Gallimard.

Thomas, H., Sen, S., Behnam, H. A. M. and Ligabue, G. 1981. New discoveries of vertebrate fossils in the "Bakhtiari formation", Injana area, Hemrin South, Iraq. *Journal of the Geological Society of Iraq*, **14**: 43–53.

Thomason, J. J. (ed.) 1995. *Functional Morphology in Vertebrate Paleontology*. Cambridge University Press.

Thompson, d'A.W. 1917. *On Growth and Form*, 2 vols. Cambridge University Press.

Thompson, R. S. and Fleming, R. F. 1996. Middle Pliocene vegetation: reconstructions, inferences, and boundary conditions for climate modelling. *Marine Micropaleontology*, **27**(1/4): 27–50.

Thon, A. (ed.) 1983. *Miscellanea Micropalaeontologica*. Hodonin.

Thulborn, T. 1990. *Dinosaur Tracks*. Kluwer.

Tibert, N. E. and Leckie, R. M. 2004. High-resolution estuarine sea level cycles from the Late Cretaceous: amplitude constraints using agglutinated foraminifera. *Journal of Foraminiferal Research*, **34**(2): 130–43.

Tobias, P. V. 1984. *Dart, Taung and the 'Missing Link'*. Witwatersrand University Press.

Toomey, D. F. and Nitecki, M. H. (eds.) 1985. *Paleoalgology*. Springer-Verlag.

Torrens, H. 2003. Wrestling with muds as strata: William Smith and the first ordering of mudrocks in Britain. *Abstracts, William Smith Meeting ('Wrestling with Mud')*, Geological Society, London.

Torsvik, T. H. and Cocks, L. R. M. 2004. Earth geography from 400–250 Ma: a palaeomagnetic, faunal and facies review. *Journal of the Geological Society, London*, **161**: 555–72.

Traverse, A., 1988. *Paleopalynology*. Unwin Hyman.

Tresise, G. and Sarjeant, W. A. S. 1997. *The Tracks of Triassic Vertebrates: Fossil Evidence from North-West England*. Stationery Office/National Museums and Galleries on Merseyside.

Trewin, N. H. and McNamara, K. J. 1995. Arthropods invade the land: trace fossils and palaeoenvironments of the Tumblagooda sandstone (?late Silurian) of Kalbarri, Western Australia. *Transactions of the Royal Society of Edinburgh (Earth Sciences)*, **85**: 177–210.

Trewin, N. H., Fayers, S. R. and Kelman, R. 2003. Subaqueous silicification of the contents of small ponds in an Early Devonian hot spring complex, Rhynie, Scotland. *Canadian Journal of Earth Sciences*, **40**: 1697–712.

Trinkaus, E. and Shipman, P. 1993. *The Neandertals*. Jonathan Cape.

Trueman, A. E. 1933. A suggested correlation of the coal measures of England and Wales. *Proceedings of the South Wales Institute of Engineers*, **1933**: 1–32.

Trueman, C. N., Benton, M. J. and Palmer, M. R. 2003. Geochemical taphonomy of shallow marine vertebrate assemblages. *Palaeogeography, Palaeoclimatology, Palaeoecology*, **197**(3–4): 151–69.

Truett, J. C. and Johnson, S. R. (eds.) 2000. *The Natural History of an Arctic Oil Field: Development and the Biota*. Academic Press.

Tschinkel, W. R. 2003. Subterranean ant nests: trace fossils past and future. *Palaeogeography, Palaeoclimatology, Palaeoecology*, **192**(1–4): 321–34.

Tsuchi, R. (ed.) 1994. *Pacific Neogene Events in Time and Space*, Tokai University Press.

Tucker, M. W. 1996. *Sedimentary Rocks in the Field*. 2nd edn. John Wiley.

Tudge, C. 1995. *The Day before Yesterday: Five Million Years of Human History*. Jonathan Cape.

Tunnicliffe, V. J. 1992. Hydrothermal vent communities of the deep sea. *American Scientist*, **80**(4): 336–49.

Tunnicliffe, V. J., McArthur, A. G. and McHugh, D. 1998. A biogeographical perspective of the deep-sea hydrothermal vent fauna. *Advances in Marine Biology*, **34**: 353–442.

Turner, A. and Anton, M. 1997. *The Big Cats and their Fossil Relatives*. Columbia University Press.

2004. *Evolving Eden: An Illustrated Guide to the Evolution of the African Large-Mammal Fauna*. Columbia University Press.

Turner, C. E. and Peterson, F. 2004. Reconstruction of the Upper Jurassic Morrison formation extinct ecosystem: a synthesis. *Sedimentary Geology*, **167**(3–4): 309–55.

Twitchett, R. J. 2004. Diachronous ecosystem recovery after the Late Permian mass extinction event. *Palaeontological Association Newsletter*, **57**(48th Annual Meeting Abstracts): 182.

Tzedakis, P. C., Andrieu, V., de Beaulieu, J.-L., *et al.* 1997. Comparison of terrestrial and marine records of changing climate of the last 500 000 years. *Earth and Planetary Science Letters*, **150**: 171–6.

Uchman, A., Drygant, D., Paszkowski, M., Porebski, S. J. and Turnau, E. 2004. Early Devonian trace fossils in marine to non-marine redbeds in Podolia, Ukraine: palaeoenvironmental implications. *Palaeogeography, Palaeoclimatology, Palaeoecology*, **214**(1–2): 67–83.

Ulfstrand, S. 2002. *Savannah Lives: Animal Life and Human Evolution in Africa*. Oxford University Press.

Underhill, J. R. 1998. *Development, Evolution and Petroleum Geology of the Wessex Basin*, Special Publication No. 133. Geological Society.

Ungar, P. and Williamson, M. 2000. Exploring the effects of tooth wear on functional morphology: a preliminary study using dental topographic analysis. *Palaeontologia Electronica*, **3**(1), http://www.palaeo-electronica.org.toc.htm.

Utescher, T., Mosbrugger, V. and Ashraf, A. R. 2000. Terrestrial climate evolution in northwest Germany over the last 25 million years. *Palaios*, **15**: 430–49.

Vail, P. R., Mitchum, R. M., Jr, Thompson, S., III, *et al.* 1977. Seismic stratigraphy and global changes of sea-level. IV. Global cycles of relative sea-level. *American Association of Petroleum Geologists Memoir*, **26**: 49–212.

Vahrenkamp, V. C. 1996. Carbon isotope stratigraphy of the upper Kharaib and Shuaiba formations: implications for the Early Cretaceous evolution of the Arabian Gulf region. *American Association of Petroleum Geologists Bulletin*, **80**(5): 647–62.

Valentine, J. W., Jablonski, D. and Erwin, D. H. 1999. Fossils, molecules and embryos: new perspectives on the Cambrian explosion. *Development*, **126**: 851–9.

Valicenti, V. H., Benson, J. M., Petrie, H. S. P., Pringle, A. A. and McMillan, I. K. 1991. Correlation of depositional systems tract boundaries within the Albian 14A sequence, Bredasdorp basin, South Africa, using condensed sections. *Abstracts, 1er Colloque de Stratigraphie et de Paléogeographie des Bassins Sédimentaires Ouest-Africains/11e Colloque Africain de Micropaléontologie*, Libreville, Gabon.

Van Andel, T. H. 1994. *New Views on an Old Planet: A History of Global Change*, 2nd edn. Cambridge University Press.

Van Buchem, F. S. P., Razin, P., Homewood, P. W., *et al.* 1996. High-resolution sequence stratigraphy of the Natih formation (Cenomanian/Turonian) in northern Oman: distribution of source rocks and reservoir facies. *GeoArabia*, **1**(1): 65–91.

Van Buchem, F. S. P., Pittet, B., Hillgartner, H., *et al.* 2002a. High-resolution sequence stratigraphic architecture of Barremian/Aptian carbonate systems in northern Oman and the United Arab Emirates (Kharaib and Shuaiba formations). *GeoArabia*, **7**(3): 461–500.

Van Buchem, F. S. P., Razin, P., Homewood, P. W., Oterdoom, W. H. and Philip, J. 2002b. Stratigraphic organization of carbonate ramps and organic-rich intrashelf basins: Natih formation (Middle Cretaceous) of northern Oman. *American Association of Petroleum Geologists Bulletin*, **86**(1): 21–53.

Van den Bold, W. A. 1974. Ostracod associations in the Caribbean Neogene. *Verhandlungen Naturforschungs Geseuschaft, Basel*, **84**(1): 214–21.

Vandenbroucke, T. R. A. 2004. Chitinozoans from historical type areas and key sections in the UK: towards a biozonation for the Upper Ordovician on Avalonia. *Palaeontological Association Newsletter*, **57**(48th Annual Meeting Abstracts): 184.

Van Morkhoven, F. P. C. M., Berggren, W. A. and Edwards, A. 1986. Cenozoic cosmopolitan deep-water benthic foraminifera. *Centre des Recherches Exploration–Production Elf-Aquitaine Mémoires*, **11**.

Vaughan, A. 1905. The palaeontological sequence in the Carboniferous Limestone of the Bristol area. *Quarterly Journal of the Geological Society, London*, **61**: 181–307.

Vecoli, M. 2000. Palaeoenvironmental interpretation of microphytoplankton diversity trends in the Cambrian–Ordovician of the northern Sahara Platform. *Palaeogeography, Palaeoclimatology, Palaeoecology*, **160**: 329–46.

Vecoli, M. and le Herisse, A. 2004. Biostratigraphy, taxonomic diversity and patterns of morphological evolution of Ordovician acritarchs (organic-walled microphytoplankton) from the northern Gondwana margin in relation to palaeoclimatic and palaeogeographic changes. *Earth-Science Reviews*, **67**: 267–311.

Velichko, A. A. (ed.) 1984. *Late Quaternary Environments of the Soviet Union*. Longman.

Vennemann, T. W. and Hegner, E. 1998. Oxygen, strontium and neodymium isotope composition of fossil shark teeth as a proxy for the palaeoceanography and palaeoclimatology of the Miocene northern Alpine Paratethys. *Palaeogeography, Palaeoclimatology, Palaeoecology*, **142**: 107–21.

Vennin, E., van Buchem, F. S. P., Joseph, P., *et al.* 2003. A 3D outcrop analogue model for Ypresian nummulitic carbonate reservoirs: Jebel Ousselat, northern Tunisia. *Petroleum Geoscience*, **9**: 145–61.

Vermeij, G. J. 1978. *Biogeography and Adaptation*. Harvard University Press.

1991. Anatomy of an invasion: the trans-Arctic interchange. *Paleobiology*, **17**: 335–56.

Vidal, G. and Knoll, A. H. 1982. Radiations and extinctions of plankton in the late Proterozoic and Early Cambrian. *Nature*, **297**: 57–60.

Vidal, G. and Moczydlowska Vidal, M. 1997. Biodiversity, speciation and extinction trends of Proterozoic and Cambrian phytoplankton. *Paleobiology*, **23**: 230–46.

Videtich, P. E., McLimans, R. K., Watson, H. K. S. and Nagy, R. M. 1988. Depositional, diagenetic, thermal, and maturation histories of Cretaceous Mishrif formation, Fateh field, Dubai. *American Association of Petroleum Geologists Bulletin*, **72**(10): 1143–59.

Vignaud, P., Duringer, P., Mackaye, H. T., *et al.* 2002. Geology and palaeontology of the Upper Miocene Toros-Menalla hominid locality, Chad. *Nature*, **418**: 152–5.

Vink, A. 2004. Calcareous dinoflagellate cysts in South and equatorial Atlantic surface sediments: diversity, distribution, ecology and potential for palaeoenvironmental interpretation. *Marine Micropaleontology*, **59**(1–2): 43–88.

Vinks, V. J. 1988. Biostratigraphy of the Jurassic deposits of the south of the Lesser Caucasus according to foraminifers. *Revue de Paléobiologie, Volume Spéciale*, **2**: 213–18.

Vinogradov, A. 1969. *Atlas of the Lithological–Paleogeographical Maps of the USSR*. Ministry of Geology of the USSR/Academy of Sciences of the USSR. (in Russian)

Vinther, J. and Nielsen, C. 2004. The Lower Cambrian *Halkieria* is a mollusc. *Palaeontological Association Newsletter*, **57** (48th Annual Meeting Abstracts): 189.

Vishnevskaya, V. S., de Wever, P., Baraboshkin, E. Y., *et al.* 1999. New stratigraphic and palaeogeographic data on Upper Jurassic to Cretaceous deposits from the eastern periphery of the Russian platform (Russia). *Geodiversitas*, **21**(3): 347–64.

Vivas, V., Macsotay, O., Furrer, M. and Alvarez, E. 1988. Inyectitas clasticas asociados a desplomes en sedimentitas batiales del Cretaceo Superior de Venezuiela nor-oriental. *Boletín de la Sociedad Venezolana de Geología*, **34**: 3–33.

Vizcaino, S.F., Farina, R.A., Zarate, M.A., Bargo, M.S. and Schultz, P. 2004. Palaeoecological implications of the Mid-Pliocene faunal turnover in the Pampean region (Argentina). *Palaeogeography, Palaeoclimatology, Palaeoecology*, **213**(1–2): 101–13.

Vogel, K. 1975. Endosymbiotic algae in rudists? *Palaeogeography, Palaeoclimatology, Palaeoecology*, **17**: 327–32.

Von Bitter, P.H., Scott, S.D. and Schenk, P.E. 1992. Chemosynthesis: an alternative hypothesis for Carboniferous biotas in bryozoan/microbial mounds, Newfoundland, Canada. *Palaios*, **7**(4): 466–84.

Vorren, T.O., Bergsager, E., Dahl-Stamnes, O.A., *et al.* (eds.) 1993. *Arctic Geology and Petroleum Potential*, Norwegian Petroleum Society Special Publication, No. 2. Elsevier.

Vrba, E.S., Denton, G.H., Partridge, T.C. and Burckle, L.H. (eds.) 1995. *Paleoclimate and Evolution (with Emphasis on Human Origins)*. Yale University Press.

Vrsaljko, D. 1999. The pannonian palaeoecology and biostratigraphy of molluscs from Kostanek-Medvednica Mt., Croatia. *Geologia Croatica*, **52**(1): 9–27.

Waggoner, B. 2003. The Ediacaran biotas in space and time. *Integrative and Comparative Biology*, **43**(1): 104–13.

Wagner, R.H., Hill, C.R. and El-Khayal, A.A. 1985. *Gemmelitheca gen. nov.*, a fertile pecopterid fern from the Upper Permian of the Middle East. *Scripta Geologica*, **79**: 51–74.

Wagner, R.H., Winkler Prins, C.F. and Granados, L.F. (eds.) 1996. *The Carboniferous of the World*, vol. 3, *The Former USSR, Mongolia, Middle Eastern Platform, Afghanistan and Iran*. Instituto Tecnológico Geomínero de España/Nationaal Natuurhistorische Museum.

Wagreich, M. (ed.) 2002. *Aspects of Cretaceous Stratigraphy and Palaeobiogeography*. Osterreichischen Akademie der Wissenschaften.

Walker, G. 2003. *Snowball Earth*. Bloomsbury.

Wallace, A.R. 1858. On the tendency of species to form varieties, and on the perpetuation of varieties and species by natural means of selection. *Proceedings of the Linnean Society, Zoology*, **3**(9): 53–62.

Wallace, D.R. 2004. *Beasts of Eden: Walking Whales, Dawn Horses, and Other Enigmas of Mammal Evolution*. University of California Press.

Walliser, O.H. (ed.) 1995. *Global Events and Event Stratigraphy in the Phanerozoic*. Springer-Verlag.

Walter, M.R. (ed.) 1976. *Stromatolites*. Elsevier.

Walter, M.R. and Heys, G.R. 1985. Links between the rise of the Metazoa and the decline of the Stromatolites. *Precambrian Research*, **19**: 149–74.

Walter, R.C., Buffler, R.T., Bruggemann, J.H., *et al.* 2000. Early human occupation of the Red Sea coast of Eritrea during the last interglacial. *Nature*, **405**: 65–9.

Wang, Y. Boucot, A.J., Rong, J.-Y. and Xang, X.-C. 1987. Community paleoecology as a geologic tool: the Chinese Ashgillian–Eifelian (latest Ordovician through early Middle Devonian). *Geological Society of America Special Publication*, **211**: 1–100.

Ward, P. 2000. *Rivers in Time: The Search for Clues to Earth's Mass Extinctions*. Columbia University Press.

Warny, S.A., Bart, P.J. and Suc, J.-P. 2003. Timing and progression of climatic, tectonic and glacioeustatic influences on the Messinian salinity crisis. *Palaeogeography, Palaeoclimatology, Palaeoecology*, **201**(1–2): 59–66.

Warren, A. and Turner, S. 2004. The first stem tetrapod from the Lower Carboniferous of Gondwana. *Palaeontology*, **47**(1): 151–84.

Waters, J.A. and Maples, C. (eds.) 1997. *Echinoderm Paleobiology*, Short Courses in Paleontology No. 10. Paleontological Society.

Watkins, R. and Wilson, E.C. 1989. Paleoecologic and biogeographic significance of the biostromal organism *Palaeoaplysina* in the Lower Permian McCloud Limestone, Eastern Klamath Mountains, California. *Palaios*, **4**: 181–92.

Watkins, T. 2004. *Origins of Agriculture in the Near East*. Routledge.

Watkinson, D. (ed.) 1987. *First Aid for Finds*. RESCUE/United Kingdom Institute for Conservation Archaeology Section.

Webby, B.D., Paris, F., Droser, M.L. and Percival, I.G. (eds.) 2004. *The Great Ordovician Biodiversification Event*. Columbia University Press.

Wefer, G., Heinze, P.-M. and Berger, W.H. 1994. Clues to ancient methane release. *Nature*, **369**: 282.

Weinberg, S. 1999. *A Fish Caught in Time*. HarperCollins.

Weishampel, D.B., Dodson, P. and Osmolska, H. (eds.) 1990. *The Dinosauria*. University of California Press.

Wellnhofer, P. 1991. *The Illustrated Encyclopaedia of Pterosaurs*. Crescent Books/Salamander.

Wells, J. 1963. Coral growth and geochronometry. *Nature*, **197**: 948–50.

Wells, S. 2002. *The Journey of Man: A Genetic Odyssey*. Penguin.

Wenban-Smith, F.F. and Hosfield, R.T. (eds.) 2001. *Palaeolithic Archaeology of the Solent River*, Occasional Paper No. 7. Lithic Studies Society.

Wendler, J. and Willems, H. 2004. Pithonelloid wall-type of the Late Cretaceous calcereous dinoflagellate cyst genus *Tetratropis*. *Review of Palaeobotany and Palynology*, **129**(3): 133–40.

Wendt, J. 1993. Steep-sided carbonate mud mounds in the Middle Devonian of the eastern Anti-Atlas, Morocco. *Geological Magazine*, **130**(1): 69–83.

Wendt, J., Belka, Z., Kaufmann, B., Kostrewa, R. and Hayer, J. 1997. The world's most spectacular carbonate mud mounds (Middle Devonian, Algerian Sahara). *Journal of Sedimentary Research*, **67**(3): 424–36.

Wescott, W.A., Krebs, W.N., Sikora, P.J., Boucher, P.J. and Stein, J.A. 1998. Modern applications of biostratigraphy in exploration and production. *The Leading Edge*, **(Sept.)**: 1204–10.

West, R.G. 2000. *Plant Life of the Quaternary Cold Stages: Evidence from the British Isles*. Cambridge University Press.

Westermann, G.E.G. 2000. Marine faunal realms of the Mesozoic: review and revision under the new guidelines for biogeographic classification and nomenclature. *Palaeogeography, Palaeoclimatology, Palaeoecology*, **163**: 49–68.

Whatley, R. and Maybury, C. (eds.) 1990. *Ostracoda and Global Events*. Chapman & Hall.

White, E.I. 1950. The vertebrate faunas of the lower Old Red Sandstone of the Welsh borders. *Bulletin of the British Museum (Natural History)*, **A1**: 57–67.

White, M.E. 1996. *The Flowering of Gondwana*. Princeton University Press.

White, T.D., Suwa, G. and Asfaw, B. 1994. *Australopithecus ramidus*, a new species of early hominid from Aramis, Ethiopia. *Nature*, **371**: 306–12.

Whittaker, J.E., Jones, R.W. and Banner, F.T. 1998. *Key Mesozoic Benthic Foraminifera of the Middle East*. Natural History Museum.

Whittaker, J.E., Horne, D.J. and Jones, R.W. 2003. Micropalaeontology in the service of archaeololgy: advances in Quaternary biostratigraphy and palaeoenvironmental analysis using foraminifera and ostracods. *Abstracts, Micropalaeontological Society Annual General Meeting Presentations*.

Whittington, H.B. 1992. *Trilobites*. Boydell Press.

Whitton, B.A. and Potts, M. (eds.) 2000. *The Ecology of Cyanobacteria: Their Diversity in Time and Space*. Kluwer.

Whybrow, P.J. (ed.) 2000. *Travels with the Fossil Hunters*. Cambridge University Press/Natural History Museum.

Whybrow, P.J. and Hill, A. (eds.) 1999. *Fossil Vertebrates of Arabia*. Yale University Press.

Wicander, R. and Monroe, J.S. 2000. *Historical Geology*, 3rd edn. Brooks/Cole and Thomson Learning.

Wierbowski, H. 2004. Carbon and oxygen composition of Oxfordian–early Kimmeridgian belemnite rostra: palaeoenvironmental implications for Late Jurassic seas. *Palaeogeography, Palaeoclimatology, Palaeoecology*, **203**(1–2): 153–68.

Wilby, P.R., Williams, M., Purnell, M., *et al.* 2003. Deep marine Lower Palaeozoic mudrocks of Scotland and Wales: a 'window' for preserving non-mineralised fossil animal tissues. *Abstracts, William Smith Conference ('Wrestling with Mud')*, Geological Society, London.

Wilgus, C., Hastings, B.S., Kendal, C.G.St.C., *et al.* (eds.) 1988. *Sea-Level Changes: An Integrated Approach*. Special Publication No. 42. American Association of Petroleum Geologists.

Williams, M., Siveter, D.J. and Peel, J.S. 1996. *Isoxys* (Arthropoda) from the Early Cambrian Sirius Passet Lagerstatte, north Greenland. *Journal of Paleontology*, **70**(6): 947–54.

Williams, M., Wilkinson, I.P., Leng, M., *et al.* 2003. Ostracods cross the Rubicon: colonising non-marine habitats during the Early Carboniferous. *Palaeontological Association Newsletter*, **54**(47th Annual Meeting Abstracts): 160.

Williamson, P.G. 1981. Palaeontological documentation of speciation in Cenozoic molluscs from the Turkana basin. *Nature*, **293**: 437–43.

Willis, K.J. and McElwain, J.C. 2002. *The Evolution of Plants*. Oxford University Press.

Willmer, P. 1990. *Invertebrate Relationships*. Cambridge University Press.

Wilson, L. and Ratcliffe, P. 2003. Three-dimensional phosphatic preservation of hyolith guts from the Montagne Noire: insights into hyolith ontogeny and phylogeny. *Palaeontological Association Newsletter*, **54**(47th Annual Meeting Abstracts): 190.

Winchester, S. 2001. *The Map that Changed the World*. Penguin.

Winter, A. and Siesser, W.G. (eds.) 1994. *Coccolithophores*. Cambridge University Press.

Wnuk, C. and Pfefferkorn, H.W. (eds.) 1996. Paleozoic paleophytogeography. *Review of Palaeobotany and Palynology*, **90**(1–2): 1–153.

Woillard, G.M. 1978. Grande Pile peat bog: a continuous pollen record for the last 140 000 years. *Quaternary Research*, **9**: 1–21.

Wood, A., Stedman-Edwards, P. and Mang, J. (eds.) 2000. *The Root Causes of Biodiversity Loss*. Earthscan.

Wood, C.J. 1988. The stratigraphy of the Chalk of Norfolk. *Bulletin of the Geological Society of Norfolk*, **38**: 3–120.

Wood, R. 1999. *Reef Evolution*. Oxford University Press.

2004. Palaeoecology of a post-extinction reef: Famennian (Late Devonian) of the Canning basin, north-western Australia. *Palaeontology*, **47**(2): 415–45.

Wood, R., Zhuravlev, A.Y. and Debrenne, F. 1992. Functional biology and ecology of the Archaeocyatha. *Palaios*, **7**: 131–56.

Woodburne, M.O. (ed.) 2004. *Late Cretaceous and Cenozoic Mammals of North America: Biostratigraphy and Geochronology*. Columbia University Press.

Woodcock, N. and Strachan, R. (eds.) 2000. *Geological History of Britain and Ireland*. Blackwell.

Woodman, P., McCarthy, M. and Monaghan, N. 1993. The Irish Quaternary fauna project. *Quaternary Science Reviews*, **16**: 129–59.

Woods, C.A. (ed.) 1989. *Biogeography of the West Indies: Past, Present and Future*. Sandhill Crane Press.

Worsley, D. and Aga, O.J. 1986. *The Geological History of Svalbard: Evolution of an Arctic Archipelago*. Statoil.

Wray, G.A., Levinton, J.S. and Shapiro, L.H. 1996. Molecular evidence for deep Precambrian divergences among metazoan phyla. *Science*, **274**: 568–73.

Wrenn, J.H., Suc, J.-P. and Leroy, S.A. (eds.) 1999. *The Pliocene: Time of Change*. American Association of Stratigraphic Palynologists Foundation.

Wright, A.J., Young, G.C., Talent, J.A. and Laurie, J.R. (eds.) 2000. *Palaeobiogeography of Australasian Faunas and Floras*, Memoir No. 23. Association of Australasian Palaeontologists.

Wright, T. 2001. *Bolivinoides* saves Ipswich from flooding disaster! *Newsletter of Micropalaeontology*, **65**: 17.

Wymer, J.J. 1988. Palaeolithic archaeology and the British Quaternary sequence. *Quaternary Science Reviews*, **7**: 79–98.

Xiangjun Sun, Xu Li, Yinli Luo and Kudong Chen. 2000. The vegetation and climate at the last glaciation on the emerged continental shelf of the South China Sea. *Palaeogeography, Palaeoclimatology, Palaeoecology*, **160**: 301–16.

Xiaoqiao Wan, Wignall, P.B. and Wenjin Zhao 2003. The Cenomanian–Turonian extinction and oceanic anoxic event: evidence from southern Tibet. *Palaeogeography, Palaeoclimatology, Palaeoecology*, **199**(3–4): 283–98.

Yalden, D. 1999. *The History of British Mammals*. Poyser.

Yanagisawa, Y. and Akiba, F. 1998. Refined Neogene diatom biostratigraphy for the northwest Pacific around Japan, with an introduction of code numbers for selected diatom biohorizons. *Journal of the Geological Society of Japan*, **104**: 395–414.

Young, J., Geisen, M., Cros, L., *et al.* 2003. A guide to extant coccolithophore taxonomy. *Journal of Nannoplankton Research, Special Issue*, **1**: 1–125.

Zalasiewicz, J.A., Rushton, A.W.A., Hutt, J.E. and Howe, M. (eds.) 2000. *Atlas of Graptolite Type Specimens*, Folio 1. Palaeontographical Society/British and Irish Graptolite Group.

Zaporozhets, N.I. 1999. Palynostratigraphy and dinocyst zonation of the Middle Eocene–Lower Miocene deposits of the Belaya river (Northern Caucasus). *Stratigraphy and Geological Correlation*, **7**(2): 161–78.

Zazzo, A., Mariotti, A., Lecuyer, C. and Heintz, E. 2002. Intratooth variations in Late Miocene bovid enamel from Afghanistan: paleobiological, taphonomic and climatic implications. *Palaeogeography, Palaeoclimatology, Palaeoecology*, **186**: 145–61.

Zempolich, W.G. and Cook, H.E. (eds.) 2002. *Paleozoic Carbonates of the Commonwealth of Independent States (CIS): Subsurface Reservoirs and Outcrop Analogs*, Special Publication No. 74. Society of Economic Paleontologists and Mineralogists.

Zhizhchenko, B.P. 1959. *Atlas of the Middle Miocene Fauna of the Northern Caucasus and Crimea*. Gostoptekhizdat. (in Russian)

Zhou Chuanming, Brasier, M.D. and Xue Yaosong. 2001. Three-dimensional phosphatic preservation of giant acritarchs from the terminal Proterozoic Doushantuo formation in Guizhou and Hubei provinces, south China. *Palaeontology*, **44**(6): 1157–78.

Zhuravlev, A.Y. and Riding, R. (eds.) 2001. *The Ecology of the Cambrian Radiation*. Columbia University Press.

Zhuravlev, A.Y. and Wood, R. 1996. Anoxia as the cause of the mid-Early Cambrian (Botomian) extinction event. *Geology*, **24**: 311–14.

Ziegler, M.A. 2001. Late Permian to Holocene paleofacies evolution of the Arabian plate and its hydrocarbon occurrences. *GeoArabia*, **6**(3): 445–503.

Ziegler, W. and Sandberg, C. 1990. The Late Devonian standard conodont zonation. *Courier Forschungs-Institut Senckenberg*, **121**: 1–115.

Zubakov, V.A. and Borzenkova, I.I. 1990. *Global Palaeoclimate of the Late Cenozoic*. Elsevier.

Index

Page numbers set in italics refer to figures.